Studies in Surface Science and Catalysis 160

# CHARACTERIZATION OF POROUS SOLIDS VII

Studies in Surface Science and Catalysis

**Advisory Editors:** B. Delmon and J.T. Yates
**Series Editor:** G. Centi

**Vol. 160**

# CHARACTERIZATION OF POROUS SOLIDS VII

Proceedings of the 7th International Symposium on the Characterization of Porous Solids (COPS-VII), Aix-en-Provence, France, 26-28 May 2005

Edited by

**P.L. Llewellyn**
*Université de Provence, France*

**F. Rodriquez-Reinoso**
*Universidad de Alicante, Spain*

**J. Rouqerol**
*Université de Provence, France*

**N. Seaton**
*University of Edinburgh, United Kingdom*

Amsterdam – Boston – Heidelberg – London – New York – Oxford – Paris
San Diego – San Francisco – Singapore – Sydney – Tokyo

ELSEVIER

Elsevier
Radarweg 29, PO Box 211, 1000 AE Amsterdam, The Netherlands
The Boulevard, Langford Lane, Kidlington, Oxford OX5 1GB, UK

First edition 2007

Notice
No responsibility is assumed by the publisher for any injury and/or damage to persons
or property as a matter of products liability, negligence or otherwise, or from any use
or operation of any methods, products, instructions or ideas contained in the material
herein. Because of rapid advances in the medical sciences, in particular, independent
verification of diagnoses and drug dosages should be made

**Library of Congress Cataloging-in-Publication Data**
A catalog record for this book is available from the Library of Congress

**British Library Cataloguing in Publication Data**
A catalogue record for this book is available from the British Library

ISBN-13: 978-0-444-52022-7
ISBN-10: 0-444-52022-8
ISSN (Series): 0167-2991

For information on all Elsevier publications
visit our website at books.elsevier.com

Printed and bound in The Netherlands

07 08 09 10 11   10 9 8 7 6 5 4 3 2 1

Working together to grow
libraries in developing countries

www.elsevier.com | www.bookaid.org | www.sabre.org

ELSEVIER    BOOK AID
            International    Sabre Foundation

# Foreword

The The 7th International Symposium on the *Characterization of Porous Solids* (COPS-VII) was held in the Congress Centre in Aix-en-Provence between the 25th – 28th May 2005. This conference was the seventh in the series launched by the former IUPAC Commission of Colloid and Surface Chemistry and which began in 1987 in Bad Soden, Germany .

We welcomed around 230 guests from 27 countries. It is noteworthy that 31% of the participants were female and 12% of participants were from Industry. This industrial participation shows the interest of porous materials to processes. There were 36 oral presentations and 166 posters were given.

An intensive program over the three days was devoted to recent results of fundamental and applied research on the characterization of porous solids. Papers relating to characterization methods such as gas adsorption and liquid porosimetry, X-ray techniques and microscopic measurements as well as the corresponding molecular modelling methods were given. These characterization methods were shown to be applied to all types of porous solids such as clays, carbons, ordered mesoporous materials, porous glasses, oxides, zeolites and metal organic frameworks. A large part of this symposium was devoted to the use computational methods to characterise these porous solids.

I would like to express my thanks to the organising committee (J. Rouquerol, F. Rodriguez-Reinoso, N. A. Seaton) for their help in composing a wide spectrum of presentations for an outstanding program of scientific quality. Many thanks also go to the local organising committee and all the young volunteers who ensured the smooth running and friendly atmosphere so typical of this conference series.

Finally I would like to acknowledge the financial support of Micromerticis S.A., Quantchrome, Region Provence Alpes Côte d'Azur, Rubotherm GMBH, Thermoelectron and Université de Provence who enabled the reduced fees for 36 students.

It has been decided that COPS-VIII will be held in Edinburgh, United Kingdom.

*P. L. Llewellyn*                    *Marseille, France*

## Scientific Committee

P. L. Llewellyn, CNRS - Université de Provence, Marseille, France
J. Rouquerol, CNRS - Université de Provence, Marseille, France
F. Rodriquez-Reiniso, Universidad de Alicante, Spain
N. A. Seaton, University of Edinburgh, United Kingdom

## Organising Committee

R. Denoyel, CNRS-Université de Provence, Marseille, France
M. Fiori, CNRS-Université de Provence, Marseille, France
F. Rouquerol, CNRS-Université de Provence, Marseille, France

## Financial Support

The organisers gratefully acknowledge the financial support of the following sponsors.

Micromerticis S.A., Verneuil-en-Halatte, France
Quantchrome Instruments, Boynton Beach, USA
Region Provence – Alpes – Côte - d'Azur, France
Rubotherm GMBH, Bochum, Germany
Thermoelectron, Courtaboeuf, France
Université de Provence, Marseille, France

## Surface Science and Catalysis
## Proceedings of COPS VII
## Editors : P. L. Llewellyn, J. Rouquerol, F. Rodrigues-Reinoso, N. A. Seaton

F. Porcheron, P. A. Monson and M. Thommes

Studies in Surface Science and Catalysis 160
*P.L. Llewellyn, F. Rodriquez-Reinoso, J. Rouqerol and N. Seaton (Editors)*

1

# Effect of pore morphology and topology on capillary condensation in nanopores: a theoretical and molecular simulation study

**R. J.-M. Pellenq[a], B. Coasne[b], R. O. Denoyel[c], J. Puibasset[d]**

[a] CRMC-N, CNRS, Campus de Luminy, 13288 Marseille cedex 09, France.
pellenq@crmcn.univ-mrs.fr

[b] LCPMC, CNRS-Université de Montpellier II, Place Eugène Bataillon, 34095 Montpellier cedex 5, France.
[c] MADIREL, CNRS-Université de Provence, Centre de Saint-Jérôme,13397, Marseille cedex 20, France.
[d] CRMD, CNRS-Université d'Orléans, 45071 Orléans cedex 02, France.

## 1. ABSTRACT

We report a theoretical and simulation study of the temperature dependence of adsorption hysteresis for porous matrices having different morphologies and topologies. We used off-lattice Grand Canonical Monte Carlo (GCMC) simulations and two Density Functional Theories (DFT): we used the standard DFT in the non local approximation for cylindrical pores and the coarse-grained lattice DFT developed by Kierlik *et al.* [7] for disordered porous materials. We aim at gaining some insights on the concept of critical hysteresis temperature defined as the temperature at which the adsorption/desorption isotherm becomes reversible.

## 2. INTRODUCTION

Capillary condensation occurs when a fluid is confined within nanopores. It is analoguous to the usual gas-liquid transition but displaced towards lower pressure because of confinement. In fact, capillary condensation concerns pores large enough so that the transition can occur due to cooperative interactions between adsorbed molecules. In contrast, such a phenomenon is not expected in micropores of a few angstroms. The status of capillary condensation in nanoporous adsorbents as being or not a first order transition is still the subject of intense research. Theoretical works for slit pores have demonstrated the existence of a true first order transition with a so-called capillary critical point characterized by a critical temperature of the confined fluid $T_{cc}$ that is lower that of the bulk $T_c^{3D}$. In a pioneering work on the criticality of fluids between plates [1,2], Fisher and Nakanishi showed that the critical behavior is affected by the finite thickness of the adsorbed film and its interaction with the wall. This scaling theory predicts that pore condensation and the related hysteresis loop disappear at a temperature, $T_{cc}$, that is below the bulk critical temperature, $T_c^{3D}$. The shift in critical temperature $\Delta T_{cc} = T_c^{3D} - T_{cc}$ is dictated by the ratio of the pore width $D$ and the correlation length $\xi_0$ of the density fluctuations in the bulk fluid:

$$\frac{\Delta T_{cc}}{T_c^{3D}} = C \left[ \frac{\xi_0}{D} \right]^{1/\nu} \qquad (1)$$

In the 3D-Ising model, $\nu = 0.63$ and $\xi_0$ is of the order of magnitude of the molecular diameter $\sigma$. $C$ is a constant that takes the universal value of 1.658 for zero fluid/wall interactions or 3.432 for strong fluid/wall interactions. In the classical mean field van der Waals fluid theory, $\nu = 0.5$ and $\xi_0$ is again of the order of magnitude of $\sigma$. $C$ takes the universal constant value of 1.645 for zero fluid/wall interactions, which is close to the value found in the Ising approach, or takes the value of 4.284 for strong fluid/wall interactions. Critical-point shifts in a slit-like geometry can be rationalized by the facts that the fluid is between a 3D and a 2D states and that $T_c^{2D}$ is much smaller than $T_c^{3D}$. Evans *et al.* [3] used the Density Functional Theory (DFT) and derived another expression for the shift in the critical temperature for a fluid confined in small pores:

$$\frac{\Delta T_{cc}}{T_c^{3D}} \approx \frac{1}{\lambda D} \qquad\qquad (2)$$

where $1/\lambda$ is the range of the fluid-fluid interaction, which is equal to the adsorbate size in reference [4]. Equation (2) is compatible with the 2D-Ising fluid theory since $\nu$ equals 1.

An adsorption/desorption isotherm below $T_{cc}$, exhibits a hysteresis loop. In between condensation and evaporation metastable pressures, there is an equilibrium pressure for which both the gas and liquid phases coexist. In the case of cylindrical pores of any dimensions, there are theoretical arguments to indicate that no first order transition can exist because at the critical point, the correlation length can diverge only in the direction of the pore axis; a cylindrical pore can be considered as a one-dimensional system whatever its diameter. However, DFT [4], simulations [5] and experiments suggest that fluids in both confining geometries (slit and cylinders with size of several nm) behave similarly as far as condensation and evaporation are concerned; the hysteresis loop shrinks as temperature increases and eventually disappears. We note that in a van der Waals picture of gas-liquid transitions in such simple systems, the temperature of hysteresis disappearance is the capillary critical temperature, i.e., $T_{cc}$.

Over the last decade, significant theoretical advances have been achieved regarding the understanding of fluids confined in a disordered porous materials; it is now clear that fluids in a network of pores having topological and morphological disorders (such as Vycor or CPG) strongly affects capillary condensation as compared to that in independent pores of simple geometries [6]. The first effect is a flattening of the adsorption isotherm branch, which therefore does not exhibit any discontinuity as in a slit or a cylindrical pore. The desorption branch remains vertical and defines a hysteresis loop that shrinks and disappears with increasing temperature. In a random matrix, Kierlik *et al.* [7] used a coarse-grained lattice gas theory (see below) and showed that, for large values of the ratio of the surface-fluid to fluid-fluid energies (parameter *y*), capillary condensation cannot be a first order transition when averaging over matrix disorder. The disorder generates a complex free energy landscape, with a large number of local minima (i.e., metastable states), in which the system remains trapped without finding the true equilibrium state; capillary condensation is then described as an out-of-equilibrium phenomenon. Detcheverry *et al.* [8,9] used the same coarse-grained theory for fractal porous materials and found evidence for a first order capillary condensation that depends on the porosity; it is first order in aerogels with 98% porosity but no longer in aerogels with 87 % porosity. Woo and Monson [10] used again this on-lattice mean field theory for Vycor glasses (mean pore diameter 4-5 nm) and found evidence of a true first-order transition with a critical point at a temperature lower than $T_{cc}$ (the shift of the real critical temperature compared to $T_{cc}$ increases with the coupling parameter *y*). However, the mapping

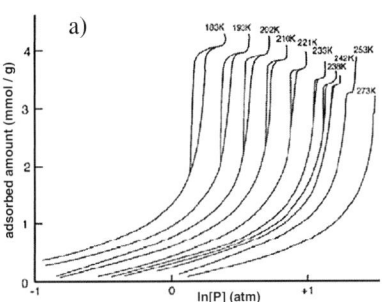

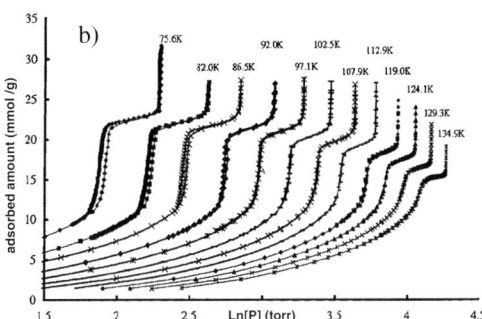

Fig. 1a. Xe Adsorption isotherms in Vycor at different temperatures. From Ref. [25].

Fig. 1b. Temperature dependence of Ar adsorption isotherms in MCM-41 (D = 4.4 nm), $T_{cc}$= 102.5 K. adapted from Ref. [26].

of these on-lattice results on more realistic situations attainable with atomistic off-lattice techniques [11] remains difficult (in particular concerning the dependence of on-lattice results with the lattice spacing and the absolute value of *y*). We note that Monte Carlo simulations on a single (small) periodic Vycor also suggest the existence of a first order transition [12,13,14], when analyzing adsorption data for Ar and $H_2O$ over a wide range of temperatures. However, such atomistic simulations do not consider any averaging over the matrix disorder. The extent of the pore topological disorder therefore seems to have some influence on the nature of capillary condensation. More recently, Coasne and Pellenq [15] showed the important effect on condensation and evaporation of extended morphological defects within a single cylindrical pore such as a constriction. Upon adsorption, the dense phase nucleates in the constrictions and subsequently propagates in the larger cavities (with a meniscus) until a gas bubble is formed and eventually collapses. Upon desorption, evaporation can follow either the so-called pore blocking mechanism (i.e., the larger pore region remains filled until the constricted region empties) or a cavitation process (i.e., the liquid in the largest pore regions undergoes a liquid-gas transition while the constricted regions remain filled), depending on the constriction size and the ratio of the fluid-fluid to fluid-wall energies [16]. These findings confirmed those obtained for constricted slit [17,18,19] and cylindrical [20] pores. Independent but constricted cylindrical pores give rise to adsorption/desorption isotherms, whose shape closely resembles that observed for connected porous networks, in agreement with the recent experiments by van der Voort *et al.* on plugged MCM-41 pores with silica clusters [21]. A constriction can therefore be viewed as a connection between two pores and a constricted single pore can be considered as the simplest model for connected porous materials. Up to now, we have only described results for infinite (periodic) systems, i.e. with pores that are not directly connected to the gas phase through the external surface of the material. If a simple cylindrical pore is connected to the external gas reservoir, then it was shown that desorption occurs at equilibrium [18]. A pore with a close end have a reversible adsorption isotherm with a sudden increase upon condensation (decrease upon evaporation) in the adsorbed amount at a pressure equal to the equilibrium pressure of the equivalent infinite pore (which corresponds to the desorption pressure of the open ended pore of the same diameter or width). Therefore, open ended pores or infinite pores of the same size will have the same hysteresis critical temperature. In disordered porous systems, the status of $T_{cc}$ as being or not relevant to a true critical point is not clear as it seems to depend on the type and degree of disorder (random or fractal porous networks or correlated structures such as Vycor).

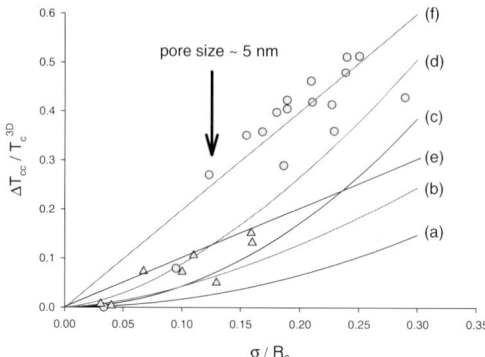

Fig. 2. Experimental hysteresis critical temperatures for various confined fluids. $T_c^{3D}$ is the bulk critical temperature and $\Delta T_c = [T_c^{3D} - T_{cc}]$. Experimental data from [26,22,27,28]. (a) 3D-classical fluid theory with zero wall-fluid interaction, (b) 3D-Ising fluid theory with zero wall-fluid interaction, (c) 3D-classical fluid theory with strong wall-fluid interactions, (d) 3D-Ising fluid theory with zero wall-fluid interactions, (e) 2D-Ising fluid theory with $1/\lambda = \sigma$, (f) same as (e) with $1/\lambda = 2\sigma$.

After these theoretical considerations, one should turn to experimental results. Materials made of unconnected pores are for instance MCM-41 [23] and porous silicon for instance [24]. A standard nanoporous disordered material (with connected pores, and a mean pore size ~ 5 nm) is Vycor glass. Experimental data of $T_{cc}$ in these porous systems are rather scarce as it requires adsorption/desorption isotherms over a wide range of temperatures. This was done for Vycor by Burgess *et al.* in the case of Xe and $CO_2$ [25]. Very recently, Morishige *et al.* were able to measure a complete set of adsorption isotherms for several simple adsorbates such Ar, $N_2$, $O_2$, etc. in MCM-41 materials of several pore diameters [26]. Figure 1 presents experimental data for Xe in Vycor and Ar in MCM-41 (D = 4.4 nm). Gathering all available data together, one can plot the reduced shift of $T_{cc}$ (with respect to the bulk fluid critical temperature) with the reduced pore dimension (with respect to the adsorbate diameter) as originally proposed by Thommes *et al.* [27] and Sing *et al.* [28] using experimental data of simple fluids in CPG and MCM-41 samples. This is presented in Figure 2 for a collection of fluids including Ar, Xe, $N_2$, $O_2$, $SF_6$, $CO_2$, $C_2H_2$ in various MCM-41 and Vycor. Clearly, none of the 3D (classical or Ising) theories is able to describe the data for MCM-41 with pore diameter smaller than 5 nm. Data for larger MCM-41 pores and disordered materials are reasonably well described using the 3D-Ising theory with strong wall-fluid interactions. To our knowledge, there is no available $T_{cc}$ data for fluids confined in disordered porous materials with mean pore size smaller than 5 nm. Data for MCM-41 pores with diameter smaller than 5 nm are adequately described using the 2D-Ising fluid theory with $1/\lambda = 2\sigma$. It seems that the mean pore size matters more than the degree of disorder as data for large MCM-41 mix with those for disordered systems.

In this work, we aim at gaining insights at a molecular level on the hysteretic behavior of adsorption/desorption isotherms of several fluids confined in a rather large collection of pore models that includes single pores of various geometries (cylinders, ellipsoids, constricted pores and Vycor).

## 3. COMPUTATIONAL DETAILS

We used the standard Grand Canonical Monte Carlo (GCMC) technique to simulate capillary condensation and evaporation of molecular fluids such as Ar and $H_2O$ in connected and independent silica nanopores over a large range of temperatures; we calculated for each system a set of adsorption/desorption isotherms to determine $T_{cc}$. Details about the procedure used to generate realistic models of the porous materials (cylinders of various diameters, ellipsoidal pores, contricted cylinders, and Vycor) and intermolecular potentials can be found in Refs. [12,15,29,30]. The reference to the pore dimension corresponds either to the diameter in the case of cylinders or to the mean size of the largest cavities as given by the pore length distribution [13] since these are the location of condensation/evaporation. We also tested different mean field approaches: *(i)* the non local DFT in the case of Ar and $N_2$ confined to perfectly smooth silica cylindrical pores as reported by Ravikovitch and Neimark [31], *(ii)* the coarse-grained lattice theory (CGLT) originally developed by Kierlik *et al.* [7]. In order to compare precisely our data to other simulation or theoretical works on smooth structureless pore models, one should subtract 2.2 Å to the value of the pore dimension of atomistic systems as it corresponds to the OH bond length plus the hydrogen van der Wall radius.

## 4. RESULTS AND DISCUSSION

We first discuss GCMC results for Ar and water on the effect of the pore morphology (cylindrical, ellipsoidal, and constricted pores) and topology (Vycor) on the temperature dependence of capillary condensation hysteresis in nanoporous materials. In Figure 3, we plot the relative decrease of the hysteresis critical temperature, $T_{cc}$, with respect to the bulk critical temperature, $T_c^{3D}$, as a function of the pore size $R_0$ (reduced to the size of the adsorbate molecule taken as the tabulated kinetic diameter $\sigma$). We also report data taken from the literature for Xe, $\sigma = 0.39$ nm, in a Vycor sample having a largest cavity of 5 nm [32], and for water, $\sigma = 0.28$ nm, in cylindrical pores of a diameter 2.4 and 4 nm [33]. The first important result is that GCMC simulations (this work and others) do follow the 2D Ising behavior $\Delta T_{cc}/T_c^{3D} = 2\sigma/R_0$ for small independent nanopores ($R_0 < 5.0$ nm) in agreement with experiments (see Figure 2). This scaling law seems to be independent *(i)* of the nature of the adsorbate since both water data and Ar data for argon fall on this master curve, *(ii)* of the shape (cylinders, ellipsoids, presence of a constriction) of the (independent) pore. For MCM-41 materials with diameter larger than 5 nm, the relative shift of $T_{cc}$ with respect to $T_c^{3D}$ is closer to the 3D Ising model, again with no sensitivity to the adsorbate as both data for Ar, Xe and water are reported. Data for independent pores can be rationalized as a 3D to 2D transition. The locus of this transition, $2\sigma/R_0 \sim 0.27$ (see Figure 3) corresponds to the point above which (small pores) the number of adsorbate particles under the influence of the pore-wall potential (*i.e.* at a distance of approximately two adsorbate diameters) after capillary condensation is larger than that of particles in the pore core; the system adopts a 2D behavior. Reciprocally, below this point (large pores), there is a larger number of molecules in the core of the pore rather than the number of particles that feel the wall potential. If we now consider data for Vycor materials, it is clear that systems with large pores behave as large MCM-41. Again, this behavior seems to be independent of the nature of the adsorbate. Surprisingly, Vycor with smaller pores do not follow the scaling behavior as found for MCM-41 of smaller pore dimension. The reasons for such a behavior in the case of Vycor samples with large cavities is not yet clear. This can be interpreted as a strong effect of the pore disorder in

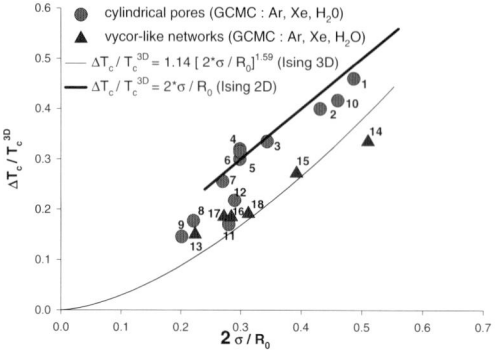

1 : this work, Ar, cyl-2.8 nm
2 : this work, Ar, cyl-3.6 nm
3 : this work, Ar, cyl-4.4nm
4 : this work, Ar, cyl-5.0 nm
5 : this work, Ar, elli-5.0-2.5 nm
6 : this work, Ar, cyl-5.0 nm + constriction
7 : this work, Ar, cyl-5.6 nm
8 : this work, Ar, cyl-6.1 nm
9 : this work, Ar, cyl-7.2nm
10 : Brovchenko *et al*, $H_2O$, cyl-2.4 nm [33]
11 : Brovchenko *et al*, $H_2O$, cyl-4.0 nm [33]
12 : Gelb *et al*, Xe, cyl-5.0 nm [33]
13 : this work, $H_2O$, Vycor-5.0 nm
14 : this work, Ar, Vycor-2.5 nm
15 : this work, Ar, vycor-3.47 nm
16 : Pelleng *et al*, Ar, Vycor-5.0 nm [13]
17 : Gelb *et al*, Xe, Vycor-5.5 nm [32]
18 : Gelb *et al*, Xe, CPG-5.5 nm [32]

Fig. 3. Hysteresis critical temperature as a function of the ratio $2\sigma/R_0$ : GCMC results. the number of the plot reads: n°: author, adsorbate, susbstrate-pore,dimension.

networked materials. For such systems, we expect the correlation length on the condensed fluid to be able to grow to larger values, on average, so that the depression of the pore critical point would be less.

In Figure 4, we report DFT results for argon and nitrogen in cylindrical pores of various diameters. As the pore wall are structureless (perfectly smooth pore surface), DFT in the non local density approximation produces a spurious multi-layering of the fluid close to the pore wall resulting in steps in the adsorption isotherm *prior* to condensation that are not observed experimentally. The adsorbate-fluid (Lennard-Jones) potential parameters are taken from the literature and reasonably mimic the experimental adsorption and condensation/evaporation of nitrogen and argon in MCM-41 materials [34]. It is striking that DFT follows rather closely a 3D-Ising behavior far away from the experimental data in the case of small MCM-41. This result may be attributed to the genuine mean-field character of DFT. As in the case of GCMC and experimental results, DFT calculations are nearly independent of the nature of the confined fluid.

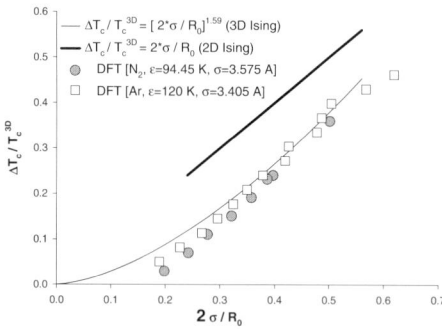

Fig. 4. Hysteresis critical temperature as a function of the ratio $2\sigma/R_0$ : DFT results

Finally, we employed the CGLT of Kierlik *et al.* [7], which is a simplified DFT (nearest neighbor), to model adsorption in our 5 nm Vycor sample. We used the adsorbate-substrate energy grid that was used in the GCMC simulations [13] with variable lattice spacing; all negative values (in-pore locations) of this 3D grid were set to zero, while positive values were set to unity at the beginning of each CGLT calculations. We first tested our CGLT code for a bulk fluid and verified that the critical temperature is simply $k_B T_c^{3D}/w_{ff} = c/4$, where $w_{ff}$ is the fluid-fluid energy parameter and $c$ the number of neighbors for a given lattice symmetry. Of course, this result is independent of the lattice spacing as expected for a bulk system. For a porous system, one adds a characteristic length that corresponds to the pore size. In a nearest neighbor approximation, a lattice theory will be equivalent to off lattice methods when the lattice spacing is chosen so that there is one molecule per site. In their work, Kierlik *et al.* chose a coarse-grained lattice in which a site is no longer one molecule but several. In other words, site occupancy is no longer zero or unity but any value in-between allowing to consider very large disordered porous systems. This implies the use of rather large lattice spacings (15 Å for instance) [10]; for instance, a pore width in a lattice model of Vycor is described by three lattice sites only. In Figure 5, we show that for a given set of fluid-fluid, fluid-wall energy parameters, and a given lattice symmetry, the value of $T_{cc}$ strongly depends on the lattice spacing. Increasing the lattice spacing is equivalent to artificially increase the fluid-wall interaction hence decrease the fluid-fluid interaction rending the CGLT approach difficult to be mapped out onto off-lattice GCMC results or experiment, unless restraining the theory to a classical version allowing one molecule per lattice site (Figure 5a and b).

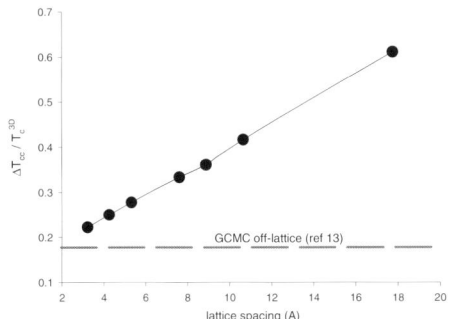 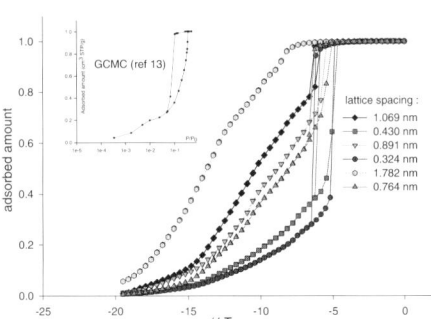

Fig. 5a. Hysteresis critical temperature as a function of the lattice spacing in CGLT. Lattice spacing values are the one given in Fig. 5b.

Fig. 5b. Adsorption/desorption isotherms for Ar in Vycor-5 nm at T= 77 K, $\varepsilon_{ff}$=120 K, $\varepsilon_{fw}$=330 K, simple cubic lattice for various lattice spacings. Inset : comparison with off-lattice GCMC results.

## 5. CONCLUSION

We report a theoretical and simulation study of the temperature dependence of adsorption/desorption hysteresis for porous matrices having different morphologies and topologies. We used the off-lattice Grand Canonical Monte Carlo (GCMC) simulation technique combined with two Density Functional Theory approaches: the first in the non local approximation and the second being a coarse-grained lattice theory. We gained some insights on the concept of critical hysteresis temperature, $T_{cc}$, defined as the temperature at which the adsorption/desorption isotherm becomes reversible. GCMC-simulated $T_{cc}$ for unconnected cylindrical, ellipsoidal, and constricted pores follow the experimental scaling law $\left[T_c^{3D}-T_{cc}\right]/T_c^{3D}=2\sigma/R_0$ established experimentally for MCM-41 silica materials with pore radius $R_0$

lower that 5 nm (σ is the adsorbate diameter). In contrast, GCMC-obtained $T_{cc}$ for larger cylindrical pores and Vycor samples with various mean pore sizes obeys a different relationship, in agreement with experiments. The determination of $T_{cc}$ (for a given the adsorbate in a given porous matrix) thus appears to be a reliable tool to characterize the mean pore size. We infer that it can replace all Kelvin-equation based procedures that rely on many unknown properties of the confined fluids such as surface, tension, density, etc... We also found that the genuine mean-field nature of DFT gives a rather poor description of the $T_{cc}$ for systems such as Ar and $N_2$ in cylindrical silica pores. We show that providing that the lattice spacings is chosen close to the adsorbate size, the coarse-grained DFT lattice theory gives acceptable results compared to GCMC simulations of critical hysteresis temperature. Since it is computationally less expensive that off lattice methods, it can be used as a quantitative predicting tool to study large systems: recently in the condition of small lattice spacing, Salazar and Gelb [35] proposed an improved version of CGLT that allows taking into account pair correlation effects rending CGLT even more effective.

# REFERENCES

[1] H. Nakanishi, M. Fisher, *J. Chem. Phys.*, **78**, 3279 (1983)
[2] A. de Kreizer, T. Michalski, G. H. Findenegg, *Pure & Appl Chem.*, **63**, 1495 (1991).
[3] R. Evans, U. Marini Bertollo Marconi, P. Tarazona, *J. Chem. Phys.*, 84, 2376 (1986).
[4] P. C. Ball and R. Evans, *Langmuir*, **5,** 714 (1989)
[5] A.Z. Panagiotopoulos, *Mol. Phys.,* **67**, 813 (1987)
[6] L. D. Gelb, K E Gubbins, R Radhakrishnan, M Sliwinska-Bartkowiak, *Rep. Prog. Phys.* **62**, 1573 (1999).
[7] E. Kierlik, P. A. Monson, M. L. Rosinberg, L. Sarkisov, G. Tarjus, *Phys. Rev. Lett.* **87**, 055701 (2001).
[8] Detcheverry, F.; Kierlik, E.; Rosinberg, M. L.; Tarjus, G. *Phys. Rev. E*, **68**, (2003) 61504.
[9] F. Detcheverry, E. Kierlik, M. L. Rosinberg, G. Tarjus, Langmuir, **20**, 8006 (2004).
[10] H. J. Woo, P. A. Monson, *Phys. Rev. E*, **67**, 041207 1-17 (2003).
[11] L. D. Gelb , *Mol. Phys.*, **100**, 2049–2057 (2002).
[12] R. J.-M. Pellenq, B. Rousseau, P. Levitz, *Phys. Chem. Chem. Phys.*, **3**, 1207 (2001).
[13] R. J.-M. Pellenq, P. E Levitz, *Molecular Physics*, **100**, 2049 (2002).
[14] J. Puibasset, R. J.-M. Pellenq, *J. Chem. Phys.*, **122**, 94714 (2005).
[15] B. Coasne and R. J.-M. Pellenq, *J. Chem. Phys.*, **121**, 3767 (2004).
[16] B. Coasne, K. E. Gubbins, R. J-.M. Pellenq, *Part. Part. Sys. Charact.*, **21**, 149 (2004).
[17] B. Libby, P. A. Monson, *Langmuir*, **20**, 4289 (2004).
[18] L. Sarkisov and P. A. Monson, *Langmuir* **17**, 7600 (2001).
[19] A. P. Manaloski, F. van Swol, *Phys. Rev. E*, **66**, 041603 (2002).
[20] P. I. Ravikovitch, A. V. Neimark, *Langmuir*, **18**, 9830 (2002).
[21] P. Van der Voort, *et al*, *J. Phys. Chem B.*, **106**, 5873 (2003).
[22] K. Morishige, H. Fujii, M. Uga and D. Kinukawa., *Langmuir*, **13**, 3494 (1997).
[23] M. Kruk, M. Jaroniec, A. Sayari, *Langmuir*, **13**, 6267 (1997).
[24] B. Coasne, A. Grosman, N. Dupont-Pavlovsly, C. Ortega, M. Simon, *Phys. Chem. Chem. Phys.*, **3**, 1196 (2001).
[25] Burgess, C.G.V., D.H. Everett, and S. Nuttal, *Langmuir*, **6**, 1734 (1990).
[26] K. Morishige, Y Nakamura, *Langmuir*, **20**, 4503 (2004).
[27] M. Thommes, G. H. Findenegg, *Langmuir*, **10**, 4270 (1994).
[28] K. S. W. Sing *et al*, Proceedings of the COPS- IV Conference, ed. B. Mc Enaney, F. Rodriguez-Reinoso, J. Rouquerol, K. S. W. Sing, K. K. Unger, The Royal Society of Chemistry, Pub., Cambridge, UK, 1997.
[29] R. J.–M. Pellenq, D. Nicholson, *J. Phys. Chem.*, **98**, 13339 (1994).
[30] J. Puibasset, and R.J.–M. Pellenq, *J. Chem. Phys.*, **119**, 9226 (2003).
[31] P. I. Ravikovitch, A. Vishnyakov, A. V. Neimark, *Phys. Rev. E*, **64**, 011602 (2001)
[32] L. D. Gelb, K.E. Gubbins, *Fundamentals of Adsorption* 7, 333 (2002).
[33] I. Brovchenko, A. Geiger and A. Oleinikova, *J. Chem. Phys.*, **120**, 1958 (2004).
[34] P. I. Ravikovitch, G. L. Haller, A. V. Neimark, Advances in Colloid and Interface Science. **76-77**, 203 (1998).
[35] R. Salazar, L. D. Gelb, *Phys. Rev. E*, **71**, 041502 (2005).

Studies in Surface Science and Catalysis 160
*P.L. Llewellyn, F. Rodriquez-Reinoso, J. Rouqerol and N. Seaton (Editors)*

# Density Functional Theory Model of Adsorption on Amorphous and Microporous Solids

Peter I. Ravikovitch[*] and Alexander V. Neimark

Center for Modeling and Characterization of Nanoporous Materials
TRI/Princeton, 601 Prospect Ave., Princeton, NJ 08540, USA

We suggest a model of adsorption in pores with amorphous and microporous solid walls, named the quenched solid non-local density functional theory (QSNLDFT) model. We consider a multicomponent non-local density functional theory (NLDFT), in which the solid is treated as a quenched component with a fixed spatially distributed density. Drawing on several prominent examples, we show that QSNLDFT model produces smooth isotherms of mono- and polymolecular adsorption, which resemble experimental isotherms on amorphous surfaces. The model reproduces typical behaviors of $N_2$ isotherms on micro- mesoporous materials, such as SBA-15. QSNLDFT model offers a systematic approach to the account for the surface roughness/heterogeneity in pore structure characterization methods.

## 1. INTRODUCTION

Recent progress in the theory of adsorption on porous solids, in general, and in the adsorption methods of pore structure characterization, in particular, has been related, to a large extent, to the application of the density functional theory (DFT) of inhomogeneous fluids [1]. DFT has helped qualitatively describe and classify the specifics of adsorption and capillary condensation in pores of different geometries [2-4]. Moreover, it has been shown that the non-local density functional theory (NLDFT) with suitably chosen parameters of fluid-fluid and fluid-solid interactions quantitatively predicts the positions of capillary condensation and desorption transitions of argon and nitrogen in cylindrical pores of ordered mesoporous molecular sieves of MCM-41 and SBA-15 types [5,6]. NLDFT methods have been already commercialized by the producers of adsorption equipment for the interpretation of experimental data and the calculation of pore size distributions from adsorption isotherms [7-9].

At the same time, current implementations of NLDFT have a significant drawback: theoretical adsorption isotherms in the region of polymolecular adsorption exhibit multiple steps associated with layering transitions. Experimentally, stepwise layering transitions are observed only for fluids confined to very smooth surfaces, such as mica

---

[*] ravikovi@triprinceton.org; [§] aneimark@triprinceton.org

or graphite. In most experimental systems, however, layering transitions are hindered due to inherent energetic and geometrical heterogeneities of real surfaces. The layering transition steps on the theoretical isotherms cause artificial gaps on the calculated pore size distributions [9,10], because the computational schemes attribute the layering transition step to the capillary condensation/pore filling step in the pore of a certain size. For example, in the case of nitrogen at 77K on graphite, the monolayer transition in NLDFT occurs at the same pressure as the pore filling in ~1 nm slit pore, which results in a prominent false gap on the pore size distribution histograms [9].

Another crucial problem is a proper account for the surface roughness and microporosity, which are inherent to a large number of novel polymer-templated silica and organosilica materials, such as, e.g. SBA-15 [11-14]. For many structures, especially those synthesized at low temperatures, the micropore volume cannot be reliably estimated from the comparison plot due to a competition between micropore filling and multilayer adsorption on the rough surface.

Several attempts were made in the literature to introduce the effects of surface heterogeneities and to improve upon the models of smooth walls. Olivier [8] used adsorption energy distributions to describe adsorption on heterogeneous surfaces using NLDFT. Maddox et al [15] performed MC simulations of adsorption in cylindrical pores of MCM-41 with a radial and axial dependent potential. Gelb and Gubbins [16], Coasne and Pellenq [17], Kuchta et al [18] simulated adsorption within energetically and geometrically heterogeneous pores. Ravikovitch et al [19] modeled toluene adsorption on MCM-41 with randomly distributed attractive and repulsive sites on the pore walls.

In order to account for the surface heterogeneity in the DFT calculations, one has to consider at least two- or, better, three-dimensional spatial dependence of the solid-fluid interaction potential. In the case of NLDFT, multidimensional calculations become very expensive [20], and the main advantage of NLDFT, namely its computational efficiency, is lost. Technically, the local DFT (LDFT) produces smooth isotherms in the region of multilayer adsorption [2]. However, due to the lack of a realistic description of short-range molecular packing, LDFT is unable to provide even a qualitatively correct description of the fluid structure. Recently, Ustinov and Do [21] suggested an original modification to the Tarazona's version of NLDFT [22], which is capable of generating smooth adsorption isotherms.

In this paper, we suggest a systematic approach that extends the applicability of NLDFT models to heterogeneous surfaces of amorphous and microporous solids. The main idea is to use a multicomponent NLDFT, in which the solid is treated as one of the components with a fixed spatially distributed density. The model, named quenched solid non-local density functional theory (QSNLDFT), is an extension of the quenched-annealed DFT model of systems with hard-core interactions recently proposed by Schmidt and coworkers [23,24]. Drawing on several prominent examples, we show that the proposed model produces smooth isotherms in the region of multiplayer adsorption. Moreover, the effects of wall microporosity can be naturally incorporated into the model. Although the parameters of the model have not been yet optimized to describe quantitatively a particular experimental system, the model generates adsorption isotherms which are in qualitative agreement with experimental isotherms of $N_2$ or Ar adsorption on amorphous silica materials.

## 2. THEORY

In the conventional NLDFT description of a single component fluid in a pore [1], the grand thermodynamic potential of the system, $\Omega[\rho(\mathbf{r})]$, is written as a functional of the fluid density distribution, $\rho(\mathbf{r})$. The confinement induced by the pore walls is described as an external potential field $V_{ext}(\mathbf{r})$:

$$\Omega[\rho(\mathbf{r})] = F_{id}[\rho] + F_{ex}[\rho] + \frac{1}{2}\iint d\mathbf{r} d\mathbf{r}' \rho(\mathbf{r})\rho(\mathbf{r}')u_{ff}(|\mathbf{r}-\mathbf{r}'|) - \int d\mathbf{r} \rho(\mathbf{r})[\mu - V_{ext}(\mathbf{r})] \qquad (1)$$

Here $F_{id}[\rho]$ and $F_{ex}[\rho]$ are the ideal and excess components of the Helmholtz free energy of the hard sphere (HS) fluid; and $u_{ff}(|\mathbf{r}-\mathbf{r}'|)$ is the attractive part of the fluid-fluid intermolecular potential, treated in a mean field fashion using, for example, the WCA scheme [25]; $\mu$ is the chemical potential. To determine the fluid density distribution, the functional is minimized with respect to $\rho(\mathbf{r})$.

Typically, NLDFT models operate with fluids confined to pores of simple geometries (slit, cylinder, sphere) with smooth structureless walls, which exert an external potential $V_{ext}(r)$ being a function of one coordinate, e.g. the distance from the pore wall, or the radial coordinate in the case of cylindrical or spherical pores. The consequence of this simplification is that it induces a strong layering of the fluid at the walls, especially at low temperatures [6], which causes artificial layering transitions discussed above.

Let us consider a NLDFT model, in which solid and fluid are treated as a two-component system in a fashion similar to the quenched-annealed DFT introduced by Schmidt [23,24] for the systems with hard-core interactions. The solid is represented as a "quenched" component with a given density distribution $\rho_S(\mathbf{r})$, and the fluid as a "mobile" (or "annealed") component, which density $\rho(\mathbf{r})$ has to be found from the functional minimization.

The grand potential functional of the fluid-solid system with attractive walls is given in the following form:

$$\Omega[\rho_S;\rho] = F_{id}[\rho_S;\rho] + F_{HS}[\rho_S;\rho] + \frac{1}{2}\iint d\mathbf{r} d\mathbf{r}' \rho(\mathbf{r})\rho(\mathbf{r}')u_{ff}(|\mathbf{r}-\mathbf{r}'|) +$$
$$\iint d\mathbf{r} d\mathbf{r}' \rho(\mathbf{r})\rho_S(\mathbf{r}')u_{sf}(|\mathbf{r}-\mathbf{r}'|) - \mu \int d\mathbf{r} \rho(\mathbf{r}) \qquad (2)$$

Similarly to the fluid-fluid intermolecular potential, we split the solid-fluid intermolecular potential into repulsive hard-sphere and attractive interactions. Here $F_{HS}[\rho_S;\rho]$ is the excess free energy of the solid-fluid HS mixture, for which we employ Rosenfeld fundamental measure functional [26] with the recent modifications that give an accurate Carnahan-Starling equation of state in the bulk limit [27,28]; $u_{sf}(|\mathbf{r}-\mathbf{r}'|)$ is the attractive part of the solid-fluid intermolecular potential. Since the solid-solid attraction interaction is not included, the solid is effectively modeled as a compound of hard spheres interacting with "soft sphere" fluid molecules via $u_{sf}\left(|\mathbf{r}-\mathbf{r'}|\right)$. Unlike for the fluid-fluid interactions, where the WCA scheme is proven to give a reasonably good description of the bulk and interfacial properties of simple fluids, the problem of implementation of the solid-fluid potential is not straightforward. We explored several different schemes of splitting the solid-fluid intermolecular potential into attractive and repulsive interactions. These results will be presented elsewhere. The calculations presented below were performed using the WCA scheme for the division of the solid-fluid LJ potential:

$$u_{sf}(r)=\begin{cases}-\varepsilon_{sf}, & r<2^{1/6}\sigma_{sf}\\ 4\varepsilon_{sf}\left[\left(\sigma_{sf}/r\right)^{12}-\left(\sigma_{sf}/r\right)^{6}\right], & r\geq 2^{1/6}\sigma_{sf}\end{cases} \tag{3}$$

The model (2)-(3) is symmetric with respect to the solid and fluid densities. At a given "quenched" solid density distribution, minimization of the functional with respect to the fluid density distribution leads to the Euler-Lagrange equation in the form:

$$\rho(\mathbf{r})=\Lambda^{-3}\exp\left\{-\beta\frac{\delta F_{HS}[\rho_S;\rho]}{\delta\rho(\mathbf{r})}-\beta\int d\mathbf{r'}\rho(\mathbf{r})u_{ff}\left(|\mathbf{r}-\mathbf{r'}|\right)-\beta\int d\mathbf{r'}\rho_S(\mathbf{r})u_{sf}\left(|\mathbf{r}-\mathbf{r'}|\right)+\beta\mu\right\} \tag{4}$$

where $\beta=1/(k_B T)$, $\Lambda$ is the de Broglie wavelength.

The model (2)-(4) is referred to as the quenched solid non-local density functional theory (QSNLDFT). There are several advantages in considering the solid as a quenched component of the system rather than a source of the external field. On the one hand, this approach offers flexibility in the description of the fluid-solid boundary by varying the solid density and the thickness of the diffuse solid surface layer. On the other hand, it retains the main advantage of NLDFT computational efficiency because even a one-dimensional solid density distribution can include the effects of surface roughness and heterogeneity. For example, the solid density distribution can be taken from simulations of amorphous silica surfaces [29,30].

It is worth noticing that QSNLDFT provides a unified model of adsorption on both smooth and heterogeneous surfaces. QSNLDFT describes the behavior of LJ fluid near a smooth hard wall by employing a uniform solid density distribution with a packing density approaching unity. The behavior of LJ fluid near a smooth attractive wall is approximated by using an appropriate division of the solid-fluid potential into repulsive and attractive parts. Below we consider several prominent examples, which demonstrate that QSNLDFT is capable of describing experimental adsorption isotherms on amorphous and microporous solids.

## 3. RESULTS AND DISCUSSION

As the first example, we consider Ar adsorption in a slit pore with attractive walls made of amorphous hard sphere (HS) solid. Two different cases are considered. In the first case, the solid is modeled using a sharp-kink approximation for the solid density profile (Figure 1). The solid density is taken near the HS closed-packing density, which does not permit penetration of fluid molecules into the solid matrix. In the second case, the solid density distribution is taken in the form that approximates simulations of amorphous silica surfaces [30]:

$$\rho_s(z) = \frac{\rho_0}{2}\left(1 - \tanh\left(\frac{2(z-z_0)}{\delta}\right)\right) \tag{5}$$

Here $\rho_0$ is the solid density, the same as in the first case; $z_0 = 0$ is the position of the solid boundary. The parameter $\delta$ characterizes the surface roughness. The solid surface is diffuse ($\delta$=0.8 nm in Eq. 5) that permits additional adsorption on the wall indentations as compared to adsorption on the wall with the sharp, impermeable boundary ($\delta\to0$). The intermolecular parameters for Ar were taken from [9]. The parameters of the solid-fluid LJ potential were $\varepsilon_{sf}/k_B$=368 K, $\sigma_{sf}$=0.3 nm, and the diameter of solid HS was $d_S$=0.3 nm.

Figure 1 (left) shows the solid and fluid density profiles for the sharp and diffuse wall surfaces. In the case of HS wall, the fluid exhibits layering, qualitatively similar to the layering near smooth LJ walls discussed above. However, the position of the fluid density peaks is shifted closer to the solid due to the surface roughness of HS solid, and a tendency for fluid molecules to fill in the cusps between the hard spheres. In the case of the diffuse HS wall, the fluid density distribution is smooth and does not exhibit layering, although the fluid structure can be clearly seen.

Figure 1 (right) shows the adsorption isotherms for the two slit pores – one pore with sharp HS walls and the other with diffuse HS walls. The adsorption-desorption isotherms are hysteretic with abrupt capillary condensation and evaporation steps located at the vaporlike and liquidlike spinodals. The lines of equilibrium transition calculated from the condition of equality of grand potentials on the adsorption and desorption branches are shown in the middle. It can be seen that although the desorption branches are almost identical, and the positions of the spinodal and equilibrium transitions are quite close, the adsorption branches differ significantly in the multilayer region. The adsorption branch for the pore with sharp HS walls exhibits some layering, while the adsorption branch for the pore with diffuse walls is smooth. Also, the adsorption per unit area on the rough surface is significantly higher. This situation is typical for adsorption isotherms on SBA-15, FDU-1 and other polymer-templated silicas, which often exhibit similar total pore volumes and pore diameters, but drastically different surface areas. This behavior of adsorption isotherms can be explained by different degree of the surface roughness.

In the second example, we consider adsorption of $N_2$ in a slit pore with diffuse and microporous walls, mimicking microporous amorphous silica surface. The intermolecular parameters for $N_2$ are taken from our previous publication [9]. Parameters of nitrogen-oxygen atom interactions are calculated from combining rules using oxygen-oxygen interaction parameters reported for amorphous silicas [31] - $\varepsilon_{OO}/k_B$=230 K, $\sigma_{OO}$=0.27 nm. Both Si and O atoms are modeled as hard spheres of 0.27 nm in diameter, but only interactions with O are considered. The solid surface density distribution is taken in the form of Eq. (5), $\delta$=0.118 nm, which approximates amorphous silica surface [30]. Microporosity of the walls is modeled by using twice lower density of the solid, $\rho_0 = 1.1$ g/cm$^3$ (the density of amorphous silica is 2.2 g/cm$^3$). The thickness of the pore wall is $10\sigma_{ff}$, and the system is periodic in z-direction.

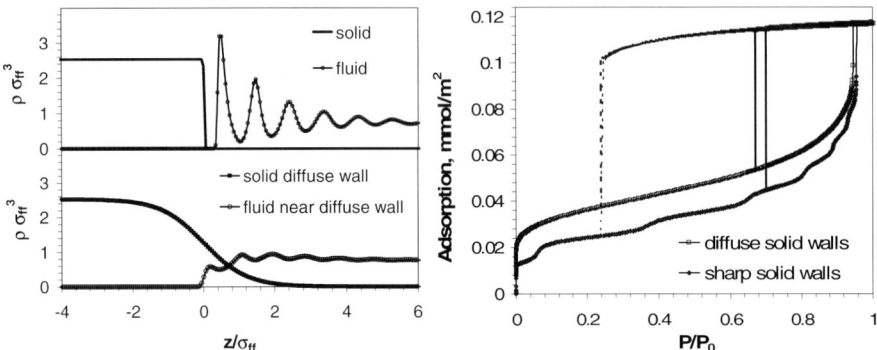

Figure 1. Ar adsorption at 87 K in two $20\sigma_{ff}$ slit pores with attractive walls made of hard spheres. The first pore has a sharp surface. The second pore has a diffuse surface. The solid density is about the close-packing density. Solid and fluid density distributions (left) and adsorption isotherms (right). Equilibrium capillary condensation is shown as vertical lines and metastable desorption branches as dotted lines.

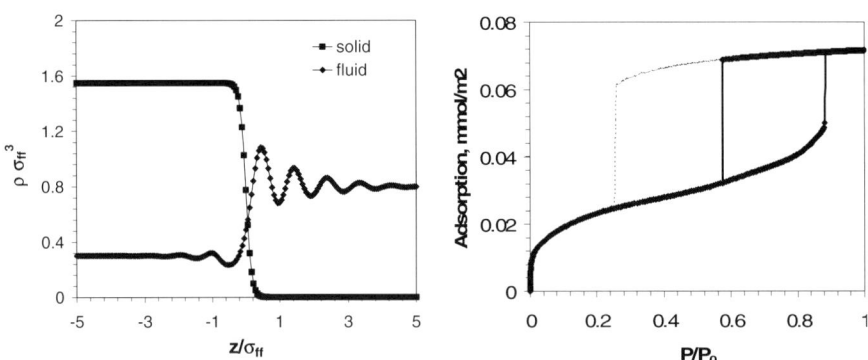

Figure 2. Adsorption of $N_2$ at 77 K in $10\sigma_{ff}$ slit pore with microporous silica walls. Solid and fluid density distributions (left) and adsorption isotherm (right). The solid density in the wall is twice lower than the density of amorphous silica. The pore wall surface is diffuse. Fluid molecules penetrate into the solid matrix mimicking the micropore filling. Equilibrium capillary condensation is shown as a vertical line and metastable desorption branch is shown as a dotted line.

The low solid density allows for the penetration of fluid molecules into the solid matrix mimicking the micropore filling, as can be seen from the density profile (Figure 2, left). The density profile shows the fluid structure at the solid-fluid interface. At the same time, the isotherm (Figure 2, right) is smooth in the region of polymolecular adsorption. The overall shape of this adsorption isotherm is qualitatively similar to the shape of adsorption isotherms on micro-mesoporous polymer-templated silicas, such as SBA-15. Depending on the synthesis temperature, these materials may possess rough and micropous walls due to the interpenetration of silica and polyethylene oxide chains during templating [12]. Template removal leads to an amorphous low-density structure of the pore walls. The example in Figure 2 clearly demostrates that QSNLDFT model captures the main factors determining adsorption on SBA-15 type materials. Note, that the width of the calculated hysteresis loop, which is somewhat wider than typical experimental hysteresis loops, will shrink when the calculations are performed in cylindrical rather than in the slab geometry.

QSNLDFT has immediate implications for the practical problems of pore structure characterization. The NLDFT models, which are routinely employed to study various materials, can now be extended to provide an assessment of the surface roughness and microporosity.

## 4. CONCLUSIONS

We presented a novel quenched solid non-local density functional (QSNLDFT) model, which provides a realistic description of adsorption on amorphous surfaces without resorting to computationally expensive two- or three-dimensional DFT formulations. The main idea is to consider solid as a quenched component of the solid-fluid mixture rather than a source of the external potential. The QSNLDFT extends the quenched-annealed DFT proposed recently by M. Schmidt and coworkers [23,24] for systems with hard core interactions to porous solids with attractive interactions. We presented several examples of calculated adsorption isotherms on amorphous and microporous solids, which are in qualitative agreement with experimental measurements on typical polymer-templated silica materials like SBA-15, FDU-1 and others. Introduction of the solid density distribution in QSNLDFT eliminates strong layering of the fluid near the walls that was a characteristic feature of NLDFT models with smooth pore walls. As the result, QSNLDFT predicts smooth isotherms in the region of polymolecular adsorption. The main advantage of the proposed approach is that QSNLDFT retains one-dimensional solid and fluid density distributions, and thus, provides computational efficiency and accuracy similar to conventional NLDFT models.

## ACKNOWLEDGMENTS

The work was supported by the TRI/Princeton exploratory research program and Quantachrome Instruments. A part of this research was performed at Princeton University, where one of us (AVN) worked in 2004-2005 as a Guggenheim Fellow. AVN gratefully acknowledges the support of the John Simon Guggenheim Memorial Foundation and thanks George Scherer for his hospitality and fruitful discussions.

## REFERENCES

[1] R. Evans In Fundamentals of Inhomogeneous Fluids; D. Henderson, Ed.; Marcel Dekker: New York, 1992; Chapter 5.

[2] R. Evans, U.M.B. Marconi and P. Tarazona, Journal of the Chemical Society-Faraday Transactions Ii 82 (1986) 1763.

[3] B.K. Peterson, K.E. Gubbins, G.S. Heffelfinger, U. Marini, B. Marconi and F. Van Swol, J. Chem. Phys. 88 (1988) 6487.

[4] P.B. Balbuena and K.E. Gubbins, Langmuir 9 (1993) 1801.

[5] P.I. Ravikovitch, S.C. O'Domhnaill, A.V. Neimark, F. Schuth and K.K. Unger, Langmuir 11 (1995) 4765.

[6] P.I. Ravikovitch, A. Vishnyakov and A.V. Neimark, Phys. Rev. E 64 (2001) 011602.

[7] C. Lastoskie, K.E. Gubbins and N. Quirke, J. Phys. Chem. 97 (1993) 4786.

[8] J.P. Olivier, J. Porous Mater. 2 (1995) 217.

[9] P.I. Ravikovitch, A. Vishnyakov, R. Russo and A.V. Neimark, Langmuir 16 (2000) 2311.

[10] J.P. Olivier, Carbon 36 (1998) 1469.

[11] M. Kruk, M. Jaroniec, C.H. Ko and R. Ryoo, Chem. Mat. 12 (2000) 1961.

[12] A. Galarneau, H. Cambon, F. Di Renzo and F. Fajula, Langmuir 17 (2001) 8328.

[13] P.I. Ravikovitch and A.V. Neimark, J. Phys. Chem. B 105 (2001) 6817.

[14] V.B. Fenelonov, A.Y. Derevyankin, S.D. Kirik, L.A. Solovyov, A.N. Shmakov, J.L. Bonardet, A. Gedeon and V.N. Romannikov, Microporous Mesoporous Mat. 44 (2001) 33.

[15] M.W. Maddox, J.P. Olivier and K.E. Gubbins, Langmuir 13 (1997) 1737.

[16] L.D. Gelb and K.E. Gubbins, Langmuir 14 (1998) 2097.

[17] B. Coasne and R.J.M. Pellenq, J. Chem. Phys. 120 (2004) 2913.

[18] B. Kuchta, P. Llewellyn, R. Denoyel and L. Firlej, Colloid Surf. A-Physicochem. Eng. Asp. 241 (2004) 137.

[19] P.I. Ravikovitch, A. Vishnyakov, A.V. Neimark, M. Ribeiro Carrott, P. Russo and P.J.M. Carrott, Langmuir, in press, 2005.

[20] L.J.D. Frink and A.G. Salinger, J. Comput. Phys. 159 (2000) 407.

[21] E.A. Ustinov, D.D. Do and M. Jaroniec, J. Phys. Chem. B 109 (2005) 1947.

[22] P. Tarazona, Phys. Rev. A 31 (1985) 2672.

[23] M. Schmidt, Phys. Rev. E 68 (2003) 021106.

[24] M. Schmidt and J.M. Brader, J. Chem. Phys. 119 (2003) 3495.

[25] J.D. Weeks, D. Chandler and H.C. Andersen, J. Chem. Phys. 54 (1971) 5237.

[26] Y. Rosenfeld, Phys. Rev. Lett. 63 (1989) 980.

[27] R. Roth, R. Evans, A. Lang and G. Kahl, J. Phys.-Condes. Matter 14 (2002) 12063.

[28] Y.X. Yu and J.Z. Wu, J. Chem. Phys. 117 (2002) 10156.

[29] A. Roder, W. Kob and K. Binder, J. Chem. Phys. 114 (2001) 7602.

[30] M. Rarivomanantsoa, P. Jund and R. Jullien, J. Phys.-Condes. Matter 13 (2001) 6707.

[31] A. Brodka and T.W. Zerda, J. Chem. Phys. 104 (1996) 6319.

Studies in Surface Science and Catalysis 160
P.L. Llewellyn, F. Rodriquez-Reinoso, J. Rouqerol and N. Seaton (Editors)

17

# Thickness of Adsorbed Nitrogen Films in SBA-15 Silica from Small-Angle Neutron Diffraction

A. Schreiber[a], I. Ketelsen[a], G.H. Findenegg[a] and E. Hoinkis[b]

[a]Stranski Laboratorium für Physikalische und Theoretische Chemie, Technische Universität Berlin, Straße des 17. Juni 135, D-10623 Berlin, Germany
[b]Hahn-Meitner-Institut, Glienicker Straße 100, D 14109 Berlin, Germany

Pore filling of SBA-15 silica with nitrogen at 77 K was studied by combining gas adsorption measurements with small-angle neutron scattering. The Bragg peaks resulting from the 2D-hexagonal packing of the cylindrical mesopores exhibit a characteristic dependence of the peak intensity on the amount of adsorbed gas, increasing or decreasing as the thickness of the adsorbed film grows. This modulation of the peak intensities can be reproduced by a structural model in which the cylindrical pores and the adsorbed nitrogen film are taken into account by a form factor $F(R, r_f)$ of cylindrical objects, with $R$ the pore radius and $r_f$ the radius of the vapor core of the pores. For our SBA-15 sample it is found that the thickness of the adsorbed film $t = R - r_f$ is a nearly linear function of the gas pressure, although the adsorption isotherm exhibits a pronounced high-affinity region at low pressures. This behaviour is attributed to adsorption into a microporous corona of the cylindrical mesopores at low pressures. A simple geometrical model is presented which accounts for the observed dependence of the film thickness on the pressure.

## 1. INTRODUCTION

Nitrogen adsorption represents the most widely used method for the determination of pore size in the micropore and lower mesopore range [1,2]. In the case of mesopores, the method relies on a knowledge of the thickness of the physisorbed nitrogen film at the pore wall, which has to be added to the Kelvin radius to obtain the pore radius $R$. For open, flat surfaces the thickness of adsorbed films can be determined directly by ellipsometry [3], X-ray reflectometry [4], or the quartz-crystal microbalance technique [5], but none of these techniques can be adapted to measure the film thickness in a narrow pore. A method to estimate the thickness of nitrogen adsorbed films at the walls of mesoporous silicas by small-angle neutron scattering (SANS) was proposed by Smarsly and coworkers [6]. It is based on an analysis of the chord-length distribution that can be extracted from the SANS data and is, in principle, applicable to any micro- or mesoporous material. In the present work a new method for determining the thickness of nitrogen adsorbed films in pores is presented which is applicable specifically to materials with periodic pore structures. Our method, which represents an extension of the formalism proposed by Imperor-Clerc et al. [7] for the characterization of the SBA-15 pore matrix, is based on an analysis of the intensity of the Bragg peaks resulting from the regular arrangement of the cylindrical pores.

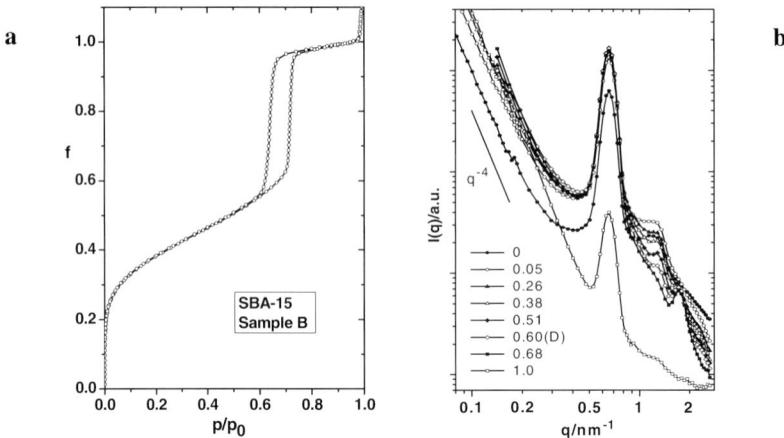

Fig. 1. Nitrogen adsorption in SBA-15 (77K): (a) Adsorption isotherm expressed as fractional pore filling $f$ vs. relative pressure $p/p_0$; (b) SANS curves $I(q)$ for a series of relative pressures $p/p_0$

## 2. EXPERIMENTAL

**Sample preparation and characterization.**  SBA-15 was synthesized by the method reported by Zhao et al. [8], using Pluronic P123 (BASF USA, Mount Olive, NJ) as the structure-directing agent and tetraethyl orthosilicate (TEOS) as the silica source. A 19.6 mL volume of HCl (36%) was added to a solution of 4g P123 in 105 mL Milli-Q water, and 9.2 mL of TEOS was added under vigorous stirring at 35°C. The polymer–silica composite formed as a fine precipitate was kept in the reaction solution at this temperature for 24 h under constant stirring, and then aged for 48 h at 80°C. The product was filtered, washed and dried, first at 105°C and then at 180°C for several hours and finally calcined at 550°C in air.

Structural features of the sample are summarized in Table 1. Nitrogen adsorption at 77 K was measured using a Micromeritics Gemini 2375 instrument. An adsorption isotherm, expressed by the fractional pore filling $f = v/v_p$ as a function of the relative pressure $p/p_0$, is shown in Fig. 1a. The pore radius $R$ was estimated from the ascending branch of the pore condensation, $(p/p_0)_a = 0.715$, and the Dollimore-Heal (DH) prescription of the film thickness, $t = 0.43[-5/\ln(p/p_0)_a]^{1/3}$. The resulting pore radius is significantly greater than the hydraulic radius of a cylinder $R_h = 2v_p/a_s$ as derived from the total pore volume $v_p$ (from the adsorbed amount at $p/p_0 = 0.95$) and the specific surface area $a_s$ (standard BET method). The parameter $\gamma = R/R_h$ represents a measure of the microporosity of the pore walls (see Discussion).

Small-angle x-ray diffraction was used to determine the lattice parameter $a_0$ of the 2D hexagonal pore system of the sample. A geometrical pore radius [2] $R_x$ estimated from the relation $R_x = (\sqrt{3}/2\pi)^{1/2}a_0[\rho_m v_p/(1+\rho_m v_p)]^{1/2}$ (with the matrix density $\rho_m = 2.16$ g cm$^{-3}$ from helium density measurements) is also given in Table 1.

Table 1.
Characterization of the SBA-15 sample by nitrogen adsorption and X-ray diffraction

| $v_p$ / cm$^3$ g$^{-1}$ | $a_s$ / m$^2$ g$^{-1}$ | $R$ / nm | $\gamma = Ra_s/2v_p$ | $a_0$ / nm | $R_x$ / nm |
|---|---|---|---|---|---|
| 1.098 | 993 | 3.91 | 1.77 | 10.74 | 4.7 |

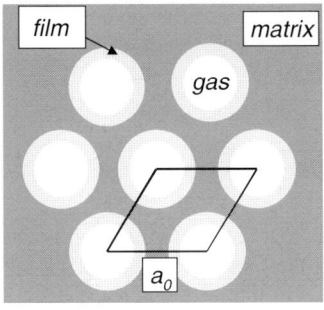 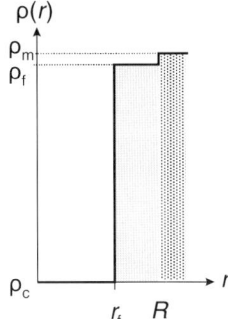

Fig. 2. 2D hexagonal packing of cylindrical pores coated with an adsorbed film (left); box model of the radial density profile $\rho(r)$ with core density $\rho_c$, film density $\rho_f$ and matrix density $\rho_m$ (right)

**Small-angle neutron scattering.** Measurements were made at instrument V4 of the Berlin Neutron Scattering Center (BENSC) using the SANSADSO set-up [9], which combines a standard gas adsorption experiment with small-angle neutron scattering. SANS measurements were made as outlined in ref. 6. Neutrons of a wave length $\lambda=0.605$ nm and a wavelength distribution $\Delta\lambda/\lambda = 0.12$ (fwhm), and sample-to-detector distances of 2, 4, and 16 m were used to cover a range of scattering vectors $q$ from 0.03 to 2.8 nm$^{-1}$. Scattering curves $I(q)$ for SBA-15 at a series of nitrogen pressures are shown in Fig. 1b.

## 3. THEORETICAL BACKGROUND

To analyse the scattering data we adopt a model of cylindrical mesopores on a 2D hexagonal lattice as sketched in Fig. 2. For such a system the scattering intensity can be written as

$$I(q) = |F(q)|^2 S(q) + I_s(q),$$ (1)

where the first term represents the Bragg scattering from the 2D ordered array of cylinders, with the form factor $F(q)$ and the structure factor $S(q)$, and the term $I_S(q)$ accounts for the diffuse scattering resulting from micropores and other inhomogeneities of the powder. For a perfect 2D hexagonal lattice the spherically averaged structure factor constitutes a sum of terms of type $m_{hk}L_{hk}(q)/q^2$, where $m_{hk}$ is the peak multiplicity ($m_{hk}=6$ for $h0$ and $hh$, and $m_{hk}=12$ otherwise), and $L_{hk}(q)$ is the delta function of the Bragg peaks at positions $q_{hk}$

$$q_{hk} = \frac{4\pi}{a\sqrt{3}}\sqrt{h^2 + k^2 + hk} \ .$$ (2)

The form factor of an infinite cylindrical pore of radius R imbedded in a uniform matrix is

$$F(q) = 2\pi(\rho_m - \rho_c)\frac{J_1(qR)}{qR},$$ (3)

where $\rho_m$ and $\rho_c$ represent the scattering length densities of the matrix and cylinder, and $J_1$ is the Bessel function of order 1. We assume that the nitrogen film adsorbed at the pore wall has a uniform thickness $t = R - r_f$ and uniform density $\rho_f$, and we take $\rho_c = 0$ for the vapour in the core of the cylindrical pores (Fig. 2). For this slab model the form factor is that of a core-shell cylinder with core radius $r = R - t$ and outer radius $R$, i.e.,

$$F(q) = 2\pi(\rho_m - \rho_c)\left[\alpha r^2 \frac{J_1(qr)}{qr} + (1-\alpha)R^2 \frac{J_1(qR)}{qR}\right], \tag{4}$$

with $\alpha = (\rho_f - \rho_c)/(\rho_m - \rho_c)$. In this work we assume that the scattering length density of the film is equal to that of the matrix ($\rho_f = \rho_m$), based on the fact that the scattering length density of liquid nitrogen at 77 K ($3.22 \times 10^{10}$ cm$^{-2}$) is similar to that of the silica matrix ($\rho_m = 3.43 \times 10^{10}$ cm$^{-2}$). With this simplification we have $\alpha = 1$ and thus Eq. (4) reduces to Eq. (3) with $R$ now replaced by $r_f$. The form factor $F(q)$ of a cylinder as given by Eq. (3) exhibits a series of sharp minima at positions $q_n r_f \approx n\pi + 0.6947$, and these minima are shifted to greater $q$ as $r_f$ decreases due to an increase of $t$. Whenever the position of one of these minima coincides with a Bragg peak, the intensity of this peak is strongly reduced. Accordingly, the intensity of the individual Bragg peaks is expected to vary in a characteristic way with the film thickness.

## 3. RESULTS

The adsorption isotherm in Fig. 1a exhibits characteristic features of the pore filling of well-ordered SBA-15 materials, viz., a steep initial increase of the adsorption isotherm, followed by an extended multilayer adsorption region and a sharp pore condensation step with a H1 type hysteresis loop. The SANS curves for the sample in the evacuated state and after equilibration with nitrogen at several gas pressures (Fig. 1b) exhibits the pronounced *(10)* peak at $q = 0.675$ nm$^{-1}$ and weaker peaks or shoulders in the $q$ regions corresponding to the *(11)* and *(20)* peaks (1.17 and 1.35 nm$^{-1}$), and of the *(21)* and *(30)* peaks (1.79 and 2.03 nm$^{-1}$), but these higher peaks are not resolved individually due to wave length broadening of the neutrons. Porod scattering, $I(q) \sim q^{-4}$, from the outer surface of the silica particles is dominating at low $q$ ($< 0.2$ nm$^{-1}$). As a general trend one observes that for given $q$ the scattering intensity increases with the filling fraction $f$ at low $q$ but decreases with increasing $f$ at the high-$q$ end of the experimental $q$ range (0.03 - 2.8 nm$^{-1}$). This inversion of the scattering intensity arises from the different length scales monitored in different $q$ ranges: In the Porod regime at low $q$, $I(q) \sim \rho_p^2$, where $\rho_p$ is the mean scattering length density of the entire particle, which increases as the pores are gradually filled with nitrogen. Scattering at high $q$ arises mainly from the inner pore walls. Adsorption of a nitrogen film at the pore walls reduces the effective internal area (due to the near-contrast match of liquid nitrogen with the silica matrix) and thus leads to a reduction of the intensity $I(q)$ with increasing loading. At complete pore filling ($f = 1$) the intensity of the *(10)* peak is reduced by almost two orders of magnitude but the peak is still clearly visible, indicating that the scattering length density of the pore liquid is not completely matching the scattering length density of the silica matrix.

    The scattering data were analysed to extract the intensities of the Bragg peaks as a function of the degree of pore filling. This was done by subtracting a smooth background from the experimental scattering curves, viz., $I_B(q) = I(q) - I_S(q)$, assuming that the background intensity $I_S(q)$ represents a linear function in a semi-logarithmic representation of the scattering data ($\log(I)$ *vs. q*), and that the width of the peaks is independent of pore filling. Fig. 3a shows the resulting Bragg peak functions $I_B(q)$ for all experimental pressures. Due to the wavelength broadening the pairs of Bragg peaks *(11)*&*(20)*, *(21)*&*(30)*, and *(22)*&*(31)* could

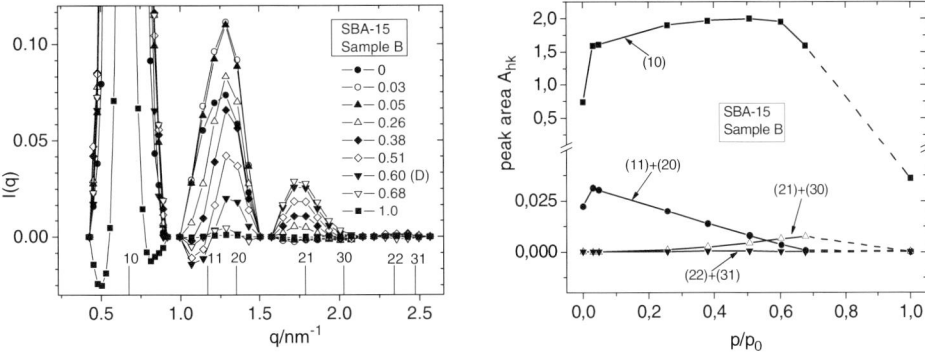

Fig. 3. (a) Bragg peaks for SBA-15 in vacuum and with adsorbed nitrogen film at a series of relative pressures $p/p_0$ (left); (b) peak areas $A_{hk}$ of Bragg peaks or pairs of peaks as a function of $p/p_0$ (right)

not be separated. However, as shown in Fig. 3a, the shape of the combined $(11)$&$(20)$ peak changes in a systematic manner with increasing gas pressure, in a way that the $(11)$ reflex decreases more strongly than that of the $(20)$ peak as the pressure is increased. This trend is consistent with the simulation of the peak intensities as a function of film thickness. The occurrence of negative values of $I_B(q)$ is an artefact arising from the choice of the background function $I_S(q)$. However, the fact that negative values appear only on the low-q side of the combined peak and only at higher relative pressures brings out the systematic nature of the changes in the peak shape with pressure. The combined $(21)$&$(30)$ peak exhibits a pronounced increase in intensity with increasing pressure. Fig. 3a shows that this peak is mostly due to the $(21)$ Bragg reflex while the intensity of the $(30)$ reflex is very low, again in agreement with the simulated peak intensities based on Eqs. (2) and (3).

The dependence of the integrated peak area $A_{hk}$ on the relative pressure is shown in Fig. 3b for the $(10)$ peak and the pairs of higher peaks $(11)+(20)$, $(21)+(30)$ and $(22)+(31)$. It is seen that the leading peaks exhibit a pronounced increase in intensity from the evacuated sample to $p/p_0 = 0.03$, i.e., in the pressure range in which the degree of pore filling increases sharply to $f = 0.26$. This effect can be explained by an increased scattering contrast $(\rho_m - \rho_c)$ due to adsorption of nitrogen into the microporous corona of the mesopores (see below). Above $p/p_0 = 0.03$ the peak area $A_{10}$ shows a further weak increase for pressures up to $p/p_0 = 0.6$, and then a decrease as the pore condensation pressure $p/p_0 = 0.715$ is approached.

Fig. 4a shows the *relative* peak areas (*relative intensities*) $A_{hk}/A_{10}$ plotted as a function of the pressure $p/p_0$ for the pairs $(11)+(20)$, $(21)+(30)$ and $(22)+(31)$. The relative intensity of the $(11)+(20)$ peak exhibits a monotonic decrease, while the relative intensity of the $(21)+(30)$ peak shows a monotonic increase with increasing $p/p_0$. The relative intensity of the $(22)+(31)$ peak is very small in the entire pressure range, but exhibits a weak maximum at $p/p_0 \approx 0.4$. Fig. 4b shows calculated values of the relative peak intensities as a function of the thickness $t$ of the adsorbed nitrogen film, based on Eqs. (1)-(3) with a pore radius $R = 4$ nm and the assumption of equal scattering length densities of matrix and adsorbed film. It is seen that this simulation reproduces the main trends of the experimental data of Fig. 4a. A comparison of the two sets of data on a quantitative basis allows us to determine the film thickness $t$ as a function of the pressure $p/p_0$. The resulting film thickness $t$ as derived from the experimental Bragg intensities on the basis of the simple model is shown in Fig. 5.

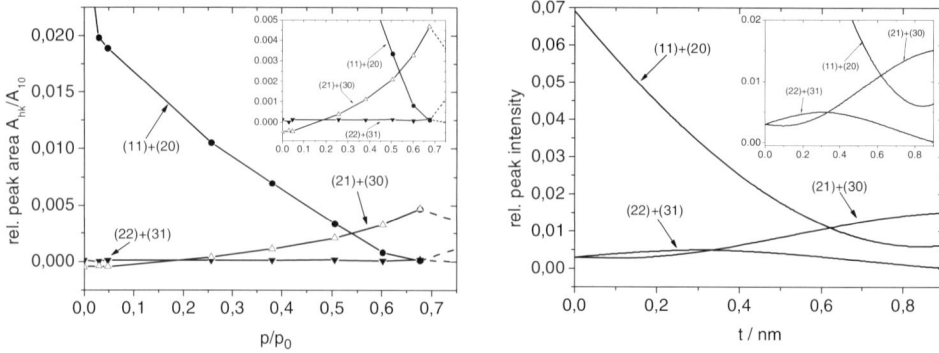

Fig. 4. Relative peak areas $A_{hk}/A_{10}$ for the pairs of peaks indicated in the graphs: Experimental results (a, left) and results of the simulation based on the form factor model with equal densities of matrix and pore liquid (b, right). The inset in the graphs shows a magnification of the region of low peak areas

## 4. DISCUSSION

For our SBA-15 sample it is found that the film thickness increases with pressure in an almost linear way over nearly the entire pressure range below pore condensation ($0 < p/p_0 < 0.65$). This behaviour contrasts with that of adsorbed films on smooth silica surfaces, at which one statistical monolayer ($t \approx 0.35$ nm) is reached already at low pressures while a further increase of $t$ occurs more gradually. The increase of $t$ above one statistical monolayer is commonly represented by $t$-layer models such as the Harkins-Jura (HJ) and the Dollimore-Heal (DH) relation which are shown in Fig. 5 by dashed curves. For nitrogen in SBA-15 these models clearly do not give a good representation of our experimental data.

The nearly linear increase of $t$ in a wide pressure range found in the SANS study contrasts with the steep increase of the filling fraction $f$ at relative pressures below $p/p_0 = 0.05$ (Fig. 1a). For cylindrical pores with smooth (uncorrugated) pore walls the film thickness is expected to increase as $t = R[1 - (1 - f)^{1/2}]$. For ideal cylindrical pores of radius 4 nm a fractional pore filling $f \approx 0.26$, as it is found for our sample at $p/p_0 = 0.03$, would correspond to a layer thickness $t \approx 0.54$ nm. Clearly this value is much larger than the layer thickness $t = (0.1 \pm 0.1)$ nm extracted from the Bragg intensities at $p/p_0 = 0.03$ (Fig. 5).

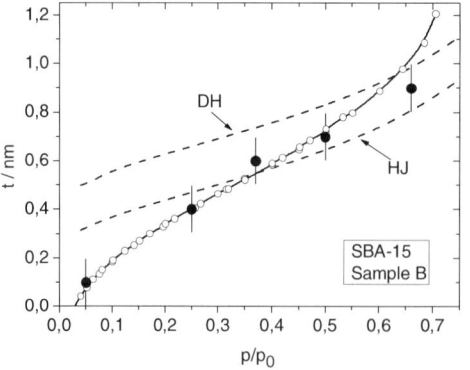

Fig. 5. Thickness $t$ of the adsorbed nitrogen film as derived from SANS (full symbols) and from the filling isotherm on the basis of Eq. (7) (open symbols and full line). The predictions of the Dollimore-Heal (DH) and Harkins-Jura (HJ) models are indicated by dashed lines

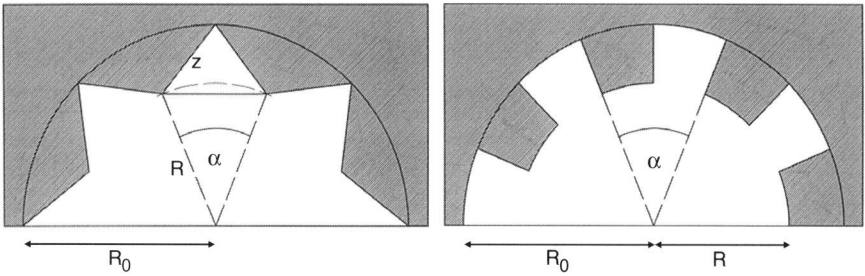

Fig. 6. Geometric models of corrugated pore walls of SBA-15: Saw-tooth profile (left) and rectangular profile (right): $R$ is the pore radius and $\Delta R = R_0 - R$ the profile depth of the surface (corona thickness)

The results for the film thickness can be reconciled with the experimental adsorption isotherm by assuming that initially nitrogen is adsorbed only into a microporous corona surrounding the cylindrical mesopores, and that an adsorbed film in the pores is formed only after complete filling of the corona. Based on a recent NMR study of the inner surface structure of SBA-15 [10] we may model a porous corona by corrugated pore walls. We adopt simple geometric models for a corona with linearly increasing mean density (saw-tooth profile), or a corona of uniform mean density (rectangular profile), as sketched in Fig. 6, where $\alpha$ is the centri-angle and $R_0$ the outer radius of the corona (profile depth of the corrugated pore wall). The model parameters $R_0$ and $\alpha$ (or the number of teeth $n = 360/\alpha$) are chosen such that the ratio of specific surface area $a_s$, pore radius $R$, and overall pore volume $v_p$, viz. the quantity $\gamma = a_s R / 2 v_p = R/R_h$, agrees with the respective value derived from the adsorption isotherm (Table 1). The model parameters are related to $\gamma$ by

$$\gamma = \frac{\sqrt{1 - 2(R_0 / R)\cos(\alpha / 2) + (R_0 / R)^2}}{[(R_0 / R) - \cos(\alpha / 2)]\sin(\alpha / 2) + (\sin \alpha)/2} \qquad \text{(saw-tooth profile)} \qquad (5)$$

$$\gamma = \frac{1 + (1 + n/\pi)\Delta R / 2R}{1 + \Delta R / R + 2(\Delta R / 2R)^2} \qquad \text{(rectangular profile)} \qquad (6)$$

where $\Delta R = R_0 - R$. To obtain a quantitative estimate of $R_0$ and $\alpha$ (or n) we assume that initially adsorption occurs only into the cavities of the pore wall until these cavities are filled at a fractional pore filling $f_0$. With $R = 4.0$ nm, $\gamma = 1.77$ (Table 1) and $f_0 = 0.264$ (corresponding to the overall filling fraction at $p/p_0 = 0.03$), Eq. (5) yields $\Delta R = 1.42$ nm with $\alpha \approx 19.5°$ (n = 18.44) for the saw-tooth profile, and Eq. (6) gives $\Delta R = 1.21$ nm with n = 25.6 for the rectangular profile. The latter value of $\Delta R$ agrees with the corona thickness obtained in the SAXD study of Imperor-Clerc et al. [7] for a similar SBA-15 sample, where a value $\Delta R = 1.2$ nm is reported for the uniform density model. Our assumption that up to a filling fraction $f_0$ adsorption leads solely to the filling of the porous corona is supported by the observed increase in peak intensities of the leading Bragg peaks at pressures up to $p/p_0 = 0.03$ (Fig. 3), since the filling of the voids will cause an increase of the mean scattering density of the corona, and thus an increased contrast against the pores. Only after complete filling of the cavities at fractional pore filling $f_0$ a film will develop at the pore wall. In terms of this model the film thickness will increase with the fractional pore filling as

$$t = R\left(1 - \sqrt{\frac{1-f}{1-f_0}}\right). \tag{7}$$

Fig. 5 shows that the film thickness $t$ derived from the gas adsorption data on the basis of this relation agrees with the experimental values of $t$ over nearly the entire range of relative pressures up to pore condensation.

In conclusion, the present work shows that the thickness $t$ of nitrogen adsorbed films and its dependence on the vapor pressure in cylindrical pores of a periodical mesoporous silica can be determined from the intensities of the leading Bragg peaks. Using neutron scattering for such studies has the advantage that the silica matrix and the adsorbed nitrogen film have similar scattering length densities, which facilitates the data analysis. On the other hand, the broad wavelength distribution of the neutron beam implies that not all Bragg peaks can be resolved separately by SANS. This is a serious disadvantage which limits the accuracy of the film thickness determination by neutron scattering. For the present sample the error in $t$ is estimated to ca. $\pm 0.1$ nm, but larger error bars ($\pm 0.2$ nm) were found for other samples. X-ray scattering is much preferable for such studies from this point of view, although it implies that nitrogen has to be replaced by a gas of higher electron density, such as Kr, Xe, dibromomethane or perfluorinated hydrocarbons. Recently, an X-ray small-angle diffraction study of the pore filling of SBA-15 with krypton has been performed [11] which supports the main conclusions of the present study, but also reveals significant differences in the pore filling of templated mesoporous silica with nitrogen and krypton.

**Acknowledgement**: This work was supported by the Deutsche Forschungsgemeinschaft through the Sonderforschungsbereich "Mesoscopically organized composites" (SFB 448).

## REFERENCES

[1]  S.J. Gregg, K.S. Sing, *Adsorption, Surface Area and Porosity*, Academic, London 1982
[2]  M. Kruk, M. Jaroniec, A. Sayari, *Langmuir* **13** (1997) 6267
[3]  W.H. Lawnik, U.D. Goepel, A.K. Klauk, G.H. Findenegg, *Langmuir* **11** (1995) 3075
[4]  W. Prange, W. Press, M. Tolan, C. Gutt, *Eur. Phys. J. E* **15** (2004) 13
[5]  V. Panella, J. Krim, *Phys. Rev. E* **49** (1994) 4179
[6]  B. Smarsly C. Göltner, M. Antonietti, W. Ruland, E. Hoinkis, *J. Phys. Chem. B* **105** (2001) 831
[7]  M. Impéror-Clerc, P. Davidson, A. Davidson, *J. Amer. Chem. Soc.* **122** (2000) 11925
[8]  D. Zhao, Q. Huo, J. Feng, B.F. Chmelka, G.D. Stucky, *J. Amer. Chem. Soc.* **120** (1998) 6024
[9]  E. Hoinkis, *Langmuir* **12** (1996) 4299
[10] I.G. Shenderovich, G. Buntkowsky, A. Schreiber, E. Gedat, S. Sharif, J. Albrecht, N.S. Golubev, G.H. Findenegg, H.-H. Limbach, *J. Phys. Chem. B* **107** (2003) 11924
[11] T. Hoffmann, D. Wallacher, P. Huber, R. Birringer, K. Knorr, A. Schreiber, G.H. Findenegg, *Phys. Rev. B*, accepted

Studies in Surface Science and Catalysis 160
P.L. Llewellyn, F. Rodriquez-Reinoso, J. Rouqerol and N. Seaton (Editors)

# Strong light scattering upon capillary condensation in silica aerogels

G. Reichenauer[a,b], J. Manara[b], H. Weinläder[b]

[a]Physikalisches Institut, Universität Würzburg, Am Hubland, 97074 Würzburg, Germany

[b]Bavarian Center for Applied Energy Research, Am Hubland, 97074 Würzburg, Germany

## 1. INTRODUCTION

Within the past decade several authors reported on the observation of light scattering in silica aerogels upon capillary adsorption or desorption of gases and vapours [1-5]. The effect has been observed for $LN_2$ at 125 K [1,2] as well as for sorption of 2-propanol at room temperature [5]. X-ray scattering of a high porosity (98%) silica aerogel partially filled with [4]He at 3.6 K indicates an increase of the size of the scattering entities with the amount of [4]He condensed; the length scale of the scatterers was found to be up to 60 nm [4]. In contrast, light scattering of a denser aerogel (porosity 87%) upon sorption revealed scattering entities on the order of 100 nm in addition to scattering from the outer sample surface [5].

The goal of this paper is to elucidate the mechanisms responsible for the scattering by analyzing the optical data in more detail.

## 2. EXPERIMENTAL

### 2.1. Sample

The silica aerogel investigated was a one-step base-catalyzed supercritically dried silica gel (TMOS). For details of the synthesis see e.g. references [5,6]. To increase the stiffness of the as-prepared sample and thus to avoid a significant volume change due to capillary forces [2] we sintered the silica aerogel to a density of about 0.29 $g/cm^3$ (initial density 0.26 $g/cm^3$).

For the characterization of the sample with optical and X-ray scattering methods a 1 mm thick slice was cut from the aerogel with a diamond saw. The adsorption was performed by placing the cuvette for the optical measurement holding the sample next to a reservoir of 2-propanol for several hours. The amount of 2-propanol adsorbed was determined by weighing. After each sorption step the sample was sealed in the cuvette.

### 2.2. Characterization of the porosity and the texture

The sample was characterized by nitrogen sorption at 77 K (ASAP 2000, software 2010, Micromeritics) to determine its total mesoporosity, the average pore size and the average size of the particles forming the silica network. Complementary, (ultra) small angle X-ray scattering ((U)SAXS) data were taken at the beam lines BW1 (JUSIFA) and BW4 (German synchrotron facility HASYLAB, Hamburg). The elastic modulus of the aerogel was determined via an ultrasonic run time measurement.

### 2.3. Optical characterization

The optical properties of the silica aerogel were investigated in a light scattering set-up (see [6]) that allows the detection of the intensity of the scattered light as a function of the scattering angle. Hereby the incident beam was unpolarized light with a wavelength of 543 nm. In addition, the direct-hemispherical transmittance and reflectance of the sample was measured with a VIS-IR spectrometer (Lambda 9 by Perkin-Elmer) combined with an integrating sphere [7, 8].

## 3. RESULTS

### 3.1. Structural properties of the aerogel investigated

The nitrogen sorption and U(SAXS) data are shown in Figs.1 and 2. In the SAXS-plot (Fig. 2) also the light scattering data are included to extend the range in scattering vector $q$. The nitrogen isotherm (Fig.1) shows a typical type IV shape with a H1 hysteresis [9]. The specific surface area, the total porosity and the average pore size (BJH) determined from the experimental data are 300 m$^2$/g, 3 cm$^3$/g and 55 nm, respectively.

The scattering data in Fig.2 exhibit different scattering regimes that also have been found for similar aerogels by other authors (e.g. [10]). Besides the characteristic scattering at $q >$ $10^{-2}$ Å$^{-1}$ that can be interpreted in terms of fractal clusters [11], an additional increase of the scattered intensity is visible at small $q$-values. Within the framework of this paper it is important to prove that this feature is not due to a significant number of large pores:

The porosity $\Phi$ as a function of the length scale $L$ can be expressed in terms of the integral over the intensity $I(q)$ times $q^2$ according to

$$\Phi(L) \ \propto \ \int_{q=1/L}^{q_{ref}=1/L_{ref}} q^2 \cdot I(q)dq \tag{1}$$

with $q_{ref}$ the scattering vector corresponding to an arbitrary reference length scale at which the sample possesses a porosity $\Phi_{Ref}$ (e.g. $L_{ref}$ =2Å corresponding to about $q_{ref}$ =0.5Å$^{-1}$). Since Fig.2 shows that the contribution to the integral (Eq.(1)) is negligible for scattering vectors $q \le$ $3 \cdot 10^{-3}$Å$^{-1}$ (corresponding to a length scale larger than about 300 Å) the porosity on a length scale of 300 Å and larger does not change any more. It can therefore be concluded that the forward scattering is not an indication of large pores, but can rather be explained in terms of small density fluctuations caused by the arrangement of large clusters (100-200 nm) with an aerogel substructure (see also AFM-images [10]).

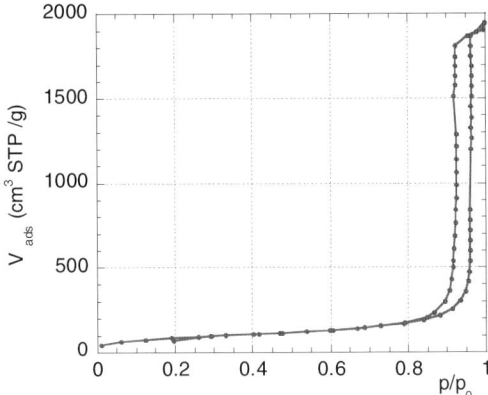

Fig.1. Nitrogen sorption isotherm taken at 77 K.

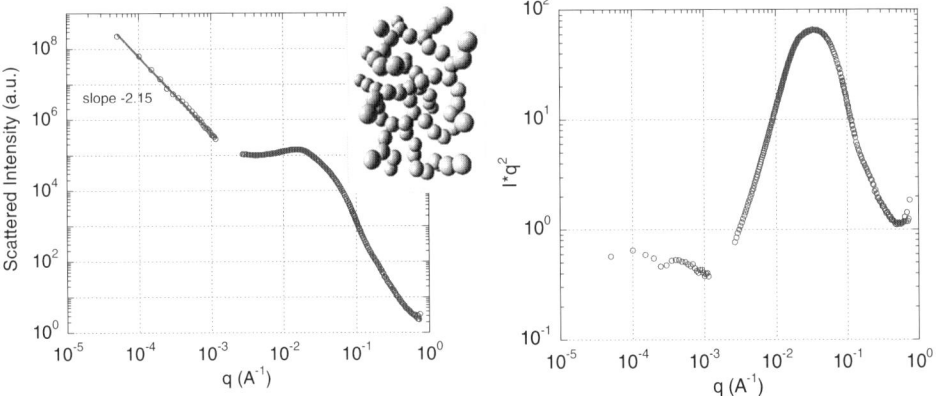

Fig.2. Left: Scattered intensity versus scattering vector; the data plotted represent the results from light scattering ($q < 10^{-3}$ Å$^{-1}$) and (U)SAXS. Right: Scattered intensity times $q^2$ vs. $q$- representation. The cartoon in the middle represents the model of an aerogel skeleton.

### 3.2. Optical measurements

The light scattering as a function of the scattering angle and the wavelength dependence of the direct-hemispherical transmittance of the silica aerogel are shown in Fig.3 for various amounts of 2-propanol absorbed.

The light scattering data $I_{light}$ were evaluated by a superposition of two "2-phase media" [12] contributions to account for scattering from the bulk (B) and the external surface (S) of the sample:

$$I_{light}(\theta) = [I(\theta, L_S) + I(\theta, L_B)] \cdot (e_{II} + e_{\perp} \cdot \cos^2(\theta)) \text{ , with} \tag{3a}$$

$$I(\theta, L_{S,B}) = a_{S,B} / (1 + (L_{S,B} \cdot q(\theta))^2)^2 \text{ and} \tag{3b}$$

$$q = 4\pi \cdot \sin(\theta/2)/\lambda , \tag{3c}$$

with $q$ the scattering vector. $\lambda$ and $\theta$ are the wavelength of the incident laser beam and the scattering angle, respectively. The parameter $a$ is a factor accounting for the intensity of the incident beam and the total scattering volume; $e_{II}$ and $e_\perp$ denote the polarization of the scattered light. $L_S$ and $L_B$ are the mean chord lengths characterizing the surface (roughness of the outer surface) and the bulk of the sample. For a statistically isotropic medium, the correlation length $L_B$ is a representative measure for of the extension of the inhomogeneities within a sample consisting of two phases with different electron density or dielectric constant; these properties are assumed to be constant within each phase. The results derived via a fit of the light scattering data (Fig.3) with a superposition of two 2-phase media models (Eq. (3a)) are given in Table 1.

Within the framework of this investigation the directional-hemispherical reflectance and transmittance were measured instead of the corresponding directional-directional quantities because directional-hemispherical data also contain information about the scattering of the incoming radiation. This is reflected by the fact that the directional-directional transmittance $T_{dd}$ decreases with $\exp(\tau_0)$, whereas the directional-hemispherical transmittance $T_{dh}$ is proportional to $\tau_0*^{-1}$ for optically thick samples; hereby is $\tau_0*$ the effective optical thickness. The effective values of the optical thickness $\tau_0^*$ and the albedo $\omega_0^*$ are related to the quantities $\tau_0$ and $\omega_0$ (that can be derived from direct-direct measurement) via the anisotropy factor $g$ [13]:

$$\tau^*(\lambda) = \tau(\lambda) \cdot (1 - \omega_0(\lambda) \cdot g), \qquad \omega_0^*(\lambda) = \frac{\omega_0(\lambda) \cdot (1-g)}{1 - \omega_0(\lambda) \cdot g} , \tag{4a}$$

$$\text{with } g = \int_{4\pi} \cos\vartheta' \cdot p(\vartheta', \vartheta) d\Omega \tag{4b}$$

the probability for radiation coming in at an angle $\vartheta$' to be scattered in the direction defined by the angle $\vartheta$. The values of the anisotropy factor $g$ can range from $-1$ (in case of pure backward scattering) to $+1$ (in case of a delta function like forward scattering). The effective scattering and the effective extinction coefficients $S^*$ and $E^*$, respectively, can be determined from $\tau_0^*$ and $\omega_0^*$ via

$$E^*(\lambda) = \tau_0^*(\lambda) \cdot D \quad and \quad S^*(\lambda) = \omega_0^*(\lambda) \cdot E^* , \tag{5}$$

with $D$ the sample thickness. The spectral variation of the effective scattering coefficient $S^*$ is affected by the pore structure of the silica aerogel, i.e. the size and shape of the pores and porosity $\Pi$ of the sample. The effective scattering coefficient

$$S^*(\lambda) = \frac{3}{2} \cdot \Pi \cdot \frac{I_{scattered}(\lambda)}{I_{incident} \cdot d} \quad . \tag{6}$$

can be calculated via the Mie-theory [15]. The effective scattering coefficient $S^*$ determined from the experimental data (Fig.4) was fitted with an equation derived from Mie-theory [14-17] assuming individual spherical scatters [7] and two different indices of refraction, i.e. $n \sim 1.45$ for silica and 2-propanol and $n = 1$ for the air in the pores, respectively. Hereby we presumed that the volume of the scattering entities within a logarithmic diameter interval $\Delta \ln d$ is distributed according to a logarithmic normal distribution [14]:

$$f(d)\Delta \ln d = \frac{1}{\sqrt{2\pi}\sigma_g} \exp\left[ -\frac{(\ln d - \ln d_M)^2}{2\sigma_g^2} \right] \Delta \ln d \tag{7}$$

The results of the fit are given in Table 1. It has to be noted that the scattering coefficient (Fig.4) shows an offset in the range of low absorption ($\lambda < 2000$ nm). This part has not been included in the fit. It is likely to originate from the larger scattering entities which have been found via the evaluation of the light scattering profile (Table 1).

Table 1
Results derived from the light scattering data (vs. scattering angle) and the direct-hemispherical spectra as described in the text. $L_{surface}$ and $L_{bulk}$ are the mean chord lengths of the entities causing scattering from the outer surface and the bulk of the sample, respectively. (* different sample, however taken from the same batch)

| | from light scattering | | from direct-hemispherical data |
| --- | --- | --- | --- |
| pore filling (pore vol %) | $L_{surface}$ (nm) | $L_{bulk}$ (nm) | average diameter $d$ (nm) |
| 0 | nd | 119* | nd |
| 4 | 517 | 94 | 13 |
| 6 | 640 | 99 | 20 |
| 30 | 641 | (61) | 30 |

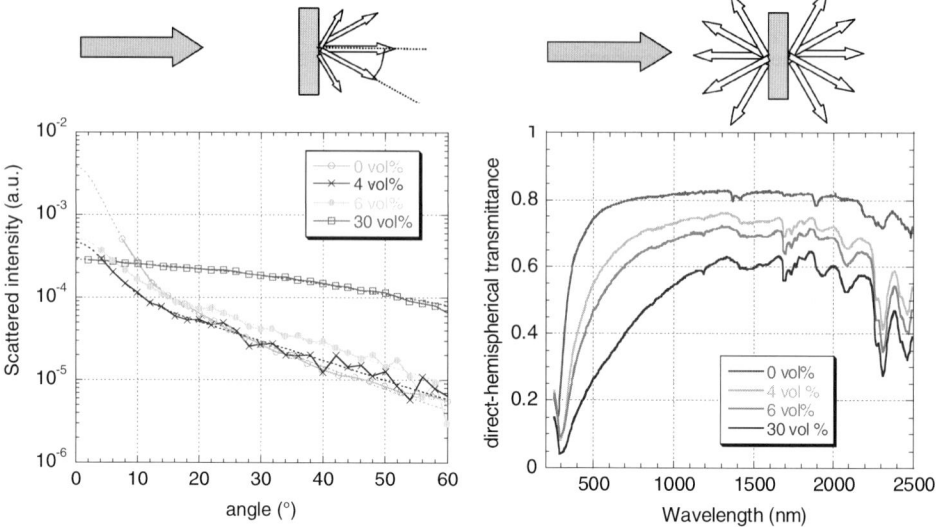

Fig.3. Left: Scattered intensity versus scattering angle for the silica aerogel with different amounts of 2-propanol adsorbed (incident beam: λ = 543 nm). The lines represent fit of the data by a two-phase model [5]. Right: direct-hemispherical transmittance for the same set of samples. The transmittance decreases with increasing amount of 2-propanol adsorbed. The two different experimental configurations are indicated by the sketches above the plots.

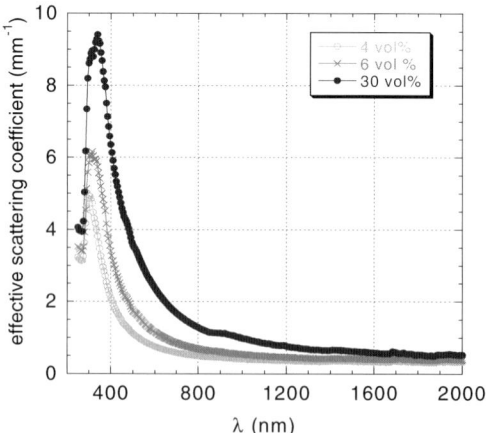

Fig.4. Effective scattering coefficient for the silica aerogel with different amounts of 2-propanol adsorbed.

## 4. DISCUSSION

The light scattering data for the silica aerogel with different amounts of 2-propanol adsorbed reveal three characteristic length scales (Table 1):
- according to Wang et al. [6] the largest scattering entities ($L_{surface}$~ 500 nm) can be attributed to scattering from the rough outer surface of the sample (caused by the sawing of the sample)
- the two other length scales of about 100 and 10 to 30 nm are related to scattering from the bulk of the silica aerogel.

The mean chord length $L_{bulk}$ is almost constant except for the sample with 30 vol.% filling. In the latter case, however, the scattering is so strong that significant multiple scattering effects are to be expected, which will yield a broadening of the scattering profile vs. scattering angle and thus mock a decrease of $L_{bulk}$. If we therefore do not include the 30 vol. % value in our discussion, we have to conclude that $L_{bulk}$ does not vary significantly with the amount of 2-propanol adsorbed.

The average diameter $d$, in contrast, clearly increases with increasing filling of the pores; this effect is qualitatively in accordance to the observations of Lurio et al. for the X-ray scattering of a LHe filled silica aerogel [4] with a higher porosity. One effect, however, is critical for a quantitative interpretation of our data: since the sample was not continuously kept at constant temperature, part of the optical measurements might have been taken from a sample that is in the process of desorption rather than of adsorption yielding larger scattering entities [4]. However, we believe that the experiments still provide a first glimpse at the effects involved, that is two types of scatterers being present in the bulk:
- small entities increasing with pore filling that can be interpreted as (partially) filled pores or pores of increasing size being filled
- large scatterers that do not significantly change in size but that still contribute increasingly to the extinction with increasing pore filling; these entities can be related to large aerogel clusters whose scattering contrast is increasing in the early stage of capillary condensation with the amount of 2-propanol adsorbed.

## 5. CONCLUSIONS

Although the experiments presented are crude in terms of controlling the sorption of 2-propanol within the sample, they indicate that the light scattering of a silica aerogel partially filled with condensed vapour has two components, i.e. scattering of small density fluctuations on a length scale of about 100 nm and scattering of liquid filled pores.
Currently, a new temperature controlled set-up is being tested that allows for simultaneous measurements of sorption and light scattering under well controlled conditions.

# REFERENCES

[1] A.P.Y. Wong, S.B. Kim, W.I. Goldburg, M.H.W. Chan, Phys. Rev. Lett. 70 (1993) 954.

[2] G.W. Scherer, D. M. Smith, D. Stein, J. Non.-Cryst. Solids 186 (1995) 314.

[3] A. Bisson, E.Rodier, A. Rogacci, D. Lecomte, P. Achard, J. Non.-Cryst. Solids 350 (2004) 230.

[4] L.B. Lurio, N. Mulders, M. Paetkau, M. Lee, S.G.J. Mochri, M.H.W. Chan, J. of Low Temperature Physics, 121 (2000) 591.

[5] G. Reichenauer, J. Fricke, J. Manara, J. Henkel, J. Non.-Cryst. Solids 350 (2004) 364.

[6] P. Wang, W. Körner, A. Emmerling, A. Beck, J. Kuhn, J. Fricke, J. Non.-Cryst. Solids 145 (1992) 141.

[7] J. Manara, R. Caps, F. Rather, J. Fricke, Optics Communications 168 (1999) 237.

[8] J. Manara, R. Caps, J. Fricke, International Journal of Thermophysics 26 (2005) 531.

[9] F. Rouquerol, J. Rouquerol, K. Sing, Adsorption by Powders and Porous Solids (Academic Press 1999).

[10] C. Marliere, F. Despetis, P. Etiennne, T. Woignier, P. Dieudonne, J. Phalippou, J. Non.-Cryst. Solids 285 (2001)     148

[11] A. Emmerling, R. Petricevic, A. Beck, P. Wang, H. Scheller, J. Fricke, J. Non- Cryst. Solids 185 (1995) 240.

[12] P. Debye, H. .R. Anderson, H. Brumberger, J. of Appl. Phys. 285 (1957) 679.

[13] B.H.J. McKellar and M.A. Box, J. Atmosph. Sci. 38 (1981) 1063.

[14] G. Mie, Annalen der Physik 25 (1908) 377.

[15] C.F. Bohren, D.R. Huffmann, Absorption and Scattering of Light by Small Particles (John Wiley & Sons 1983).

[16] M. Kerker, The Scattering of Light and Other Electromagnetic Radiation (Academic Press 1969).

[17] H.C. van de Hulst, Light Scattering by Small Particles (Dover Publications Inc.1981).

Studies in Surface Science and Catalysis 160
P.L. Llewellyn, F. Rodriquez-Reinoso, J. Rouqerol and N. Seaton (Editors)

# Characterisation of porous solids from nanometer to micrometer range by capillary condensation

**R. Denoyel, M. Barrande and I. Beurroies**

MADIREL, CNRS-Université de Provence, Centre de St Jérôme, 13397 Marseille cedex 20, France

A new experimental procedure based on the isothermal desorption of vapour is proposed to extend the domain of characterisation of porous solids and powders by capillary condensation until the macropore range. The set-up is based on the use of a Tian-Calvet type microcalorimeter that insures a full control of temperature gradients around the sample and allows the desorption isotherm to be determined very close to the saturation pressure. The principle of the experiment is described and the first results obtained for water desorption are compared to measurements based on gravimetry as well as to pore size distributions obtained by mercury porosimetry.

## 1. INTRODUCTION

The determination of pore size distributions by gas adsorption-desorption methods is probably the most commonly used method but it is admitted to be reliable only in the micro-mesopore size range (0,3-50nm) [1]. For larger pore sizes, mercury porosimetry is a useful complementary method, notably because of its unique ability to determine pore size distributions at several scales from nm to several hundred micrometers. Unfortunately, this method is destructive for many samples that are more or less damaged by the high pressures that are required by such an experiment. Moreover, the sample, even if not damaged, cannot be recovered in many cases and this technique will be probably faced to more and more drastic rules in the future for environmental reasons. Another technique, thermoporometry, can also be used for pore size characterisation, with more or less the same range of pore size as gas adsorption [2]. Its advantage relies on its possible use for samples that cannot be dried without damage. Its disadvantages are mainly due to the fact that melting-solidification in confined medium is less well understood than capillary condensation, despite clear similarities [3]. Indeed, calculation procedures and models are not as well established and useful parameters are not directly measurable. Clearly there is a need for characterisation methods which cumulate the advantages of gas adsorption (adsorbate with low interfacial tension, well defined models with measurable parameters), mercury porosimetry (pore size determination at several scales) and thermoporometry (starting from wet state). Because of the advantages quoted above, a method based on capillary condensation, i.e. adsorption or desorption of a vapour, would be interesting provided it could be applied in a larger range of pore width. The main difficulty is that for the largest accessible pore sizes (above 50nm) the capillary condensation or evaporation occurs at pressures that are very close to the saturation pressure of the adsorbate. This leads to technical difficulties that have to be overcome, notably by controlling as much as possible the temperature gradients in the experimental system.

In this paper, we present a new vapour desorption set-up based on a microcalorimeter that provides a full control of temperature gradients. We demonstrate that is possible to carry out capillary condensation studies in the nanometer-micrometer range. In a first part the set-up and the procedure established to study the desorption of a vapour from a porous solid that is initially immersed in an excess of liquid are described. In the second part a few results are shown and compared to existent techniques.

## 2. EXPERIMENTAL ASPECTS

The major difficulty with standard apparatus is that working close to the saturation pressure may lead to instabilities because of a lack of temperature control. This problem is illustrated in table 1, where the relative pressure at which capillary evaporation should occurs is calculated by the Kelvin equation as a function of Kelvin radius. This calculation is made for water at 25°C. From the relative pressure, the equilibrium pressure can be calculated. From the equilibrium pressure the temperature at which this equilibrium pressure becomes saturating is also shown. If the data for a pore size around 10μm are considered, one can observe that the equilibrium pressure should be 23.754 torrs, which corresponds to a saturation temperature of 24.988°C. It means that a temperature fluctuation of 0.012°C is sufficient to create a cold point in the system leading to condensation outside the pore system. It means also that a temperature control better than $10^{-4}$K is needed for a safe measurement of the relative pressure in that range with a sensitivity better than $10^{-4}$.

| Radius | P/P° | P (torrs) | T=f(P) (°C) | T°-T (°C) |
|--------|------|-----------|-------------|-----------|
| 10nm   | 0,9002 | 21,386  | 23,315      | 1,711     |
| 100nm  | 0,98955 | 23,508 | 24,814      | 0,185     |
| 1μm    | 0,99895 | 23,731 | 24,972      | 0,027     |
| 10μm   | 0,99989 | 23,754 | 24,988      | 0,012     |

Table 1: relative pressure P/P° is calculated by the Kelvin equation; P is the equilibrium pressure; corresponding saturation temperature T is obtained from tabulated data (Handbook of Physical Chemistry), T°=25°C. P°=23,775 torrs at 25°C.

### 2.1 Set-up
The order of magnitude of temperature gradients that are maintained in the isothermal block of Tian-Calvet type calorimeters is suitable for the requirements quoted above. Indeed, even if the absolute temperature of the isothermal block may shift with time, the difference of temperature between the reference and the sample cell that are inside the oppositely connected thermopiles is better than 0.0001K. The schema of the set-up is presented in figure 1. The porous solid or the powder are in the sample cell. The reference cell contains the pure liquid. The two cells are connected through two valves (1 and 2) to a manifold equipped with an absolute pressure transducer (100torrs Baratron MKS). The manifold is connected to a vacuum pump through a capillary. Controlled leak valves may also be used. Between the

tubes coming from the two cells, a differential pressure transducer (±1torrs Baratron MKS) allows to determine the difference of pressure between the reference and the sample cell. The calorimeter is at 25°C whereas all the other parts of the system are in a box thermostated at 35°C. The temperature must regularly increase from the calorimeter block to the manifold. The signal coming from the thermopiles is recorded as a function of time together with the total and the differential pressures.

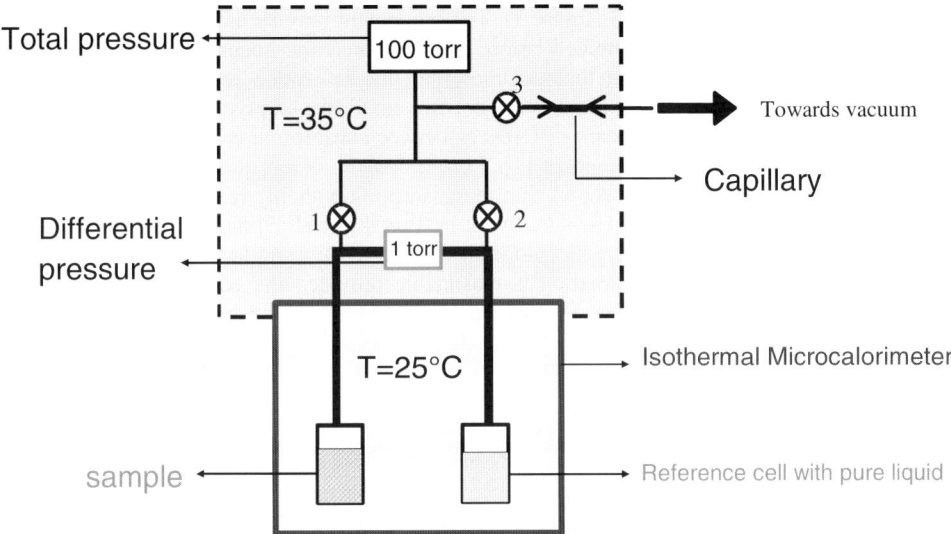

Figure 1: simplified schema of the set-up used for studying vapour desorption

## 2.2 Procedure

A continuous quasi-equilibrium procedure is used. In the initial state, the sample cell contains the solid with an excess of liquid, whereas the reference cell contains the pure liquid. The amount of solid needed is generally lower than 50mg. Both cells are at atmospheric pressure and valves 1 and 2 are open. The system is slowly evacuated through the capillary. The total pressure is recorded versus time. When the saturation pressure is reached the total pressure stays constant as a function of time (see figure 2). In this first step vaporisation is observed both in reference and sample cell whereas air is eliminated from the system. The valve 2 above the reference cell is then closed. Total pressure over the sample, differential pressure and heat flow are then recorded as a function of time. From this time, indicated by an arrow in figure 2, a stationary state is reached in the system which corresponds to vaporisation of the excess liquid inside the sample cell, whereas nothing occurs in the reference cell. There is a step clearly visible on the differential transducer recording which comes from the fact the sample cell is slightly colder than the reference cell due to endothermic vaporisation of the liquid. This stationary state is chosen to determine the value of the transducer signals that corresponds to $P/P°=1$ inside the sample cell. When the excess of liquid is evaporated the pressure decrease above the sample corresponds to the desorption of the vapour from the pore structure. The differential transducer (1torr full scale) allows determining relative pressures that are very close to 1 (between 1 and 0.95 for water at 25°C), whereas the absolute transducer gives access to the full range 0-1 but is less accurate in the high relative pressure part. Notably fluctuations of absolute pressure may be observed due to a variation of the calorimeter absolute temperature, whereas the differential pressure stays constant because the temperature gradients are negligible inside the isothermal block.

Simultaneously, the heat flow that is recorded as a function of time is used to calculate the amount of water that was eliminated from the sample cell. By integrating the heat flow versus time and dividing the result by the vaporisation enthalpy of the liquid the amount or the mass desorbed can be easily calculated (figure 3). The agreement may be as good as 1% with the value determined by weighting the cell before and after the experiment, despite the fact that this method is an approximation that assumes that the desorption energy is equal to the vaporisation energy of the liquid. If pores are not two small, the emptying of the pore is close to a vaporisation process. Significant differences between adsorption energy and liquefaction energy are observed only for nanopores [4]. Concerning the desorption from the multilayer, adsorption calorimetry generally shows that the adsorption enthalpy reaches the liquefaction energy around the monolayer whereas the work by Partyka et al [5] has shown that after 1.5 layer, the adsorbed film has energetic properties that are very similar to those of the bulk liquid. Consequently the approximation is reasonable provided the equilibrium pressure is above 0.2 (just above the monolayer for many systems). More refined analysis will be done in the future for lower pressures.

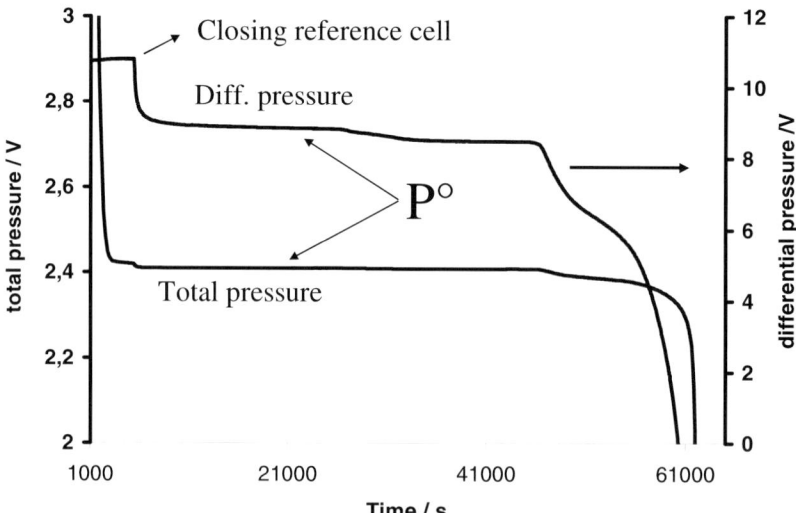

Figure 2: recordings of the signal of total and differential pressure transducers versus time

Finally, knowing mass loss and quasi-equilibrium relative pressure versus time, the desorption isotherm can be plotted as shown in figure 4 for a typical porous powder.

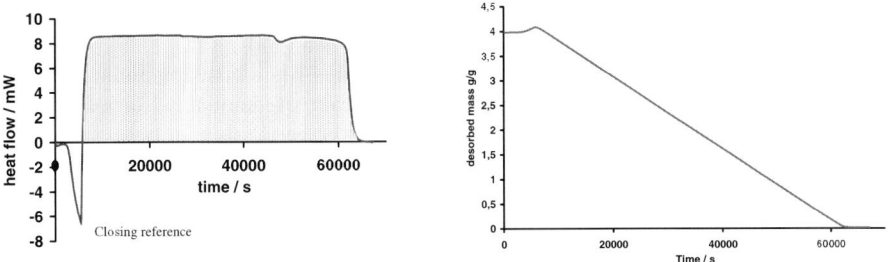

Figure 3: heat flow versus time (left) that allows the calculation of desorbed mass versus time (right)

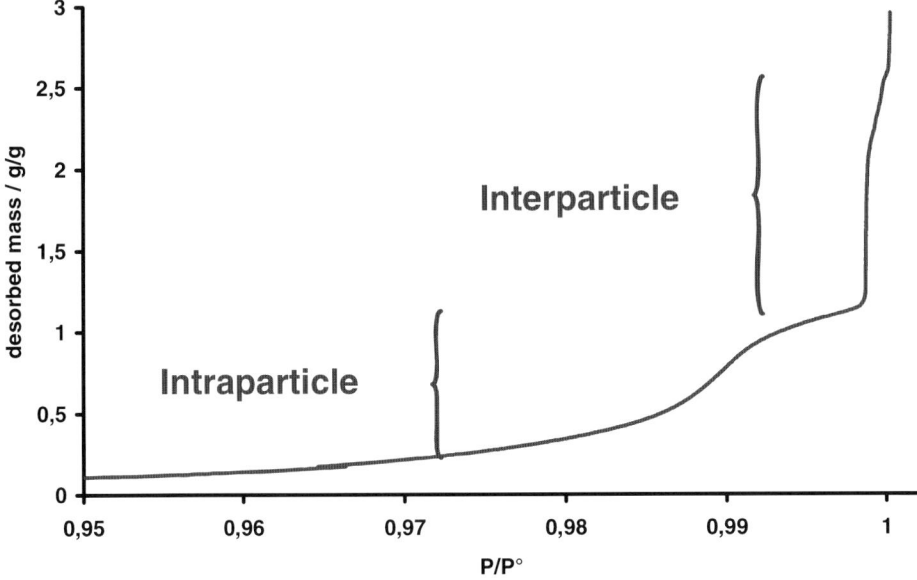

Figure 4: typical desorption isotherm of water from a porous powder

## 3. RESULTS AND DISCUSSION

Characterisation methods based on water desorption methods already exist. For example thermogravimetry [6] or Controlled Rate Thermal Analysis [7] have been developed but the range of pore size they measure is at most mesopores. In the present paper examples are only given for water desorption, but any pure liquid can be used. The desorption isotherm of figure 4 is typical of what was obtained with many porous particles, including soft gels. It is very similar to what is commonly observed with mercury porosimetry: a firs step corresponding to interparticle void volume emptying (instead of filling for mercury) followed by a step corresponding to the emptying of the pores that are inside the particles. In general, a reasonable agreement is obtained between the pore volumes (inter or intra-particles) obtained by mercury porosimetry and water desorption. The inter-particle volume may differ due to different compacting behaviours under the action of the two fluids.

This good agreement at the level of pore volume analysis is a first important result which shows that the experiment is quantitative on the scale of amount adsorbed and allows porosities to be determined. Nevertheless, if a characterisation in term of pore size analysis is wished, the experiment must be also quantitative on the scale of pressure, i.e. this is the equilibrium desorption isotherm that is actually measured. In figure 5, we compare the desorption isotherm of water from a mesoporous silica powder (particle size 12µm, mean pore size 6nm) with that obtained with a standard gravimetric apparatus [8]. The agreement is very good in the range of relative pressures that are accessible to the gravimetric apparatus, which validates our technique. Moreover, as shown in the zoom of figure 5, this new technique gives also an analysis of the pores corresponding to the bed of particles. The pore size distribution calculated by using the BJH equation [9] is shown in figure 6.

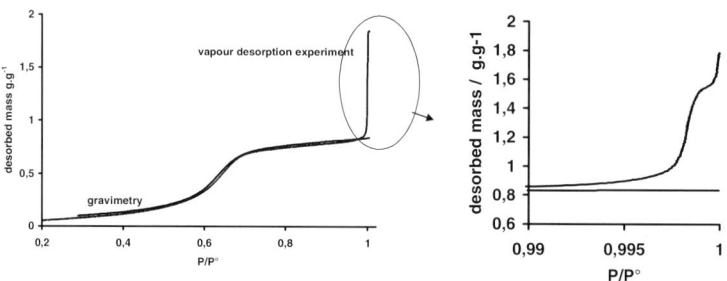

Figure 5: comparison between water desorption experiment and gravimetry

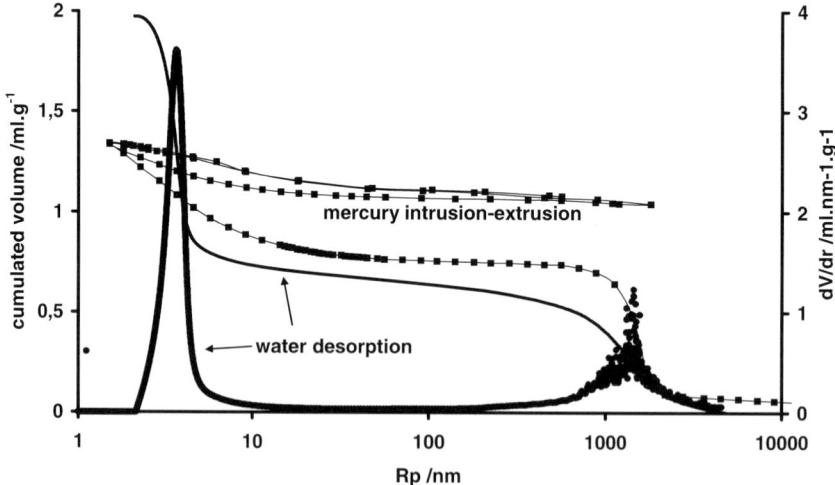

Figure 6: pore size distributions obtained by water desorption compared with intrusion-extrusion of mercury.

The equation used for the multilayer versus pressure was of the Halsey type [10] and fitted with data of water desorption in a range of pressure where no pore emptying is observed. This figure shows a pore size distribution from the nanometer to the micrometer range based on the Kelvin equation. The comparison with mercury intrusion, which is presented in the same figure, shows a very good agreement for the inter-particle range, whereas some probems are encountered with mercury in the nanometer range where the sample is probably damaged. In the case of the Kelvin equation, $72.3mJ/m^2$ was used for the water surface tension value, whereas $485mJ.m^2$ were used for mercury in the Washburn equation. In figure 7 another example is given for a silica sample that is made of porous particles with macropores around 200nm and an average particle size above 40µm. The agreement is good between the two methods for the intra-particle pores, whereas the inter-particle pore size seems to be underestimated by the desorption method whatever the desorption rate. Indeed, the larger are the pores the closer is the relative pressure from 1 and the larger is the difference between measured and equilibrium pressure at a given desorption rate. This problem is now under study in order to find procedures that are quantitative as well as completed in a reasonable duration. The experiments shown here are obtained for desorption rates around 1mg per hour. In these conditions the limit of the technique in term of pore size measurement is around 5µm.

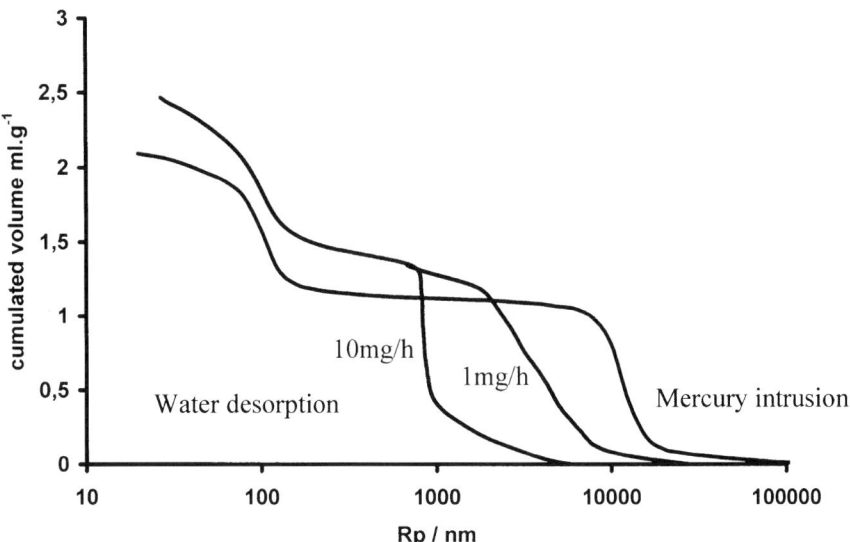

Figure 7: comparison between mercury porosimetry and water desorption at two different desorption rates for a silica powder with particle sizes in he range 40-100µm and pore radius around 100nm.

## 4. CONCLUSION

The new experiment presented here allows the desorption isotherm of a pure liquid from a porous solid to be determined from P/P°=1 to 0. The agreement with other techniques is very good in the ranges where this comparison is possible. The quantitative determination of porosities is possible for particles until 100µm, whereas pore size distributions can be calculated for pore sizes below 5µm. This method open a number of new possible studies for porous solids that are two weak to be studied by mercury porosimetry but may accept the capillary stress due to common liquids. The wettability of powders could be also studied by comparing the desorption of liquids having different surface tensions. On a fundamental point of view, it will open the opportunity to test models at several scales.

**Acknowledgements**: this work is supported by funding under the Sixth Research Framework Programme of the European Union, Project AIMs, No. NMP3-CT-2004-500160.

## REFERENCES

[1] F. Rouquerol, J. Rouquerol and K.S.W. Sing, *Adsorption by Powders and Porous Solids*, Academic Press, London, **1999**.
[2] M. Brun, A.Lallemand, J.F. Quinson, C. Eyraud, Thermochimica Acta, 21 (1977) 59-88.
[3] I. Beurroies, R. Denoyel, P. Llewellyn and J. Rouquerol Thermochimica Acta, 421 (2004) 11-18 .
[4] N. Tanchoux, P. Trens, D Maldonado, F. Di Renzo and F. Fajula, Colloids and Surfaces A: Physicochemical and Engineering Aspects, 246 (2004) 1-8.
[5] S. Partyka, F. Rouquerol and J. Rouquerol, Journal of Colloid and Interface Science, 68 (1979) 21-31.
[6] J. Goworek, W. Stefaniak, Instrumentation Science & Technology, 25 (1997) 97-105.
[7] V. Chevrot, P. L. Llewellyn, F. Rouquerol, J. Godlewski and J. Rouquerol, Thermochimica Acta, 360 (2000) 77-83.
[8] J. Rouquerol and L. Davy, Thermochimica Acta, 24 (1978) 391-397.
[9] E.P. Barrett, L.G. Joyner and P.H. Hallenda, J. Am. Chem. Soc., 73 (1951) 373
[10] G. D. Halsey, J. Chem. Phys., 16 (1948) 93.

Studies in Surface Science and Catalysis 160
P.L. Llewellyn, F. Rodriquez-Reinoso, J. Rouqerol and N. Seaton (Editors)
© 2007 Elsevier B.V. All rights reserved

# Characterization of zeolite membrane quality by using permporosimtrey

**Kazuyuki Nakai[a], Kaori Nakamura[a], Jun Kaneshiro[b] and Masahiko Matukata[b,c]**

[a]BEL Japan Inc. 2-11-27 Shin-kitano Yodogawaku, Osaka 532-0025, Japan

[b]Department of Applied Chemistry, Waseda University, Okubo, Shinjuku-ku, Tokyo 169-8555, Japan

[c]Advanced Research Institute for Science and Engineering, Waseda University

Silicalite membranes, supported on porous alumina tube, have been prepared by two-step hydrothermal synthesis. The pore structure and zeolite of membrane layer has been characterized by XRD, SEM and nano-permporosimetry technique. The permeabilities of several vapors have been measured together with vapor mixture separation behavior.

## 1. INTRODUCTION

Zeolite membrane is useful for separation of hydrocarbon isomers, water/organic solvent mixtures, etc [2]. These separations are performed through inherent micropores of zeolite (typically 0.3-0.7 nm). Checking quality of membrane such as cracks, intercrystalline boundaries and defects of crystal is important for these applications. Pore volumes of these defects are usually insufficient to determine their pore sizes by conventional nitrogen gas adsorption and mercury porosimetry technique. One of useful methods for this characterization is the permporosimetry, of which the principle is to evaluate pore size by the Washburn equation on the basis of the difference in permeabilities between dry and wet samples during pressure rising of gas. This method allows us to evaluate pore sizes in the range of 50 nm to 300 $\mu$ m (pressure applied is from 0 to 10 bar). On the other hand, in the research of zeolite membrane we need to evaluate pores in the micro- and mesopore regions (<50 nm). We will primarily focus on a technique called nano-permporosimetry that allows measurement of transport through the inherent pores of zeolite in membrane, as well as quantification of bypass flow through defects.

## 2. EXPERIMENTAL

### 2.1. Hydrothermal synthesis of silicalite membrane

Tetrapropylammoniumhydroxide(TPAOH, Aldrich, 1.0M solution) 50ml has added NaOH 13g and stirred under 80°C and 10min. And then 13g of fumed silica has add into this solution. After 10min. stir, this solution was aged for 80°C and 2 hour. This mixture molar composition is $10SiO_2:2.4TPAOH:1.0NaOH:110H_2O$. Hydrothermal synthesis were carried out by heating at 135°C, 4 hour and 24 rpm. After crystallization of silicalite seed (Fig.1) were separated by centrifugation at 10000 rpm, 1.5 hour. This separation was carried two times. The slurry of seed crystal has prepared 2.3 wt% by adding distilled water.

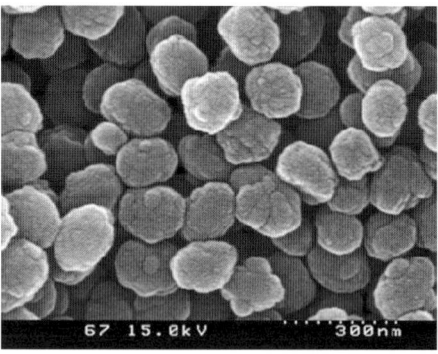

Fig. 1. SEM photo of silicalite seed

$\alpha$-alumina support tube (O.D. 10 mm, Length 30 mm, pore size 0.1 $\mu$ m) was immersed into this slurry, and then silicalite seed crystal dip coated on the alumina surface at 30 min and room temperature. After that slurry surface was moved 3.0 cm s$^{-1}$ speed for coating. After coating, tube has dried at 100°C. These dip coating procedure has carried 3 times.

Separation layer synthesis (2$^{nd}$ crystal growing up) were carried out by heating the synthesis mixture ($25SiO_2:3TPAOH:1500H_2O$) with the support in Teflon lined stainless steel autoclaves at different temperature at 125°C, 24 hours. After crystallization of Silicalite membranes, the template of TPAOH was removed by calcinated at 525°C. We got two different separation membrane A and B.

### 2.2. Characterization of membrane

Silicalite membrane were characterized XRD (RINT2000, Cu target, Rigaku Co), SEM (CC-SEM, Hitachi science systems Co.). XRD measurement has done from vertical direction of Silicalite layer of tube.

### 2.3. Mixed vapor permeation measurement

Mixed vapor permeation has measured at 100, 200 and 300°C by Fig.2 apparatus. Carrier gas has used Ar with 3%CH$_4$. Analysis vapor (n-hexane, 2-Metyl pentane and 2,2-Dimetyl butane) has dosed liquid feed pump and then vaporized. Each solvent has mixed same volume. And then each partial pressure was c.a. 17kPa. Permeated vapor has analyzed FID GC.

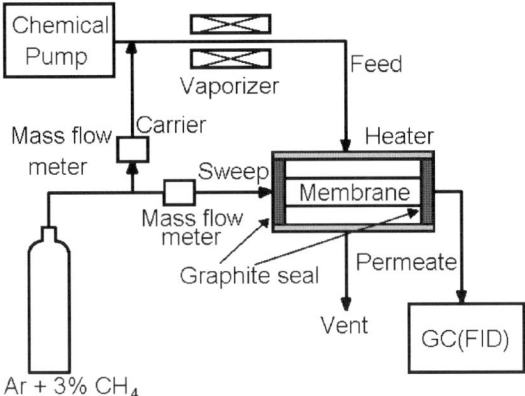

Fig.2 Schematic of mixed gas permeation measurement

## 2.4. Nano-permporosimetry

The porosimetry technique works by measuring the permeation of a non-condensable gas such as He through a membrane as the pores of a membrane are progressively filled by increasing the partial pressure of a condensable vapor carried in the helium [3,4].

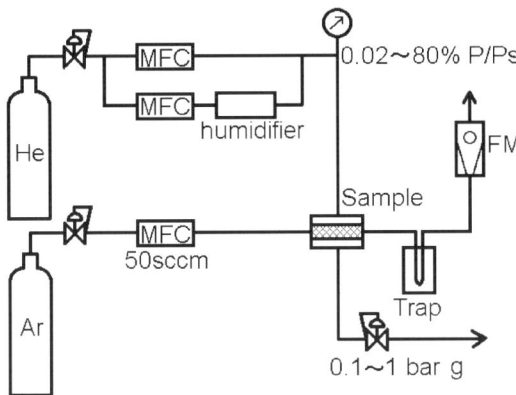

Fig.3 Schematic of nano-permporosimetry

In this experiment setup, condensable vapor of hexane is introduced through a bubbler, because of silicalite has hydrophobic surface character. By mixing together a stream of He flowing through the bubbler and a pure He stream, different concentration of vapor partial pressure can be achieved. Total flow (Carrier Ar + Permeate He gas flow) was measured electronically sensing soap flow meter after passing zeolite trap for condensable vapor.

Permeate He flow has calculate subtract from complete dry sample total flow which was measured after pretreatment. Every sample was pretreated at 300°C under dry Ar and He gas flow before measurement. Analysis Ar and He gas has increased purity from water chemical trap before MFC. Permeation driving force can adjustable backpressure regulator up to 1 bar. Hexane permporosimetry was measured at 306K.

Pore size is calculated from Kelvin equation for cylindrical pore model:

$$r_k = \frac{-2\gamma V \cos\theta}{\ln(P/Ps)RT} \qquad (1)$$

where $P/Ps$ is the relative vapor pressure of the condensable gas, $\gamma$ is the surface tension of liquid state of condensable gas, $V$ is the molar volume of the condensable gas, $\theta$ is contact angel and $r_k$ is Kelvin radius, which is in fact the meniscus of the gas liquid interface, $R$ is the gas constant and $T$ is the measurement temperature. In this work, we characterized silicalite membrane by using n-Hexane. The silicalite has hydrophobic surface and then n-Hexane can easy adsorb on the surface of zeolite. A meniscus of capillary condensation will make the film of n-Hexane layer, so we can assume contact angle is 0. For a cylindrical pore model, real pore diameter size ($P_d$) and the Kelvin radius ($r_k$) is as follows:

$$P_d = 2 \times (r_k + t) \qquad (2)$$

where t is the thickness of the "t-layer" formed on the inner surface of the pore. In this work, we used t = 0.4 nm which is model size on flat adsorption of molecule. Table 1 show hexane concentrations and pore size relation.

Table 1
Pore size from Kelvin equation in case of n-hexane nano-permporosimetry

| $P/Ps$ | 0.002 | 0.01 | 0.1 | 0.3 | 0.5 | 0.7 | 0.8 | 0.9 |
|--------|-------|------|-----|-----|-----|-----|-----|-----|
| $P_d$ / nm | 1.4 | 1.6 | 2.3 | 3.8 | 5.9 | 10.8 | 16.8 | 34.6 |

## 3. RESULTS AND DISCUSSION

Two silicalite membrane tubes have been prepared in same procedure at experiment section. Prepared silicalite membrane observed by SEM (Fig.4, 5). Sample A has 18.4 $\mu$ m, Sample B has 4.5 $\mu$ m thickness of silicalite layer on the outer surface of $\alpha$-Alumina support. By SEM observation, it is very difficult to find defect of membrane quality.

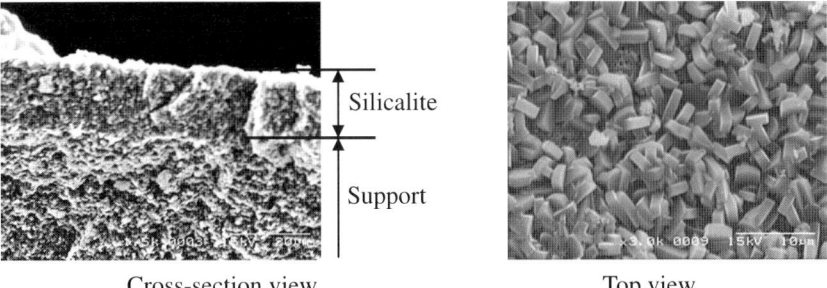

Cross-section view                                    Top view

Fig.4 SEM of Sample A

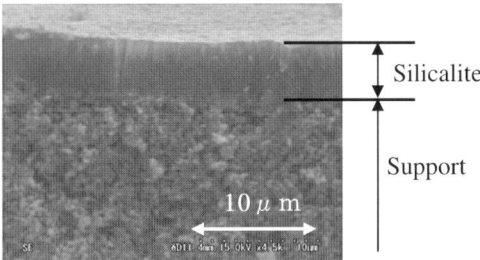

Fig.5 SEM of a cross-section Sample B

XRD pattern can assign to MFI type peak and α-alumina peak over 25° (Fig.6). Peak ratio of MFI and alumina, sample B show much higher than sample B. It indicate silicalite membrane layer of sample B thinner than Sample A. (101), (002) and (004) intensity ratio to (010) MFI peak of sample A is much higher than sample B. It shows sample A is highly oriented (001) and (101) face to the top during crystal growing up (evolutionary selection mechanism) [1]. MFI hexagonal crystal can't observe in SEM top view of sample A in Fig. 6. This also support XRD results.

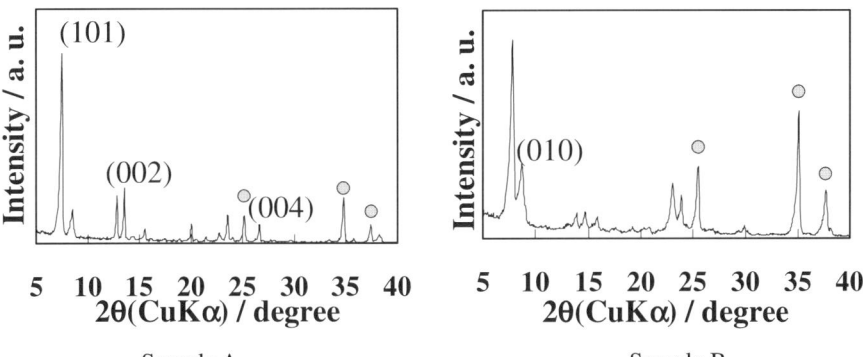

Sample A                                    Sample B

Fig.6 XRD pattern of Sample A and B ( ○: α-Alumina peak)

Mixed gas permeation is shown in Fig.7. Sample A has high permeation rate of n-Hexane compare than others. Minimum cross-sectional molecular dimension of n-Hexane is 0.4 – 0.45 nm. So it can easy permeate through zeolitic pore (ca. 0.58 nm). But sample B doesn't show any separation results for n-Hexane (n-Hex), 2-Metyl Pentane (MP) and 2,2-Dimetyl Butane (DMB) vapor. It indicates that zeolite membrane has some defect and large pore.

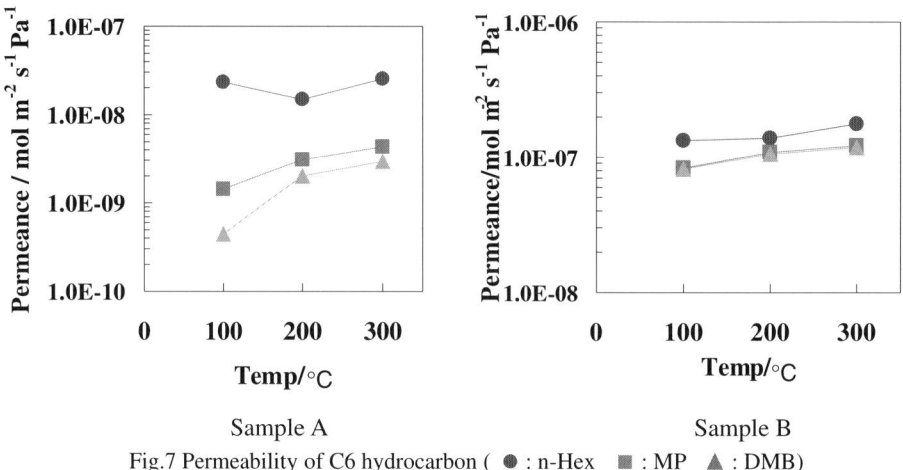

Sample A                              Sample B

Fig.7 Permeability of C6 hydrocarbon ( ● : n-Hex    ■ : MP    ▲ : DMB)

Fig.8 show separation factor of branched C6 hydrocarbon. Sample A has high separation factor of MP and DMB from n-Hex at 100°C. Increase the temperature, separation factor decrease. Sample B doesn't have any selectivity of separation.

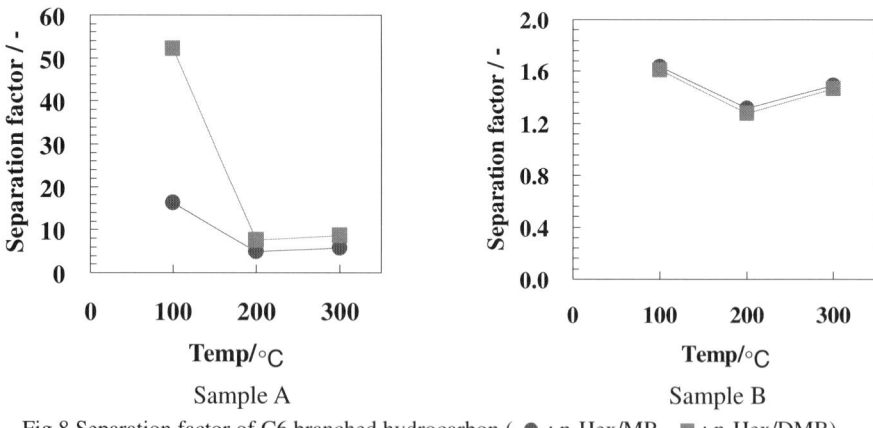

Sample A                              Sample B

Fig.8 Separation factor of C6 branched hydrocarbon ( ● : n-Hex/MP    ■ : n-Hex/DMB)

Hexane nano-permporosimetry is shown in Fig. 9. Permeation rate of sample A and B decrease from increase concentration of n-Hexane. Both curve show drastically drop at initial concentration increase (0 – 0.2%) due to zeolitic micropore has plugged by hexane molecule. After that only sample B has decrease permeation until 80%. From this result sample A has complete silicalite membrane layer. Sample B has some mesopore in the layer.

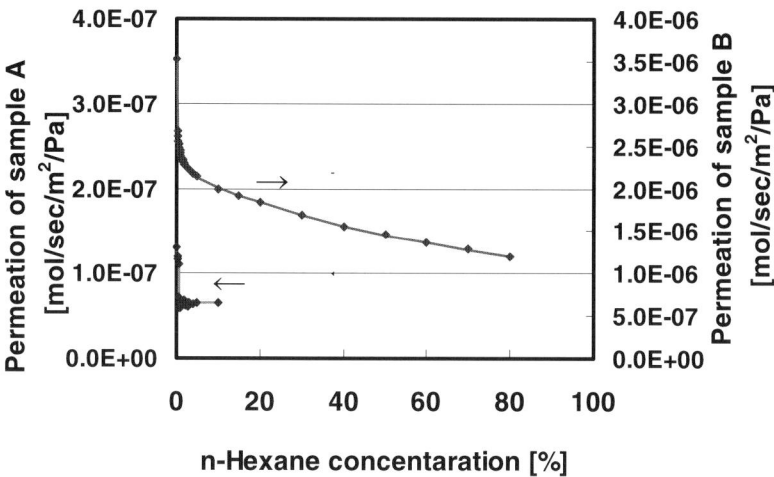

Fig.9 Hexane nano-porosimetry of two different Silicalite membrane at 306K.

Fig. 10 show pore size distribution of sample B from nano-permporosimetry. It shows Sample B has 1 to 7 nm pore size distribution. It will come imperfect growing up zeolite layer.

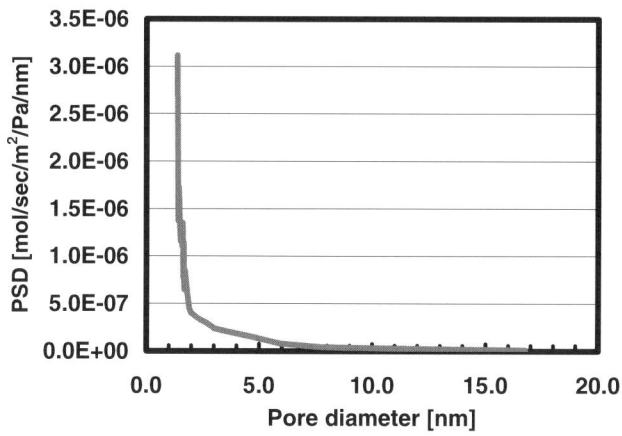

Fig. 10 Pore size distribution of silicalite membrane (Sample B)

## 4. CONCLUSION

Silicalite membrane useful for separation of n-Hexane and branched hydrocarbon (MP, DMB) at 100 – 300°C. 2 step hydrothermal synthesis (Evolutionary selection mechanism) of zeolite membrane has improved to create perfect zeolite layer. Hexane nano-permporosimetry technique determined pore size range 1 to 20 nm. The instrument was also found to be useful tool to confirm the membrane quality.

## REFERENCE

[1] M.C. Lovallo and M. Tsapastsis, *AIChE J.* **42**, (1996) 3020.

[2] S. Kallus, P. Langlois, G.E. Romanos, Th. Steriotis, E.S. Kikkinides, N.K. Kanellopoulos and J.D.F. Ramsay, Studies in Surface Science and Catalysis, **128** (2000) 467.

[3] F.P. Cuperus, D. Bargeman and C.A. Smolders, *J. Membrane Sci.,* **71** (1992) 57.

[4] R. Gallaher and P.K.T. Liu, *J. Membrane Sci.,* **92** (1994) 29.

Studies in Surface Science and Catalysis 160
P.L. Llewellyn, F. Rodriquez-Reinoso, J. Rouqerol and N. Seaton (Editors)

# Is the BET equation applicable to microporous adsorbents?

## J. Rouquerol, P. Llewellyn and F. Rouquerol

MADIREL Laboratory, CNRS-Université de Provence, Centre de St Jérôme, 13397 Marseille Cedex 20, France, e-mail : jean.rouquerol@up.univ-mrs.fr

## 1. INTRODUCTION

Although most scientists are aware that the application of the BET method to microporous adsorbents is essentially erroneous, we must also recognize that most of us find it convenient, year after year, to use this popular method even when we know or suspect the existence of micropores. Its success probably has much to do with the fact that it eventually provides us with an "area", which is a quantity easily available to our imagination, especially since it is usually expressed in m$^2$, an order of magnitude which we can daily see, touch and imagine. Now, since we all expect science to do something more than to feed imagination, *is it a rigorous and sound way to apply the BET method in the case of microporous materials*? This is the issue we wish to address hereafter. For that purpose, after recapping the basic limitations of the BET method, we propose to examine whether and how it can still provide reproducible, meaningful and useful information about microporous adsorbents.

## 2. WHY NOT SIMPLY USE THE LANGMUIR EQUATION WHEN THE ADSORBENT IS KNOWN TO BE MICROPOROUS ?

Micropores tend to provide Type I gas adsorption isotherms (in the IUPAC classification for *physisorption* isotherms [1]), similar to the Langmuir isotherm. It therefore looks natural and simple to apply the Langmuir equation to the above isotherms. Now, this is not really suitable for the following reasons:
1/ The Langmuir equation was clearly established for the specific case of *chemisorption in a single layer* in free contact with the gas phase; this is quite different from adsorption in micropores, where adsorption is not necessarily limited to adsorption sites (simple filling can occur) and where most of the adsorbed phase is not in contact with the gas phase.
2/ *If the material is purely microporous,* its isotherm is perfectly Type I. In this case, *the horizontal plateau exactly provides the micropore capacity.* There is no need for any further assumption and therefore no need for applying the Langmuir equation.
3/ *Most often, the isotherms obtained for microporous materials are composite, i.e.* mixing either Types I and II (resulting from to micropores and external surface, respectively) or Types I and IV (resulting from micropores and mesopores, respectively). Now, the Langmuir equation does not apply to composite isotherms, simply because the Langmuir theory was established for a single phenomenon (chemisorption in a single layer). It follows that, for composite isotherms, the calculated "Langmuir monolayer" is never completed within the pressure range in which it is calculated (as usual for true Langmuir isotherms), but at a higher pressure, which is a sign of inconsistency.
We therefore conclude that something else than the Langmuir equation is needed to interpret the adsorption isotherms for microporous materials. Is the BET equation the good answer?

## 2. WHY IS THE BET METHOD ALSO LIMITED IN CASE OF MICROPOROUS MATERIALS?

The BET method can be considered, essentially, as a mathematical means to analyse the adsorption isotherm in order to derive a "monolayer capacity" and then a surface area. This analysis requested a number of assumptions, of which we just remind the main ones [2]:

1/ Adsorption takes place on a *uniform surface* and the energies of adsorption of all molecules in the first layer are identical

2/ *Each molecule adsorbed* in a layer *is itself a potential adsorption site* for the next layer

3/ There is *no* steric *limitation to the thickness* of the multilayer

4/ It is *only for the first layer* that the differential energy of physical adsorption $E_1$ is higher than the energy of liquefaction $E_l$

5/ *Interactions between molecules adsorbed in the same layer do not play any part* in the adsorption equation

6/ The second and further layers start to build up before the completion of the first one

If the application of the BET equation was to be limited to the type of adsorbent assumed above (an energetically uniform surface and no pores), there would probably not be many people to remember it to-day. In reality, most interesting adsorbents are either heterogeneous from the viewpoint of adsorption energy, or porous, or both. Finally, assumptions 1/ and 2/ are exceptionally fulfilled, if ever, assumption 3/ does not hold for porous adsorbents, assumption 4/ is an acceptable approximation, assumption 5/ is incorrect, and, finally, only assumption 6/ is usually right … except for those ultra-micropores whose width cannot accommodate more than two molecules. Moreover, at the time of deriving a surface area from the monolayer content, three other assumptions are used:

7/ All molecules in the monolayer cover the same area $\sigma$

8/ The arrangement of these molecules is assumed to be a hexagonal close-packing

9/ The molecular cross-sectional area $\sigma$ can be derived from the density of the adsorptive in the bulk liquid state

The above three assumptions again raise a number of questions, especially when the BET method is applied to adsorbents containing micropores (in the presence or not of larger pores). It follows that the questions we wish to address in this paper are the following:

1/ Irrespective of its meaning, is the BET "monolayer capacity" a *reproducible* quantity?

2/ Is *adsorption calorimetry* on microporous adsorbents helpful to evaluate the BET energetic assumptions (1/, 4/ and 5/)?

3/ Since assumptions 3/,6/,7/,8/ and 9/ do not hold for a microporous adsorbent, what meaning can we then give to the "*BET surface area*"? Should we complement this concept by another one to make it useful for industrial applications of microporous adsorbents?

## 3. IS THE BET "MONOLAYER CAPACITY" OF A MICROPOROUS MATERIAL A REPRODUCIBLE QUANTITY?

One could think that the problem of reproducibility of the BET "monolayer content" is a simple matter of experimental accuracy. In relality, the problem does not lie there anymore since, with modern gas adsorption equipment and with careful and reproducible outgassing [3,4], one can obtain reliable and satisfactory data. The issue is that once a reliable adsorption isotherm is obtained it may still lead to several values of the BET "monolayer content". This is due to the fact that the BET equation must be applied to a straight part of the BET plot, but that *various portions of that plot may look linear*, so that, in the absence of any other criteria, their choice can be purely subjective.

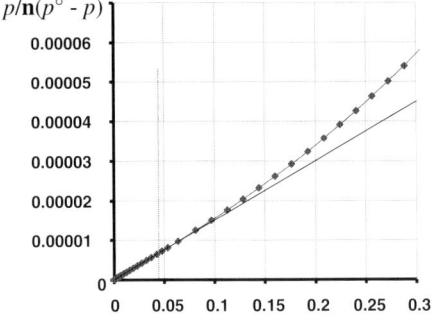

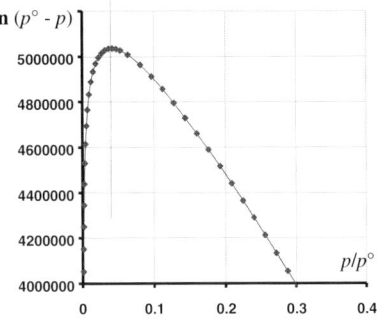

Fig. 1. BET plot for Ar on Zeolite 13 X at 87 K    Fig. 2. Plot of the term $n^a$ $(p^\circ - p)$ *vs.* $p/p^\circ$

The simplest way to avoid subjectivity could be to decide for ever on a pressure range in which to carry out the calculation. This was proposed by Brunauer with a relative pressure range from to 0.05 to 0.35 [1] which is indeed well suited, in general, in the absence of micropores or of any other type of strongly adsorbing sites, *i.e.* for adsorption isotherms of type II or IV. Now, in the case of materials containing micropores, such a broad pressure range largely exceeds the real range of linearity of the BET plot, as illustrated by the plot given in Fig 1 and obtained for the adsorption of argon on a zeolite 13X at 87 K.

This leads us to the selection of an *appropriate pressure range for each adsorbent* with the problem that the above "linearity criterion", which is essential, is unfortunately not self-sufficient, since several portions of the BET plot can fulfil this requirement. One can indeed consider, in Fig 1, several relative pressure ranges in which the plot is reasonably linear: for instance 0.01-0.2, 0.02-0.05 and 0.05-0.15, the latter being the most commonly used. The resulting monolayer contents are 4.0, 4.5 and 5.2 mmol.g$^{-1}$, respectively, i.e. a variation of up to 30%! Hence the need of other *criteria leading to the "objective choice" of a single linear portion of the BET plot*. Those two selected hereafter are aimed to provide such a single choice:

1/ The straight portion selected should have a *positive intercept on the ordinate, i.e.* no negative value of "C", which would be meaningless. In Fig 1, one sees that this criterion eliminates the upper half portion of the BET plot represented there

2/ *The term $n^a$ $(p^\circ - p)$ should continuously increase* together with $p/p^\circ$ ; if not, the pressure range should be narrowed. In Fig 2, derived from the same data as Fig 1, one sees that this criterion now eliminates the whole relative pressure range above 0.04. This criterion can be considered as a self-consistency criterion for the modified BET equation proposed by Keii *et al.* [5] and used by Parra *et al.* in the case of microporous carbons [6].

Athough the above criteria look sensible and consistent with each other, we don't claim that they allow to reach "the" actual surface area of the sample (which, furthermore is never really accessible by the BET method in the case of microporous materials, as developed later on in this paper). We simply think that these criteria, easy to apply and to introduce in the automatic processing of data (especially in the software of modern adsorption equipment) can make easier and safer the comparison between BET results obtained in different places. We have been using them systematically, for years, and never found any problem in their application [7,8]. Furthermore, the two checks of consistency hereafter were systematically applied and we never saw them failing:

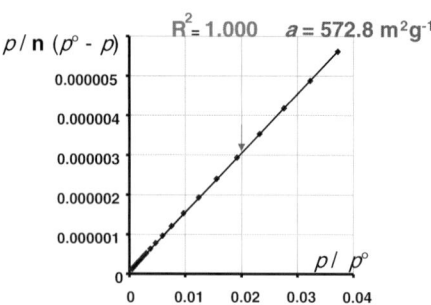

Fig. 3. BET plot in the finally selected pressure range for Ar on Zeolite 13 X at 87 K

a/ The calculated BET monolayer capacity, when reported on the adsorption isotherm, should correspond to a relative pressure $p/p°_m$ located within the pressure range selected for the calculation

b/ Alternatively, the relative pressure $p/p°_m$ for the monolayer capacity can be recalculated from the value of "C", through the BET equation, after stating $n^a = n^a_m$ . The calculated $p/p°_m$ and the experimental one (read on the isotherm, like in the previous check) should not be apart from each other by, say, more than 10%.

Figure 3 shows the portion of the BET plot finally selected with help of these criteria for our Ar/13 X system. The calculated BET monolayer capacity is now 4.8 mmol.g$^{-1}$, whereas it previously ranged between 4.0 and 5.2 mmol.g$^{-1}$.

## 4. IS THE BET APPROACH REASONABLY SUPPORTED BY CALORIMETRY?

For non-porous adsorbents, it was early shown, by adsorption calorimetry, especially by Isirikyan and Kiselev [9] and by Beebe [10] that, on an energetically homogeneous surface like graphitized carbon, the part of adsorbate-adsorbate interactions can hardly be ignored. It was also seen, especially on common oxides like silicas [11] or titanias [12], that the surface of these technological adsorbents is energetically heterogeneous so that they do not lend themselves to the assumption of a constant energy of adsorption on the first layer. Most fortunately, the effects of the adsorbate-adsorbate interactions and of the surface heterogeneities upon the energies of adsorption partly compensate each other, which therefore supports, in some respect, the simultaneous use of these simplifying assumptions in the BET theory.

Now, what happens for microporous adsorbents? We shall illustrate the state of affairs by four examples. For each of them, we report a figure providing simultaneously the adsorption isotherm and the curve of adsorption enthalpy, both determined during the same experiment.

Fig. 4 gives the results for methane on Silicalite at 77 K. Like in the figures to follow, the adsorption isotherm starts from the bottom left, with its initial and steep portion merging within the ordinate axis. The calorimetric curve starts from the bottom right and makes use of the same ordinate, which provides the amount adsorbed, whereas the differential energy of adsorption is indicated in the scale on top of the figure. One should first notice the high and constant value of the enthalpy of adsorption over most of the experiment : -17 kJ.mol$^{-1}$, more than twice higher than the enthalpy of liquefaction. In spite of the appearance, this does not support the BET assumption of a constant adsorption energy on the first layer: we know indeed Silicalite to be microporous, so that what occurs is not the completion of a monolayer but the filling of micropores in volume. Since most of the filling (up to 4 mmol.g$^{-1}$) takes

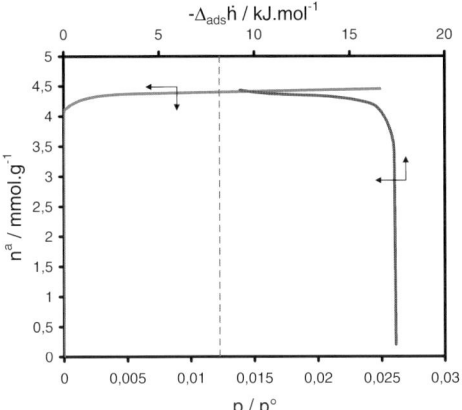

Fig.4. Methane on Silicalite at 77 K: adsorption isotherm (left) and calorimetric curve (right)

place under a pressure lower than $10^{-2}$ mbar, it should also be pointed out that only direct calorimetric experiments are able to assess the adsorption enthalpies, whereas the isosteric method would not be here reliable at all (because it is too sensitive to small errors in pressure or to small amounts of less adsorbed impurities like nitrogen or oxygen). Finally, one is struck by the symmetrical shape of the two curves, which shows that the enthalpy of adsorption drops just once the plateau is reached on the Type I isotherm, *i.e.* once the micropores are filled. Applying the BET equation with the criteria given in section 3 leads here to an apparent monolayer content of 4.2 mmol.$g^{-1}$, corresponding, as usual, to the "knee" of the isotherm.

The calorimetric curve shows that *this apparent BET monolayer content exactly corresponds to the most strongly adsorbed portion of the adsorbate.*

Fig. 5 also reports results obtained with methane adsorption at 77 K on a micron-sized ZSM-48 sample whose external surface is approximately 50 m$^2$ g$^{-1}$ and whose pore width is 5-6 nm. In spite of the still ordered porous structure, the picture is quite different from above, both for the isotherm and for the calorimetric curve. The adsorption isotherm is clearly composite (Type I + Type II), which is easily explained by the existence of an appreciable external .

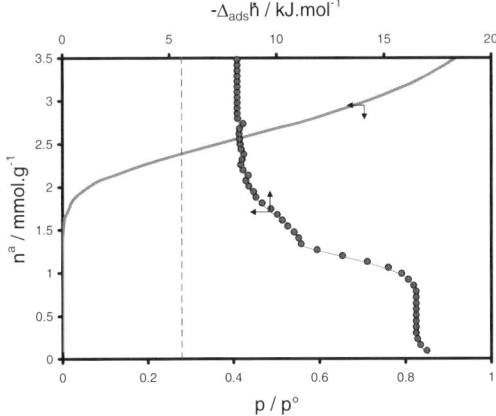

Fig.5. Methane on Zeolite ZSM-48 at 77 K: adsorption isotherm (left) and calorimetric curve (right) surface, adding its effect to that of micropores.

The calorimetric curve is also more complicated than previously, since now three steps can be distinguished:

-   step 1 ( the initial vertical part, in the bottom right ) corresponds to the filling of the micropores with the same high enthalpy of adsorption as for Silicalite. The small "tail" just before reveals the existence of a small amount of defects or heterogeneities which were not visible (and probably did not exist) in the case of Silicalite
-   step 2 (starting around 12 kJ.mol$^{-1}$ and continuing close to the enthalpy of liquefaction) corresponds to the formation of the statistical monolayer (*i.e.* simultaneous formation of first and upper layers)on the external surface
-   step 3 (final vertical part, at the level of the enthalpy of liquefaction) exclusively corresponds to the formation of the upper layers (the "multilayer")

Here again, the apparent BET monolayer content (1.9 mmol.g$^{-1}$) exactly corresponds to the most strongly adsorbed portion of the adsorbate.

Fig. 6 reports results for the adsorption of nitrogen on a microporous carbon (charcoal 26, from Sutcliffe Speakman, obtained from NPL) at 77K. We see some features in common with the previous example: composite Type I + Type II isotherm and at least two steps in the calorimetric curve. Now, the initial decay of the adsorption enthalpy is much stronger and longer than on the previous example, in part because of the higher heterogeneity of the activated carbon and in part because the permanent quadrupole moment of the nitrogen molecule makes it more sensitive to heterogeneities than the methane molecule.

Fig. 7 shows the results obtained for the adsorption of argon on a microporous silica at 77 K. We get again a composite isotherm, whereas the steadily decreasing enthalpy of adsorption indicates a broad range of micropores, starting with ultramicropores (as it was the case for the activated carbon).

## 5. FINALLY, IN THE PRESENCE OF MICROPORES, CAN THE BET EQUATION BE MEANINGFUL AND USEFUL?

Everybody agrees on the fact that *the sole concept of "BET monolayer content" is inadequate* in the presence of micropores where a monolayer has no clear physical or theoretical meaning. Now, if we do not want to lose, in the case of microporous materials, the benefits of the BET equation (recapped in the introduction and in section 3), we need to introduce another concept.

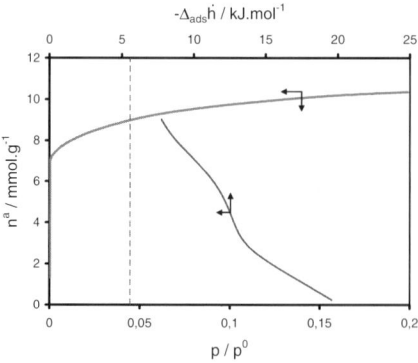

Fig.6. Nitrogen on microporous carbon at 77 K: adsorption isotherm (left) and calorimetric curve (right)

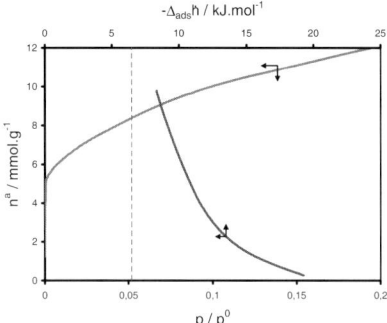

Fig.7. Argon on microporous silica (Davison 950) at 77 K: adsorption isotherm (left) and calorimetric curve (right)

This is fortunately offered by the calorimetric experiments which suggest that the "BET monolayer content" physically corresponds to an energetically strong retention. This quantity, provided by the BET equation, could therefore be called safely the "*BET strong retention capacity*". This quantity includes two parts, which are the "*micropore capacity*" and the "*monolayer content*"on the non-microporous portions of the surface. The latter, which provides the "external (*i.e.* non-microporous) surface area" is easily assessed by means of the $\alpha_S$ or *t* methods, without even requesting the very low part of the adsorption isotherm. The $\alpha_S$ method is to be preferred when one wishes to carry out a more detailed analysis of the micropores and when the low pressure range of the adsorption isotherm is available. Conversely, if one only wishes to assess a reliable external surface area, he will probably find it simpler to use the *t* method: this can indeed easily be done in a software, after introducing the appropriate multilayer equation, like the Harkins and Jura *t*-curve equation [13]. The recommended succession of calculations is therefore:

1/ Calculation of the "*BET strong retention capacity*", $n^a_{m(BET)}$, with help of the BET equation and of the criteria given in section 3

2/ Calculation of the "*external surface area*", $a_{ext}$, with help of the $\alpha_S$ or *t* methods, and of the corresponding external monolayer content, $n^a_{m(ext)}$

3/ Derivation of the "*micropore capacity*", by substracting $n^a_{m(ext)}$ from $n^a_{m(BET)}$

(It may be worth pointing out that the micropore *capacity* (for a given adsorptive) is highly meaningful and close to physical reality, which is not the case for the micropore *volume* whose calculation relies on the unknown packing of the adsorbate in the micropores)

4/ Along the same lines and always for a given adsorptive, determination of the "*saturation capacity*" from the amount adsorbed at saturation (*i.e.* on the final plateau, usually beyond $p/p^o = 0.9$)

The four quantities above can be determined systematically and automatically by a software, whatever the adsorbent: in the presence of micropores, they should all be meaningful, whereas in their absence the derived "*micropore capacity*" should simply be negligible. It is only in the latter case that the concept of "BET monolayer content" can be restored and used without ambiguity.

## 6. CONCLUSIONS

1. Because it was not devised for that, the BET method *should not be applied blindly* to adsorbents containing micropores (and the Langmuir equation even less).

2. *Beyond the "linearity criterion" of the BET plot, two other criteria are found necessary,* especially in the presence of micropores, to draw the specific advantage of the BET equation, *i.e.* to reach a single and reproducible value for the so-called "monolayer content".

3. *Calorimetric data* for adsorption on microporous adsorbents confirm the fact that the BET monolayer content, calculated with the above criteria, mostly corresponds to the adsorbate in energetical interaction with the surface.

3. For adsorbents containing micropores, the concept of "BET monolayer content" is misleading and could well be replaced by that of *"BET strong retention capacity"*. This concept includes the adsorbate present in the micropores together with the content of the statistical monolayer on the non-microporous portion of the surface.

4 Finally, we suggest that instead of the"absolute" concepts of BET surface area (inadequate for micropores) or microporous volume (which is never correctly assessed, because of the unknown packing of the adsorbate), one should more safely use the concepts of *"BET strong retention capacity"*, *"external surface area"*, *"micropore capacity"* and *"saturation capacity"* which are close to physical reality and therefore suitable for sound interpretation and practical application.

## REFERENCES

[1] K.S.W. Sing, D.H. Everett, R.A.W. Haul, L. Moscou, R.A. Pierotti, J. Rouquerol and T. Siemieniewska, Pure Appl. Chem., 57 (1985) 603.
[2] S. Brunauer, The Adsorption of Gases and Vapours, Oxford University Press, Oxford, 1945.
[3] F. Rouquerol, J. Rouquerol and K. Sing, Adsorption by powders & porous solids. Principles, methodology and applications, Academic Press, London, 1999, pp79-83.
[4] O. Toft Sorensen and J; Rouquerol (eds), Sample Controlled Thermal Analysis. Origin, Goals, Multiple forms, Applications and Future, Kluwer Academic Publishers, Dordrecht, 2003, pp 167-170.
[5] T. Keii, T. Tagaki and S. Kanataka, Anal. Chem., 33 (1961) 1965.
[6] J.B. Parra, J.C. de Sousa, R.C. Bansal, J.J. Pis and J.A. Pajares, Adsorption Sci. Tech., 11 (1994) 51.
[7] ] F. Rouquerol, J. Rouquerol and K. Sing, Adsorption by powders & porous solids. Principles, methodology and applications, Academic Press, London, 1999,pp 166-168.
[8] F. Rouquerol, L. Luciani, P. Llewellyn, R. Denoyel and J. Rouquerol, Techniques de l'ingénieur, 2003, P 1050.
[9] A.A. Isirikian and A.V.Kiselev, J. Phys. Chem., 65 (1961) 601.
[10] R.A. Beebe, J. Biscoe, W.R. Smith, and Wendell, J. Am. Chem. Soc., 69 (1947) 95.
[11] J. Rouquerol, F. Rouquerol, C. Pérès, Y. Grillet and M. Boudellal, *in* Characterization of Porous Solids, (S.J. Gregg, K.S.W. Sing and H.F. Stoeckli, eds), Society of Chemical Industry, London, 1979, p. 107.
[12] D.N. Furlong, F. Rouquerol, J. Rouquerol and K.S.W. Sing, J. Chem. Soc., Faraday Trans. I, 76 (1980) 774.
[13] W.D. Harkins and G. Jura, J. Chem. Phys., 11 (1943) 431.

Studies in Surface Science and Catalysis 160
*P.L. Llewellyn, F. Rodriquez-Reinoso, J. Rouqerol and N. Seaton (Editors)*

# A new classification of pore sizes

**T.J. Mays**[1]

Department of Chemical Engineering, University of Bath, Bath BA2 7AY, United Kingdom

## 1. INTRODUCTION

A new classification of pore sizes is proposed. It is based on prefixes defined by the Bureau International des Poids et Mesures under Le Système International d'Unités (SI) [1] (in particular *nano-*, *micro-* and *milli-*), unlike the current classification scheme defined by the International Union of Pure and Applied Chemistry (IUPAC) [2]. Thus the new classification is also consistent with other common scientific terms based on SI prefixes such as *nanotechnology* [3]. Further advantages are that unlike the IUPAC scheme – which is derived from physical adsorption phenomena in pores narrower than 50 nm – the new classification is entirely decoupled from any physico-chemical system or process and is not biased towards small pores. However, the proposed new scheme is more complicated than the current IUPAC one, especially regarding sub-divisions of the main pore size classes. Also, the term micropore occurs in both schemes, which makes them incompatible, at least over the micropore size range as defined in the new classification.

## 2. BACKGROUND

Pores in solids have many properties such as shape, location, connectivity and surface chemistry. Perhaps the easiest property of a pore to visualise (though not necessarily to define) is its size, i. e., its extent in one spatial dimension. This is probably one reason why size is often the first or main property used to characterise a pore. However, another more important reason is that pore size has, arguably, the greatest or widest influence on the properties (and hence uses) of solids, compared to other parameters such as pore shape. It is therefore unquestionably useful and convenient to use pore size (or pore size distribution) as a means to characterise and compare different porous solids.

The current classification of pore size recommended by IUPAC is as follows [2]:

| | |
|---|---|
| micropore | pore width smaller than 2 nm |
| mesopores | pore width between 2 and 50 nm |
| macropore | pore width greater than 50 nm |

---

[1]  Tel: +44 (0)1225 386528, Fax: +44 (0)1225 385713, E-mail: t.j.mays@bath.ac.uk

where the term 'width' refers to the limiting size of a pore, taken to be its smallest dimension. The idea is that once a definition of pore size (or width) is made or adopted (see for example [4]), then any pore or set of pores may be classified using the scheme above.

The history of the current IUPAC scheme [5-7] is interesting, and sets the scene for the new scheme proposed below.

Brunauer [8] gave an early definition of micropores and macropores as pores smaller or larger than 10 nm respectively[2], though these terms appear to have been used informally before. The 10 nm boundary was taken to be a transition from bulk or continuum properties of fluids inside pores to where the molecular sizes of fluid molecules approached the pore size. In the late 1940s and early 1950s Dubinin and co-workers developed these ideas in papers on adsorption and porosity originally published in Russian. The main themes of these were eventually published by Dubinin in English in 1955 [9], where Brunauer's terms and broad meanings of micropore and macropore were retained, with the addition of transitional pores to account for a gradual rather than abrupt transition between bulk and molecular properties of fluids inside pores. While precise boundaries for these pore size classes were not given, it is clear that micropores were rather smaller than 10 nm and macropores rather larger than 10 nm, with transitional pores in between. In a seminal paper, published under the auspices of IUPAC, Everett later suggested the term mesopore rather than transitional (or intermediate) pore, and that the micropore-mesopore and mesopore-macropore boundaries should be 2 and 50 nm respectively [10], with the general idea (after Brunauer and Dubinin) that fluids in macropores behave as on free or external surfaces, capillary condensation occurs in mesopores and that adsorption in micropores occurs by a filling process partly influenced by fluid molecules being about the same size as the pores. The term mesopore appears to have been used earlier to refer to pores 10s of μm in size in soil [11, 12], though it is not clear whether Everett knew that. Gregg and Sing [13] subsequently adopted Everett's classification, and reinforced Dubinin's ideas that micropores were also characterised by enhanced adsorption due to the overlap of interaction potentials from opposite pore walls. The IUPAC confirmed this scheme in 1985 [14] and recommended it formally in 1994 [2]. While Mikhail *et al.* [15] suggested the term ultramicropore for the smallest micropores and later Dubinin [16] used the term supermicropore to refer to pores around the micropore-mesopore boundary, the recommended pore size classification has remained the IUPAC one [2].

While the IUPAC scheme has served the scientific community well it has some drawbacks. The most obvious of these is that the term micropore is unrelated to the SI unit micrometer, μm. While the origins of the IUPAC scheme discussed above preceded common usage of SI prefixes, those new to this classification are often initially confused that micropore has nothing to do with SI. Also, the terms mesopore and macropore have no links with SI or any other conventions. This lack of conformity also means that confusion can arise when common SI-related terms such as nanotechnology [3] are applied to porous solids. The link between the IUPAC classification and adsorption processes is also a problem, for other equally important processes depend on pore size in different ways and not all of these processes are dominated or even affected by small pores.

---

[2]    Though at the time the common unit used for small distances was Å not nm.

Given these drawbacks, the present author considered that it was timely to consider a new pore size classification that would be consistent with SI and related terms, and that would replace (rather than be used in parallel with) the current IUPAC scheme. This new classification is presented below.

## 3. NEW PORE SIZE CLASSSIFICATION

The new pore size classification is sumarised in Fig. 1. It is based on a logarithmic scale (to base 10), and all pores of interest are considered to be no smaller than 0.1 nm and no larger than 0.1 m (though in principle the new scheme could be extended beyond these boundaries). The three main pore size ranges are:

nanopore            pore size between 0.1 and 100 nm
micropore           pore size between 0.1 and 100 μm
millipore           pore size between 0.1 and 100 mm

Each range covers three decades of pore size, with the lower boundary of the middle decade being the reference pore size for the range (1 nm, 1 μm and 1 mm for nanopores, micropores and millipores respectively). None of these terms appears to have been formally defined before, though Sing and Schüth [6] referred to 0.1 to 100 nm as a 'nanoscale range' in relation to pore size, and this range is now broadly accepted as that covered by nanotechnology [3]. Unlike the IUPAC classification, the new scheme clearly conforms to the spirit of SI, is entirely decoupled from any physico-chemical process occurring in or associated with pores, and is not biased to either small or large pores. It also covers most pore sizes of interest in the characterisation of porous solids.

While the new scheme appears to have many advantages, one problem is that it uses the term micropore, which is also used in the current IUPAC classification. Therefore the two schemes are incompatible, at least over the micropore size range in the new scheme. A further problem is that the three orders of magnitude of pore size covered in each range may be rather broad to distinguish many porous solids. To resolve this last issue, further divisions of each pore size range in the new classification are proposed as follows:

sub-nanopore        pore size between 0.1 and 1 nm
inter-nanopore      pore size between 1 and 10 nm
super-nanopore      pore size between 10 and 100 nm

sub-micropore       pore size between 0.1 and 1 μm
inter-micropore     pore size between 1 and 10 μm
super-micropore     pore size between 10 and 100 μm

sub-millipore       pore size between 0.1 and 1 mm
inter-millipore     pore size between 1 and 10 mm
super-millipore     pore size between 10 and 100 mm

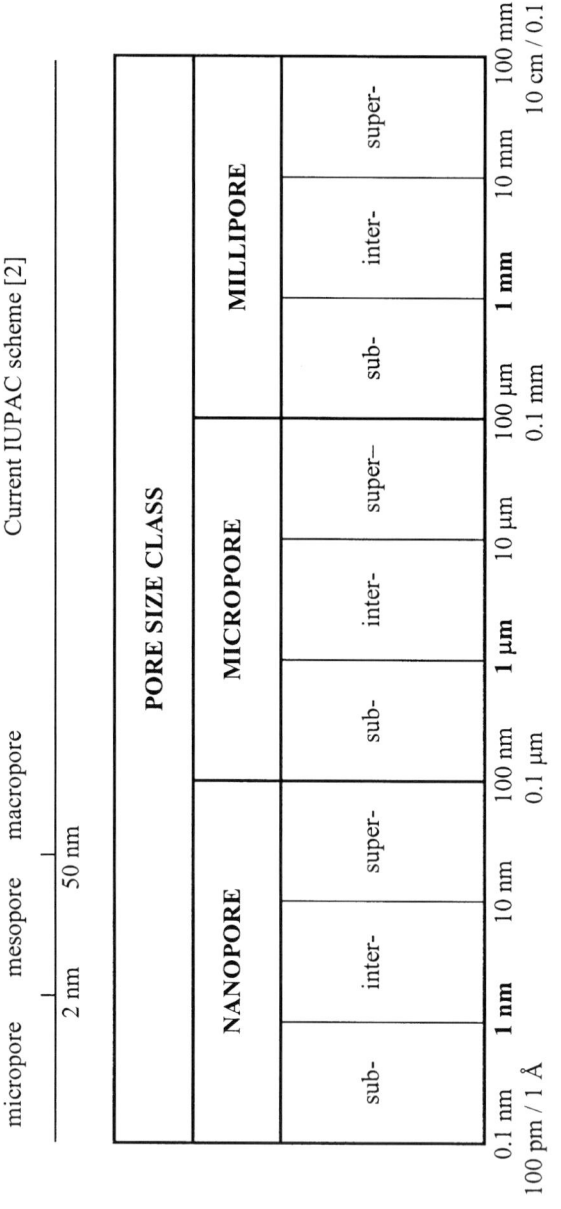

Fig. 1 New pore size classification compared with the current IUPAC scheme [2].
Figures in **bold** in the new scheme refer to the reference points for each main pore size class.

The prefixes sub-, inter- and super- (it is suggested that these should be used with hyphens, unlike the prefixes in the main pore size classes) refer to the lower, middle and highest decade of pore size in each main class. All of these prefixes are derived from Latin [17]: sub- means under, inter- means between, among or amid and super- means above. The derived terms sub-nanoporosity, sub-nanoporous, etc., follow naturally. It may be noted that the upper boundary of each main or sub-class of pore size is the same as the lower boundary of the class above it (in mathematical terms, all the classes are closed intervals). While the sub-classification is rigorous, it makes the new scheme more complicated than the IUPAC one.

## 4. CONCLUDING REMARKS

A new pore size classification is proposed that unlike the current IUPAC scheme [2] is consistent with SI prefixes [1], and terms based on these such as nanotechnology [3], is de-coupled from any physico-chemical processes occurring in or associated with pores, and is not biased towards small pores. While the new scheme covers pores in the size range 0.1 nm to 0.1 m, it can in principle be easily extended beyond these boundaries. However, the new scheme is more complicated than the current IUPAC one, especially regarding sub-divisions of the main pore size classes. Also, the term micropore occurs in both schemes, which makes them incompatible, at least over the micropore size range as defined in the new classification (0.1 to 100 $\mu$m).

The new scheme is proposed with the idea that it should replace the IUPAC classification entirely, though of course this will require general agreement that this appropriate. At least the term nanopore and its sub divisions may initially be adopted as this might overcome the confusion that those new to the characterisation of porous solids feel when they encounter the term micropore.

It is interesting to note that the use of the term nanopore (or derivative terms) has, in fact, been increasing in recent years. Fig. 2 is a plot of the number of journal papers published in the years 1990 to 2004 inclusive with the strings 'nanopor' and 'micropor' in their titles. The data in Fig. 2 were obtained from an electronic search of the Science Citation Index expanded database. Fig. 2 shows that while there appears to be continued interest in micropores, interest in (or reference to) nanopores has increased rapidly, with no sign of slowing down, and to an extent that references to nanopores exceeded those to micropores in 2003. The earliest references to nanopores appeared in the late 1980s / early 1990s, though the term nanopore has never been officially or formally defined.

## ACKNOWLEDGEMENTS

The author is grateful to Professor K.S.W. Sing for valuable discussions and to the Engineering and Physical Sciences Research Council in the UK for financial support.

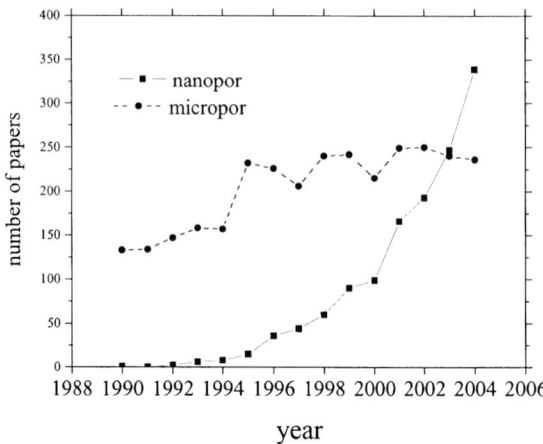

year

Fig. 2 Journal papers published with the strings 'nanopor' or 'micropor' in their titles
in 1994 to 2004 inclusive (data from Science Citation Index).

## REFERENCES

[1]    Bureau International des Poids et Mesures, The International System of Units (SI), (7th ed.),
       1998.
[2]    J. Rouquerol, D. Avnir, C.W. Fairbridge, D.H. Everett, J.H. Haynes, N. Pernicone, J.D.F.
       Ramsay, K.S.W. Sing and K.K. Unger, Pure. Appl. Chem., 66 (1994) 1739.
[3]    A. Franks, J. Phys. E: Sci. Instrum. 20 (1987) 1442.
[4]    P.E. Levitz, In: Handbook of Porous Solids, Vol. 1 (F. Schüth, K.S.W. Sing and J. Weitkamp,
       eds.), Wiley, Weinheim, 2002, pp. 37-80.
[5]    B.H. Davis and K.S.W. Sing, In: Handbook of Porous Solids, Vol. 1 (F. Schüth, K.S.W. Sing
       and J. Weitkamp, eds.), Wiley, Weinheim,, 2002, pp. 3-23.
[6]    K.S.W. Sing and F. Schüth, In: Handbook of Porous Solids, Vol. 1 (F. Schüth, K.S.W. Sing and
       J. Weitkamp, eds.), Wiley, Weinheim, 2002, pp. 24-33.
[7]    F. Rouquerol, J. Rouquerol and K.S.W. Sing, Adsorption by Powders and Porous Solids,
       Academic Press, London, 1999.
[8]    S. Brunauer, The Adsorption of Gases and Vapors. Vol. 1: Physical Adsorption, Princeton
       University Press, Princeton and Oxford, Oxford University Press, 1943.
[9]    M.M. Dubinin, Q. Rev. Chem. Soc., 9 (1955) 101.
[10]   D.H. Everett, Pure Appl. Chem., 31 (1972) 579.
[11]   A. Jongerius, In: Proc. Int. Symp. Soil Structure, Ghent, Belgium, 1958, p.206.
[12]   R. Brewer, Fabric and Mineral Analysis of Soils, Wiley, London, 1964.
[13]   S.J. Gregg and K.S.W. Sing, Adsorption, Surface Area and Porosity (2nd ed.), Academic Press,
       London, 1982,
[14]   K.S.W. Sing, D.H. Everett, R.A.W. Haul, L. Moscou, R.A. Pierotti, J. Rouquerol and T.
       Siemieniewska, Pure Appl. Chem., 57 (1985) 603.
[15]   R.Sh. Mikhail, S. Brunauer and E.E. Bodor, J. Colloid and Interface Sci., 26 (1968) 45.
[16]   M.M. Dubinin, Prog. Surf. Membr. Sci., 9 (1975) 1.
[17]   Oxford English Dictionary (2nd ed.), Oxford University Press, Oxford, 1989.

Studies in Surface Science and Catalysis 160
P.L. Llewellyn, F. Rodriquez-Reinoso, J. Rouqerol and N. Seaton (Editors)

63

# Characterization of nanoporous carbons

**T.X. Nguyen and S.K. Bhatia***

Division of Chemical Engineering, The University of Queensland, St. Lucia, QLD 4072,
Australia.
E-mail: sureshb@cheque.uq.edu.au

Hitherto, adsorption has been traditionally used to study only the porous structure in disordered materials, while the structure of the solid phase skeleton has been probed by crystallographic methods such as x-ray diffraction. Here we show that for carbons density functional theory, suitably adapted to consider heterogeneity of the pore walls, can be reliably used to probe features of the solid structure hitherto accessibly only approximately even by crystallographic methods. We investigate a range of carbons, including commercial activated carbons, activated carbon fibres as well as coal chars prepared in our laboratory, and determine pore wall thickness distributions using argon adsorption. The results are corroborated by x-ray diffraction studies, which yield a parallelism indicator that correlates well with its counterpart determined from the adsorption-based pore wall thickness distributions. Further, characterization of progressively heated carbons using the proposed approach shows clear evidence of thickening of pore walls, confirmed by xrd.

## 1. INTRODUCTION

The use of nonlocal density functional theory (NLDFT) for modeling adsorption isotherms of Lennard-Jones (LJ) fluids in porous materials is now well-established [1-5], and is central to modern characterization of nanoporous carbons as well as a variety of other adsorbent materials [1-3]. The principal concept here is that in confined spaces the potential energy is related to the size of the pore [6], thereby permitting a pore size distribution (PSD) to be extracted by fitting adsorption isotherm data. For carbons the slit pore model is now well established, and known to be applicable to a variety of nanoporous carbon forms, where the underlying microstructure comprises a disordered aggregate of crystallites. Such slit width distributions are then useful in predicting the equilibrium [1-5] and transport behavior [7,8] of other fluids in the same carbon.

Numerous investigations of the effect of pore wall thickness, carbon density, $\rho_s$, interlayer spacing, $\Delta$, and pore-pore correlation have shown [5,9-12] that the pore wall thickness and carbon density have the most significant impact on adsorption behavior such as pore filling pressure and adsorbed density profile. However, defects related to lower carbon density may be small in activated carbons during carbonization and activation processes due to high reactivity of the resulting free radicals. Accordingly, other defects arising from formation of highly-strained rings having number of carbon atoms different from six carbon atoms may be more likely. However, such defects may lead to crumpled sheets rather than to lower density of the sheets. Thus, it is inferred that expected deviation of the number density in carbon sheets, $\rho_s$, from that of graphite is minor. The minor deviation can be imbedded in the solid-fluid interaction parameter, $\varepsilon_{sf}$, determined by fitting experimental adsorption isotherm on non-porous carbonaceous material, since $\rho_s$ and $\varepsilon_{sf}$ occur as a product in the theory.

Despite these findings it is still common to arbitrarily assume infinite pore wall thickness in using the slit pore model, and the associated pore size dependent Steele 10-4-3 potential [6] is then employed for estimating the potential energy profile in a pore of any size. The inappropriateness of this assumption has recently been demonstrated in our laboratory, where it has been shown [5] that in typical nanoporous carbons having surface area in the range important for practical application (>800 m²/gm) the pore walls must actually be rather thin, and comprised of only a very small number (2-3) of graphene layers. For such small wall thicknesses the adsorption potential is much weaker than that obtained for the infinitely thick wall, and the adsorbed amounts can be lower by factors of 2 or more, particularly at low pressures where fluid-solid interactions dominate [5].

Here we present a novel technique whereby both pore size and pore wall thickness distributions can be simultaneously determined. We illustrate the method for a variety of nanoporous carbons, showing good correspondence of the results with x-ray diffraction. In addition, we also find that pore-pore correlation, involving interaction between fluid molecules in adjacent pores separated by thin walls, has a negligible effect on results of pore size distribution (PSD) and pore wall thickness distribution (PWTD) obtained from interpretation of argon adsorption 87K using our current approach. In the remaining part of this paper, we further apply the proposed model to characterize progressively heated carbons and predict supercritical adsorption of simple gases, such as ethane, on BPL carbon.

## 2. THEORY

In our technique, the Tarazona [13] NLDFT is applied to calculate the local density profile at the single pore level, with the confining potential taken as the sum of the potentials [6,12], $\phi_{wf}(z)$, from opposite walls of the pore:

$$\phi_{wf}(z) = 2\pi\rho_s \varepsilon_{cf} \sigma_{wf}^2 \left\{ \sum_{i=0}^{n-1} \left[ \frac{2}{5} \left( \frac{\sigma_{cf}}{z+i\Delta} \right)^{10} - \left( \frac{\sigma_{cf}}{z+i\Delta} \right)^4 \right] \right\}, \; z > 0 \qquad (1)$$

Here $n$ is the number of graphene layers in the pore wall, $\Delta$ is the interlayer spacing and $\rho_s$ is the number of carbon atoms per unit area in a single sheet. Following Steele [6] we use the values $\Delta = 0.335$ nm and $\rho_s = 0.3817$ atoms.Å$^{-2}$ for carbon. Considering the adsorption of an LJ fluid in a carbon having slit pore width distribution $f(H)$, and random pore wall thickness characterized by the probability distribution $p(n)$, we obtain the overall isotherm

$$\hat{\rho}(P) = \sum_{m=1}^{\infty} p(m) \sum_{\ell=1}^{\infty} p(\ell) \int_0^{\infty} \rho_{lm}(P,H) f(H) dH \qquad (2)$$

where $\rho_{lm}(P,H)$ is the local isotherm in a pore of slit width $H$, with left wall having $l$ graphene layers and right wall having $m$ layers, obtained from the NLDFT. Here we assume that the thicknesses of the two opposing walls of a pore are uncorrelated, and that the interaction potential between adsorbed molecules in neighboring pores is negligible in comparison to the fluid-solid potential energy for small molecules such as nitrogen [5], justifying the latter assumption. In the calculations we used the potential parameters $\sigma_f = 0.3375$ nm, $\varepsilon_f / k = 110.2$ K for argon, and $\sigma_c = 0.3349$ nm, $\varepsilon_c / k = 30.5474$ K.

Earlier attempts at characterization considering finite pore walls have been largely *ad hoc*, utilising arbitrary assumptions regarding the thicknesses of the confining walls [11,14], for example that one wall is infinitely thick while the other has exactly one graphene layer [14]. Such assumptions have resulted in unrealistic surface areas, and have therefore not led to a viable solution. Our use of the probability distribution $p(n)$ eliminates such arbitrary assumptions. Nevertheless, given the ill-posed nature of the problem [15,16] the determination of $p(n)$ is fraught with much uncertainty, with a multitude of possible solutions. The problem is, however, made tractable by the relation between the mean number of graphene layers in a pore wall, $\gamma$, and the surface area

$$\gamma = \frac{2N_o(1 - w_a)}{\rho_s M_c S} \tag{3}$$

which is a modification of that derived recently [5] based on mass balance over a pore wall, accounting also for the mass fraction of inorganic impurities (ash), $w_a$, in the carbon. Here $N_o$ is the Avogadro number, $M_c$ is atomic weight of carbon and and $S$ is the specific surface area.

Eq. (3) provides an important constraint to be satisfied by the probability distribution $p(n)$, since $\gamma = \sum_{n=1}^{\infty} np(n)$, and the pore wall thickness and pore size distributions are then correlated, which serves to restrict the solution space. In practice for $n \geq 5$ the potential energy is insensitive to $n$, and essentially matches that for an infinite wall. Consequently we lump the combined probability for all pore walls having five or more layers into $p(4+)$, leading to

$$\sum_{n=1}^{4+} p(n) = 1 \quad , \qquad \sum_{n=1}^{4} np(n) + 5p(4+) \leq \gamma \tag{4}$$

yielding constraints to be satisfied by $p(n)$, $n = 1,2,3,4,4+$. The solution of the distributions $f(H)$ and $p(n)$ is now accomplished by fitting the predicted isotherm to its experimental counterpart using Tikhonov regularization [16], combined with a genetic algorithm to search the solution space. In solving for $\hat{\rho}(P)$ the summations in eq. (2) are taken over the domain [1,4+], since walls having five or more layers are essentially infinitely thick.

## 3. EXPERIMENTAL

The above approach has been applied to data for argon adsorption at 87 K, for a variety of commercial carbons. The former were BPL, F-400 and PCB carbons manufactured by Calgon Corpn., RB-2, ROW-0.8, ROX-0.8, Norit R1 Extra (R1E) and Sorbonorit B4 (B4) carbons produced by Norit, and activated carbon fibre ACF-15 from American Kynol Corpn. Heat treatment of the R1E carbon was conducted at 1100°C for one hour (R1E1H), two hours (R1E2H), and six hours (R1E6H) under nitrogen atmosphere. All the carbons were degassed at 300 °C overnight prior to argon adsorption. A Micromeritics ASAP 2010 volumetric adsorption analyzer was used to obtain argon adsorption data at 87 K. The ash content of each carbon was determined gravimetrically following direct combustion in air at 873 K in a

Lindberg box furnace. X-ray diffraction patterns of the carbons were also obtained using Cu Kα radiation ($\lambda = 0.15406$ nm).

## 4. RESULTS AND DISCUSSION

Figure 1. illustrates the pore wall thickness distribution obtained for activated carbon BPL, with insets depicting the PSD and argon isotherm in each case. Similar results were also obtained for the other carbons, as shown in Table 1. From this table it is seen that surface area, S, calculated from the proposed model is slightly greater than that obtained if infinite wall thickness is assumed. Further, a shift of fitted pore size distribution to slightly larger pore size can be seen in the insets in Fig. 1. This has also been noted in the earlier study using arbitrarily sized pore walls [14], with some reduction in the S-shaped deviation between fitted and experimental isotherms, prominent in isotherm fits based on the infinitely thick wall assumption, when the opposing walls differ in thickness. In the inset in Fig. 1 the S shaped deviation in the isotherms is completely eliminated by the current approach. In the case of pore volume, the infinitely thick pore wall model underestimates the value by only a small amount, as seen in Table 1. This is largely due to the slightly lower density of the condensed argon in the pores with finite thickness walls.

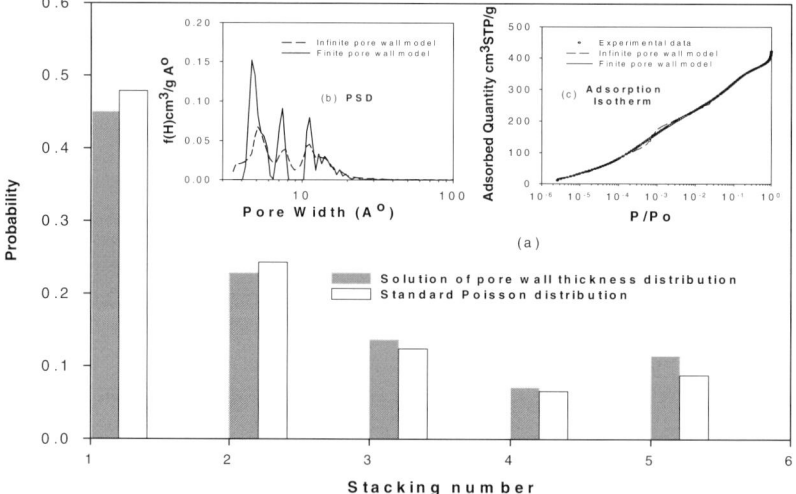

Fig. 1. Pore wall thickness distribution of BPL activated carbon determined by adsorption. Left inset depicts a comparison of the pore size distribution using the present finite wall thickness method and that using the conventional infinitely thick wall assumption. Right inset depicts fitted and experimental argon adsorption isotherms, illustrating the S shaped deviation using the infinitely thick wall assumption, in the $P/P_o$ range lying between $10^{-4}$ and $10^{-3}$.

Table 1 also lists the mean number of graphene sheets, $\gamma$, in a pore wall for each carbon, determined from the pore size distribution and the mineral matter content [15]. The latter was estimated from the residue obtained upon burning the carbon to complete conversion. The mean value $\gamma$ is found to be very low in general, and in the range of about 1.5-2, underscoring the inappropriateness of the assumption of infinitely thick walls. Besides highlighting the small wall thicknesses the pore size distributions also suggest that the carbon nanopores occur at discrete sizes or widths. This was evident from the sharp peaks in the distributions at discrete sizes from the application of the present approach, as opposed to the broad distribution peaks obtained from the infinite wall thickness approach. Furthermore, an average distance between the first three major peaks in pore size distribution of the investigated carbons from present approach is 3.23 Å respectively, while the corresponding values from the infinite wall thickness model are invariant at 3.04 Å. The former agrees well with the interlayer spacing of activated carbons.

Table 1
Comparison between results obtained for activated carbons based on the present technique and those obtained assuming infinitely thick pore walls

| Carbon | Infinite wall thickness model | | Finite wall thickness model | | |
|---|---|---|---|---|---|
| | $S_g$ $(m^2/g)$ | $Vp$ $(cm^3/g)$ | $S_g$ $(m^2/g)$ | $Vp$ $(cm^3/g)$ | $\gamma$ |
| ACF15 | 1446 | 0.52 | 1745 | 0.54 | 1.51 |
| BPL | 1023 | 0.51 | 1173 | 0.53 | 2.17 |
| F-400 | 1012 | 0.55 | 1158 | 0.57 | 2.14 |
| PCB | 1178 | 0.55 | 1382 | 0.57 | 1.80 |
| ROW | 1034 | 0.59 | 1198 | 0.61 | 2.06 |
| ROX | 1161 | 0.65 | 1348 | 0.66 | 1.95 |
| RB-2 | 1125 | 0.47 | 1294 | 0.49 | 1.94 |
| R1E | 1241 | 0.61 | 1461 | 0.64 | 1.62 |
| B4 | 1206 | 0.61 | 1408 | 0.63 | 1.77 |

While yielding the PSD and mean number of graphene sheets in the pore walls, the approach also provides the pore wall thickness distribution $p(n)$, as shown in Fig. 1 for the BPL carbon. In these figures $p(5)$ represents the lumped probability for all the walls having five or more layers, given above as $p(4+)$. The large proportion of single layer walls (45%) is evident from the figure, and was found to be the case for all the carbons examined. While one may expect details of the pore wall thickness distribution to depend on the manufacturing process, the predominance of single sheet walls appears to be a common feature for moderate and high surface area carbons. Further, it was noted that the pore wall thickness obtained from the isotherm fit could be well approximated by a generalized Poisson distribution [17] for all the carbons, as seen in Fig. 1. In particular for the carbon fiber ACF-15 it corresponds well to the limiting case of the Poisson distribution, corresponding to maximum randomness. Thus, the activated carbon fiber is expected to have a highly disordered structure.

As an independent characterization of the solid phase the parallelism indicator, $R$, was determined from the (002) peak of the x-ray diffractograms as discussed above. This parameter has been used by Dahn et al. [18] to estimate the fraction of graphene sheets that have no parallel neighbor. The (002) peak comes from constructive inference between X-rays scattered from parallel stacked graphene sheets. As the proportion of graphene layers with parallel neighbors increases, so will $R$. In terms of the probability distribution $p(n)$ the fraction of graphene sheets in walls having more than one sheet is $[\gamma - p(1)]/\gamma$, which may be expected to have a direct correspondence with the x-ray parallelism indicator $R$. Fig. 2. depicts this relationship for the various carbons studied in our laboratory, illustrated in terms of a plot of $R$ versus $[\gamma - p(1)]/\gamma$. In this figure the open triangles represent the data for the various coal-based activated carbons in Table 1, indicating a strong and nearly linear relationship between the two quantities. Given the somewhat diffuse nature of the (002) peak the parallelism indicator is not expected to be quantitative.precise. The good correlation between the adsorption and xrd-derived quantities, despite the approximate nature of $R$, is therefore representative of a strong correspondence between the two quantities for the carbons derived from a common precursor. The result for ACF-15 is seen to deviate from the correlation for the other carbons, probably due to the different precursor and manufacturing process.

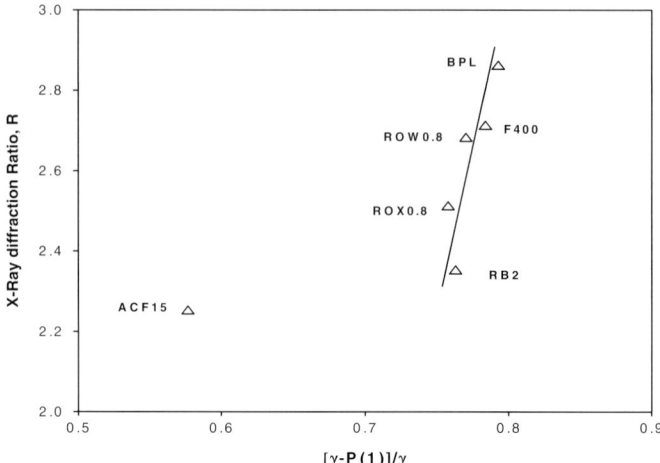

Fig. 2. Correlation between x-ray diffraction based parallelism indicator, $R$, and its counterpart, $[\gamma-p(1)]/\gamma$ determined by adsorption, for various carbons.

From the results above, it is evident that our current approach is more versatile than the infinitely thick pore wall model, and provides good correspondence with x-ray diffraction. However, pore-pore correlation may lead to enhancement of the local isotherm that may affect results of PSD and PWTD obtained from the current approach. Such correlation, as noted earlier [5], has negligible enhancement of local isotherm of small molecules like nitrogen. In the present work, it was found that pore-pore correlation has negligible effect on results of PSD and PWTD obtained from interpretation of argon adsorption at 87 K using the current approach [19].

Further validation of our current approach, through characterization of progressively heated carbons was conducted. From Fig. 3a and its inset, significant decrease in intensity of the first major peak in the PSD of these carbons, with increasing time of heat treatment, can be observedl. Such decrease in association with increasing proportions of thick wall containing at least 3 graphene layers, as seen in Fig. 3b., indicates significant variation of wall structures as well as strong interaction between the opposite walls of pores smaller than 6Å., leading to coalescence of thin pore walls to form thicker ones during heat treatment.

The last issue addressed here is prediction of supercritical adsorption of gases in porous carbons using the characteristic parameters of their internal structure probed by gas adsorption. The issue is not new but has not been well resolved due to mainly poor knowledge of carbon structure. In order to address this issue, we applied the results of the PSD and PWTD of BPL, obtained from interpretation of argon adsorption at 87 K using our proposed approach, to predict supercritical adsorption of simple gases such as methane, ethane. The results were then directly compared with those obtained from the use of PSD calculated by infinitely thick pore wall model and experimental data, and demonstrated the better capability of our approach. As an example Fig. 4 below depicts the results for ethane adsorption on BPL carbon, illustrating the improved predictions by the proposed approach incorporating the PWTD.

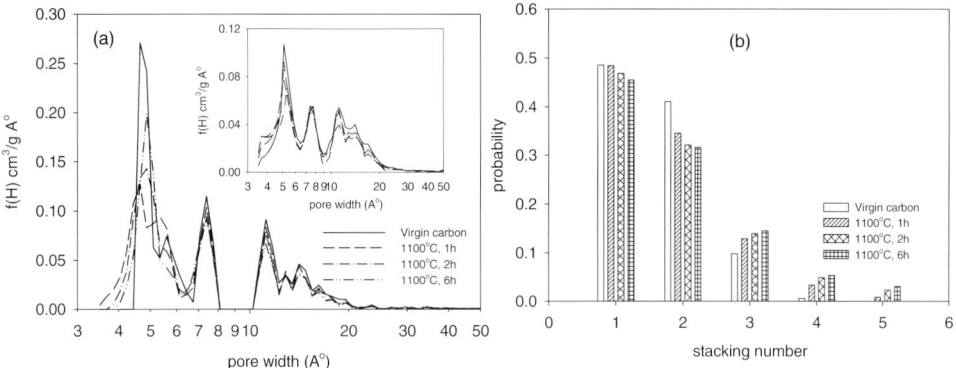

Fig. 3. PSD and PWTD of progressively heated carbons: (a) FWT based PSDs and the inset depicts IWT based PSDs; (b) FWT based pore wall thickness distribution.

Further application of the proposed approach to predict different sets of adsorption experimental data of methane in BPL carbon, reported in the existing literature, showed that the approach overpredicts some sets of experimental data of methane in BPL carbon such as those obtained by Gusev *et al.* [22]. From this observation, it was inferred that such overprediction was due to finite pore connectivity that leads to occurrence of regions in the BPL carbon that are inaccessible to methane, even though the smallest pore size of the BPL carbon is larger than the methane molecule. Accordingly, we proceeded to introduce the fraction of accessible carbon volume into the generalized adsorption isotherm, as earlier proposed by Davies and Seaton [23]. It was found that the current approach predicts correctly the sets of adsorption experimental data of methane in BPL carbon at different temperatures with a constant fraction of accessible volume.

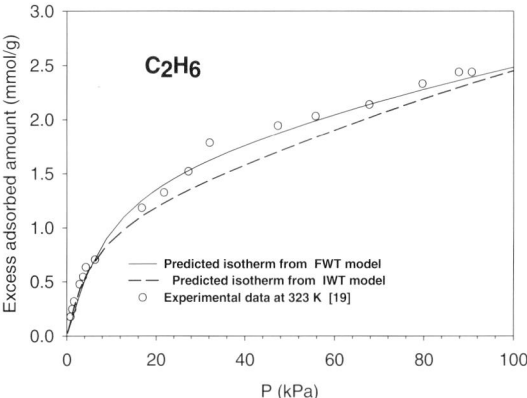

Fig. 4. Predicted and experimental adsorption isotherms of ethane in BPL at 323 K. Solid line depicts predicted isotherm obtained from finitely wall thick model and dash line represents that using finitely thick wall model. Open circles are experimental data up to 1 bar taken from Russel and LeVan [21].

## 5. CONCLUSIONS

The above results demonstrate that adsorption can be reliably used to probe the physical structure of the solid phase in carbons, conventionally studied by x-ray diffraction. The good correspondence with xrd, combined with the sharp PSD peaks, suggests that the narrow micropores are defects within the same crystallite, most likely arising from missing portions of neighboring graphene sheets. Nevertheless, some influence of intercrystalline pore space, particularly in the larger pores beyond about 1 nm may be expected. Further, pore-pore correlation has been shown to have a negligible effect on results of PSD and PWTD obtained from interpretation of argon adsorption at 87 K using our current approach. Finally, it has been illustrated that our current charaterisation approach is superior to the infinitely thick wall model, and enables one to accurately predict supercritical adsorption of other gases on the same carbon.

## 6. ACKNOWLEDGEMENT

This research has been supported by the Australian Research Council. Use of the High Performance Computing facilitility at the University of Queensland, and the supercomputing facilities at the National Centre for Advanced Computing, are gratefully acknowledged.

## REFERENCES

[1]  N. A. Seaton, J.P.R.B. Walton and N. Quirke, Carbon, 27 (1989) 853.
[2]  C. Lastoskie, K.E. Gubbins, and N. Quirke, Langmuir, 9 (1993) 2693.
[3]  A. Vishnyakov, P.I. Ravikovitch and A.V. Neimark, Langmuir, 15 (1999) 8736.
[4]  S.K.Bhatia, Langmuir, 14 (1998) 6231.
[5]  S.K.Bhatia, Langmuir, 18 (2002) 6845.
[6]  W.A. Steele, Surf. Sci., 36 (1973) 317.
[7]  S.K. Bhatia and D. Nicholson, Phys. Rev. Lett., 90 (2003) 016105.
[8]  O.G. Jepps, S.K. Bhatia and D.J. Searles, Phys. Rev. Lett., 91 (2003) 0126102.
[9]  Y.F. Yin, B. McEnaney and T.J. Mays, Carbon, 36 (1998) 1425.
[10] T. Suzuki, K. Kaneko, N. Setoyama, M. Maddox and K. E. Gubbins, Carbon, 34, (1996) 909.
[11] B. McEnaney, Carbon, 26 (1988) 267.
[12] T. J. Mays, Fundamentals of Adsorption 5, Kluwer Academic, Boston, 1996.
[13] P. Tarazona, Phys. Rev. A 1985, 31, 2672; 1985, 32, 3148. Tarazona P.; Marini Bettolo Marconi, U.; Evans, R. Mol. Phys. 1987, 60, 573.
[14] P.I. Ravikovitch, J. Jagiello, D. Tolles and A.V. Neimark. Carbon'01, International Conference on Carbon, Lexington, 2001.
[15] T.X. Nguyen and S.K. Bhatia, Langmuir, 20 (2003) 3532.
[16] A.N.Tikhonov, Dokl. Akaf. Nauk, SSSR, 49 (1963) 153.
[17] P.C. Consul, Generalized Poisson Distributions: Properties and Applications; Marcel Dekker: NewYork, 1989.
[18] J.R. Dahn, W. Xing and Y. Gao, Carbon, 35 (1997) 830.
[19] T.X. Nguyen, and S.K. Bhatia, J. Phys. Chem. B, 108 (2004) 14032.
[20] T.X. Nguyen and S.K. Bhatia, Carbon, 43 (2005) 775.
[21] B.P. Russel and M.D. LeVan, Ind. Eng. Chem. Res., 36 (1997) 2380.
[22] V.Y. Gusev and J.A. O'Brien, Langmuir, 13 (1997) 2815.
[23] G.M. Davies and N.A. Seaton, Langmuir, 15 (1999) 6263.
[24] T.X. Nguyen, S.K. Bhatia and D. Nicholson, Langmuir, 21 (2005) 3187.

Studies in Surface Science and Catalysis 160
P.L. Llewellyn, F. Rodriquez-Reinoso, J. Rouqerol and N. Seaton (Editors)

# Adsorption and neutron scattering studies: a reliable way to characterize both the mesoporous MCM-41 and the filling mode of the adsorbed species

**N. Floquet[a], J.P. Coulomb[a], P.L. Llewellyn[b], G. André[c] and R. Kahn[c]**

[a]Centre de Recherche en Matière Condensée et Nanosciences, CNRS, Campus de Luminy, Case 901, 13288 Marseille Cedex 9 - France.

[b]MADIREL - UMR 6121, Centre de S[t] Jérome, 13397 Marseille Cedex 20 - France

[c]Laboratoire Léon - Brillouin, CEA - Saclay, 91191 Gif - sur - Yvette, Saclay - France.

Numerous studies concern the MCM-41 material, and yet the MCM-41 porosity description and especially the MCM-41 pore diameter (Ø) characterization is still open for discussion. Here, we report on adsorption and extensive neutron diffraction analyses for the hydrogen sorption in the mesoporous material MCM-41. The type IV isotherm for hydrogen sorption in MCM-41 is similar to many species sorption isotherms such as $N_2$, Ar, Kr. The hydrogen isotherm particularity is that the two distinctive parts of the isotherm (first vertical uptake at the low relative pressure $p/p_0 \leq 0.1$ and second uptake at the high relative pressure $p/p_0 > 0.1$) are equally developed. Thus, the neutron scattering experiments realized during the hydrogen adsorption gave accurate data for each adsorption mechanism associated to these two main uptakes, and consequently for MCM-41 porosity structural characterisation. The main findings are the low density of the MCM-41 silica walls (20% voids) and the fine description of $D_2$ adsorption mechanism : at the low relative adsorption pressure $D_2$ fills the wall voids and forms a layer on the rough wall surface. Then at the relative pressure of the second uptake, $D_2$ fills the whole free MCM-41 mesopores as expected (capillary condensation phenomenon). Even the solid capillary phase does not grow layer by layer on the inner pore walls but is growing up along the pore axis.

## 1. INTRODUCTION

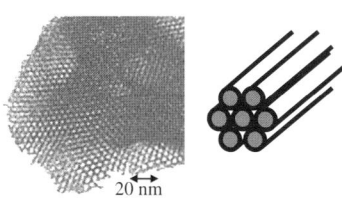

Fig. 1. Transmission electron micrograph of our MCM-41 samples and illustration of the hexagonal network of the cylindrical MCM-41 mesopores.

MCM-41 is the well known model mesoporous material containing extended hexagonal arrays of parallel tubular pores separated by amorphous silica (Fig. 1.). This is a common description, on the base of most studies to interpret adsorption phenomenon results. However, the structural and chemical parameters of the host porous material have a crucial influence on the adsorption mechanism. Actually each synthesis produces MCM-41 sample with its own silica wall and porosity. Our MCM-41 (19 Å < Ø < 40 Å) samples have been extensively characterized by sorption isotherm, microcalorimetry and neutron diffraction measurements of many adsorbates such as Ar, $N_2$, $O_2$, $C_2$, $H_2$ and $CO_2$ [1-7]. However, although these MCM-41 samples were produced by different laboratories, our adsorption studies proved that the ideal MCM-41 porosity could not be the base to interpret all our experimental results coming from different and coupled measurements. Actually, they led us to consider a certain roughness of the MCM-41 walls or a model of low density silica walls that is rarely reported [8-9]. Note that uncoupled experimental data such as adsorption and X-ray data could produce uncertain characterizations as underlined previously [10]. The present work reports on extensive neutron diffraction analyses for the hydrogen sorption in the mesoporous material MCM-41 ($\varnothing \cong 25$ Å). The main goal of the present paper is to outline how our experimental and analytical approach is an illustrative way both to give detailed structural data of the porous structure of the MCM-41 and to characterize the filling mode of the adsorbed species in the MCM-41 [7].

## 2. EXPERIMENTAL SECTION

The neutron diffraction experiment has been performed at the laboratory Léon Brillouin (C.E.A. - Saclay) on the G4-1 diffractometer. The MCM-41 ($\varnothing \cong 25$ Å) sample used was synthesized at the laboratory of K.K. Unger (Mainz - Germany). Prior to the experiment it was outgazed at T = 573K under vacuum (P < $10^{-6}$ torr) during around 12 hours. Fourteen diffractograms have been measured at different loadings for deuterated hydrogen confined phase in MCM-41 ($\varnothing \cong 25$ Å) at T = 16.4K. A calibration sorption isotherm is measured during the neutron diffraction experiment.

## 3. RESULTS AND DISCUSSION

### 3.1. Adsorption isotherm characterization: filling process of the MCM-41 mesopores.

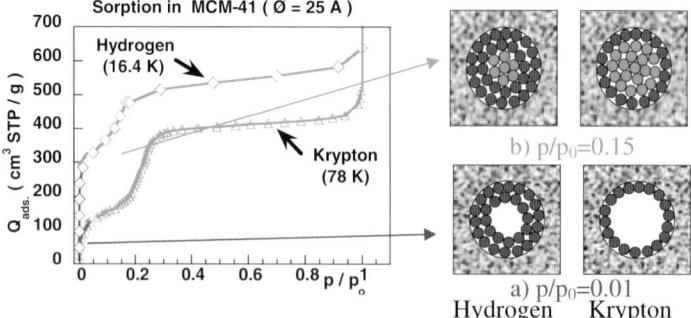

Fig.2. Sorption isotherms of $D_2$ ( $\diamond$ ) at T = 16.4K and Kr ( $\triangle$ ) on MCM-41 ($\varnothing \cong 25$ Å) at T = 77.4K and illustration of the two adsorption steps; at p/p0 = 0.01: pore wall physisorption; at p/p0 = 0.15 : capillary condensation

Most species adsorption on MCM-41 gives rise to type IV isotherms (Fig. 2). Two steps are well observed. The relative extension of the two parts depends on the chemical nature of the confined molecules [1,11]. It is very high in the case of hydrogen and low in the case of krypton. The first uptake at this very low relative pressure ($P/P_0 \leq 0.1$) corresponds to the formation of a film of uniform thickness on the pore walls: up to two layers in the case of hydrogen, one layer in the case of krypton ($P_0$ is the saturated vapor pressure of the bulk sorbate). The second sharp increase at higher pressure (P/ $P_0$ > 0.1) is due to the formation of the so called "capillary phase". This analysis of the adsorption process in the mesoporous MCM-41 is currently prevailing. But for a more precise description of the adsorption process, detailed structural characterization of both the host MCM-41 and the confined phase will be required.

### 3.2. Neutron diffraction characterization: MCM-41 structure and mesopore filling signatures.

MCM-41 samples and the adsorption stages of the confined molecules in MCM-41 have been structurally characterized by neutron diffraction (Fig. 3.). The hexagonal network of

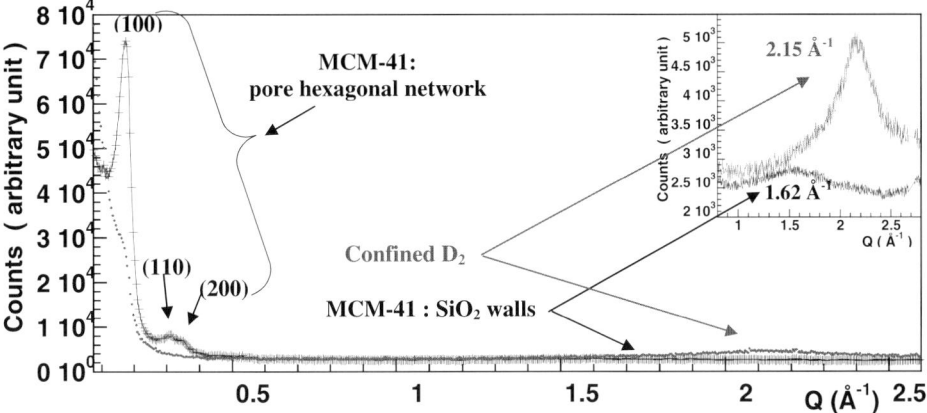

Fig. 3. Neutron diffractograms of MCM-41 (Ø ≅ 25 Å) measured at T=16.4 K : empty (black plot) and for a $D_2$ loading of 95% (red plot).

MCM-41 produces diffraction peaks at low diffraction angle. One intense peak (100) and weak peaks are generally well observed. The MCM-41 walls made of the amorphous silica

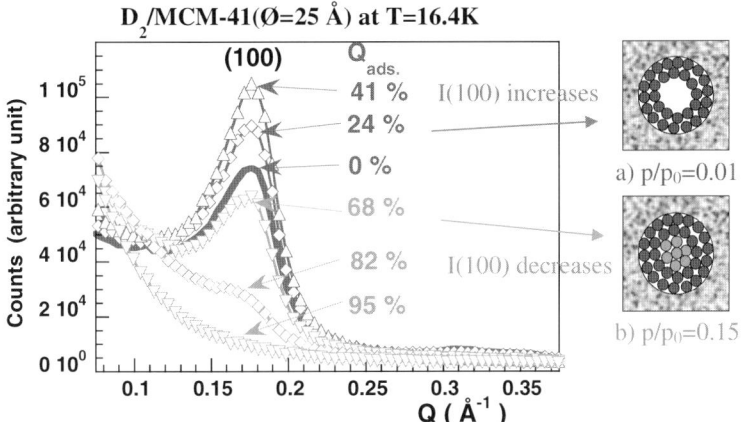

Fig.4. Neutron diffractograms of MCM-41 (Ø ≅ Å Å) measured at T=16.4 K : empty and for increasing $D_2$ loading ; illustration of the two adsorption steps; at p/p0 = 0.01: layer by layer filling mode; at p/p0 = 0.15 : capillary filling mode

give a very large bump near 1.6 Å$^{-1}$. And the confined molecules would diffract in this range between 1 to 3 Å$^{-1}$. Intensity modifications of the MCM-41 diffraction peaks tell about the location of the confined molecules within the MCM-41 mesopores [7]. It allows to identify the two adsorption stages observed in the isotherms. In the case of deuterium (neutron coherent scattering length = 6.67 10$^{-13}$ cm), we observe very large intensity modifications for the first strong (100) peak during the adsorption (Fig. 4). During the first stage of adsorption, up to $Q_{ads.}$ = 41%, which corresponds to the pore wall physisorption in a layer by layer filling mode, its intensity is increasing. Then, during the second stage, when occurs the capillary filling mode, the (100) intensity is decreasing.

### 3.3. Diffraction pattern analysis from the MCM-41 idealised structure.

To validate this usual scenario of the two filling modes, the first attempt was a modelling of the diffraction patterns from a model structure for MCM-41, that is an hexagonal pore, amorphous silica walls with a silica density of 2.2. Both the MCM-41 wall atoms and the sorbate atoms are supposed to be located on a discrete hexagonal lattice. The MCM–41 hexagonal network [7] is defined by the pore diameter (Ø) and the pore wall thickness ($t_w$). Then, the unit cell parameter is $a_h$ = Ø + $t_w$. The layer by layer filling mode, and the capillary filling mode are defined from the sorbate positions within the pore. Usual diffraction peak intensity is calculated for the atomic sorbate positions and at the different sorbate loadings corresponding either to the layer by layer filling mode, either to the capillary filling mode.

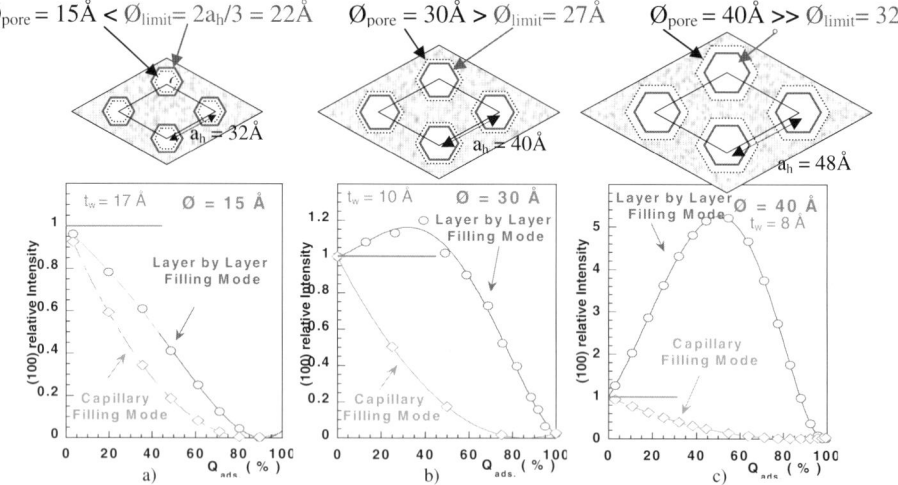

Fig. 5. Calculated relative intensity $I_r$ of the MCM-41 (100) peak as function of $D_2$ loading in the two filling modes, the layer by layer mode and the capillary mode, the density of the walls is $d_{SiO2}$ = 2.2 g.cm$^{-3}$, MCM-41 pore diameter (Ø) and wall thickness ($t_w$) are either with Ø < $2t_w$ (Ø < $2a_h/3$) in the case a) or with Ø > $2t_w$ (Ø > $2a_h/3$) in the case b and c).

Three idealised MCM-41 samples have been tested (Fig. 5). They are characterized by three different increasing pore sizes (Ø = 15, 30 and 40 Å) and three different decreasing wall thicknesses $t_w$ = 17, 10 and 8 Å . In the case of the capillary filling mode, whatever is the MCM-41 structure, the (100) intensity is continuously decreasing. In the case of the layer by layer filling mode, there is an intensity increase for the two MCM-41 samples having large Ø pores and thin $t_w$ walls ($\varnothing > 2t_w$ or $\varnothing > 2a_h/3$). For the other sample with $\varnothing < 2t_w$ (small pores and thick $t_w$ walls), intensity is continuously decreasing as in the case of the capillary filling mode. Thus, in conclusion, the simulation outlines that intensity modifications depend both on the filling mode but also on the MCM-41 pore diameter $\varnothing$ and wall thickness $t_w$. An intensity increase of the MCM-41 (100) diffraction peak would correspond to: a MCM-41 having large pores and thin walls ($\varnothing > 2t_w$ or $\varnothing > 2a_h/3$), and to an adsorption mode starting by a film physisorption.

In this idealised structural description of MCM-41, the quantity of adsorbed $D_2$ at $Q_{ads.} = 100\%$ corresponds to a pore diameter $\varnothing = 30$ Å. Fig. 6. compares the experimental and the computed intensity $I_r$ of the (100) peak corresponding to the two filling modes of MCM-41 hexagonal pores having a wall density $d_{SiO2} = 2.2$ g.cm$^{-3}$. The modelling would not be satisfying. The computed intensity for a layer by layer mode has its maximum at $Q_{ads.} = 43\%$, but it is 30% lower than the experimental one. There is two ways in this modelling to increase the calculated intensity. The first way is an increase of the pore size, but it would not correspond to the experimental adsorbed quantity. The second way is a decrease of the silica density of the walls. Actually a best fit was obtained for a pore diameter $\varnothing = 30$ Å, a wall thickness $t_w = 10$Å and a density of the walls $d_{SiO2} = 1.62$ g.cm$^{-3}$. The calculated intensity is shown in Fig. 7. The layer by layer mode ends at $Q_{ads.} = 43\%$, it corresponds to a $D_2$ film thickness $t_{film} = 1.25$ layer. Then starts the capillary filling mode up to the full loading. A misfit between the calculated and experimental data well appears during the first adsorption stage. The calculated intensity is larger than the experimental one. It means that the decrease of the silica density from 2.2 g.cm$^{-3}$ to 1.6 g.cm$^{-3}$ could be micropores inside the walls and that deuterium adsorbs this wall microporosity at the beginning of the adsorption. Such findings are expected, a previous work [8, 12] on structure and dynamics of hydrogen sorption in MCM-41 led to similar conclusions.

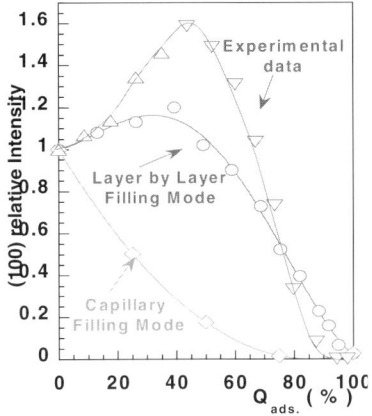

Fig.6. Experimental and calculated relative intensity $I_r$ of the MCM-41 (100) peak as function of $D_2$ loading in the two filling modes, the layer by layer mode and the capillary mode, case of a MCM-41 pore with diameter $\varnothing = 30$ Å and a density of the walls $d_{SiO2} = 2.2$ g.cm$^{-3}$.

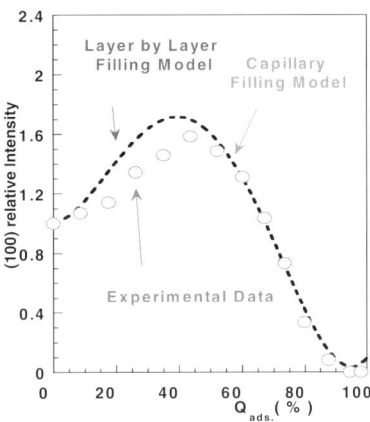

Fig. 7. Experimental (O) and calculated (dashed line) relative intensity $I_r$ of the MCM-41 (100) peak as function of $D_2$ loading. Successive filling mode modelling, the layer by layer mode up to 43%, then the capillary mode up to full loading; case of a MCM-41 pore diameter $\varnothing = 30$ Å and a density of the walls $d_{SiO2} = 1.62$ g.cm$^{-3}$.

### 3.4. Diffraction pattern analysis by Rietveld method.

In view of the above findings, the Rietveld method was performed for the diffractogram refinement, using a more realistic structural model for MCM-41 [7]. First, we have checked the MCM-41 pore shape : cylindrical pore and hexagonal pores in two different orientations have been tested. Each intensity of the weak (110), (200) and (210) peaks is modified. It is the hexagonal pore oriented as usual (see Fig. 5.) which gives rise to the highest intensity of these peaks and best fits our experimental data. Then we have checked the influence of the MCM-41 pore size. Two pore sizes of $\varnothing$ = 30 Å and 25 Å have been simulated. The intensity of the three weak peaks is higher in the case of $\varnothing$ = 30 Å. But our experimental data would be between these both cases. Finally we have simulated the two successive filling modes corresponding to our experimental diffraction patterns. The layer by layer mode up to a hydrogen filling of 41%. Then, the capillary mode up to the full hydrogen loading. Diffraction patterns have been calculated for the three pore sizes : $\varnothing$ = 25, 30 and 32 Å. For $\varnothing$ = 25 Å pore size, the experimental diffraction patterns are not reproduced. There is no intensity increase at

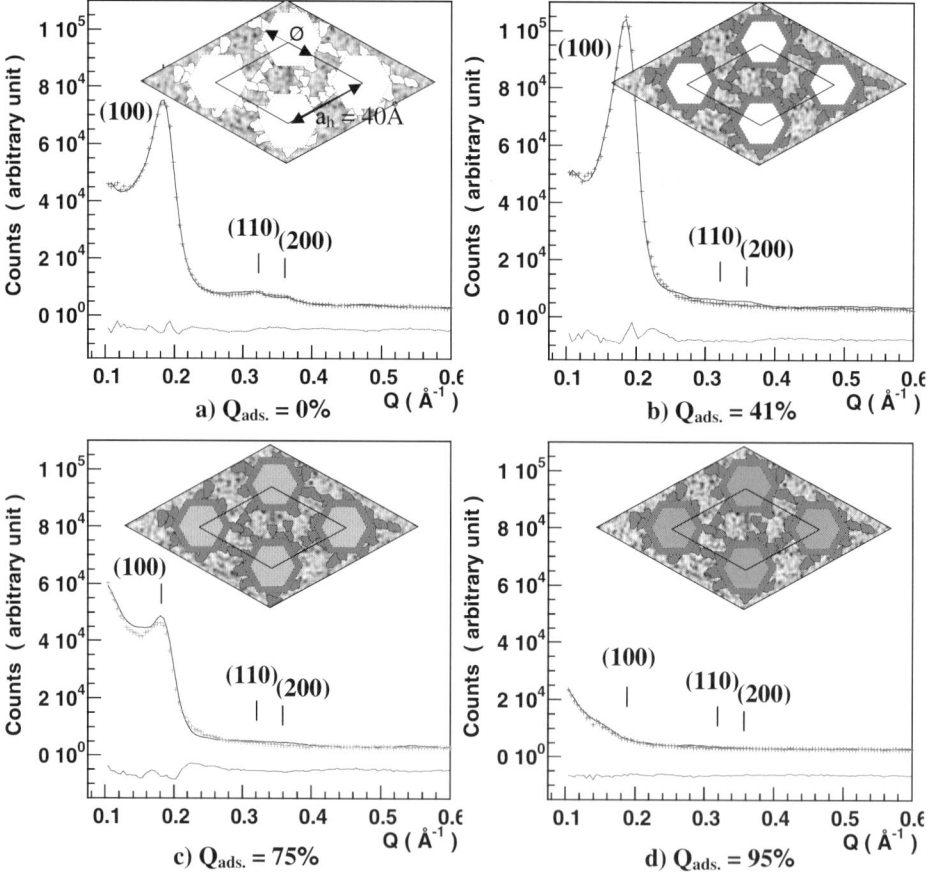

Fig.8. Rietveld refinement of the MCM-41 (hk0) peaks for a) empty MCM-41 and b) 41%, c) 75% and d) 95% $D_2$ loadings and illustrations of the MCM-41 porosity filling.

any hydrogen loading. For Ø = 30 Å pore size, the intensity increase is too low and its decrease is too strong. For Ø = 32 Å pore size, the intensity increase is almost reproduced, but its decrease is still stronger. These simulations confirm our previous findings. By taking a 2.2 density for the silica walls of our MCM-41 sample, simulation does not reproduce the diffraction data. Thus, the next step of the diffraction pattern refinement was to adjust a microporosity within the silica walls and its filling with hydrogen molecules. This refinement

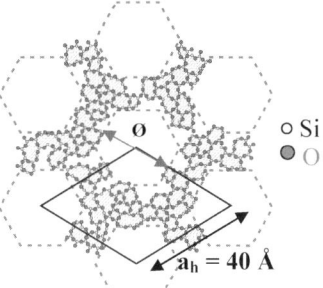

o Si
● O

Fig. 9. Illustration of the MCM-41 structural model as a result of the Rietveld analysis of our neutron diffractograms: the hexagonal network of the MCM-41 mesopores ($a_h$ = 40 Å) have quasi hexagonal mesopores Ø = 27 Å and rough amorphous silica walls ($SiO_2$ density = 1.76 g.cm$^{-3}$)

was successful. The full set of diffraction patterns was reproduced (Fig. 8.).
The MCM-41 structural model was fitted to a pore size of Ø = 27 Å, with microporous voids in the walls (wall roughness as represented in Fig. 9) corresponding to a density of d = 1.76 g.cm$^{-3}$ instead d = 2.2 g.cm$^{-3}$ that is a 20% of free space in the wall. This MCM-41 adsorbs hydrogen in two steps (Fig. 10). In the first step, up to a filling of $Q_{ads.}$ = 48%, hydrogen adsorbs both microporosity within the wall and the first wall layer. In the capillary step, from a filling of $Q_{ads.}$ = 48% to full loading, hydrogen fills the MCM-41 central mesopore. It follows that the limit at $Q_{ads.}$ = 48% between the two steps corresponds to the inflexion point that is the usually accepted limit between the two steps.

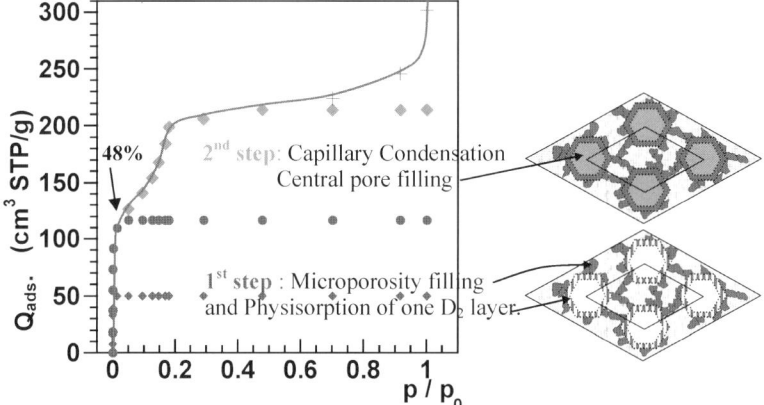

Fig.10. Illustration of the MCM-41 complex porosity and its filling modes by $D_2$, interpreting the $D_2$ adsorption isotherm in MCM-41 and resulting from the Rietveld refinements of our neutron diffractograms: 1) $D_2$ fills the microporosity of the MCM-41 walls (♦ purple) and physisorbs the inner surface of the MCM-41 walls (purple ●) up to 48% of loading, 2) $D_2$ fills the MCM-41 central mesopore (orange ♦).

## 4. CONCLUSION

An accurate analysis of experimental diffractograms recorded during the adsorption is a successful way both to give detailed structural data of the porous structure of the MCM-41 sample and to characterize the growth mode of the adsorbed species in the MCM-41 sample. Actually from a simple structural modelling, it appears that, in the case of adsorption of most gas, an initial increase of the diffracted intensity of the first intense (100) peak gives information on the pore diameter Ø and or the roughness and microporosity of the walls. It means that there is gas adsorption in the MCM-41 region r and $a_h/3 < r < (1-a_h/3)$. In other words, it means either that there is physisorption of a $t_{film}$ thickness film on the inner MCM-41 walls and $(Ø-2t_{film}) > 2a_h/3$, and or there is microporosity within the MCM-41 wall, and adsorption occurs in the MCM-41 wall porosity. Such experimental data could be obtained in most cases and would help greatly to the characterization of the porous material and to the adsorption phenomenon analysis.

## REFERENCES

[1] P.L Llewellyn, Y. Grillet, F. Schüth, H. Reichert, K.K. Unger, Microporous Materials 3 (1994) 345.
[2] P. L. Llewellyn, F. Schüth, Y. Grillet, F. Rouquerol, J. Rouquerol and K. K. Unger, Langmuir, 11 (1995) 574.
[3] N. Floquet, J.P. Coulomb, S. Giorgio, Y. Grillet, P.L. Llewellyn, Studies in Surface Science and Catalysis, 117 (1998) 583.
[4] N. Floquet, J.P. Coulomb, C. Martin, Y. Grillet, P.L. Llewellyn and G. André, Proceedings of the 12th Int. Zeolite onference, Ed. M.J. Treacy et al., Material Research Society (1999) 659
[5] J.P. Coulomb, N. Floquet, Y. Grillet, P.L. Llewellyn, R. Kahn and G. André, Studies in Surface Science and Catalysis, 128 (2000) 235.
[6] J.P. Coulomb, N. Floquet, C. Martin, R. Kahn, Eur. Phys. J. E 12 (2003) 25
[7] N. Floquet, J.P. Coulomb and G. Andre, Microp. Mesopor. Mater. 72 (2004) 143
[8] K. J. Edler, P. A. Reynolds,' P.J. Branton,F. R. Trouw and J. W. White, J. Chem. Soc., Faraday Trans., 93 (1997) 1667.
[9] L. A. Solovyov, S. D. Kirik, A. N. Shamkov and V. N. Romannikov, Microporous and Mesoporous Materials 44-45 (2001) 17.
[10] M. Kruk and M. Jaroniec, Chem. Mater., 11 (1999) 492.
[11] J. P Coulomb, C. Martin, Y. Grillet, P. Llewellyn and G. André, Studies in Surface Sciences and Catalysis, 105 (1997) 1827.
[12] K. J. Edler, P. A. Reynolds, J. W. White and D. Cookson, J. Chem. Soc., Faraday Trans., 93 (1997)199

Studies in Surface Science and Catalysis 160
P.L. Llewellyn, F. Rodriquez-Reinoso, J. Rouqerol and N. Seaton (Editors)

# Absolute assessment of adsorption-based microporous solid characterisation methods

**M.J. Biggs[†], A. Buts, Q. Cai and N.A. Seaton**

Institute for Materials and Processes, University of Edinburgh

Kenneth Denbigh Building, King's Buildings, Mayfield Road, Edinburgh, UK, EH9 3JL.

## 1. INTRODUCTION

Although adsorption is by far the most widely used means of microporous solid characterisation, it is not without its problems [1]; these may be broadly described in terms of correctness, consistency (*i.e.* is the parameter purely related to what it purports to represent or does it 'include' more), and meaningfulness (*e.g.* what does 'surface area' mean in a microporous solid) [2]. Much effort has been directed towards addressing these concerns using *relative assessment* in which data obtained from two or more methods for a solid are compared. This approach is rarely satisfying for a variety of reasons including, amongst others, the difficulty faced in understanding any observed differences.

An alternative to relative assessment is to use a solid whose characteristics are exactly known and for which the interstitial fluid behaviour can be probed in detail. Whilst such an *absolute assessment* process is (perhaps) experimentally feasible for solids like zeolites, it is clearly not for ill-defined solids such as carbons, which are most in need of assessment. We have, therefore, developed and applied a molecular simulation based methodology for the absolute assessment of adsorption-based characterisation methods. A summary of this methodology with examples of its application are given here; greater details may be found elsewhere [2, 3].

## 2. ABSOLUTE ASSESSMENT METHODOLOGY

Figure 1 illustrates the absolute assessment methodology. Grand Canonical Monte Carlo (GCMC) simulation is used to determine the adsorption and desorption isotherms for a model fluid in an *in silico* solid for which measures of the characteristics are known exactly. The adsorption or, if appropriate, desorption isotherm is then submitted to the method to be assessed and estimates obtained. These estimates are compared with the corresponding exactly known measures and conclusions are drawn regarding the correctness (closed loop in Figure 1) and, if appropriate, meaningfulness and consistency of the methods for the particular model system. Reasons for lack of correctness can be identified and assessment of meaningfulness and consistency can be made by probing the adsorption process at the

---

† Author for correspondence. E-mail: m.biggs@ed.ac.uk. Fax: +44-131-650-5891. Phone: +44-131-650-6551.

molecular level; such analysis can be used to suggest improvements to the characterisation method (feedback loop in Figure 1) or an entirely new method that can in turn be assessed.

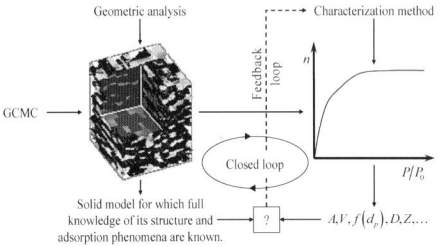

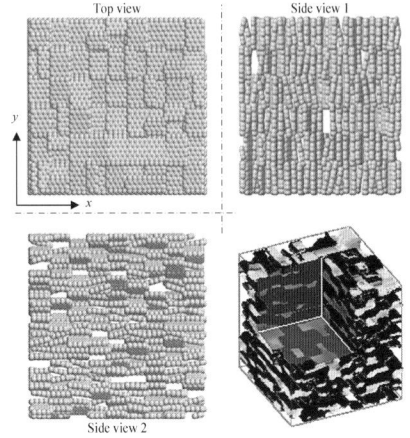

Figure 1. Absolute assessment methodology.

Figure 2 (right). Model carbon used in study of size (in angstroms) $88.42 \times 76.51 \times 80.50$.

## 3.    STUDY DETAILS

### 3.1.    Solid model

A variety of model solids have been used by us in assessing adsorption-based characterisation methods ranging from simple slit pore models with various surfaces (*e.g.* basal, armchair) through to complex solids such as that shown in Figure 2 – the work presented here is based on this particular model. This model is made up of small *basic structural units* (BSUs) of $m = 1-3$ parallel evenly spaced graphene domains of size $9.82 \times 12.76$ Å$^2$ that are randomly tilted up to $\pm 15°$ about the $x$ and $y$ axes and arranged in such a way that microporosity and surfaces of highly irregular character are created between the BSUs [4]. The complex microporosity effectively decouples the rigid link between pore size and energy that exists in simpler models. Whilst this model does not match any specific carbon, previous work [4] has demonstrated that it and similar models can produce a wide range of isotherm shapes and heats of adsorption loading dependencies that match, at least qualitatively, those observed experimentally.

A variety of direct methods have been used to determine a range of characteristics for this solid against which those derived from adsorption-based methods can be compared. Many are based on three geometric sets [5]: (i) those cubes and tetrahedra of a tessellation that are wholly located in the pore space, $\varsigma$, (ii) those cubes and tetrahedra of the tessellation that are wholly located in the solid, $\Sigma$, and (iii) those planes that define the interface between these two sets, I. A number of them are also based on the subdivision of the pore space into a set of individual pores, $\Pi$, and associated pore throats, T, that are assumed to be defined by those planes in the pore space that represent significant barriers to diffusing molecules [6]. Those quantities relevant here are:

1. **Pore volume.** The *total pore volume* is equal to that of $\varsigma$ for a fine tessellation (*i.e.* it is independent of $d$). The *pore volume accessible to fluid-i in the absence of pore network effects* is equal to that part of the total that is covered by the fluid molecules in a GCMC simulation at saturation. The *pore volume accessible to a fluid-i in the presence of pore network* effects is equal to $\Sigma_j(\delta\varsigma_j)$ for a fine tessellation that is covered by fluid molecules in a GCMC simulation at saturation, for all pores-$j$ visited by the fluid in a non-equilibrium MD simulation (see below).

2. **Pore surface area.** The *geometric surface area* is equal to the area of I for a fine tessellation. The *covered surface area*, a more meaningful quantity for solids dominated by pores that cannot accommodate more than one fluid layers, is determined from the number of fluid molecules that cover the surface at saturation [2].

3. **Fractal dimension.** This is derived by the box counting method, which uses the surface area of I as a function of the size of the cubic tessellation.

4. **Pore size distribution (PSD).** This is derived from the volume and size of the pores in $\Pi$, where the former is defined by the volume of the sets $\delta\Sigma_i$ and the latter by the size of the largest hard spherical probe that can be accommodated within a pore.

5. **Mean coordination number.** Two measures are used: (i) the mean number of pores emanating from junctions in T; and (ii) the *dynamic mean coordination number*, which is the mean coordination number of the network defined by the pore throats in T crossed by diffusing molecules of a fluid during a non-equilibrium MD simulation (see below).

## 3.2. Fluid model

The fluids are modelled by a truncated and shifted Lennard-Jones (LJ) pair potential [7]. The potential parameters and equations of state of the fluids are given in Table 1 with the parameters for carbon; cross-interaction parameters are obtained by the Lorentz-Berthelot rules. The cut-off radius for simulations in the solid is $2.5\sigma_{ff}$, whilst that used in generating the single pore isotherms for determining the PSDs is $5\sigma_{ff}$, inline with standard procedure.

## 3.3. Simulation details

Adsorption is modelled by Grand Canonical Monte Carlo (GCMC); full details of the method and simulation protocols may be found elsewhere [2]. By applying Grand Canonical molecular dynamics (GCMD) [8] to the model solid in an initially evacuated state, the pore throats in T passed through and pores in $\Pi$ accessed by the fluids are also identified.

| Species | Potential Parameters | | | Pore resolution | | EOS |
|---|---|---|---|---|---|---|
| | $\varepsilon/k_b$ (K) | $\sigma$ (Å) | Ref. | $w_l$ (Å) | $w_u$ (Å) | Ref. |
| $N_2$ | 95.2 | 3.75 | [9] | 4.10 | | [10] |
| $CH_4$ | 149.9 | 3.73 | [11] | 3.94 | | [11] |
| $CF_4$ | 152.5 | 4.70 | [11] | 4.45 | | [11] |
| $SF_6$ | 200.9 | 5.51 | [11] | 4.95 | | [11] |
| C | 28.0 | 3.40 | [12] | - | - | - |

Table 1. LJ parameters, $\varepsilon$ and $\sigma$, lower and upper pore sizes resolvable by fluid, $w_l$ and $w_u$, and bulk phase equation of state used. Pore size is distance between surface carbons less 3.4 Å.

## 4.    RESULTS AND DISCUSSION

### 4.1.    Adsorption isotherms

The adsorption isotherms for nitrogen at 77 K up to the saturation pressure, $P_0$, and the remaining fluids at the supercritical temperature of 258 K for pressures up to 1 bar are show in Figure 3. As the inserts in this figure show, the $N_2$ and $SF_6$ isotherms are clearly Type I as expected. The rate of take-up with pressure for the three super-critical gases increases inline with the energy of the fluid-solid interactions. The rate of take-up of $SF_6$ is similar to that of $N_2$ despite the higher temperature because of its greater size and affinity for carbon.

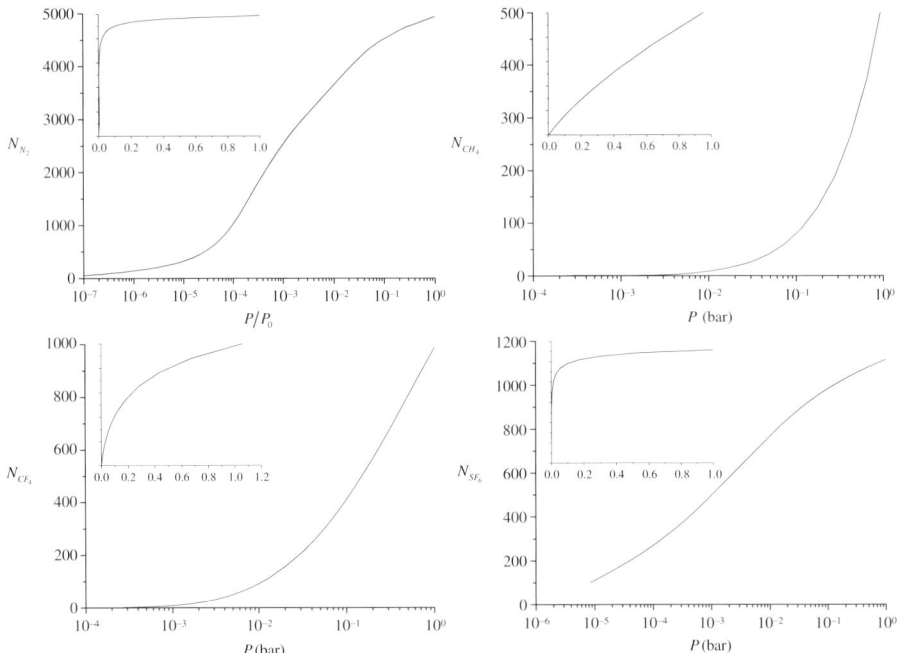

Figure 3. Adsorption isotherms for $N_2$ at 77 K and $CH_4$, $CF_4$ and $SF_6$ at 258 K on the model solid as determined by GCMC simulation; $N_i$ are the number of molecules.

### 4.2.    Pore size distribution

PSDs obtained by inversion of the adsorption integral equation as per [11]

$$N(P) = \int_{0}^{\infty} \rho(P,w) f(w) dw \qquad (1)$$

using the adsorption isotherms in Figure 3 and GCMC derived local adsorption isotherms, $\rho(P,w)$, are shown in Figure 4(a)-(c), whilst Figure 4(d) shows the PSD obtained by combining those of the three supercritical gases and taking into account their window of reliability, $(w_l, w_u)$; the PSDs in all cases have been normalised to unity area to facilitate comparison. The character of the PSDs obtained from the supercritical isotherms are similar to those of López-Ramón *et al.* [11] (their pore sizes can be converted to the pore size definition used here by subtracting the size of the carbon atom from their values), with the PSD based on methane, Figure 4(b), presenting two peaks at the lower end of the pore size range, whilst those derived from $CF_4$ and $SF_6$, Figure 4(c), each have a peak to the right of these and a second peak centred around 17 Å. A third peak centred around 28 Å is also present in the $SF_6$-based PSD; as this is outside the relevant window of reliability, it is omitted from the composite PSD shown in Figure 4(d), which is assembled following [11].

The PSD obtained from direct analysis of the solid, which is also shown in Figure 4(a)-(d), reveals that the smallest dimension of the model solid's pores are distributed between ~5 Å to ~14 Å. Whilst each of the adsorption-based PSDs contain significant porosity somewhere between 5 and 14 Å, there appears to be little correspondence between these PSDs on the one hand and the direct PSD on the other in terms of both the number and position of the peaks.

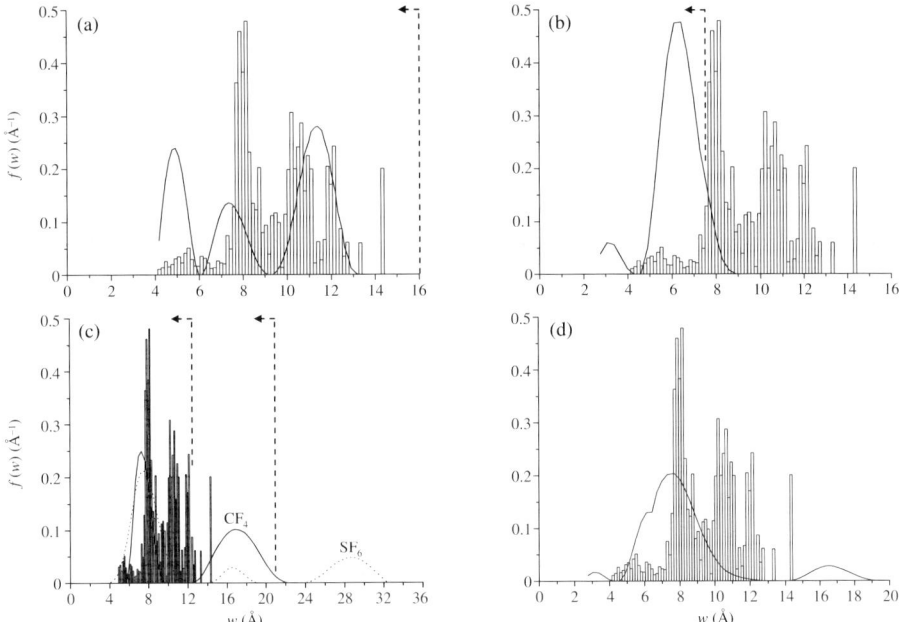

Figure 4. Normalised PSDs from direct analysis (bars) and inversion of the adsorption isotherms of Figure 3 (lines): (a) $N_2$, (b) $CH_4$, and (c) $CF_4$ and $SF_6$ isotherms; (d) combining the PSDs of (b) and (c) within their respective windows of reliability, indicated by the dashed arrows.

Comparing the distribution of the total energy experienced by the molecules within the model solid as determined during the GCMC simulations with the average energies associated with the slit pores used in the inversion of the adsorption integral equation indicates that the peaks in the PSDs are in fact related to the energy distribution. This is perhaps not surprising as various relationships between adsorption energy and pore size have been advanced over the decades. However, consideration of the energy associated with individual molecules in the model solid show that there is only a weak link between 'pore size' and adsorption energy. For example, the volume associated with the smallest pores in the $N_2$ and $CH_4$ based PSDs are in fact attributable to high energy nooks and crannies in the pore space rather than 'pores' of the indicated size. Such features of the pore space are better described in terms of fractals.

### 4.3.    Mean coordination number

The only adsorption-based method proposed to date for determining the mean coordination number of microporous solids [11] is based on the premise that the difference between the PSDs obtained from molecules of different size is due to inaccessible porosity arising from pore network effects. Application of this approach to the model solid using the PSDs for $CF_4$ and $SF_6$ in Figure 4(c) leads to a mean coordination number of $Z = 2.26$. This value suggests that the pore space is percolating to $SF_6$, which is inline with the observation that molecules of this fluid diffuse freely through the solid under a chemical potential gradient. This mean coordination number would, however, normally be interpreted as indicating that some of the porosity is inaccessible to the $SF_6$ because of network effects.

Encouragingly, the adsorption-based mean coordination is inline with the mean number of pores that actually emanate from the pore junctions as determined by direct analysis of the pore space. However, analysis of where the $SF_6$ molecules diffuse during non-equilibrium MD simulations show that whilst not all the pore throats are open to the fluid, they do have access to *all* the pores; this is reflected in the dynamic coordination number for the pore network defined by $SF_6$ at 256K of 3.6. Thus, the assumption that underpins the method of López-Ramón *et al.* [11] – *i.e.* that the difference between the PSDs of the two molecules is due to some pores being inaccessible to the latter because of network effects – is not met here. Inspection of the fluid structure obtained from the GCMC simulations shows that the difference between the lower end of the PSDs in Figure 4(c) is in fact caused by $SF_6$ not being able to access all the 'nooks and crannies' of individual pores that are accessible to the $CF_4$ rather than entire pores – such variation in accessible pore volume as a function of the molecule size is in fact the province of fractals.

### 4.4.    Fractal dimension

A wide range of different adsorption-based methods have been proposed for the determination of the fractal dimension. The earliest involves exploiting the well known scaling $N_m = \sigma^{-D}$, where $\sigma$ is the size of the adsorbate molecule, and $N_m$ the number of these molecules required to cover the surface, which is normally equated to the monolayer coverage obtained from a suitable method [13]. Changes in slope at specific values of $\sigma$ also reveal the range of the fractal(s). This *multi-isotherm approach* was assessed here using the Langmuir monolayer coverage derived from sub-critical isotherms for a series of molecules that vary only in size from 1 to 1.8 times that of nitrogen. This approach yields a fractal dimension for the model solid that well exceeds the upper limit of 3. This poor result despite the energy of interaction and molecule shape being identical is due to a number of factors. The errors in the monolayer coverage are a particular problem as it makes accurate fitting of the data almost impossible given the relatively narrow range over which it can be obtained for microporous solids.

A raft of methods have been proposed since the early 1980s for determining the fractal dimension from a single isotherm. One major group of such methods is based on the inversion of the adsorption integral equation, Eq. (1), using a local isotherm with the 'fractal pore size distribution'. The earliest such approach, the fractal FHH equation [14] and its various elaborations (*e.g.* [15]), have already been shown to be less than satisfactory when applied to microporous solids [16] and our analysis confirms these findings, including the absence of any clear linear region from which an unambiguous fractal dimension can be derived. This poor performance lead Terzyk *et al.* [16] to propose for microporous solids a six parameter *non-linear* fractal FHH equation. Both the non-linearity and number of parameters means fitting this model to obtain the fractal dimension amongst other things is extremely challenging. Indeed, whilst attempts to fit the $N_2$ isotherm in Figure 4(a) using a genetic algorithm [17] have lead to isotherm shapes that are satisfactory, none of the parameter value sets obtained have been sensible despite considerable effort. Forcing the model to take the parameters derived from the direct analysis of the model solid leads to a very poor fit indeed, suggesting that the isotherm is unsuitable for determining the fractal dimension, at least in this instance.

A second major group of single isotherm methods extend the BET analysis to fractal surfaces. Once again, there are a variety of incarnations of this approach. One of these [18] has produced very promising results in the study reported here. Whilst the match between this model and the entire isotherm is poor if the exact parameter values are used, fitting of the model to that part of the isotherm which is linear in the traditional BET coordinates is very good indeed. In particular, using the procedure outlined below, average values from 10 fits for the monolayer coverage and fractal dimension are within 5% of the actual values of 4282 molecules and 2.92 respectively, whilst the range of fractality is within less than 10% of the actual value of 12.45 Å. Whilst the limited nature of the current study precludes a definitive statement on the optimal protocol for this method, several have been tried and the following yields consistently good results when applied to the region of the isotherm that is linear in the traditional BET coordinates (in all cases, the fractal dimension is constrained to lay between 2 and 3 inclusive):

1. Fit the BET series to the experimental isotherm to obtain values for the number of adsorbate layers, $n$, the energy parameter, $C$, and the monolayer coverage, $W_m$.

2. Fix $n$ and $C$ at the values obtained from the BET analysis (the latter was always obtained even when left free to change), set bounds on the monolayer coverage to be roughly $\pm20\%$ around that obtained from the BET analysis, and then fit the model to the experimental isotherm several times to obtain an average for the range of fractality, $r^*$.

3. Finally, further fix the value of $r^*$ at the average obtained from the second step and then carry out several further fits to obtain averages for the remaining free variables (*i.e.* the fractal dimension and monolayer coverage).

## 5. CONCLUSIONS

An absolute assessment process based on molecular simulation has been briefly outlined. This process is a useful tool in determining the correctness or methods, and the consistency and meaningfulness of quantities. These capabilities have been demonstrated here by application of the absolute assessment process to methods for determining the pore size distribution, mean coordination number of the pore network, and the fractal dimension.

The study shows that the pore size distribution is just another way of presenting the energy distribution, and that the concept of 'pore size' is only weakly linked to actual pore size. The study also reveals that whilst the only method currently available for determining the mean coordination number of microporous solids does yield a quantity with meaning (it is related to the mean number of pores emanating from pore junctions), it does not relate to the coordination number relevant to transport in the pore network, which is substantially higher. Finally, many of the methods considered here for the determination of the fractal dimension do not appear to be satisfactory when applied to microporous solids with one exception, which consistently gives good results provided the protocol outlined in this paper is followed.

## ACKNOWLEDGEMENTS

The Engineering and Physical Science Research Council (EPSRC) of the UK is thanked for its support of this research.

## REFERENCES

1.  J. Rouquerol, D. Avnir, C.W. Fairbridge *et al.*, Pure & Appl. Chem., 66 (1994) 1739.
2.  M.J. Biggs, A. Buts and D. Williamson, Langmuir, 20 (2004) 7123.
3.  A number of publications are currently in preparation that will report assessment of the methods covered in this paper and others relating to the energetics of adsorption.
4.  M.J. Biggs, A. Buts and D. Williamson, Langmuir, 20 (2004) 5786.
5.  A tessellation of cubes sized, $d$, laid on the volume is initially split into three sets: (i) those cubes wholly in the pore space, $\varsigma$, (ii) those cubes wholly in the solid, $\Sigma$, and (iii) the remainder, which contain the interface between the solid and pore space, I. The position of the interface within each cube is then identified using tetrahedral decomposition [A. Doi and A. Koide, IEICE Trans. Commun., 74 (1991) 214] and the sets $\varsigma$ and $\Sigma$ are augmented with those tetrahedra wholly in the pore space and solid respectively, whilst the members of set I are replaced with those surfaces defining the interface. Cubes and tetrahedra are considered to lay wholly within a phase if all their vertices lay in the phase; a vertex is considered to lay in the solid if the potential energy of a test molecule placed on it exceeds an energy that defines an accessible pore width (*i.e.* in which the centre of a molecule can be placed) of $2.5\sigma_{ff} - 2\sigma_{fs}$ for an infinite-extent graphite basal plane slit pore whose surface carbons are $2.5\sigma_{ff}$ apart.
6.  The tetrahedra and the cubes of set $\Sigma$ are eroded layer by layer, starting at the fluid-solid interfaces. As the erosion process progresses, the percolating cluster defining the pore space gradually becomes disconnected and increasing numbers of isolated clusters are formed. Once an isolated cluster is formed, its erosion ceases. This continues until all the pore space is occupied by isolated clusters – each cluster at this stage defines the kernel of a pore in the set of all pores, $\Pi$. The sets of cubes that define each pore, $\delta\Sigma_i$, are then determined by growing the clusters uniformly layer by layer until they can no longer grow due to intersection with the pore walls and adjacent clusters – the set of interfaces between clusters, T, are considered to be pore throats.
7.  M.P. Allen and D.J. Tildesley, Computer Simulation of Liquids, Clarendon Press, Oxford, 1989.
8.  R.F. Cracknell, D. Nicholson N. and Quirke, Phys. Rev. Lett., 74 (1995) 2463.
9.  J.P.R.B. Walton and N. Quirke, Mol. Simul., 2 (1989) 361.
10. B. Smit, J. Chem. Phys., 96 (1992) 8639.
11. M.V. López-Ramón, J. Jagiełło, T.J. Bandosz and N.A. Seaton, Langmuir, 13 (1997) 4445.
12. WA. Steele, The Interaction of Gases with Solid Surfaces, Pergamon, Oxford, 1974.
13. P. Pfeifer and D. Avnir, J. Chem. Phys., 79 (1983) 3558; see also Avnir *et al.*, *ibid.*, 3565.
14. P. Pfeifer, Y.J. Wu, M.W. Cole and J. Krim, Phys. Rev. Lett., 62 (1989) 1997.
15. D. Avnir and M. Jaroniec, Langmuir, 5 (1989) 1431.
16. A.P. Terzyk, P.A. Gauden, G. Rychlicki and R. Wojsz, Colloids Surf. A, 152 (1999) 293.
17. D. Djurdjevic and M.J. Biggs, unpublished.
18. R. Segars and L. Piscitelle, Mat. Res. Soc. Symp. Proc., 407 (1996) 349.

Studies in Surface Science and Catalysis 160
*P.L. Llewellyn, F. Rodriquez-Reinoso, J. Rouqerol and N. Seaton (Editors)*

# Molecular Modeling of Mercury Porosimetry

**F. Porcheron[a], P. A. Monson[a] and M. Thommes[b]**

[a] Department of Chemical Engineering, University of Massachusetts Amherst, MA 01003-9303, USA
[b] Quantachrome Corporation, 1900 Corporate Drive, Boynton Beach, FL 33426

## Abstract

We review a recently developed molecular-based approach for modeling mercury porosimetry. This approach is built on the use of a lattice model of the porous material microstructure and the use of mean-field density functional theory (MF-DFT) calculations and Monte Carlo simulations to calculate the three-dimensional density distribution in the system. The lattice model exhibits a symmetry between the adsorption/desorption of a wetting fluids and intrusion/extrusion of a nonwetting fluid. In consequence, macroscopic approaches used previously to transform mercury porosimetry curves into gas adsorption isotherms are essentially exact in the context of the model. We illustrate the approach with some sample results for intrusion and extrusion in Vycor and controlled pore glass (CPG).

## 1.    Introduction

Probing the pore structure by gradually filling/draining the material with a fluid as a function of an increasing/decreasing external pressure represents the most important approach to porous material characterization. This idea was originally built on a simple pore analysis where a thermodynamic expression connects the pore radius $r$ to the pressure at which a liquid/vapor interface can exist in the pore. If the fluid wets the solid surface (contact angle<90°) then this relation is provided by the well-known Kelvin equation and the corresponding experiments are gas (usually $N_2$) adsorption/desorption in the solid material. On the other hand, if the fluid is nonwetting (contact angle>90°) the appropriate relation is the Washburn equation which yields [1]:

$$P_h r = -2\gamma \cos\theta \tag{1}$$

where $P_h$ is the external hydraulic pressure, $\gamma$ is the liquid surface tension and $\theta$ is the contact angle between the solid and the liquid. Mercury porosimetry experiments are consequently designed based on this equation as mercury is a nonwetting liquid with respect to most of the solid materials [2-4]. The experiments are conducted by imposing a hydraulic pressure above the bulk saturation pressure $P_0$, which forces the liquid to intrude into the solid material. By simply monitoring the quantity of liquid in the material as a function of $P_h$, one can directly invert the obtained isotherms using the Washburn equation and obtain a pore size distribution of the material. However, since this methodology is built on a simple pore analysis, the material is view as a bundle of capillary tubes and the pore size distribution provides only a fingerprint of the material which might be totally uncorrelated to the real pore structure. The mercury porosimetry has been particularly used in the characterization of mesoporous materials since the method can probe over a wide range of pore sizes from about 4 nm to 400 nm. Usually a mercury porosimetry curve displays several characteristic features. For instance, the existence of a hysteresis loop between the intrusion and extrusion curve is similar to the one observed in gas adsorption. Due to the high pressure constraint on the material, the mercury porosimetry can also sometimes lead to a partial alteration or destruction of the pore network [5]. A more problematic feature is the entrapment of mercury appearing at the end of a cycle as the extrusion curve does not close the intrusion one. This quantity of entrapped mercury can sometimes reach up to 90% of the total intruded mercury [3]. Nevertheless this method has had a great deal of success to characterize mesoporous materials and is an important complement with gas adsorption experiments.

It is therefore quite surprising that few theoretical works have addressed the modeling of mercury porosimetry experiments. Most of the theoretical treatments rely on macroscopic assumptions like contact angle hysteresis [6, 7], energy barrier [8] or percolation assumption [9-19] and there is a lack of molecular-based approaches. In a series of recent papers [20-22] we model mercury porosimetry experiments with a molecular approach based on a lattice model which was successfully employed for gas adsorption modeling [24-29]. The lattice model is a very useful compromise that allows us to sample the pore structure over reasonably large length scales while remaining computationally tractable. In this paper we present brief review of this approach and we illustrate it using some results for the cases of Vycor and CPG.

## 2.    Models and Methods

### 2.1    Lattice Hamiltonian

To model mercury porosimetry experiments we discretize the porous space employing a face-centered cubic lattice. Each lattice site has therefore 12 nearest neighbors and can either be occupied by a solid particle ($\zeta=0$) or available for a fluid occupation ($\zeta=1$). Each fluid site can then either be occupied and the fluid occupation variable $n_i=1$ or vacant and then $n_i=0$. The Hamiltonian of the lattice model can be written as:

$$H = -\varepsilon \sum_{<ij>} \zeta_i \zeta_j n_i n_j - \mu \sum_i \zeta_i n_i - \alpha\varepsilon \sum_{<ij>} [n_i \zeta_i (1-\zeta_j) + n_j \zeta_j (1-\zeta_i)] \qquad (2)$$

where $\varepsilon$ is the fluid-fluid interaction parameter, $\alpha$ is the ratio of the wall-fluid to the fluid-fluid interaction parameter and $\mu$ is the chemical potential. The double summations in the first and third terms of Eq. 2 run over all the distinct nearest neighbor site pairs.

This lattice Hamiltonian presents a symmetry with respect to the bulk liquid-vapor coexistence chemical potential $\mu_0$ and with respect to $\alpha=1/2$ [20]. This symmetry can easily be seen in the Ising description of the lattice model (where $s_i=2n_i-1$ is the lattice spin) and gives

$$H(1/2 + \delta\alpha, \mu_0 + \delta\mu, s_i, \zeta_i) = H(1/2 - \delta\alpha, \mu_o - \delta\mu, -s_i, \zeta_i) \qquad (5)$$

that is the Hamiltonian of a state of a fluid where $\delta\mu<0$ and $\delta\alpha>0$ (i.e. gas adsorption/desorption of a wetting fluid) has the same value as that for the fluid with $\delta\mu>0$ and $\delta\alpha<0$ and all the fluid occupancies reversed (i.e. liquid intrusion/extrusion of a nonwetting fluid).

In the present example the configuration of the solid sites $\{\zeta_i\}$ is build to model the mesoporous structure of a porous glass. Each sample of the glass material is obtained with the Gaussian random field method [30]. During a calculation, we use periodic boundary conditions in all directions of space. An illustration of a Vycor glass sample obtained with the Gaussian random field is reported on Fig. 1. We use the same procedure for CPG.

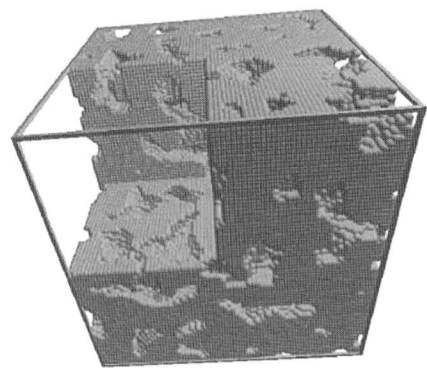

Fig. 1 : An illustration of a mesoporous Vycor glass built on the lattice model with the Gaussian random field method. The parameters used are $p=0.3$, $a=15$Å, $L_x=L_y=L_z=80$. For CPG we use $p=0.6$, $a=15$Å, $L_x=L_y=L_z=180$.

## 2.2   Mean-Field Density Functional Theory

Mean-field density functional theory (MF-DFT) calculations have been widely used for the modeling of gas adsorption experiments in mesoporous materials [24-27]. For our lattice model, the mean-field expression of the grand potential is

$$\beta\Omega[\zeta_i]=\sum_i[\rho_i\ln\rho_i+(1-\rho_i)\ln(1-\rho_i)-\beta\mu\rho_i]-\beta\varepsilon\sum_{<ij>}[\rho_i\rho_j+\alpha\rho_i(1-\zeta_j)+\alpha\rho_j(1-\zeta_i)] \quad (6)$$

where $\beta=1/k_bT$ and $\rho_i$ is the average density of the site $i$. The minimization of the grand potential under the constraint of fixed chemical potential yields the mean-field equation for the local density at a site $i$

$$\rho_i = \frac{n_i}{1 + \exp\left[-\beta\mu - \beta\varepsilon\sum_{j\in i}[\rho_j + \alpha(1-\zeta_j)]\right]} \quad\quad (7)$$

The equations are solved by an iterative scheme and we consider that the algorithm has converged when for an iteration $k+1$ we obtain $\sqrt{\sum_i([\rho_i^{k+1} - \rho_i^k]^2)} < 10^{-4}$.

## 3. Application to porous glasses

### 3.1 Transformation of Intrusion/Extrusion Curve into Adsorption/Desorption Isotherm

Previous experimental works have attempted to make a connection between liquid porosimetry and gas adsorption by proposing transformations between the respective isotherms based upon macroscopic considerations [31-33]. We have shown that the Hamiltonian symmetry contained in our model leads to an exact transformation between gas adsorption and liquid porosimetry curves [20]. The integration of the Gibbs-Duhem equation expressed in terms of activity leads to

$$\frac{\lambda_l}{\lambda_o} = \exp(P_h V_l / RT) \quad\quad (8)$$

where $\lambda_l$ is the activity of the liquid phase, $\lambda_o$ is its value at saturation and $V_l$ is the molar volume of the bulk liquid. In this treatment the liquid phase is considered to be incompressible and since the hydraulic pressure in mercury experiments is very large we therefore neglect the contribution from the saturation pressure. The symmetry of our lattice model Hamiltonian then yields,

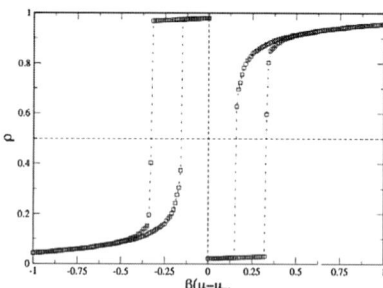

Fig. 2 : MF-DFT adsorption/desorption curves of a wetting fluid ($\alpha$=1.0, left part) and intrusion/extrusion curves of a nonwetting fluid ($\alpha$=0.0, right part) in our model of CPG. $T/Tc$=0.5. The curves are symmetric with respect to $\mu$=$\mu_0$ and $\rho$=1/2.

$$\frac{\lambda_v}{\lambda_0} = \frac{\lambda_0}{\lambda_l} = \exp(-P_h V_l / RT) \tag{9}$$

where $\lambda_v$ is the activity of the vapor phase. This transforms the hydraulic pressure scale for a liquid intrusion experiment into a relative activity scale for a gas adsorption experiment. If we further assume that the bulk gas in a gas adsorption experiment is ideal then we can rewrite Eq. (9) as

$$\frac{P_v}{P_o} = \exp(-P_h V_l / RT) \tag{10}$$

and we now have an expression relating the hydraulic pressure scale for the porosimetry experiment to a relative pressure scale for a gas adsorption experiment. In reference 20 we illustrated this transformation for the case of our model of Vycor. Here we do the same for CPG.

In Fig. 2 we plot the liquid intrusion/extrusion ($\alpha$=0) and gas adsorption/desorption ($\alpha$=1) isotherms obtained for CPG by MF-DFT. The curves are obtained for a single realizations of a CPG sample of dimension $L_x = L_y = L_z = 180$ and $a = 15$Å at a relative temperature $T/T_c = 0.5$ where $T_c$ is the bulk critical temperature. We clearly observe a symmetry around $\mu_0$ and $\rho = 1/2$ between the nonwetting and the wetting fluid curves. The symmetry of the lattice model dictates that the origin of the hysteresis in the wetting and nonwetting cases is identical and in the first instance hysteresis in mercury porosimetry for disordered porous materials has the same origin as hysteresis in gas adsorption/desorption experiments in the same materials. Woo and Monson [26] showed that the appearance of a very large number of local minima of the grand free energy is the direct origin of the hysteresis in mesoporous glass materials.

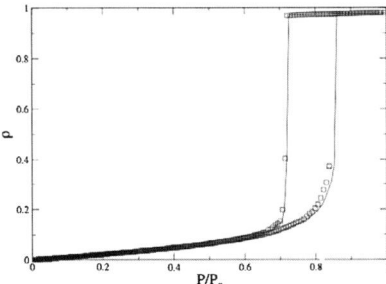

Fig. 3 : MF-DFT adsorption/desorption curves of a wetting fluid ($\alpha$=1.0, solid line) and transformed intrusion/extrusion curves of a nonwetting fluid ($\alpha$=0.0,   ). $T/T_c$=0.5.

By using Eq. (10) and the hole-particle symmetry of the lattice model we can transform the liquid porosimetry isotherm into an adsorption/desorption isotherm of density versus relative pressure for the wetting fluid. This is shown in Fig. 3 together with the gas adsorption results and the agreement is excellent.

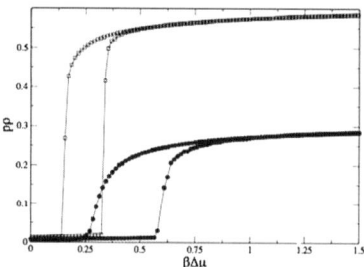

Fig. 4: Intrusion extrusion curves from MF-DFT for CPG (open symbols) and Vycor (closed symbols) for $T/T_c=0.5$. This plot gives the density multiplied by the porosity versus the chemical potential (relative to the value for the saturated liquid). Notice the transition from type I hysteresis (CPG) to type II hysteresis (Vycor) between the two systems.

## 3.2    Intrusion and Extrusion behavior

In figure 4 we show intrusion and extrusion curves for our models of mercury in Vycor and in CPG. We see that the amount of liquid intruded is much higher for the CPG reflecting the higher porosity in this case. We also note the transition from type I hysteresis (CPG) to type II hysteresis (Vycor) between the two systems. Our calculations capture the trends seen in experimental mercury porosimetry studies of these systems [24].

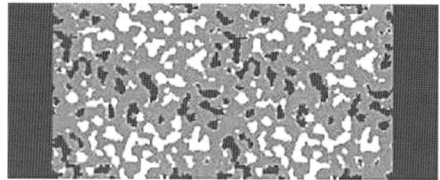 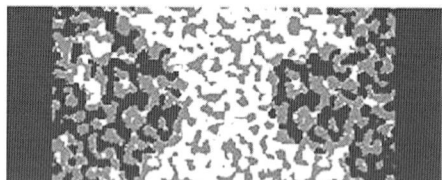

Fig. 5: Computer graphics visualization of the density distribution for a state during intrusion for Vycor (left) and CPG (right). $T/T_c=0.5$. We show a section through the sample of the porous material in each case.

The MF-DFT calculations yield the density distribution within the porous material and it is of interest to make computer graphics visualizations of this distribution during intrusion and extrusion. Figure 5 shows visualizations of the density distributions for states during intrusion for both Vycor and CPG. We see a significant difference in the density distribution in the two cases. For the CPG we see that intrusion resembles a percolation process with some nucleation of liquid drops ahead of the percolation front. For the Vycor the intrusion is also like a percolation process but we find more significant nucleation of liquid ahead of the percolation front. This nucleation of droplets leaves some pore unfilled while the liquid moves deeper into the Vycor. We may emphasize that this mechanism is symmetrically similar to the one obtained in gas desorption [34].

Figure 6 shows visualizations of the density distributions for states during extrusion for the two systems. In this case an important feature of the behavior shown by the density distribution is the fragmentations of the liquid. This is particularly pronounced for the Vycor system but is also seen for the CPG. The significance of this fragmentation is that it indicates that vapor transfer must play a significant role in the kinetics of extrusion. This leads to a much lower flux of mercury during extrusion than would be the case if extrusion simply involved the withdrawal of liquid without fragmentation. We believe that this fragmentation process provides a key mechanism for mercury entrapment. Calculations using dynamic Monte Carlo simulations provide additional support for this hypothesis [21, 23].

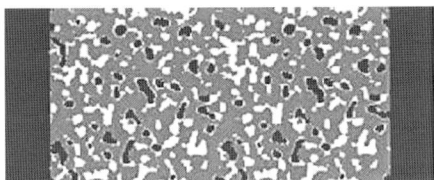 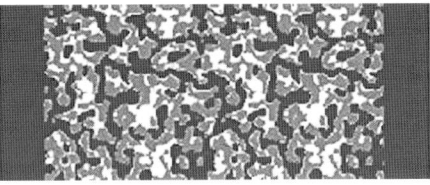

**Fig. 6:** Computer graphics visualization of the density distribution for a state during extrusion for Vycor (left) and CPG (right). $T/T_c$=0.5. We show a section through the sample of the porous material in each case.

## 4.    Conclusions

We have reviewed a recently developed approach to modeling mercury porosimetry experiments using statistical mechanics. The approach is based on a lattice model which allows studying large system sizes while maintaining computational tractability. The Vycor and CPG samples studied here are generated by the Gaussian random field method which provides a realistic description of the pore network. We first show that the symmetry of the lattice Hamiltonian leads to a symmetry between gas adsorption/desorption isotherms and liquid intrusion/extrusion curves. The transformations between gas adsorption and mercury porosimetry isotherms proposed in previous experimental works and based upon macroscopic considerations are essentially exact for our model. Mean-field density functional theory calculations display most of the characteristics of the experimental curves as a hysteresis loop develops above $P_0$ for a nonwetting fluid (liquid porosimetry) or below $P_0$ for a wetting fluid (gas adsorption). The present approach therefore provides a unified framework for modeling gas adsorption and mercury porosimetry in mesoporous materials.

Visualizations of the density distributions in the porous materials reveal insights into the mechanisms of intrusion and extrusion in the systems. We find that intrusion resembles an invasion percolation process with a front of liquid gradually moving into the porous material from the external surface. However vapor transport and liquid condensation ahead of this fromt may also contributed to the mechanism. An important aspect of extrusion is the fragmentation of the liquid. This means that the transport processes in extrusion involve vapor transport. This provides a mechanism for mercury entrapment. We have made dynamics studies of these effects in our recent work [21, 23].

**Acknowledgement**

This work was supported by the National Science Foundation (CTS-0220835).

**REFERENCES**

[1] E. W. Washburn, Proceedings of the National Academy of Sciences of the United States of America, 7 (1921) 115.
[2] S. Lowell and J. E. Shields (3$^{rd}$ ed.) Powder Surface Area and Porosity, Chapman and Hall, New York, 1991.
[3] C. Leòn y Leòn, Adv. Colloid Interface Sci., 76-77 (1998) 341.
[4] J. Vanbrakel, S. Modry and M. Svata, Powder Technol., 29 (1981) 1.
[5] R. Pirard, S. Blacher, F. Brouers and J. P. Pirard, J. Mater. Res., 10 (1995) 2114.
[6] S. Lowell, Powder Technol., 25 (1980) 37.
[7] S. Lowell and J. E. Shields, J. Colloid Interface Sci., 80 (1982) 192.
[8] H. Giesche, Mat. Res. Soc. Symp. Proc., 80 (1982) 192.
[9] J. C. Orr, Powder Technol., 3 (1970) 117.
[10] S. P. Rigby, J. Colloid Interface Sci., 224 (2000) 382.
[11] S. P. Rigby and L. F. Gladen, Chem. Eng. Sci., 224 (2000) 382.
[12] S. P. Rigby, S. P. Fletcher and S. N Riley, J. Colloid Interface Sci. 240 (2001) 190.
[13] S. P. Rigby, S. P. Fletcher ans S. N. Riley, Ind. Eng. Chem. Res., 41 (2002) 1205.
[14] S. P. Rigby, S. P. Fletcher and S. N. Riley, Applied Catalysis A, 8526 (2003) 1.
[15] S. P. Rigby and K. J. Edler, J. Colloid Interface Sci., 250 (2002) 175.
[16] G. P. Matthews, C. J. Ridgway and M. C. Spearing, J. Colloid Interface Sci. 171 (1995) 8.
[17] C. Salmas and G. Androutsopoulos, J. Colloid Interface Sci., 239 (2001) 178.
[18] G. Zgrablich, S. Mendioroz, L. Daza, J. Pajares, V. Mayagoitia, F. Rojas and W. C. Conner, Langmuir, 7 (1991) 779.
[19] A. M. Vidales, E. Miranda, M. Nazzaro, V. Mayagoitia, F. Rojas and G. Zgrablich, Europhys. Lett., 36 (1996) 259.
[20] F. Porcheron, P. A. Monson and M. Thommes, Langmuir, 20 (2004) 6482.
[21] F. Porcheron and P. A. Monson, Langmuir, 21 (2005) 3179.
[22] F. Porcheron, P. A. Monson and M. Thommes, Adsorption, 11 (2005) 325.
[23] F. Porcheron, P. A. Monson and M. Thommes, (to be published).
[24] H. J. Woo, L. Sarkisov and P. A. Monson, Langmuir, 17 (2001) 7472.
[25] L. Sarkisov and P. A. Monson, Phys. Rev. E, 65 (2001) 011202.
[26] H. J. Woo and P. A. Monson, Phys. Rev. E, 67 (2003) 041207.
[27] E. Kierlik, P. A. Monson, M. L. Rosinberg, L. Sarkisov and G. Tarjus, Phys. Rev. Lett., 87 (2001) 055701.
[28] L. Sarkisov and P. A. Monson, Langmuir, 17 (2001) 7600.
[29] L. Sarkisov and P. A. Monson, Langmuir, 16 (2000) 9857.
[30] J. W. Cahn, J. Chem. Phys., 42 (1965) 93.
[31] L. Moscou and S. Lub., Powder Technol., 29 (1981) 45.
[32] K. L. Murray, N. A. Seaton and M. A. Day, Langmuir, 15 (1999) 8155.
[33] S. Lowell and J. E. Shields, Powder Technol., 29 (1981) 225.
[34] H. J. Woo, F. Porcheron and P. A. Monson, Langmuir, 20 (2004) 4743.

Studies in Surface Science and Catalysis 160
*P.L. Llewellyn, F. Rodriquez-Reinoso, J. Rouqerol and N. Seaton (Editors)*

# Predicting ambient temperature adsorption of gases in active carbons

**M.B. Sweatman[a], N. Quirke[b] and P. Pullumbi[c]**

[a]Department of Chemical and Process Engineering, University of Strathclyde, Glasgow, G1 1XQ, Scotland, United Kingdom.

[b]Department of Chemistry, Imperial College, South Kensington, London, SW7 2AY, United Kingdom.

[c]Air Liquide, Centre de Recherche Claude-Delorme, Les Loges-en-Josas, 78354, Jouy-en-Josas, France.

## ABSTRACT

We describe procedures, based on the slit pore model and Monte Carlo simulation, for predicting the adsorption of pure gases in active carbons given only a single carbon dioxide 'probe' adsorption isotherm. Predictions are made at ambient temperature up to quite high pressure for methane, ethene, ethane, propene and propane. The key development in our work concerns our method for calibrating gas – surface interactions, i.e. we calibrate these interactions to a reference active carbon rather than a low surface area carbon as in most other work of this type. Our predictions highlight limitations in our surface model and experiments.

## 1. INTRODUCTION

The aim of our research is to create a fast, quantitatively accurate method for predicting gas mixture adsorption in active carbons and other adsorbents. Activated carbons are used for a variety of purposes, including separation of one fluid component from another on an industrial scale. To design a separation process it is necessary to understand how gas mixtures are adsorbed by a given material. Recently[1-5], we described how the adsorption of some simple gas mixtures (involving carbon dioxide, nitrogen, methane and hydrogen) at near-ambient temperatures and up to moderately high pressure can be predicted quite accurately with a fast 'slab-DFT' in combination with characterisation of the active carbon in terms of the polydisperse ideal independent slit-pore model. The first step in this approach concerns development of accurate molecular models for each gas of interest[6]. One of the key developments in our molecular modelling approach, which is of central importance in producing results that are quantitatively accurate, concerns the calibration of gas-surface interactions. In our approach, gas-surface interactions are calibrated separately for each gas to reproduce adsorption on a reference *active* carbon (codenamed PNC in our dataset). This represents a significant departure from all previous work of this type (for example, see references [7-13]) where these interactions are instead calibrated to reproduce adsorption on a *low surface area* carbon (such as graphite or Vulcan 3G[14]). The implication here is that the surfaces of active carbons could be more similar to each other than to low surface area active carbons. If this finding is corroborated by independent work then we are required to explain this difference. Suggested possible causes include effective 3-body interactions (surface – gas – surface) that are not present when gas is adsorbed on a single isolated surface, and the difference in chemistry and curvature between the surfaces of active and low surface area carbons.

This paper describes our attempt to extend this molecular modelling and calibration protocol to other gases, namely ethene, ethane, propene and propane. We have generated new molecular models for these gases using Gibbs ensemble Monte Carlo simulation[15] and calibrated their respective gas-surface interactions as before[1]. We make predictions for adsorption of these gases at ambient conditions up to pressures close to saturation on a variety of active carbons. We find that our predictions are accurate for some active carbons, but less accurate for others. Some of the discrepancies are attributed to experimental error; the remainder highlight potential failings in our approach. Much of this work is based on methods identical to that in other work. So here we can afford to be brief and we direct readers to this other work, and comprehensive reviews[1,4,5], for more information.

## 2. MOLECULAR MODELS

In this work we employ the polydisperse independent ideal slit-pore model to model the surface of active carbons. The pore-size distribution (PSD) is calculated by analysis of a single experimental carbon dioxide adsorption isotherm at 293 K. Adsorption isotherms for any gas are then predicted using the appropriate kernel with the same PSD. Our previous work[1] has shown that this is a useful procedure for modelling the adsorption of some simple gases (carbon dioxide, methane, nitrogen and hydrogen) at ambient temperatures up to about 50 bar. However, some of the gas species that we are concerned with here, including propene and propane, are slightly more complex than these simple gases in respect of their less spherical molecular shape (i.e. their modelled pair-interactions) and their stronger gas-surface interactions.

We model gas molecules as rigid assemblies of Lennard-Jones and partial charge sites[1,6]. We tailor the Lennard-Jones (LJ) parameters until good agreement with experimental data[16] for saturation properties (coexisting gas and liquid densities and pressures) is obtained by Gibbs ensemble simulation[15]. Initially, we started with models defined by Bourasseau et.al.[17]. However, we employ a cutoff, described elsewhere[1,6], in gas-gas interactions to avoid interactions between periodic images, and also use a 'ramp' (starting at 0.9 times the cutoff distance) to smooth these interactions near the cutoff[1,6]. Final molecular models for our gases are presented in Table 1.

Table 1

Model parameters used in MC simulations. $\sigma_{ff}$ and $\varepsilon_{ff}$ are the Lennard-Jones interaction length and energy parameters respectively, $L$ and $\theta$ denote bond lengths and angles respectively, $q$ is the partial charge centred on a Lennard-Jones site, $r_c$ denotes the cutoff distance between molecular centres (the CH site in the case of propene), while $\varepsilon_{ww}$ is the calibrated strength of surface – surface interactions.

| Parameter | $CO_2$ | $CH_4$ | $C_2H_4$ | $C_2H_6$ | $C_3H_6$ | $C_3H_8$ |
|---|---|---|---|---|---|---|
| $\sigma_{ff}$ (nm) | C: 0.275 O: 0.304 | 0.373 | $CH_2$: 0.3483 | $CH_3$: 0.3625 | $CH_3$: 0.3561 CH: 0.3431 $CH_2$: 0.3391 | CH3: 0.3608 CH2: 0.3458 |
| $L$ (nm) | C-O: 0.1149 | 0 | $CH_2$-$CH_2$: 0.1922 | $CH_3$-$CH_3$: 0.1976 | $CH_3$: 0.1936 CH: 0 $CH_2$: 0.1896 | CH3-CH2: 0.1966 |
| $\theta$ (deg) | O-C-O: 0 | 0 | 0 | 0 | $CH_3$-CH-$CH_2$: 124.0 | $CH_3$-$CH_2$-$CH_3$: 114.0 |
| $r_c$ (nm) | 1.5 | 1.492 | 1.6 | 1.6 | 1.75 | 1.75 |
| $\varepsilon_{ff}/k_B$ (K) | C: 28.3 O: 84.2 | 151.5 | $CH_2$: 112.6 | $CH_3$: 121.3 | $CH_3$: 122.8 CH: 92.6 $CH_2$: 112 | $CH_3$: 123.4 CH2: 88.7 |
| $q$ (e) | C: 0.6512 O: -0.3256 | 0 | $CH_2$: 0 | $CH_3$: 0 | $CH_3$: 0 CH: 0 $CH_2$: 0 | $CH_3$: 0 $CH_2$: 0 |
| $\varepsilon_{ww}/k_B$ (K) | 24.0 | 26.0 | 28.0 | 25.5 | 28.0 | 26.0 |

Our slit pores are formed from two identical opposing walls, each described by the Steele potential[18]. If we use the Lorentz-Berthelot combining rules[4] then for a given gas molecular model our surface model is fully defined by definition of the surface – surface interaction length and strength parameters respectively. The length parameter is fixed at the commonly used value[1] of 0.34 nm while the strength parameter (see Table 1) is adjusted separately for each gas to achieve the best fit to experiment (for each pure gas on PNC) based on the carbon dioxide PSD for this material. The kernels for these gases are obtained by grand-canonical MC simulation[1,6]. We model the helium calibration of pore volume by defining a 'chemical' pore width[1,3], $H_c = H_p - 0.285$ nm.

## 2.1. Gas – surface interaction calibration

Fig. 1 shows the fitted isotherms for each pure gas on PNC active carbon. The carbon dioxide PSD is obtained using the procedures described in reference [1], i.e. the PSD is constrained to a sum of log-normal modes, and a downhill simplex[19] algorithm is applied to obtain the smoothest PSD that produces a fit to the experimental adsorption isotherm within experimental error. These results are presented in terms of the deviation between experiment and theory relative to the maximum amount of carbon dioxide adsorbed at the highest pressure. We see that the fit for carbon dioxide is rather good, with only occasional deviations beyond experimental error, but the fits for the alkenes and alkanes deteriorate, roughly, with increasing molecular weight to the extent that most deviations are well outside of experimental error. Also note that because adsorption is small at the lowest pressures that the deviation relative to the adsorbed amount is rather large at the lowest pressures. This calibration approach immediately informs us that our models are not able to capture some significant features of the experiments. However, also note that the deviation for nitrogen is also significant, yet in our previous work we were still able to make good predictions for nitrogen adsorbed in a wide range of active carbons.

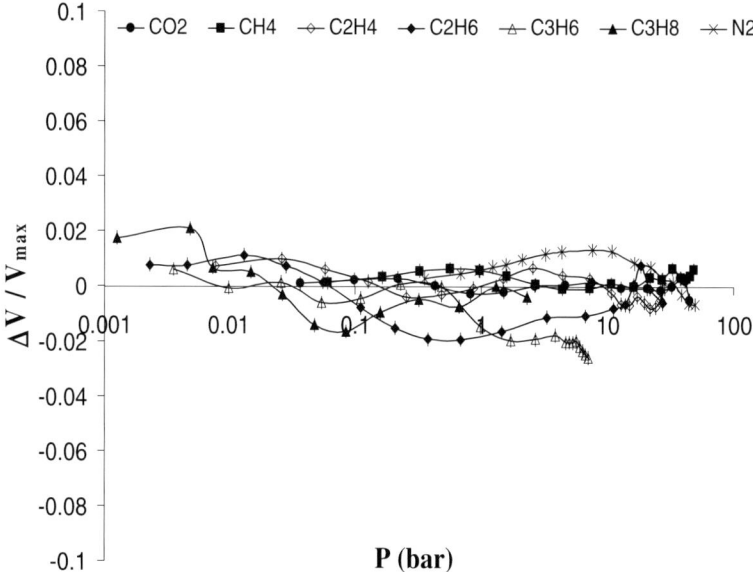

Fig.1. Calibration of gas – surface interaction strength parameters to PNC active carbon. The lines with symbols show the deviation between theory and experiment, relative to the maximum amount of carbon dioxide adsorbed ($V_{max}$). Error bars denote experimental error. Pressure is on a logarithmic scale.

## 3. PREDICTION OF PURE GAS ADSORPTION

Figs. 2a to 2f show predictions for the adsorption of methane, ethane, ethane, propene and propane at 293 K on six active carbons. These active carbons represent a wide cross-section of material types; from molecular sieves through to micro- and meso porous carbons. Fig. 2a corresponds to what might be called an 'ordinary' active carbon, i.e. it consists mostly of nano- and micropores. Predictions for all gases are made with a similar level of accuracy as the fit to carbon dioxide. This is our best result, indicating that this material is most similar to our reference (PNC). However, the poor accuracy at the lowest pressures remains. It will be seen that this is a common feature. Fig. 2b corresponds to a carbon with a wide range of pores, including significant contributions from nano-, micro- and mesopores. These predictions are acceptable (even at low pressure) and show no overall trend, although the absolute deviation for the predictions is generally significantly outside of experimental error. Figs. 2c and 2d correspond to molecular sieves. Most of the pores in these materials are less than 10 Angstroms wide. Predictions for both these materials are rather good (except at the lowest pressures). The remaining two Figures correspond to 'ordinary' active carbons with most pores in the nano- to micropore regime, similar to the material corresponding to Fig. 2a. Our predictions for both alkenes and alkanes in Fig. 2e are consistently too high. The relative error for methane at the highest pressure is about 9%, for the other gases it is about 3%. This indicates that there is a significant difference between the surface of this material and PNC that our surface model has not captured. Considering that carbon dioxide has a moderately strong electrostatic quadrupole moment, while the other gases do not, these results might be highlighting the absence of surface polarity in our surface model. The results in Fig. 2f are poor, except for methane. We cannot explain why methane adsorption is uniformly well predicted, while the adsorption of ethane, ethane, propene and propane is not. We might speculate that adsorption in this material is particularly sensitive to molecular shape and size, although there is nothing in the PSD that hints at this, and it should be recognised that carbon dioxide itself is only very slightly smaller than ethene and ethane.

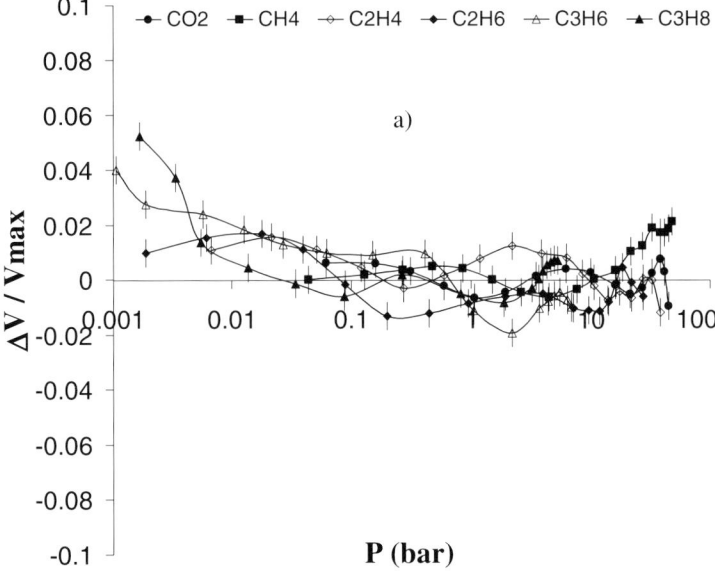

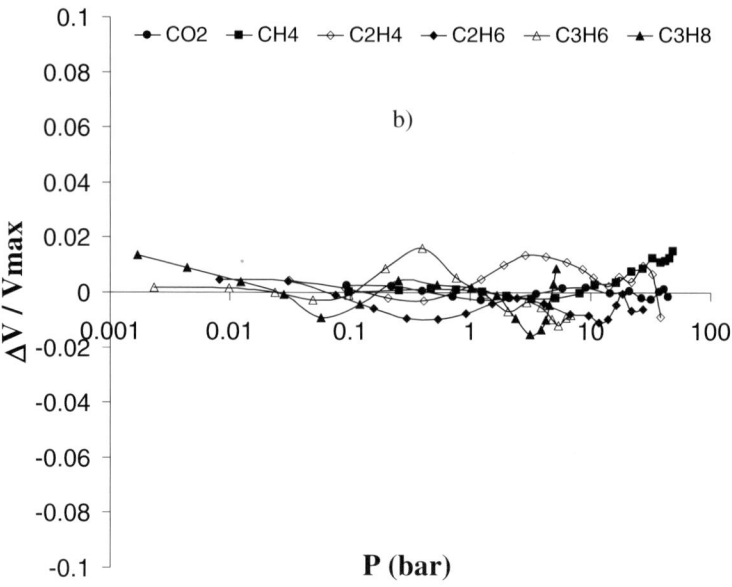

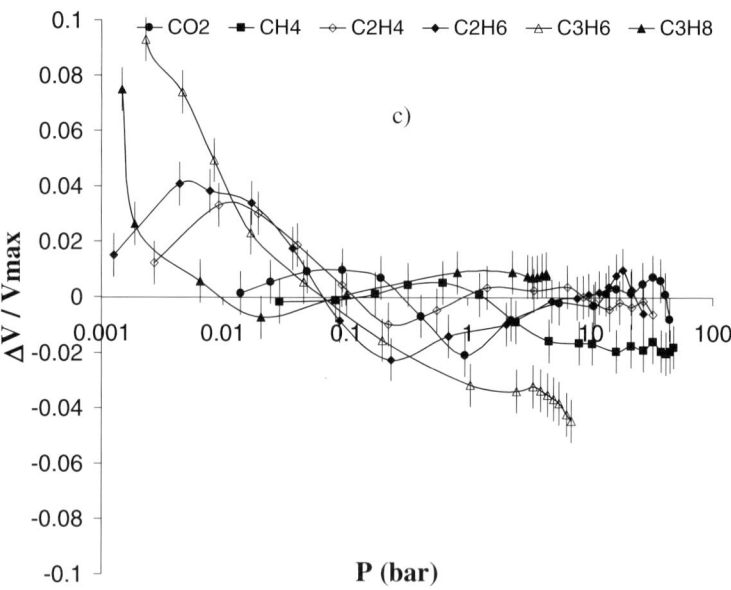

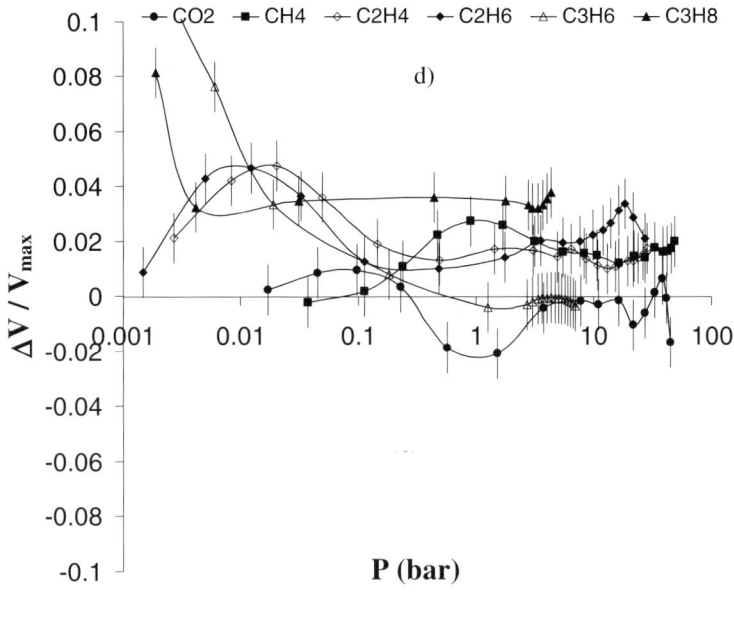

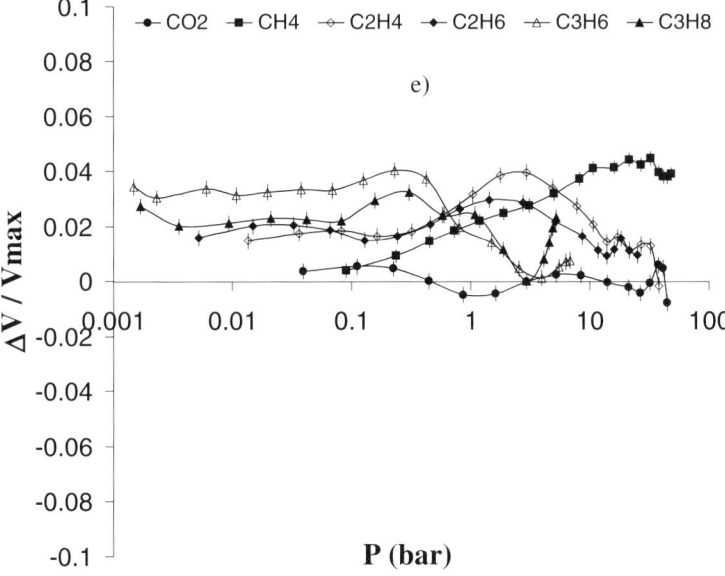

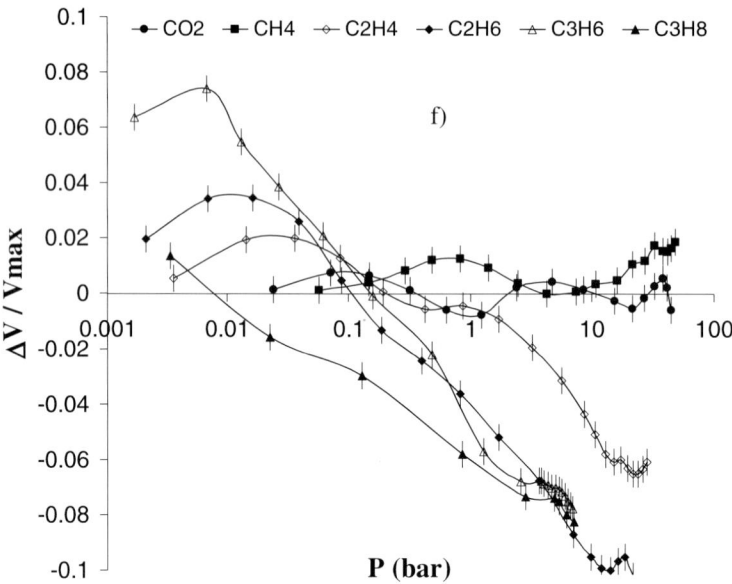

Fig. 2. As for Fig. 1 except that these are predictions for the adsorption of alkenes and alkanes in several active carbons.

## 4. DISCUSSION

While the results in our previous work and in Figs. 2a to 2d of this work are encouraging, those in 2e and particularly 2f highlight potential problems with our approach. In our view, the generally poor performance at the lowest pressures in nearly all our results in this work might be the result of experimental error – in this case systematic error caused by non-equilibrium adsorption. We understand that experimental adsorption readings are taken when diffusion reaches a lower threshold. It might be that this threshold is set too high at the lowest pressures for the heaviest gases where diffusion is slowest. Only the low pressure predictions in Fig. 2b are satisfactory, possibly because diffusion in this material, which has significant mesoporosity, is faster for the larger gas species. The errors in Fig. 2e highlight a potential inadequacy of the ideal slit pore model. However, it might be relatively straightforward to correct for this by introducing a 'surface polarity' term to account for polar gas – surface interactions. In this case two different gas probes would be needed to distinguish geometric and chemical effects. Finally, if the errors in Fig. 2f cannot be attributed to experiment, then they highlight a more fundamental inadequacy of the ideal slit pore model that, we expect, cannot be removed by any minor modifications. It might be that these results can be explained in terms of molecular sieving or the modelled uniformity of pores in the slit pore model. Recent articles[4,7,20] have addressed the development of detailed and demanding 3-D surface models in some depth.

**REFERENCES**

[1] M.B. Sweatman and N. Quirke, J. Phys. Chem. B, 109 (2005) 10381.

[2] M.B. Sweatman and N. Quirke, J. Phys. Chem. B, 109 (2005) 10389.

[3] M.B. Sweatman and N. Quirke, Langmuir, 18 (2002) 10443.

[4] M.B. Sweatman and N. Quirke, in W. Schommers, M. Rieth (Eds.), Handbook of Theoretical and Computational Nanotehnology. American Scientific Publishers, in press.

[5] M.B. Sweatman and N. Quirke, Mol. Sim., (in print).

[6] M.B. Sweatman and N. Quirke, Mol. Sim., 27 (2001) 295.

[7] T.J. Bandosz, et al, in L.R. Radovic (Ed.), Chemistry and Physics of Carbon.vol. 28, Marcel Dekker, New York, 2003.

[8] G.M. Davies and N.A. Seaton, AICHE, 46 (2000) 1753.

[9] R.F. Cracknell, et al, Adsorption, 2 (1996) 193.

[10] M. Heuchel, et al, Langmuir, 15 (1999) 8695.

[11] M.L. Lastoskie, K.E. Gubbins and N. Quirke, Langmuir, 9 (1993) 2693.

[12] P.I. Ravikovitch, et al, Langmuir, 16 (2000) 2311.

[13] S. Samios, et al, Mol. Sim., 27 (2001) 441.

[14] S. Scaife: The characterization of Porous Carbons Using Computer Simulation and Experimental Techniques, Ph.D., University of Wales, Bangor, 1999.

[15] B. Smit and D. Frenkel: Understanding molecular simulation: from algorithms to applications, Academic, New York, 1996.

[16] Thermophysical properties of fluid systems, NIST Chemistry WebBook 2005, http://webbook.nist.gov/chemistry/fluid/.

[17] E. Bourasseau, et al, J. Chem. Phys., 118 (2003) 3020.

[18] W.A. Steele, Surf. Sci., 36 (1973) 317.

[19] W.H. Press, et al: Numerical recipes in Fortran 77: The Art of Scientific Computing, Cambridge University Press, Cambridge, 1992.

[20] J.K. Brennan, et al, Colloid Surface A, 187 (2001) 539.

Studies in Surface Science and Catalysis 160
P.L. Llewellyn, F. Rodriquez-Reinoso, J. Rouqerol and N. Seaton (Editors)

# Characterisation of periodic mesoporous silicas using molecular simulation

**M. Pérez-Mendoza[a], C. Schumacher and N.A. Seaton**

Institute for Materials and Processes, School of Engineering and Electronics, University of Edinburgh, King's Buildings, Mayfield Road, Edinburgh, EH9 3JL, UK

[a] Current address: Department of Inorganic Chemistry, University of Granada, Avda. Fuentenueva, 18071 Granada, Spain

We have developed models of the pore structure of two periodic mesoporous silicas: MCM-41 and SBA-2. The MCM-41 model was generated using a kinetic Monte Carlo algorithm, which replicates the hydrothermal process used to synthesise the real material. SBA-2 was represented by a pore network model. Adsorption was simulated in both these model materials using Grand Canonical Monte Carlo simulation. By comparing the simulated and experimental adsorption we were able to determine the parameters of the structure models, giving new insights into the structure of the real materials. Predictions of adsorption in the model materials are very accurate, both for the original inorganic materials, and when the surface is modified by the addition of organic surface groups.

## 1 INTRODUCTION

We are interested in the design of periodic mesoporous silicas (PMSs), for gas storage and separation applications. In order to be able to effectively design these materials, we need good models of the pore structure, and we need to be able to relate the parameters of the models to experimental data. As we are primarily interested in adsorption applications, we use adsorption data for this purpose. In this paper, we report our work on the development of models for MCM-41, which consists of a hexagonal array of unconnected, parallel pores, and for SBA-2, which is composed of mesoporous, spherical cavities in a hexagonal, close-packed (hcp) arrangement, linked by microporous channels.

We have taken different approaches to the modelling of these two materials. Our pore model for MCM-41, which has the simpler structure, is generated using a kinetic Monte Carlo (kMC) algorithm, which mimics the hydrothermal synthesis of the real material. In principle, this method may be applied to any PMS. However, we have not yet done this with the more complex SBA-2 structure. Instead, we represent the SBA-2 structure using a pore network model, in which adsorption is considered to take place independently in the individual pores (either cavities or channels) and the interconnectedness of the pores is taken into account using percolation theory. For both materials, adsorption is computed using Grand Canonical Monte Carlo (GCMC) simulation, and the model parameters are fixed by comparing the simulated adsorption with experimental data.

## 2    KINETIC MONTE CARLO-BASED MODEL FOR MCM-41

PMSs are formed by the reaction of silica monomers in the aqueous phase of a microemulsion, in which the array of micelles has long-range order. Once the gelation is complete, the surfactant molecules which form the micelles are removed, either by extraction or calcination. We simulate both the gelation and, if applicable, calcination processes by a kinetic Monte Carlo method. In this approach, silica monomers react to form an amorphous network, obeying the relative kinetics of the real reaction. That is, the rates of, e.g., bond formation and bond breaking are in the same ratio as in the real reaction, and the temperature dependence is that of the real reaction, but the time step does not correspond to real reaction time and the absolute reaction rate is not computed. To make the simulation achievable with current computers, the non-bonding interactions between molecules are treated in a highly simplified way; the generation of each structure nevertheless requires several days of time on a fast processor. The synthesis process may be followed by calcination at an elevated temperature, in which the model micelles and the water molecules are removed. The surface may then be "functionalised" by replacing some of the silanol groups on the surface of the model material by the desired organic group. (This is an important element of the optimisation of PMSs for particular adsorption applications.)

We studied, by GCMC simulation [1], the adsorption of ethane and carbon dioxide on pure-silica MCM-41 and on MCM-41 with surface phenyl and aminopropyl groups. The fluid-fluid and fluid-solid potentials took into account dispersion and, where appropriate, electrostatic interactions. The surface groups - phenyl and aminopropyl - are modelled as flexible chain molecules. The solid-fluid potentials are "transferable"; that is, they are applicable to all the oxide materials we have studied, and are not optimised for particular materials; this is an indicator of the consistency of the approach. The silicon atoms are ignored in the simulation of adsorption. Further details of the kMC and GCMC simulation methods are given in reference [2].

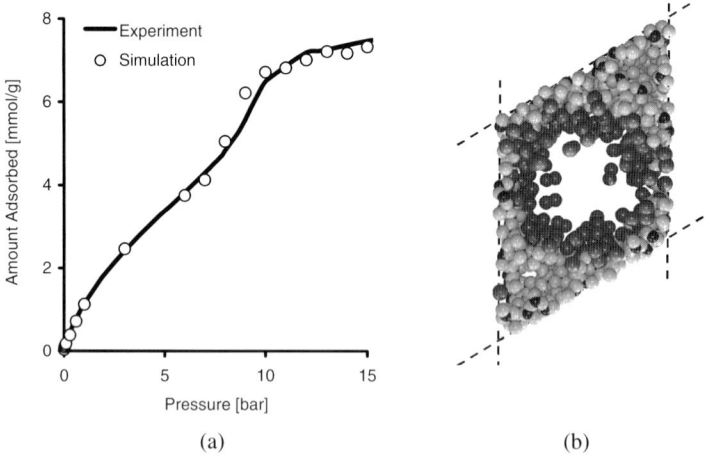

(a)                                                    (b)

Figure 1. (a) Experimental and simulated adsorption isotherms for ethane on pure-silica MCM-41 at 264 K. (b) A snapshot from the GCMC simulation; silicon atoms, oxygen atoms and ethane molecules are blue, grey and green, respectively. (The experimental data are from He and Seaton [3].)

Figure 1 shows a comparison of the simulated and experimental adsorption of ethane on the pure-silica MCM-41. The only adjustable parameters are: the porosity of the model MCM-41, to take into account the possibility of imperfections (such as the presence of closed pores) in the real material; and the pore size, which is not known precisely for the real material. Agreement with experiment is excellent. Figures 2 and 3 are GCMC predictions of adsorption on MCM-41 modified with either phenyl or aminopropyl groups. There are no adjustable parameters in these simulations, which are thus pure predictions.

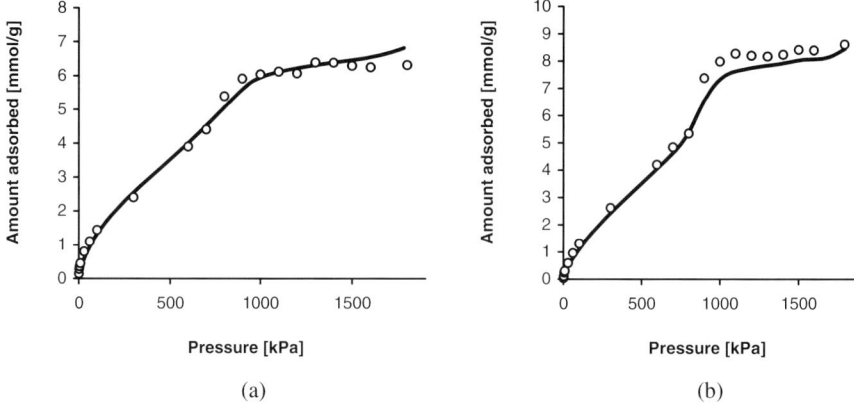

Figure 2. Experimental (line) and simulated (○) adsorption isotherms for ethane on (a) 10% phenyl-modified MCM-41 and (b) aminopropyl-modified MCM-41.

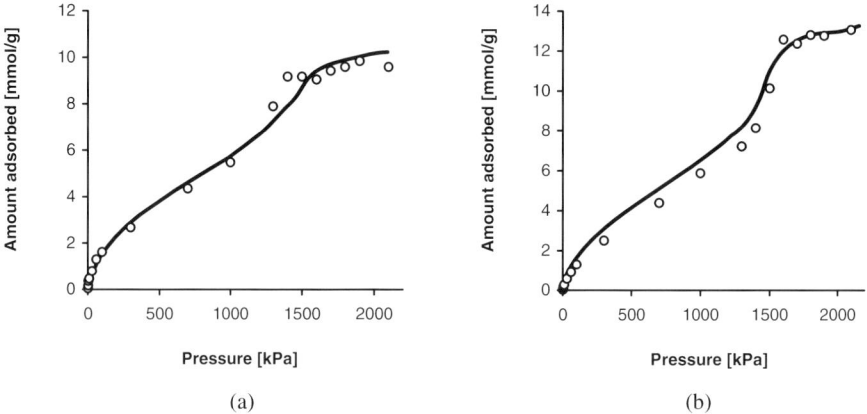

Figure 3. Experimental (line) and simulated (○) adsorption isotherms for carbon dioxide on (a) 10%-phenyl-modified MCM-41 and (b) aminopropyl-modified MCM-41.

## 3    PORE NETWORK MODEL FOR SBA-2

The pore network model of the SBA-2 material is composed of cavities, which we model as spheres bounded by an ordered array or oxygen atoms, and channels, which we model as cylinders, again bounded by an ordered array of atoms. (As in the simulation of adsorption in MCM-41, we ignore the silicon atoms.) We allow the cavities and channels to have a range of diameters. Adsorption in the individual pores (cavities or channels) is assumed to be independent of adsorption in adjacent pores. This approximation is necessary to make the computation of adsorption in the pore network as a whole tractable. As the cavities form an hcp array, each cavity has 12 neighbours. It does not necessarily follow that each cavity actually has 12 channels connecting it to neighbouring cavities. Our preliminary analysis of adsorption data, in which we assumed this to be the case, revealed that a proportion of the cavities were effectively isolated from certain adsorptives. (It is an open question whether some of the channels are in fact absent.) It is thus necessary to take into account the effect of the reduced connectivity of the cavities; we do this by carrying out a percolation analysis of the network. The effect of the presence of pores too small to permit the passage of all adsorptive species is illustrated in Figure 4, for two species – nitrogen and ethane. The main points to note are: (i) although all cavities are large enough to contain both species, some are isolated from the larger species (ethane) by small channels, which are not necessarily adjacent to that cavity; and (ii) some channels large enough to contain ethane are isolated in the same way.

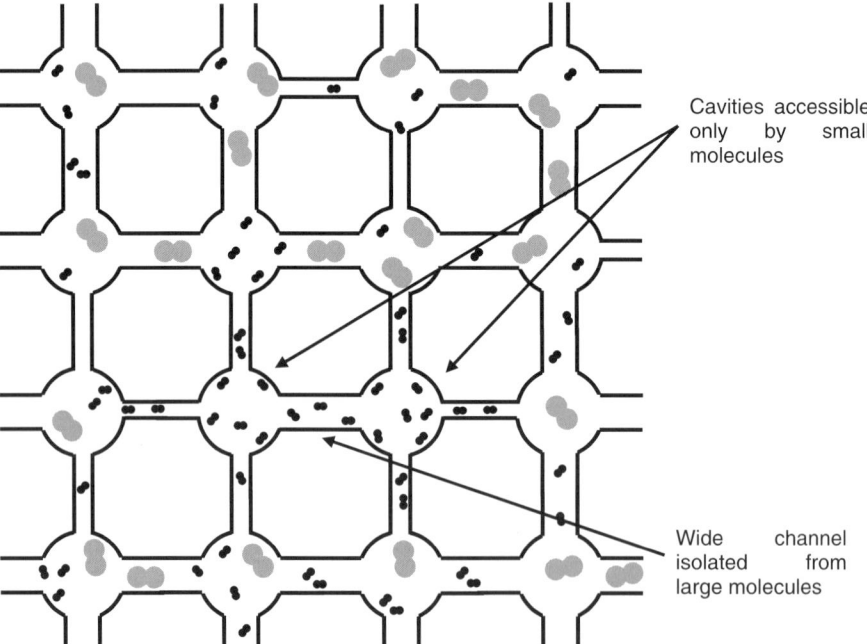

Cavities accessible only by small molecules

Wide channel isolated from large molecules

Figure 4. Schematic diagram showing connectivity in a 2D lattice for large (ethane) and small (nitrogen) molecules. The relative sizes of cavities, channels and molecules have been modified for the sake of clarity.

Adsorption was simulated using the GCMC method, as described in the previous section. These were obtained by the usual approach of inverting the adsorption integral equation, using isotherms generated by molecular simulation as the kernel of the integral; we used the method of Davies and Seaton [4] to fit to the experimental ethane isotherm at 263 K. The PSD, which consists of two peaks – one for the channels and one for the cavities, is shown in Figure 5. The fit to experimental data at 263 K, and the predicted adsorption at 293 K, are shown in Figure 6. Both the fit at 263 K and the prediction at 293 K are in good agreement with experiment, demonstrating the realism of the model. Further details of the PSD analysis are given in reference [5].

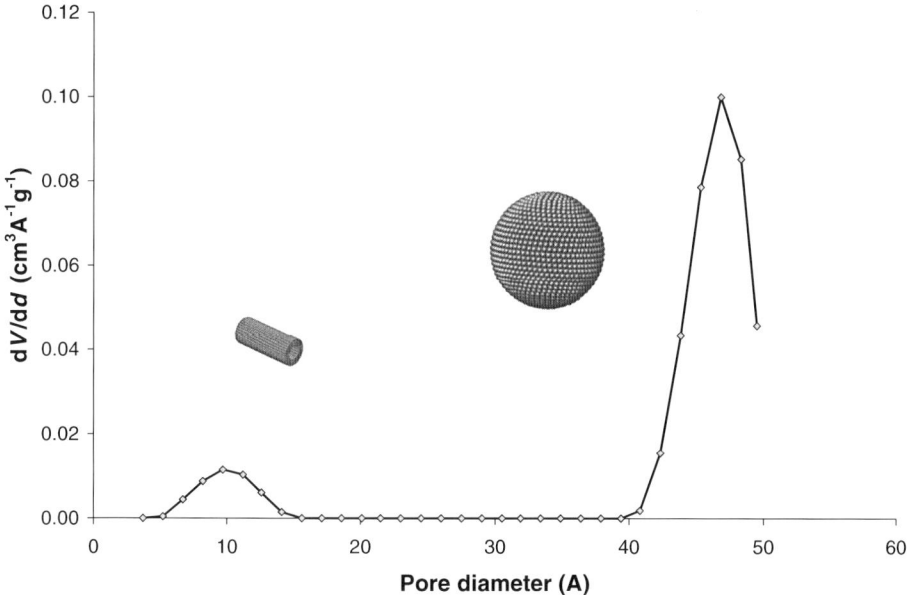

Figure 5. PSD obtained from the analysis of ethane adsorption at 263 K. The peak at smaller sizes corresponds to the channels, and the peak at larger sizes corresponds to the cavities.

The percolation analysis is described in detail elsewhere [6]. The main outcome of the analysis is that the effective coordination number of the cavities, defined as the average number of channels per cavity that are large enough to allow nitrogen (the smaller of our adsorptives) to pass, is 4.9, much less than the theoretical maximum value of 12. Taking into account only the smaller set of channels large enough to admit ethane, the effective coordination number is 1.8, just above the percolation threshold of the network. The different accessibility of the pore network to the two adsorptives is illustrated in Figure 7. Figure 8 shows a comparison between simulated and experimental nitrogen adsorption isotherms, with and without taking into account the lower network connectivity experienced by ethane, demonstrating the need to take into account pore network connectivity in modelling adsorption in SBA-2. The picture provided by percolation theory is thus one of a sparsely connected array of cavities, with different subnetworks available to each species, and with ethane in particular moving through a tortuous subnetwork of poorly connected pores.

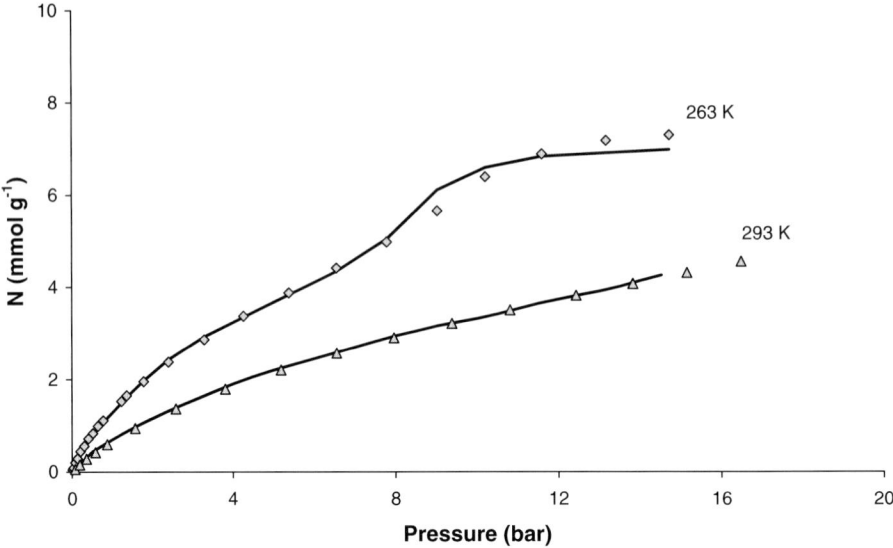

Figure 6. Fit of the simulated isotherm to experimental data at 263 K, and comparison of the predicted isotherm at 293 K with the corresponding experimental data. Symbols: experimental data. Solid line: GCMC simulation.

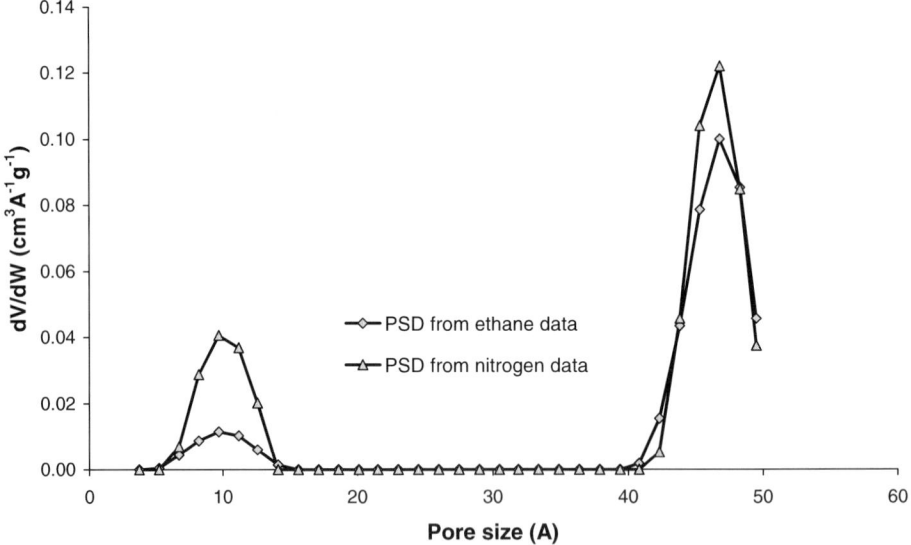

Figure 7. Comparison of PSDs accessible to ethane and nitrogen.

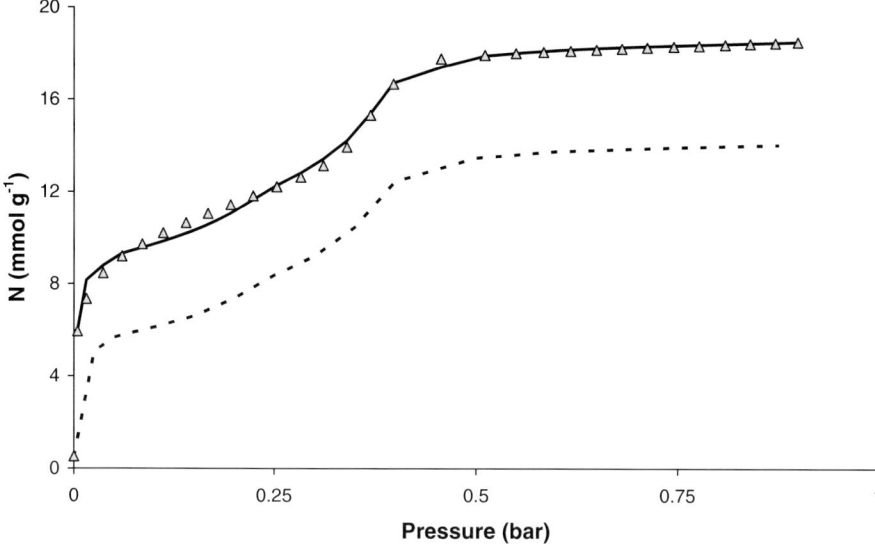

Figure 8. Experimental (symbols) and simulated (lines) adsorption of nitrogen at 77 K. Dashed line: prediction from the PSD obtained from the analysis of ethane adsorption. Solid line: isotherm obtained by fitting the PSD to nitrogen adsorption data.

## 4    DISCUSSION

We presented two different types of model for PMSs: a model based on kMC simulation of the synthesis process, which we applied to MCM-41; and a pore network model, which we applied to SBA-2. The parameters of the kMC-based model, which is the more fundamental of the two, were fitted to ethane adsorption data. The model was then used to predict adsorption in materials functionalised with surface organic groups, with excellent results. The pore network model for SBA-2 allowed us to determine the PSD of the cavities and the channels, and to obtain information on the pore network connectivity, and in particular the coordination numbers of the subnetworks accessible to different adsorptives. Taken together, these two studies have improved our knowledge of the structure of these materials, and have established that we can use these methods as tools for the design of PMSs as adsorbents and catalyst supports.

**ACKNOWLEDGEMENTS** The authors are grateful to Paul Wright and Jorge Gonzales for providing samples, and to the UK Engineering and Physical Sciences Research Council for financial support.

## REFERENCES

1   D. Frenkel, and B. Smit, Understanding Molecular Simulation, Academic Press, London (1996).
2   C. Schumacher, PhD thesis, University of Edinburgh, 2005.
3   Y. He and N.A. Seaton, *Langmuir* **19**, 10132 (2003)
4   G.M. Davies and N.A. Seaton *Langmuir* **15**, 6263 (1999).

5  M.J. Pérez-Mendoza, J. Gonzalez, P.A. Wright and N.A. Seaton, *Langmuir* **20**, 7653 (2004).
6  M. Pérez-Mendoza, J. Gonzalez, P.A. Wright and N.A. Seaton, The structure of the mesoporous silica SBA-2, determined by a percolation analysis of adsorption, *Langmuir* **20**, 9856 (2004).

Studies in Surface Science and Catalysis 160
P.L. Llewellyn, F. Rodriquez-Reinoso, J. Rouqerol and N. Seaton (Editors)

# Structural characterization of porous carbonaceous materials using high-pressure adsorption measurements

**Y. Belmabkhout, M. Frère (\*), G. De Weireld.**

Faculté polytechnique de Mons - Thermodynamics department, 31 bd Dolez 7000 Mons – Belgium

The aim of this work is to present a new method for porous carbonaceous solids characterization: the pore size distribution function (PSDF) of the adsorbent is determined by a theoretical treatment applied to adsorption data. The measurements are performed with different adsorbates at different temperatures in a wide pressure range. The theoretical model is based on the concept of integral adsorption equation (IAE). The main assumptions of the model are: (1) The PSDF is considered as an intrinsic property of the adsorbent. (2) The slit shaped pore model is used to describe the geometric configuration of the porous structure. (3) The adsorbent-adsorbate interactions are described by a Lennard-Jones potential model. (4) The pore wall surface is considered to be energetically homogeneous and the adsorbed phase is monolayer. (5) Both the gas phase and the adsorbed phase are supposed to be efficiently described by a Redlich-Kwong type equation of state. (6) We assumed a priori the analytical form of the pore size distribution function. We used adsorption isotherms of nitrogen, oxygen, argon and methane on different carbonaceous solids (four activated carbons and one molecular sieve) at 283 K, 303 K and 323 K and for pressures up to 2200 kPa.

## 1. INTRODUCTION

Characterization of heterogeneous micropourous carbonaceous solids from theoretical treatments of adsorption data is a complex issue. The large number of existing characterization methods is an evidence of the constant evolution in the understanding of the way adsorption properties and sorbents structural characteristics are related [1-13]. During the last decades, the disposal of efficient calculation means [11] has led to significant improvements in this matter and the major challenges which were clearly identified a few tens years ago are about to be overcome. These challenges may be classified as follows:

- the analytical form of the pore size distribution function should not be assumed a priori;
- the geometrical structure of the pores should be well defined;
- the interaction models (adsorbate-adsorbate and adsorbate-adsorbent) should be accurate;
- the local adsorption model should be efficient in the whole pore range (micro, meso and macropores);
- The energetic heterogeneity of the pore walls should be taken into account.

All these aspects have been partially or thoroughly treated in previous works so that their relative importance for the assessment of a reliable pore size distribution function can be evaluated [1,10]. On such a basis, it should be interesting to develop an efficient model which would be able to simulate adsorption data in a wide range of temperature and pressure conditions and for any adsorbate. Such a model should remain as simple as possible so that it could be used for adsorption process simulation.

The purpose of this paper is to present such a model which can be regarded as a synthesis of fundamental considerations resulting from the activities in porous sorbents characterization and classical thermodynamics developments. This model is able to represent in a correct way adsorption data of oxygen, argon, nitrogen and methane on a given carbonaceous adsorbent using a unique pore size distribution function whatever the adsorbate. The procedure was applied on four different activated carbons and on a carbon molecular sieve. The adsorption isotherms were measured at 283 K, 303 K and 323 K and for pressures up to 2200 kPa.

## 2. THEORITICAL DEVELOPMENT

### 2.1. General approach

Our theoretical approach is based on the Integral Adsorption Equation (IAE) concept. The IAE allows the generalization of statistic models initially developed for homogeneous surfaces to heterogeneous ones. Such surfaces are characterized by an adsorption energy distribution function $F(\varepsilon)$ for a given adsorbate. Considering porous carbonaceous adsorbents, it is generally accepted that the apparent energetic heterogeneity results from pore size heterogeneity. The concept of Integral Adsorption Equation may then be used for such adsorbents if $F(\varepsilon)$ may be correlated to the pore size distribution function $F_V(H)$. The basic expression of the Integral Adsorption Equation applied to geometrically heterogeneous adsorbents is:

$$m(T,P) = \int_{Bi}^{Bs} M\Gamma(T,P,H)F_A(H)dH \qquad (1)$$

In which:

- $m(T, P)$ is the adsorbed mass at temperature T and pressure P;
- $M$ is the molar mass of the adsorbate;
- $\Gamma(T, P, H)$ is the molar adsorbate surface concentration on the walls of the micropores of diameter H at temperature T and pressure P;
- $F_A(H)$ is the pore size distribution function defined on a surface basis;
- $B_i$ and $B_s$ are the integration limits.

Using Eq. (1) for the calculation of adsorbed masses requires the knowledge of both $\Gamma(T, P, H)$ and $F_A(H)$. $F_A(H)$ may be related to the pore size distribution function defined on a volume basis $F_V(H)$ for a given pore shape. $\Gamma(T, P, H)$ is called the local adsorption isotherm model. Considering the pore walls as energetically homogeneous surfaces, classical models derived from statistical thermodynamics developments may be used once the correlation between the adsorption energy $\varepsilon$ and the pore size $H$ is known. The analytical form of such a correlation depends on the pore shape on the one hand and on the adsorbate-pore wall interaction energy function on the other hand.

### 2.2. Geometrical and energetic modelling of the porous structure of carbonaceous adsorbents

Classical carbonaceous adsorbents (activated carbons and carbon molecular sieves) are composed of amorphous and graphitic carbon. Such materials exhibit a very important microporous structure due to missing graphitic layers in the graphite crystallites. The disordered arrangement of these crystallites is responsible for the mesoporous and macroporous structure. The presence of heteroatoms in the raw materials used for the synthesis of such adsorbents leads to a chemical heterogeneity of the pore wall surface. As far as gas adsorption at supercritical temperatures is concerned, it is generally assumed that the adsorption process occurs mainly in micropores of molecular dimensions of which the walls may be considered as energetically homogeneous [13]. As a consequence, we chose the simplified slit-shaped model to describe the porous structure of our adsorbents. The pore diameter $H$ is defined as the distance between the parallel graphite walls. The interactions between the adsorbate molecules and the pore wall are assumed to be well described by the Lennard-Jones model. The adsorption energy in a pore of diameter H is calculated by summing the contribution of each wall of the pore. The resulting correlation between the adsorption energy and the pore diameter $\varepsilon(H)$ is used to develop the local model $\Gamma(T, P, H)$ on the basis of classical existing models defined on a energetic basis $\Gamma(T, P, \varepsilon)$.

In this work, we set the mathematical form of the pore size distribution function $F_V(H)$; we opted for a bimodal distribution:

$$F_V(H) = \frac{V_{1\,pores}}{\sqrt{2\pi}\sigma_{1Hpores}}\exp\left(\frac{-\left(H - m_{1Hpores}\right)^2}{2\sigma_{1Hpores}^2}\right) + \frac{V_{2\,pores}}{\sqrt{2\pi}\sigma_{2Hpores}}\exp\left(\frac{-\left(H - m_{2Hpores}\right)^2}{2\sigma_{2Hpores}^2}\right) \qquad (2)$$

In which:
o   $V_{1\,pores}$ and $V_{2\,pores}$ are the micropores volumes;
o   $m_{1Hpores}$ and $m_{2Hpores}$ are the averages pore diameters;
o   $\sigma_{1Hpores}$ and $\sigma_{2Hpores}$ are the standard deviations.

Given the slit-like shape of the pores, the pore size distribution function defined on a surface basis $F_A(H)$ is related the corresponding function defined on a volume basis $F_V(H)$ by Eq. (3):

$$F_A(H) = \frac{x\,F_V(H)}{H} \qquad (3)$$

In which $x$ is equal to 1 or 2 according to the number of monomolecular layer formed within the pore.

**2.3. Local adsorption isotherm model**
$\Gamma(T, P, \varepsilon)$ and hence $\Gamma(T, P, H)$ are obtained by expressing the equality of the chemical potentials of the adsorbate in both the gas and adsorbed phases. The chemical potentials are derived from statistical thermodynamics developments [14].

*2.3.1. Gas phase modeling*
The canonical partition function of a system composed of $N_g$ molecules in a volume $V_g$ at temperature T is given by:

$$Q_g = \frac{1}{N_g!}q_{gtrans}{}^{N_g}q_{grot}{}^{N_g}q_{gvibi}{}^{N_g}q_{ge}{}^{N_g}\exp\left(-\frac{U'_{int}}{2kT}\right) \qquad (4)$$

In which:

• $q_{g\,trans}$ is the translational contribution to the partition function (3 degrees of freedom of the centre of mass of the molecule);
• $q_{g\,rot}$ is the rotational contribution to the partition function;
• $q_{g\,vibi}$ is the internal vibrational contribution to the partition function;
• $q_{g\,e}$ is the electronic contribution to the partition function;

- $U'_{int}$ is the potential energy of interaction between any molecule and all the others in the system;
- $k$ is the Boltzmann constant.

$q_{g\,trans}$ may be expressed by:

$$q_{gtrans} = \frac{V_g - N_g b}{\Lambda^3} \qquad (5)$$

with:

$$\Lambda^{-1} = \left[\frac{2\pi M_{ml} kT}{h^2}\right]^{1/2} \qquad (6)$$

In which:
- $\Lambda$ is the thermal de Broglie wavelength;
- $h$ is the Planck constant;
- $M_{ml}$ is the mass of the molecule;
- $b$ is the volume of the molecule calculated using the Redlich-Kwong approach [15].

The molecular chemical potential of the gas phase is obtained by differentiating $\ln Q_g$ (given by Eq. (4) with respect to $N_g$ at constant temperature and volume:

$$\mu_g = -kT\left(\frac{\partial \ln Q_g}{\partial N_g}\right)_{T,V_g} \qquad (7)$$

Using Eq. (4) and Eq. (7) and the Redlich-Kwong approach for the calculation of $U'_{int}$ [16], we obtain:

$$\mu_g = -kT\left[\begin{array}{l} \ln kT \dfrac{q_{grot} q_{gvibi} q_{ge}}{\Lambda^3} - \ln P + \ln\dfrac{PV_g}{kTN_g} - \ln\dfrac{V_g}{V_g - N_g b} - \\[2ex] \dfrac{N_g b}{V_g - N_g b} + \left[\dfrac{a}{bkT\sqrt{T}} \ln\left(\dfrac{V_g + N_g b}{V_g}\right)\right] + \left[\dfrac{N_g b}{V_g + N_g b}\dfrac{a}{bkT\sqrt{T}}\right] \end{array}\right] \qquad (8)$$

In which: $a$ is the molecular parameter of the Redlich-Kwong eos [15] for the adsorbate in the gas phase.

Eq. (8) may also be written:

$$\mu_g = \mu_g^0 + kT \ln P + kT \ln \Phi' \qquad (9)$$

With:

$$\mu_g^0 = -kT \ln kT \frac{q_{grot} q_{gvibi} q_{ge}}{\Lambda^3} \qquad (10)$$

$$\ln \Phi' = -\ln\frac{PV_g}{kTN_g} + \ln\frac{V_g}{V_g - N_g b} + \frac{N_g b}{V_g - N_g b} - \frac{a}{bkT\sqrt{T}}\left[\ln\left(\frac{V_g + N_g b}{V_g}\right) + \frac{N_g b}{V_g + N_g b}\right] \qquad (11)$$

$\Phi'$ is the fugacity coefficient which takes into account the non-ideality of the gas phase.

*2.3.2. Adsorbed phase modeling*

We consider a system of $N_s$ molecules adsorbed on a surface A; the molecules interact with each other. The same kind of developments as the ones presented for the gas phase may be achieved for the adsorbed phase. They lead to the expression of the chemical potential of the adsorbate in the adsorbed phase by using the corresponding two dimensional equation Redlich-Kwong equation of state:

$$\mu_s = -kT\left[\begin{array}{l}\left[\ln kT\,\dfrac{q_{svib}\,q_{srot}\,q_{se}\,q_{svibi}}{\Lambda^2}\right]-\dfrac{U_0}{kT}-\ln\Pi+\ln\dfrac{\Pi}{kT}\dfrac{1}{\Gamma_{ml}}-\ln\dfrac{1/\Gamma_{ml}}{1/\Gamma-b_s}-\\[2mm]\dfrac{b_s}{1/\Gamma_{ml}-b_s}+\dfrac{a_s}{b_s kT\sqrt{T}}\left(\ln\left(\dfrac{1/\Gamma_{ml}+b_s}{1/\Gamma_{ml}}\right)+\dfrac{b_s}{1/\Gamma_{ml}+b_s}\right)\end{array}\right] \quad (12)$$

In which:
- $U_0$ is the energy of adsorption. It can be calculated from the Lennard-Jones expression of the potential energy of interaction of an adsorbate molecule in a pore;
- $\Pi$ is the spreading pressure;
- $\Gamma_{ml}$ is the surface molecular concentration $\Gamma_{ml}=\dfrac{N_s}{A}$;
- $a_s$ and $b_s$ are the Redlich-Kwong molecular parameters of the adsorbate in the adsorbed phase;
- $q_{s\,vib}$ is the contribution to the partition function due to the vibration of the molecule perpendicularly to the adsorption surface.

$q_{s\,trans}$ and $q_{s\,vib}$ may be calculated respectively by:

$$q_{strans}=\dfrac{A-N_s b_s}{\Lambda^2}\quad (13)\qquad\text{and}\qquad q_{svib}=\dfrac{\exp\left(-\dfrac{h\nu_z}{2kT}\right)}{1-\exp\left(-\dfrac{h\nu_z}{kT}\right)}\quad (14)$$

In which:
- $\nu_z$ is the vibration frequency of the molecule perpendicularly to the surface. It can be calculated from the Lennard-Jones expression of the potential energy (adorbate-pore).

Eq. (12) may also be written:

$$\mu_s=\mu_s^0+kT\ln\Pi+kT\ln\Phi \quad (15)$$

With:

$$\mu_s^0=-kT\left[\ln kT\,\dfrac{q_{svib}\,q_{srot}\,q_{se}\,q_{svibi}}{\Lambda^2}\right]+U_0 \quad (16)$$

And

$$\ln\Phi=-\ln\dfrac{\Pi}{kT}\dfrac{1}{\Gamma_{ml}}+\ln\dfrac{1/\Gamma_{ml}}{1/\Gamma_{ml}-b_s}+\dfrac{b_s}{1/\Gamma_{ml}-b_s}-\dfrac{a_s}{kT\sqrt{T}b_s}\left[\ln\left(\dfrac{1/\Gamma_{ml}+b_s}{1/\Gamma_{ml}}\right)+\dfrac{b_s}{1/\Gamma_{ml}+b_s}\right] \quad (17)$$

$\phi$ is the fugacity coefficient of the adsorbate in the adsorbed phase; it takes into account of the non-ideality of the adsorbed phase.

### 2.3.3. Expression of equilibrium

The equality of the chemical potentials of the adsorbate in both phases calculated respectively by eq. (8) and Eq. (12) lead to the final expression of the local adsorption isotherm model. The electronic, the internal vibrations and the rotational behaviour of the molecules are supposed to undergo no change when the molecules pass from the gas to the adsorbed phase. Using molar quantities instead of molecular quantities leads to:

$$\Gamma\left[\dfrac{1}{1-\Gamma b_{smol}}\exp\left(\dfrac{\Gamma b_{smol}}{1-\Gamma b_{smol}}\right)\left[1+\Gamma b_s\right]^{-\frac{N_0 a_{smol}}{RT\sqrt{T}b_{smol}}}\exp\left(-\dfrac{N_0\Gamma a_{smol}}{RT\sqrt{T}}\dfrac{1}{1+\Gamma b_{smol}}\right)\right]=\dfrac{hN_0}{RT\sqrt{2\pi MRT}}\dfrac{\exp\left(-\dfrac{N_0 h\nu_z}{2RT}\right)}{1-\exp\left(-\dfrac{N_0 h\nu_z}{RT}\right)}\quad(18)$$

$$\exp\left(\dfrac{-N_0 U_0}{RT}\right)P\left[\dfrac{RT}{Pv}\dfrac{v}{v-b_{mol}}\exp\left(\dfrac{b_{mol}}{v-b_{mol}}\right)\left[\dfrac{v+b_{mol}}{v}\right]^{-\frac{N_0 a_{mol}}{RT\sqrt{T}b_{mol}}}\exp\left(-\dfrac{N_0 a_{mol}}{RT\sqrt{T}}\dfrac{1}{v+b_{mol}}\right)\right]$$

In which:

- $v$ is the molar volume of the adsorbate in the gas phase; it is calculated as a function of temperature and pressure using the Redlich-Kwong equation of state;
- $R$ is the ideal gas constant;
- $N_0$ is the Avogadro number
- $a_{mol}$, $b_{mol}$, $a_{smol}$, $b_{smol}$, $\Gamma$ are the molar values corresponding respectively to $a$, $b$, $a_s$, $b_s$, $\Gamma_{ml}$.

### 2.4. Global model

The use of Eq. (1), (2), (3) and eq. (18) allows the calculation of the adsorbed mass as a function of temperature and pressure for a given adsorbate and a given carbonaceous adsorbent. The unknown parameters are:

- $a_{smol}$ and $b_{smol}$ (for each adsorbate).
- $V_{1\ pores}$ and $V_{2\ pores}$, $m_{1Hpores}$ and $m_{2Hpores}$ and $\sigma_{1Hpores}$ and $\sigma_{2Hpores}$, the parameters of the bimodal pore size distribution function

### 3. RESULTS AND DISCUSSION

The theoretical developments presented in the previous section were applied to experimental adsorption isotherms. We studied four activated carbons (Centaur, BPL, F30/470, WS42) and one molecular sieve (CMS1).

High pressure adsorption isotherm measurements of $N_2$, Ar, $CH_4$, and $O_2$ (at three different temperatures) were carried out with a volumetric device described earlier by Belmabkhout et al [17]. These experimental data have also been published [18]. By minimizing the discrepancies between the experimental and calculated results, it was possible to determine the adjustable unknown parameters of the model. Fig. 1 shows an example of the experimental and theoretical isotherms on Centaur (AC).

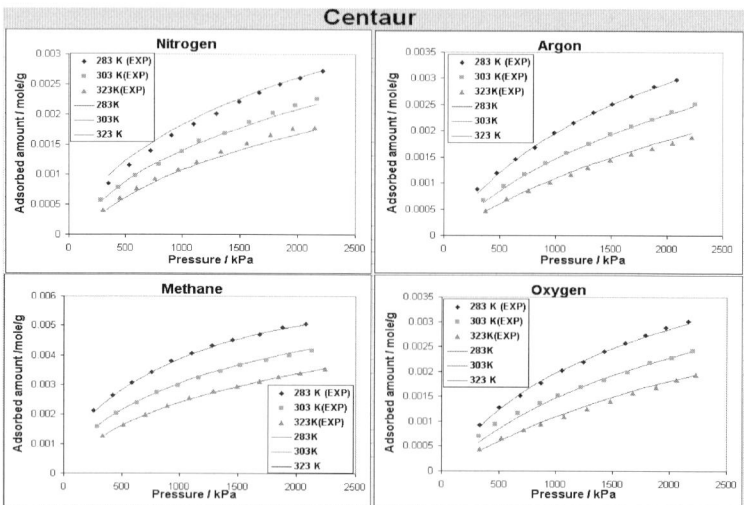

Fig. 1. Experimental and theoretical adsorption isotherms of $N_2$, Ar, $CH_4$ and $O_2$ on Centaur

For all the studied systems, a perfect fitting is noticed between experiment and theory (discrepancies are within the experimental error on the adsorbed mass: $5*10^{-3}$ g/g)

Tables 1 and 2 present the values of the fitted parameters for all the studied systems.

Table 1
Fitted parameters of the pore size distribution function for each adsorbent

| Sample | $m_{1Hpores}/Å$ | $\sigma_{1Hpores}/Å$ | $V_{1pores}/$ m$^3$/kg | $m_{2Hpores}/Å$ | $\sigma_{2Hpores}/Å$ | $V_{2pores}/$ m$^3$/kg | $Vtotalmicr/$ m$^3$/kg |
|---|---|---|---|---|---|---|---|
| *Centaur* | 6.98 | 0.49 | 9.75E-05 | 10.46 | 2.49 | 1.98E-04 | 2.95E-04 |
| *BPL* | 6.73 | 0.47 | 1.74E-04 | 10.77 | 2.35 | 1.10E-04 | 2.84E-04 |
| *F30/470* | 6.27 | 0.66 | 2.26E-04 | 8.51 | 3.28 | 1.70E-04 | 3.97E-04 |
| *WS42* | 6.74 | 0.57 | 1.65E-04 | 9.62 | 1.52 | 1.29E-04 | 2.93E-04 |
| *CMS1* | 6.00 | 0.87 | 3.13E-04 | 13.04 | 2.53 | 3.32E-09 | 3.13E-04 |

Table 2
Fitted energetic and geometric bidimensional Redlich-Kwong parameters for each adsorbent-adsorbate system ($a_{smol}$ / J m$^2$ mole$^{-2}$ , $b_{smol}$ / m$^2$ mole$^{-1}$)

| Sample | $a_{smol}(N_2)$ | $b_{smol}(N_2)$ | $a_{smol}(Ar)$ | $b_{smol}(Ar)$ | $a_{smol}(CH_4)$ | $b_{smol}(CH_4)$ | $a_{smol}(O_2)$ | $b_{smol}(O_2)$ |
|---|---|---|---|---|---|---|---|---|
| *Centaur* | 8.10E+09 | 7.88E+04 | 9.80E+08 | 3.42E+04 | 4.18E+09 | 4.45E+04 | 2.18E+09 | 4.25E+04 |
| *BPL* | 9.43E+09 | 7.59E+04 | 7.15E+08 | 3.24E+04 | 5.83E+09 | 4.27E+04 | 1.29E+09 | 3.69E+04 |
| *F30/470* | 8.14E+09 | 6.74E+04 | 2.50E+09 | 4.64E+04 | 4.93E+09 | 3.96E+04 | 1.09E+09 | 3.67E+04 |
| *WS42* | 6.52E+09 | 5.76E+04 | 9.91E+08 | 2.88E+04 | 2.96E+09 | 3.13E+04 | 1.53E+09 | 3.32E+04 |
| *CMS1* | 8.66E+09 | 5.94E+04 | 2.26E+09 | 3.87E+04 | 4.20E+09 | 2.89E+04 | 2.23E+09 | 4.19E+04 |

In Fig. 2 we plot the derived PSDF representative for each adsorbent. The PSDF of CMS1 seems to reduce to a Gaussian which points out that the range of pore size of CMS1 is very small. The values of the standard deviations $\sigma_{1Hpore}$, $\sigma_{2Hpore}$ and the volumes $V_{1pore}$ and $V_{2pore}$ for F30/470 point out that the degree of the structural heterogeneity of this adsorbent is the highest as far as micropores are concerned. Centaur, WS42, and BPL seem to have a relatively intermediate degree of microporous structural heterogeneity.

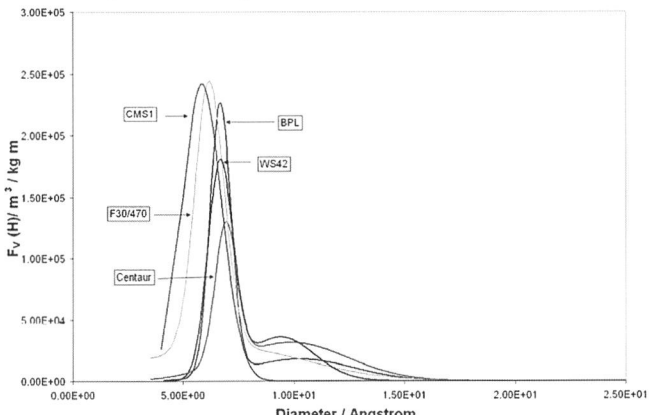

Fig. 2. PSDF of the studied adsorbents.

From Table 2, it appears that the value of the fitted energetic and geometric bidimensional Redlich-Kwong parameters $a_{smol}$ and $b_{smol}$ seem to be physically realistic; they vary weakly from one adsorbent to another. The variation of these bidimensional parameters can be explained by the difference in the surface chemistry for the different adsorbents which is not taken into account in our model.

## 4. SUMMARY AND CONCLUSIONS

In this paper, we have presented and tested a model which allows the calculation of adsorption isotherms for carbonaceous sorbents. The model is largely inspired of the characterization methods based on the Integration Adsorption Equation concept. The parameters which characterize the adsorbent structure are the same whatever the adsorbate. In comparison with the most powerful characterization methods, some reasonable hypothesis were made: the pore walls of the adsorbent are assumed to be energetically homogenous; the pores are supposed to be slit-like shaped and a simple Lennard-Jones model is used to describe the interactions between the adsorbate molecule and the pore wall; the local model is obtained considering both the three-dimension gas phase and the two-dimension adsorbed phase (considered as monolayer) described by the Redlich-Kwong equation of state; the pore size distribution function is bimodal. All these hypotheses make the model simple to use for the calculation of equilibrium data in adsorption process simulation. Despites the announced simplifications, it was possible to represent in an efficient way adsorption isotherms of four different compounds at three different temperatures on a set of carbonaceous sorbents using a unique pore size distribution function per adsorbent.

## REFERENCES

1. M. Kruk, M. Jaroniec, and J. Choma, Carbon., 36 (1998)1447.
2. K. Kaneko, C. Ishii and t. Rybolt. J.Rouquerol,F.Rodrigues-Reinoso, K.S.W.Sing and K. K. Unger (eds.), Proceeding of Characterisation of porous solids III.. Elsevier, Amsterdam, (1994), 583.
3. L.S. Cheng and R.T Yang., Chemical engineering science., 49 (1994) 2599.
4. P. I. Ravikovitch, A. Vishnyakov, R. Russo, and A.V.Neimark, Langmuir., 16 (2000) 2311.
5. K. Wang and Do .D.D, Langmuir., 13 (1997) 6226.
6. C. Lastoskie, and al, J.phys.Chem., 97 (1993) 4786.
7. C. Dapeng and al, Carbon., 40 (2002) 2359.
8. J. P. Olivier, Carbon., 36 (1998)1469.
9. L. P. Ding, and S. K. Bathia, Carbon., 39 (2001) 2215
10. P. I. Ravikovitch, and A. V. Neimark, J.Phys.Chem.B., 102 (2001) 6817.
11. P. Kowalczyk, and al., Carbon., 41 (2003) 1113.
12. Z. Tan, and K. E. Gubbins, J.Phys.Chem., 94 (1990) 6061.
13. J. Jagiello, T. Bandosz, K. Putyera, and J. Schwarz. Rouquerol J, Rodriguez-Reinoso F., Sing K.S.W.and K.K.Unger (eds.), Proceeding of Characterisation of porous solids III., Elsevier Science,Amsterdam,1994, 679.
14. T. L. Hill, An introduction to statistical thermodynamics, Addison-Wesley Publishing company, London, 1962.
15. J. M. Pruasnitz, R. N. Lichtenthaler, E.G de Azevedo, Second edition, Molecular thermodynamics of fluid-phase equilibria, Prentice-Hall, Inc, New Jersey, 1986.
16. J. H. Vera, J. M. prausnitz, The chemical engineering journal., 3 (1972) 1.
17. Y Belmabkhout, M Frère and Guy De Weireld., Meas.Sci.Technol., 15 (2004) 848.
18. Y Belmabkhout, M Frère and Guy De Weireld., J.Chem.eng.data., 49 (2004) 1379.

Studies in Surface Science and Catalysis 160
P.L. Llewellyn, F. Rodriquez-Reinoso, J. Rouqerol and N. Seaton (Editors)

# Microcalorimetric Characterization of Hydrogen Adsorption on Nanoporous Carbon Materials

Akihiko Matsumoto[a], Kazumasa Yamamoto[a] and Tomoyuki Miyata[b]

[a] Department of Materials Science, Toyohashi University of Technology, Tempaku-cho, Toyohashi 441-8580, JAPAN
[b] New Material Center, Osaka Science and Technology Center, 8-4 Utsubohommachi 1 Chome, Nishi-ku, Osaka 550-0004, JAPAN[+]

An apparatus for adsorption microcalorimetry at low temperature under high-pressure conditions was set up. Adsorption isotherms and differential energy of adsorption of hydrogen on activated carbon fibers ACF of different pore widths and single-wall carbon nanotube SWNT were measured in the range of equilibrium pressure from 0 to 10 MPa at isothermal condition, 203, 223, 243, and 298 K. The adsorption capacity of hydrogen depended upon adsorption temperature and micropore size of ACF: the ACF with narrower pore size (pore diameter: 0.5 nm) exhibited higher sorption capacity than that with wider pore size (1.7 nm), and the uptakes at lower temperature were higher than those at higher temperature. The SWNT showed the highest adsorption uptakes among the samples. The differential energy of adsorption was calorimetrically evaluated as $-10 \sim -5$ kJ/mol, which well agreed with the isosteric enthalpy of adsorption evaluated by Clausius-Clapeyron equation.

## 1. INTRODUCTION

Hydrogen is one of encouraging alternatives to fossil fuels, and harnessing hydrogen as a future energy has been studied especially in the fields of motor vehicle industry and fuel cell development. A significant challenge in practical use of hydrogen is safety and high-density storage of the gas, in which adsorption plays an important role.

Nanoporous carbon materials have been focused attention as hydrogen storage materials, since possibility of hydrogen storage on the materials, such as single-wall carbon nanotubes (SWNT) [1], graphite nano-fibers [2,3] and carbon nanohorns [4] has been reported in the latter half of the last decade. Microcalorimetric characterization of hydrogen adsorption is important to understand the adsorption mechanism on these materials. However, the adsorption capacity and the differential adsorption energy of hydrogen on the carbon materials would be far less than those of condensable vapors near room temperature and atmospheric pressure because of weak intermolecular force between carbon and hydrogen [5].   Therefore, the hydrogen adsorption should be measured under the special conditions, at high pressure and low temperatures.   The development of an adsorption apparatus for hydrogen adsorption at low temperature under high-pressure conditions was conducted in this study.   The hydrogen adsorption on nanoporous carbon materials at high-pressure conditions was also characterized microcalorimetrically by use of the adsorption apparatus.

## 2. EXPERIMENTAL

### 2.1. Nanoporous carbons
Tow kinds of activated carbon fibers (ACF, Ad'all Co., Japan) of different micropore sizes, designated as A7 and A20, and single wall carbon nanotubes (SWNT, Carbon-Nanotechnologies, Inc, USA) were used in this study as adsorbents.   The ACFs have slit-shaped micropores with relatively uniform pore widths and are often used as model substances in adsorption researches.   The SWNT contains 8 wt% of iron.   Each sample was used without further chemical treatments.

### 2.2. Physicochemical characterization
The nitrogen adsorption isotherms were measured volumetrically at 77 K using an automatic adsorption apparatus (Quantachrome, Autosorb-1-MP).   Each sample was outgassed at 423 K and 1 mPa for 12 h before the adsorption measurements.

### 2.3. Hydrogen adsorption
The high-pressure adsorption isotherms and differential energy of adsorption of hydrogen were measured at isothermal condition, 203, 223, 243 and 298 K, respectively, by a twin conduction type microcalorimeter (Setarum, BT-2.15) equipped with a volumetric adsorption apparatus.   The apparatus was made of stainless steel tubing being operated under high pressure up to 20 MPa.   Each adsorption isotherm was consecutively measured with microcalorimetric measurements.   The hydrogen gas of 99.99999 % purity (Nippon Sanso) was used in the research.   Sample was evacuated at 1 mPa and 423 K for 12 h before each measurement.

## 3. RESULTS AND DISCUSSION

### 3.1. Pore characteristics of carbon materials
Nitrogen adsorption isotherms of ACFs and SWNT are shown in Fig. 1.   The isotherm of A20 was of type Ib in the IUPAC classification suggesting microporous character of the material with wide distribution of pore size.   The isotherm of A7 exhibited a sharp knee at very low relative pressure region (P/Po < 0.02) and a long plateau, which is a typical shape

for adsorption by primary micropore filling. The isotherm of SWNT is rather a combination of type I and II although it had a sharp knee at a low P/Po region. Particles of SWNT are consisted by bundles of SWNTs, and the interspaces of nanotubes would not wide enough to diffuse nitrogen molecules there. Therefore, the isotherm would show the pore characteristics of the aggregate structures. Pore characteristics was evaluated by density functional theory (DFT method) and shown in Table 1.

## 3.2. Hydrogen adsorption

The adsorption came to equilibrium within 50 minutes after hydrogen dose. The adsorption isotherms on ACF at different temperatures were shown in Fig. 2. The ordinate in the figure is represented in molecules per unit area to clarify the effect of surface against adsorption capacity or surface excess concentration. Each adsorption isotherm was of Henry type regardless of adsorption temperature at the fugacity ranges of less than ca. 5 MPa; the uptakes increased linearly with fugacity, and they gradually increased over 5 MPa. The adsorption at lower temperature tended to show higher surface excess concentration. For example, the surface excess concentration at ca. 8 MPa varied from 0.9 to 1.3 molecules/nm$^2$ (from 2.4 to 3.4 mmol/g) for A20, and 2.4 to 2.6 molecules/nm$^2$ (from 3.4 to 3.7 mmol/g) for A7 with decreasing temperature from 298 K to 203 K.

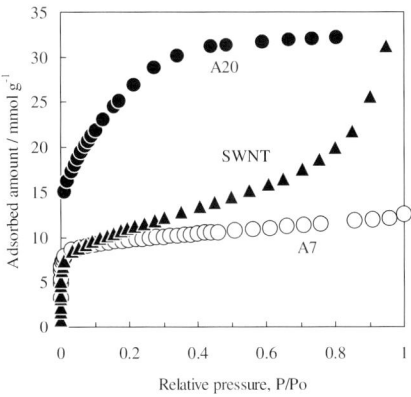

Fig. 1.   Adsorption isotherms of nitrogen on ACF.

Table 1   Pore characteristics of samples determined by DFT method.

| Sample | Pore width, mode (nm) | Surface area (m$^2$/g) | Pore volume (mL/g) |
|--------|------------------------|-------------------------|---------------------|
| A7 | 0.7 | 856 | 0.35 |
| A20 | 1.5 | 1586 | 1.03 |
| SWNT | 0.6 | 723 | 0.53 |

It is interesting to note that the surface excess concentration on A7 were higher than those on A20 at the same sorption temperature. In the case of vapor adsorption in ultramicropore, the micropore filling takes place at a very low pressure, which is associated with enhanced adsorbent-adsorbate interactions caused by overlapping of potential fields on facing pore walls [6]. Although hydrogen is a supercritical gas under the temperature range 203~298 K, the enhancement of the adsorbent-adsorbate interactions would affect the sorption capacity. Activated carbon fibers used in the present study have slit-shaped micropores [7,8]. Regarding a slit-shaped micropore being composed of graphite layers with the pore width of $2d$, the total interaction potential $\Phi$ between a hydrogen molecule in the pore and the facing

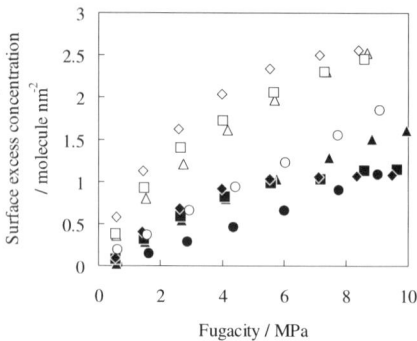

Fig. 2. Adsorption isotherms of hydrogen on ACF at different adsorption temperatures. Keys: circle, adsorbed at 298 K; triangle, 243 K; square, 223 K; diamond, 203 K; opened, A7; solid, A20.

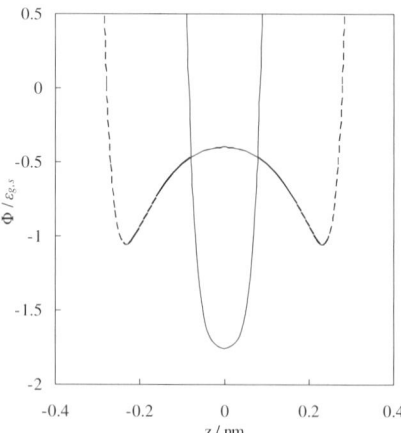

Fig. 3. Profiles of the interaction potential between a hydrogen molecule and pore walls. Keys: solid line, A7; broken line, A20.

graphite layers is a function of the distance between the surface of pore wall and the middle of the micropore $z$, and represented as,

$$\Phi = \Phi_{g,s}(d + z) + \Phi_{g,s}(d - z).$$

Here $\Phi_{g,s}$ is the potential at the distance $r$ from the center of carbon atom on the surface;

$$\Phi_{g,s} = A\left\{\frac{2}{5}\left(\frac{\sigma_{g,s}}{r}\right)^{10} - \left(\frac{\sigma_{g,s}}{r}\right)^{4} - \frac{\sigma_{g,s}^4}{3\Delta(r + 0.61\Delta)^3}\right\}$$

$$A = 2\pi\rho_s\varepsilon_{g,s}\sigma_{g,s}^2\Delta$$

where $r$ is the distance between the center of a hydrogen molecule and graphite surface, and energy, $\varepsilon_{g,s}$, and size, $\sigma_{g,s}$, parameters of Lennard-Jones potential between a carbon atom and a hydrogen molecule can be calculated as 30 K and 0.32 nm, respectively, by Lennard-Jones parameters of hydrogen ($\varepsilon_{g,g} = 33.3/k$ K and $\sigma_{g,g} = 0.30$ nm. $k$: Boltzman constant) and carbon atoms ($\varepsilon_{s,s} = 28$ K and $\sigma_{s,s} = 0.34$ nm) using Lorentz-Bertherot rule [9,10]. The solid density of graphite $\rho_s$ and the distance between graphite layers $\Delta$ are 114 nm⁻¹ and 0.335 nm, respectively. Thus, the potential energy profiles of A7 and A20 were evaluated as shown in Fig. 3. There are two minima in the potential curve of A20, but the minima merge to give a minimum in the curve of A7. The ordinate is a measure of the enhancement of the energy of adsorption in micropore as compared with that on an open surface, which means that the potential minima for A20 were almost identical to that of an open surface. On the other hand, the potential curve for A7 had the minimum which was 1.66-fold deeper than those for A20. This result indicates that hydrogen molecules interact stronger in the micropores of A7 than those of A20. Therefore, the higher adsorption capacity of A7 would be due to the strong

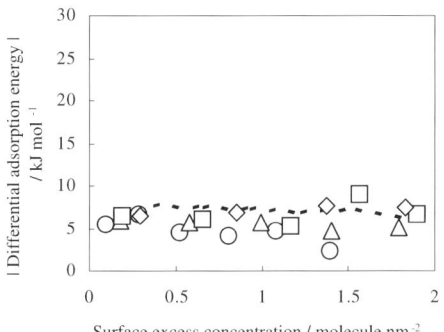

Fig. 4. Differential heats of adsorption of hydrogen on A7. Keys: circle, adsorbed at 298 K; triangle, 243 K; square, 223 K; diamond, 203 K; broken line, isosteric enthalpy of adsorption.

interaction by the enhanced potential field.

The differential energy of adsorption $\Delta_{ads}u$ is calorimetrically evaluated as the differential heats of adsorption $q_{diff}$ (>0), those are related by the expression: $\Delta_{ads}u = -q_{diff}$ [6]. The differential energy of adsorption, $\Delta_{ads}u$, of hydrogen on A7 was shown in Fig. 4. It should be noted that an ordinate in Fig.4 shows an absolute value which corresponds to $q_{diff}$. The differential energy was –8 ~ –4 kJ/mol and they varied widely with increasing the adsorption uptakes. The unevenness of the energy values would be due to experimental error caused by rather smaller increase of uptakes. The differential adsorption energy tended to be lower at higher adsorption temperature. $\Delta_{ads}u$ is written by the differential enthalpy of adsorption $\Delta_{ads}h$ as: $\Delta_{ads}u = \Delta_{ads}h + RT$ (R: gas constant, T: temperature) [6]. According to this relationship, the differential energies at 298 K would be –0.8 kJ/mol higher than those at 203 K, which qualitatively coincides with the difference in the differential energies between these temperatures shown in Fig.3. Isosteric enthalpy of adsorption $\Delta_{ads}h$ were estimated from the adsorption isotherms by the Clausius-Clapeyron equation as –7 ~ –6 kJ/mol. These values agreed with the experimental results and those for activated carbon [11]. In the adsorption on A20, the differential energies at initial stage were –19 ~ –15 kJ/mol but they immediately increased to ca. –5 kJ/mol, as shown in Fig. 5. These adsorption energies was comparable to the activation energy of desorption for single-walled carbon nanotubes, 19.6 kJ/mol [1]. However, the differential energies of adsorption would include considerable error at the initial stage of adsorption, because the adsorption uptake on A20 was smaller comparing to that on A7. We will confirm the initial heat carefully. The differential energies of adsorption on A20 were comparable to those on A7. Although the adsorption potential for A7 is higher than that for A20 as already mentioned, significant differences in the adsorption energies were not observed.

The adsorption isotherm on SWNT is shown in Fig. 6. The surface excess concentration on SWNT is the highest among all samples. Judging from the differential energies being

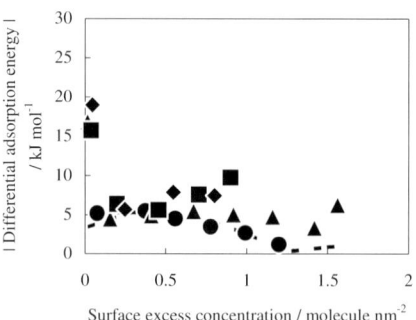

Fig. 5. Differential heats of adsorption of hydrogen on A20. Keys: circle, adsorbed at 298 K; triangle, 243 K; square, 223 K; diamond, 203 K; broken line, isosteric enthalpy of adsorption.

discussed later, there would not be any special interactions between the SWNT surface and hydrogen molecules. Therefore, the high surface excess concentration might be due to diffusion and adsorption of hydrogen on the interspaces of SWNT bundles where nitrogen molecule cannot access. The enhancement of the adsorption potential would also take place in the interspaces and gave rise to higher sorptivity. The surface excess concentration increased with decreasing the adsorption temperature as well as the cases in ACFs.

The differential energies of adsorption on SWNT were similar to those of ACFs as shown in Fig. 7; the energy values attained −10 ~ −5 kJ/mol at the initial stage of adsorption and gradually increased with increasing the uptakes. Significant changes in the differential energies of adsorption by the potential enhancement were not detected. The differential energies at higher adsorption temperature were lower, as observed in the cases of ACFs. These results indicate that the hydrogen sorption mechanism for SWNT is rather similar to

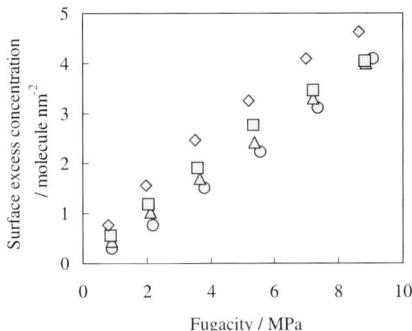

Fig. 6.   Adsorption isotherms of hydrogen on SWNT at different adsorption temperatures. Keys: circle, adsorbed at 298 K; triangle, 243 K; square, 223 K; diamond, 203 K.

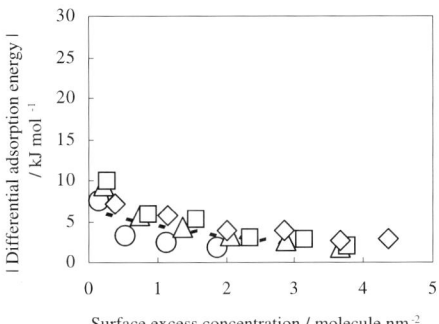

Fig. 7.   Differential heats of adsorption of hydrogen on SWNT.   Keys: circle, adsorbed at 298 K; triangle, 243 K; square, 223 K; diamond, 203 K; broken line, isosteric enthalpy of adsorption.

those of ACFs.    More detailed measurements of hydrogen sorption on microporous materials will be further studied.

## 4. CONCLUSION

An apparatus for adsorption microcalorimetry at low temperature under high-pressure conditions is set up.    Adsorption isotherms were of Henry type in the fugacity range $0 \sim 5$ MPa at isothermal condition, 203, 223, 243 and 298 K.    The surface excess amount increased with decreasing the adsorption temperature.    The ACF with narrower pore size A7 exhibited higher sorption ability than that with wider pore size A20, suggesting the enhancement of the interaction potential between pore wall and hydrogen molecule in A7. The differential energies of adsorption by the adsorption microcalorimetry were $-7 \sim -5$ kJ/mol for A7 and A20, which well agreed with those evaluated by Clausius-Clapeyron equation.    The adsorption capacity of SWNT was the highest among these three samples. The high adsorption capacity would due to adsorption in voids of the bundles where the enhancement of the adsorption potential may take place.

## ACKNOWLEDGEMENTS

This work was supported by New Energy and Industrial Technology Development Organization (NEDO) through Project 03015, Development for Safe Utilization and Infrastructure of Hydrogen.

## REFERENCES

[1]    A.C. Dillon, K.M. Jones, T.A. Bekkedahl, C.H. Kiang, D.S. Bethune and M.J. Heben, Nature, 386(1997)377.
[2]    A. Chambers, C. Park, R.T.K. Baker and N.M. Rodriguez, Phys. Chem. B, 102(1998) 4253.
[3]    C. Park, P.E. Anderson, A. Chambers, C.D. Tan, R. Hidalgo, N.M. Rodriguez, J. Phys. Chem. B, 103(1999)10572.
[4]    K. Murata, K. Kaneko, H. Kanoh, D. Kasuya, K. Takahasi, F. Kokai, M. Yudasaka and S. Iijima, J. Phys. Chem. B, 106(2002)11132.
[5]    J. Israelachvili, Intermolecular and Surface Forces, Academic Press, London, 1991.
[6]    For example, S.J. Gregg and K.S. W. Sing, Adsorption Surface Area and Porosity, $2^{nd}$ Ed, Academic Press, London, 1982, F. Rouquerol, J. Rouquerol and K. Sing, Adsorption by Powders and Porous Solids, Academic Press, London, 1999.
[7]    K. Kaneko, R.F. Cracknell, D. Nicholson, Langmuir, 10(1994)4606.
[8]    A. Matsumoto, J. Zhao, K. Tsutsumi, Langmuir, 13(1997)496.
[9]    W.A. Steele, Surf. Sci., 36(1973)317.
[10]  O.J. Hirschfelder, C.F. Curtiss and R.B. Bird, Molecular Theory of Gases and Liquids, Wiley, New York, Chapter 3, 1954.
[11]  P. Bénard and R. Chahine, Langmuir, 17(2001)1950.

Studies in Surface Science and Catalysis 160
P.L. Llewellyn, F. Rodriquez-Reinoso, J. Rouqerol and N. Seaton (Editors)
© 2007 Elsevier B.V. All rights reserved

# Effect of thermal treatments on the surface chemistry of oxidized activated carbons

**J. Ruiz-Martínez, E.V. Ramos-Fernández, A. Sepúlveda-Escribano and F. Rodríguez-Reinoso**

Laboratorio de Materiales Avanzados. Departamento de Química Inorgánica, Universidad de Alicante, Apartado 99, E-03080 Alicante, Spain.

## 1. INTRODUCTION

Activated carbons are widely used as adsorbents in either the gas or the liquid phase, and also as catalysts and catalyst support. In some cases their adsorption behaviour depends basically on their textural characteristics, i.e., porous structure and pore volume. As an example carbon molecular sieves (CMS) are a kind of activated carbons that make use of a narrow pore size distribution, of a few angstroms in diameter, to selectively separate gas mixtures [1], such as $N_2/O_2$, $CO_2/CH_4$, $C_3H_6/C_3H_8$ and some others. But also the surface chemistry can condition in many cases [2, 3] the adsorption behaviour, as well as the activity as catalyst and catalyst support [4, 5]. This means that the adsorption properties of the activated carbons cannot be easily explained only on the basis of textural characteristics (surface area and pore size distribution); the nature of the chemical surface must also be taken into account. Therefore, for a complete characterization of the activated carbon surface, textural and chemical characteristics must be assessed.

Oxygen surface groups are the most important chemical groups in carbons. They are responsible of many physico-chemical and surface properties of the carbons, such as surface acidity, cation exchange capacity, and adsorption of polar and non polar gases and vapours. It is a normal practise to increase the amount of oxygen surface groups by treatment with air or oxidizing aqueous solutions without substantially modifying the porous texture of the carbon [6]. The controlled heat treatment of the oxidized carbon under an inert atmosphere produces the selective decomposition of the oxygen groups [7], this modifying its surface chemistry not only because of the removal of the oxygen functionalities but also the creation of new active sites on the surface. But the characteristics of the active sites remaining depend on the gaseous atmosphere in which such treatment is carried out [8,9]. When the activated carbon is treated under $H_2$, not only oxygen surface groups are removed, but also some of more reactive carbon atoms (e.g., as $CH_4$). Furthermore, some of the active sites created can be stabilized by hydrogen chemisorption. In contrast, treatment in an inert gas removes oxygen but leaves behind many highly active sites.

The goal of this work was the study of the effect the surface chemistry of a carbon molecular sieve (Takeda 5A) in the adsorption of propane and propylene. The CMS was oxidized by two different agents, and thermal treatments under vacuum and different atmospheres were used to selectively remove the oxygen surface groups and the highly active sites.

## 2. EXPERIMENTAL

The starting material (CMS) was a commercial carbon molecular sieve, Takeda 5A. It was oxidized with two different agents in aqueous solution: $HNO_3$ 6M at 353 K for 1 h (CMS-$HNO_3$), and $H_2O_2$ 6M at room temperature for 24 h (CMS-$H_2O_2$). Then, the carbons were washed with distilled water until the oxidizing solution had been completely removed, and dried at 383 K during 24 h.

Textural properties of the samples were evaluated from the $N_2$ and $CO_2$ adsorption isotherms at 77 K (after out-gassing for 4 h. at 523 K), using a Coulter Omnisorp 610. The apparent surface areas were determined by means of the BET equation, and the micropore volumes were calculated by applying the Dubinin-Radushkevich (DR) equation.

The amount and type of surface oxygen complexes were determined by temperature-programmed decomposition (TPD) experiments. The samples (0.1 g) were placed into a differential flow reactor coupled to a quadrupole mass spectrometer (GSD 301), and then heated at 10 K/min up to 1323 K under a He flow of 50 ml/min.

The enthalpies of immersion of the carbons into hexane and 1-hexene were determined at 303 K in a Tian-Calvet type differential microcalorimeter (Setaram, C80D). Prior to the experiments, the samples (0.1 g) were out-gassed under vacuum at 523 and 773 K for 4 h, as described previously [10].

A Tian–Calvet microcalorimeter (model BT 2.15, Setaram, France) was used to measure the enthalpies of adsorption of propane and propylene at room temperature. The samples (0.1 g) were treated under different conditions: (i) vacuum at 523 K, (ii) vacuum at 773 K, (iii) He at 1073 K and (iv) $H_2$ at 1073 K, all for 4h. Then, they were sealed into a Pyrex RMN tube in pure He and placed into the microcalorimetric cell. A conventional manometric system coupled to the microcalorimeter was used to measure the amount adsorbed employing a MKS (type 660) manometer with a precision of 0.001 Torr. The maximum apparent leak rate of the manometric system (including the calorimetric cells) was $10^{-5}$ Torr·min$^{-1}$, in a volume of about 60 cm$^3$.

## 3. RESULTS AND DISCUSSION

The adsorption isotherms of $N_2$ at 77 K (fig. 1) show that the carbons exhibit a narrow pore size distribution, and the contribution from meso- and macroporosity is rather small. After oxidation both with $HNO_3$ and $H_2O_2$, the apparent BET surface area slightly decreases (table1), and this could be related to a partial blocking of the pore mouth by the oxygen functionalities created. In fact, $CO_2$ adsorption at 273 K hardly decreases with the oxidation treatment, this indicating that the blocking effect and the subsequent diffusion limitations suffered by nitrogen at 77 K do not apply for $CO_2$ at 273 K. Anyway, the liquid phase oxidation treatments have no significant impact on the texture of the carbon material, as already reported in the literature [11].

Table 1
Apparent BET surface area and micropore volume of the carbons.

| Carbon | $S_{BET}$, $m^2 \cdot g^{-1}$ | $V_{D-R}$ ($N_2$), $cm^3 \cdot g^{-1}$ | $V_{D-R}$ ($CO_2$), $cm^3 \cdot g^{-1}$ |
|---|---|---|---|
| CMS | 743 | 0,33 | 0.26 |
| CMS-$H_2O_2$ | 469 | 0,22 | 0.25 |
| CMS-$HNO_3$ | 514 | 0,25 | 0.25 |

Fig. 2a shows the $CO_2$ evolution profiles of the starting CMS and the two oxidized samples. The $CO_2$ evolution profile of CMS-$HNO_3$ is similar to others reported in the literature for activated carbon oxidized with $HNO_3$ [12,13]. The maximum rate of $CO_2$ evolution is at 570 K, with two shoulders close to 700 and 1000 K that tails up to around 1200 K. This profile is indicative of the occurrence of chemically different complexes and/or the same oxygen complex existing on energetically different sites [13]. The first $CO_2$ peak may be attributed to carboxylic acid groups, also called low-temperature $CO_2$ groups, while the higher temperature $CO_2$ peaks, that show a shoulder at 700 and 1000 K may result from lactones and carboxylic anhydrides respectively (usually named "high-temperature $CO_2$ groups"). The carboxylic anhydride groups that can be formed from two adjacent carboxyl

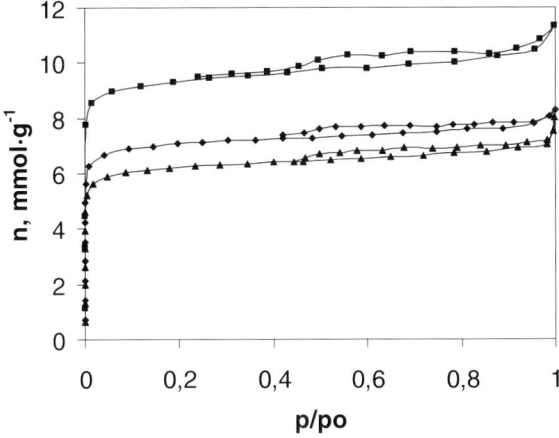

Fig. 1. $N_2$ adsorption isotherms at 77 K on carbons CMS
(■), CMS-$H_2O_2$ (▲) and CMS-$HNO_3$ (♦).

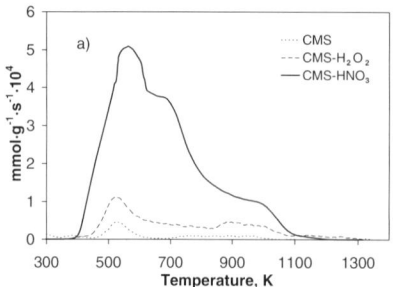 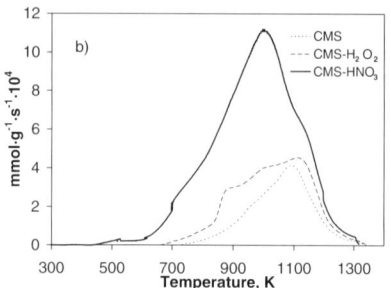

Fig. 2. $CO_2$ (a) and CO (b) evolution profiles for the parent and oxidized CMS´s

groups during the TPD runs, are more stable than carboxylic acid groups and, therefore, will evolve as $CO_2$ and CO above 570 K [13]. In the CMS-$H_2O_2$ sample, the maximum rate of $CO_2$ evolution is shifted to a lower temperature, 530 K, thus indicating that the thermal stability of the carboxylic acid groups on this sample is lower than in the sample CMS-$HNO_3$. After this maximum there is a smaller peak around 900 K and a tail up to around 1300 K, possibly due to carboxylic anhydride groups. Finally, sample CMS shows a $CO_2$ evolution profile that is similar to CMS-$H_2O_2$, although the peak at higher temperature begins at lower temperature than the later.

Fig. 2b shows the CO evolution profiles from the carbons in TPD experiments. Substantial differences can be found among them. CO evolution from sample CMS-$HNO_3$ has a small plateau around 570 K that does not appear in the other samples, the maximum rate of CO evolution being at 1000 K with a small shoulder at 1125 K. Sample CMS-$H_2O_2$ has the maximum at 1125 K, with two shoulders at 890 and 1000 K. Therefore, oxidation with $H_2O_2$ generates CO-evolving oxygen groups that are more stable than those created by treatment with $HNO_3$. The shoulder at 1000 K in the evolution profile of sample CMS-$H_2O_2$ is located at the same temperature than the maximum in CMS-$HNO_3$ that can be due to anhydride groups (CO + $CO_2$ evolution), overlapping with the CO from other types of structures (phenol, quinone, etc.). Finally, CO evolution from sample CMS only shows one peak at 1100K, a lower temperature as compared to oxidized sample CMS-$H_2O_2$.

The amount of $CO_2$ and CO evolved from all the samples upon TPD has been determined by integration of the area under the curves of the profiles and are listed in Table 2. The results indicate that the amount of the oxygen surface groups is greater in sample CMS-$HNO_3$ than sample CMS-$H_2O_2$, and the $CO/CO_2$ ratio shows that treatment with $HNO_3$ creates more oxygen groups that evolve as $CO_2$ than the treatment with $H_2O_2$.

Table 2
Amounts of CO and $CO_2$ evolved in TPD experiments up to 1323 K

| Carbon | $CO_2$, mmol·$g^{-1}$ | CO, mmol·$g^{-1}$ | $CO/CO_2$ |
|---|---|---|---|
| CMS | 0.052 | 0.598 | 11.5 |
| CMS-$H_2O_2$ | 0.187 | 0.885 | 4.7 |
| CMS-$HNO_3$ | 0.940 | 2.098 | 2.2 |

Table 3 reports the enthalpy of immersion ($-\Delta H_i$, J·g$^{-1}$) of the carbons in hexane and in 1-hexene at 303 K, after out-gassing under vacuum at 523 and 773 K for 4 h. The areal enthalpies (mJ·m$^{-2}$) calculated taking into account the surface area obtained for $N_2$ adsorption are also reported. For all the samples, the enthalpy of immersion in 1-hexene is higher than in hexane. Comparing the areal enthalpies of immersion in the two liquids, the values increase with the degree of strength of oxidation.

Table 3
Enthalpies of immersion of the carbons into hexane and 1-hexene after out-gassing at different temperatures

| Sample | $-\Delta H_i$, J·g$^{-1}$ Hexane | $-\Delta H_i$, J·g$^{-1}$ 1-Hexene | $-\Delta H_i$, mJ·m$^{-2}$ Hexane | $-\Delta H_i$, mJ·m$^{-2}$ 1-Hexene |
|---|---|---|---|---|
| CMS 523 K | 73.3 | 106.4 | 98.6 | 143.2 |
| CMS 773 K | 78.3 | 98.8 | - | - |
| CMS-H$_2$O$_2$ 523 K | 66.0 | 87.8 | 140.7 | 187.2 |
| CMS-H$_2$O$_2$ 773 K | 86.4 | 144.8 | - | - |
| CMS-HNO$_3$ 523 K | 81.7 | 106.0 | 158.9 | 206.2 |
| CMS-HNO$_3$ 773 K | 104.2 | 124.5 | - | - |

Fig. 3a shows the differential heats of propane adsorption as a function of the amount adsorbed in samples CMS and CMS-H$_2$O$_2$, after out-gassing at 773 K. The heat of adsorption decreases continuously as the coverage increases. In both cases, the initial heat of adsorption is 58 kJ·mol$^{-1}$, and the profiles are very similar. Fig. 3b compares the adsorption heat of propane of the same carbons but out-gassed at different temperatures (CMS at 523 K and CMS-H$_2$O$_2$ at 773 K). Again, very similar profiles have been obtained. The adsorption isotherms of propane at room temperature and the micropore volumes obtained from them indicate that sample CMS-H$_2$O$_2$ out-gassed at 773 K adsorbs the highest amount of gas (0.38 cm$^3$·g$^{-1}$ vs. 0.32 cm$^3$·g$^{-1}$ for CMS); this can be explained by the progressive removal of the oxygen groups, that has two effects: (i) the elimination of the oxygen groups that block the microporosity and (ii) slight activation of the carbon (the oxygen atoms leave the surface together with carbon atoms, to form $CO_2$ and CO) [14]. Therefore, the oxidation and thermal treatments only affect the amount of propane adsorbed, but the modification of the surface chemistry does not affect the heats of adsorption.

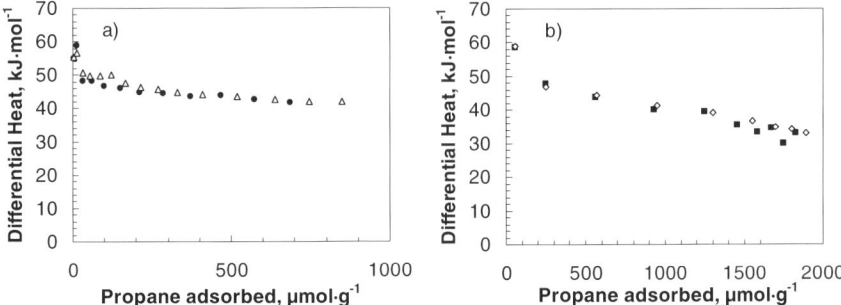

Fig. 3. Differential heat of propane adsorption at 298 K: a) on CMS (■) and CMS-H$_2$O$_2$(Δ) out-gassed at 773 K; b) on CMS at 523 K and CMS-H$_2$O$_2$ at 773 K.

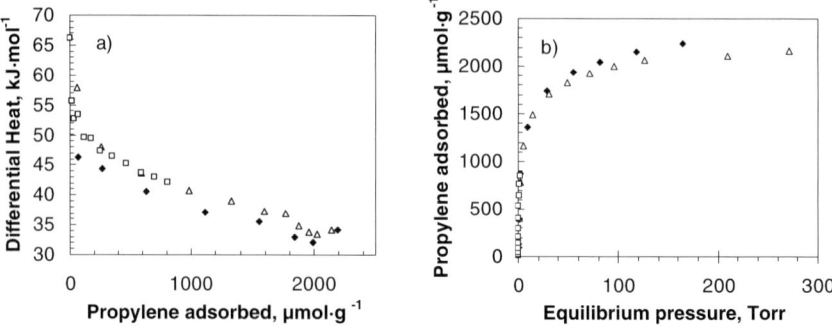

Fig. 4. Propylene adsorption at 298 K on CMS (♦), CMS-H$_2$O$_2$ (Δ) and CMS-HNO$_3$ (□), differential heat plot (a) and adsorption isotherm (b).

Fig. 4 plots the differential heats of adsorption (a) and the adsorption isotherm (b) of propylene for the carbons out-gassed at 527 K. In the three samples, the tendencies are similar. The initial heats of adsorption are 47, 58 and 66 kJ·mol$^{-1}$ for CMS, CMS-H$_2$O$_2$ and CMS-HNO$_3$ respectively. Sample CMS has the lower heat of propylene adsorption in all the coverage range, although the difference decreases with increasing propylene uptake. Oxidized carbons have similar differential heats of adsorption profiles, but sample CMS-HNO$_3$ has a higher initial heat of adsorption. Similar results are obtained when the carbons are out-gassed at 773 K. The adsorption isotherms of propylene show that the amount adsorbed in CMS and CMS-HNO$_3$ are practically the same, but it is a bit lower in the case of CMS-H$_2$O$_2$. An important conclusion for these results is that the adsorption strength of propylene in the oxidized carbons is stronger than in the non-oxidized counterparts, and this strength does not depend on the out-gassing temperature under vacuum.

Microcalorimetric measurements of propylene adsorption in the oxidized carbons treated at 1073 K under He and under H$_2$ are plotted in Fig. 5. For the samples treated under He, the initial heats of adsorption are 64 and 58 kJ·mol$^{-1}$ for CMS-HNO$_3$ and CMS-H$_2$O$_2$, respectively. For the samples treated under H$_2$, the initial heats of adsorption are 53 kJ·mol$^{-1}$ for both carbons. The samples treated in H$_2$ have a weaken adsorption strength of propylene, while samples treated in He give the same results than when out-gassed at 527 and 773 K under vacuum.

Treatment in He at 1073 K removes essentially all oxygen groups of the carbons, leaving unsaturated carbon atoms at the edge of the basal planes which are highly unstable [15], thus making the surface very reactive. When the samples are treated under H$_2$, surface stabilization is achieved, hydrogen treatment removes the less saturated carbon atoms ("dangling" carbons) from the edges (probably as methane) and also form stable C-H bonds with carbon active sites [16,17]. When the oxidized carbons are treated under vacuum at 527 and 773 K the differential heat profiles are the same; therefore, the active sites that are formed when the low-temperature CO$_2$ groups are removed are those mainly contributing to the strong adsorption of propylene. The results could also be explained by assuming a similar effect of the heat of adsorption on propylene of both the oxygen groups and the active sites formed when they are removed by thermal treatments under vacuum or inert atmosphere.

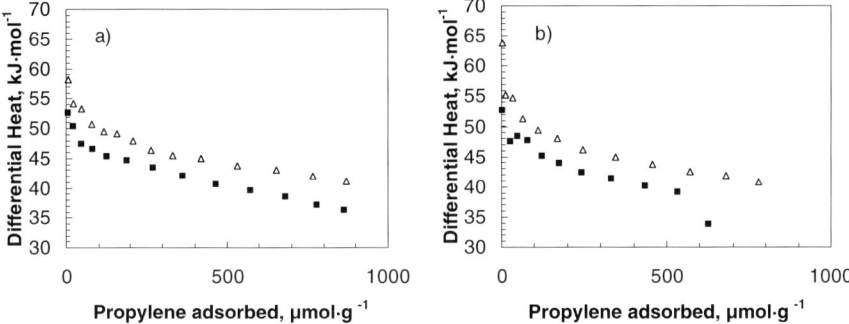

Fig. 5. Differential heat of propylene adsorption at 298 K on CMS-$H_2O_2$ (a) and CMS-$HNO_3$ (b) treated in He($\Delta$) and $H_2$ (■) at 1073 K.

## 4. CONCLUSIONS

In this study, the surface chemistry of a carbon molecular sieve (Takeda 5A) was modified by oxidation with $HNO_3$ and $H_2O_2$ solutions. Different thermal treatments under vacuum or under different gases (He and $H_2$) were used to selectively remove oxygen surface groups, and the adsorption of propane and propylene was studied by microcalorimetry.

Adsorption microcalorimetry of propane reveals that the surface chemistry of the carbons has no relevance in the heat of adsorption. However, the amount of propane adsorbed can vary in the oxidized carbon with the progressive removal of the oxygen groups that may block part of the microporosity.

In the case of propylene, the adsorption strength in the oxidized carbons is higher than in the non oxidized CMS, although it does not depend on the out-gassing temperature used. The adsorption isotherms follow the same tendency with the thermal treatments in both adsorbates. When the oxidized samples are treated at high temperature (1073 K) in $H_2$ the surface is stabilized, and therefore, a heats of adsorption profile very similar to the non oxidized carbon is obtained. These finding suggest that the active sites that are formed when the low-temperature $CO_2$-groups are removed are the most important in conditioning the propylene heat of adsorption, although the possibility that active sites and oxygen groups provide very similar features in the strong adsorption strength of propylene may also be taken into account.

**REFERENCES**

[1] Y. Euchi. J. Jpn. Petrol Inst., 13:105-11, 1970.

[2] C.A. León y León, and L.R. Radovic. Chemistry and Physics of Carbon, Vol. 25, p. 213. Marcel and Dekker, New York, 1994.

[3] F. Rodríguez-Reinoso. Introduction to carbon technologies, p. 55, Secretariado de Publicaciones, Universidad de Alicante (Spain), 1997.

[4] M. Domingo-García, I. Fernández-Morales, C. Moreno-Castilla, M. Pérez-Mendoza. Langmuir, 15, 3226, 1999.

[5] M. Domingo-García, I. Fernández-Morales, F.J. López-Garzón. Appl. Catal. A 112, 75, 1994.

[6] B.R. Puri, P. L. Jr Walker, Chemistry and Physics on Carbon, Ed.; Marcel Dekker: New York, Vol. 6,1970

[7] G. Tremblay, F. J. Vastola, P. J. Walker. Carbon, 16, 35, 1978.

[8] J. A. Menéndez, J. Phillips, B. Xia, L. R. Radovic. Langmuir, 12, 4404, 1996.

[9] J. A. Menéndez, L. R. Radovic, B. Xia, J. Phillips. J. Phys. Chem., 100, 17243, 1996.

[10] M. T. Gonzaléz, A. Sepúlveda-Escribano, M. Molina-Sabio, F. Rodríguez-Reinoso. Langmuir. 11, 2151, 1995.

[11] J. S. Noh, J. A. Schwarz. Carbon. 28, 1990.

[12] M. Molina-Sabio, M. A. Muñecas, F. Rodríguez-Reinoso. Characterization of Porous Solids, 96, 1992, 2707-13.

[13] C. Moreno-Castilla, M. A. Ferro-Garcia, J. P. Joly, I. Bautista-Toledo, F. Carrasco-Marín, J. Rivera-Utrilla. Langmuir, 11, 4386, 1995.

[14] F. Rodríguez-Reinoso, M. Molina-Sabio, M.A. Muñecas. J. Phys. Chem., 96, 1992.

[15] G. Herzberg, K. F. Herzfeld, E. J. Teller. J. Phys. Chem. 41, 1937.

[16] J. A. Menéndez, J. Philips, B. Xia, L. R. Radovic. Langmuir. 12, 1996.

[17] M. C. Román-Martínez, D. Cazorla-Amorós, A. Linares-Solano, C. Salinas-Martínez de Lecea. Carbon. 31, 6, 1993.

Studies in Surface Science and Catalysis 160
P.L. Llewellyn, F. Rodriquez-Reinoso, J. Rouqerol and N. Seaton (Editors)

# Digital reconstruction of silica gels based on small angle neutron scattering data.

**J.D.F. Ramsay[a], M. Kainourgiakis[b], Th. A. Steriotis[b*] and A.K. Stubos[b]**

[a] Institut Européen des Membranes, IEM / CNRS - 1919 Route de MENDE - 34293 Montpellier Cedex 5, France
[b] National Center for Scientific Research "Demokritos", 15310 Ag. Paraskevi Attikis, Athens, Greece

[*] Corresponding author; E-mail address: tster@chem.demokritos.gr (Th. Steriotis)

## 1. ABSTRACT

Digitally reconstructed porous domains contain statistical information well beyond the pore level and thus constitute accurate models that can be used to relate the structure of porous solids to their macroscopic properties. In this work, we focus on two different porous silica gels and, based on adsorption and SANS data, propose two different processes for their digital reconstruction. The first gel was formed from a sol in which the primary particles exist as aggregates that give a high porosity open gel structure, which was simulated by a diffusion-limited aggregation procedure. For the second gel the uniform discrete sol particles pack efficiently during gelation and produce a low porosity mesoporous structure, which was reconstructed through a Monte Carlo ballistic sphere deposition procedure. In both cases the reconstructed materials possess similar statistical properties with the original gels and can thus serve as models for the simulation of sorption and transport phenomena.

## 2. INTRODUCTION

The performance of nano-porous solids in various fields of environmental, technological and pharmaceutical interest is a sensitive function of the topology of the pore matrix. A common structural feature of real porous materials is a certain degree of disorder, spanning over some or several length scales. In general there is a great difficulty in representing adequately the natural systems, thus the relation of their macroscopical behaviour, with the geometrical properties of the porous network is a challenging task. Recent advances in theoretical techniques and computational resources have led to computer-based tools for the representation of the actual pore structure [1,2,3] and the reliable assessment of several properties [4,5]. A key point has been the development of methods that generate binary images based on certain statistical properties (usually porosity and two-point correlation function). Experimental information on these properties can be obtained either directly from SEM or TEM images, or indirectly by small angle scattering (SAS) measurements [6], which can efficiently probe details of the nano-scale inhomogeneities of matter [7].

The aim of this work is to present process-based digital reconstruction techniques, which based on adsorption and Small Angle Neutron Scattering (SANS) data are used for the 3D representation of the disordered structure of two different nano-porous silica gels. These "virtual" solids can be further used as accurate models for process simulations. The potential of this approach is revealed by an example of sorption simulations in one of the gels. In this case the spatial distribution of adsorbate is determined and compared with experimental data.

## 3. MATERIALS

The silica gels used in the present work were prepared through a sol-gel process from different silica sols as described previously [8,9]. The first gel (S1) was formed from a sol in which the primary spherical particles exist as aggregates. These give a very open structure on conversion to a gel, with a high porosity ($\varepsilon \sim 0.8$; pore radius, $r_p$ =14 nm, determined by $N_2$ adsorption at 77 K) despite the small size of the primary particles (diameter $\sim$ 10 nm), as has been described previously [9]. For the second gel (S4) the uniform spherical particles (diameter $\sim$ 30 nm) exist as discrete units in the sol. These pack much more efficiently in the transformation to a gel to give a low porosity ($\varepsilon \sim 0.3$) mesoporous structure ($r_p$ = 2.9 nm).

## 4. SMALL ANGLE NEUTRON SCATTERING

### 4.1 Theory

Small-angle neutron scattering (SANS) is a powerful, non-destructive technique [10] that can provide information about the geometry and morphology of inhomogeneities with size between 1 - 200 nm. Since porous solids commonly contain pores that fall well within this range, the technique has become an essential tool for their investigation [11,12,13,14]. In a SANS experiment the sample is exposed to the neutron beam, and the scattering intensity, $I(Q)$, is measured as a function of the scattering vector, $Q$ (=$4\pi\sin\theta/\lambda$, where $\lambda$ is the wavelength and $2\theta$ is the scattering angle). For an isotropic, stationary, two-phase (solid-pores) system, with porosity, $\varepsilon$, the spherically averaged intensity is given by [15,16]:

$$I(Q) = 4\pi\left(\eta_s - \eta_p\right)^2 V_{sample}\varepsilon(1-\varepsilon)\int_0^\infty u^2 R_Z(u)\frac{\sin Qu}{Qu}\,du \qquad (1)$$

where $u$ is the real space variable, $V_{sample}$ is the volume of the sample and $\eta_s$, $\eta_p$ are the scattering length densities of the solid and the pores, respectively (the difference is known as "contrast"). For an evacuated porous medium, $\eta_p$ is effectively zero. $R_Z(u)$ is the autocorrelation function, which is a measure of the medium mass distribution correlations at different scales and can be deduced by using Fourier-inversion of the $Q$, $I(Q)$ data. The scattering intensity from an assembly of particles can be expressed more simply as $I(Q) = N_s V_s^2 \eta_s^2 P(Q)S(Q)$, where $N_s$ is the particle number density and $V_s$ is the volume of each particle. $P(Q)$ is the single particle form factor, which depends on its size and shape and describes the decay of the intensity at low $Q$ (Guinier region). The structure factor, $S(Q)$, describes the effects of interference in the scattering from particles that are in close separation. For the limiting case of high $Q$, $S(Q)$ tends to unity. This is the outer part of a scattering curve where if the inequality $Qd_{sc}\gg 1$ holds ($d_{sc}$ is the size of the scattering object), $I(Q)$ may be approximated by Porod's law: $I(Q) = 2\pi\eta_s^2 AQ^{-4}$ ($A$ is the surface area) [17].

Many porous structures can be described as mass or surface fractals [18,19]. For mass fractals, the mass, $M$, and the corresponding radius of gyration, $R_g$, obey the scaling relationship: $M \propto R_g^{D_m}$ where $D_m$ is the mass fractal dimension ($1 \leq D_m \leq d$, d is the dimension of the space, in which the object is embedded). On the other hand, the surface area, $A$, of surface fractals follows the scaling law $A \propto R_g^{D_s}$ where $D_s$ is the surface fractal dimension ($d - 1 \le D_s < d$). The structure of both systems can be quantitatively determined by SANS experiments. For mass fractals $I(Q)$ shows a power law behavior given by $I(Q) \propto Q^{-D_m}$, while for surface fractals, $I(Q)$ follows the power law $I(Q) \propto Q^{D_s-2d}$ [20]. In practical terms, for a three-dimensional object mass fractality will produce a slope between $-1$ and $-3$, while surface fractality between $-3$ and $-4$ in a log-log $Q$-$I(Q)$ plot.

SANS in conjunction with in-situ contrast matching (CM) adsorption, can highlight structural details of porous materials and also related physical processes (i.e. micropore filling, multilayer growth, capillary condensation, etc.) [14,21]. In practice, an appropriately selected compound, having the same scattering length density with the solid, is introduced in the sample before its exposure to neutrons, which cannot distinguish the solid and the adsorbate and thus the spectra of the "new" material (solid and adsorbate) is obtained. The use of neutrons is very convenient in this context, since different isotopes of an element have different scattering length densities, allowing thus the appropriate adjustment of the contrast of a certain system, without affecting its thermodynamic properties. For example, hydrogen and deuterium have scattering length densities of opposite sign and thus mixtures of e.g. $H_2O/D_2O$ or $C_6H_6/C_6D_6$, can match almost any given scattering length density.

### 4.2 SANS spectra of silica gels

The SANS spectra of dry (evacuated) S1 and S4 silica gels were measured on the V4 SANS instrument at HMI, Berlin. Experimental details can be found elsewhere [22]. The SANS results for the high porosity silica gel (S1) (Fig. 1, left) reveal characteristics of a mass fractal structure giving a slope of -1.8 ($D_m$=1.8) at the $Q$ region 0.08-0.15 nm$^{-1}$. In the high $Q$ region, $I(Q)$ scaling occurred as $Q^{-3.5}$, corresponding to a fractally rough surface with $D_s$=2.5.

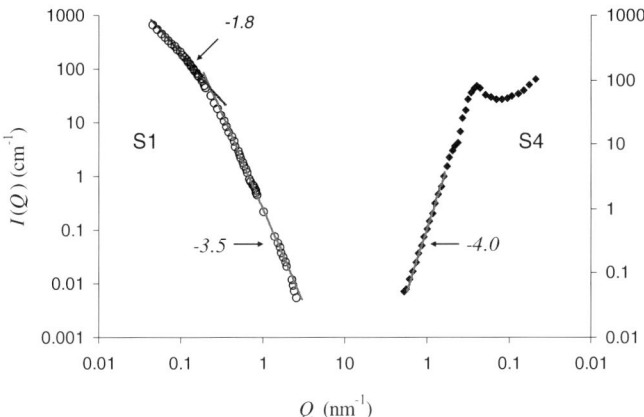

Fig. 1: SANS curves of dry S1 (left, open circles) and S4 (right, filled diamonds).

The scattering curve of dry S4 (Fig. 1, right) is in marked contrast to that of S1, being characteristic of a structure formed by packing of nearly mono-dispersed spherical particles. The maximum at $Q$=0.025 Å arises from the interference in the scattering from a partially ordered structure, where the interparticle separation is given approximately by $2\pi/Q$ (~25 nm). The inflexion at $Q=0.045$ Å results from the form factor, $P(Q)$, of the spherical particles. For mono-dispersed spheres of radius $R$, $P(Q)$ has a primary maximum at $Q \cdot R=5.9$. Although the feature is smeared, it corresponds to a particle diameter of 26 nm, in agreement with electron microscopy and other SANS measurements on dilute sols. Beyond this, $I(Q)$ decays linearly with $Q^{-4}$ in accord with the Porod law, i.e. the surface of the S4 is Euclidian ($D_s=2$).

## 5. DIGITAL REPRESENTATION OF THE STRUCTURE OF BIPHASIC MEDIA

Starting from rather simplified representations of the porous structure through serial or parallel pore models or pore networks [23], the evolution of experimental and numerical techniques allowed the development of novel approaches that are now used for the production of binary images of porous matrices. Stochastic reconstruction methods [1,24,25,26] generate binary arrays that respect a number of statistical properties of the actual porous medium. The required experimental information can be obtained either directly from serial-tomography, SEM or TEM images or indirectly by SAS techniques [2,15]. Process-based methods [27,28], where the computational procedure tries to imitate the physical processes that take place during the formation of the medium, comprise interesting alternatives. These methods, although more sound from a physical point of view, frequently suffer from computational requirements. In the present work, two different process-based reconstruction approaches are employed in order to adequately reproduce the structural features of the silica gels.

The spatial distribution of matter in a porous medium can be represented by the phase function, $Z(\mathbf{x})$, which is equal to 1 if $\mathbf{x}$ belongs to the pore space, or 0 if $\mathbf{x}$ belongs to solid ($\mathbf{x}$ is the position vector from an arbitrary origin). A reliable 3D representation of a porous medium should possess the same statistical properties as those determined in a single two-dimensional section, properly reflected by the various moments of $Z(\mathbf{x})$ [6] and in most cases, the first two moments are considered sufficient. The one-point correlation function, $S_1$, is the probability that any point lies in the pore space and is defined as $S_1=<Z(\mathbf{x})>$ ($<\cdot>$ indicates spatial average, hence, $S_1$ is equal to the porosity, $\varepsilon$). On the other hand, the two-point correlation function, $S_2(\mathbf{u})$, is the probability that two points a specified distance $\mathbf{u}$ apart, both lie in the pore space and is defined as $S_2(\mathbf{u})=<Z(\mathbf{x}) \cdot Z(\mathbf{x}+\mathbf{u})>$. $S_2(\mathbf{u})$, is directly related to the autocorrelation function, $R_z(\mathbf{u})$ ($R_z(\mathbf{u})=[S_2(\mathbf{u})-\varepsilon^2]/(\varepsilon-\varepsilon^2)$). For an isotropic medium, $R_z(\mathbf{u})$ is only a function of $u=|\mathbf{u}|$ [1] and thus degenerates to the one-dimensional autocorrelation function, $R_z(u)$, of Eq. (1).

### 5.1. Process based reconstruction of the silica gels

For the digital reconstruction of the S1 gel, the following procedure was used: In a cubic lattice of size $L^3$, $N$ seed particles were placed in random positions ($L=100$ and $N=2000$ for this study). The size of every particle was equal to that of the voxel and thus each of those randomly chosen voxels was considered occupied. New particles undergoing random walk were generated sequentially in arbitrary unoccupied positions of this lattice. Each of those particles was allowed to travel until it arrived at an unoccupied site adjacent to an occupied one. The random walk was then terminated and the current site was considered occupied. During the simulation procedure, periodic boundary conditions were applied (when a particle reaches one of the lattice edges it is reinserted in the opposite edge), while the fraction of the total number of unoccupied sites over the total number of voxels was set to be 80%, i.e. equal to the porosity. This procedure resembles closely to the diffusion limited (DL) cluster-cluster aggregation (CCA) model [29], the main difference being that in our case the number of growing clusters is pre-fixed (*N*), while the clusters are not diffusing. This simplified version of the DLCCA model is more efficient computationally, while still adequately capturing the process of gel formation. The reconstructed material is depicted in Fig. 2A.

$R_z(u)$ of the reconstructed material (Fig. 2D, left, circles) is in good agreement with the experimental $R_z(u)$ (Fig. 2D, left, solid line). However, our lattice model implies that the voxel size is equal to the silica primary particles and thus short-range (comparable to the primary particle size) correlations cannot be captured. In order to represent the finer structural details of this length scale, off-lattice aggregation of diffusing monodisperse spheres was employed in a similar manner as above, by using a single seed sphere (Fig. 2B). Based on this more accurate model (the radius of each primary particle is 10 voxels) it was possible to extract the short length scale information (i.e. the initial part of the total autocorrelation function). In such a case our multi-cluster growth process was simplified to single-cluster one, i.e. a diffusion-limited-aggregation (DLA) model [29]. There is however a marked difference from typical DLA simulations, where particles start diffusing outside the growing cluster until they are "deposited" on it. In order to stay as close as possible to the original gelation process, in our simulations, particles can start diffusing inside the void spaces of the growing cluster.

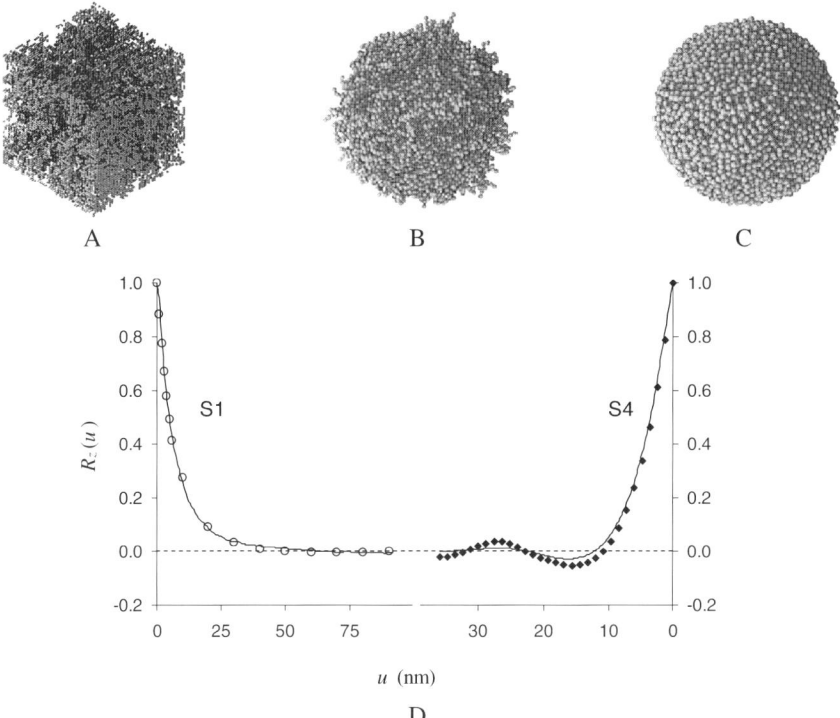

Fig. 2: A. Multi-cluster CCA-type lattice reconstruction of S1 (each point represents a primary silica sphere) B. Single-Cluster DLA-type off-lattice reconstruction of S1 structure (Sphere DLA) C. Center attraction ballistic deposition reconstruction of S4 D. Comparison of the autocorrelation functions of reconstructed S1 (left) and S4 (right) silica gels (points), with the experimental ones (solid lines).

For the digital reconstruction of the porous matrix of S4, a ballistic Monte Carlo deposition procedure [30] was used. The first sphere of the stack is placed at the origin while the second at random position, in contact with the former. Then, $N$ "test" balls are dropped but only the nearest to the centre of attraction becomes a part of the stack. For a large enough $N$ (typically $N>10^5$), random sphere packs (Fig 2C) with the same structural properties found by the more rigorous deposition algorithms can be generated (in the present case, packs consist of 1000 spheres and have a porosity of 0.35). The random sphere packs are then digitised (the size of the domains was $100^3$) and the phase function is derived. The $R_z(u)$ (Fig. 2D, right), obtained from the reconstructed sample is in agreement with the experimental $R_z(u)$ of S4 (Figure 2D, right, solid line) as obtained from the scattering curve.

## 6. DETERMINATION OF SORPTION PROPERTIES (S4 gel)

The simultaneous calculation of the amount adsorbed and the distribution of adsorbate within the pores of a reconstructed solid can be implemented after the application of a Density Functional Theory (DFT) mean field model, which is particularly suited for on-lattice simulations on digitised structures [31]. In a lattice model the spatial distribution of adsorbate can be described at each site by the local density function. The equilibrium density profile for a given matrix realization and chemical potential $\mu$ can then be determined by minimising the grand potential $\Omega\{\rho\}$ with respect to the fluid density on each lattice site, $\rho(\mathbf{x})$, leading to:

$$\rho(\mathbf{x}) = \frac{Z(\mathbf{x})}{1 + exp\left\{-\frac{1}{k_B T}\left[\mu + \sum_{y/x}\left(w_{ff}\rho(\mathbf{y}) + w_{mf}(1 - Z(\mathbf{x}))\right)\right]\right\}} \qquad (2)$$

where, $w_{ff}$ and $w_{mf}$ is the fluid-fluid ($ff$) and matrix-fluid ($mf$) interactions respectively, $T$ is the temperature and $k_B$ is the Botlzmann constant. The sum is over all nearest to $\mathbf{x}$ neighbours. Equation (2) is solved iteratively and if $L$ is the size of the domain the overall fluid density is:

$$\rho_f = \frac{\sum \rho(\mathbf{x})}{\varepsilon L^3} \qquad (3)$$

In this work we have used SANS measurements combined with CM in-situ sorption and applied the above-described methodology to study sorption on S4. The in-situ measurements have been carried out by means of a specially designed apparatus (SANSADSO) mounted on the V4 instrument at HMI, Berlin. The adsorbate was an isotopic mixture of benzene (59% $C_6D_6$, 41% $C_6H_6$), having the same scattering length density with $SiO_2$ (details in [22]). Two scattering curves have been studied, i.e. S4 equilibrated at $p/p_0= 0.6$ and 0.67 (referring to relative volumes $V_s$=0.2 and 0.5; $V_s$ is the pore volume fraction occupied by adsorbate). $V_s$, is equal to the corresponding $\rho_f$ and the transformation of the DFT density field into a binary field where the adsorbed phase was treated as solid phase, in analogy with the SANS experiments, was done by a relatively simple thresholding technique [32]. The isotherm cannot be quantitatively predicted since the voxel size of reconstructed silica gel is larger than $C_6H_6$. Therefore, we have used the method as a tool to extract the spatial distribution of the adsorbate by varying the ratio of $w_{ff}$ over $w_{mf}$ until the experimental $R_z(u)$ was matched. Fig. 3A (left) shows the $R_z(u)$ of the 20% loaded ($V_s$=0.2) material and the reconstructed analogue and reveals good agreement between the simulation and the experiment. As expected the autocorrelation function of the loaded material is similar to that of "dry" S4, since at such a low loading the adsorbate is considered to form a film on the pore walls and the material retains its "sphere pack" characteristics. This situation is confirmed by our simulations and well observed in the visualization pictures (Fig. 3B). Fig. 3A (right) illustrates the simulation and experimental $R_z(u)$ for S4 at $V_s$=0.5. The agreement is again very good. In this case "consolidation" of the non-void phase (solid and adsorbate) due to the initialisation of capillary condensation at narrow contact zones, is observed (Fig. 3C). As a result $R_z(u)$ changes from that of a random sphere pack system to that of interpenetrating spheres.

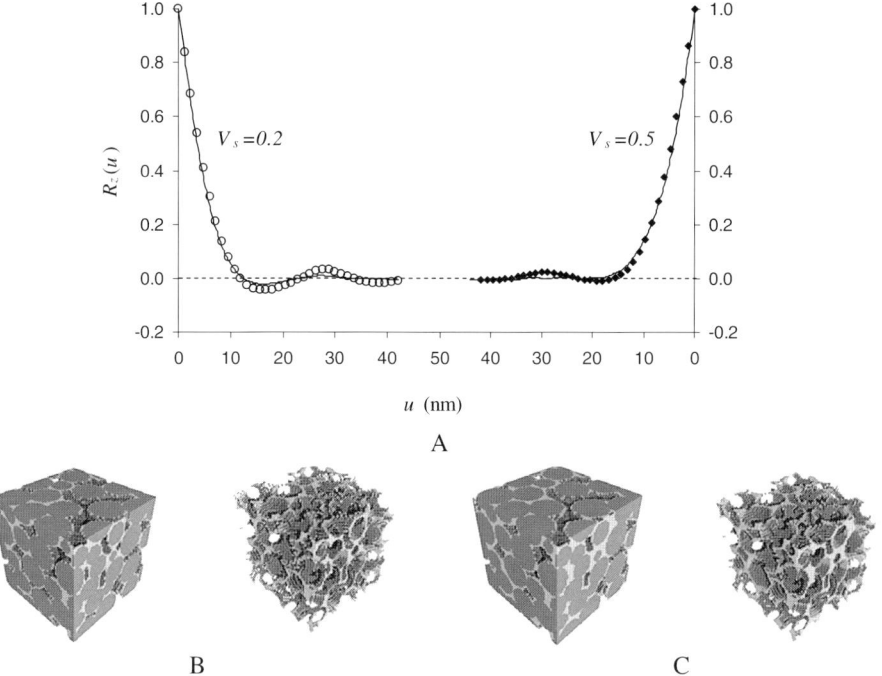

Fig. 3: A. Auto correlation functions of S4 (solid line) and sphere pack (points) loaded with benzene ($V_s$=0.2, left, circles and $V_s$=0.5, right, filled diamonds). B. 3D images showing part of the S4 reconstructed domain and the adsorbate spatial distribution at $V_s$ =0.2 C. 3D images showing part of the S4 reconstructed domain and the adsorbate spatial distribution at $V_s$ =0.5

## 7. CONCLUSIONS

A methodology based on the combination of SANS with process-based reconstruction techniques is developed in an effort to represent in a sufficiently accurate manner the structure of two different silica gels. Appropriate mathematical transformation of SANS spectra provides valuable information about the disordered geometry of the gels, in terms of the corresponding $R_z(u)$. The resulting reconstructed domains reveal similar statistical properties with the actual materials and can be further used for the simulation of equilibrium (sorption) and dynamic (diffusion and permeation) processes. The spatial distributions of an adsorbed phase filling the pore space is determined by the use of a DFT-based methodology for one of the silica gels (S4). Similar work on the second silica gel (S1) is currently underway. Additionally, typical diffusion limited cluster-cluster aggregation simulations mimicking the gelation of S1 are being performed for comparison.

## REFERENCES

[1] P.M. Adler, Porous Media: Geometry and Transports, Butterworth, London, 1992.
[2] P. Levitz, D. Tchoubar, J. Phys. I 2 (1992) 771.
[3] S. Torquato, Random Heterogeneous Materials: Microstructure and Macroscopic Properties, Springer-Verlag, New York, 2002.
[4] J.-F. Thovert, F. Yousefian, P. Spanne, C.G. Jacquin, P.M. Adler, PHYSICAL REVIEW E 63 (2001) 061307.
[5] S. Bekri, J. Howard, J. Muller, P.M. Adler, Transport in Porous Media, 51 (2003) 41.
[6] E.S. Kikkinides, M.E. Kainourgiakis, K.L. Stefanopoulos, A.Ch. Mitropoulos, A.K. Stubos, N.K. Kanellopoulos, J. Chem. Phys. 112 (2000) 9881.
[7] J.D.F. Ramsay, E. Hoinkis, Phys. B: Condensed Matter 248 (1998) 322.
[8] J.D.F. Ramsay, B.O. Booth, J. Chem. Soc. Faraday Trans. (1) 79 (1983) 173.
[9] J.D.F. Ramsay, in: D.J. Wedlock Ed.., Controlled Particle, Droplet and Bubble Formation, Butterworth Heinemann, 1994, p. 1.
[10] J.C. Dore, A.N. North, J.S. Rigden, Radiation Physics and Chemistry 45 (1995) 413.
[11] M.H. Stacey, Stud. Surf. Sci. Catal. 62 (1991) 165.
[12] L. Auvray, S. Kallus, G. Golemme, G. Nabias, J.D.F. Ramsay, Stud. Surf. Sci. Catal. 128 (2000) 459.
[13] W. Gille, O. Kabisch, S. Reichl, D. Enke, D. Fürst, F. Janowski, Microporous and Mesoporous Materials 54 (2002) 145.
[14] J. D. F. Ramsay, Advances in Colloid and Interface Science 76-77 (1998) 13.
[15] P. Debye, H.R. Anderson, H. Brumberger, J. Appl Phys. 28 (1957) 679.
[16] G. Porod, in: O. Glatter, O.Kratky (Eds.), Small Angle X-Ray Scattering, Academic Press, London, 1982, p. 17.
[17] G. Porod, in Small Angle X-Ray Scattering, O. Glatter, O.Kratky (Eds.), Academic Press, London, 1982, p. 17-51.
[18] J. Teixeira, J. Appl. Cryst.. 21, (1988) 781
[19] Keefer K D and Schaefer D W Phys. Rev. Lett. 56 (1986) 2376; Keefer K D Mat. Res. Soc. Symp. Proc. 73 (1986) 295
[20] H.D. Bale and P.W. Schmidt, Phys. Rev. Lett., 53 (1984) 596.
[21] E. Hoinkis, Advances in Colloid and Interface Science 76-77 (1998) 39.
[22] J. D. F. Ramsay, E. Hoinkis, Journal of Non-Crystalline Solids 225 (1998) 200.
[23] D. Nicholson, J.H. Petropoulos, J. Chem. Soc. Farad. Trans. I 80 (1984) 1069.
[24] J. A. Quiblier, J. Colloid Interface Sci. 98 (1986) 84.
[25] C.L.Y. Yeong, S. Torquato, Phys. Rev. E 57 (1998) 495.
[26] P. Levitz, V. Pasquier, I. Cousin, in: B. Mc Enaney, T. J. Mays, J. Rouqerol, F. Rodriguez-Reinoso, K. S. W. Sing, K. K. Unger (Eds.) Characterization of Porous Solids IV, Bath, UK, 1996, p. 135.
[27] S. Bakke, P.E. Ören, SPE Journal 2 (1997) 136.
[28] P.E. Ören, S. Bakke, O.J. Arntzen, SPE Journal 3 (1998) 324.
[29] T. Vicsek, Fractal Growth Phenomena", World Scientific, Singapore, 1989
[30] Th. Steriotis, E. Kikkinides, M. Kainourgiakis, A. Stubos, J. D. F. Ramsay, Coll. Surf. A 241 (1-3) (2004) 231.
[31] E. Kierlik, P.A. Monson, M.L. Rosinberg, L. Sarkisov, G. Tarjus, Phys. Rev. Lett. 87 (2001) 055701.
[32] E.S. Kikkinides, M.E. Kainourgiakis, A.K. Stubos, Langmuir 19 (2003) 3338.

Studies in Surface Science and Catalysis 160
*P.L. Llewellyn, F. Rodriquez-Reinoso, J. Rouqerol and N. Seaton (Editors)*

# The impact of mesoporosity on microporosity assessment by $CO_2$ adsorption, revisited

**S. Brouwer,[*] J.C. Groen, and L.A.A. Peffer**

DelftChemTech, Delft University of Technology, Julianalaan 136, 2628 BL Delft, The Netherlands
[*] e-mail: s.brouwer@tnw.tudelft.nl

The use of both $CO_2$ and $N_2$ physisorption measurements provide complementary information on the determination of micropore characteristics. The presence however of a substantial amount of external surface area, as e.g. present in hierarchically structured porous materials, can lead to an overestimation of the Dubinin-Radushkevich micropore volume as derived from $CO_2$ physisorption. A simplified method of correction to the already known isotherm subtraction approach is proposed here. This proposed correction is proportional to the amount of external surface area determined by $N_2$ adsorption and is based on correction of the DR-results. Advantages of this method are the ease of determining the correction factor and the simplicity of correction the obtained micropore volume under normal conditions using only sub-atmospheric $CO_2$ and $N_2$ adsorption measurements.

The results of our method show a good similarity to the isotherm subtraction approach in the obtained micropore volumes. Novel adsorption models such as Density Functional Theory can only properly take into account the contribution of the external surface area if this has been actually measured during the analysis, which requires high-pressure $CO_2$ adsorption.

## 1. INTRODUCTION

The use of carbon materials is widely spread throughout industry and research due to its wide availability, the relatively low production costs, and interesting properties as a good chemical stability and a high specific surface area. Typical applications involve separation of gases, storage of gases and liquids, detoxification, and catalysis. Burn-off and activation processes can be used to tune the surface properties of the carbon [1]. The majority of the surface area, which yields up to 2000 $m^2$/g, is located in micropores. The purely microporous character is not an issue for most storage and separation applications, but for catalysis purposes it makes these materials less efficient. The limited accessibility to the active sites is often responsible for a reduced activity, induces enlarged contact time, and consequently leads to undesired follow-up reactions [2].

Recent developments on various materials focus therefore on more hierarchically ordered structures with defined mesoporosity supplementary to microporosity [2,3]. The advantage of the hierarchical architecture of porosity is the high surface area of a typical microporous system coupled to an improved accessibility by enhanced molecular transport properties in the mesopores. This combination would result in a more efficient use of the material. Phenolic-resin-derived activated carbons are a typical example of carbonaceous materials with a hierarchical architecture of porosity [4]. The synthesis of these carbons leads to well-controlled mesopores and even the amount of impurities in the material is much less compared to conventionally prepared carbons.

Since the use of carbons is widely spread, their characterization is also practiced intensively. Textural characterization of these materials is generally done by physisorption of carbon dioxide at close to ambient temperatures (273 K or 298 K) in combination with nitrogen physisorption at 77 K [1,5]. $N_2$ adsorption enables the measurement of pores between approximately 5 Å and 2000 Å, while pores with diameters between approximately 4 Å and 15 Å can be measured when using $CO_2$ up to an absolute pressure of 100 kPa. The fact that $CO_2$ adsorption enables the assessment of smaller pores is a result of its smaller kinetic diameter and the higher kinetic energy in comparison with $N_2$. As a result of this partially overlapping range between the two techniques, the micropores (pores ≤ 20 Å) can be divided into three ranges as presented in Fig. 1. Pores that can only be determined using $CO_2$ adsorption will be referred to as ultra-micropores; super-micropores can only be measured with $N_2$ adsorption; the intermediate micropores can be assessed by both techniques.

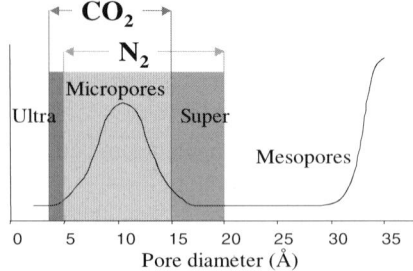

Fig. 1. Classification of micropores based on $N_2$ and $CO_2$ physisorption.

Calculation of the micropore volume from the $CO_2$ adsorption data can be done using the commonly applied Dubinin-Radushkevich (DR) equation, while the *t*-method is used to derive this information from the $N_2$ adsorption data. Expression of the micropore contribution should preferably be done as a micropore volume. The reason for this is because the micropore surface area has no physical meaning if micropore filling is taking place and the cross sectional area of the adsorbed molecule cannot unambiguously be assigned.

In case of combined micro- and mesoporous materials, the $N_2$ adsorption isotherm additionally provides information on mesopores. The impact of the presence of this non-micropore surface area on the characterization of microporosity is however often overlooked. The non-micropore area, which is called external surface area, is responsible for an overestimation in the obtained micropore volume because at low relative pressure not only micropore filling occurs but also monolayer adsorption takes place on the external surface area. In the case of purely microporous materials the effect is negligible but when the amount of external surface increases, the effect gets more pronounced. This phenomenon has previously been observed with different adsorptives and adsorbents [6,7]. Dubinin also pointed it out for carbons in his earlier work [8,9]. More recently, Moreno-Castilla and co workers refined some of Dubinins work [10]. They used the Dubinin-Stoekli (DS) equation corrected with the Dubinin-Zaverina (DZ) equation. This method comprehends the measurement of hydrocarbon adsorption on microporous and non-microporous carbons and subsequent isotherm subtraction to eliminate the effect of external surface on the determination of the micropore volume. The DS-equation is an extended form of the commonly used DR-equation and applicable to less homogenous surfaces. Disadvantages of the DS-equation are however the non-straightforward utilization and the large amount of variables that leads to an enhanced uncertainty in the calculated micropore volume. Moreover, the DR-equation is easily available to most routine users because and therefore more commonly used.

In this contribution, a simplified procedure is proposed to correct for the contribution of $CO_2$ adsorption by external surface area on the micropore volume as derived from the DR-equation. This approach will be supported by experimental evidence on both carbon and siliceous materials.

## 2. EXPERIMENTAL

### 2.1. Materials

The carbon materials used in this study were obtained from MAST Carbon Ltd. and are phenolic resin derived activated carbons, NOVACARB™ [4]. The voids between the primary microporous particles result in a controlled mesopore size. In this way combined micro- and mesoporous materials are obtained. The different letter in the sample name of the carbon samples refers to a different percentage burn-off or a different precursor is used in preparation. The number in the sample name refers to the temperature (K) at which the samples have been thermally treated after burn-off.

Two purely mesoporous MCM-41 materials were prepared as described by Beck et al. [11].

### 2.2. Characterization

$N_2$ and $CO_2$ adsorption isotherms were measured using an Autosorb-6B from Quantachrome Inc. $N_2$ adsorption and desorption was performed at 77 K up to a relative pressure ($p/p_0$) of 1, while $CO_2$ adsorption and desorption has been measured at 273 K up to 0.03 $p/p_0$ (absolute pressure 100 kPa.). The applied densities for adsorbed $N_2$ and $CO_2$ at corresponding temperatures are 0.808 and 1.044 $g/cm^3$, respectively [12]. Calculation methods used are available with various commercial instruments; BJH-method, BET-method, *t*-method, and DR-equation.

Low-pressure adsorption isotherms of $N_2$ at 77 K were measured on an ASAP 2010 from Micromeritics, equipped with a 0.1 kPa pressure transducer. The micropore size distribution has been deconvoluted using the Saito-Foley (SF-) method incorporating cylindrical pore shape geometry. This particular pore shape geometry is used since the micropores in the

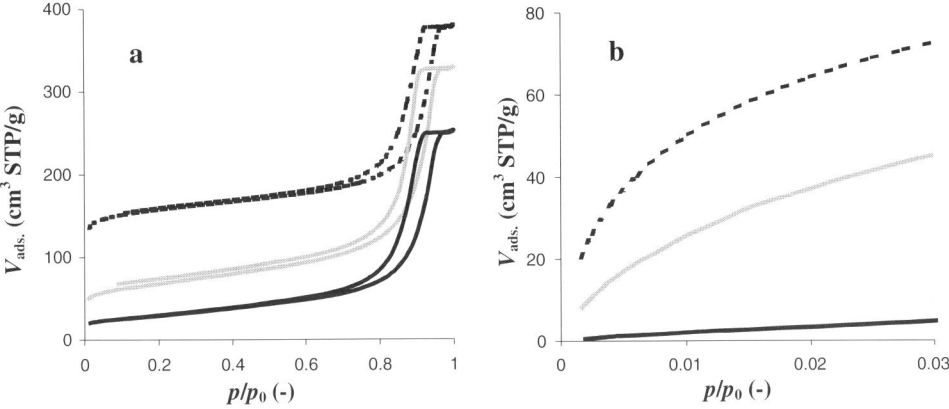

Fig. 2. $N_2$ (a) and $CO_2$ (b) isotherms of sample A (dashed line), A-1500 (gray line), A-2000 (solid line).

phenolic resin derived activated carbons are known to be cylindrical [4], in contrast to conventionally prepared carbons that have slit shaped micropores.

## 3. RESULTS AND DISCUSSION

### 3.1. $CO_2$ and $N_2$ adsorption on carbons

$CO_2$ and $N_2$ physisorption isotherms were measured of various carbon materials and some isotherms are displayed in Fig. 2. The $N_2$ isotherm of material A-2000 shows a type IV, typically for mesoporous materials [12]. Material A and A-1500 show a combination of type I and IV, pointing out the combined micro- and mesoporous character of these samples. The $CO_2$ adsorption shows the measured uptake at absolute pressures up to 100 kPa.

The BJH pore size distribution derived from the $N_2$ isotherm points out narrow maxima between 100 and 200 Å. The corresponding external surface area ($S_{external}$), micropore volume ($V_{micro, t}$) and micropore volume from the DR-equation obtained by $CO_2$ adsorption ($V_{micro, DR}$) are given in Table 1. Despite the fact that the DR-equation is in theory only applicable to materials with homogeneous surfaces, the correlation between the adsorption points in the DR-plot is linear over a wide pressure range.

Table 1
$N_2$ and $CO_2$ physisorption results of various carbons.

| Sample name | $N_2$ physisorption | | $CO_2$ physisorption | Difference |
|---|---|---|---|---|
| | $S_{external}$ $(m^2/g)$ | $V_{micro, t}$ $(cm^3/g)$ | $V_{micro, DR}$ $(cm^3/g)$ | $V_{micro}$ $(cm^3/g)$ |
| A | 101 | 0.20 | 0.24 | -0.04 |
| A-1200 | 140 | 0.15 | 0.25 | -0.10 |
| A-1500 | 134 | 0.05 | 0.17 | -0.12 |
| A-1800 | 97 | 0.01 | 0.03 | -0.02 |
| B | 242 | 0.51 | 0.43 | 0.08 |
| B-1800 | 235 | 0.15 | 0.16 | -0.01 |
| B-2000 | 219 | 0.05 | 0.07 | -0.02 |
| C | 311 | 0.51 | 0.52 | -0.01 |
| D | 102 | 0.63 | 0.54 | 0.09 |
| A-2000 | 104 | 0.00 | 0.02 | -0.02 |
| C-2500 | 254 | 0.00 | 0.06 | -0.06 |

The above results show a relative large discrepancy between the micropore volumes obtained from $N_2$ and $CO_2$ adsorption for most samples. Based on these results it can be suggested that all materials that show a larger micropore volume determined with $CO_2$ have ultra-micropores, which cannot be measured by $N_2$. This is however a premature conclusion because the samples contain additional mesopores, pointed out by the high external surface area, which contribute to the DR-outcome. Consequently the results need to be corrected. This overestimation of the micropore volume by $CO_2$ adsorption on the external surface area is also confirmed with the two purely mesoporous materials A-2000 and C-2500. These two samples have had a thermal treatment by which all micropores have been removed, resulting in a purely mesoporous material.

Dubinin proposed a method for correction of the adsorption on the external surface area that is based on isotherm subtraction [8-10]. This method separates the total amount of adsorption ($a_{total}$) into a contribution due to the micropores and external surface area. The amount of adsorption on the external surface area is subsequently related to the quantity of external surface area, and the pressure related interaction of the adsorptive ($\gamma$).

$$a_{total} = a_{micro} + a_{external} = a_{micro} + \gamma\, S_{external} \qquad (1)$$

The value for $\gamma$ can be approximated theoretically or determined experimentally by using organic vapor adsorption on a non-microporous reference sample. When Eq. 1 is combined with the DR-equation and homogeneous surfaces are presumed, it can be rewritten to Eq. 2. This equation performs a correction on the outcome of the DR-equation. The assumption of a homogeneous surface is physically not substantiated for bi-modal distributions but practically applicable since the correlation between the adsorption points in the DR-plot is linear.

$$W_0\,/\,\rho = V_{micro,\,DR} = V_{micro,\,DR,\,correct} + V_{external} = V_{micro,\,DR,\,correct} + \alpha\, S_{external} \qquad (2)$$

Here $W_0$ is the total uptake calculated with the DR-equation in grams, which is converted into a micropore volume using the liquid density ($\rho$) of the adsorbate. This obtained micropore volume is the sum of the actual micropore contribution ($V_{micro,\,DR,\,correct}$) and the contribution of the external surface area ($V_{external}$). The latter term is proportionally related to the extent of external surface area via the pressure independent factor $\alpha$ (in $cm^3/m^2$). Advantages of this methodology are the ease at which the correction can be applied on new but also readily available results. Secondly, the external surface area is quantified by $N_2$ physisorption rather than adsorption of organic vapors like benzene. This enables the use of low-pressure instruments and requires no special adaptations from $CO_2$ adsorption.

It is important for the determination of the factor $\alpha$ that the used reference materials are non-microporous. The two purely mesoporous carbons A-2000 and C-2500 are such materials. They show a linear relation between $S_{external}$ and $V_{external}$ through the origin according to Eq. 2, resulting in $\alpha = 2.2 \cdot 10^{-4}\ cm^3/m^2$. This indicates that 1 square meter of external surface area leads to an overestimation of the micropore volume of $2.2 \cdot 10^{-4}\ cm^3$. The experimentally determined $\alpha$ is used to calculate the corrected micropore volumes for the other carbons, which results are given in Table 2.

Table 2
$N_2$ and corrected $CO_2$ physisorption results of various carbons.

| Sample name | $N_2$ physisorption | $CO_2$ physisorption | Difference |
|---|---|---|---|
| | $V_{micro,\,t}$ $(cm^3/g)$ | $V_{micro,\,DR,\,correct}$ $(cm^3/g)$ | $V_{micro}$ $(cm^3/g)$ |
| A | 0.20 | 0.21 | -0.01 |
| A-1200 | 0.15 | 0.22 | -0.07 |
| A-1500 | 0.05 | 0.14 | -0.09 |
| A-1800 | 0.01 | 0.01 | 0.00 |
| B | 0.51 | 0.38 | 0.13 |
| B-1800 | 0.15 | 0.11 | 0.04 |
| B-2000 | 0.05 | 0.02 | 0.03 |
| C | 0.51 | 0.44 | 0.07 |
| D | 0.63 | 0.52 | 0.11 |

All samples contain a substantial amount of external surface area and the applied correction induces therefore a significant decrease in the corrected value of the $V_{micro,\,DR}$. Since the two techniques do not fully overlay in measurement range, as described in the introduction, discrepancies in micropore volume remains to exist between $V_{micro, t}$ and $V_{micro, DR, correct}$. The samples B, C, and D show super-micropores while the samples A-1200 and A-1500 contain ultra-micropores. Sample A shows similar micropore volumes and these micropores can therefore fully be quantified using both techniques.

To confirm the observations in Table 2, micropore size distributions were measured on selected materials using low-pressure $N_2$ adsorption. The Saito-Foley (SF) pore size distribution of the selected materials is displayed in Fig. 3. Micropore size distributions as calculated by classical models like SF are also affected by monolayer adsorption on the external surface [7]. For this reason the obtained pore size distributions are corrected for mesoporous content.

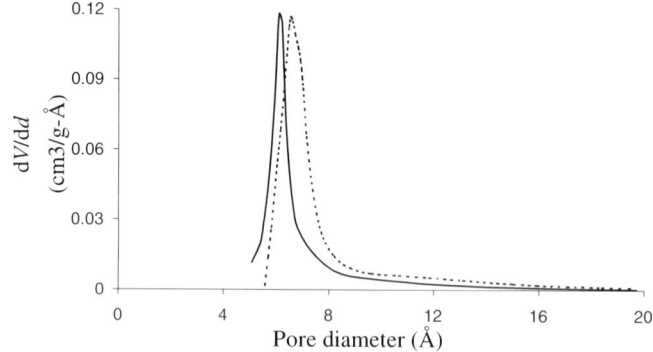

Fig. 3. SF-micropore size distribution of material A (dashed line) and A-1200 (solid line) by low-pressure $N_2$ adsorption.

The micropore size distribution of material A-1200 shows porosity close to the lower limit of measurement using $N_2$ as adsorptive. It can therefore be anticipated that not all pores are measured and consequently pores will also be present in the ultra-micropore region for this material. The distribution of sample A however shows a maximum in the distribution at larger pore size and no pores are present around the lower limit. Material A is therefore expected to lack the presence of ultra-micropores. These observations show a good correlation with the corrected data in Table 2, and the presumed ultra-microporosity in material A (Table 1) was indeed caused by the external surface mesopores.

### 3.2. $CO_2$ and $N_2$ adsorption on Siliceous MCM-41

In addition to the evaluation of the combined micro and mesoporous carbons, also siliceous materials were analyzed. More recently is there also a growing interest in the characterization of siliceous materials with $CO_2$ adsorption and the application of the DR-equation [13]. The application of the DR-equation and the proposed correction on this material will provide further insight on the universal applicability of this approach. To this end the $N_2$ and $CO_2$ physisorption characteristics were measured on two purely mesoporous MCM-41 materials. The BJH pore size distribution ($N_2$) of these mesoporous materials shows a narrow pore size distribution around 23 Å and 32 Å respectively. The corresponding specific surface area ($S_{BET}$), total pore volume ($V_{total}$), and results from the DR-equation obtained by $CO_2$ adsorption are given in Table 3.

Table 3
Textural properties of two MCM-41 materials as derived from $N_2$ and $CO_2$ physisorption.

|  | | $N_2$ | | $CO_2$ |
| --- | --- | --- | --- | --- |
|  | Pore size (Å) | $S_{BET}$ (m²/g) | $V_{total}$ (cm³/g) | $V_{micro, DR}$ (cm³/g) |
| MCM-41 A | 23 | 715 | 0.44 | 0.13 |
| MCM-41 B | 32 | 718 | 0.77 | 0.13 |

The results of these samples also show an apparent ultra microporosity by the application of the DR-equation on the $CO_2$ data, despite the fact that these materials are known to be purely mesoporous [14]. Application of Eq. 2, using $\alpha = 2.2 \cdot 10^{-4}$ cm³/m² as determined for carbon, results in a slight overestimation of the correction for these two materials, leading to $V_{micro, DR, correct} = -0.03$ cm³/g. The $\alpha$ for siliceous MCM-41 is therefore expected to be smaller compared to carbon, pointing out a lower surface coverage of the external surface. This lower coverage indicates a smaller interaction of the $CO_2$ on the external surface of the MCM-41 in comparison to the external surface of carbon. Although it is known from multilayer adsorption that the degree of curvature of the surface also influences the amount of adsorption [15], this does not seem to have a significant influence here; the two MCM-41 materials different in pore size by nearly 40 %, while their ration of external surface area to $V_{micro, DR}$ is equal.

### 3.3. Comparison of other methodologies

The utilization of the proposed correction has been compared to modern calculation methods like Density Functional Theory (DFT), as this methodology is in principle applicable to both micro- and mesoporous materials. To this end the DFT-pore size distribution of the two non-microporous carbons has been calculated (Kernel: $CO_2$ adsorption on Carbon [16]). These pore size distributions however showed significant microporosity in the range 4-10 Å. $N_2$ as adsorptive however showed no contribution in this region. It can therefore be concluded that not only the classical methods fail to describe the contribution of the external surface by $CO_2$ adsorption at low relative pressure but also modern methods cannot correct properly. It is expected that the DFT-method can only take into account the presence of mesopores if these pores have actually been quantified (this requires high pressure $CO_2$ adsorption).

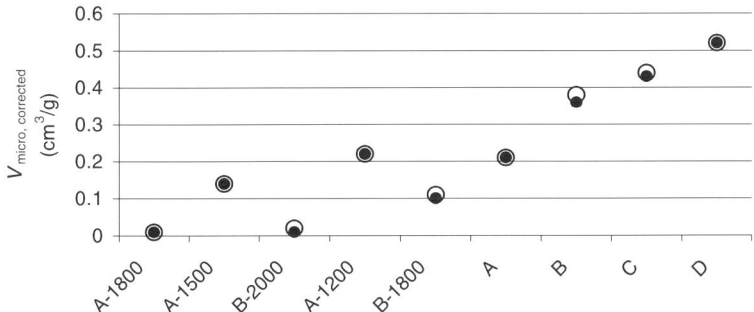

Fig. 4. Comparison of corrected $V_{micro}$ using proposed method (solid symbol) and isotherm subtraction (open symbol).

The correlation between this proposed method and the method of Dubinin using isotherm subtraction prior to using the DR-equation is fairly good for the investigated carbons as can be observed in Fig. 4. Difference in the corrected micropore volume by either technique did not exceed 0.02 $cm^3/g$, which also proves that the method of correction was applicable.

## 4. CONCLUSIONS

The presence of a significant external surface area, as e.g. present in hierarchically structured porous materials, can lead to an overestimation of the Dubinin-Radushkevich micropore volume as derived from $CO_2$ physisorption. The method proposed here corrects for this apparent ultra-micropore contribution by subtraction of the proportional contribution from the DR-results. Advantages of this latter method are the ease of determining the correction factor and the simplicity of correction the obtained micropore volume under normal conditions using sub-atmospheric $CO_2$ and $N_2$ adsorption measurements.

A good correlation has been established between our method and the isotherm subtraction approach. The use of different materials also pointed out the potentials of the method to other materials than carbon although the correction is material depending. Novel adsorption models such as Density Functional Theory can only properly take into account the contribution of the external surface area if this has been actually measured during the analysis, which requires high pressure $CO_2$ adsorption.

**Acknowledgements**
The authors thank MAST Carbon Ltd. for providing the *Novacarb* mesoporous carbon.

## REFERENCES

[1]  J. Garrido, A. Linares-Solano, J.M. Martin-Martinez, M. Molina-Sabio, F. Rodriguez-Reinoso, Langmuir 3 (1997) 76.
[2]  S. Gheorghiu, M.O. Coppens, AIChE J. 50 (2004) 812.
[3]  J.C. Groen, L.A.A. Peffer, J.A. Moulijn, J. Perez-Ramirez, Micro- Mesoporous Mater. 69 (2004) 29.
[4]  S.R. Tennison, Appl. Catal. A 173 (1998) 289.
[5]  D. Lozano-Castello, D. Cazorla-Amoros, A. Linares-Solano, Carbon 42 (2004) 1233.
[6]  S. Storck, H. Bretinger, W.F. Maier, Appl. Catal. A 174 (1998) 137.
[7]  Z. Shan, W. Zhou, J.C. Jansen, C.Y. Yeh, J.H. Koegler, Th. Maschmeyer, Nanoporous Materials III, Stud. Surf. Sci. Catal. 141 (2002) 635.
[8]  M.M. Dubinin, O. Kadlec, Carbon 13 (1975) 263.
[9]  M.M. Dubinin, Carbon 23 (1985) 373.
[10]  C. Moreno-Castilla, F. Carrasco-Marin, M.V. Lopez-Ramon, Langmuir 11 (1995) 247.
[11]  J.S. Beck, J.C. Vartuli, W.J. Roth, M.E. Leonowicz, C.T. Kresge, K.D. Schmitt, C.T-W. Chu, D.H. Olson, E.W. Sheppard, S.B. McCullen, J.B. Higgins, J.L. Schlenker, J. Am. Chem. Soc. 114 (1992) 10834
[12]  S.J. Gregg, K.S.W. Sing, Adsorption, Surface Area and Porosity, Academic Press, New York, 2nd edn. 1982.
[13]  J. Garćia-Martínez, D. Cazorla-Amorós, A. Linares-Solano, Stud. Surf. Sci. Catal. 128 (2000) 485
[14]  P.J. Branton, P.G. Hall, M. Treuguer, K.S.W. Sing, J. Chem. Soc., Chem. Commun. 16 (1993) 1257.
[15]  J.C.P. Broekhoff, J.H. de Boer, J. Catal. 9 (1967) 8.
[16]  P.I. Ravikovitch, A. Vishnyakov, R. Russo, A.V. Neimark, Langmuir 16 (2000) 2311.

Studies in Surface Science and Catalysis 160
P.L. Llewellyn, F. Rodriquez-Reinoso, J. Rouqerol and N. Seaton (Editors)

# A Monte Carlo study of capillary condensation of krypton within realistic models of templated mesoporous silica materials

**Francisco R. Hung[a], Benoit Coasne[a,†], Keith E. Gubbins[a], Flor R. Siperstein[b], Matthias Thommes[c] and Malgorzata Sliwinska-Bartkowiak[d]**

[a] Center for High Performance Simulation and Department of Chemical and Biomolecular Engineering, North Carolina State University, Raleigh, NC 27695-7905, USA

[b] Departament d'Enginyeria Quimica, Universitat Rovira i Virgili, Campus Sescelades, Avenida dels Països Catalans 26, 43007, Tarragona, Spain

[c] Quantachrome Instruments, 1900 Corporate Drive, Boynton Beach, FL 33426, USA

[d] Institute of Physics, Adam Mickiewicz University, Umultowska 85, 61-614 Poznan, Poland

## 1. ABSTRACT

We report molecular simulations of Kr adsorption at 87 and 100 K in three atomistic silica mesopores with an average pore diameter of 6.4 nm: (a) a pore that keeps the morphological features of a MCM-41 mesoscale model, generated from lattice Monte Carlo simulations mimicking its fabrication process; (b) a smooth, regular cylindrical pore; and (c) a cylindrical pore with constriction. Surface roughness and structural defects significantly affect Kr adsorption: marked differences were observed in the adsorption isotherms, isosteric heat curves and pore filling mechanisms for the three pore models. Our results suggest that the molecular-level surface disorder for the first pore model is too high, but its roughness at larger length scales (10-50 Å), as determined from simulated SANS spectrum, is in agreement with experimental results. The dense phase of Kr inside the three pore models exhibits a liquid-like global structure, even though the temperatures considered are well below the bulk triple point.

## 2. INTRODUCTION

Templated mesoporous silica material MCM-41 [1] consists of hexagonal arrays of cylindrical pores with diameters between 1.5 and 20 nm, narrow pore size distributions and negligible pore networking. These properties make these materials ideal for fundamental studies aimed at determining the effect of surface forces, confinement and reduced dimensionality on the phase behavior of host molecules. The features of MCM-41 materials [2] make them suitable for a number of applications in catalysis, adsorption, optics, as low dielectric constant materials to insulate integrated circuits, and as host materials for polymers, nanoparticles and enzymes [2]. The gas-liquid transition of adsorbates in templated mesoporous silica materials has been extensively studied by experiment, theory and molecular simulation [3]. From a molecular simulation viewpoint, a number of silica pore models have

---

[†] Present address: Laboratoire de Physicochimie de la Matière Condensée (UMR CNRS 5617), Université de Montpellier II, Place Eugène Bataillon, 34095 Montpellier, Cedex 5, France.

been used recently [4-7] to study adsorption of different pure substances and mixtures on MCM-41 type materials. Most of these pore models exhibit a regular cylindrical geometry, but it is unclear whether those models are realistic representations of MCM-41, due to the lack of conclusive experiments regarding its pore surface roughness and morphology. Some experimental results [8,9] attribute some surface roughness to the pore walls, whereas others [10,11] suggest a smooth surface at molecular scales (3-7 Å), with roughness and other morphological defects (constrictions, tortuosity) at larger length scales (20-50 Å).

One possible approach to modeling porous solids is to mimic the synthesis process of the real material using simulations, a strategy used in the past [12] to develop realistic models for Vycor and controlled pore glasses (CPG). Recently, Siperstein and Gubbins [13] used lattice Monte Carlo simulations to study the behavior of surfactant-inorganic oxide-solvent systems and mimic the synthesis of templated mesoporous silica materials. Low values of the surfactant/inorganic oxide concentration ratio led to formation of hexagonally ordered porous structures resembling that of MCM-41 material. Pores with an important degree of surface roughness and structural defects were obtained from this mesoscale simulation protocol. Recently, Coasne *et al.* [14,15] developed a fully-atomistic model based on the morphological features of this mesoscale model. In doing this "downscaling" process, atomic details can be included, more accurate potentials can be used for the adsorbate-wall interactions, and the effect of structural defects on the adsorbate phase behavior can be assessed. In this paper we summarize our methodology used to generate atomistic silica mesopores. Three pore models were considered in this study: 1) a pore that keeps the morphological features of the mesoscale model of Siperstein and Gubbins [13]; 2) a pore with regular cylindrical shape; and 3) a cylindrical pore with constrictions. We performed Grand Canonical Monte Carlo (GCMC) simulations of krypton adsorption at 100 K and 87 K, for which we report adsorption isotherms and isosteric heats of adsorption. Kr adsorption at low temperatures (77, 87 K) has been used in the past for characterization of porous materials with very low surface areas and volumes [16,17]. We discuss the effect of surface roughness and structural defects, and we compare with experimental results available in the literature. Similar results for argon and xenon on atomistic silica mesopores have been presented in other publications [14,15].

## 2. SIMULATION DETAILS

A detailed description of the simulation protocol used to generate and characterize atomistic silica mesopores has been reported elsewhere [14,15]. Similarly, the mimetic simulation procedure to obtain mesoscale, lattice models of MCM-41 type materials has been described in detail in previous publications [13]. The lattice positions of the inorganic oxide segments from the mimetic simulations were scaled to obtain a material with a pore diameter $D = 6.4$ nm and a length $L = 28$ nm. We isolated one of these pores and carved out its morphology from an atomistic box with a number of unit cells of cubic cristobalite. Following that, silicon atoms that are in an incomplete tetrahedral environment are removed, as well as oxygen atoms with two dangling bonds; oxygen atoms with one dangling bond are saturated with hydrogen atoms. This procedure ensures that all silicon atoms have no dangling bonds, all oxygen atoms have at least one saturated bond with a silicon atom, the atomistic simulation box has no net electrical charge and the pore surface is modeled in a realistic way. To mimic an amorphous silica surface, all the O, Si and H atoms were sightly displaced a random distance, and in a final step, the structure was further relaxed by performing a *NVT* Monte Carlo simulation using suitable potentials and parameters [15]. The generated atomistic silica mesopore (model A) has an average pore radius of 3.2 nm with a dispersion of

±1 nm, and its roughness at length scales between 10 Å and 50 Å was determined from simulated small angle neutron scattering spectra (SANS) [15]. This procedure to generate atomistic silica pores was originally proposed by Pellenq and Levitz [18] to model Vycor glass. We used a similar simulation protocol to generate two additional silica pores: a regular cylindrical pore with $D = 6.4$ nm (model B), and a cylindrical pore of diameter $D = 8.2$ nm with a constriction of $D = 4.1$ nm (model C). Similar constricted pores were considered in previous simulation studies [5,7,19,20]. The dimensions of pore model C are such that model B is an equivalent regular cylindrical pore with the same length, volume, and on average, the same degree of confinement. Both pore models B and C have a length of 15 nm, and periodic boundary conditions were applied in the axial direction for all three pore models. A schematic of the procedure used to generate the atomistic pores, as well as front views and cross sections of the three pore models used in this study are shown in Fig. 1.

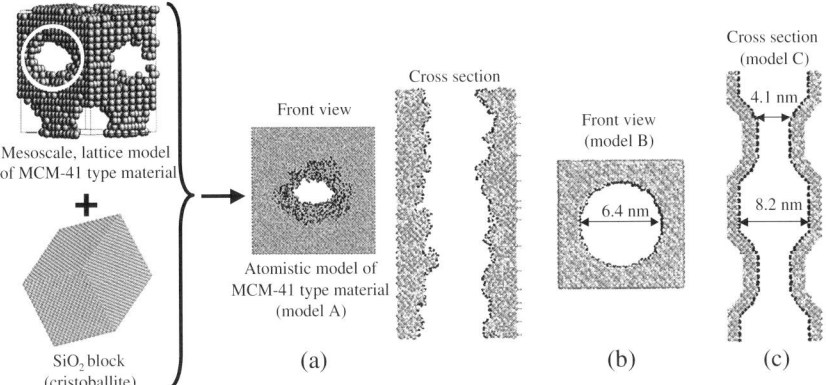

Fig. 1. Scheme of the procedure used to generate atomistic silica pore models used in this study. Front views and/or cross sections of the three pore models are also shown. Oxygen, silicon and hydrogen atoms are depicted in white, grey and black, respectively. For model C, we have represented two simulation boxes aligned in the axial direction $z$.

Krypton was modeled as a Lennard-Jones (LJ) fluid with parameters taken from Ref. [18]: $\sigma_{ff} = 0.369$ nm, $\varepsilon_{ff}/k_B = 170$ K. The interaction of Kr atoms with O, Si and H atoms in the substrate was modeled using the PN-TraZ potentials and parameters [18]. We performed Grand Canonical Monte Carlo (GCMC) simulations of Kr adsorption at $T = 100$ K and 87 K. The ideal gas equation and the properties of the LJ fluid at coexistence [21] were used to calculate chemical potentials $\mu$ from relative pressures $P/P_0$. To speed up our simulation runs, we have used an energy grid [18] with an elementary cube size of about 1 Å$^3$ to compute the adsorbate-wall potential energy. Thermodynamic properties were averaged over a minimum of $10^5$ MC steps per particle (typical systems had up to 15000 particles); however, much longer runs were considered near the phase transitions. We also determined the isosteric heat of adsorption $q_{st}$ [22], the local density profile $\rho$ and the "renormalized" 3D positional pair correlation function $g(R)$ (where $R$ is the in-pore distance) following the procedure reported in our previous work [23].

## 3. RESULTS

Kr adsorption isotherms are presented in Fig. 2 for the three atomistic silica mesopore models. The adsorbed amounts were normalized with respect to the quantity of adsorbate when each pore is filled. At a fixed $T$, the adsorption curves exhibit different features and are significantly affected by the pore morphology. For pore model B there is one vertical jump due to capillary condensation, whereas two discontinuities were observed for model C. The adsorption isotherm for pore model A also exhibits two small jumps that are not as sharp as those observed in models B and C. The surface area of pore model A is larger than that of model C, which in turn is larger than that of model B. In addition, pore model A exhibits an important degree of surface roughness and therefore there are more sites where molecules can get preferentially adsorbed. These factors cause pore model A to exhibit a lower capillary condensation pressure, a larger value of $N/N_0$ at any given $P/P_0$, and an adsorption curve that looks smoother than those for the other pore models. Pore model C fills at a lower value of $P/P_0$ than model B, and at a fixed value of $P/P_0$, model C exhibits a larger value of $N/N_0$ than that for model B. Pore model A shows an asymmetrical hysteresis loop, which is typical of adsorption on materials with interconnected pores, e.g. Vycor [7,12,18], but were also observed in unconnected pores with morphological defects [5,7].

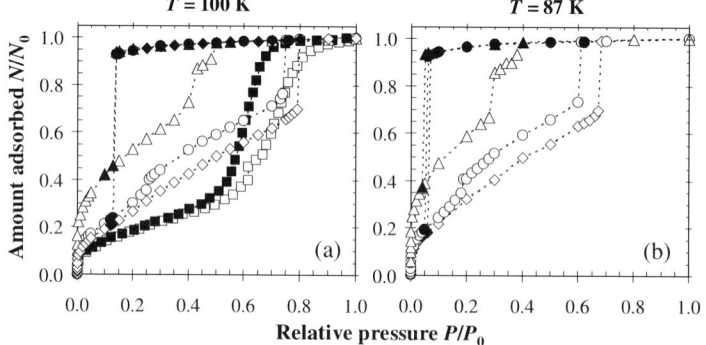

Fig. 2. Adsorption-desorption isotherms for Kr at (a) $T = 100$ K, and (b) $T = 87$ K. Results for pore models A, B and C are depicted by triangles, diamonds and circles, respectively; squares represent experimental results on SBA-15 with a similar mean pore diameter (6.4 nm) [24]. Open and closed symbols represent adsorption and desorption, respectively.

Evaporation takes place approximately at the same value of $P/P_0$ in the three pore models. Due to the use of periodic boundary conditions, there is no interface with the bulk and evaporation occurs via nucleation of a gas bubble (cavitation) in the center of the pore. The mean pore sizes in term of Kr molecular diameters ($\sigma_{Kr} = 0.369$ nm) are such that the desorption process does not seem to be affected by the pore morphology. Our adsorption-desorption simulation study of Ar ($\sigma_{Ar} = 0.34$ nm) on pore model A [15] shows a desorption curve similar to that of Kr, but for Xe ($\sigma_{Xe} = 0.410$ nm) it was found that the desorption curve is composed of reversible paths and two small jumps in the adsorbed amount [14,15]. Moreover, for Ar adsorption-desorption on cylindrical pores of $D = 6.0$ nm with a constriction of $D = 4.0$ nm [7], the isotherms exhibit the same features observed in this study for Kr on pore model C; however, for Ar on cylindrical pores of $D = 5.0$ nm with a constriction of $D = 2.5$ nm [25], a two-step desorption process was observed. These results suggest that the desorption mechanism in pores with morphological defects strongly depends on pore geometry and adsorbate size. For a given pore model, as temperature increases the amount adsorbed at a certain value of $P/P_0$ decreases, capillary condensation and evaporation take place at higher values of $P/P_0$, the hysteresis loop shrinks, and the jumps in the adsorption isotherms are less marked.

In Fig. 2 we also compare our simulation results at 100 K with experiments for SBA-15 with the same mean pore diameter (6.4 nm) from Morishige *et al.* [24]. The shape of the experimental adsorption isotherm resembles that of pore model A, and the step due to capillary condensation is steep but not vertical. This can be due to the presence of a pore size distribution, as well as morphological and topological features (changes in pore shape, surface roughness, energetic heterogeneity, microporosity, interconnected pores). On the other hand, the experimental capillary condensation pressure is similar to what was obtained for pore model B, and $N/N_0$ is overestimated by all pore models. The lack of quantitative agreement between simulations and experiments could be due to at least two reasons. First, the intermolecular potential adsorbate-wall can be overestimating the attractive energy: for simulated Ar adsorption in regular cylindrical silica pores, it has been reported that a reduction of 3% in selected parameters can lead to quantitative agreement in film thickness (*t*-plot) with experimental results [7]. Second, the degree of surface disorder for pore model A is too high at the molecular level, leading to overestimation of $N/N_0$ and underestimation of the capillary condensation pressure. These conclusions are corroborated by our results for the isosteric heat of adsorption $q_{st}$ for Kr at 100 K, which are presented in Fig. 3. The experimental value of $q_{st}$ at low coverages is 15 kJ/mol [18], in agreement with results for pore models B and C. At low pore filling fractions ($N/N_0 < 0.3$), $q_{st}$ for pore model A largely overestimates the values of models B and C [Fig. 3(a)], since the adsorbate-wall contribution to $q_{st}$ for pore model A is significantly larger than those for the other pore models [Fig. 3(b)]. This explains why $N/N_0$ at a given pressure prior to capillary condensation is much larger for pore model A, and suggests that the degree of surface disorder for pore model A is too high at the molecular level. This conclusion is also supported by our results for Ar and Xe adsorption [14,15] and their comparison with available experimental data. Nevertheless, simulated SANS results for pore model A [15] indicates that its roughness at length scales between 10 Å and 50 Å is in agreement with experimental measurements for MCM-41 type materials [10,11].

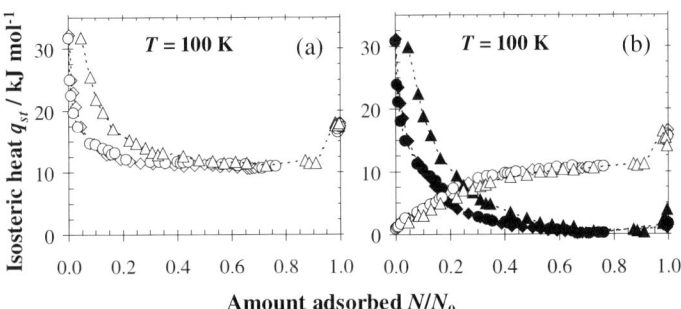

Fig. 3. Isosteric heat of adsorption $q_{st}$ as a function of coverage fraction $N/N_0$, for Kr adsorption on atomistic silica mesopores at 100 K: (a) total $q_{st}$, and (b) adsorbate-wall (filled symbols) and adsorbate-adsorbate (open symbols) contributions to $q_{st}$. Symbols as in Fig. 2.

We have used our pore models to discuss the effect of surface roughness and structural defects on the adsorption mechanism and on the nature of the dense phases. In Fig. 4 we present plots of the local density profile $\rho(r, z)$, as well as representative simulation snapshots for the three pore models at different values of $P/P_0$ and $T = 100$ K. For pore model A, $\rho$ also depends on the angular coordinate $\theta$; nevertheless, a plot of $\rho(r, z)$ can provide a suitable measure of the local state of the confined phase. As $P/P_0$ increases, the pore walls are covered by an adsorbate film whose thickness increases gradually with $P/P_0$, until it reaches a point when there is formation of a condensate "bridge" between low density regions of adsorbate in the center of the pore [$P/P_0 = 0.40$, Fig. 4(a)]. Most of the low density regions condense suddenly at $P/P_0 = 0.43$ [Fig. 4(a)], which is also signaled by a jump in the adsorption isotherm (Fig. 2). Finally, at $P/P_0 = 0.50$ the pore is filled with a condensed phase and a smaller jump in the isotherm is also observed (Fig. 2). A similar behavior has been observed in past studies for adsorbents with distinct chemical and morphological heterogeneities [7,18-20,26-28], and is similar to the filling mechanism observed for the constricted pore [Fig. 4(c)]. As $P/P_0$ increases, we see again an increase in the adsorbed film thickness until it becomes unstable and condensation takes place in the constrictions [$P/P_0 = 0.26$, Fig. 4(c)]. The size of the constriction ($D = 4.1$ nm) is comparable to the diameter at the narrowest part of pore model A ($D = 5.4$ nm), where the liquid-like "bridge" first forms [$P/P_0 = 0.40$, Fig. 4(a)]. This leads to the formation of hemispherical gas-liquid interfaces that coexist with condensed regions, as was observed for pore model A. These gas-like regions slightly shrink as $P/P_0$ increases, until the main cavity is suddenly filled with a condensed phase [Fig. 4(c)]. The adsorption isotherm also exhibits two jumps when these condensation processes take place (Fig. 2). In contrast, for pore model B [Fig. 4(b)], the adsorbed film thickness increases with $P/P_0$ until it reaches its stability limit [$P/P_0 = 0.79$, Fig. 4(b)]. A slight increase in $P/P_0$ makes the pore to be suddenly filled with a condensed phase and a vertical jump is observed in the adsorption isotherm [Fig. 4(b) and Fig. 2]. The adsorption mechanisms at $T = 87$ K for the three pore models are similar to those described at $T = 100$ K.

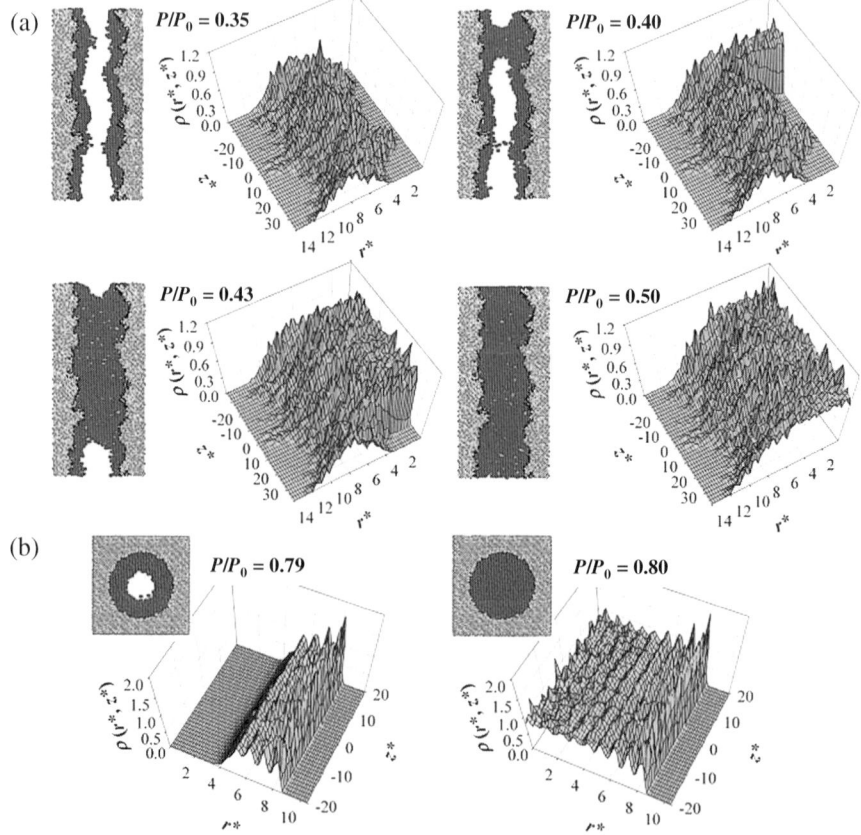

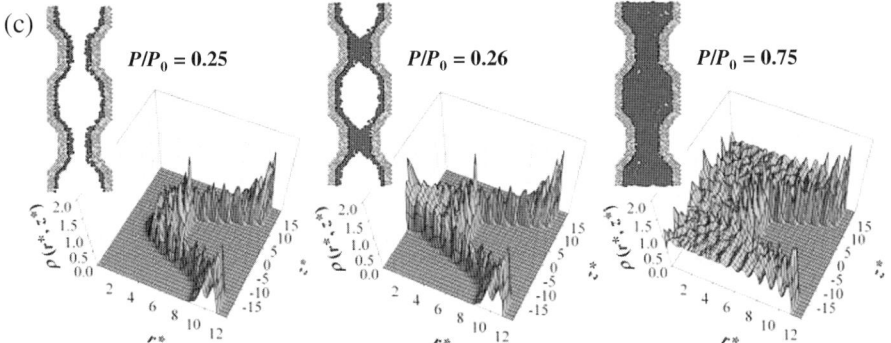

Fig. 4. Local density profile $\rho(r, z)$, and front views or cross sections of representative simulation snapshots of the confined phase inside (a) pore model A, (b) pore model B, and (c) pore model C, for different values of $P/P_0$ at $T = 100$ K. Two simulation boxes aligned in the axial direction $z$ are represented in the snapshots for pore model C, to help visualization of the confined phase features.

In Fig. 5 we show the 3D positional pair correlation function $g(R)$ for Kr inside the three pore models at $T = 87$ K and $P/P_0 = 1$. These functions suggest that the three pore models are filled with dense Kr with a liquid-like structure, even though the temperature is well below the bulk triple point (116 K). The maximum in the first peak of $g(R)$ is slightly larger for pore model B as compared to the other models, with Kr inside pore model A exhibiting the lowest value for this first peak. An investigation of the global and local structure of Kr freezing within these pores is currently in progress.

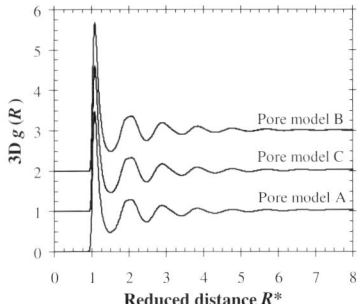

Fig. 5. 3D positional pair correlation function $g(R)$ for Kr at $T = 87$ K and $P/P_0 = 1$.

## 4. CONCLUSIONS

We performed GCMC simulations of Kr adsorption on three atomistic silica mesopores: (a) a pore (model A) that keeps the morphological features of the MCM-41 mesoscale model of Siperstein and Gubbins [13]; (b) a regular cylindrical pore (model B); and (c) a cylindrical pore with constriction (model C). Adsorption isotherms, isosteric heats, density profiles and 3D positional pair correlation functions were obtained at $T = 100$ K and 87 K, below the bulk triple point of Kr (116 K), and compared with experimental data. Our results suggest that the degree of surface disorder for pore model A is too high at the molecular level. However, simulated SANS spectrum shows that its roughness at larger length scales (10-50 Å) is in agreement with MCM-41 experimental results [10,11]; a detailed account of these results are presented in ref. 15. We are currently improving our pore model A so that its molecular roughness agrees with experimental data. Surface roughness and structural defects significantly affect Kr adsorption: marked differences were observed in the adsorption isotherms, isosteric heat curves and pore filling mechanisms. The dense phase of Kr inside the three pore models exhibit a liquid-like global structure; a detailed investigation of the freezing behavior of Kr within these pores is currently in progress.

We are grateful to R. J.-M. Pellenq (CNRS-Marseille) for helpful discussions. This work was supported by grants from NSF (CTS-0211792), KBN (2P03B 014 24) and NATO (PST.CLG.978802). This research used supercomputing resources from HPC-NCSU, SDSC (NSF/MRAC CHE050047S) and NERSC (DOE DE-FGO2-98ER14847).

# REFERENCES

[1] C. T. Kresge, M. E. Leonowicz, W. J. Roth, J. C. Vartuli and J. S. Beck, Nature, 359 (1992) 710.
[2] For recent reviews, see P. Selvam, S. K. Bhatia and C. G. Sonwane, Ind. Eng. Chem. Res., 40 (2001) 3237; G. J. de A. A. Soler-Illia, C. Sanchez, B. Lebeau and J. Patarin, Chem. Rev., 102 (2002) 4093; F. Schüth and W. Schmidt, Adv. Mater., 14 (2002) 629.
[3] For a review, see L. D. Gelb, K. E. Gubbins, R. Radhakrishnan and M. Sliwinska-Bartkowiak, Rep. Prog. Phys., 62 (1999) 1573.
[4] M. W. Maddox, J. P. Olivier and K. E. Gubbins, Langmuir, 13 (1997) 1737.
[5] B. Coasne, A. Grosman, C. Ortega and R. J. M. Pellenq, Studies in Surface Science and Catalysis, 144 (F. Rodríguez-Reinoso, B. McEnaney, J. Rouquerol and K. Unger, eds.), Elsevier, Amsterdam (2002) 35.
[6] Y. He and N. A. Seaton, Langmuir, 19 (2003) 10132.
[7] B. Coasne and R. J.-M. Pellenq, J. Chem. Phys., 121 (2004) 3767; 120 (2004) 2913.
[8] V. B. Fenelonov, A. Y. Derevyankin, S. D. Kirik, L. A. Solovyov, A. N. Shmakov, J.-L. Bonardet, A. Gedeon and V. N. Romannikov, Micropor. Mesopor. Mat., 44-45 (2001) 33.
[9] A. Berenguer-Murcia, J. García-Martínez, D. Cazorla-Amorós, A. Martínez-Alonso, J. M. D. Tascón and A. Linares-Solano, Studies in Surface Science and Catalysis, 144 (F. Rodríguez-Reinoso, B. McEnaney, J. Rouquerol and K. Unger, eds.), Elsevier, Amsterdam (2002) 83.
[10] K. J. Edler, P. A. Reynolds and J. W. White, J. Phys. Chem. B, 102 (1998) 3676.
[11] C. G. Sonwane, S. K. Bhatia and N. J. Calos, Langmuir, 15 (1999) 4603.
[12] L. D. Gelb and K. E. Gubbins, Langmuir, 14 (1998) 2097; Mol. Phys., 96 (1999) 1795.
[13] F. R. Siperstein and K. E. Gubbins, Mol. Sim., 27 (2001) 339; Langmuir, 19 (2003) 2049.
[14] B. Coasne, F. R. Hung, F. R. Siperstein and K. E. Gubbins, Ann. Chim-Sci. Mat., in press (2005).
[15] B. Coasne, F. R. Hung, R. J.-M. Pellenq, F. R. Siperstein and K. E. Gubbins, Langmuir, submitted (2005).
[16] S. J. Gregg and K. S. W. Sing, Adsorption, Surface Area and Porosity, 2nd edition, Academic Press, London, 1982.
[17] T. Takei and M. Chikazawa, J. Ceram. Soc. Jpn., 106 (1998) 353.
[18] R. J.-M. Pellenq and P. E. Levitz, Mol. Phys., 100 (2002) 2059.
[19] L. Sarkisov and P. A. Monson, Langmuir, 17 (2001) 7600.
[20] A. Vishnyakov and A. V. Neimark, Langmuir, 19 (2003) 3240.
[21] R. Agrawal and D. A. Kofke, Mol. Phys., 85 (1995) 43.
[22] D. Nicholson and N. G. Parsonage, Computer Simulation and the Statistical Mechanics of Adsorption, Academic Press, London, 1982.
[23] F. R. Hung, B. Coasne, K. E. Gubbins, E. E. Santiso, F. R. Siperstein and M. Sliwinska-Bartkowiak, J. Chem. Phys., 122 (2005) 144706.
[24] K. Morishige, K. Kawano and T. Hayashigi, J. Phys. Chem. B., 104 (2000) 10298.
[25] B. Coasne, K. E. Gubbins and R. J.-M. Pellenq, Part. Part. Syst. Charact., 20 (2004) 1.
[26] H. Bock and M. Schoen, Phys. Rev. E, 59 (1999) 4122.
[27] J. Puibasset, J. Chem. Phys., 122 (2005) 134710.
[28] F. Detcheverry, E. Kierlik, M. L. Rosinberg, and G. Tarjus, Phys. Rev. E, 68 (2003) 061504.

Studies in Surface Science and Catalysis 160
P.L. Llewellyn, F. Rodriquez-Reinoso, J. Rouqerol and N. Seaton (Editors)
© 2007 Elsevier B.V. All rights reserved

# Using molecular simulation to characterise metal-organic frameworks and judge their performance as adsorbents

**Tina Düren[a] and Randall Q. Snurr[b]**

[a] Institute for Materials and Processes, School of Engineering and Electronics, University of Edinburgh, King's Buildings, Edinburgh, EH9 3JL, UK

[b] Department of Chemical and Biological Engineering, Northwestern University, 2145 Sheridan Road, Evanston, IL 60208, USA

Porous metal-organic frameworks (MOFs) have recently gained much attention as promising materials for gas adsorption. These materials are generated in a directed-assembly process from corner units and linker molecules. As a result of the building block approach, these materials offer the possibility to tune host / guest interactions and therefore to tailor them rationally for specific adsorption separation or storage tasks. In this paper, molecular simulations are used to study methane adsorption in isoreticular metal organic frameworks (IRMOFs). The simulations revealed three distinct regimes: at low loading, the amount adsorbed is proportional to the heat of adsorption, at medium loading to the accessible surface area and at high loading to the free volume. A detailed analysis of adsorption in interpenetrated structures revealed that interpenetration leads to larger surface areas and stronger sorbate – framework interactions but also to lower free volumes, higher framework densities and smaller pores all of which can have a negative impact on the adsorption performance.

## 1. INTRODUCTION

Recent advances in supramolecular chemistry and metal-based directed-assembly chemistry have yielded a large variety of metal-organic materials. These materials are generated in a self-assembly process and generally consist of metal vertices interconnected by organic linker molecules as shown schematically in Fig. 1 a. The large variety of possible linker and corner units results in a large number of metal-organic frameworks. Over 13000 crystalline, extended metal-containing frameworks are catalogued in the Cambridge Structure Database [1] and more and more porous metal-organic materials are emerging that might be used as adsorbents for gas storage or gas separation tasks [2-4]. The modular building process allows systematic tailoring of the physical and chemical properties of the cavities. Yet, in order to choose the appropriate building blocks for a given application it is essential to understand how the characteristics of the building blocks and the resulting material influence the adsorption behaviour.

An ideal class of materials to study in this context are the isoreticular metal-organic frameworks synthesised by Yaghi and co-workers [5, 6], which consist of oxide-centered $Zn_4O$ tetrahedra linked by six dicarboxylates as illustrated in Fig. 1 b. The resulting materials feature extended, open, three-dimensional frameworks. Because all IRMOFs are built from the same corner unit and have the same network topology and therefore only vary in the linker molecules, they allow the systematic study of the influence of the linker molecules on adsorption.

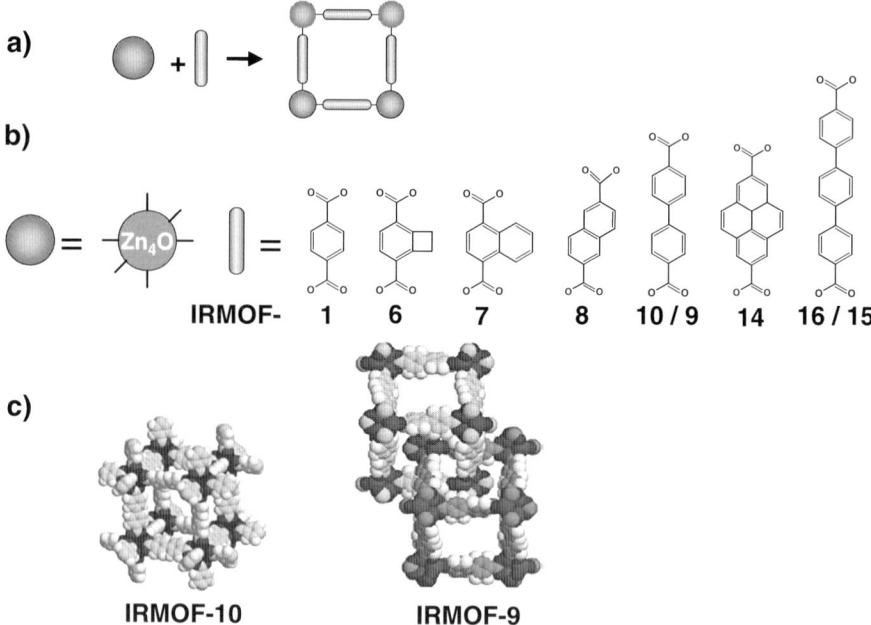

**IRMOF-    1        6        7        8      10/9     14    16/15**

**IRMOF-10            IRMOF-9**

Fig. 1. a) Schematic representation of the self-assembly process of metal-organic materials from corner and linker units. b) Building blocks for different IRMOF materials. IRMOF-9 and IRMOF-15 are the interpenetrated forms of IRMOF-10 and IRMOF-16, respectively. c) Examples of the resulting materials. For clarity, the carbon atoms in one of the networks in IRMOF-9 are shown in dark grey.

The materials included in this study are listed in Fig. 1 b. Two of the materials studied consist of two interpenetrated networks: IRMOF-9 and IRMOF-15 are the interpenetrated forms of IRMOF-10 and IRMOF-16 respectively (see Fig. 1 c). Interpenetration results in two opposite effects for adsorption, namely a gain in surface area and stronger energetic inter-actions between the sorbate molecules and the framework but also a loss of porosity and an increase in the crystal density which affects the gravimetric uptake negatively.

In this paper, we use molecular simulations to study methane adsorption in different IRMOFs. Molecular simulations provide a powerful tool to investigate adsorption as they not only allow quantitative predictions of methane adsorption in IRMOFs as we have shown before [7] but they also give a detailed picture on the molecular scale that allows, for example, a thorough analysis of the siting of the methane molecules in the cavities.

## 2. SIMULATION DETAILS

Adsorption isotherms were simulated with grand canonical Monte Carlo simulations [8] using our multipurpose simulation code Music [9]. For the sorbate molecules as well as for the IRMOF frameworks, atomistic models were employed. Methane – methane and methane – IRMOF interactions were modelled by the Lennard Jones potential using the Lorentz Berthelot mixing rules to calculate mixed parameters. Interactions beyond 12.8 Å were neglected. The potential parameters for methane were taken from Goodbody et al. [10] ($\sigma_{CH4}$ = 3. 73 Å, $\varepsilon_{CH4}/k_B$ = 98 K), whereas the Lennard Jones parameters for the IRMOF frameworks were taken from the DREIDING force field [11]. All simulations were carried out at 298 K.

The output of a GCMC simulation is the absolute amount adsorbed, whereas experimentally the excess amount adsorbed is measured. The excess number of molecules, $N_{ex}$, is related to the absolute number of molecules, $N_{abs}$, by

$$N_{ex} = N_{abs} - V_{free}\rho_g \qquad (1)$$

where $V_{free}$ is the pore volume of the adsorbent determined according to reference [12] and $\rho_g$ is the molar density of the bulk gas phase calculated with the Peng Robinson equation of state. The isosteric heat of adsorption at low loading, $Q_{st}$, was calculated from [13]. Pore size distributions were determined according to the method of Gelb and Gubbins [14]. The accessible surface area defined by the centre of a methane molecule rolling on the surface was calculated by a simple Monte Carlo integration [7].

## 3. RESULTS

Fig. 2 shows the simulated methane adsorption isotherms for a variety of IRMOF materials. In order to judge the performance of the materials, both the amount adsorbed per volume and per mass are shown. The materials with the best volumetric performances are the ones with the smallest linker molecules and therefore the smallest cavities (see Fig. 1 and Table 1). Within the group of the smaller IRMOFs, materials that consist of linker molecules with more carbon atoms (e.g. the dicarboxylate naphthalene linker of IRMOF-7 compared to the dicarboxylate benzene linker of IRMOF-1) show a better performance, as the additional carbon atoms result not only in an increased surface area but also in stronger interactions between the methane molecules and the cavities. Fig. 2 a also illustrates that the interpenetrated IRMOFs, which have comparable (in the case of IRMOF-15) or even smaller cavities (IRMOF-9), show a similar performance. At high pressures the differences between the absolute and the excess amount adsorbed become apparent. Whereas the absolute adsorption isotherms start to level off, the excess adsorption isotherms show a maximum, which is expected for any gas above its critical temperature when the increase of the bulk gas density is larger than the increase in the density of the adsorbate (see Eq. (1)).

Table 1
Properties of IRMOFs investigated

|  | $d_{cavity}$ Å | $\rho_{crys}$ g cm$^{-3}$ | $Q_{st}$ kJ mol$^{-1}$ | $S_{acc}$ m$^2$ cm$^{-3}$ | $V_{free}$ cm$^3$ g$^{-1}$ | $V_{free}$ % |
|---|---|---|---|---|---|---|
| IRMOF-1 | 10.9/14.3 | 0.59 | 11.6 | 2099 | 1.37 | 81 |
| IRMOF-6 | 9.1/14.5 | 0.65 | 13.1 | 1966 | 1.18 | 77 |
| IRMOF-7 | 10.1 | 0.71 | 13.6 | 2230 | 1.05 | 74 |
| IRMOF-8 | 12.5/17.1 | 0.45 | 11.4 | 1974 | 1.91 | 86 |
| IRMOF-9 | 4.5/6.3/8.1/10.7 | 0.66 | 16.2 | 2610 | 1.15 | 76 |
| IRMOF-10 | 16.7/20.2 | 0.33 | 10.9 | 1637 | 2.69 | 89 |
| IRMOF-14 | 14.7/20.1 | 0.37 | 11.6 | 1821 | 2.38 | 89 |
| IRMOF-15 | 7.3/9.5 | 0.41 | 10.7 | 2562 | 2.04 | 84 |
| IRMOF-16 | 23.3 | 0.21 | 10.5 | 1268 | 4.49 | 92 |

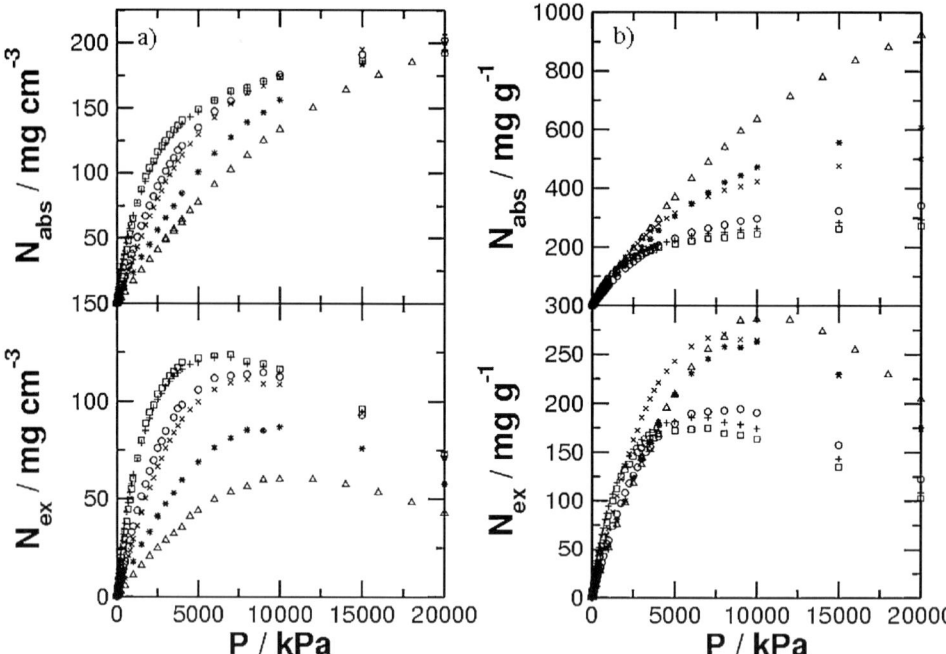

Fig. 2. Absolute (top) and excess (bottom) adsorption isotherms for methane in different IRMOFs (○ IRMOF1, □ IRMOF7, + IRMOF9, * IRMOF10, × IRMOF15, △ IRMOF16) at 298 K. a) adsorption per volume, b) adsorption per mass.

At higher pressure, the gap in performance between the larger IRMOFs, such as IRMOF-10 or IRMOF-16, and the smaller IRMOFs becomes smaller when the absolute amount adsorbed is considered. The reason for this is that at higher pressure the capacity of the materials plays an important role. Yet, for the excess amount adsorbed, the large bulk gas volume, $V_{free}$, of the large IRMOFs results in a low performance. Looking at the amount adsorbed per mass (Fig. 2 b), it becomes apparent that the additional carbon atoms in the linker molecules come with a price. Materials with lower framework densities show better performances at higher pressure. In general, the smaller IRMOFs show the best performance at lower pressure. However, because of the lower crystal density (Table 1), the performance of IRMOF-1 is better than of IRMOF-7 and at pressures larger than 7,500 kPa IRMOF-16, the material with the lowest framework density, shows the best performance.

From this argumentation it becomes clear that there are three main factors that influence adsorption: at low pressure, the energetic interactions between the sorbate molecules and the framework dominate and materials with small pores and many atoms in the framework show the best performance. At higher pressure, the accessible surface area and, when the capacity of the pores is reached, the free volume or the porosity play an important role.

The strength of the interaction between the sorbate molecules and the framework can be measured by the isosteric heat of adsorption at low loading. As illustrated by Fig. 3 a, the amount of methane adsorbed at 10 kPa depends linearly on the isosteric heat of adsorption for the IRMOFs investigated.

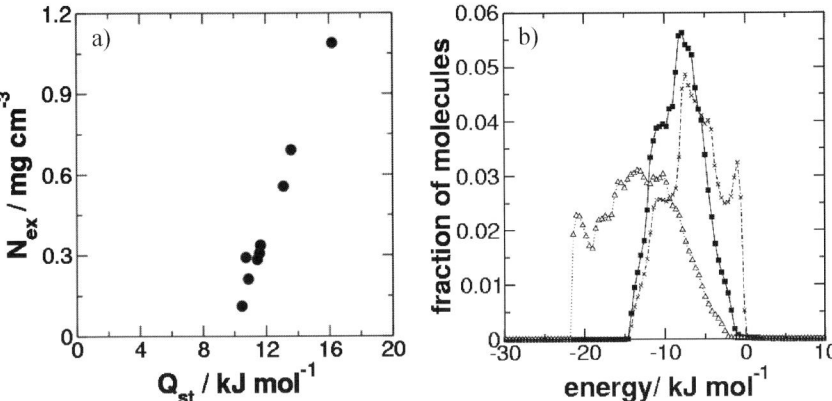

Fig. 3. a) Excess amount adsorbed as function of the isosteric heat of adsorption at 10 kPa. b) Corresponding distribution of methane / framework potential energies for individual methane molecules (■ IRMOF-1, △ IRMOF-9, × IRMOF-10). The lines were added to guide the eye.

A histogram of the energies between individual methane molecules and the framework elucidates the processes on the molecular level. The plots for the two non-interpenetrated structures IRMOF-1 and IRMOF-10 (Fig. 3 b) show a distinct plateau between -11 and -10 kJ mol$^{-1}$ as well as a peak around -7 kJ mol$^{-1}$. The plateau corresponds to the energetically favourable corner regions of the frameworks. Yet, as the size of these places is limited and the molecules move freely inside the cavity, methane molecules can also be found at the less favourable linker sites – and in the case of IRMOF-10, which has larger cavities than IRMOF-1 – also in the centre of the cavities where the energy is close to zero. The energies that a methane molecule experiences in the interpenetrated IRMOF-9 framework are much more favourable as illustrated by Fig. 3 b. The reason for this is that the smaller pores result in stronger interaction as the potential energies of opposite linker molecules overlap. This stronger interaction is the reason why IRMOF-9 shows the highest uptake of methane at 10 kPa (1.1 mg cm$^{-3}$).

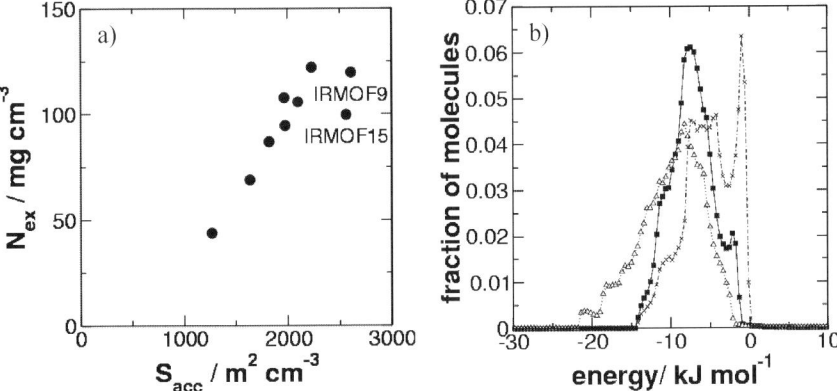

Fig. 4. a) Excess amount adsorbed as function of the accessible surface area at 5,000 kPa. b) Corresponding distribution of methane / framework energies for individual methane molecules (■ IRMOF-1, △ IRMOF-9, × IRMOF-10). The lines were added to guide the eye.

At 5,000 kPa, the amount adsorbed is proportional to the accessible surface area for the non-interpenetrated structures as illustrated by Fig. 4 a. Note, that the accessible surface area, reported here for $CH_4$ molecules, is not an inherent property of the material but depends on the specific sorbate molecule. Smaller sorbate molecules might be able to access more of the framework than larger molecules, e.g. if rather bulky functional groups on the linker molecules block access or if interpenetration gives rise to pores that are too small for larger molecules. The two interpenetrated structures, IRMOF-9 and IRMOF-15, have rather large surface areas because they consist of two interpenetrated frameworks. Yet, in contrast to the non-interpenetrated structures where the scaffold structure of the framework allows molecules to adsorb on both sides of the linker molecules, the cavities are too small to accommodate two layers of methane molecules. (To accommodate two layers of methane molecules, the cavities should have a diameter of at least 7.46 Å.) Therefore, their performance on a volumetric basis is worse than for non-interpenetrated structures with comparable surface areas as illustrated in Fig. 4 a. The energy distributions in Fig. 4 b show that the position of the peaks is the same as at 10 kPa (see Fig. 3 b). In comparison to the situation at 10 kPa, the relative importance of the different sites is shifted towards the sites at the linker molecules.

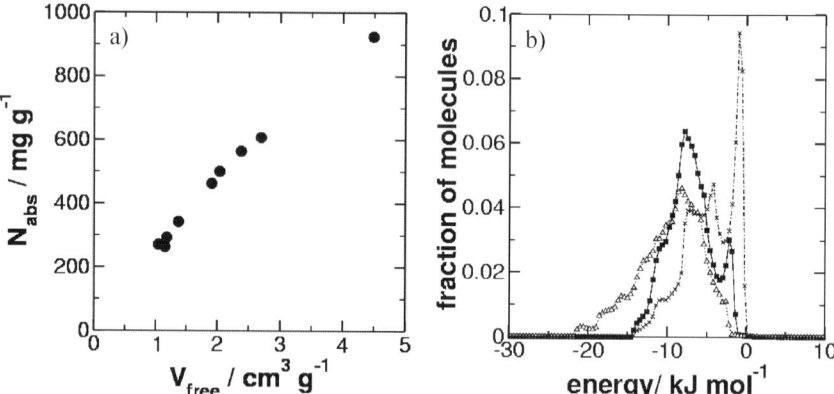

Fig. 5. a) Absolute amount adsorbed as function of the porosity at 20,000 kPa. b) Corresponding distribution of methane / framework energies for individual methane molecules (■ IRMOF-1, △ IRMOF-9, × IRMOF-10). The lines were added to guide the eye.

At very high pressure, when the absolute adsorption isotherms level off (compare Fig. 2) and the capacity of the cavities is reached, the absolute amount adsorbed (i.e. the number of methane molecules present in the pores) is proportional to the free volume or porosity of the materials as illustrated by Fig. 5 a. Note, that this relationship only holds true for the absolute amount adsorbed and not the excess amount adsorbed as the excess adsorption isotherms show a maximum at high pressure. The energy distribution (Fig. 5 b) shows that energetically less favourable sites within the pores such as the centre of the cavities (denoted by the sharp peak around -1 kJ $mol^{-1}$ in IRMOF-10) gain in importance as every available space within the cavities is occupied by methane molecules.

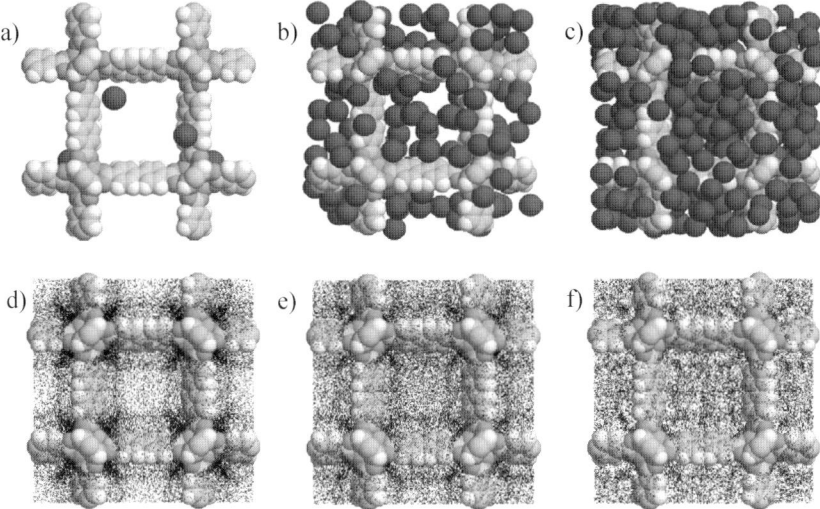

Fig. 6. Snapshots of methane molecules (dark spheres) adsorbed in IRMOF-10 a) 10 kPa, b) 5,000 kPa, c) 20,000 kPa. Density distributions of methane molecules in IRMOF-10 d) 10 kPa, e) 5,000 kPa, f) 20,000 kPa. Each dot represents the position of a methane molecule during the simulation. For each distribution, 75,000 configurations were used.

The importance of different sites at different pressures can also be seen in the snapshots and density distributions given in Fig. 6 for the example of IRMOF-10. At 10 kPa, only a few molecules are present in the pores (Fig. 6 a) and the dark patches around the corners in the density distribution (Fig. 6 d) indicate that the molecules are located preferentially at the energetically favourable corner sites. Yet, it can be also seen that molecules are located along the linker and to a much lesser extent in the centre of the cavities, which is in line with the energy distribution (Fig. 3 b). At 5,000 kPa, many more molecules are present, but there is still plenty of space especially in the centre of the cavity (Fig. 6 b). This is also confirmed by the darker areas along the linker molecules in the corresponding density distribution shown in Fig. 6 e. The snapshot at 20,000 kPa shows that the framework is completely occupied by methane molecules (Fig. 6 c) and the corresponding density distribution (Fig. 6 f) does not exhibit any sign for preferential sites. Rather, the molecules are distributed evenly throughout the cavities. Note that there is faint evidence for layering indicated by three black stripes in the centre cavity. Yet, as the cavities are structurally much more complex than, for example, slit pores, layers cannot evolve distinctly.

## 4. CONCLUSIONS

In this paper, molecular simulations were used to study methane adsorption in IRMOF materials and to examine which properties of the materials influence methane uptake. The materials were carefully characterised using Monte Carlo methods to get pore size distributions, accessible surface areas and pore volumes. Depending on the pressure, three distinct regimes can be distinguished. At low loading (10 kPa), the amount adsorbed is proportional to the isosteric heat of adsorption which is a measure for the strength of the interaction between the methane molecules and the framework. Here, the methane molecules are located preferentially in the energetically favourable corner sites of the materials. Materials with small cavities (i.e. small linker molecules) and with linker molecules consisting of many carbon atoms show the highest methane uptake. With increasing pressure, the number of molecules in the pores increases and the corner sites become fully occupied and the energetically less favourable sites along the linker molecules gain in importance. At medium loading (5,000 kPa), the linker molecules are covered by methane molecules so that the amount adsorbed depends linearly on the accessible surface area. If the pressure is increased further to 20,000 kPa, the absolute adsorption isotherms start to show saturation and level off. Here, the absolute amount adsorbed is proportional to the free volume of the materials. Note that this relationship does not hold for the excess amount adsorbed, as the excess adsorption isotherms show a maximum at such high pressures.

It was postulated that interpenetrated MOF would be ideal candidates to store small molecules as the interpenetration leads to stronger interaction between the sorbate molecules and the framework [15]. Our simulations show that the energetic interaction between methane molecules and an interpenetrated structure is indeed higher. Yet, the interpenetration comes with a price: the smaller pores might result in pore spaces that cannot accommodate two layers of methane molecules between opposite linker molecules. Therefore, the volumetric uptake is lower than in non-interpenetrated structures where methane molecules can adsorb on both sides of a linker molecule. Furthermore, the crystal density of an interpenetrated structure is about twice as large as the density of the equivalent non-interpenetrated structure which lowers the performance on a gravimetric basis.

In general, we have shown that molecular simulations are a powerful tool to analyse adsorption in MOFs. Although the results presented here are for methane adsorption in one particular group of MOFs, namely IRMOFs, these results can be extended to other materials and guide the choice of the linker molecule.

## ACKNOWLEDGEMENTS

We thank Houston Frost for helpful discussions and the U.S. Department of Energy, the Alexander von Humboldt foundation and the Nuffield foundation for financial support.

## REFERENCES
[1] J.L.C. Rowsell and O.M. Yaghi, Micropor. Mesopor. Mat., 73 (2004) 3.
[2] S. Kitagawa, R. Kitaura and S. Noro, Angew. Chem. Int. Edit., 43 (2004) 2334.
[3] S.L. James, Chem. Soc. Rev., 32 (2003) 276.
[4] C. Janiak, Dalton T., (2003) 2781.
[5] M. Eddaoudi, J. Kim, N. Rosi, D. Vodak, J. Wachter, M. O'Keefe and O.M. Yaghi, Science, 295 (2002) 469.
[6] H. Li, M. Eddaoudi, M. O'Keeffe and O.M. Yaghi, Nature, 402 (1999) 276.
[7] T. Düren, L. Sarkisov, O.M. Yaghi and R.Q. Snurr, Langmuir, 20 (2004) 2683.
[8] D. Frenkel and B. Smit, Understanding of Molecular Simulation: from Algorithms to Applications, Academic Press, San Diego, 2002.
[9] A. Gupta, S. Chempath, M.J. Sanborn, L.A. Clark and R.Q. Snurr, Mol. Simul., 29 (2003) 29.
[10] S.J. Goodbody, K. Watanabe, D. Macgowan, J.P.R.B. Walton and N. Quirke, J. Chem. Soc. Faraday T., 87 (1991) 1951.
[11] S.L. Mayo, B.D. Olafson and W.A. Goddard, J. Phys. Chem., 94 (1990) 8897.
[12] A.L. Myers and P.A. Monson, Langmuir, 18 (2002) 10261.
[13] R.Q. Snurr, A.T. Bell and D.N. Theodorou, J. Phys. Chem., 97 (1993) 13742.
[14] L.D. Gelb and K.E. Gubbins, Langmuir, 15 (1999) 305.
[15] B. Kesanli, Y. Cui, M.R. Smith, E.W. Bittner, B.C. Bockrath and W.B. Lin, Angew. Chem. Int. Edit., 44 (2005) 72.

Studies in Surface Science and Catalysis 160
*P.L. Llewellyn, F. Rodriquez-Reinoso, J. Rouqerol and N. Seaton (Editors)*
169

# Stability of Porous Carbon Structures Obtained from Reverse Monte Carlo using Tight Binding and Bond Order Hamiltonians

**S. K. Jain[1], J. Fuhr[2], R. J-M Pellenq[2], J. P. Pikunic[3], C. Bichara[2], K. E. Gubbins[1]**

[1] Center for High Performance Simulation and Department of Chemical and Biomolecular Engineering, North Carolina State University at Raleigh, Box 7905, Raleigh, NC 27695-7905, U.S.A.

[2] Centre de Recherche en Matière Condensée et Nanosciences, Campus de Luminy, 13288 Marseille, cedex 09, France.

[3] Department of Biochemistry, University of Oxford, South Parks Road, Oxford OX1 3QU, UK.

The constrained Reverse Monte-Carlo (RMC) technique [1,2] was used to generate atomic configurations of disordered microporous carbons in a previous work. However, a carbon structure obtained from RMC is a result of the fitting to some structural data such as obtained from X-ray diffraction; it does not guarantee the stability of the resulting models when a realistic interatomic potential is used. In the present work, we studied the stability of these RMC structures using canonical Monte-Carlo simulations. Two different descriptions of the carbon-carbon and carbon-hydrogen interactions are used, both encompassing the bonding processes characteristic of carbon chemistry. The first approach is based on a bond-order potential while the second considers a tight binding model. We found that the structures obtained from RMC simulations undergo local structural changes upon relaxation, however the porous structure of the models remains intact.

## 1. INTRODUCTION

Porous carbons are disordered materials with heterogeneous pore structures. They are widely used in industry for separation and purification of gases and liquid mixtures [3], because of their excellent adsorption properties. The link between texture of porous carbons and diffusion and adsorption properties, is far from being clear at the present time. Thus suitable microscopic scale models for the internal structure of these materials are required to understand and predict the adsorption and diffusion properties. To that purpose, models that describe in a realistic way the pore structure in porous carbons have to be developed. Models that have been proposed in the last few years have been recently reviewed [4].

Reverse Monte Carlo [5] is a reconstruction method, which is normally used to produce molecular models that match the experimental structure data and some other constraints. The constraints are usually based on the knowledge of the material being studied and also can be obtained from experimental evidence. However, no intermolecular potential is used in this methodology, so the stability of the models produced is not guaranteed. In previous works [1,2], a reconstruction procedure based on constrained Reverse Monte Carlo (RMC) was used to generate atomistic models that quantitatively match experimental (diffraction and small angle) properties of two real porous saccharose cokes CS400 (density ~

1.2 g/cm$^3$) and CS1000 (density ~ 1.5 g/cm$^3$), which were heat treated at 400° and 1000°, respectively. In this work, we study the stability of those models by relaxing them using a realistic interatomic potential. We relax the models in two different ways. In the first case we use an empirical interatomic potential developed by Brenner and in the other case we use a Tight Binding potential for carbon. Finally, we discuss the structural differences between the RMC models and the subsequent relaxed models by comparing the result we obtained in this work to that obtained in a previous work [2].

## 2.  COMPUTATIONAL METHODOLOGY

### 2.1  The Reactive Empirical Bond Order (REBO) potential
The empirical interatomic potential used in this work was developed by Brenner [6], the REBO potential for carbons. REBO is an empirical potential that is based on Tersoff's covalent bond formalism [7]:

$$E_{ij} = V_{ij}^R(r_{ij}) + b_{ij}V_{ij}^A(r_{ij})$$

It has a pair repulsive, $V_{ij}^R$, a pair attractive, $V_{ij}^A$, potential terms and a bond order term, $b_{ij}$, which weights the attractive part of the potential with respect to the repulsive part. The bond order term is a many body term, which depends on the local environment of atoms $i$ and $j$. A variety of chemical effects that affect the strength of the covalent bonding interaction are accounted for in this term. Coordination numbers, bond angles and conjugation effects contribute to the strength of a particular bonding interaction in the REBO potential. REBO is a highly parameterized potential that has been optimized for different properties of diamond, graphite and small hydrocarbons.

### 2.2  The Tight Binding Formalism
Among the available techniques for studying electronic structure and atomic cohesion, the tight-binding (TB) method has found favour for its simplicity and its versatility. The TB approach consists in replacing the many-body Hamiltonian operator by a parameterized Hamiltonian matrix and  solving the Schrödinger equation using an atomic orbital basis set. This is the minimal model that enables one to calculate not only the atomic geometry but also electronic structure, e.g., density of electronic states, ionization energy and band energy. This method has been applied to many materials, including *sp* bonded systems, *pd* bonded systems and transition metals; the TB approach has been demonstrated to work very well for covalently bonded systems (C, Si, Ga, Ge, In, etc.) and transitional d-band metals. The explicit introduction of pairwise atomic repulsion makes it possible to implement the TB method within molecular-dynamics simulations (TBMD), which can be employed for calculating characteristics properties of both crystaline and amorphous solids, as well as atomic clusters.

In this work we have used a minimal *sp* model, with one *s* orbital per hydrogen atom and one *s* orbital and three *p* orbitals per carbon atom. The tight-binding Hamiltonian and the pair potential were modeled using the Winn *et al.* parameterization [8], which gives reasonable energies and bond lengths for hydrocarbon molecules and hydrogenated surfaces. Self-consistency is modeled by the addition of a Hubbard-type term in the Hamiltonian with U = 10 eV, which reduces unphysical charge transfer. For the O(N) TB algorithm, we used the kernel polynomial method [9] with a smearing Fermi function (700 Chebyshev polynomials, and a spatial truncation of 5.5 Angstrom for the moments).

## 3. RESULTS AND DISCUSSION

In previous works Pikunic *et al.* [1,2] developed molecular models of CS400 and CS1000 based on RMC and using three body constraints for porous carbons (see Fig. 1).

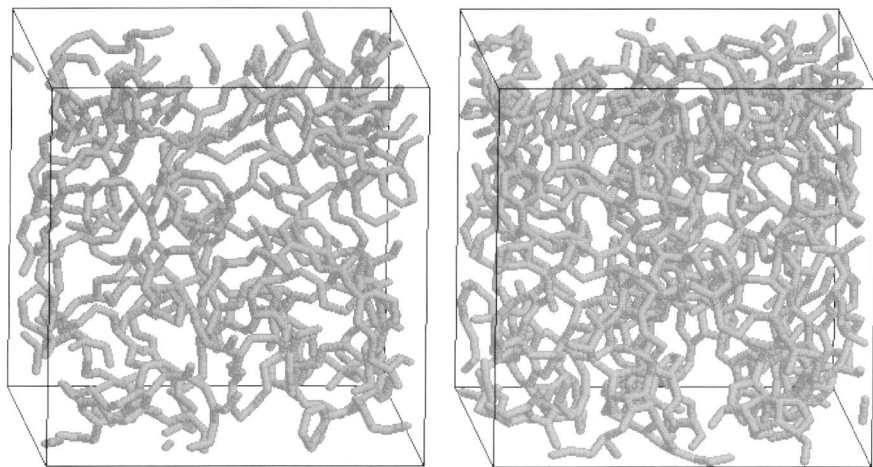

Fig. 1. RMC snapshots of CS400 (left) and CS1000 (right)). The size of the box is 25 Å [2].

The simulated radial distribution functions and high resolution TEM images were found to be in very good agreement with the experimental data [2]. Pikunic *et al.* also performed adsorption of $N_2$ and Ar at 77 K in these structures and found that the models correctly capture the energetic heterogeneity of the real material, which was reflected from the very good agreement with the isosteric heat of adsorption [10].

The models obtained using Constrained Reverse Monte Carlo result from the fit to experimental radial distribution functions along with a set of constraints [1]. On the other hand, there is no interatomic potential in the simulation method. Thus, the stability of the models obtained is not guaranteed. In this work, we study the stability of these models in the presence of a realistic interatomic potential. Moreover, the models developed by Pikunic *et al.* [1,2] contain only carbon atoms; the hydrogen atoms are included implicitly by assuming that each carbon atom is $sp^2$ hybridised, *i.e.* has 3 nearest neighbors. However, we need to take hydrogen into account before doing the relaxation. Hence, we started with the final configuration obtained from RMC and added hydrogen atoms to carbon atoms with less than 3 neighbors to make it $sp^2$ hybridized. We then relaxed the resulting models using Monte Carlo simulation in the canonical ensemble at 300 K.

### 3.1 Results for CS400 (density : 1.2 g/cm³)
As we can see from Fig. 2, the TB and REBO relaxations give similar results for the radial distribution function. The relaxation of the RMC models results in an intensity increase along with a decrease in width of the first peak. These results arises from the different bond length constraints used in RMC and the REBO or TB potential. In RMC, it is assumed that two atoms can have a minimum distance of approach of 1.2 Å and that two atoms are considered bonded if the distance between them is less than 1.6 Å. This definition is obtained from the

end points of the first peak in the experimental radial distribution function, g(r). This discrepancy in the first peak is due to the fact that in RMC, we place the atoms so that g(r) matches the experimental structural data. It does not take into account the interaction between carbon atoms. The height of the second peak also increased upon relaxation with the REBO or TB potentials. This suggests that the short range order is enhanced upon relaxation. One can note that the first three peaks occur approximately at the same distance. The features beyond the third peak are lost upon relaxation, presumably due to the short range nature of the potential that does not have any dispersion (long range) interactions.

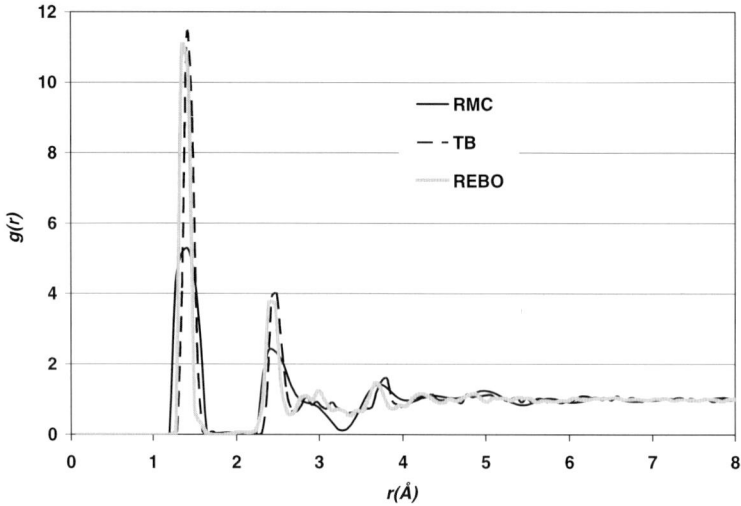

Fig. 2. Radial distribution functions comparing RMC [2], REBO relaxed and TB relaxed models.

We calculated the bond angle and neighbor distributions of the RMC and relaxed structures to further study the changes in the local environment. As seen in Fig. 3, the RMC and the relaxed structures, both have a bond angle distribution centered at 120°. It can be seen that the structures obtained from REBO, as well as from TB relaxation, have a broader distribution as compared to the RMC model. This is due to the use in RMC of a bond angle constraint that minimizes the deviation of the bond angles from 120° [1]. However, during relaxation, the bond angle also depends on the local environment of an atom and its neighbors. This is the reason why the distribution is slightly broader upon relaxation. From Fig. 4, we see that the neighbor distribution of the relaxed as well as the RMC structure remains nearly the same, especially when relaxed with the REBO potential. This shows that there is not a dramatic breaking or creation of C-C bonds upon relaxation. However, the structure obtained after relaxation using TB has a slight increase in the number of atoms having 3 neighbors and a slight decrease in the number of atoms having 2 neighbors. This suggests that the TB potential is slightly more attractive as compared to the REBO potential and thus results in the formation of a few new bonds.

It has been reported in the literature that the structures obtained from RMC contain some 3 and 4 membered rings, which are high energy structures and are thus unphysical. These are believed to be an artifact of the RMC method [11]. To test this we calculated the ring statistics of the RMC and the relaxed structures using the method of Franzblau [12].

From Fig. 5, we see that the RMC structure of CS400 contains some 4 member rings, which is decreased upon relaxation with the REBO potential; they are completely removed by the TB potential. Moreover, the number of 5, 6, 7 member rings remain mostly the same upon relaxation with the REBO potential, while an increase is observed upon relaxation with the TB potential. This again reflects the attractive nature of the TB potential as compared to the REBO potential. The distribution however is centered at 6 member rings in both the RMC and relaxed cases.

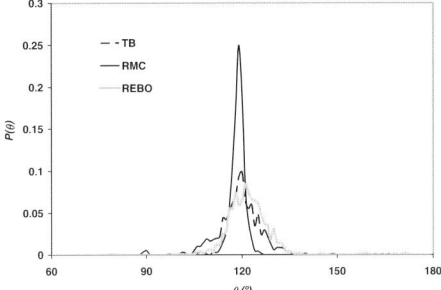
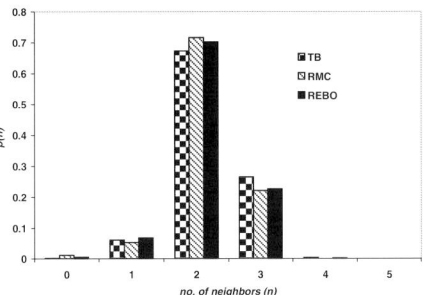

Fig. 3. Bond angle distributions (C-C-C bond angle, i.e. all the atoms involved in a bond angle are carbon atoms) comparing RMC, REBO relaxed and TB relaxed models.

Fig. 4. Neighbor distributions (only carbon atom neighbors) comparing RMC, REBO relaxed and TB relaxed models

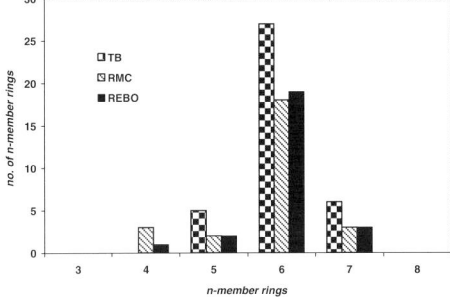
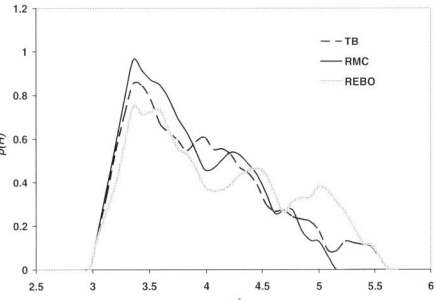

Fig. 5. Ring statistics (rings contain only carbon atoms) comparing RMC, REBO relaxed and TB relaxed models.

Fig. 6. Pore size distribution comparing RMC [2], REBO relaxed and TB relaxed models.

The results shown above reflect the change in the carbon skeleton, *i.e.* the change in the local environment or arrangement in the carbon walls. We calculated the porosity of the RMC and relaxed models using Ar as a test particle, to study the relaxation effect on the pore structure. The parameters for argon-carbon are the same as those used by Jain *et al.* [13]. The porosity of the RMC, REBO relaxed and TB relaxed models was found to be 0.1209, 0.1386 and 0.1405 respectively. This shows that the overall porosity of the RMC structure is increased by ~ 9 % upon relaxation. This can be attributed to the fact that the REBO and TB potential are reactive in nature and bring the carbon atoms close together. We calculated the pore size distribution (PSD) using the method suggested by Gelb and Gubbins [14]. The shape of the PSD for the RMC model changes slightly upon relaxation (see Fig. 6) and the PSD of the relaxed models is slightly inclined towards larger pores as compared to the models from RMC. This reveals that some of the pores are slightly enlarged upon relaxation which is consistent with the fact that the porosity of the relaxed models is slightly higher than that of the RMC models. On the other hand, we do not expect to create any new pores since there is not a significant amount of breaking of C-C bonds upon relaxation.

## 3.2   Results for CS1000 (density : 1.5 g/cm³)
We present results of the relaxation of CS1000 using REBO potential only. We show the radial distribution function (Carbon-Carbon) of RMC and relaxed models in Fig. 7.

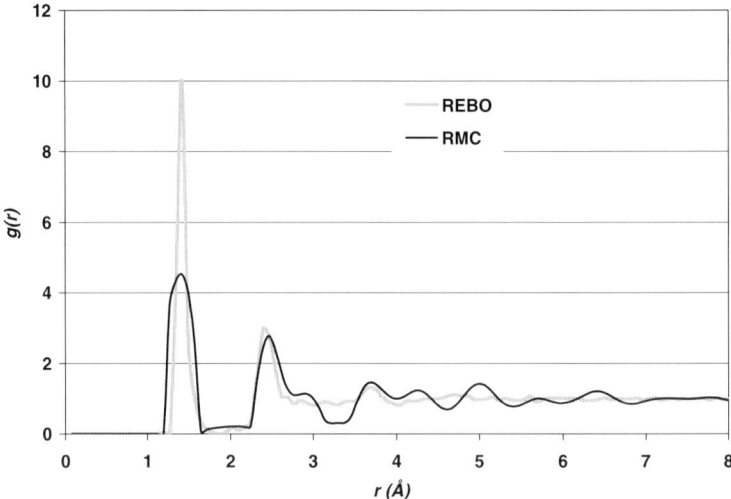

Fig. 7. Radial distribution function comparing RMC [2] and REBO relaxed models for CS1000.

As with CS400, relaxation results in an increase in the height and decrease in width of the first peak. Moreover, the second peak does not change as in the case of CS400, where the height of the second peak was increased. The first three peaks are approximately at the same distance. Again, the features beyond the third peak are lost upon relaxation. This may be due to the short range nature of the potential, which does not have any dispersion interactions that accounts for the long range part of the potential. We further study the structural changes by calculating the bond angle distributions, nearest neighbor distributions and ring statistics as done for CS400.

Fig. 8 shows that the bond angle distribution of the relaxed model is centered at 120° as for the RMC model, but the distribution changes. RMC contains small peaks at 60 and 90 degrees which are eliminated upon relaxation. These peaks are due to the 3 and 4 member carbon rings (Fig. 10). The number of carbon atoms having 1 or 2 carbon neighbors increases upon relaxation while the number of carbon atoms having 3 neighbors decreases, as can be seen from the neighbor distribution (Fig. 9). This shows that relaxation actually breaks some of the carbon-carbon bonds. Moreover from Figure 10 we see that CS1000 has a number of 3 and 4 member rings which are destroyed during relaxation. This explains the breaking of some of the carbon-carbon bonds. This also shows that the 3 and 4 member rings are high energy structures in the REBO framework and are thus eliminated. The number of 6 and 7 member rings is almost the same for both RMC and relaxed models. However, the number of 5 member rings decreases upon relaxation, which indicates that relaxation removes some of the defects having rings of 5 carbon atoms.

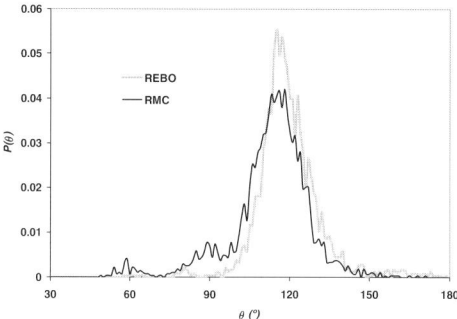 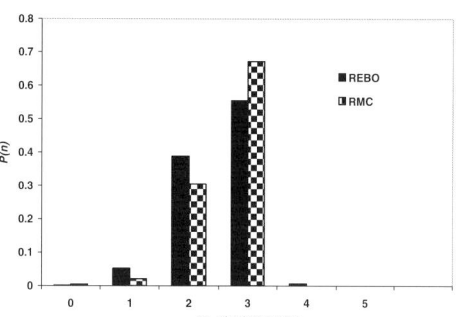

Fig. 8. Bond angle distribution, comparing RMC and REBO relaxed models for CS1000.

Fig. 9. Neighbor distribution, comparing RMC and REBO relaxed models for CS1000.

We also calculated the porosity and pore size distribution for the RMC and relaxed model of CS1000 using Argon as a probe particle. The porosity of the RMC and relaxed model was found to be 0.0964 and 0.0968, respectively. This shows that the overall porosity is not changed upon relaxation, although the local environment of the carbon atoms is rearranged. Figure 11 suggests that there is a slight modification of the PSD after relaxation, which is to be expected as some of the 3 and 4 member rings are eliminated and the bond lengths are relaxed. However the range of PSD is the same in the RMC and relaxed models.

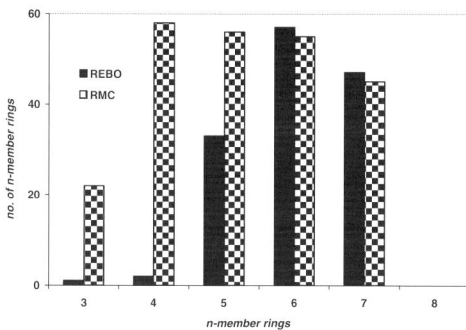 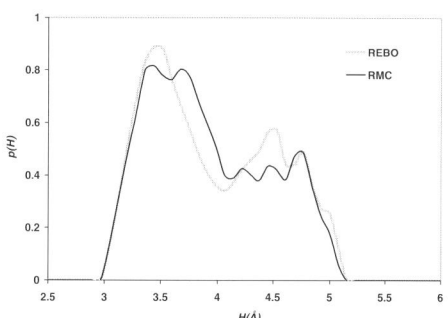

Fig. 10. Ring statistics, comparing RMC and REBO relaxed models for CS1000.

Fig. 11. Pore size distribution, comparing RMC [2] and REBO relaxed for CS1000.

## 4. CONCLUSIONS

We have studied the structural changes on RMC models brought about by relaxation using two different approaches. We found that both TB and REBO relaxation give similar results. This shows that these two potentials describe the local environment (or interaction between C-

C, C-H and H-H) in a very similar way. The RMC models undergo structural changes upon relaxation, *i.e.* relaxation modifies the local environment of carbon atoms as can be seen from the bond angle and neighbor distributions. This is particularly evident for CS1000, where relaxation removes most of 3 and 4 member rings that are present in the RMC model. The first three peaks of *g(r)* are approximately at the same location, which shows that the main features of the RMC models remain stable upon relaxation. On the other hand, we find that the discrepancy in the first peak is a result of different bond length constraints used in RMC and the relaxing Hamiltonians. The long range features are lost since the relaxing Hamiltonians does not have dispersion (long range) interactions. Stuart *et al.* [15] have developed a method that includes the dispersion interaction in the REBO framework. Including such dispersion interactions may give better long range agreement. The porous structure, however, remains intact and does not collapse. This shows that the assumption of RMC ($sp^2$ hybridization) is valid for these carbons. We have developed a methodology in which we replace the constraints in the RMC procedure by an energy penalty term. The energy term in the RMC procedure has been used before [16]. Such an energy penalty reduces the probability of having 3 and 4 member rings, which are high energy structures, and captures the correct local environment or short range features of carbon atoms.

## 5. ACKNOWLEDGEMENTS

We thank Isabelle Rannou (CRMD, Université d'Orléans, France) for providing us the X-ray results, Nathalie Cohaut and Jean Michael Guet (CRMD, Université d'Orleans, France) for providing us with SAXS and density results. We also thank Dr. S. J. Stuart (Clemson University, Department of Chemistry, South Carolina, US) for providing us with the REBO code. This work was funded by the National Science Foundation, through the grant CTS-0211792. This research used supercomputing resources from HPC-NCSU, SDSC (NSF/MRAC CHE050047S) and NERSC (DOE DE-FGO2-98ER14847).

## 6. REFERENCES

1.      J. Pikunic, C. Clinard, N. Cohaut, K. E. Gubbins, J. M. Guet, R. J-M Pellenq, I. Rannou, J. N. Rouzaud, Stud. Surf. Sci. Catal., 144 (2002) 19.
2.      J. Pikunic, C. Clinard, N. Cohaut, K. E. Gubbins, J. M. Guet, R. J-M Pellenq, I. Rannou, J. N. Rouzaud, Langmuir, 19(20) (2003) 8565.
3.      S. Sircar, T. C. Golden, M. B. Rao, Carbon, 34(1), (1996) 1.
4.      T. J. Bandosz, M. J. Biggs, K. E. Gubbins, Y. Hattori, T. Liyama, K. Kaneko, J. Pikunic, K. Thomson, in: L.R. Radovic (Eds.), 28, pp. 137-199 Marcel Dekker: New York, 2003.
5.      R. L. McGreevy, and L. Pustzai, Mol. Simul., 1 (1988) 359.
6.      D. W. Brenner, Phys. Rev. B, 42(15), (1990) 9458.
7.      J. Tersoff, Phys. Rev. Lett., 61 (1988) 2879.
8.      M.D. Winn, M. Rassinger, and J. Hafner, Phys. Rev. B, 55 (1997) 5364.
9.      S. Goedecker, Rev. Mod. Phys. 71 (1999) 1085.
10.     J. Pikunic, P. Llewellyn, R. J-M. Pellenq, K. E. Gubbins, Langmuir, 21 (2005) 4431.
11.     G. Opletal, T. Petersen, B. O'Malley, I. Snook, D. G. McCulloch, N. A. Marks, I. Yarovsky, Mol. Simul., 28(10-11), (2002) 927.
12.     D. S. Franzblau, Phys. Rev. B, 44(10), (1991) 4925.
13.     S. K. Jain, J. P. Pikunic, R. J-M Pellenq and K. E. Gubbins, Adsorption, 11 (2005) 355.
14.     L. D. Gelb, and K. E. Gubbins, Langmuir, 15 (1999) 305.
15.     S. J. Stuart, A. B. Tutein and J. A. Harrison, J. Chem. Phys., 112, (2000), 6472.
16.     T. Petersen, I. Yarovsky, I. Snook, D. G. McCulloch, G. Opletal, Carbon , 41 (2003) 2403.

Studies in Surface Science and Catalysis 160
P.L. Llewellyn, F. Rodriquez-Reinoso, J. Rouqerol and N. Seaton (Editors)

# Simulation of mercury porosimetry using MRI images of porous media

**Matthew J. Watt-Smith[a], Sean P. Rigby\*,[a], J. A. Chudek[b], and Robin S. Fletcher[c]**

[a]Department of Chemical Engineering, University of Bath, Claverton Down, Bath, BA2 7AY, U.K.

[b]Division of Physical and Inorganic Chemistry, School of Life Sciences, University of Dundee, Dundee, DD1 4HN, U.K.

[c]Johnson Matthey Catalysts, P.O. Box 1, Belasis Avenue, Billingham, Cleveland, TS23 1LB, U.K.

**Abstract**
Models of the pore structure of pellets taken from a batch of sol-gel silica spheres have been constructed from magnetic resonance images of the macroscopic ($\sim$0.04-1 mm), spatial distribution of spin density and spin-spin relaxation time within the material. Simulations of mercury porosimetry on these models gave rise to good predictions for the point of deviation of the intrusion and retraction curves, and the level of mercury entrapment, in agreement with those found by experiment. This finding suggested that mercury intrusion and retraction within the pellets are determined by the macroscopic structure of the material, as detected using MRI.

## 1. INTRODUCTION

Knowledge of the mechanisms of the entrapment of non-wetting fluids within porous media is important in a number of fields of study. The mercury extrusion curve in porosimetry potentially contains useful information on the nature of a porous structure. However, in order to extract that information it is necessary to have an understanding of how the entrapment of mercury arises. Mercury porosimetry is also often used in the oil industry to evaluate reservoir rock cores. This is because the mercury recovery efficiency is expected to provide an indication of the oil recovery efficiency in a strongly water-wet system.

Previous experimental work [1] studying the mechanisms of mercury entrapment in glass micro-models has suggested that larger scale heterogeneities in the spatial distribution of pore size give rise to entrapment. If isolated regions containing large pores are located within a continuous network of smaller pores then, on commencing retraction of mercury from a fully imbibed pore space, mercury first withdraws from the smaller pores. However, at the stage where the pressure has been reduced below the threshold for emptying the clusters of larger pores, these have already become disconnected by snap-off and extensive residual mercury is retained. More recently [2-4], some experimental evidence has suggested that the same mechanism may give rise to the mercury entrapment observed in commercial mesoporous, catalyst-support pellets. For various sol-gel silica and alumina pellets it was found that mercury entrapment did not occur when the porosimetry experiment was run on fragmented samples (powder size ~30-100 μm) rather than whole pellets (size ~3-10 mm). This is what would be expected if the particle size of the fragmented material was smaller than the characteristic dimension of the isolated regions of larger pores within the materials. For transparent sol-gel silica materials, it was observed [3] that the spatial distribution of entrapped mercury within whole pellets was heterogeneous, and the isolated regions in which mercury was entrapped were macroscopic in size (0.01-1 mm).

These findings suggest that, for macroscopically heterogeneous materials, the only information that should be required to predict levels of mercury entrapment is a knowledge of the spatial distribution of porosity and pore size. This type of information can be obtained using magnetic resonance imaging (MRI) techniques [5,6], and macroscopic heterogeneities in the spatial distribution of porosity and pore size have been observed within catalyst pellets with MRI. It is the purpose of this paper to directly test the above theory of mercury entrapment within certain mesoporous catalyst support pellets by predicting the level of mercury entrapment using models of the porous media derived from MR images. MRI pore characterisation techniques are only suitable for application to model porous media which are relatively chemically homogeneous, such as pure silica and alumina pellets. However, if the theory above is shown to be correct then the form of the mercury extrusion curve and level of mercury entrapment could potentially be used to deduce information about the spatial distribution of pore sizes for relevant materials not amenable to MRI.

## 2. THEORY

### 2.1. MRI

The use of MRI to probe the structure of porous media is based upon the phenomenon that the NMR relaxation rate of fluids imbibed within pores is enhanced due to molecular interactions with the pore walls. A thin layer of liquid, typically ~1-2 molecular layers thick, in contact with the walls has restricted motion, and thus an increased relaxation rate. This surface layer is typically in fast diffusional exchange with the bulk fluid in the rest of the pore. Hence the observed relaxation rate is the volume-weighted mean value of the surface and bulk phases. This model is known as the two-fraction fast-exchange model [7] and the overall spin-spin (say) relaxation rate, $T_2$, in a pore of surface area $S$ and volume $V$ is given by:

$$\frac{1}{T_2} = \left(1 - \frac{\lambda S}{V}\right)\frac{1}{T_{2B}} + \frac{\lambda S}{V}\frac{1}{T_{2S}} \approx \frac{\lambda S}{V}\frac{1}{T_{2S}} \qquad (1)$$

where $\lambda$ is the thickness of the surface layer, the subscripts $B$ and $S$ refer to the bulk and surface layer, respectively. In general $T_{2B} \gg T_{2S}$, and hence the measured $T_2$ is directly proportional to the volume-to-surface area ratio of a pore (= radius/2 for a cylindrical pore). In order to obtain a characteristic pore dimension an assumption concerning the pore geometry must be made. However, in this work, it is only necessary to assume that $T_2$ is directly proportional to the characteristic dimension(s) of a pore that determine(s) the pressure at which mercury intrusion and extrusion occur during porosimetry. The MRI technique allows this measurement to be made spatially resolved over length-scales $>\sim0.01$ mm. More detail concerning MRI can be found elsewhere [5]. MRI, using spin-spin relaxation time pre-conditioning, has been used here to simultaneously obtain maps of the spatial distribution of both spin density (proportional to porosity) and $T_2$.

The $T_2$ maps obtained using MRI have been analysed using the fluctuation auto correlation function, $C(s)$, which measures the degree of correlation between $f(x_n)$ values at successive data points. Explicitly, defining $\delta f_n$ as:

$$\delta f_n = f(x_n) - \langle f \rangle \tag{2}$$

then:

$$C(s) = \frac{\langle \delta f_n \delta f_{n+s} \rangle}{\langle \delta f_n^2 \rangle} \tag{3}$$

where the averages denoted by the brackets $\langle\rangle$ are over the data set $\{x_n\}$. In the context of the images, $f(x_n)$ is the characteristic $T_2$ value in image pixel $x_n$. For Eq. (3) successive annular shells at a distance $s$ from each pixel are considered for each pixel in turn. The characteristic values of this function are the value of $C(s=1)$ and the value of $s$ when $C(s)=0$. The first value characterises the degree of correlation, while the second value is known as the correlation length ($\xi$) and characterises the linear extent of that correlation.

## 2.2. Simulation of mercury porosimetry

The MRI spin density and $T_2$ images were used to construct a model of the porous pellet. The structural representation consisted of a lattice site model where each site corresponded to a voxel in the MR images. The mechanisms of mercury intrusion and retraction were the same as those employed in previous work [2]. As mentioned above it was assumed that the characteristic pore dimension determining the pressure at which mercury intrudes or retracts from a given region of the sample was proportional to the value of $T_2$ in the image voxel volume corresponding to that region of the sample. In order to simulate mercury intrusion, the value of a cutoff in $T_2$, above which mercury was deemed to be able to penetrate, was gradually lowered in small steps to mimic the stepwise increase in pressure in a real experiment. The cutoff initially commenced at a value of $T_2$ above any value present in the image, and was subsequently lowered until all sites (voxels) in the model were penetrated with mercury. The intruded pore volume in any model site (voxel) was taken as being proportional to the spin density for the corresponding image voxel. In simulations of mercury retraction the value of the $T_2$ cutoff was raised steadily in small steps. Mercury was allowed to retract back from a given model site if the $T_2$ value in the corresponding image voxel was below the cutoff value, and a path existed between the site under consideration and the edge of the model which involved only 'stepping' on intermediate sites still filled with mercury. The retraction continued until the initial value of the $T_2$ cutoff was re-reached.

### 2.3. Analysis of raw, experimental mercury porosimetry data

In general, mercury porosimetry data is analysed using the Washburn equation [8]. In addition, raw porosimetry data is also generally characterised by the presence of hysteresis. This hysteresis is generally acknowledged [1] to arise from either, or both, contact angle hysteresis and structural hysteresis. Recently, building on earlier work on controlled pore glasses [9,10], Rigby and Edler [2] proposed semi-empirical alternatives to the Washburn equation for the analysis of raw mercury porosimetry data that deconvolves the contact angle component of the hysteresis from the structural. This allows the structural contribution to hysteresis to be used to characterise the pore structure of a particular porous material. For mercury intrusion the pore radius (nm) is given by:

$$r = \frac{302.533 + \sqrt{91526.216 + 1.478p}}{p}, \tag{4}$$

where $p$ is the applied pressure (in MPa), while for mercury retraction the pore radius is given by:

$$r = \frac{68.366 + \sqrt{4673.91 + 471.122p}}{p}. \tag{5}$$

Since Equations (4) and (5) are empirical in origin, then their use leads to an experimental error in the pore sizes obtained, which is estimated [10] to be ~4-5 %.

### 3. EXPERIMENTAL

The material studied in this work is a batch of sol-gel silica spheres, denoted G2, with a typical pellet diameter of ~3 mm and a BET surface area of ~99 $m^2.g^{-1}$.

### 3.1. MRI

Samples were prepared by impregnation with de-ionised water under ambient conditions for 24 h. The values of the specific pore volume obtained independently from the ultimate intruded mercury volume (below), and gravimetrically following water impregnation, were found to be identical, within experimental error. This finding suggests that mercury and water probe the same void space features. MRI experiments were carried out on a Bruker AVANCE NMR System with a static field strength of 7.05 T, yielding a resonance frequency of 300.05 MHz. All samples were placed within a 10 mm Birdcage coil. Spin-spin relaxation time ($T_2$) and spin density maps (which probe porosity) were acquired together using the Bruker sequence "*m_msme*", and employed 90° selective and 180° non-selective pulses. A $T_2$ pre-conditioned imaging sequence with an echo time of 7 ms was used. 3D images were acquired using the "*m_se3d*" sequence. Data acquisition, initial data transformation, two and three dimensional data processing, and workup was handled on an Aspect X32 workstation, running the Paravision© suite of software (Bruker Analitische Messtechnik Gmbh, Karlsruhe, Germany). The in-plane pixel resolution was 40 µm and the slice thickness was 250 µm. 3D spin-echo data was subsequently worked up using AMIRA software.

### 3.2. Mercury porosimetry

Mercury porosimetry experiments were performed using a Micromeritics Autopore IV 9420. Samples typically consisted of ~3-4 pellets. The sample was first evacuated to a pressure of 6.7 Pa in order to remove physisorbed gases from the interior of the sample. The standard equilibration time used in the experiments was 15 s. Following the experiments, the pellets were left for several days under ambient conditions, and no further mercury was observed to leave the sample. X-ray images revealed that, even several days after the experiments, mercury entrapped within the central core region of the pellets was still there (further details of x-ray imaging experiments of entrapped mercury will be described in a subsequent publication).

## 4. RESULTS AND DISCUSSION

### 4.1. Image analysis

Figure 1 shows examples of two $T_2$-contrasted images of different slices through the centre of a spherical pellet from batch G2. Separate images were obtained of each of the slices in the stack from the top to the bottom of an individual pellet sample taken from batch G2. The individual $T_2$-contrasted images of slices through pellets taken from batch G2 were analysed using the auto-correlation function. The variation of the values of $C(s=1)$ and $\xi$ with position within the pellet, for images taken of different slices through a sample of a typical pellet taken from batch G2, is shown in Figure 2. It can be seen that the values of both parameters of the correlation function peak within the central region of the pellet. As shown schematically in Figure 3, this is the result that would be expected if the distribution of $T_2$ through the pellet possessed spherical symmetry arising from a correlated structure. It is supposed that the pellet consists of a structure where higher $T_2$ values, say, are more concentrated towards the centre of the sphere and decrease in concentration in weakly-defined bands located progressively further from the centre of the pellet. If a 2D slice MR image were taken of this structure, such that the plane of the image passed through the central zone of the pellet, then, as shown in Figure 3, the image would slice through several different bands and would detect the correlation in $T_2$ values. Hence, the correlation function would show a relatively high value of $C(s)$ at shorter distances (corresponding to one pixel). However, if the plane of the 2D MR image were taken closer to the top ('pole') of the pellet, then it would slice through only one or two bands. The $T_2$ values within one particular band are envisaged to be closer together than those in different bands, and hence the image taken nearer the top of the image would be relatively more dominated by the (unavoidable) noise in the image. Hence, the correlation function would be of a form closer to that expected for a completely random arrangement of pixel intensities (a horizontal line along $C(s) = 0$). The experimental data shown in Figure 2 is consistent with this scenario. Therefore, it is proposed that the structure of pellets from batch G2 has some similarities to the type of structure shown schematically in Figure 3, and possesses some sort of spherical symmetry. Hence, it is reasonable, in this case, to use a 2D model, constructed from a single central MR image, for the structure of pellet G2.

### 4.2. Simulations of mercury porosimetry

Simulations of mercury intrusion and retraction, according to the mechanisms described above, were performed on 2D structural models constructed from spin density and $T_2$-weighted images of central slices through pellets taken from batch G2. An example of a typical set of data

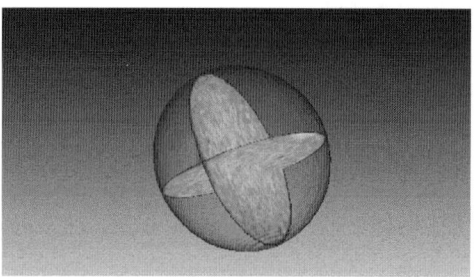

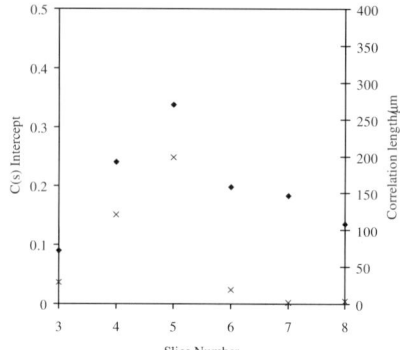

Fig. 1. $T_2$ images of perpendicular 2D slices through the centre of a pellet from batch G2. The pixel resolution is 40 μm and the slice thickness is 250 μm.

Fig. 2. Variation with position of the correlation length (◆) and degree of correlation (x) for $T_2$ maps of parallel slices of a pellet from batch G2.The central region of the pellet corresponds to slice 5.

for one particular model is shown in Figure 4. Figure 4 shows a plot of the fractional occupied volume for the extrusion curve against the fractional occupied volume for the intrusion curve at the same $T_2$ value obtained from simulations of the mercury porosimetry experiment on a model created from MR images. The choice of variables for this Figure makes it very clear when the deviation between intrusion and extrusion commences in the simulations, independent of the form of the pore size probability density function of $T_2$ values (or critical pore sizes). It can be seen that the curves separate at an occupied volume fraction of ~0.65, and the entrapment is ~40 %. Data from simulations on models constructed from different images of separate pellet samples are shown in Table 1. Previous simulations on abstract model grids [2] have shown that different spatial arrangements of pore sizes lead to different combinations of the point of deviation of the intrusion and extrusion curves, and level of entrapment. It is noted that the level of mercury entrapment for the models generated from MR images is significantly different to the value (of 50 %) expected for a completely random arrangement of pixel intensities [2].

### 4.3. Experimental mercury porosimetry data

Figure 5 shows the mercury intrusion and retraction curves, analysed using Eqs. (4) and (5), respectively, for a sample of pellets from batch G2. It can be seen that at smaller pore sizes, within the error in eqs. (4) and (5), the intrusion and extrusion curves overlay each other, while the point of separation of the intrusion and retraction curves occurs at a fractional occupied volume of ~0.67, and that the final level of mercury entrapment is ~40 %. Table 1 shows the mercury entrapment levels for several samples taken from batch G2. From Table 1, it can be seen that the experiments are repeatable, and the value of entrapped mercury agrees well with that predicted from the models derived from MR images above.

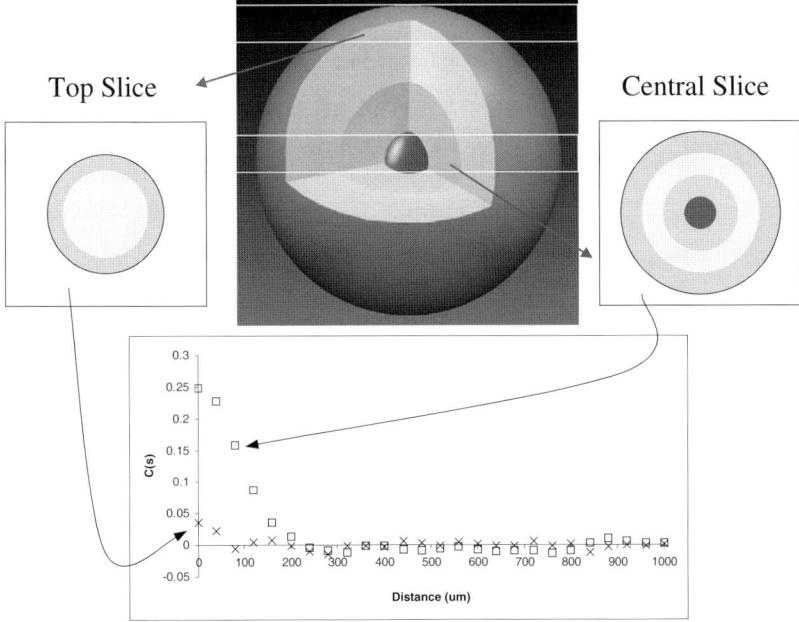

Fig. 3. Schematic diagram illustrating the expected variation of the form of the correlation function with the position of the $T_2$ image slice if the pellet possessed a spherically symmetric spatial distribution of $T_2$.

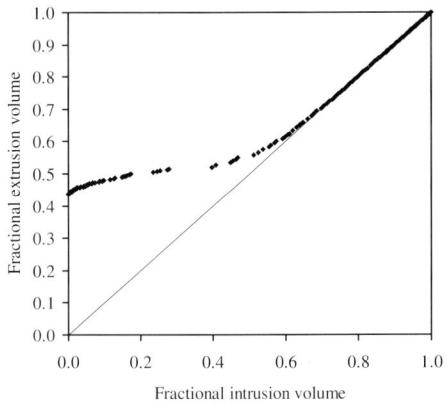

Fig. 4. Variation in fractional occupied volume on the extrusion curve against the corresponding fractional occupied volume on the intrusion curve for simulations of porosimetry on a typical model created from MR images of a pellet from batch G2.

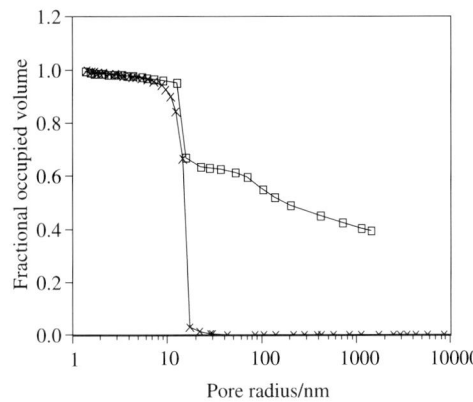

Fig. 5. Experimental mercury intrusion (x) and extrusion (□) curves, analysed using eqs. (4) and (5), for a sample from batch G2. The solid lines are to guide the eye.

Table 1
Point of separation for the mercury intrusion and extrusion curves, and level of entrapment, for both simulations on image-derived models and experimental data for batch G2. (N.B. The data shown is for several samples but image and porosimetry sample numbers do not correspond).

| Sample | Simulation | | Experiment | |
|---|---|---|---|---|
| | Entrapment (%) | Fractional occupied volume at point of separation of curves | Entrapment (%) | Fractional occupied volume at point of separation of curves |
| 1 | 43.8 | 0.70 | 41.9 | 0.66 |
| 2 | 43.9 | 0.69 | 39.4 | 0.67 |
| 3 | 45.6 | 0.71 | 42.4 | 0.71 |

## 5. CONCLUSIONS

It has been found that models for the structure of sol-gel silica spheres, constructed from spin density and spin-spin relaxation time images, give rise to good predictions for the point of separation of mercury intrusion and retraction curves, and the level of mercury entrapment, found for experimental porosimetry data for the same material. This finding suggests that the MR images contain sufficient information to determine the level of mercury entrapment. Hence, this result supports the view that mercury intrusion and retraction within this material is determined by macroscopic (0.01-1 mm) heterogeneities in the spatial distribution of porosity and pore size.

## REFERENCES

[1]    N.C. Wardlaw and M. McKellar, Powder Technol. 29 (1981) 127.
[2]    S.P. Rigby and K.J. Edler, J. Colloid Interface Sci. 250 (2002) 175.
[3]    S.P. Rigby, R.S. Fletcher, and S.N. Riley, Appl. Catal. A 247 (2003) 27.
[4]    S.P. Rigby, R.S. Fletcher, and S.N. Riley, Chem. Engng Sci. 59 (2004) 41.
[5]    M.P. Hollewand and L.F. Gladden, J. Catal. 144 (1993) 254.
[6]    S.P. Rigby and L.F. Gladden, Chem. Engng Sci. 51 (1996) 2263.
[7]    K.R. Brownstein and C.E. Tarr, J. Magn. Reson. 26 (1977) 17.
[8]    E.W. Washburn, Phys. Rev. 17 (1921) 273.
[9]    A.A. Liabastre and C. Orr, J. Colloid Interface Sci. 64 (1978) 1.
[10]   J. Kloubek, Powder Technol. 29 (1981) 63.

Studies in Surface Science and Catalysis 160
P.L. Llewellyn, F. Rodriquez-Reinoso, J. Rouqerol and N. Seaton (Editors)
© 2007 Elsevier B.V. All rights reserved

# Adsorption and microcalorimetric measurements on activated carbons prepared from Polyethylene Terephtalate

M. C. Almazán-Almazán[a], M. Domingo-García[a], I. Fernández-Morales[a], F. J. López-Garzón[a], I. Rodríguez-Ramos[b], A. Guerrero-Ruíz[b] and A. Martínez-Alonso[c].

[a]Departamento de Química Inorgánica. Facultad de Ciencias. 18071. Granada. Spain.

[b]Instituto de Catálisis y Petroleoquímica. CSIC. Madrid. Spain.

[c]Instituto Nacional del Carbon. CSIC. Oviedo. Spain.

## 1. ABSTRACT

This paper deals with the characterization of activated carbons obtained from Polyethylene Terphtalate (PET). This has been carried out by using several techniques. Among them immersion calorimetry of several organic vapours (n-hexane, benzene, cyclohexane and 2,2-DMB) and adsorption of the same vapours. Nice agreement is found between the textural characteristics determined by both techniques.

## 2. INTRODUCTION

Immersion calorimetry is a useful technique for the characterization of porous materials. The heat evolved in the immersion process is directly related to the integral enthalpy of adsorption if the experiment is carried out at constant pressure and temperature [1]. The experimental data, which are obtained by immersion measurements, are normally used to determine the textural characteristics of the adsorbent, i.e. the micropore volume or the surface areas accessible to the wetting liquid. In relation to the former Stoeckli et al [2] consider that a thermodynamic consequence of the Dubinin theory is the equation 1.

$$-\Delta H_i = \frac{\beta E_o V_o (1 + \alpha T)\sqrt{\pi}}{2 V_m} \qquad (1)$$

Where $\Delta H_i$ is the enthalpy of immersion of a microporous solid into organic liquids, $\beta$ is the affinity coefficient, $V_o$ is the micropore volume of the solid, $V_m$ is the molar volume of the liquid filling the micropore system, $\alpha$ is the thermal expansion coefficient and $E_o$ is the characteristic energy of the adsorbent. In general the experimental immersion enthalpy, $\Delta H_{i(exp)}$, includes the contribution of micropores, $\Delta H_{i(mic)}$, and of the external surface, $S_e$ [3]:

$$\Delta H_{i(exp)} = \Delta H_{i(mic)} + h_i S_e \qquad (2)$$

where $h_i$ is the specific enthalpy of immersion of an open surface. Nevertheless, for microporous carbons the external surface is usually very small and its contribution to the experimental enthalpy of immersion is negligible compared to the contribution of micropore surface. Thus the micropore volume filled by the wetting liquid, $V_o$, can be related with the experimental enthalpy of immersion by equation 1.

Denoyel et al [4] proposed a different approximation focused on the determination of the surface area of porous materials from immersion calorimetry measurements. It is based on geometrical considerations so that they conclude that the immersion enthalpy is proportional to the surface accessible to the liquid probe and that it is independent of the shape and size of the pores. The areal enthalpy of immersion, $h_i$ ($J.m^{-2}$), for a non-porous carbon black into each wetting liquid is used as reference to determine the surface area wetted by the liquid probe in a porous carbon.

The aim of this paper is to determine the textural characteristics of several activated carbons by using immersion calorimetry measurements and to relate these results with the information obtained from other experimental techniques. The textural characteristics of the adsorbents obtained from the immersion enthalpies by the application of the several approaches, already commented, are discussed.

## 3. EXPERIMENTAL

Polyethylene terephtalate (PET) is the raw material used for the preparation of the activated carbons. The experimental procedure has been already reported [5]. Two series of samples A, obtained at 1073 K, and B, prepared at 1223 K, have been used. Mercury porosimetry up to 4200 $kg\cdot cm^{-2}$ was used to determine the pore size distribution of pores wider than 3.6 nm. The measurements were carried out in a Quantachrome commercial system (Model Autoscan 60). The adsorption of nitrogen and carbon dioxide were carried out at 77 and 273 K, respectively. The former was measured in an automatic volumetric apparatus (ASAP 2010, Micromeritics) in the range of relative pressure between $10^{-6}$ and 0.99, and the latter in a conventional gravimetric system. The chemical surface groups of the samples were analysed by FTIR and selective neutralization using the method proposed by Boehm [6]. This uses bases of different strength ($NaHCO_3$, $Na_2CO_3$ and $NaOH$) to titrate the acid groups (carboxyls, lactones and phenols) and $HCl$ to titrate the basic groups. Pellets of KBr containing about 0.5 % of sample were used to obtain the transmission FTIR spectra in the range 4000-450 $cm^{-1}$.

Table 1
Physical properties of the molecular probes

|             | $d_{min}$ (nm) | $\alpha\ 10^3 (1/K)$ | $\beta$ | $V_m$(cc mol$^{-1}$) | $A\ 10^{19}$ (m$^2$) |
|-------------|----------------|---------------------|---------|----------------------|----------------------|
| benzene     | 0.37           | 1.24                | 1.00    | 88.9                 | 4.2                  |
| n-hexane    | 0.43           | 1.4                 | 1.22    | 130.5                | 4.8                  |
| cyclohexane | 0.49           | 1.0                 | 1.04    | 108.1                | 5.1                  |
| 2,2-DMB     | 0.59           | 1.4                 | 1.12    | 132.8                | 5.4                  |

$d_{min}$: minimum molecular dimension, $\alpha$ : Thermal expansion coefficient, $\beta$ : Affinity coefficient, $V_m$: Liquid molar volume, A: Molecular area

The immersion enthalpies of the four activated carbons and of a graphitised carbon black (V3G, surface area 59 $m^2g^{-1}$), which is used as reference to obtain the accessible surface area by the method proposed by Denoyel at al [4], were measured into benzene, n-hexane, cyclohexane and 2,2-dimethyl butane (2,2-DMB). The experiments were carried out at 303 K in an isothermal calorimeter of the Tian-Calvet type, Setaram C-80. Prior to the immersion experiments the samples were out-gassed at 383 K for 12 hours. A complete explanation of the experimental procedure is reported elsewhere [7,8]. The immersion enthalpies were corrected for the breaking of the bulb and the evaporation of the liquid to fill the empty volume of the bulb. The experimental procedures were repeated several times in order to have an error smaller than 5%. The physical properties of the adsorbates are collected in Table 1.

## 4. RESULTS AND DISCUSSION

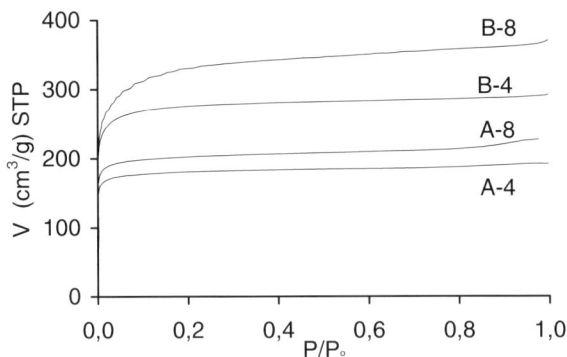

Figure 1. Nitrogen adsorption isotherms.

The nitrogen adsorption isotherms are shown in Figure 1. The four isotherms are of type I and the adsorbed amounts seems to be close in the case of samples A-4 and A-8 while those of samples B-4 and B-8 are clearly different in such a way that nitrogen is adsorbed in much larger extent on the sample obtained after 8 hours of treatment. The knee of the four isotherms is sharp although this of sample B-8 is more rounded. This suggests that samples of A-series and B-4 have homogeneous and narrow microporosity while B-8 should have a wider and more heterogeneous distribution of micropores.

The DR micropore volumes calculated from nitrogen, $V_o(N_2)$, and from carbon dioxide, $V_o(CO_2)$, are collected in Table 2. In the samples of A-series both volumes are very similar and the change with the activation time is negligible. Nevertheless, for samples of B-series $V_o(N_2)$ is clearly larger than $V_o(CO_2)$. Moreover $V_o(N_2)$ increases with activation time. This difference between $V_o(N_2)$ and $V_o(CO_2)$ is usually attributed to the existence of super-micropores i.e. micropores larger than 0.7 nm in width. While $N_2$ at 77 K is adsorbed in super-micropores and even small mesopores the upper limit of micropore size on which $CO_2$ is adsorbed at 273 K and sub-atmospheric pressure is around 1 nm [9]. The increase in $V_o(N_2)$ with the activation time is related with the widening of the micropore size and with a more heterogeneous micropore network of sample B-8 (Figure 2). The $N_2$ and $CO_2$ adsorption data together with the porosimetry measurements (Table 2) suggest that the porous network of the samples mainly consists of micropores. In fact the meso, $V_2$, and macro-porous, $V_3$, volumes determined from this experimental technique are very small compared to the micropore volumes determined from nitrogen and carbon dioxide adsorption measurements (Table 2).

Table 2
Textural characteristics of the samples

| | $N_2$ | | $CO_2$ | | | |
|---|---|---|---|---|---|---|
| | $V_o(N_2)^*$ $(cm^3 g^{-1})$ | $E_o^*$ $(kJ\ mol^{-1})$ | $V_o(CO_2)^*$ $(cm^3 g^{-1})$ | $E_o^*$ $(kJ\ mol^{-1})$ | $V_2$ $(cm^3 g^{-1})$ | $V_3$ $(cm^3 g^{-1})$ |
| A-4 | 0.29 | 27.3 | 0.25 | 31.3 | 0.02 | 0.01 |
| A-8 | 0.32 | 26.7 | 0.27 | 30.3 | 0.02 | 0.01 |
| B-4 | 0.44 | 22.4 | 0.34 | 29.5 | 0.02 | 0.01 |
| B-8 | 0.53 | 19.1 | 0.34 | 33.0 | 0.01 | 0.01 |

\* Parameters derived from the application of DR equation to $N_2$ and $CO_2$ adsorption.
$V_2$, mesopore and $V_3$, macropore volumes obtained by mercury porosimetry

The DFT calculation has been applied to the nitrogen adsorption isotherms and the results are shown in Figure 2. Both samples of A-series have narrow microporosity and almost unimodal distribution which extends up to 1nm. Activated samples of B series are different, sample B-4 has a narrow micropore distribution with a maximum at 0.6nm but also has larger micropores in the region of super-micropores, sample B-8 has a more heterogeneous and poly-modal micropore distribution which extends from 0.5 up to 2.1 nm.

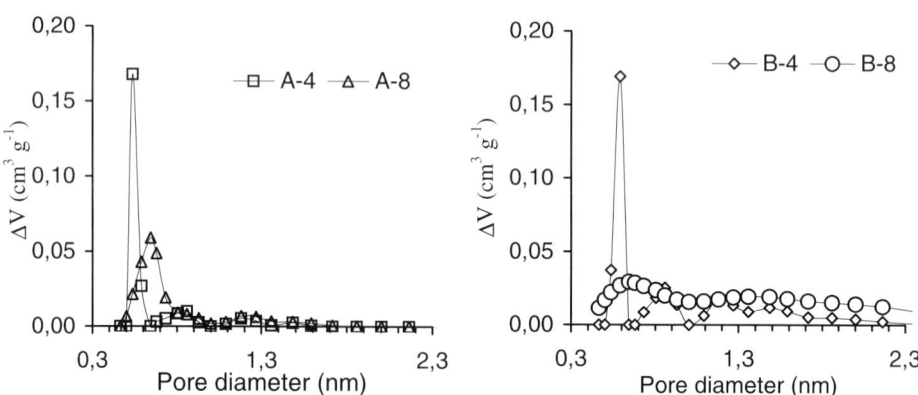

Figure 2. Micropore distributions obtained by DFT.

The results of the chemical surface groups analysis are collected in Table 3. This shows that basic groups are the most abundant on the surface of these samples, while the acid group content is small. In particular carboxyls amount is negligible in all cases. The FTIR spectra show that almost no chemical surface groups exist. This means that the basic groups (Table 3) mainly consist on $\pi$ electrons of the graphite structures [10].

Table 3
Chemical surface groups content (meq. g$^{-1}$)

|       | Carboxyls | Phenols | Lactones | Basic |
|-------|-----------|---------|----------|-------|
| A-4   | n.d       | 0.043   | 0.072    | 0.271 |
| A-8   | n.d       | 0.074   | 0.043    | 0.228 |
| B-4   | n.d       | 0.049   | 0.040    | 0.327 |
| B-8   | n.d       | 0.045   | 0.034    | 0.377 |

The immersion heats of the activated carbons and of the graphitised carbon black (V3G) in the organic liquids are collected in Table 4. The values of $\Delta H_i$ could be useful to show the trends in the molecular sieving properties of each carbon as the four molecular probes used have different shapes and sizes. Only for non-porous surfaces the values of $\Delta H_i$ depend on the molecular weight and on the polarizability of the probes and not to large extent on the molecular shape [11,12]. The values of immersion heats for B-series are in all cases, in absolute value, larger than those of A-series. The variation represents a change of 36% and 38% in B-4 and B-8, respectively. Nevertheless, the change is 79% and of 72% for samples A-4 and A-8, respectively. These data suggest that samples of B-series have more open microporous networks than the samples of A-series, in agreement with the differences in the nitrogen and carbon dioxide adsorption volumes (Table 2) and with the micropore size distributions obtained from $N_2$ adsorption measurements (Figure 2). The largest values of enthalpy are produced by the immersion in n-hexane and benzene in all samples. This is probably due in the former case to the several conformations that n-hexane can adopt. In the latter this could be related to the contribution of specific interactions to the adsorption. This could be caused because the $\pi$ electrons of the benzene aromatic ring interact with the $\pi^*$ orbital of the graphitic planes [8, 12]. Moreover the specific interactions could be produced because the basic groups on the carbon surface (Table 2) interact with the $\pi^*$ orbital of the benzene ring. Of the two possibilities the first one seems to be the most plausible if the nature of the basic groups ($\pi$ electrons), already commented, is considered.

Table 4
The enthalpies of immersion of the carbon samples in the organic liquids at 303 K

| Sample | $-\Delta H_i$ (J g$^{-1}$) | | | |
|--------|-----------|------------|-----------|---------|
|        | $C_6H_6$  | n-$C_6H_{14}$ | $C_6H_{12}$ | 2,2-DMB |
| A-4    | 91.8      | 93.4       | 34.1      | 19.0    |
| A-8    | 86.7      | 59.5       | 34.7      | 23.8    |
| B-4    | 139.2     | 127.8      | 102.1     | 91.3    |
| B-8    | 126.8     | 104.5      | 88.0      | 92.8    |
| V3G    | 6.8       | 7.1        | 6.5       | 6.3     |

It has been reported [4] that the immersion heats can be used to determine the surface area of porous materials. For this purpose it is necessary to measure the immersion heat in a non-porous adsorbent of known surface area, as V3G, which is used as reference. Thus it is accepted that the immersion enthalpy of the porous adsorbent in a liquid is simply proportional to the surface area which is accessible to the liquid whatever shape and size of the micropores are. The enthalpies obtained by the immersion of V3G into the four organic liquids are also collected in Table 4. These values and the immersion heats of the activated carbons have been used to determine the surface of the adsorbents which is accessible to the four probes. The results together with the nitrogen BET surface area, $S_{BET}$, and with the surface area obtained using the $\alpha_s$-method, $S_\alpha$, are collected in Table 5.

Table 5
Comparison of the surface areas deduced from immersion and adsorption measurements.

| Sample | $S_{BET}(m^2 g^{-1})$ | $S_\alpha (m^2 g^{-1})$ | $C_6H_6$ | $n\text{-}C_6H_{14}$ | $C_6H_{12}$ | 2,2-DMB |
|--------|--------|--------|------|------|------|------|
|        |        |        | \multicolumn{4}{c}{$S (m^2 g^{-1})$} | | | |
| A-4 | 718 | 783 | 796 | 776 | 309 | 178 |
| A-8 | 797 | 864 | 752 | 495 | 315 | 223 |
| B-4 | 1063 | 1068 | 1208 | 1062 | 927 | 855 |
| B-8 | 1219 |  | 1102 | 864 | 800 | 871 |

$S_{BET}$= from nitrogen adsorption, $S_\alpha$= from $\alpha_s$ plots, S= from immersion.

The values of $S_{BET}$ and $S_\alpha$ for samples A-4, A-8 and B-4 are very similar which means that the nitrogen adsorption is produced in micropores with an average size of two molecular dimension. Moreover, $S_{BET}$ values for these samples are also close to S of benzene. This could be expected as the molecular sizes of both adsorbates are similar and this means that the adsorption of benzene is similar to that of nitrogen. The fact that in two cases (A-4 and B-4) the surface areas measured by benzene are larger than $S_{BET}$ could be due to the partial specific interaction of benzene, already commented, or to the different temperatures at which $N_2$ and benzene are adsorbed: benzene at 303 K reaches narrow microporosity which is not accessible to $N_2$ at 77 K. The results for B-8 are rather different as $S_\alpha$ is not included in Table 5. This is because the extrapolation of the linear part of the $\alpha$-plot yields a negative value. This means that the nitrogen adsorption is produced by a secondary or cooperative micropore filling mechanism in pores of sizes which are from 2 to 5 times the adsorbate molecular dimension [13]. For this reason the $S_{BET}$ is larger than the benzene surface area. The surface areas obtained by the rest of the adsorbates are smaller (except for one case with n-hexane due to the possibility of several conformations) than these measured by $N_2$. In addition, these surface areas decrease as the molecular size increases (Table 1). This is particularly evident in A-series and means that both samples have molecular sieve effect for the adsorption of the largest molecules. Nevertheless as the molecular dimensions of these adsorbates are smaller than the mean value of the pore size distributions (Figure 2) it is plausible to think that there are constrictions at the entrance of the micropores which hinder the access of these molecules. For this reason these organic molecules can only access to a fraction of the microporosity which is smaller as larger the molecular size is. Similar results have been obtained by the adsorption of these adsorbates, already reported [5].

Nitrogen and carbon dioxide adsorption and porosimetry measurements have shown that the activated carbons used in this work are essentially microporous with very low values of meso, $V_2$, and macropores, $V_3$ (Table 2). In addition the external surface, $S_{ext}$, is negligible (between 12 and 28 $m^2 g^{-1}$) compared to the micropores surface. For this reason the contribution of the external surface to the immersion enthalpy is also negligible compared to this of the micropores. Under this circumstance Stoeckli at al have proposed equation 1 (see introduction section) which relates the immersion enthalpy, $\Delta H_i$, of a solid in a liquid with the micropore volume of the solid, $V_o$. This equation points out that the immersion enthalpy depends on both the micropore volume and the characteristic energy, which means that $\Delta H_i$ is related to the micropore size distribution [2]. Then if $E_o$ is obtained from adsorption data and $\Delta H_i$ is known from immersion measurements, the volume of micropores, $V_o$, can be determined by using that equation. $E_o$ is usually determined from the adsorption data of small molecules which can access to most of the porosity. Nitrogen and benzene values of $E_o$ can be used as references. The $V_o$ values obtained using nitrogen as reference are collected in Table 6. Similar results are obtained if benzene is used.

Table 6
Micropore volume obtained by using equation 1

| | $V_o$ (cm$^3$ g$^{-1}$) | | | |
| --- | --- | --- | --- | --- |
| Sample | $C_6H_6$ | n- $C_6H_{14}$ | $C_6H_{12}$ | 2,2-DMB |
| A-4 | 0.25 | 0.29 | 0.12 | 0.06 |
| A-8 | 0.24 | 0.19 | 0.12 | 0.08 |
| B-4 | 0.45 | 0.48 | 0.41 | 0.38 |
| B-8 | 0.48 | 0.46 | 0.43 | 0.45 |

The trend in $V_o$ is similar to that found in the surface values in Table 5, as it is seen a general decrease as the adsorbate molecular size increases. This means that the samples have restrictions which partially hinder the access of the adsorbates to the porosity, in particular for the largest molecules in A-series. In general benzene and n-hexane can "see" similar micropore volume than nitrogen (Table 2) while cyclohexane and 2,2-DMB only reach 40 and 23% of the nitrogen micropore volume in A-4, and 50 and 35% in A-8. Again the micropore distribution of these samples (Figure 2) does not support the molecular sieve behaviour as the molecular dimensions (Table 1) are smaller than the micropore width. As already commented, the existence of some constrictions in the micropores should be accepted to explain this behaviour. 2,2-DMB occupies 60 and 73 % of the microporosity in B-series which is probably due to a lack of packing.

It is useful to compare the values of micropore volumes, $V_o$, obtained by equation 1 with those determined by adsorption experiments. These are plotted in Figure 3 which shows a good agreement between the micropore volume measured by the two methods. The trend is the same if nitrogen is used to determine $E_o$ or if benzene is chosen. This supports the use of equation 1 to determine micropore volumes. In spite of this it should be mentioned that, although the physical meaning of Denoyel and Stoeckli approximations are different, the results of both methods in terms of the accessibility or molecular sieve effects are similar.

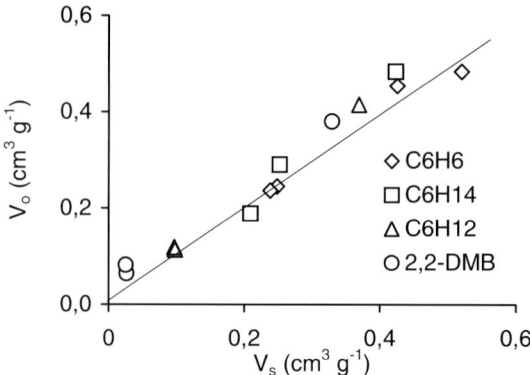

Figure 3. Micropore volumes obtained by adsorption and from equation 1.

## CONCLUSIONS

The results of the characterization of the activated carbons by immersion calorimetry are coincident with those obtained by vapours adsorption. The variation of the surface areas and of the micropore volumes obtained from calorimetry data clearly reflects the characteristics of the adsorbents.

## ACKNOWLEDGEMENTS

This work has been supported by Ministerio de Educación y Ciencia (Projects CTQ2004-07698-CO2-01/PPQ and BQU2001-2936-C02-02), Ministerio de Medio Ambiente (expediente 033/2004/3) and Junta de Andalucía. M. C. A. A. also acknowledges the financial support of Junta de Andalucía.

## REFERENCES

[1]. K.S.W. Sing, Advances in Colloid and Interface Science, 76 (1998) 3.
[2]. H. F. Stoeckli and F. Kraehenbuehl, Carbon 19 (1981) 353.
[3]. H. F. Stoeckli and F. Kraehenbuehl, Carbon 22 (1984) 297.
[4]. R. Denoyel, J. Fernández-Colinas, Y. Grillet and J. Rouquerol, Langmuir 9 (1993) 515.
[5]. I. Fernández-Morales, M. C. Almazán-Almazán, M. Pérez-Mendoza, M. Domingo-García and F. J. López-Garzón, Microporous and Mesoporous Materials 80 (2005) 107.
[6]. H.P. Boehm, Carbon 32 (1994) 759.
[7]. M.T. González, A. Sepúlveda-Escribano, M. Molina-Sabio and F. Rodríguez-Reinoso, Langmuir 11 (1995) 2151.
[8]. E. Castillejos-López, D.M. Nevskaia, V. Muñoz, I. Rodríguez-Ramos and A. Guerrero-Ruíz, Langmuir 20 (2004) 1013.
[9]. A. Vishnyakov, P.I. Ravikovitch and A.V. Neimark, Langmuir 15 (1999) 8736.
[10]. F. Rodríguez-Reinoso, Carbon 36 (1998) 159.
[11]. D. Atkinson, A. I. McLeod and K.S.W. Sing, Carbon 20 (1987) 319.
[12]. M. Domingo-García, F.J. López-Garzón, C. Moreno-Castilla and M. Pyda, J. of Phys. Chem. B, 101 (1997) 8191.
[13]. N. Setoyama, F. Suzuki, K. Kaneko, Carbon 36 (1998) 1459.

Studies in Surface Science and Catalysis 160
P.L. Llewellyn, F. Rodriquez-Reinoso, J. Rouqerol and N. Seaton (Editors)

# Compressing some sol-gel materials reduces their stiffness: a textural analysis

C.J. Gommes, N. Job, S. Blacher and J.-P. Pirard

Université de Liège, Laboratoire de Génie Chimique B6a, B-4000 Liège, Belgium.

The mechanical behaviour of two series of silica and of resorcinol xerogels is analyzed by mercury porosimetry. The data are expressed as pressure-density curves, which enables textural information to be obtained. In particular, it is shown that some of the analyzed samples exhibit a marked lowering of their mechanical stiffness upon compression. This observation is analyzed in terms of the collapse of the sample's porosity and of the heterogeneity of the microstructure.

## 1. INTRODUCTION

The general problem of textural analysis is to extract information about the microscopic structure of a material out of a macroscopic measurement. Many methods exist to characterize the microstructure of porous materials, for instance adsorption, small angle scattering, mercury porosimetry, thermoporometry, and so forth [1]. On the other hand, some measurements aim at assessing directly a useful property of the analyzed materials. This is usually the case for mechanical or electrical measurements. Textural measurements and useful property measurements can however be analyzed at once using appropriate models to link the macroscopic property to the microscopic structure.

Mechanical properties are sometimes measured to obtain textural information. For example, rheological measurements are common to gain some insight into the texture of polymer solutions and gels [2]. In principle, the same type of method could be applied to characterize the texture of dry porous materials. A convenient way to do this is to use mercury porosimetry, which is generally used to assess the pore size distribution of porous materials. However, the samples must be strong enough to withstand high mercury pressure without deforming. When it is applied to fragile materials, such as those synthesized by sol-gel process, the samples are crushed by mercury rather than intruded [3, 4]. As a consequence, the information obtained by this technique is purely mechanical. It conveys some information on the way in which the microstructure of the material resists an isostatic pressure. Although the absence of mercury intrusion renders the usual models useless for interpreting the measurements, the collected data are not devoid of textural relevance. Many models exist that relate the mechanical strength of porous materials to their microstructure [e.g. 5-7], and some have been specifically developed to analyze the compaction of porous materials in mercury porosimeters [4].

In the present study, the pressure-density curves of two series of silica and of organic xerogels, obtained by mercury porosimetry, are analyzed at the light of their microstructure. A special emphasis is given to the fact that the stiffness of some of these materials decreases significantly when the samples are compressed, which can also find a textural interpretation.

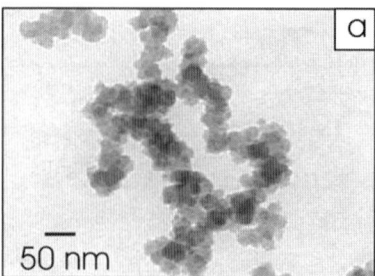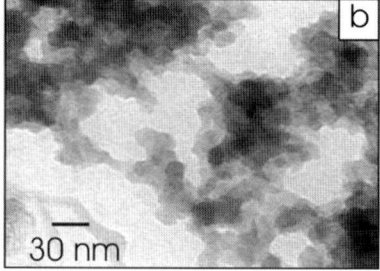

Figure 1. Typical transmission electron micrographs of fragments of samples ET2.5 (a) and of ET20 (b).

## 2. EXPERIMENTAL SECTION

### 2.1 Analyzed samples

A series of phase-separated silica xerogel samples was synthesized using tetraethoxysilane (TEOS) as main silica precursor and 3-(2-aminoethyleamino)propyltri-methoxysilane (EDAS) as additive, as described elsewhere [8]. **The ratios of water/silica precursors and of ethanol/silica precursors are the same for all samples,** and only the relative amount of TEOS and EDAS was modified. After gelation, the samples were aged for one week at 60°C, and afterwards dried under vacuum. The samples are hereafter labelled ETX where X is the molar ratio EDAS/TEOS in %.

The preparation of resorcinol-formaldehyde xerogels was performed according to a previous study [9]. The **resorcinol/formaldehyde and water/(resorcinol + formaldehyde) ratios** were fixed at 0.5 and 5.7 respectively. The pH of the starting solution was adjusted to the desired value with NaOH. After gelation and ageing at 85°C (72h) the samples were dried under vacuum. These samples are labelled RFX, where X is pH of the solution.

Typical transmission electron micrographs of fragments of the first and last samples of each series are displayed Figures 1 and 2. These micrographs were obtained by crushing the samples into a very fine powder. The global microstructure of the analyzed materials therefore corresponds to a random piling of such fragments. The overall appearance of the samples is typical of many sol-gel materials, *i.e.* a sponge-like structure supported by columns. Globally, the microstructure of these materials becomes denser throughout both series.

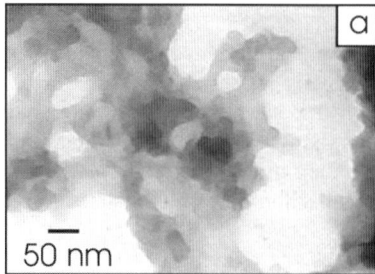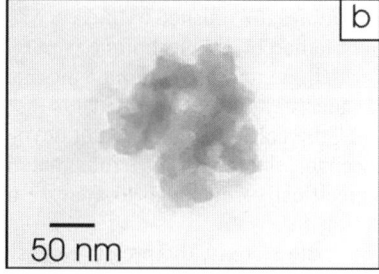

Figure 2. Typical transmission electron micrographs of fragments of samples RF-5.45 (a) and of RF-7.35 (b).

Table 1. Textural characteristics of the samples.

|  | $\rho_b$ (g/cm³) | $\rho_s$ (g/cm³) | $S_{BET}$ (m²/g) | $V$ (cm³/g) | $\phi$ (-) | $l_P$ (nm) | $l_S$ (nm) |
|---|---|---|---|---|---|---|---|
| ET2.5 | 0.30 | 2.65 | 165 | **3.33** | 0.11 | 71.7 | 9.1 |
| ET4 | 0.32 | 2.13 | 179 | **3.13** | 0.15 | 59.3 | 10.5 |
| ET6 | 0.34 | 2.10 | 215 | **2.94** | 0.16 | 45.9 | 8.9 |
| ET10 | 0.37 | 2.04 | 243 | **2.70** | 0.18 | 36.4 | 8.1 |
| ET15 | 0.39 | 1.97 | 256 | **2.56** | 0.20 | 32.1 | 7.9 |
| ET20 | 0.50 | 1.91 | 327 | **2.00** | 0.26 | 18.1 | 6.4 |
| RF5.8 | 0.63 | 1.54 | 435 | **1.59** | 0.41 | 8.6 | 6.0 |
| RF6 | 0.75 | 1.53 | 475 | **1.33** | 0.49 | 5.7 | 5.5 |
| RF6.25 | 0.85 | 1.52 | 505 | **1.18** | 0.56 | 4.1 | 5.2 |
| RF6.5 | 1.01 | 1.53 | 510 | **0.99** | 0.66 | 2.6 | 5.1 |
| RF7.35 | 1.19 | 1.53 | 470 | **0.84** | 0.78 | 1.6 | 5.6 |

**The bulk and skeletal densities estimated from mercury and helium pycnometry,** $\rho_b$ **and** $\rho_s$**,** and specific surface area **estimated from nitrogen adsorption,** $S_{BET}$, of these samples are reported in Table 1, and were already published elsewhere [9, 10]. The specific volume is estimated as $V = 1/\rho_b$, and the volume density as $\phi = \rho_b/\rho_s$. The pore and solid chord lengths $l_P$ and $l_S$, were estimated from these textural data as discussed previously [10]. The meaning of these two parameters is as follows. Let a random straight line be drawn across the porous material, the average length of intersection of the line with the pore or solid phase would be the corresponding chord length.

### 2.2 Mercury porosimetry measurements

Mercury porosimetry was performed on monolithic samples outgassed down to 0.01 Pa for at least 2 h at room temperature. The samples were transferred to a Carlo Erba Pascal 140 porosimeter on which the mercury pressure was raised from *ca* 0.01 MPa to 0.1 MPa, and afterwards to a Carlo Erba Pascal 240 porosimeter on which the mercury pressure is further raised to 200 MPa. A blank curve was subtracted from the raw data to correct for the compressibility of mercury.

**Whenever mercury intrusion occurs in a silica sample, this is manifest through a change of colour of the sample being measured, as it becomes greyish. For both silica and organic samples, intrusion is accompanied by a slight increase of the weight of the sample. For most of the analysed samples, however, no mercury intrusion occurs [3, 4]. The monoliths of these samples shrink uniformly during the measurement and their shape remains almost unchanged.**

### 3. RESULTS

The mercury porosimetry data are displayed in Figure 3. The curves of samples ET2.5 to ET10 exhibit a two stage phenomenon, already reported for other low density materials [11]: at low pressure the samples are compressed but not intruded by mercury, and mercury enters the remaining pores of the crushed material only above a given pressure threshold. The change of mechanism corresponds to the riser in the curves, also marked by an arrow on Figure 3. For the samples synthesized with the largest amount of EDAS and for all the RF samples, no intrusion occurs, and only the compression is observed. It can also be noted that

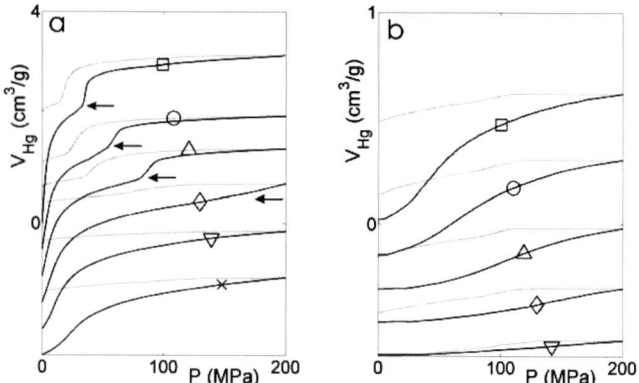

Figure 3. Mercury porosimetry curves measured (a) on silica xerogels: ET2.5 (□); ET4 (○), ET6 (Δ), ET10 (◊), ET15 (∇), and ET20 (×), and (b) of resorcinol xerogels: RF5.8 (□); RF6 (○), RF6.25 (Δ), RF6.5 (◊), RF7.35 (∇). The full and dotted lines are the pressure increase and decrease respectively. The curves are arbitrarily shifted vertically. In Figure a, the arrows highlight the onset of mercury intrusion.

intrusion is reversible as the riser is also observed during the decrease of pressure, but the compression is mainly irreversible.

In the present paper, we shall only focus on the compression of the samples. In this context the data are conveniently expressed as the pressure needed to compress the samples to reach a given volume density $\phi$. At any given pressure, the density $\phi(P)$ is estimated as $1/(\rho_s V(P))$, where the specific volume of the skeleton has been estimated as $1/\rho_s$, and $V(P)$ is the specific volume of the material at pressure $P$. Since no mercury intrusion occurs, $V(P)$ is derived from the mercury porosimetry data as $V(P)=1/\rho_b-V_{Hg}(P)$. The $P$ vs. $\phi$ curves obtained from the data of Figure 3, and from the densities of Table1, are plotted in Figure 4 in logarithmic scales. For all samples, when the pressure is initially increased, the volume fraction remains first unchanged, and compaction only begins when a given yield pressure is reached. For both series, the yield pressure increases with the initial volume fraction of the material.

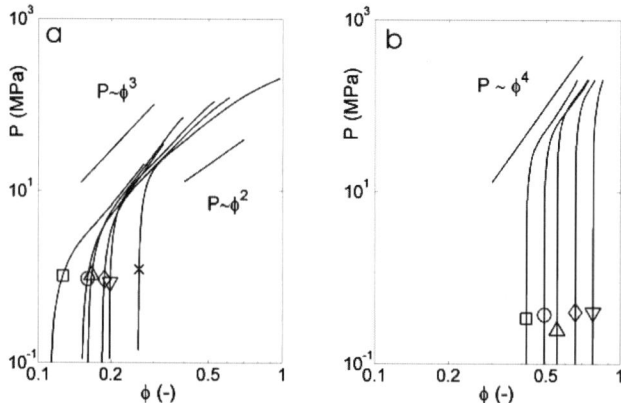

Figure 4. Pressure density curves of (a) on silica xerogels: ET2.5 (□); ET4 (○), ET6 (Δ), ET10 (◊) ET15 (∇), and ET20 (×), and (b) of resorcinol xerogels: RF5.8 (□); RF6 (○), RF6.25 (Δ), RF6.5 (◊) RF7.35 (∇).

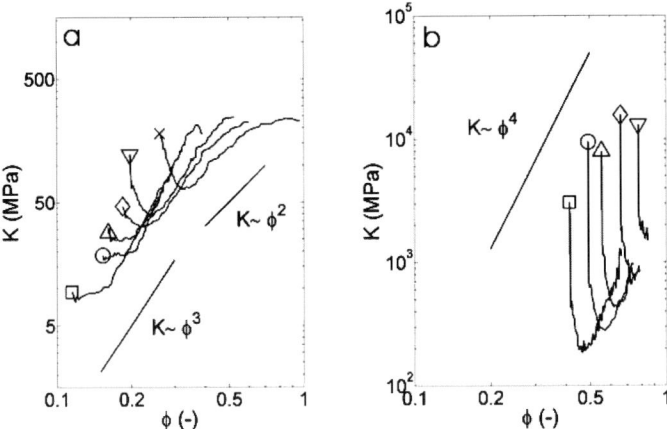

Figure 5. Pressure density curves of (a) on silica xerogels: ET2.5 (□); ET4 (○), ET6 (Δ), ET10 (◊), ET15 (∇), and ET20 (×), and (b) of resorcinol xerogels: RF5.8 (□); RF6 (○), RF6.25 (Δ), RF6.5 (◊), RF7.35 (∇).

**At pressures higher than the yield, the compaction of the materials follows a power law of the form $P \sim \phi^{\alpha}$, corresponding to the scaling lines plotted in Figure 4. Sol-gel materials are generally characterized by exponents $\alpha$ between 3 and 4 [*e.g.* 3]. It is not totally clear yet, however, whether the value of the exponent is typical of the morphology of the materials being compressed [5], or of the mechanism of deformation of a material with a given morphology [7].**

In order to analyze in more details the behaviour of the samples near the yield pressure, it is useful to estimate their compression modulus, defined as $K = \phi \, dP/d\phi$. This quantity is plotted in Figure 5 for all samples. For samples ET2.5 to ET6, $K$ exhibits a plateau at the beginning of the compression and $K$ increases continuously afterwards according to the same power law as $P$ (see Figure 4). An interesting feature that is shared by all the other samples is that the beginning of compression is accompanied by a marked decrease of $K$, which means that these samples initially become softer upon compression.

## 4. DISCUSSION

A first observation that can be made from Figures 4 and 5 is that the compaction of RF xerogels and that of silica xerogels ET15 and ET20 does not resemble that of an aggregate of colloidal particles [*e.g.* 7, 12]. For such a structure, the mechanical strength of the links between particles is weaker than that of the particles themselves. At low pressure, the compaction of colloidal aggregates is therefore expected to occur mainly via a reorganization of the particles, with destruction of the links between particles. This mechanism can only last until the density approaches that of a dense packing ($\phi = 0.64$ for monodisperse spheres). Further compaction can only occur through the destruction of the particles themselves, which should only occur at a significantly higher pressure. This would lead to a riser in the $P$ *vs.* $\phi$ curves near $\phi = 0.64$. No such riser is detected for any of the studied samples (see Figure 4). In the case of ET20 for instance, $P$ increases smoothly until very high value of $\phi$ are reached (Figure 4a).

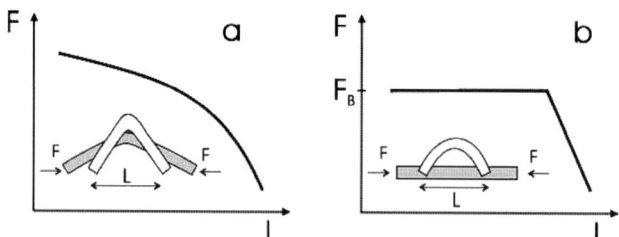

Figure 6. Deformation of the struts that support the porous network of sol-gel materials, according (a) to Gross *et al.* [12] and (b) to Pirard *et al.* [4], and corresponding force-length relations *F(L)*.

From TEM observations, the ET2.5 to ET6 seem to be particulate, as is also visible in Figure 1a. For these samples however, mercury intrusion does occur at high pressure. Since the compaction of samples ET2.5 to ET6 ceases before high values of $\phi$ are reached, no conclusion on the particulate nature of these materials can be drawn from the present analysis.

Another striking observation is the lowering of $K$ for some samples when they are compressed (see Figure 5). This phenomenon has been known for long for similar systems. For instance Gross *et al.* observed that the velocity of sound propagation in some sol-gel materials decreases when they are compressed [13]. The explanation given by these authors for the phenomenon is summarized in Figure 6a. The microstructure of many sol-gel materials is a sponge-like structure, supported by elongated struts similar to those visible in Figure 1a. Because the struts often have knee-like defects, they are expected to become progressively weaker when they are compressed. The force-length relation $F(L)$ of the struts becomes progressively flatter when $L$ is decreased or $F$ increased. Therefore, the macroscopic lowering of $K$ may be directly related to the lowering of the microscopic stiffness $k = dF/dL$ of the individual struts.

It can be noted that the knee model of the struts does predict their collapse when the force is increased. Indeed, the change in length $dL$, resulting from an increase of the force $dF$, increases very rapidly for increasing values of $F$ (see Figure 6a). In another model of the struts, Pirard *et al.* [4, 11] neglects the influence of the knees and suggests that the collapse of the pores be related to the buckling of the struts (Figure 6b). This geometrical approximation leads to a well defined value for the force at which buckling occurs $F_B$. According to Euler's law, the critical force at which buckling occurs is related to the length $L$ of the beam by $F_B \sim 1/L^2$ [14]. One therefore expects that the pressure at which the porosity of a material collapses be inversely related to the size of its pores. This general conclusion is also expected to hold qualitatively for the knee model. In the ambit of Pirard's buckling model, the quantitative relation is $P \sim 1/L^4$ [4, 11]. From the curves of Figure 4, it is clear that the yield pressure of the analyzed materials increases continuously from ET2.5 to ET20 and from RF5.8 to RF7.35, which is in qualitative agreement with the values of $l_P$ reported in Table 1.

In addition to the mechanism suggested by Gross *et al.* [13], the very collapse of the pores can explain the lowering of $K$ of the ET samples upon compression. The change in density $d\phi$ of a sample upon increasing the pressure by a small amount $dP$ can be written as:

$$d\phi = d\phi_E + d\phi_C \tag{1}$$

where $d\phi_E$ is the elastic contribution and $d\phi_C$ is the contribution from the collapse of pores.

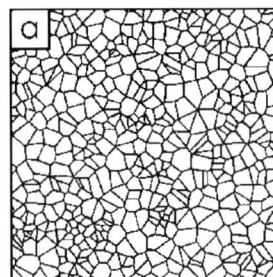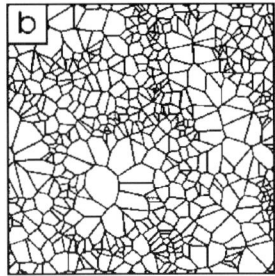

Figure 7. Pictorial examples of materials whose porosity is supported by elongated struts and whose microstructure is (a) homogeneous or (b) heterogeneous at a scale much larger than the struts.

Recalling the definition of the compression modulus, $K = \phi \, dP/d\phi$, Eq. (1) can be written equivalently as:

$$\frac{1}{K} = \frac{1}{K_E} + \frac{1}{K_C} \tag{2}$$

where the individual compression moduli were defined as $K_{E/C} = \phi \, (dP/d\phi)_{E/C}$. The compaction of a material which microstructure would be similar to that of Figure 7a, is expected to be the following. Since the length distribution of the struts is narrow, all the pores collapse at the same pressure, say $P_B$. For $P < P_B$, the struts are deformed elastically, and according to Eq. (2), ones has $K = K_E$. When $P$ approaches $P_B$, the largest pores begin to collapse. Since the volume loss resulting from the collapse of a pore is much larger than any possible elastic deformation of the same pore, collapse can only lead to a lowering of $K$ near $P = P_B$. For $P > P_B$, there are no pores left to collapse and the crowding of the microstructures is such that $K$ increases again. This succession of events could easily explain the shape of the $K$ vs. $\phi$ curve of sample ET20 (Figure 5). Furthermore, the fact that the minimum of $K$ shifts towards lower values of $\phi$ and of $P$ from ET20 to ET2.5 compares well with the size of the pores, $l_P$, in these samples (Table 1).

The collapse of the porosity cannot be responsible for the lowering of $K$ in the RF samples, as for these samples it is initially not accompanied by any significant increase of $\phi$ (Figure 5b). In this case, the origin of the lowering of $K$ lies in the heterogeneity of the samples. In a sample such as in Figure 7b, the largest struts that would buckle at low pressure are surrounded by denser regions of shorter struts. When the pressure exerted on the material increases, so does the load supported by each individual strut. The largest struts begin to buckle when their yield load is reached, but the collapse of the pores is prevented by the surrounding smaller pores. Since the load supported by a buckled strut remains constant and it cannot exceed $F_B$ (see Figure 6b), any subsequent increase in the pressure concentrates the additional load in the not yet buckled shorter struts. The net effect of an increase in pressure is therefore to share the total load among a smaller number of structures without significantly reducing the volume, by which the macroscopic stiffness can only decrease. It should be stressed that a qualitatively similar conclusion could be reached with the knee model of the struts. The collapse of the porosity only occurs at pressures sufficiently high for the network of buckled structures to percolate throughout the macroscopic solid.

## 5. CONCLUSIONS

The texture of two series of silica and organic xerogels was analyzed by mercury porosimetry. The samples are mainly compressed but not intruded by mercury in the porosimeter, and the information obtained is therefore purely mechanical. A textural information can however be extracted from the data by analyzing the way in which a given microstructure should resist a compressive stress.

In the case of the analyzed silica xerogels, the lowering of the stiffness when some samples are compressed can be interpreted as the collapse of a homogeneous porosity. In the case of the organic xerogels, the lowering of $K$ stems from the buckling of the microstructures of the materials without any significant collapse, which is likely to occur only for heterogeneous samples.

**Acknowledgements** It is a pleasure for the authors to acknowledge stimulating discussions with Dr. René Pirard of the University of Liège. C.J. Gommes is grateful to the National Fund for Scientific Research (FNRS, Belgium) for a Ph.D. research fellow position. The authors thank the Ministère de la Région Wallonne (DGTRE) and the Ministère de la Communauté Française de Belgique (Action de recherche concertée 00/05-265) for their financial support.

## REFERENCES

[1] K. Kaneko, J. Membrane Sci. 96 (1994) 59.
[2] H.A. Barnes, J.F. Hutton, K. Walters, An Introduction to Rheology, Elsevier: Amsterdam, 1989, Chap 6.
[3] G.W. Scherer, D.M. Smith, X. Qiu, J.M. Anderson, J. Non-Cryst Solids 194 (1996) 283.
[4] R. Pirard, S. Blacher, F. Brouers, J.-P. Pirard, J. Mater. Res. 10 (1995) 2114.
[5] A.P. Roberts, E.J. Garboczi, J. Mech. Phys. Solids 50 (2002) 33.
[6] A.A. Potanin, W.B. Russel, Phys. Rev. E 53 (1996) 3702.
[7] R. Botet, B. Cabane, Phys. Rev. E 70 (2004) 031403.
[8] C. Gommes, S. Blacher, B. Goderis, R. Pirard, B. Heinrichs, C. Alié, J.-P. Pirard, J. Phys. Chem. B 108 (2004) 8983.
[9] N. Job, R. Pirard, J. Marien, J.-P. Pirard, Carbon, 42 (2004) 619.
[10] C.J. Gommes, S. Blacher, M. Basiura, B. Goderis, J.-P. Pirard, Chem. Mat. submitted for publication.
[11] R. Pirard, C. Alié, J;-P. Pirard, Handbook of Sol-Gel Technology, In Characterization of Sol-Gel Materials and Products, Sakka, S. Ed.; Kluwer: Boston, 2005; Vol 2, pp 211-33.
[12] G.M. Channel, K.T. Miller, C.F. Zukoski, AIChE J. 46 (2000) 72.
[13] J. Gross, J. Fricke, R.W. Pekala, L.W. Hrubesh, Phys. Rev. B 45 (1992) 12774.
[14] L.D. Landau, E.M. Lifshiftz, Theory of Elasticity, Butterworth-Heinemann, 1986.

Studies in Surface Science and Catalysis 160
P.L. Llewellyn, F. Rodriquez-Reinoso, J. Rouqerol and N. Seaton (Editors)

# Characterisation of new Pd / hierarchical macro-mesoporous ZrO$_2$, TiO$_2$ and ZrO$_2$-TiO$_2$ catalysts for toluene total oxidation

H. L. Tidahy[1], S. Siffert*[1], J.-F. Lamonier[1], E.A. Zhilinskaya[1], A. Aboukaïs[1],
Z. Y. Yuan[2], A. Vantomme[2] and B.-L. Su[2], X. Canet [3], G. Deweireld [3], M. Frère [3]

[1]Laboratoire de Catalyse et Environnement E.A. 2598, Université du Littoral Côte d'Opale, 145 avenue Schumann, 59140 Dunkerque (France)
*Tel : (33) 03 28 65 82 68/56 ; siffert@univ-littoral.fr
[2]Laboratoire de Chimie des Matériaux Inorganiques (CMI), Université de Namur, (Belgium)
[3] Laboratoire de Thermodynamique Physique, Faculté polytechnique de Mons (Belgium)

Amorphous (macro)-mesoporous ZrO$_2$, TiO$_2$ and ZrO$_2$-TiO$_2$ have been synthesised and characterised for volatile organic compounds (VOCs) catalytic oxidation. These solids used as catalytic supports present high surface areas. The stability of the porous structure after the calcination at 400°C is observed for ZrO$_2$-TiO$_2$ and TiO$_2$ but partial breakdown of the structure occurred for ZrO$_2$ by the crystallisation to tetragonal phase. All these Pd impregnated solids are found to be powerful catalysts for total oxidation of toluene. Pd/TiO$_2$ presents the highest catalytic potential. The lowest toluene adsorption enthalpy, the low coke content observed after the catalytic test and the highest pore diameter leading to palladium particles more accessible and reducible should explain the interesting catalytic behaviour of Pd/TiO$_2$.

## 1. INTRODUCTION

Volatile Organic Compounds (VOCs) emitted from industrial processes and automobile exhaust emissions represent a serious environmental problem. An effective way of VOC removal is complete catalytic oxidation to harmless products such as H$_2$O and CO$_2$. Noble-metal-based (e.g. Pt and Pd) catalysts show the highest activity for the oxidation of volatile organic compounds (VOCs) [1-3]. However, it was found that the support nature plays an important role in improvement of the efficiency of the catalyst, particularly in oxidation reaction. Indeed, recently many nanostructured mesoporous oxides with high surface area and uniform pore size distribution were used as support for multiple catalyst applications [4,5]. For example, nanostructured mesoporous zirconia and titania supports for vanadia and gold catalysts have provided excellent catalytic properties for complete benzene oxidation [4]. Practical approaches require the preparation of mesoporous materials having hierarchical porous structures at different length scales in order to achieve highly organized functions [6], since hierarchical materials with multiple-scaled porosity can be expected to combine reduced resistance to diffusion and high surface areas for yielding improved overall reaction and adsorption/separation performances. We have thus synthesised hierarchical bimodal mesoporous-macroporous zirconium oxide, titanium oxide and binary zirconium and titanium oxide by a simple method in the presence of a single surfactant [7,8]. These new supports are then impregnated with palladium and characterised for VOCs oxidation.

## 2. SYNTHESIS AND CHARACTERISATIONS

### 2.1. Mesoporous materials preparation

The preparations of the meso-macroporous materials are processed by the method described elsewhere [6-8].

The first step to the preparation of the mesoporous titania oxide samples is adding dropwise 0.05 moles titanium ethoxide in 60ml aqueous acidic solution (pH = 2). After further stirring (100rpm) for 1h, the mixture is transferred into a teflon-lined autoclave, and heated at 60°C for 48 hours. The final samples are dried at 60°C in vacuum.

For zirconia preparation 15wt% micellar solution of cetyltrimethylammonium bromide (CTMABr) is prepared by dissolving CTMABr in an aqueous acidic solution (pH = 2) at 40°C under stirring for at least 3h. A appropriated content of zirconium propoxide ($Zr(OC_3H_7)_4$) is added dropwise into the above solution with a surfactant/Zr molar ratio of 0.33. After further stirring for 1h, the mixture is transferred into a Teflon-lined autoclave, and heated at 60°C for two days. The product is filtered by Soxhlet extraction with ethanol for at least 30h in order to remove the surfactant species, and then dried at 60°C in vacuum. The binary zirconium and titanium oxide is synthesised by the same way using a mixture of zirconium propoxide and titanium isopropoxide with 50/50 metal to metal molar ratio.
The samples are calcined in air at 400°C for 4h.

### 2.2. Catalysts preparation

Pd supported by zirconia catalysts are prepared by aqueous impregnating method using palladium nitrate (Pd $(NO_3)_2·XH_2O$, Johnson Matthey). The impregnated powders are dried at 100°C overnight and calcined in air at 400°C for 4h.
$Pd/ZrO_2$, $Pd/TiO_2$ and $Pd/ZrO_2-TiO_2$ catalysts are prepared by impregnating palladium with 0.5wt% Pd content on calcined mesoporous supports.

### 2.3. Characterization techniques

The structures of solids are analysed by powder X-ray diffraction (XRD) technique at room temperature with a Bruker diffractometer using Cu-$K_\alpha$ radiation scanning 2θ angles ranging from 10 to 80°.

The nitrogen adsorption analysis is performed on Sorptomatic 1990 apparatus at -196 °C and the specific surface areas of the solids are determined by BET method.

The Scanning electron microscopy (SEM) is carried out on a Philips XL-20 microscope, and the transmission electron microscopy (TEM) is performed on a Philips Tecnai-10 microscope at 100 kV. The specimens for TEM observation are prepared by embedding the samples in epoxy resin and ultramicrotoming, and mounting on a copper grid.

Thermal analysis measurements are performed using a Netzsch STA 409 equipped with a microbalance differential analysis (DTA) and a flow gas system. The dried catalysts are treated under air; the temperature is raised at a rate of 5°C.min$^{-1}$ from room temperature to 650°C.

The temperature programmed reduction experiments are carried out in an Altamira AMI-200 apparatus. The TPR profiles are obtained by passing a 5% $H_2$/Ar flow (30 ml.min$^{-1}$) through the calcined sample (about 100 mg). The temperature is increased from -40 to 300°C at a rate of 5 °C.min$^{-1}$. The hydrogen concentration in the effluent is continuously monitored by a thermoconductivity detector (TCD).

The palladium content is determined by inductive coupling plasma optical emission spectroscopy and mass spectroscopy (ICP/OES/MS Thermo Jarell Ash) after dissolution of sample on a mixture of HF and $HNO_3$ solution. This analysis shows closed to about 0.5wt.% palladium for all the studied samples.

Toluene oxidation is carried out in a conventional fixed bed microreactor and studied between 25 to 300°C (1°C.min$^{-1}$). The reactive flow is composed of air and 1000ppm of gaseous toluene. The analysis of combustion products is performed evaluating the toluene conversion and the $CO/(CO+CO_2)$ molar ratio from a Perkin Elmer autosystem chromatograph equipped with TCD and FID. Before the catalytic test, the solid (100mg) is calcined under a flow of air (2L/h) at 400°C (1°C.min$^{-1}$) and reduced under hydrogen flow (2L.h$^{-1}$) at 200°C (1°C.min$^{-1}$).

The electron paramagnetic resonance (EPR) measurements are performed at -196°C and 25°C on a Bruker EMX spectrometer. A cavity operating with a frequency of 9.5 GHz (X band) is used. Precise g values are determined from simultaneous precise frequency and magnetic field values.

The adsorption studies are achieved to measure the Henry constant at low partial pressure for toluene for temperature ranging from 175 to 350°C with the pulse chromatography technique [9]. From these results, the adsorption enthalpy $\Delta H$ is determined using Eq. 1.

$$K' = K'_0 \exp(\frac{-\Delta H}{RT})$$  (Eq. 1)

K' is the Henry constant (mol/kg/Pa), K'$_0$ the Van't Hoff pre-exponential factor (mol/kg/Pa) and $\Delta H$ the adsorption enthalpy (kJ/mol)

## 2.3. Structural characterisation of supports
### 2.3.1. Electron microscopy
The zirconia meso-macroporous particles used as catalyst support in this work have a size of around 10μm [7,8]. The synthesized zirconia particles contain regular arrays of macropores having a diameter range from 300 to 500 nm. The hollow macrochannels are always orthogonal to the face of the monolithic particle. Moreover, macroporous walls are mesostructured with a disordered wormhole like assembly of meso/micropores into macroporous framework.

The titania particles have diameters ranging from 1μm to 10μm [5]. Moreover, a disordered wormlike assembly of mesopores on each titania particle is observed. The zirconia meso-macroporous particles used as catalyst support in this work have a size of around 10μm.

The binary zirconia-titania oxide particles contain arrays of macropores having a diameter range from 600 to 3000nm [6]. The macroporous frameworks are composed of nanoparticules with accessible and interconnected mesopores of a disordered wormhole-like array.

### 2.3.2. X-ray diffraction
The X-ray diffraction patterns of samples are shown in fig.1. Untreated $ZrO_2$ and $ZrO_2$-$TiO_2$ (not shown) framework possess a global amorphous structure and untreated $TiO_2$ possesses a normal anatase lines with a low intensity. After calcination, the crystallisation of amorphous zirconia to a tetragonal phase is observed, intense line appeared at $2\theta = 30.2°$ (fig.1a). Titania support (fig.1c) indicates a deeper crystallisation to anatase phase after calcination. For the mixed oxide only a low broad band located in the range of $25° < 2\theta < 35°$ appears after calcination at 400°C, then this material stays almost amorphous (fig.1d).

*2.3.3. Thermal analysis*

The DTA curves obtained with the different mesoporous materials are shown in fig.2. For all the samples, an endothermic peak appeared at about 80°C corresponding to the loss of water adsorbed at the surface, the second exothermic peak at 277°C for $ZrO_2$, 222°C for $TiO_2$ and at 260°C for $ZrO_2$-$TiO_2$ is attributed to the combustion of the organic substances which are not removed during extraction proceedings. Moreover, for $ZrO_2$ (fig 2b), an exothermic peak appears at 450°C without mass loss. This peak is attributed to the crystallisation from amorphous $ZrO_2$ to the tetragonal phase [10] observed by XRD (fig. 2a).

For $TiO_2$, an exothermic broad peak with a small intensity is observed at 363°C and assigned to a slower crystallisation of the rest of amorphous $TiO_2$ phase to anatase phase (XRD).

According to XRD result, $ZrO_2$-$TiO_2$ stays amorphous after calcination at 400°C and then the exothermic peak centred at 400°C (fig 2a) and accompanied by a loss mass of about 3% should not be attributed to a change of phase but rather to a second combustion of surfactant.

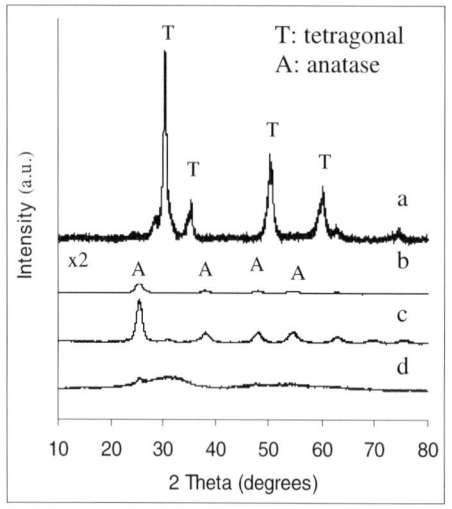
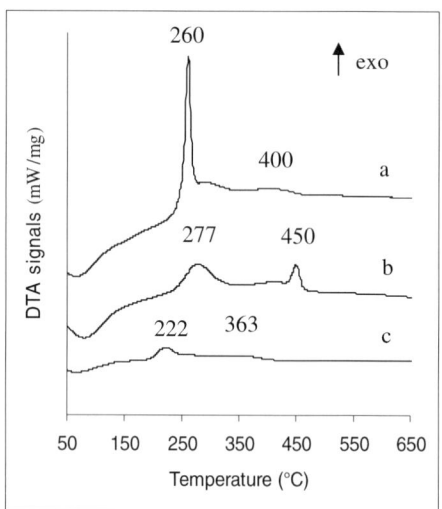

Figure 1: XRD patterns of mesoporous materials: (a) $ZrO_2$; (c) $TiO_2$; (d) $ZrO_2$-$TiO_2$ calcined at 400°C; (b) $TiO_2$ untreated

Figure 2: Thermal decomposition of supports under air (a) $ZrO_2$-$TiO_2$; (b) $ZrO_2$; (c) $TiO_2$

*2.3.4. Nitrogen adsorption analysis*

Surface areas of untreated and calcined supports and pore size distribution are summarised in table 1

Table 1

Specific surface area of oxides before and after calcination at 400°C

| sample | untreated | | After calcination at 400°C | |
|---|---|---|---|---|
| | $S_{BET}$ (m² g⁻¹) | Pore size(nm) | $S_{BET}$ (m² g⁻¹) | Pore size (nm) |
| $ZrO_2$ | 463 | 1.8 | 184 | 1.2 |
| $TiO_2$ | 222 | 2.8 | 183 | 2.8 |
| $ZrO_2$-$TiO_2$ | 542 | 1.8 | 440 | 1.8 |

For pure mesoporous titania, a low decrease of the surface area from 222 to $183 \text{m}^2\text{g}^{-1}$ after the calcination at 400°C is observed. This should be due to the crystallisation of the rest of amorphous $TiO_2$ not crystallised before the calcination (XRD data).

For pure mesoporous zirconia, the surface area is sharply decreased from 463 to $184 \text{m}^2\text{g}^{-1}$ after the calcination. The crystallisation to tetragonal phase of the walls separating mesopores leading to partial destruction of the porous structure should explain this decrease.

In the case of the mixed oxide, the amorphous character maintained after the calcination (XRD data) should explain the high surface area of the calcined sample, even though the thermal treatment leads to a decrease of the surface.

The pore sizes order of the calcined samples : $TiO_2 > ZrO_2\text{-}TiO_2 > ZrO_2$ shows larger mesopores for $TiO_2$ which could be important for the accessibility for Pd particles for the impregnation. Moreover, it is interesting to observe that the pore diameter for $ZrO_2\text{-}TiO_2$ (1.8nm) is about the average of those for $ZrO_2$ (1.2nm) and $TiO_2$ (2.8nm). However, the specific surface for $ZrO_2\text{-}TiO_2$ is not the average.

## 2.4. Characterisation of the impregnated palladium catalysts

### 2.4.1. Nitrogen adsorption analysis

Surface areas and pores sizes of the catalysts are presented in table 2. Pd incorporation leads to a low decrease of the surface areas for $Pd/ZrO_2$ and $Pd/TiO_2$, but the average pore diameters is not change. With $Pd/ZrO_2\text{-}TiO_2$ no further decrease in surface can be observed and the pore diameter remains centred at 1.8 nm.The specific surface for $Pd/ZrO_2\text{-}TiO_2$ is more than twice the value for $Pd/TiO_2$. $Pd/ZrO_2\text{-}TiO_2$ should be therefore a powerful catalyst.

Table 2
Specific surface areas and pore diameters of the catalyst after calcination at 400°C

| Catalyst | $S_{BET} (\text{m}^2\text{ g}^{-1})$ | Pore diameter (nm) |
|---|---|---|
| $Pd/ZrO_2$ | 142 | 1.2 |
| $Pd/TiO_2$ | 174 | 2.8 |
| $Pd/ZrO_2\text{-}TiO_2$ | 440 | 1.8 |

### 2.4.2. TPR measurements

The TPR profiles of calcined catalysts are displayed in Fig.3. In all cases the signals of $H_2$ consumption correspond to complete reduction of PdO to $Pd^0$ [11, 12]. The different peaks observed can be assigned to the reduction of PdO to Pd metallic corresponding to different types of PdO species. There are variations in the distribution of the palladium oxide with support composition but three peaks are found for each catalyst. A first one at near -22°C for $Pd/ZrO_2$, -20°C for $Pd/TiO_2$ and -28°C for $Pd/ZrO_2\text{-}TiO_2$ which could be attributed to the reduction of smaller PdO particles on the surface (easily reducible); the second appearing at 11°C for $Pd/ZrO_2$, -8°C for $Pd/TiO_2$ and 11°C for $Pd/ZrO_2\text{-}TiO_2$ is attributed to the reduction of bulky particles and the last one centred 102°C for $Pd/ZrO_2$, 10°C for $Pd/TiO_2$ and 90°C for $Pd/ZrO_2\text{-}TiO_2$ 20°C is related to the reduction of Pd species highly dispersed and certainly into the pores. The treatments on those compounds should lead partially to palladium particles confined in blind pores while the particle size should be probably close to the size of the pores [13]. Accordingly, it can be assumed that Pd particles particularly in this case are less accessible. However, with $Pd/TiO_2$, the total PdO reduction recorded at the lowest temperature could be explain by palladium species probably more accessible due to the higher pore size of the mesoporous material.

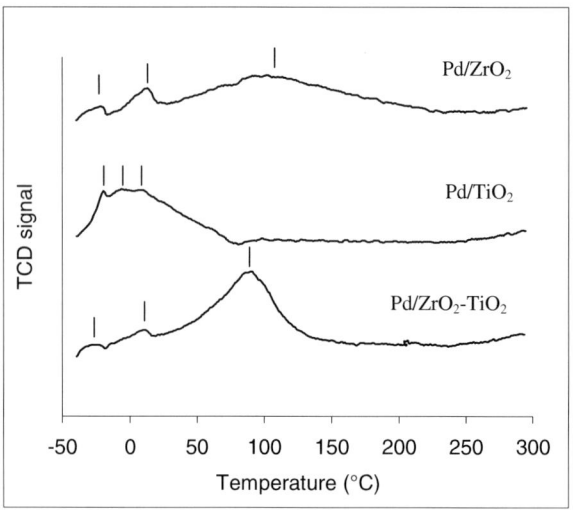

Figure 3: TPR spectra of Pd based catalysts

## 2.4.3. Catalytic activity for total toluene oxidation

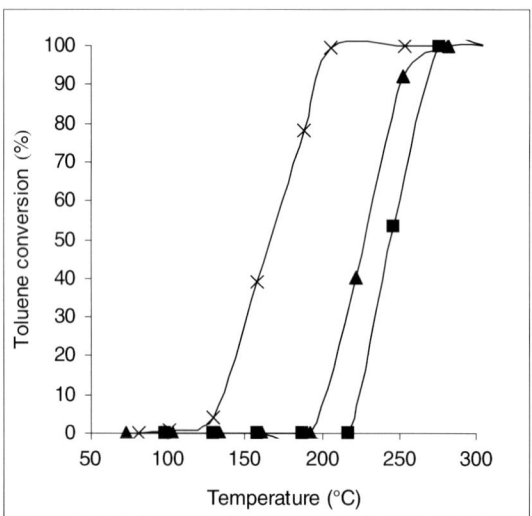

Figure 4: Toluene total conversion versus the temperature for the Pd supported catalysts: (■) Pd/ZrO₂; (x) Pd/TiO₂, (▲) Pd/ZrO₂-TiO₂

In fig.4 are compared catalytic activity data of the catalysts. The observed products are only carbon dioxide and water, indicating the complete combustion occurring during the reaction.

Pd/TiO$_2$ exhibits a very interesting catalytic activity for complete oxidation of toluene: T50[*] at about 164°C could be compared to 154°C obtained in the same conditions with 0.5wt% Pd/NaFAU [14]. Moreover, this activity is much higher than for Pd/ZrO$_2$-TiO$_2$ and Pd/ZrO$_2$. The reason of this behaviour could be related to the much easier reducibility of palladium species for Pd/TiO$_2$ than for the both other catalysts (TPR results). Pd/ZrO$_2$-TiO$_2$, with its much higher specific surface, is less active than Pd/ TiO$_2$ but better than Pd/ZrO$_2$.

### 2.4.4. Adsorption of toluene on ZrO$_2$, TiO$_2$ and TiO$_2$-ZrO$_2$

The measurement of adsorption Henry constants for toluene on the supports leads to the determination of the Van't Hoff parameters (pre-exponential factor and adsorption enthalpy) collected in table 3. Toluene is more adsorbed on TiO$_2$-ZrO$_2$ than on ZrO$_2$ and TiO$_2$ is the worst material for toluene adsorption. The lower toluene adsorption enthalpy for TiO$_2$ (53.22 kJ/mol) corresponds to the higher catalytic activity for the corresponding impregnated palladium catalyst and the two supports ZrO$_2$ and TiO$_2$-ZrO$_2$ have close adsorption Henry constants. The same behaviour is found for the toluene oxidation activity of Pd/zeolites [14].

Table 3
Adsorption Van't Hoff parameters for toluene

| Sample | K'$_0$ (mol/kg/Pa) | $\Delta$H (kJ/mol) |
|---|---|---|
| ZrO$_2$ | 2.706E-11 | 65 |
| TiO$_2$ | 1.709E-10 | 53 |
| TiO$_2$-ZrO$_2$ | 7.468E-12 | 71 |

### 2.4.5. EPR studies before and after the catalytic test

EPR measurements before and after the catalytic test (fig.5) are performed to give more information on the catalytic behaviour of all the catalysts.

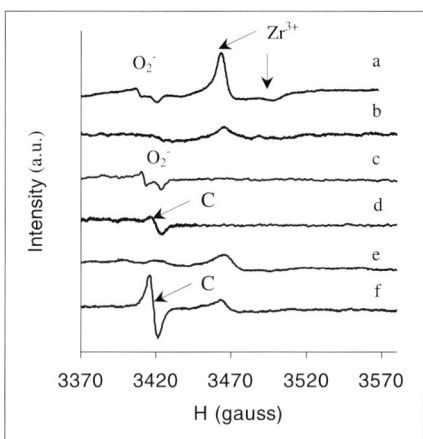

Figure 5: EPR spectra of catalysts: (a) Pd/ZrO$_2$ after calcination; (b) Pd/ZrO$_2$ after test, (c)Pd/TiO$_2$ after calcination; (d) Pd/TiO$_2$ after test, (e) Pd/ZrO$_2$-TiO$_2$ after calcinations, (f) Pd/ZrO$_2$-TiO$_2$ after test

EPR spectra of Pd/ZrO$_2$ (fig.5a and b) samples are composed of two signals. The first signal with g value g$_{xx}$= 2.025 (not shown), g$_{yy}$= 2.009, and g$_{zz}$= 2.002 is attributed to O$_2$[-]

---

[*] T50 is often used and corresponds here to 50% of toluene conversion.

species [15] and the second signal is ascribed to $Zr^{3+}$ with $g_{//} = 1.959$ and $g_{\perp} = 1.977$ [10]. $O_2^-$ signal disappears and $Zr^{3+}$ signal decreases after test. This result can be related to the intervention of these both ions during oxidation reaction.

For $Pd/TiO_2$, $O_2^-$ signal is also detected after calcination (fig.5c), this signal disappears after test due to its participation to the oxidation reaction. It is possible that carbon is deposited on the catalysts during the reaction, EPR signal at $g_{iso} = 2.003$ (C signal) is ascribed to carbon radical (fig.5d). This signal is a typical EPR spectrum of coked catalyst [16].

EPR signals of $Pd/ZrO_2\text{-}TiO_2$ (fig.5e and f) are composed of a signal of $Zr^{3+}$ less intense due to the amorphous character of this solid and a carbon radical signal (C signal) which intensity is higher than that of $Pd/TiO_2$ after test. These results are confirmed by coke combustion which represents a mass loss of 2.9 % for $Pd/TiO_2$ and 3.5 % for $Pd/ZrO_2\text{-}TiO_2$.

## CONCLUSION

The synthesised amorphous (macro)-mesoporous $ZrO_2$, $TiO_2$ and $ZrO_2\text{-}TiO_2$ used as catalytic supports present high surface areas. The stability of the porous structure after the calcination at 400°C is observed for $ZrO_2\text{-}TiO_2$ (amorphous structure maintained) and $TiO_2$ (low crystallisation to anatase phase) but partial breakdown of the structure occurred for $ZrO_2$ by the crystallisation to tetragonal phase of the walls separating mesopores. However, all these Pd impregnated solids are found to be powerful catalysts for VOCs total oxidation because they are active at low temperature and are totally selective for $CO_2$ and $H_2O$. The highest catalytic potential of $Pd/TiO_2$ for the total oxidation of toluene is explained by the lowest toluene adsorption enthalpy, the low coke content observed after the catalytic test but especially by the highest pore diameter leading to palladium particles more accessible and reducible.

## ACKNOWLEDGEMENTS
The authors would like to thank Mr Fabrice Cazier for elemental analysis and the Interreg IIIa France-Wallonie-Flandre project and Walloon Region for financial support.

## REFERENCES
[1] J. J. Spivey, Ind. Eng. Chem. Res. 26 (1987) 2165.
[2] T. Maillet, C. Solleau, J. Barbier, D. Duprez, Appl. Catal. B 14 (1997) 85.
[3] J. Carpentier, J.F. Lamonier, S. Siffert, E.A. Zhilinskaya, A. Aboukaïs, Appl. Catal. A 234 (2002) 91.
[4] V. Idakiev, L. Ilieva, D. Andreeva, J.L. Blin, L. Gigot, B.L. Su, Appl. Catal. A 243 (2003) 25.
[5] V. Idakiev, T. Tabakova, Z.Y. Yuan, B.L. Su, Appl. Catal. A 270 (2004) 135.
[6] Z.Y. Yuan, T.Z. Ren, A. Vantomme, B.L. Su, Chem. Mater. 16 (2004) 5096.
[7] J.L. Blin, A. Léonard, Z.Y. Yuan, L. Gigot, A. Vantomme, A.K. Cheetham, B.L. Su, Angew. Chem. Int. Ed. 42 (2003) 2875.
[8] Z.Y. Yuan, A. Vantomme, A. Léonard, B.L. Su, Chem. Commun. (2003) 1558.
[9] X. Canet, J. Nokerman, M. Frère, Adsorption, 11 (2005), 213-216
[10] M. Labaki, S. Siffert, J.F. Lamonier, E.A. Zhilinskaya, A. Aboukaïs, Appl. Catal. B 43 (2003) 261.
[11] S. T. Homeyer, W. M. H. Sachtler, J. Catal. 117 (1989) 91.
[12] W. J. Shen, M. Okumura, Y. Matsumura, M. Haruta, Appl. Catal. A 213 (2001) 225
[13] M. P. Kapoor, Y. Ichihashi, W.-J. Shen, Y. Matsumura, Catal. Lett. 76 (2001) 139
[14] H. L. Tidahy, S. Siffert, J.-F. Lamonier, E.A. Zhilinskaya, A. Aboukaïs, B.-L. Su, X. Canet, G. Deweireld, M. Frère, J-M Gireaudon, G. Leclercq, Paper for 4th International Conference on Environmental Catalysis (ICEC) 5-8 june 2005 Heidelberg Germany
[15] H. D. Gesser, L. Kruczynski, J. Phys. Chem. 88 (1984) 2751
[16] C.L. Li, O. Novaro, E. Munoz, J.L. Boldu, X. Bokhimi, J.A. Wang, T. Lopez, R. Gomez, Appl. Catal. A: Gen. 199 (2001) 214

Studies in Surface Science and Catalysis 160
P.L. Llewellyn, F. Rodriquez-Reinoso, J. Rouqerol and N. Seaton (Editors)
© 2007 Elsevier B.V. All rights reserved

# Characterisation of palladium supported on exchanged BEA and FAU zeolites for VOCs catalytic oxidation

H. L. Tidahy[1], S. Siffert*[1], J.-F. Lamonier[1], E.A. Zhilinskaya[1], A. Aboukaïs[1], B.-L. Su[2], X. Canet [3], G. Deweireld [3], M. Frère [3], J-M Gireaudon[4], G. Leclercq[4]

[1] Laboratoire de Catalyse et Environnement, E.A. 2598, Université du Littoral Côte d'Opale, 145 avenue Schumann, 59140 Dunkerque (France)
* tel : (33) 03 28 65 82 68/56 ; siffert@univ-littoral.fr
[2] Laboratoire de Chimie des Matériaux Inorganiques, Université de Namur (Belgium)
[3] Laboratoire de Thermodynamique, Faculté polytechnique de Mons (Belgium)
[4] Laboratoire de Catalyse de Lille, Université des Sciences de Technologies de Lille (France)

0.5wt% palladium supported on exchanged BEA and FAU zeolites have been characterised for volatile organic compounds (VOCs) catalytic oxidation. BEA and FAU zeolites have been exchanged with different cations to study the influence of alkali metal cations ($Na^+$, $Cs^+$) and $H^+$ in Pd based catalysts on propene total oxidation. Oxidation of propene depends significantly on the type of zeolites and on the compensating cation of the zeolite. According to sample, the catalytic activity is explained by the surface areas of the catalysts, the adsorption energies for VOC, the influence of the electronegativity of the compensating cation on the Pd particles, the Pd dispersion and the PdO reducibility.

## 1. INTRODUCTION

The development of active catalysts for total combustion of volatile organic compounds (VOCs) has been desired from the viewpoint of environmental protection. Noble metal which possess high activity for total oxidation are widely applied to the low temperature complete oxidation [1]. Moreover, it was shown that supports play an important role in catalytic activity and zeolites were widely used as powerful catalytic support [2-4]. Therefore, zeolites FAU and BEA exchanged with different alkali metal cations were prepared and 0.5wt% of palladium was incorporated in these supports. The catalysts obtained were calcined, characterised and tested for propene total oxidation. Some of these solids were also tested for VOCs adsorption.

The aim of this work was to study the influence of the alkali metal cations ($Na^+$, $Cs^+$) and $H^+$ in palladium based catalysts for VOCs oxidation reaction.

## 2. EXPERIMENTAL

### 2.1 Synthesis and characterisation of catalysts

HBEA, HFAU zeolites were prepared from NaBEA zeolite (framework Si/Al ratio of 25 from P.Q. Corporation) and NaFAU zeolite (framework Si/Al ratio of 2.43 from Union Carbide) respectively by exchange with a 2M solution of $NH_4NO_3$ (Acros) at 80°C. Afterward the samples were filtered, washed and dried overnight at 100°C.

CsBEA and CsFAU were prepared from the same parents zeolites by exchanging seven times with 0.5M of CsCl (Prolabo) at 60°C. After each exchange the samples were filtered, washed and heated at 350°C. Finally the solids obtained were grinded.

Calcination of all zeolite samples was performed under air flow with heating rate of 1°Cmin$^{-1}$ from ambient temperature till 500°C with a hold of 4h.

Pd/zeolite BEA catalysts were prepared by ion exchange of HBEA, NaBEA and CsBEA using an adequate dilute solution of Pd(NO$_3$)$_2$.XH$_2$O (Johnson Matthey), continuously stirred for 18h at 60°C. Afterward the samples were filtered and washed.

Pd/zeolite FAU catalysts were prepared by aqueous impregnation method using the same salt as palladium precursor.

All catalyst samples were then dried overnight at 100°C and calcined at 400°C under air flow 1°Cmin$^{-1}$ at 400°C for 4h.

## 2.2. Characterisation techniques

The palladium content was determined by inductive coupling plasma optical emission spectroscopy and mass spectroscopy (ICP/OES/MS Thermo Jarrell Ash) after dissolution of Pd/zeolite on a mixture of HF and HNO$_3$ solution. The Pd content was closed to 0.5wt%.

The structures of solids were analysed by powder X-ray diffraction (XRD) technique at room temperature with a Bruker diffractometer using Cu-K$_\alpha$ radiation scanning 2θ angles.

The nitrogen adsorption analysis was performed on Sorptomatic 1990 apparatus at -196 °C and the specific surface areas of the solids were determined by BET method.

Temperature programmed reduction experiments were carried out in an Altamira AMI-200 apparatus. The TPR profiles were obtained by passing 5% H$_2$/Ar flow (30 mL.min$^{-1}$) through the calcined sample (about 100 mg). The temperature was increased from -40 to 200°C at a rate of 5°C.min$^{-1}$. The hydrogen concentration in the effluent was continuously monitored by a thermoconductivity detector (TCD).

Pulse chemisorption measurements were performed using Micromeritics Autochem II 2920 instrument. The samples were pre-treated for 2h in a flow of hydrogen at 200°C, in order to reduce the Pd in the catalyst, then the samples were cooled to 100°C in a stream of argon. Pulse chemisorption measurements were performed at this temperature.

Electron paramagnetic resonance (EPR) measurements were performed at -196°C and 25°C on a Bruker EMX spectrometer. A cavity operating with a frequency of 9.5 GHz (X band) was used. Precise g values were determined from simultaneous precise frequency and magnetic field values.

Propene oxidation was carried out in a conventional fixed bed microreactor using 100mg of catalysts and studied between 25 to 300°C (1°C.min$^{-1}$). The reactive flow was composed of air 99mL.min$^{-1}$ and 0.6mL.min$^{-1}$ of propene. The analysis of combustion products was performed evaluating the propene conversion and the CO/(CO+CO$_2$) molar ratio from a Varian 3600 chromatograph equipped with TCD and FID.

Before the catalytic test and adsorption studies, the solid is calcined under a flow of air (2L/h) at 400°C (1°C.min$^{-1}$) and reduced under hydrogen flow (2L.h$^{-1}$) at 200°C (1°C.min$^{-1}$).

Adsorption studies were achieved on FAU zeolites by the determination of Henry constants for toluene for temperature ranging from 225 to 400°C using the pulse chromatography technique [5]. The Van't Hoff law (Eq. 1) was used to determine the adsorption energy.

$$K = K_0 \exp(\frac{-\Delta U}{RT}) \qquad \qquad \text{(Eq. 1)}$$

Where K is the dimensionless Henry constant, $K_0$ is the Van't Hoff pre-exponential factor and $\Delta U$ is the adsorption energy (kJ.mol$^{-1}$).

## 3. RESULTS AND DISCUSSIONS

Surface areas and microporous volumes of FAU and BEA zeolites are given in table 1. The effect of the exchange on the structure could be observed in the BET surface area changes. Surface areas and microporous volumes for Cs zeolites and H zeolites are lower than for Na zeolites. For H zeolite, it is well known that the dealumination processes which occur in the protonated zeolites may lead to a partial breakdown of the lattice. For Cs zeolites, Cs cation is bigger than Na and the exchange is rather difficult [6]. Therefore, several exchanges are needed and as the sample is heated many times, a partial breakdown of the zeolites could occur. The decrease is more evident with FAU zeolite, from 731m$^2$.g$^{-1}$ for NaFAU before exchange to 485m$^2$.g$^{-1}$ for CsFAU afterwards. Si/Al ratio is lower for FAU zeolite (2.43) than for BEA zeolite (25), therefore the exchange with compensating cations should be higher for FAU zeolites and then leads to a higher decrease of the BET surface area.

Figures 1 and 2 show the comparison of XRD spectra of the exchanged FAU and BEA zeolites. The crystallinity of the HBEA and CsBEA zeolites, observed by XRD (fig.2), is only slightly altered after exchange proceeding by a decrease of the intensities. For the exchanged FAU zeolites, the cristallinity seems to be more altered (XRD patterns, fig.1), due certainly to a partial breakdown of the zeolite lattice.

Table 1
Supports characterisation after calcinations at 500°C

| Sample | $S_{BET}$ (m$^2$.g$^{-1}$) | Micropore volume[a] (cm$^3$.g$^{-1}$) |
|--------|-----------------|-----------------------|
| HFAU | 547 | 0.200 |
| NaFAU | 731 | 0.291 |
| CsFAU | 485 | 0.190 |
| HBEA | 551 | 0.160 |
| NaBEA | 584 | 0.140 |
| CsBEA | 487 | 0.132 |

[a]determined by t-method

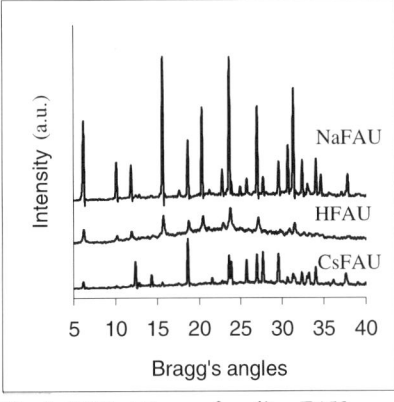

Fig. 1: XRD patterns of zeolites FAU

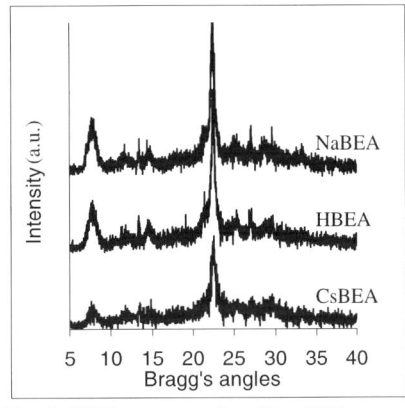

Fig. 2: XRD patterns of zeolites BEA

It is well known that the introduction of a larger cation such as $Cs^+$ in the zeolitic framework can cause a modification on the zeolite framework and therefore on the XRD pattern. A sharp decrease in intensity of the diffraction lines situated at low angle values is observed (fig.1). The crystallite shapes of starting NaFAU particles (not shown here) are multi facetted spherulites, which are not altered by ion exchange. In the case of HFAU, it seems that a partial breakdown of the zeolite lattice occurs.

As differences in the specific areas for FAU and BEA zeolites are observed, differences in the palladium dispersions on Pd/zeoliteFAU and Pd/zeoliteBEA catalysts should also be found. Palladium dispersions and particle sizes of these solids are summarised in table 2.

The palladium dispersion order: Pd/CsFAU > Pd/HFAU > Pd/NaFAU is surprising since it is the inverse than the order for the specific areas. Pd/CsFAU which has the lower specific area exhibits the higher dispersed palladium particles. The partial breakdown of the zeolite lattice during exchange should imply a more open structure of the zeolite more accessible for the palladium particles. Besides the palladium particle sizes from 2.5 to 4.8 nm determined from $H_2$ chemisorption should mean that the palladium particles are only located at the outer surface of the zeolites. Channels and pores of the both zeolites are theoretically too small to accept such palladium particles except if their lattices are more open after the exchanges. Moreover, the electronegativity of the cations could be also another factor for the dispersion [7] explaining the observed differences. For the Pd/BEA samples, lower dispersions are observed, certainly because of the lower partial breakdown of the zeolite. The palladium dispersion order: Pd/CsBEA > Pd/NaBEA > Pd/HBEA is also the inverse than the order for the specific areas and then it seems to confirm that the partial breakdown of the zeolite leads to more accessible Pd particles.

Table 2
Palladium dispersions and particle sizes of Pd/zeolite FAU catalysts after calcination at 400°C

| Catalyst | $S_{BET}$ $(m^2.g^{-1})$ | Palladium Dispersion[b] (%) | Palladium particle size[b] (nm) |
|---|---|---|---|
| Pd/HFAU | 589 | 34 | 3.2 |
| Pd/NaFAU | 714 | 22 | 4.8 |
| Pd/CsFAU | 414 | 41 | 2.5 |
| Pd/HBEA | 590 | 18 | 6.2 |
| Pd/NaBEA | 550 | 19 | 5.8 |
| Pd/CsBEA | 500 | 28 | 3.9 |

[b]determined from $H_2$ chemisorption at 100°C

The total oxidation of propene is made in order to determine the best catalyst for this reaction by comparing the zeolite structure, Si/Al ratio and the influence of the compensating cation. It should be noted that CO is not detected in the reaction products but only $CO_2$ and $H_2O$.

Fig. 3 and 4 show the propene conversion as a function of temperature over the Pd supported catalysts. First of all, FAU zeolite is more active in this reaction than BEA zeolite except for Pd/HBEA which is more active than Pd/HFAU. The activity order for FAU and BEA catalysts are inversed, respectively: Pd/CsFAU > Pd/NaFAU > Pd/HFAU and Pd/CsBEA < Pd/NaBEA < Pd/HBEA. For the Pd/FAU catalysts, propene combustion may be dependent on palladium dispersion. Pd/CsFAU presents the highest Pd dispersion and also the best activity.

However, Pd/NaFAU exhibits a lower dispersion than Pd/HFAU but is more active. Moreover, the activity order could not be related to the surface areas of catalysts. This order should then be explained also by the electronegativity of the compensating cations and the acido-basicity properties of those samples. L. Pinard et al. observed the same activity order for Pt/FAU exchanged with $Cs^+$, $Na^+$ and $H^+$ for total oxidation of chloromethane and explained the activity by the basicity which increases from the protonated Pt/HFAU to the Pt/CsFAU sample [8]. Moreover, It was shown from XPS results the presence of an electronic shift from the zeolite framework to the Pt particles which increases when increasing basicity from Pt/HBEA to Pt/NaBEA to Pt/CsBEA [9]. Nevertheless, for zeolite BEA samples the cation effect should be less important than for Pd/FAU because of the lower exchange due to the much higher Si/Al ratio. Therefore this activity order for Pd/BEA could be rather explained by the surface areas order which follows (table 4) the activity order and the reducibility of the palladium particles (see $H_2$-TPR). It is interesting to know that the same catalytic orders were observed for the total oxidation of toluene [10].

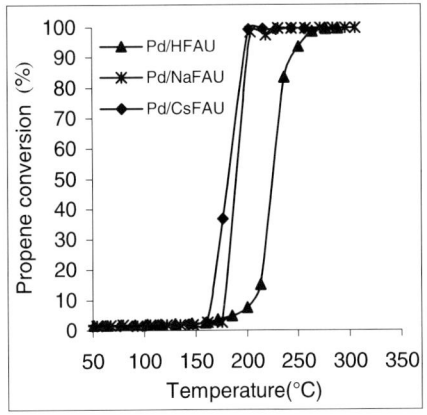 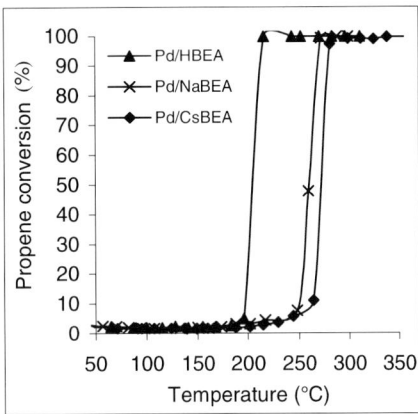

Fig.3: Propene conversion over Pd/zeolite FAU    Fig.4: Propene conversion over Pd/zeolite BEA

$H_2$-TPR is used to characterise the reducibility of palladium species which could be an important factor for the catalytic oxidation reactions. The TPR profiles of Pd/zeolite BEA and Pd/zeolite FAU catalysts are displayed in fig.5 and 6. According to literature [11, 12], hydrogen consumed under each peak observed can be assigned to the reduction of PdO to metallic Pd as PdO is an easily reducible oxide. In our samples, PdO reduction could be observed by one to three peaks corresponding to different types of PdO species.

For all the samples, PdO species are reduced under 50°C except for Pd/CsBEA which PdO reduction occurs till 120°C. If we compare the activity order for propene oxidation and the TPR profiles, it seems that the quantity of PdO species reducible at around 15°C for Pd/BEA and 25°C for Pd/FAU is correlated to the activity in propene oxidation. In fact, the quantity of PdO species reducible at around 25°C for FAU zeolite increases from Pd/HFAU to Pd/NaFAU to Pd/CsFAU whereas it is the inverse for the signal at about 15°C for BEA zeolite which increases from Pd/CsBEA to Pd/NaBEA to Pd/HBEA. This type of species could be in a special interaction with the support leading to high active Pd particles. For Pd/FAU, these species should be small PdO particles in high interaction with the support and the peak at 0°C should correspond to bigger PdO particles. These results could be correlated to the palladium particle size which increases with the intensity of the peak at 0°C. For

Pd/BEA, the species reducible at 15°C should be also small PdO particles in an higher interaction with the support than bigger PdO particles corresponding to the peak at about - 5°C. The large signal centered at 90°C could correspond to PdO particles less accessible and then more difficult to reduce.

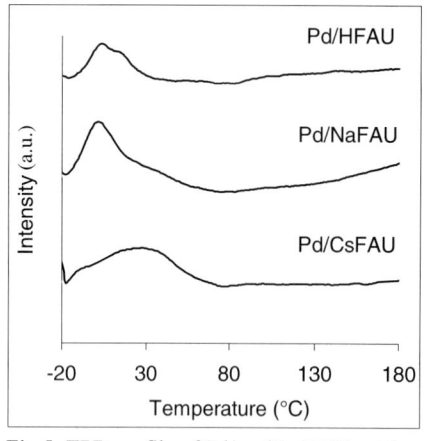

Fig 5: TPR profile of Pd/zeolite FAU catalysts

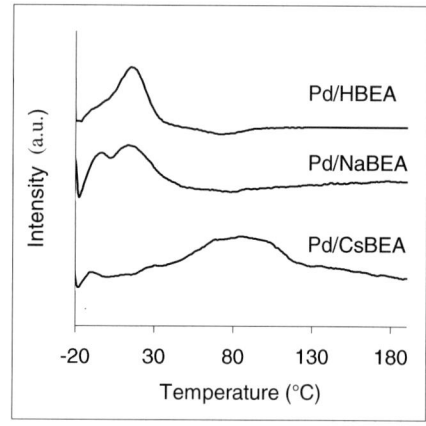

Fig 6: TPR profile of Pd/zeolite BEA catalysts

VOCs adsorption studies are done by the determination of the adsorption Henry constants for toluene (table 3) on FAU samples only. The results show an increase of the Henry constant with rising cation size. The impregnated zeolites have quite lower Henry constants than the respective original zeolites. The average reduction of the adsorption isotherm slope is 42% for Pd/HFAU, 26% for Pd/NaFAU and 36% for Pd/CsFAU. Moreover, it is interesting to observe that the activity order for VOC oxidation follows the order of the adsorption energy. The lower is the adsorption energy for toluene, the higher is the catalytic activity.

Table 3
Adsorption Energy and pre-exponential factor for toluene on FAU samples

| Sample | $K_0$ | $\Delta U$ (kJ.mol$^{-1}$) |
|---|---|---|
| HFAU | 0.00098396 | 65 |
| Pd/HFAU | 0.00032093 | 68 |
| NaFAU | 0.01159015 | 62 |
| Pd/NaFAU | 0.00407453 | 66 |
| CsFAU | 0.00932832 | 68 |
| Pd/CsFAU | 0.31553048 | 47 |

The samples are also characterised after the catalytic test. Table 4 shows a low decrease of the surface areas after test except for Pd/HFAU. The catalytic test conditions (water formation during the reaction) could lead to a partial breakdown of zeolites and agglomeration of palladium particles and also to a formation of carbonaceous compounds (coke) on the zeolites. EPR measurement (fig.7) is used to make evidence coke deposit by a signal of carbon radical observed at g=2.003 [13]. Pd/HBEA, Pd/NaBEA and Pd/HFAU present this signal after propene oxidation. Therefore, the acidic character of these zeolites should favour the coke deposition [14]. This result could be also correlated to the highest adsorption energy for VOC observed for Pd/HFAU (table 3). Iron impurities ($Fe^{3+}$ signal) and palladium ($Pd^{3+}$ signal) are also detected on our samples.

Table 4
Surface areas of catalysts

| Catalyst | $S_{BET}$ (m$^2$.g$^{-1}$) before test | $S_{BET}$ (m$^2$.g$^{-1}$) after test |
|---|---|---|
| Pd/HBEA | 590 | 564 |
| Pd/NaBEA | 550 | 520 |
| Pd/CsBEA | 500 | 450 |
| Pd/HFAU | 589 | 231 |
| Pd/NaFAU | 714 | 675 |
| Pd/CsFAU | 414 | 347 |

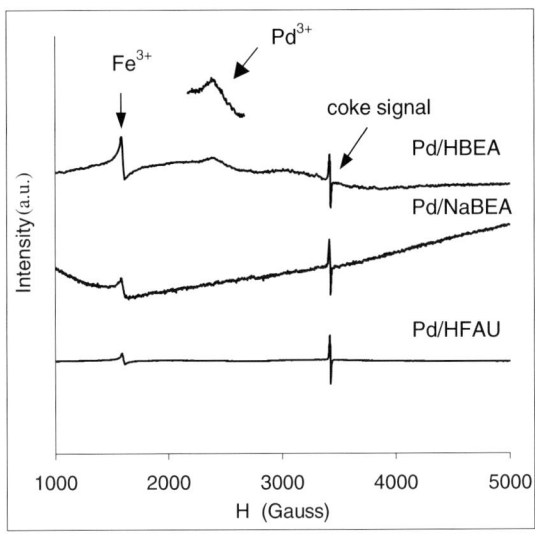

Fig 7: EPR spectra of Pd based catalysts after test obtained at 77 K

## 4. CONCLUSION

0.5wt% palladium supported on exchanged BEA and FAU zeolites are characterised for propene total oxidation. The exchanges with different cations ($Na^+$, $Cs^+$) and $H^+$ lead to a decrease of surface area and micropore volume. However, Pd/BEA and Pd/FAU solids are found to be powerful catalysts for VOCs total oxidation because they are active at low temperature and are totally selective for $CO_2$ and $H_2O$. Moreover, the total oxidation of propene depends especially on the type of zeolites and on the compensating cation of the zeolite. For the catalysts based on zeolite BEA, the following activity order PdHBEA > PdNaBEA > PdCsBEA is observed and could be explained by the surface areas of the catalysts, and the PdO reducibility (TPR measurements). Nevertheless, the activity order is the inverse for FAU solids: PdCsFAU > PdNaFAU > PdHFAU. According to these last samples, the catalytic activity is explained by the adsorption energies for VOC, the influence of the electronegativity of the compensating cation on the Pd particles, the Pd dispersion and also the PdO reducibility.

## ACKNOWLEDGEMENTS

The authors would like to thank Mr Fabrice Cazier (CCM, Dunkerque) for elemental analysis and the Interreg IIIa France-Wallonie-Flandre project and Walloon Region for financial support.

## REFERENCES

[1] J. J. Spivey, Ind. Eng. Chem. Res. 26 (1987) 2165.
[2] Y. Li, J. L. Armor, Appl. Catal. B 51 (1994) 275.
[3] J. Tsou, P. Magnoux, M. Guisnet, J. J.M.Orfao, J. L. Figueiredo, Appl. Catal. B 57 (2005) 117.
[4] D. Jaumain, B.-L. Su, J. Mol. Catal. A 197 (2003) 263.
[5] X. Canet, J. Nokerman, M. Frère, Adsorption, 11 (2005) 213.
[6] J. Tsou, P. Magnoux, M. Guisnet, J. J.M.Orfao, J. L. Figueiredo, Appl. Catal. B 51 (2004) 129.
[7] Y. Yazawa, H. Yoshida, S. Komai, T. Hattori, Appl. Catal. A 233 (2002) 113.
[8] L. Pinard, J. Mijoin, P. Magnoux, M. Guisnet, C. R. Chimie 8 (2005) 457.
[9] S. Siffert, J.-L. Schmitt, J. Sommer, F. Garin, J. Catal. 184 (1999) 19.
[10] H. L. Tidahy, S. Siffert, J.-F. Lamonier, E.A. Zhilinskaya, A. Aboukaïs, B.-L. Su, X. Canet, G. Deweireld, M. Frère, J-M Gireaudon, G. Leclercq, Paper for 4th International Conference on Environmental Catalysis (ICEC) 5-8 june 2005 Heidelberg Germany.
[11] S. T. Homeyer, W. M. H. Sachtler, J. Catal. 117 (1989) 91.
[12] W. J. Shen, M. Okumura, Y. Matsumura, M. Haruta, Appl. Catal. A 213 (2001) 225.
[13] C. L. Li, O. Novaro, E. Munoz, J. L. Boldu, X. Bokhimi, J. A. Wang, T. Lopez, R. Gomez, Appl. Catal. A 199 (2001) 211.
[14] M. Guisnet, P. Dégé, P. Magnoux, Appl. Catal. B 20 (1999) 1.

Studies in Surface Science and Catalysis 160
P.L. Llewellyn, F. Rodriquez-Reinoso, J. Rouqerol and N. Seaton (Editors)

# Comparison of transport characteristics and textural properties of porous material; the role of pore sizes and their distributions

Vladimír Hejtmánek, Petr Schneider, Karel Soukup, and Olga Šolcová

Institute of Chemical Process Fundamentals, Czech Academy of Sciences,
165 02 Praha 6, Czech Republic

## ABSTRACT

For a set of six porous materials with a range of mean pore radii from 50 to 3000 nm, and mono- or bidisperse pore structure, transport characteristics and textural properties were compared.

Two standard methods (mercury porosimetry and helium pycnometry) together with liquid expulsion permporometry (that takes into account only flow-through pores) were used for determination of textural properties. Pore structure characteristics relevant to transport processes were evaluated from multicomponent gas counter-current diffusion and gas permeation. For data analysis the Mean Transport-Pore Model (MTPM) based on Maxwell-Stefan diffusion equation and a simplified form of the Weber permeation equation was used.

It appears that for porous solids with monodisperse pore-size distribution the MTPM mean-pore radii and transport-pore distributions agree with the information from standard textural analysis. For porous solids with bidisperse pore-size distribution the MTPM mean-pore radii and transport-pore distributions are close to large pore sizes from standard textural analysis.

*Keywords:* Counter-current gas diffusion, Permeation, Transport parameters, Mean Transport-Pore Model, Maxwell-Stefan equation, Weber equation

## 1. INTRODUCTION

The structure of porous solids is traditionally characterized by textural analysis methods (e.g. high pressure mercury porosimetry, physical adsorption of inert gases – nitrogen, argon, krypton) [1–3]. Even though these methods provide consistent information on texture characteristics, they take into account all pores and partition into transport-pores (pores through which the decisive part of mass transport takes place) and blind pores (which contain the surface area but are not active in mass transport) is not feasible. Therefore, information on pore-size distributions (PSD) obtained from classic textural analyses cannot, in general, be used for rational prediction of mass (e.g. gas) transport in many processes where only the transport-pores are significant and knowledge of transport-pore size distribution is essential.

Hence, alternative experimental methods were suggested based on evaluation of simple transport processes inside the pore structure [4–6]. Usually, binary or multicomponent counter-current gas diffusion [7] or simple gas permeation [8] (or their combination) has been applied.

To obtain transport characteristics of the porous solid via evaluation of transport processes a simple model of the pore-structure and theoretical relations for transport description are required.

As a model we have used the Mean Transport-Pore Model (MTPM) [6] which assumes that the decisive part of the gas transport takes place in transport-pores that are visualized as cylindrical capillaries. The transport-pore radii are distributed around the mean value $<r>$ (first model parameter). The width of this distribution is characterized by the mean value of the squared transport-pore radii, $<r^2>$ (second model parameter). The third model parameter is the ratio of porosity, $\psi_t$, and tortuosity of transport-pores, $q_t$, $\psi = \varepsilon_t/q_t$. Pore diffusion is described by the Maxwell-Stefan diffusion equation extended to account for Knudsen transport [6]. For gas permeation the simplified form of Weber equation [8–10] is used.

In this way MTPM distinguishes between transport properties of gases (gas viscosity, binary bulk-diffusion coefficients of all gas pairs) and textural properties of porous materials characterized by the set of transport parameters ($<r>$, $<r^2>$, $\psi$). Transport parameters represent material properties of the porous solid, and, thus do not depend on temperature, pressure and the kind of used gases. The obtained transport characteristics have a wide practical use for simulation and prediction in many industrial processes (e.g. calculation of effective diffusion coefficients for any pairs of gases in automotive catalytic converter [11]).

The pore-size distribution of transport pores can be provided by the liquid expulsion permporometry (LEPP); based on Washburn equation for individual groups of pores. LEPP has been commonly used for description of textile materials, porous filtration - in general, for very thin porous materials. Recently, we have shown [12] that LEPP could be used also for other porous material (e.g. pelleted porous catalysts, adsorbents, etc).

## 2.    EXPERIMENTAL

### 1.1    Porous solids

Six porous materials in the form of cylindrical pellets were used; industrial hydrogenation catalyst (**Cherox 42–00**, Chemopetrol Litvínov, Czech Republic), catalyst for removal of hydrogen and oxygen (**G43a**, Süd-Chemie AG, Germany), two laboratory prepared porous α-alumina **A5** and **F1200**, abrasive carbide **C400** (Carborundum, Benátky nad Jizerou, Czech Republic) and knit iron powder **SP400**.

All porous materials were chosen to cover as wide as possible range of pore sizes and to represent both monodisperse and bidisperse PSD. Textural properties were determined by mercury porosimetry (AutoPore III, Micromeritics, USA) and helium pycnometry (AccuPyc 1330, Micromeritics, USA) are summarized in Table 1.

### 1.2    Gases

Argon, helium, hydrogen and nitrogen (99.9 % purity, Technoplyn Linde, Czech Republic) were selected for transport measurements. Because of their inertness the surface transport was absent.

### 1.3    Pore-filling liquid

As the liquid for filling all pores in LEPP we have used the fluorinated hydrocarbon Porofil 3 (Beckman–Coulter, USA). According to the manufacturer $\sigma \cos \theta = 0.16$ bar µm (where $\sigma$ is the surface tension, and $\theta$ the contact angle). From comparison of capillary outflow-time for water and Porofil its viscosity was estimated as $\mu = 1.25$ mPa s.

Table 1
Textural properties of porous materials

| Material | True density [g/cm³] | Apparent density [g/cm³] | Porosity [–] | Maximum of pore radii [nm] | Total intrusion volume [cm³/g] |
|---|---|---|---|---|---|
| G43a | 3.308 | 1.323 | 0.60 | 3.5/257 | 0.428 |
| Cherox 42–00 | 3.472 | 2.234 | 0.36 | 70 | 0.133 |
| F1200 | 3.689 | 1.857 | 0.50 | 375 | 0.251 |
| A5 | 3.989 | 2.685 | 0.33 | 290/2070 | 0.101 |
| SP600 | 7.703 | 6.957 | 0.10 | 1050 | 0.015 |
| C400 | 3.383 | 1.950 | 0.42 | 6600 | 0.191 |

## 1.4 Diffusion measurement

The Graham diffusion cell [7] was used for measurement of isobaric counter-current diffusion. All experiments were performed at laboratory temperature. The cell is schematically shown in Fig. 1.

The diffusion cell consists of two flow-through compartments separated by a metallic disc with cylindrical holes of slightly larger diameter than the tested porous materials. Porous pellets were forced into undersized silicon rubber tubing and the pellet-tubing assembly was then forced into holes of the metallic disc. Tightness was verified by employing nonporous (metallic) pellets.

Pure gases or their mixture flowed steadily through the both compartments until the steady state was established. After that the gas inlet and outlet in one of the cell compartment (lower compartment in Fig. 1) were stopped and the net volumetric diffusion flux was determined from the rate of movement of a soap film in a calibrated gas burette.

From the net molar diffusion flux density $N^d = N_A^d + N_B^d$, the component diffusion flux densities, $N_A^d$ and $N_B^d$, were determined from the Graham law

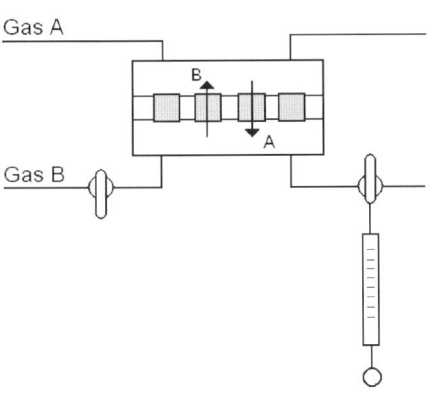

Fig. 1    Graham's diffusion cell

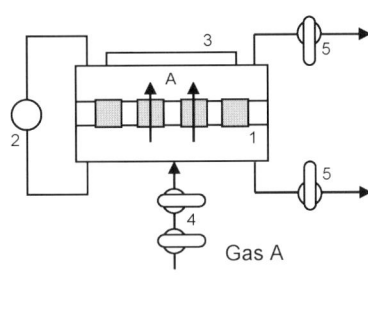

Fig. 2    Permeation cell.

$$N_A^d / N_B^d = -\sqrt{M_B / M_A}$$

where $M_A$ and $M_B$ are component molecular weights. The general form of the Graham law follows from the Maxwell-Stefan constitutive equation [5] if the requirement of constant pressure is imposed.

## 1.5    Permeation measurement

Arrangement of the permeation cell is shown in Fig. 2 [8], where (1) are metallic disc with porous membrane, (2) differential pressure gauge, (3) pressure gauge, (4) differential gas feed, (5) connections to vacuum pump, respectively. The permeation cell is divided into two parts with equal volumes ($V = 62.95 \pm 0.01$ cm$^3$) separated by a metallic disc with mounted porous pellets. The time dependent pressure difference between cell compartments were followed by a differential pressure transducer (Baratron, MKS Instruments GmbH., Germany) and logged in a computer. Experiments were performed in the pressure range from 5 to 101 kPa. All experiments were performed at laboratory temperature.

Both compartments were filled by the inert gas to an appropriate starting pressure (5 – 101 kPa). A small amount of gas was admitted into upper compartment of the cell to achieve the initial pressure difference $\Delta p_0$ between both compartments ($\Delta p_0 = 500 - 900$ Pa). The time dependence of the pressure difference was then followed. From the $\Delta p$–$t$ dependence the (pressure dependent) effective permeability coefficient, $B$, was evaluated from

$$\Delta p = \Delta p_0 \exp\left[-2S/(VL)B(p)t\right]$$

with the total cross-section, $S$; length of porous pellets, $L$; the cell volume, $V$, and the (pressure dependent) effective permeability coefficient, $B(p)$.

## 1.6    Permporometry measurement

The permporometry cell is shown schematically in Fig. 3. The metallic cylindrical cell consists of two parts (1, 2) separated by a circular partition (3) with cylindrical holes in which the liquid pre-wetted cylindrical porous pellets (4) are fixed. A metallic porous support (5) and the perforated insert (6) below the pellets guarantee also removal of the wetting liquid expelled from the pores. Pressure of nitrogen entering the upper cell part is regulated by the sensitive pressure regulator (Brooks Instruments Model 5866; accuracy $\approx 0.5\%$). Volumetric flow-rates of nitrogen leaving the bottom cell are determined either by the DryCal flowmeter (DC–2; BIOS International Corp.); or by a bubble flowmeter with electronic sensing of soap-film movement (Digital Flowmeter; Agilent Technologies).

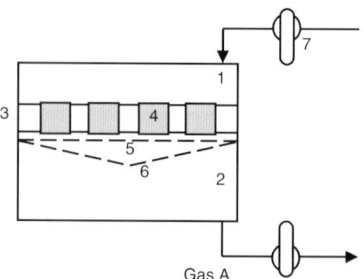

Fig. 3    Permporometry set-up

Table 2
Transport parameters from counter-current diffusion and permeation

| Porous material | Diffusion | | | Permeation | | | |
|---|---|---|---|---|---|---|---|
| | $\psi_d$ [−] | $(<r>\psi)_d$ [nm] | $<r>_d$ [nm] | $(<r>\psi)_p$ [nm] | $(<r^2>\psi)_p$ [nm²] | $<r>_p$ [nm] | $\psi_p$ [−] |
| G43a | 0.031 | 7.8 | 252 | 8.9 | 2061 | 234 | 0.038 |
| Cherox 42−00 | 0.136 | 4.9 | 36 | 4.8 | 738 | 155 | 0.031 |
| F1200 | 0.212 | 76 | 358 | 71 | 50996 | 710 | 0.10 |
| A5 | 0.116 | 195 | 1680 | 140 | 209168 | 1490 | 0.094 |
| SP600 | 0.021 | 20.5 | 980 | 38 | 34000 | 895 | 0.042 |
| C400 | 0.142 | 991 | 6980 | 584 | 8400000 | 14240 | 0.041 |

*) $<r>_d = (<r>\psi)_d / \psi_d$, **) $\psi_p = (<r>\psi)_p^2 / <r^2> \psi_p$

## 3.   RESULTS AND DISCUSSION

Optimum transport parameters obtained from diffusion and permeation measurements are summarized in Table 2 and visualized in Fig. 4. The Cherox 42-00 and F1200 samples show an excellent agreement between transport parameters $(<r>\psi)_d$ and $(<r>\psi)_p$ (deviation less than 5%; index "d" stands for diffusion and "p" for permeation). For the other pellets the agreement is slightly worse.

It appears that for porous solids with monodisperse pore-size distribution the MTPM mean-pore radii agree reasonably with the information from standard textural analysis. For porous solids with bidisperse pore-size distribution the MTPM mean-pore radii is similar to large pore sizes from standard textural analysis.

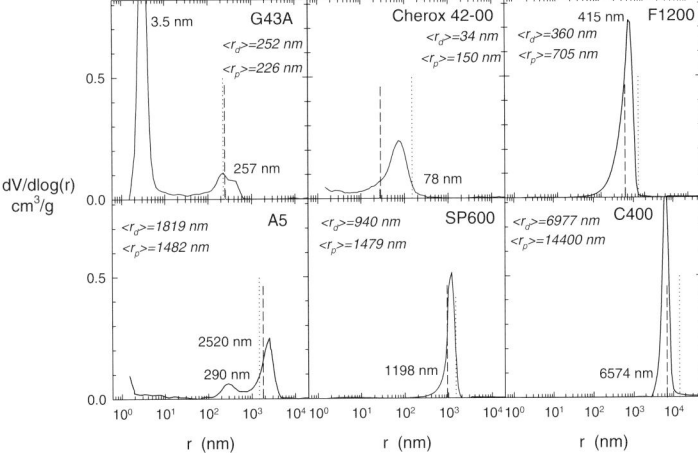

Fig. 4    Comparison of pore-size distribution (mercury porosimetry; solid line) with mean transport pore radii from diffusion (dashed line) and permeation (dotted line) measurements

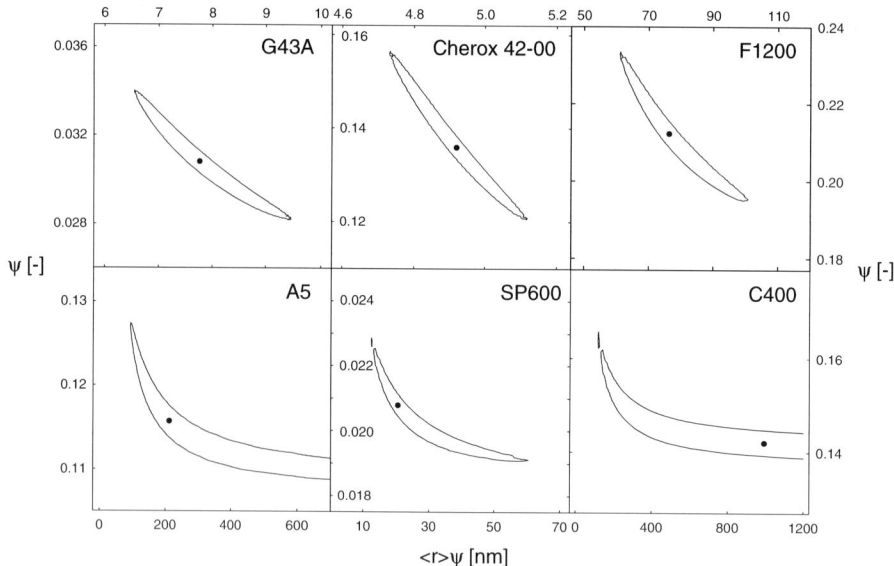

Fig. 5    95% confidence regions of calculated transport parameters for gas diffusion inside porous pellets of the various porous materials

In Fig. 4 the mean transport-pore radii from permeation measurements, $(<r>)_p$, are nearly in all cases slightly higher than from diffusion measurements, $(<r>)_d$. The reason could lie in the way how $(<r>)_p$ was determined. From directly evaluated $(<r>\psi)_p$ $(<r>)_p$ was calculated with the use of $(\psi)_p$, which followed from $(<r^2>\psi)_p$ and $(<r>\psi)_p$. This approach assumes that $<r>^2 = <r^2>$ which is true only for (infinitely) thin transport-pore radii distributions. Therefore, values of $(<r>)_p$ and, consequently, of $(\psi)_p$ have to be looked upon with substantial caution. The uncertainty of $(<r>\psi)_p$ and $(<r^2>\psi)_p$ depend on the relative role of Knudsen transport and viscous transport.

In addition, one should realize that the confidence of transport parameters depends on the amount of experimental information available as well as on the accuracy of experimental points. This is visualized in Fig. 5 where the 95% confidence regions for transport parameters from counter-current diffusion measurements are plotted. Confidence regions were determined with the use of the Beale criterion [13] which reads $SSD_{crit}/SSD_{min} = 1 + [p/(n-p)]$ $F(p,(n-p),\alpha)$ where $SSD_{min}$ is the smallest sum of squared deviations between observed and calculated results, $SSD_{crit}$ is the largest sum of deviations which is still statistically undistinguishable from $SSD_{min}$, p is the number of parameters and n is the number of experiments. $(1-\alpha)$ is the probability level (for 95% confidence region $\alpha = 0.05$) and $F()$ is the Fisher statistical distribution. Inside the confidence region combinations of transport parameters are statistically indistinguishable from optimum values (point inside the confidence regions). This makes the comparison of optimum parameter values from the two experimental methods less comprehensible.

There are two limiting diffusion mechanisms: the Knudsen-diffusion, which is significant in narrow pores and the bulk-diffusion significant in wide pores. In the transition region both cases of diffusion play a role and the contribution of each mechanism can be seen from the shape of the confidence. E.g. pores in samples C400, A5 and SP600 are wide and the

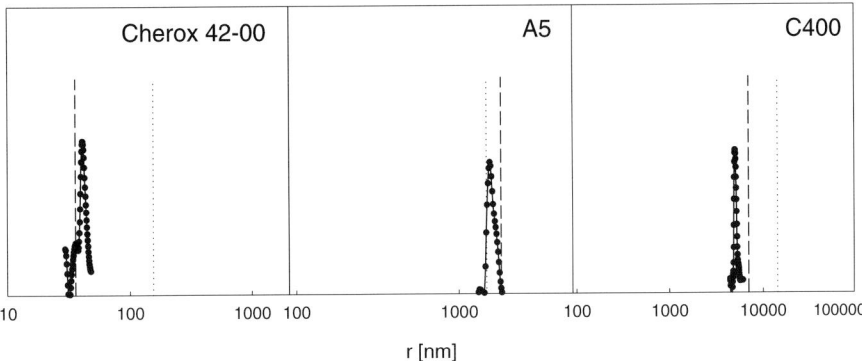

Fig. 6     The comparison of transport pore-size distribution (LEPP; solid line) and mean transport pore radii from diffusion (dashed line) and permeation (dotted line) measurements

accuracy of the transport parameter $(<r>\psi)_d$, which characterizes the Knudsen diffusion mechanism, is small. Hence, the bulk diffusion mechanism has a dominating role. On the other hand, diffusion in samples F1200, G43a and Cherox 42–00 is predominantly in the transition region: the shapes of confidence regions are a combination of both limiting shapes for Knudsen and bulk diffusion mechanisms.

Fig. 6 shows the pore size distribution of transport-pore radii from LEPP for three porous samples. When comparing transport-pore radii from diffusion and permeation measurements with these distributions the size and shape of confidence regions (Fig. 5) have to be taken into account as well as the above arguments concerning the validity of $(<r>)_p$. Therefore, the agreement of pore-size distribution of transport pores with $(<r>)_d$ is excellent and $(<r>)_p$ is a little worse.

## 4.     CONCLUSIONS

Textural properties of six porous materials with mono- and bidisperse porous structure and a range of pore radii from nanometers to microns were determined by mercury porozimetry and helium pycnometry. The obtained pore-size distributions were compared with transport characteristics obtained independently from diffusion and permeation measurements. For three chosen samples the distribution of transport-pores was obtained from LEPP.

It appears that for porous solids with monodisperse pore-size distribution the MTPM mean-pore radii and transport-pore distributions agree with the information from standard textural analysis. For porous solids with bidisperse pore-size the MTPM mean-pore radii and transport pore-sizes distributions are close to large pore sizes from standard textural analysis.

**ACKNOWLEDGEMENT**

The financial support of Grant Agency of the Czech Academy of Sciences (# A 4072404) and Grant Agency of the Czech Republic (104/04/0963, 203/03/H140) is gratefully acknowledged.

## References

[1]   P.J. Pomonis, D.E. Petrakis, A.K. Ladavos, K.M. Kolonia, C.C.Pantazis, A.E. Giannakas, A.A. Leontiou: The I-point method for estimating the surface area of solid catalysts and the variation of C-term of the BET equation. Catal. Commun. 6, 93−96 (2005).

[2]   D.D. Do, H.D. Do: Comparative adsorption of spherical argon and flexible n-butane in carbon slit pores − a GCMC computer simulation study. Colloids and Surf. A: Physicochem. Eng. Aspects 252, 7−20 (2005).

[3]   J.C. Groen, L.A.A. Peffer, J. Pérez-Ramírez: Pore size determination in modified micro- and mesoporous materials. Pitfalls and limitations in gas adsorption data analysis. Microporous and Mesoporous Mater. 60, 1−17 (2003).

[4]   E.A. Mason, A.P. Malinauskas: Gas Transport in Porous Media: The Dusty Gas Model, Elsevier, Amsterdam 1983.

[5]   P.J.A.M. Kerkhof: A Modified Maxwell–Stefan Model for Transport through Inert Membranes: The Binary Friction Model, Chem. Eng. J. 64 319−343 (1996).

[6]   P. Schneider: Multicomponent Isothermal Diffusion and Forced Flow of Gases in Capillaries, Chem. Eng. Sci. 33, 1311−1319 (1978).

[7]   P. Fott, G. Petrini, P. Schneider: Transport parameters of monodisperse porous catalysts. Coll. Czech. Chem. Commun. 48, 215−227 (1983).

[8]   V. Hejtmánek, O. Šolcová, P. Schneider: Gas Permeation in Porous Solids: Two Measurements Modes. Chem. Eng. Commun. 190(1), 48−64 (2003).

[9]   S. Weber: Dan. Mat. Phys. Medd. 28, 1−37 (1954).

[10]  P. Fott, G. Petrini: Determination of transport parameters of porous catalysts from permeation measurements. Appl. Catal., 2, 367−378 (1982).

[11]  T. Starý, O. Šolcová, P. Schneider, M. Marek: Effective Diffusivities and Pore-Transport Characteristics of Washcoated Ceramic Monolith for Automotive Catalytic Converter. Appl. Catal., B, (submitted 2005).

[12]  O. Šolcová, H. Šnajdaufová, P. Schneider: Liquid-Expulsion Perm-Porometry (LEPP) for Characterization of Porous Solids. Micropor. Mesopor. Mat. 65(2–3), 209–217 (2003).

[13]  E.M.L. Beale: Confidence Regions in Nonlinear Estimation. J. Roy. Stat. Soc. B22, 41, (1960).

Studies in Surface Science and Catalysis 160
P.L. Llewellyn, F. Rodriquez-Reinoso, J. Rouqerol and N. Seaton (Editors)

225

# Effect of noble metal deposition in zeolitic structures on their adsorption capacities

X. Canet [a], J. Nokerman [a], M. Frère [a], H.L. Tidahy [b], S. Siffert [b], B.-L. Su [c]

[a] Thermodynamics Department, Faculté Polytechnique de Mons, 31 bd Dolez 7000 Mons – Belgium

[b] Laboratoire de Catalyse et Environnement, Université du Littoral-Côte d'Opale, MREID, 145 Avenue Maurice Schumann 59140 Dunkerque – France

[c] Laboratoire de Chimie des Matériaux Inorganiques, Université de Namur (FUNPD), 61 Rue de Bruxelles 5000 Namur – Belgium

Experimental determination of Henry constants and adsorption heat for various VOCs on faujasite type zeolites (NaY, HY and CsY with a Si/Al ratio of 2.4) and their Palladium impregnated form were achieved by a pulse chromatographic technique from 448 to 623K for seven alkanes (linear, branched and cyclic probes) and from 523 to 673K for three aromatic compounds (toluene, m-xylene and chlorobenzene). The Henry constant increases with the cation size and decreases if the zeolite is impregnated.

## 1. INTRODUCTION

Volatile Organic Compounds (VOCs) released by industrial activities or by vehicles are responsible for important environmental and health problems [1]. More and more constraining regulations which aim at limiting VOCs contents in waste gases are set on an international level [Directive 2004/42/CE of the European Parliament and of the Council of 21 April 2004 on the limitation of emissions of Volatile Organic Compounds due to the use of organic solvents in certain paints and varnishes and vehicle refinishing products and amending Directive 1999/13/CE]. As a consequence, existing abatement techniques; either destructive (thermal oxidation, catalytic oxidation) or recuperative (adsorption, absorption, condensation or separation by membrane) have to be improved in order to fulfil the new requirements. It is admitted that considerable improvements of the purification rates can be achieved by coupling these methods in an efficient way.

Adsorption and catalytic oxidation may be coupled in a very simple way by using catalysts showing a high physical affinity for the compound to be eliminated which is the case when using faujasite zeolites [2]. Catalytic oxidation is even more efficient when impregnated with dispersed noble metals like platinum or palladium [3-5]. A previous work [2] presents adsorption and diffusion data on impregnated zeolites but complete studies presenting adsorption and catalytic results for this kind of materials are still rare.

In this work, we present adsorption data for very low partial pressures obtained with the pulse chromatographic method. We studied both a Na-faujasite zeolite with a Si/Al ratio of 2.43 (NaY) and two exchanged form HY and CsY. The same zeolites were impregnated with 0.5 % in weight of palladium (noted Pd-NaY, Pd-HY and Pd-CsY) as to obtain catalysts for VOCs oxidation. The experimental adsorption measurements consist of Henry constants for various alkanes and aromatics in the temperature range 448 - 673K.

## 2. MATERIALS SYNTHESIS AND CHARACTERISATION

CsY was prepared from NaY zeolite (Si/Al = 2.43 provided by Union Carbide) by cation exchange with a 0.5M solution of $CsNO_3$ (Strem Chemicals) at 333K. Afterwards the sample was filtered and washed. Calcination was performed under air flow from ambient temperature till 773K with a hold of 4h. The same protocol was carried out to obtain the $NH_4Y$ using a 10 wt% $NH_4NO_3$ solution. After each exchange, a calcination was performed at 673K for 6h at a heating rate of 100K per hour from room temperature in a flow of dry oxygen. The calcination is to decompose the $NH_4^+$ ions to give $H^+$, a small ion, in order to facilitate the next exchange. The procedure is repeated seven times.

0.5 wt% Pd/zeolite catalysts were prepared by aqueous impregnating method using palladium nitrate $Pd(NO_3)_2.xH_2O$ (Johnson Matthey). The impregnated powders were dried at 373K overnight and calcined in air at 673K for 4h.

The palladium content was determined by inductive coupling plasma optical emission spectroscopy and mass spectroscopy (ICP/OES/MS) after dissolution of the Pd/zeolite in a mixture of HF and $HNO_3$ solutions, the precision of this analysis was closed to 0.5 wt.%

The nitrogen adsorption isotherms were measured at 77K using a Quantasorb Junior apparatus. The BET surface areas and micropore volumes obtained from these measurements are given in Table 1.

Table 1: Specific surface area and micropore volume after calcination at 773K

| | Specific surface ($m^2$/g) | micropore volume[a] ($cm^3$/g) |
|---|---|---|
| HY | 547 | 0.200 |
| Pd-HY | 589 | |
| NaY | 731 | 0.291 |
| Pd-NaY | 714 | |
| CsY | 485 | 0.190 |
| Pd-CsY | 414 | |

[a]determined by t-method

The exchanges with $H^+$ and $Cs^+$ from NaY lead to a decrease of the surface areas and micropore volumes (Table 1) for HY and CsY and also to a decrease of the intensities in the XRD diffractograms (Fig. 1). For HY, the dealumination process which occurs in the protonated zeolites may lead to a partial breakdown of the lattice. For CsY, Cs cation is bigger than Na and the exchange is rather difficult [6]. Therefore, several exchanges are needed and as the sample is heated many times, a partial breakdown of the zeolites could occur. The HY and CsY samples thus contain a certain amount of amorphous material of which the adsorptive properties are not known. The proportion of amorphous material may be estimated from the specific surface area ratio considering the NaY sample as the reference. This approximation leads to a proportion of amorphous material equal to 25.1% for HY and to

33.7% for CsY. These values seem to be underestimations. Anyway, the results provided in the section 3.2 do not take into account the presence of amorphous material. They should not be considered as characteristic results of purely crystalline samples.

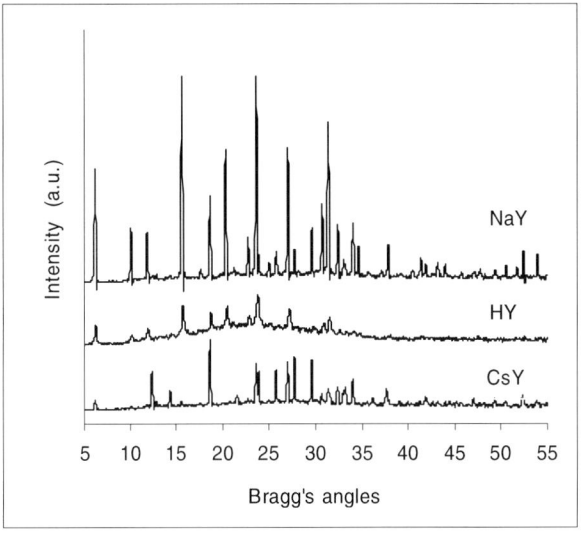

Figure 1: XRD diffractograms for NaY, HY and CsY

## 3. ADSORPTION STUDY

### 3.1 The Pulse Chromatographic Method

Faujasite zeolite-heavy hydrocarbon systems are characterised by type I isotherms (high slope in the very low pressure area followed by a plateau for higher pressures). The initial linear part of the isotherm is known as the Henry's zone for which the adsorption capacity of the adsorbent for a given adsorptive may be characterised by the Henry constant (slope of the isotherm) defined by Eq. (1):

$$\theta = K'P \tag{1}$$

In which:
-     $\theta$ is the coverage ratio (mol kg$^{-1}$)
-     K' is the Henry constant (mol kg$^{-1}$ Pa$^{-1}$)
-     P is the pressure (Pa)

Given the Henry's zone is limited to very low adsorption pressures, it is generally impossible to perform direct measurements of the coverage ratio as a function of pressure [7, 8]. The pulse chromatographic method appears as a fast and convenient technique for the direct measurement of the Henry constants which does not required any coverage ratio and pressure determination. It is based on the mathematical treatment of the concentration profile of the adsorptive in an inert carrier gas at the outlet of an adsorption column which was submitted to an inlet pulse concentration.

The column filled with the adsorbent is placed in the oven of a chromatograph equipped with a Thermal Conductivity Detector (TCD). A purificator, filled with 3A zeolite, is placed in the system to remove water eventually contained in the carrier gas. A mass flow controller (0-10Nl/h) was used to regulate the flow of carrier gas (nitrogen) which is measured at the TCD outlet with a soap film flowmeter as to equilibrate the flow in each part of the TCD. Before any measurement, the temperature was raised to 673K with a rate of 1K/min and maintained at this temperature during a 10-hour period under a flow of inert gas in order to regenerate the adsorbent. After this regeneration step, the temperature is set to the experimental temperature. Liquid adsorptive is injected with a syringe in the heated injector, passes through the column where it is adsorbed and desorbed and goes to the thermal conductivity detector. The outlet concentration profile is then recorded as a function of time on a personal computer.

The first moment μ of the signal (retention time of the adsorptive in the column) is related to the Henry constant by Eq. (2):

$$\mu = \frac{V_{column}}{Q_T}\left[\varepsilon + (1-\varepsilon)\rho_c.RTK'\right] + \frac{V_d}{Q_T} \qquad (2)$$

in which $V_{column}$ is the column volume (m$^3$), $V_d$ is the dead volume (m$^3$), $Q_T$ is the total volume flow (m$^3$.s$^{-1}$), T is the experimental temperature (K), R is the ideal gas constant (Pa.m$^3$.K$^{-1}$.mol$^{-1}$), $\varepsilon$ is the total bed porosity and $\rho_c$ is the crystal density (kg.m$^{-3}$)

$V_{column}$, $V_d$, $\rho_c$ and $\varepsilon$ may be determined prior to adsorption runs by experimental and calculation means. $Q_T$ and T are measured during the experimental run. Eq. (2) allows the determination of K'.

Eq. (2) is only valid if the column is working in the Henry's zone. This hypothesis is checked by performing several experimental runs with different injected volumes of adsorptive for which we calculate the retention times. When the retention time does not depend on the injected volume, the hypothesis is valid. Typically, the injected volume is between 0.02 and 0.2μl.

The Henry constant K' may be expressed in a dimensionless form K by Eq. (3):

$$K = \rho_c RTK' \qquad (3)$$

Eq. (4) expresses the temperature dependence of the dimensionless Henry constant:

$$K = K_0 \exp(\frac{-\Delta U}{RT}) \qquad (4)$$

in which $\Delta U$ is the potential adsorption energy (J.mol$^{-1}$) and $K_0$ is a pre-exponential constant which takes into account the reduction of the degree of freedom of the molecule when passing from the gas phase to the adsorbed phase.

Eq. (4) shows that ln K versus 1/T is a linear plot which allows the determination of both the energetic and entropic parameters $\Delta U$ and $K_0$.

## 3.2 Experimental results and Comments

The experimental and treatment procedures described in section 3.1 were applied to the determination of the Henry constants of adsorption of various alkanes and aromatics on NaY, HY and CsY and on their Pd impregnated forms for temperatures ranging from 448K to 623K for the alkanes and from 523K to 673K for the aromatic compounds (25K steps)

Fig. 2 presents the plot of the logarithm of the experimental K values versus 1/T for the three zeolites Y and their Pd impregnated forms.

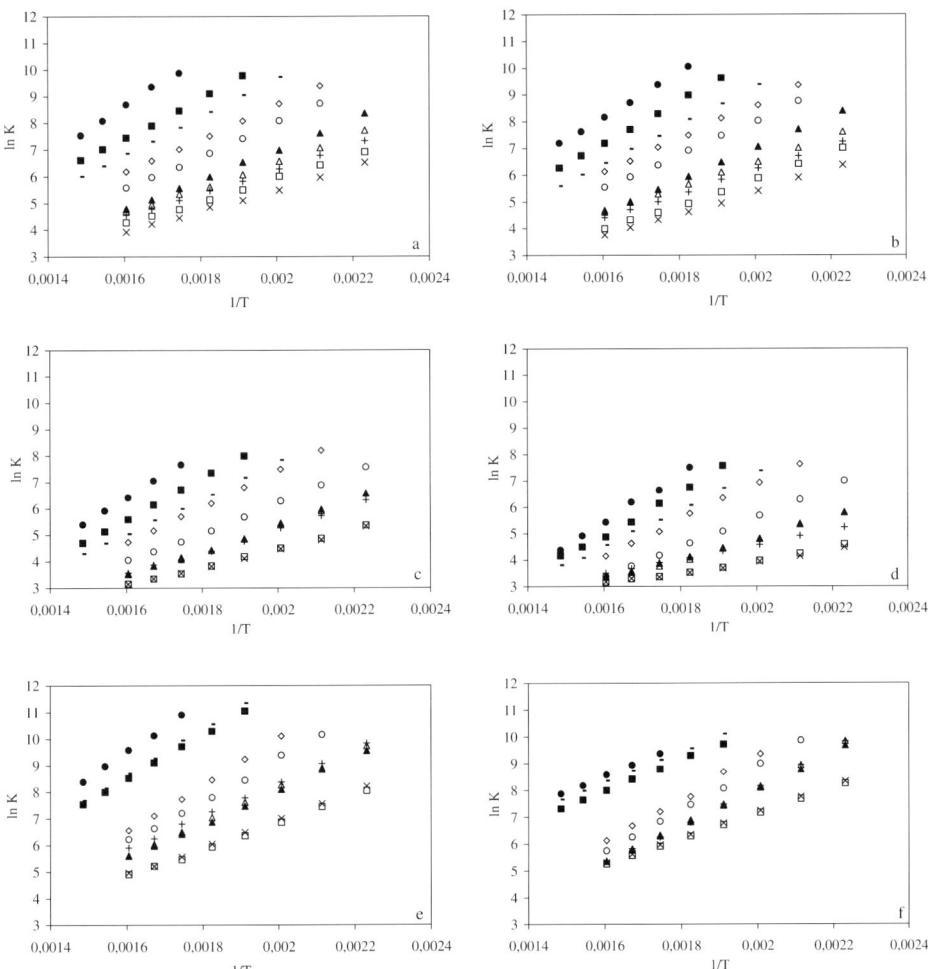

Figure 2 : Plot of ln K versus 1/T for (a) NaY (b) Pd-NaY (c) HY (d) Pd-HY (e) CsY (f) Pd-CsY
(x) 2-methylbutane, (□) n-pentane, (△) 3-methylpentane, (▲) n-hexane, (○) n-heptane,
(◊) 2,2,4-trimethylpentane, (+) Cyclohexane, (■) toluene, (●) meta-xylene and (-) chlorobenzene

The experimental values of K at 573K are given for each adsorptive and for NaY and Pd-NaY in Table 2. The deviation between the values of K for each couple of adsorbent in the whole temperature range are also provided for each adsorptive in Table 3. It is calculated by equation (5).

$$Deviation\ (\%) = \frac{K_{MY} - K_{Pd-MY}}{K_{MY}} * 100 \tag{5}$$

Where M=H, Na or Cs

Table 2: Experimental values of K at 573K for NaY and Pd-NaY

|                      | K NaY | K Pd-NaY | Deviation |
|----------------------|-------|----------|-----------|
| 2-methylbutane       | 85.76 | 75.70    | 11.7%     |
| n-pentane            | 118.7 | 99.93    | 15.8%     |
| Cyclohexane          | 166.4 | 150.0    | 9.8%      |
| 3-methylpentane      | 210.9 | 200.1    | 5.1%      |
| n-hexane             | 262.0 | 236.4    | 9.8%      |
| n-heptane            | 577.4 | 588.8    | -2.0%     |
| 2.2.4-trimethylpentane | 1121 | 1140    | -1.7%     |
| Toluene              | 4727  | 3971     | 16.0%     |
| m-xylene             | 19589 | 11797    | 39.8%     |
| Chlorobenzene        | 2525  | 1738     | 31.2%     |

Table 3: Mean deviations on the K values calculated in the whole temperature range

|                      | Pd-HY / HY | Pd-NaY / NaY | Pd-CsY / CsY |
|----------------------|------------|--------------|--------------|
| 2-methylbutane       | 28.6%      | 13.6%        | -35.0%       |
| n-pentane            | 28.1%      | 11.8%        | -38.4%       |
| Cyclohexane          | 34.1%      | 7.9%         | 26.6%        |
| 3-methylpentane      | 35.4%      | 2.6%         | 10.6%        |
| n-hexane             | 32.5%      | 3.7%         | 6.1%         |
| n-heptane            | 44.8%      | -0.1%        | 31.1%        |
| 2.2.4-trimethylpentane | 41.7%    | 3.5%         | 36.6%        |
| Toluene              | 45.0%      | 19.8%        | 48.2%        |
| m-xylene             | 62.3%      | 39.0%        | 61.2%        |
| Chlorobenzene        | 38.4%      | 31.5%        | 34.7%        |

Table 4 to 6 present the $K_0$ and $\Delta U$ values calculated from equation (4) for all zeolites and catalysts.

Table 4: *Experimental values of $\Delta U$ (kJ.mol$^{-1}$) and $K_0$ for NaY and Pd-NaY*

|  | $\Delta U$ NaY | $\Delta U$ Pd-NaY | $K_0$ NaY | $K_0$ Pd-NaY |
|---|---|---|---|---|
| 2-methylbutane | 33.91 | 34.95 | 0.072204 | 0.049002 |
| n-pentane | 35.80 | 40.05 | 0.068694 | 0.022820 |
| Cyclohexane | 37.15 | 37.94 | 0.070038 | 0.053552 |
| 3-methylpentane | 40.10 | 38.56 | 0.045394 | 0.062356 |
| n-hexane | 47.01 | 50.27 | 0.013604 | 0.006193 |
| n-heptane | 51.95 | 52.49 | 0.011294 | 0.009978 |
| 2,2,4-trimethylpentane | 52.59 | 52.41 | 0.018581 | 0.018518 |
| Toluene | 61.76 | 65.82 | 0.011639 | 0.004075 |
| m-xylene | 76.19 | 70.77 | 0.002387 | 0.004140 |
| Chlorobenzene | 59.39 | 60.08 | 0.010005 | 0.005914 |

Table 5: *Experimental values of $\Delta U$ (kJ.mol$^{-1}$) and $K_0$ for HY and Pd-HY*

|  | $\Delta U$ HY | $\Delta U$ Pd-HY | $K_0$ HY | $K_0$ Pd-HY |
|---|---|---|---|---|
| 2-methylbutane | 28.78 | 17.23 | 0.085294 | 0.808560 |
| n-pentane | 29.46 | 18.78 | 0.074743 | 0.587311 |
| Cyclohexane | 36.46 | 22.89 | 0.029177 | 0.396928 |
| 3-methylpentane | 40.15 | 32.87 | 0.013876 | 0.045978 |
| n-hexane | 40.93 | 32.43 | 0.011551 | 0.052518 |
| n-heptane | 47.37 | 47.72 | 0.005723 | 0.002914 |
| 2,2,4-trimethylpentane | 57.01 | 57.28 | 0.001861 | 0.001020 |
| Toluene | 64.97 | 67.53 | 0.000984 | 0.000321 |
| m-xylene | 72.81 | 75.83 | 0.000496 | 0.000105 |
| Chlorobenzene | 56.40 | 57.64 | 0.003010 | 0.001433 |

Table 6: *Experimental values of $\Delta U$ (kJ.mol$^{-1}$) and $K_0$ for CsY and Pd-CsY*

|  | $\Delta U$ CsY | $\Delta U$ Pd-CsY | $K_0$ CsY | $K_0$ Pd-CsY |
|---|---|---|---|---|
| 2-methylbutane | 43.67 | 40.18 | 0.028976 | 0.085273 |
| n-pentane | 42.01 | 39.71 | 0.038308 | 0.089448 |
| Cyclohexane | 52.14 | 59.38 | 0.015231 | 0.002125 |
| 3-methylpentane | 54.69 | 59.31 | 0.007003 | 0.002178 |
| n-hexane | 53.14 | 56.89 | 0.008932 | 0.003569 |
| n-heptane | 65.47 | 67.14 | 0.001496 | 0.000711 |
| 2,2,4-trimethylpentane | 73.62 | 67.28 | 0.000459 | 0.001026 |
| Toluene | 68.32 | 47.44 | 0.009328 | 0.315531 |
| m-xylene | 79.61 | 48.25 | 0.002967 | 0.472083 |
| Chlorobenzene | 71.65 | 47.33 | 0.005922 | 0.452757 |

Fig. 2 and Table 2 show that the higher values of the Henry constants are obtained for the aromatics compounds for not impregnated zeolites. The Henry constant increases with the carbon number of the molecules for both the aromatics and the alkanes. The same remarks are valid for the values of $\Delta U$ (Tables 4 to 6). The high affinity between the aromatics and the solids (high $\Delta U$ and K values) is due to the strong interactions between the electrons of the $\Pi$-orbitals of the aromatic kern and the cations of the zeolites framework.

The value of the pre-exponential parameter $K_0$ decreases when the adsorption energy increases or when the size of the probe increases for alkanes. This is due to the reduction of degrees of freedom.

Tables 2 and 3 show that the impregnation of the zeolite NaY 2.4 with palladium does not lead to a significant difference in the adsorption capacity of alkanes except for the 2-methyl butane and n-pentane, the lightest compounds considered in this study. On the other hand significant differences appear between NaY and Pd-NaY for the aromatic compounds: Pd-NaY is characterized by lower Henry constants. The maximum deviation between the two solids is observed for m-xylene.

Tables 4 and 5 show that the impregnated zeolites Pd-HY and Pd-CsY are quite less adsorbent at low coverage than the HY and CsY for nearly all probes.

The Henry constant K increases for a given adsorbate probe when changing the zeolite from HY to NaY and to CsY. The same comments can be done for the adsorption energy $\Delta U$. When rising compensating cation size (from H to Cs), the electronegativity of Sanderson [9,10] decreases and then the partial charge of both oxygen and counter-ion increase in absolute values: Interactions are then more important between probes and the zeolite framework.

## 4. CONCLUSIONS

The compensating cation has a very important effect on adsorption performances. The isotherm slope and the adsorption energy in the low partial pressure are very dependent on the size of the counter-ion. For a given probe, NaY has higher Henry constants than HY, but lower than CsY.

The impregnation of zeolites with palladium leads in most cases to a decrease of the Henry constant except for NaY and alkanes.

## 5. ACKNOWLEDGMENTS

This work was supported by the Interreg III program and by the Department for Environment and Natural Resources of the Walloon region (Belgium).

## REFERENCES

[1]     P. Le Cloirec, Les composés organiques volatils (COV) dans l'environnement, *Technique et Documentation Lavoisier*, 1998
[2]     J.F. Denayer and G.V. Baron, *Adsorption*, **3** (1997), 251-265
[3]     J.J. Spivery, Ind. Eng. Chem. Res. 26 (1987) 2165-2180.
[4]     T. Maillet, C. Solleau, J. Barbier, D. Duprez, Appl. Catal. B 14 (1997) 85-95.
[5]     J. Carpentier, J.F. Lamonier, S. Siffert, E.A. Zhilinskaya, A. Aboukaïs, Appl. Catal. A 234 (2002) 91-101.
[6]     J. Tsou, P. Magnoux, M. Guisnet, J.J.M. Orfao, J.L. Figueiredo, Appl. Catal. B 51 (2004) 129
[7]     X. Canet, J. Nokerman and M. Frère, *Adsorption, 11 (2005), 213-216*
[8]     S. Dutour, J. Nokerman, S. Limborg-Noetinger and M. Frere, *Measurement Science and Technology*, **15**, 185-194 (2004)
[9]     R.P. Sanderson, Chemical Bonds and Bond Energy, Academic Press, New-York, 1976
[10]    W.J. Mortier, Journal of Catalysis, 55 (1978), 138-145

Studies in Surface Science and Catalysis 160
*P.L. Llewellyn, F. Rodriquez-Reinoso, J. Rouqerol and N. Seaton (Editors)*

# Influence of the Bentonite/Titania ratio on the textural characteristics of incorporated ceramics for photocatalytic destruction of volatile organic compounds.

M. Yates[a], J.C. Martin[a], P. Ávila[a] and F.J. Gil-Llambias[b]

[a] Instituto de Catálisis y Petroleoquímica, C.S.I.C., c/Marie Curie 2, 28049 Madrid, Spain

[b] Departamento de Química, Universidad de Santiago de Chile, Chile

Titania in its anatase form has been shown to have a high photocatalytic activity for the degradation of volatile organic compounds at low temperature. This property is especially important when these compounds contain chlorine where catalytic oxidation or conventional combustion at higher temperatures can lead to the formation of highly toxic dioxins and furans. In systems with large volumes of gas to be treated the conformation of titania as open channelled monoliths, extrudates or sheets can ease the handling characteristics and improve the abrasion resistance. However, it is practically impossible to extrude titania without the inclusion of agglomerating agents. In this study bentonite, a natural silicate, was chosen as the agglomerating agent in order to conform the incorporated type ceramic bodies. A series of titania/bentonite materials ranging from 0 to 80% titania by weight were prepared. The inclusion of bentonite was found to increase the total specific surface area, abrasion resistance and mechanical strength of the monoliths, although causing a reduction in the total pore volume. The relationship between the surface coverage and the composition was determined by measuring the zero point charge as photocatalytic activity is a surface phenomenon since the photons can only penetrate the first 2 $\mu$m.

Key words: Titania, bentonite, composite materials, photocatalysts.

## 1. INTRODUCTION

Photocatalytic oxidation has been demonstrated as an efficient technology for the destruction of aromatic compounds [1,2,3], halocarbons [4] and alcohols [5]. Titanium dioxide in its anatase crystalline phase is commonly used for the complete mineralization of volatile organic compounds (VOCs) [6,7]. However, when large volumes of gas are to be treated there can be problems of loss of this active material when it is present as a thin supported layer. Thus, the production of composite incorporated bodies where the titania can be a majority component of a ceramic shaped material gives enhanced mechanical strength and abrasion resistance. Using ceramic monoliths with high titania content (50%) the total oxidation of chlorinated organic compounds at low temperature has been demonstrated [8]. An advantage of these incorporated systems is the simplicity of their manufacture compared to supported thin layers made by sol-gel technology that require expensive precursors and a more elaborate production process involving successive coating and firing cycles in order to maintain good adhesion between the photocatalytically active layer and the substrate. For incorporated materials since the titania is present throughout the structure any abrasion during use will reveal fresh surfaces and thus the photocatalytic activity is not affected.

In this study a series of incorporated extrudates based on mixtures of titania with bentonite, as an agglomerating agent, were produced. The textural characterisation of these materials: surface area, porosity, pore volume and pore size distribution, mechanical strength and analysis of the surface composition were undertaken in order to select the most favourable composition.

## 2. EXPERIMENTAL

### 2.1 Composite preparation

A series of ceramic materials were prepared at various bentonite/titania ratios. The titania G5 from Rhone-Poulenc had a purity of >99% while the magnesium silicate clay (Bentonite Cabañas from Tolsa SA) had a purity of 98%. After premixing of the dry powders by the careful addition of water a paste was formed with adequate rheological properties for its successful extrusion. By passing this dough through a Bonnot single screw extruder the various materials were conformed as extrudates of 1 mm diameter and approximately 5 cm length. The green ceramic bodies were allowed to dry at room temperature for two days then heated at $3°C$ $min^{-1}$ up to 500°C and maintained at this temperature for 4h. These materials were subsequently used in all of the characterisation techniques.

### 2.2 Characterisation techniques

The specific surface area ($S_{BET}$) micro and mesopore volumes and pore size distributions of the samples were determined from measurement of nitrogen adsorption/desorption isotherms at -196°C on a Carlo Erba 1800 Sorptomatic. The samples were previously outgassed overnight at 300°C to a vacuum of $\leq 10^{-2}$ Pa, to ensure a dry clean surface free from any loosely held adsorbed species. The $S_{BET}$ was calculated by application of the BET equation [9] taking the area of the nitrogen molecule as 0.162 $nm^2$ [10]. However, it should be noted that the $S_{BET}$ calculated for microporous materials may only be taken as a guideline since the adsorption in narrow micropores is a volume filling effect [11]. The micropore volumes and external surface areas, *i.e.*, the area not associated with micropore filling, were calculated using a *t*-plot analysis [12], taking the thickness of an adsorbed layer of nitrogen as 0.354 nm assuming that the arrangement of nitrogen molecules in the film is hexagonal close packed [13]. The mesopore volumes were calculated from the amount adsorbed at $p/p° = 0.96$ on the desorption branch of the isotherm, equivalent to 50 nm pore diameter, minus any micropore volume calculated from the corresponding *t*-plot. The *t*-plots used the Harkins-Jura [14] relationship between the relative pressure and the thickness of the adsorbed layer since this data was based on nitrogen adsorption over a nonporous anatase sample.

Mercury intrusion porosimetry (MIP) analyses were employed to determine the pore size distribution and pore volume over the range of approximately 100 µm down to 7.5 nm diameter, utilising CE Instruments Pascal 140/240 apparatus, on samples previously dried overnight at 150°C. The pressure/volume data were analysed by use of the Washburn Equation [15] assuming a cylindrical nonintersecting pore model and taking the mercury contact angle as 141° and surface tension as 484 mN m$^{-1}$[10]. The primary particle sizes of the two raw materials heat treated at 500°C were determined from analysis of the mercury intrusion curves in the interval related to the filling of the interparticulate pore space, assuming spherical particle geometry [16]. Combination of the results from these two techniques leads to the characterisation of the total pore volume and pore size distribution of the conformed composite materials. The axial crushing strengths of the extrudates were measured with a Chatillon LTCM Universal Tensile and Spring Tester.

The isoelectric points (IEP) of the two raw materials treated at 500°C and the zero point charge (ZPC) of the composites were determined by electrophoretic migration [17], measuring the zeter potentials as a function of the solution pH using a Zeter-Meter Inc. Model 3.0+. Experiments were determined with 30 mg of approximately 2 µm diameter particles, suspended in 300 ml of 10$^{-3}$ M KCl, adjusting the pH value with 0.2 M KOH and HCl solutions. Each curve was recorded at least twice to ensure reproducibility.

Diffuse reflectance UV-Vis spectra were obtained with a Shimadzu UV-2401PC UV-Vis spectrometer equipped with an integration sphere diffuse reflectance attachment. The samples were used as powders and a halon white (PTFE) reflectance standard used to record the baseline. The reflectance spectra were taken over the range 200 to 800 nm and converted into Kubelka–Munk function $F(R)$.

## 3. RESULTS AND DISCUSSION

### 3.1 Textural properties of the raw materials

Six different compositions were prepared, with titania contents ranging from 0 to 80 wt. %. The results of the textural characterisation after heat treatment at 500°C are shown in Table 1. The specific surface areas of the raw materials treated at 500°C were 96 and 132 m$^2$g$^{-1}$ for the titania and bentonite, respectively. Thus, on forming the composites it should be noted that the specific surface areas were higher than expected for a purely physical mixture of the two raw materials treated at the same temperature, indicating that there had been a stabilisation of the titania that originally had a specific surface area of 330 m$^2$g$^{-1}$ when dried at 110°C [18]. The results indicated that for the composites the specific surface area of the titania was 120 m$^2$g$^{-1}$. The changes observed in the total pore volume and porosity of the extrudates was directly related to the increase in the fraction of titania.

Table 1

Textural characteristics of the extrudates.

| Composition Clay/Titania (wt. %) | Area $S_{BET}$ ($m^2g^{-1}$) | External Area ($m^2g^{-1}$) | Micropore Volume ($cm^3g^{-1}$) | Mesopore Volume ($cm^3g^{-1}$) | Macropore Volume ($cm^3g^{-1}$) | Total Pore Volume ($cm^3g^{-1}$) | Porosity (%) | Density ($gcm^{-3}$) | Crushing Strength ($kgcm^{-1}$) |
|---|---|---|---|---|---|---|---|---|---|
| 100/0 | 132 | 111 | 0.009 | 0.265 | 0.079 | 0.353 | 36 | 2.41 | 0.61 |
| 80/20 | 130 | 115 | 0.008 | 0.266 | 0.262 | 0.536 | 46 | 2.07 | 0.46 |
| 65/35 | 129 | 116 | 0.006 | 0.283 | 0.412 | 0.701 | 58 | 2.45 | 0.33 |
| 50/50 | 126 | 119 | 0.003 | 0.292 | 0.572 | 0.867 | 68 | 2.78 | 0.22 |
| 35/65 | 124 | 122 | 0.002 | 0.319 | 0.646 | 0.967 | 75 | 3.40 | 0.16 |
| 20/80 | 123 | 115 | 0.001 | 0.330 | 0.725 | 1.056 | 78 | 3.64 | >0.16 |

Bentonite powder had a total pore volume of 0.85 $cm^3g^{-1}$ measured by MIP with bimodal distribution with average pore diameters of about 14 nm and 10 μm for the intraparticulate and interparticulate porosity, respectively. However, when conformed by high pressure extrusion of the paste all of the porosity above 400 nm is lost but that below this point is left unchanged since it is due to the porosity of the individual primary particles. The titania powder also gave rise to a bimodal pore size distribution with average pore sizes of about 8 nm and 400 nm with a total pore volume of 1.2 $cm^3g^{-1}$. Analysis of the interparticulate porosity of this material gave an average primary particle size of c. 800 nm.

From the total pore volume curves shown in Fig. 1 it may be seen that with the incorporation of titania all the composites displayed bimodal distributions with average pore diameters of about 14 nm and 550 nm due to the intrinsic porosity of the raw materials and the primary particle size of the titania, respectively.

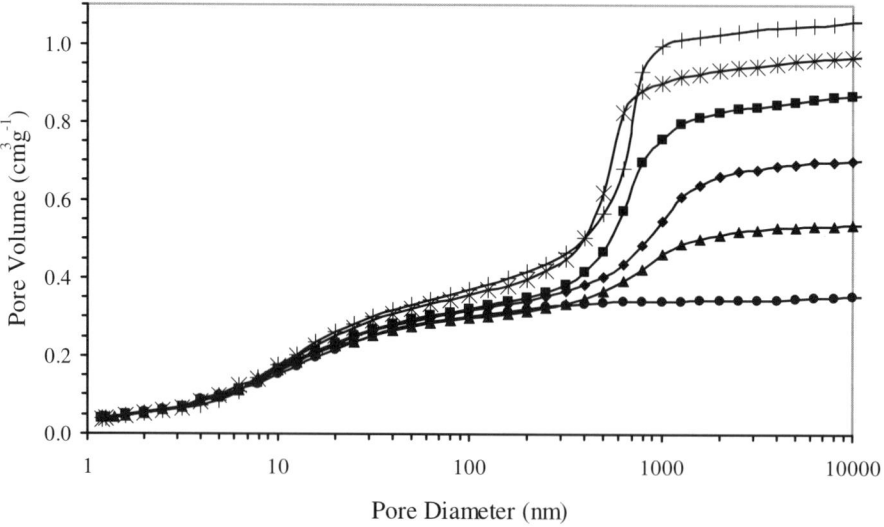

Fig.1. Total pore volume curves for extrudates: 100/0 (●), 80/20 (▲), 65/35 (♦), 50/50 (■), 35/65 (*) and 20/80 (+).

The relationship between the crushing strength and the total pore volume of the extrudates is presented in Fig. 2. From this figure it may be seen that the crushing strength for a brittle ceramic was related to the total pore volume, although the exact equation depends on the actual combination of the components used to form the ceramic, their primary particle sizes and the final heat treatment temperature. These curves are useful for estimating the expected strength for an unknown composition which falls within the measured range.

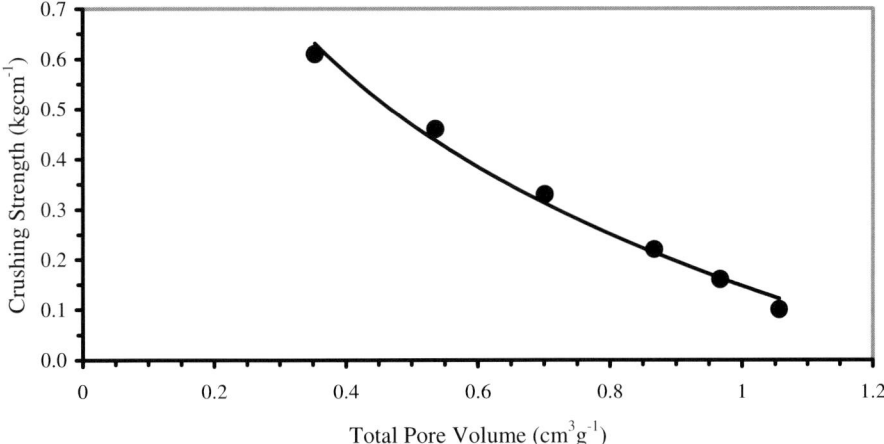

Fig. 2. Relationship between the Total Pore Volume and the Crushing Strength of the extrudates.

## 3.2 Surface properties of the raw materials

The photocatalytic activity of any material is related to the composition of the surface since the photons can only penetrate the first 2 μm. Thus, although the bulk composition of the extrudates was known the fraction of titania, which is the photocatalytically active component, exposed on the surface was unknown. Since the isoelectric points of the pure materials treated at 500°C were widely separated at pH's of 2.1 and 5.8 for bentonite and titania, respectively the determination of the zero point charge for the composite materials by electrophoretic migration should give a good indication of the surface distribution of these two components. Thus, based on the assumption that the zero point charge of a composite material should lie between the isoelectric points of the two pure materials the fraction of titania at the surface was determined by applying the following equations:

$$ZPC = X_A \, (IEP)_A + X_B \, (IEP)_B \qquad (1)$$

$$X_A + X_B = 1 \qquad (2)$$

The results from analysis of the zeta potential curves for the pure materials and the composites are presented in Fig. 3. From this curve it was obvious that in the composites the amount of titania at the surface was well below the bulk composition. Since the composites were produced on a by weight relationship these results may be explained in part by the differences in the densities and the surface areas of bentonite and titania. Thus, considering the 50/50 mixture the differences in the densities would lead to volume fractions of 0.615 and 0.385 for the bentonite and titania, respectively. The lower surface area of the titania compared to bentonite would further reduce the effective surface coverage to 34%. However, as the measured surface coverage for this material was only 16% obviously there are other criteria to be considered that lead to the titania particles being effectively covered by the clay agglomerating agent during the kneading, extruding, drying and heat treatment operations.

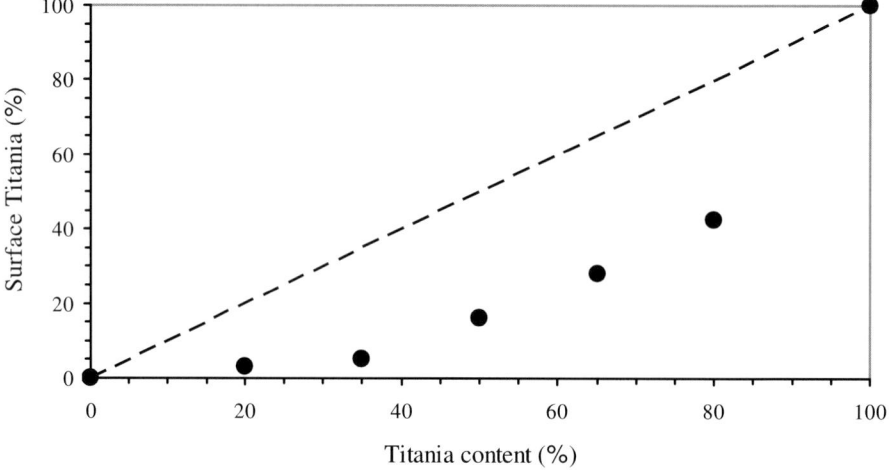

Fig. 3. Percentage of titania at the surface against bulk titania content.

The measurement of the UV-Vis spectra of the materials allows the determination of the Band Gap, that gives an idea of the energy necessary to photocatalytically activate a solid. This is the energy required to form an electron hole pair that cause the high oxidation capacity of the surface. For clarity UV-Vis spectra between 200 and 500 nm, for four of the six extrudates: 100/0, 80/20, 50/50, 20/80 are presented in Fig. 4. The UV-Vis spectrum of titania in its anatase form presents an absorption edge with an onset close to 400 nm, and an absorption maximum at 320 nm (due to the charge transfer $O^{2-} \rightarrow Ti^{4+}$ corresponding to the excitation of electrons from the O 2p valence band to the Ti 3d valence band and a shoulder at 260 nm due to the bending of the Ti 3d orbital of the Ti ions. The position of this band with its corresponding energy gap of 3eV characterises titania as an intrinsic semiconductor [11].

The spectrum for bentonite had an intense absorption at 200 nm which slowly diminished with higher wavelengths but extended into the visible region, giving rise to its characteristic tan colour. In the composite materials the spectra were very similar to that expected for the pure titania, although the absorbance between 200 nm and 300 nm was reduced and the absorption edge less well defined as the proportion of bentonite was increased.

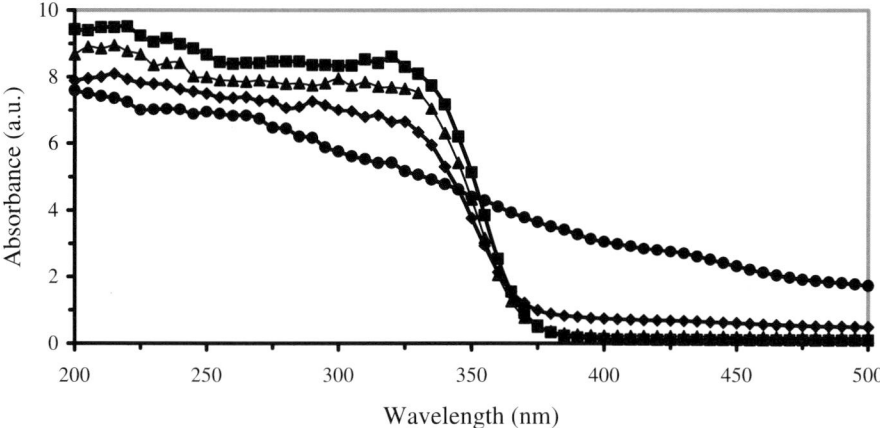

Fig. 4. UV-Vis spectra of the bentonite/titania samples 100/0 (●), 80/20 (♦), 50/50 (▲) and 20/80 (■).

## 4. CONCLUSIONS

From these results it has been shown how composite materials with high titania loadings can be successfully produced. The production of these incorporated materials has economic advantages compared to the fabrication of supported thin films due to the simplicity of the technique. Furthermore, the mechanical strength and abrasion resistance of these types of structures is adequate for their adoption in the photocatalytic degradation of VOCs in high volume effluents. Although the surface coverage of the titania was below that expected from the bulk composition analysis of the UV-Vis spectra demonstrated that the absorption edge and the band gap of the composite materials were hardly effected by the presence of the bentonite. This indicated that their photocatalytic activities should not be greatly reduced in comparison to pure titania.

**Acknowledgements**
        The authors appreciate the financial assistance received from the Spanish Ministry of Science and Technology through its projects PPQ2001-0692-CO2-O2 and CTQ2004-08232-CO2-O2.

**REFERENCES**

[1]  T.T. Ibusuki and K. Takeuchi, Atmos. Environ. 20 (1986) 1711.
[2]  J. Peral and D.F. Ollis, J. Catal. 136 (1992) 554.
[3]  J. Blanco, P. Avila, A. Bahamonde, E. Alvarez, B. Sánchez and M. Romero, Catal. Today 29 (1996) 437.

[4] L.A. Dibble and G.B. Raupp, Environ. Sci. Technol. 26 (1992) 492.

[5] L.A. Dibble and G.B. Raupp, Catal. Lett. 4 (1990) 345.

[6] M.R. Hoffmann, S.T. Martin, W. Choi and D.W. Bahnemann, Chem. Rev. 95 (1995) 69.

[7] X. Fu, W.A. Zeltner and M.A. Anderson, in Semiconductor nanoclusters, P.V. Kamat, D. Meisel (Eds.), Elsevier Science , New York, 11 (1996) 445.

[8] P. Avila, A. Bahamonde, J. Blanco, B. Sánchez, A.I. Cardona and M. Romero, Applied Catalysis B: Environmental 17 (1998) 75.

[9] S. Brunauer, P.H. Emmett and E. Teller, J. Amer. Chem. Soc., 60 (1938) 309.

[10] J. Rouquerol, D Avnir, C.W. Fairbridge, D.H. Everett, J.H. Haynes, N. Pericone, J.D.F. Ramsay, K.S.W. Sing and K.K. Unger, Pure and Appl. Chem., 66, 8 (1994) 173.

[11] M.M. Dubinin, E.D. Zaverina and D.P. Timofeeva, Zhur. Fiz. Khim., 23 (1949) 1129.

[12] B.C. Lippens and J.H. de Boer, J. Cat., 4 (1965) 319.

[13] B.C. Lippens, B.G. Linsen and J.H. de Boer, J. Cat., 3 (1964) 32.9.

[14] W.D. Harkins and G. Jura, J. Amer. Chem. Soc., 66 (1944) 1362.

[15] E.W. Washburn, Proc. Nat. Acad. Sci. U.S.A., 7 (1921) 115.

[16] R.P. Mayer and R.A. Stowe, J. Colloid Interf. Sci., 20 (1965) 893.

[17] G.A. Parks, Chem. Rev., 65 (1965) 177.

[18] P. Avila, J. Blanco, C. Knapp and M. Yates, Studies in Surface Science and Catalysis, 116 (1998) 233.

[19] J.B. Torrance, P.L. Chinnarong and R.M. Metzger, J. Solid State Chem., 90 (1991) 168.

Studies in Surface Science and Catalysis 160
P.L. Llewellyn, F. Rodriquez-Reinoso, J. Rouqerol and N. Seaton (Editors)
© 2007 Elsevier B.V. All rights reserved

# Simultaneous Determination of Intrinsic Adsorption and Diffusion of n-Butane in Activated Carbons by using the TAP Reactor

**V. Fierro**[a]**, Y. Schuurman**[b]** and C. Mirodatos**[b]

[a]Departament d'Enginyeria Quimica. Escola Tècnica Superior d'Enginyeria Química. Universitat Rovira i Virgili. Campus Sescelades 43007 Tarragona, Spain

[b]Institut de Recherches sur la Catalyse, CNRS, 2 avenue Albert Einstein, F-69626 Villeurbanne Cédex, France

The TAP reactor was used to simultaneously determine the adsorption and diffusion parameters of n-butane over chemically activated carbons in a packed bed. Measurements were performed over activated carbons prepared by phosphoric acid activation of kraft lignin with varying carbonization temperatures (400-650°C), weight ratios of phosphoric acid to lignin (P/L=1.0-1.8) and impregnation times (1 and 48h). The diffusion and adsorption in these carbons were adequately described with a simplified three-parameter model. The determined adsorption and diffusion parameters agree well with the values reported in literature. TAP pulse responses allow the study of diffusion in the microporous region.

## 1. INTRODUCTION

Activated carbons are widely used in gas separation and purification processes because of their large adsorptive capacity and low cost. Adsorption processes occurring on the surface of activated carbons are connected with their surface area, porous structure and surface heterogeneity. To model adsorption processes using activated carbons, information is needed on the basic features of adsorption isotherms and pore diffusion coefficients.

Transient kinetic investigation in a TAP reactor is an efficient tool for acquiring adsorption/desorption and diffusion parameters in porous materials and provides an excellent method for acceleration in catalyst development giving fundamental insights in reaction mechanisms [1]. The TAP reactor offers two main advantages due to the operation in ultra-high vacuum: external mass transfer limitations are completely absent and experiments are carried out at low coverages. Therefore, heat effects or interactions between molecules do not influence the values of the determined parameters and they can be referred to as the intrinsic heat of adsorption and the intrinsic diffusivity. In a previous work [2], we showed that the TAP reactor is also a powerful tool for the fast characterization of activated carbons.

This paper is focused on the determination of the diffusion and adsorption parameters of n-butane by transient experiments in several activated carbons prepared by phosphoric acid activation.

## 2. MATERIAL AND METHODS

### 2.1 Activated carbons

Lignin was mixed with varying amounts of $H_3PO_4$ in the range of 1.0 to 1.75 phosphoric acid to lignin weight ratio (P/L) on a wet basis. The slurry was left for varying impregnation times ($t_i$) from 1 to 48h at room temperature and under air atmosphere, then transferred to a furnace where carbonization was carried out under air atmosphere. The furnace was heated at $10°C$ $min^{-1}$ up to the carbonization temperatures, from 400 to 650°C, where the temperature was held for 2h. To remove the excess of $H_3PO_4$ after carbonization, the activated carbon was extensively washed with distilled water until a neutral pH was attained. Then, the samples were dried in an oven at 110°C overnight.

Surface area and pore size characterizations were performed using a Micromeritics ASAP2000 gas adsorption surface area analyzer. The specific surface area of the samples was determined from the nitrogen isotherms at 77K and by using the BET equation. Micropore volume was determined using the DR equation and the total volume of pores was calculated at a relative pressure ($p/p^0$) of 0.97. Table 1 shows the activation conditions and the structural characterization. More details are given elsewhere [3].

*Table 1.* Activation conditions, surface area and pore volumes of the activated carbons

| carbon | $T$ (°C) | P/L | $t_i$ (h) | $S_{BET}$ ($m^2/g$) | $V_{total}$ ($cm^3/g$) | $V_{micro}$ ($cm^3/g$) |
|--------|------|-----|------|------|------|------|
| C400 | 400 | 1.4 | 1 | 956 | 0.49 | 0.41 |
| C450 | 450 | 1.4 | 1 | 1047 | 0.51 | 0.41 |
| C525 | 525 | 1.4 | 1 | 1114 | 0.57 | 0.43 |
| C600 | 600 | 1.4 | 1 | 1269 | 0.64 | 0.45 |
| C650 | 650 | 1.4 | 1 | 844 | 0.45 | 0.29 |
| CR10 | 450 | 1.0 | 1 | 921 | 0.46 | 0.37 |
| CR12 | 450 | 1.2 | 1 | 955 | 0.47 | 0.38 |
| CR18 | 450 | 1.8 | 1 | 844 | 0.44 | 0.35 |
| CI48 | 450 | 1.4 | 48 | 863 | 0.42 | 0.35 |

### 2.2. TAP experiments

Transient responses experiments were performed in the TAP-2 reactor under vacuum conditions and at temperatures ranging from 423 to 498 K. An activated carbon loading of 15 mg with a particle size of approximately 0.2-0.3 mm was placed in the center of two layers of 0.2 - 0.3 mm size quartz particles. A microreactor filled with quartz only was employed to determine the Knudsen diffusion coefficients. n-Butane together with argon were introduced in the microreactor (25.4 mm in length and 4 mm in diameter) in a volume ratio of 1:1. The reactor is evacuated continuously and the transient responses of n-butane and argon were monitored by a quadrupole mass spectrometer following the signal of a single atomic mass unit (amu) per pulse as a function of time. The amu's of 43 and 40 were used to monitor butane and argon respectively. Pulse broadening reflects the diffusion between the particles, diffusion inside the particles and adsorption/desorption at the surface. The data-acquisition time to measure the entire pulse response was 1 s for argon and up to 20s for butane. By modeling the pulse responses the adsorption and diffusion parameters can be determined. For details of this technique, the reader is referred to [4].

## 2.3. Modeling

The studied carbons have a high fraction of mesopores that also contribute to the surface area, resulting in high adsorption rate constants. We assume that adsorption takes place at the carbon surface followed by diffusion into the interior of the particles.

The carbon particles are modeled as squared slabs with a characteristic length L and are considered to be symmetrical. The reversible sorption is described by an equation analogous to Henry's law:

$$C_{A,s}\big|_{z=L/2} = H'C_A \tag{1}$$

where H' is the analogous Henry coefficient ($m_g^3/m_{carb}^3$). Only the adsorbed molecules diffuse into the micropores. During TAP experiments, the concentration in the reactor remains very low and therefore the diffusion in the carbon pores is assumed to be independent of the concentration and is therefore described by Fick's law.

The reactor is divided into three zones: two inert zones of quartz beads between which the catalyst is placed. The diffusion in all three zones is described by Knudsen diffusion. In the carbon zone, the flux into the carbon particles is included in the model. Actually, the diffusion into the meso- and macropores is relatively fast and can therefore be lumped into the Knudsen diffusion coefficient according to the following equation [5]

$$D_{K,p} = \frac{D_K}{(1+(\varepsilon_p(1-\varepsilon_b))/\varepsilon_b)} \tag{2}$$

where $D_K$ is the Knudsen diffusivity in the bed ($m^2/s$), $D_{K,p}$ the Knudsen diffusivity in the pores of the particle ($m^2/s$), $\varepsilon_p$ the carbon particle porosity for meso- and macropores ($m_g^3/m_{carb}^3$) and $\varepsilon_b$ the bed porosity ($m_g^3/m_{react}^3$). This leads to the following continuity equation for the catalyst zone:

$$\varepsilon_b \frac{\partial C_A}{\partial t} = D_{K,p} \frac{\partial^2 C_A}{\delta x^2} - (1-\varepsilon_b)a_s J_{A,s} \tag{3}$$

where $C_A$ is the reactant concentration ($mol/m^3$), t is the time (s), x the reactor coordinate (m), $a_s$ is the micropore surface area per volume ($m^2/m^3$) and $J_{A,s}$ ($mol/m^2$ s) is the molar flux into the micropores as given by equation 4:

$$J_{A,s} = D_{s,eff} \frac{\partial^2 C_{A,s}}{\partial z^2} \tag{4}$$

where $D_{s,eff}$ is the effective diffusivity ($m^2/s$) in the activated carbons. The initial and boundary conditions for TAP experiments are reported in [4]. The additional boundary condition for this model is equation (1). The model has already been described in Schuurman et al. [6] and similar models to describe diffusion in microporous materials have been employed by Keipert and Baerns [7] and Nijhuis et al. [8].

Parameter estimation was performed by fitting the entire simulated response curve to the experimental one in the time domain. For each curve, the sum of the squared deviations over

5000 data points was used as the objective function which was minimized using an algorithm based on Marquardt's method [9].

For the regression analysis, a reparameterized form of the Arrhenius and Van 't Hoff equations was used. A full statistical analysis, which included the calculation of the 95% confidence intervals on the estimated parameters, was performed after regression. Initially, the response curves were recorded using a sampling time of 1 ms. For a correct statistical analysis the sampling time had to be increased to 8 ms.

## 3.  RESULTS

Fig. 1a) shows the experimental responses for butane over C450 and C650 carbons at temperatures between 423 and 498 K. Fig. 1b) shows the experimental and model responses for butane over C450 carbon at the same range of temperatures. Equally good fits were found for the rest of the carbons tested.

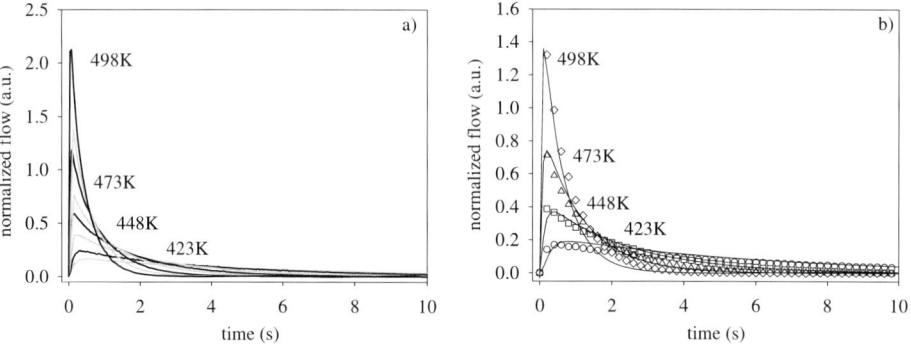

Figure 1.    a) Experimental response curves for n-butane pulses over C450 (grey) and C650 (black) and b) Experimental (symbols) and model (lines) response curves for butane pulses over C450 at temperatures from 423 to 498 K.

The data did not permit an estimation of the activation energy for the micropore diffusion significantly different from zero. Therefore, the model applied has only 3 adjustable parameters: the diffusivity in the micropores, the Henry coefficient for reversible sorption on the surface and the adsorption enthalpy. The Knudsen diffusivities for the extraparticle transport were determined from independent experiments. The other estimated parameters together with their 95% confidence interval are presented in Table 2.

At low pressures, the following relation exists between the analogous Henry coefficient, $H'$, and the Langmuir adsorption equilibrium coefficient, $K$ (Pa$^{-1}$):

$$H' = RTKq_{sat} \tag{5}$$

where $q_{sat}$ (mol m$_{carb}^{-3}$) is the butane saturation adsorption concentration. $q_{sat}$ was calculated from the BET surface area of the carbons and the molecular surface area of butane (0.321nm$^2$).

The adsorption entropy, $\Delta S$, reported in Table 2 is calculated from the Langmuir adsorption equilibrium coefficient, expressed in $atm^{-1}$, according to the following expression

$$\Delta S = R \ln(K) \tag{6}$$

Table 2. Adsorption and diffusion parameters for n-butane over different carbons.

| carbon | $D_K$ ($m^2 s^{-1}$) | $q^*_{sat}$ (mol $m^{-3}$ carb) | $\Delta S$ ($J mol^{-1} K^{-1}$) | $\Delta H$ ($kJ mol^{-1}$) | $K$ (473K) ($atm^{-1}$) | $D_{s,\,eff}$ (473K) ($m^2 s^{-1}$) |
|---|---|---|---|---|---|---|
| C400 | 12.9 | 3269 | $-116 \pm 0$ | $-49.4 \pm 1.3$ | $9.3\ 10^{-7}$ | $1.2 \pm 0.1\ 10^{-7}$ |
| C450 | 12.3 | 3807 | $-118 \pm 0$ | $-51.2 \pm 1.2$ | $6.9\ 10^{-7}$ | $9.3 \pm 0.8\ 10^{-8}$ |
| C525 | 12.0 | 3401 | $-117 \pm 0$ | $-50.7 \pm 2.5$ | $8.4\ 10^{-7}$ | $8.7 \pm 0.1\ 10^{-8}$ |
| C600 | 11.6 | 3308 | $-116 \pm 0$ | $-50.4 \pm 1.4$ | $9.1\ 10^{-7}$ | $7.4 \pm 0.8\ 10^{-8}$ |
| C650 | 12.0 | 2200 | $-120 \pm 0$ | $-53.4 \pm 1.5$ | $5.9\ 10^{-7}$ | $4.7 \pm 0.5\ 10^{-8}$ |
| CR10 | 12.6 | 3595 | $-112 \pm 0$ | $-49.1 \pm 1.2$ | $1.5\ 10^{-6}$ | $1.0 \pm 0.1\ 10^{-7}$ |
| CR12 | 12.3 | 3455 | $-117 \pm 0$ | $-51.1 \pm 1.2$ | $7.9\ 10^{-7}$ | $9.2 \pm 0.8\ 10^{-8}$ |
| CR18 | 12.6 | 3103 | $-120 \pm 0$ | $-52.5 \pm 1.9$ | $5.5\ 10^{-7}$ | $1.1 \pm 0.2\ 10^{-7}$ |
| CI48 | 13.1 | 2791 | $-110 \pm 0$ | $-49.2 \pm 1.3$ | $1.9\ 10^{-6}$ | $7.5 \pm 0.8\ 10^{-8}$ |

* $q_{sat}$, calculated from the BET surface area of the carbons and the molecular surface area of butane (0.321 $nm^2$)

## 4. DISCUSSION

### 4.1 Adsorption parameters

The adsorption parameters for the activated carbons were rather similar and no clear trends were observed with the variation of the activation conditions. An average value of approximately 116 J $mol^{-1} K^{-1}$ was determined for the adsorption entropy of n-butane in accordance with the somewhat lower value of its entropy of condensation, 90 J $mol^{-1} K^{-1}$. Adsorption enthalpies for n-butane varied between 49.1 and 53.4 kJ $mol^{-1}$. These values are very high, more than two times the vaporization heat of butane (21 kJ $mol^{-1}$). Such values would be unexpected if the dispersion forces were responsible for the interaction between the pore and the n-butane. Nevertheless, if the adsorption occurs in the Henry's law region, at very low adsorbate concentration, the molecules have large probabilities of being adsorbed on the highest energy centers, i.e. into pores of dimensions close to the molecular size [10,11]. In this case the adsorption heat is much larger, in absolute value, than the vaporization heat, because of a cooperative effect of the walls, which rises the potential. Finding adsorption enthalpies which almost do not depend on the textural characteristics of the carbons means that all the samples have micropores whose dimensions are close to the n-butane molecular size.

The adsorption enthalpies for n-butane determined in this work are in the (rather wide) range of those reported in the literature. Hu et al. [12] and Do and Do [13] determined adsorption enthalpies of approximately 16 and 35 kJ $mol^{-1}$ in the micropore and meso-macropore region, respectively. Allen et al. [14] measured isosteric heats of adsorption of 32 and 66 kJ $mol^{-1}$ for n-butane adsorption over a non-porous carbon and over a molecular sieve carbon, respectively. Liu et al. [15] determined an adsorption enthalpy of 53.9 kJ $mol^{-1}$ at zero coverage. Linders et. al. [16] found adsorption enthalpies between 45 and 50 kJ $mol^{-1}$ for two different carbons in a Multitrack apparatus a technique similar to the TAP system described here. Prasetyo et al. [17] calculated an adsorption enthalpy of 42.3 kJ mol$^{-1}$ by using the Clapeyron equation.

## 4.2. Diffusion parameters

The pore diffusivity at 473 K, $D_{s,eff}$, varied between 4.7 and 12 $10^{-8}$ m$^2$ s$^{-1}$. These values are in the range of those found by other authors. In Do et al [18], Do and Do [13] and related works by the same authors, values of surface diffusivities between 1.0 and 9.3 $10^{-8}$ m$^2$ s$^{-1}$ at 303 K are reported. Linders et al. [16] determined a value of the diffusion coefficient of 2.5 $10^{-9}$ m$^2$ s$^{-1}$, for a essentially microporous carbon, and of 7.2 $10^{-7}$ m$^2$ s$^{-1}$, for an activated carbon with a high fraction of meso- and macropores, both at 473K.

Figure 2 shows the diffusion coefficient measured from TAP experiments as a function of the mean micropore diameter of the various active carbons. The mean pore diameter has been calculated from:

$$\overline{d}_p = \frac{4\ V_{micro}}{S_{BET}} \qquad (7)$$

A linear decrease in the diffusivity can be observed with decreasing mean pore diameter. This implies that even the smallest micropores are accessible to the gas during a pulse.

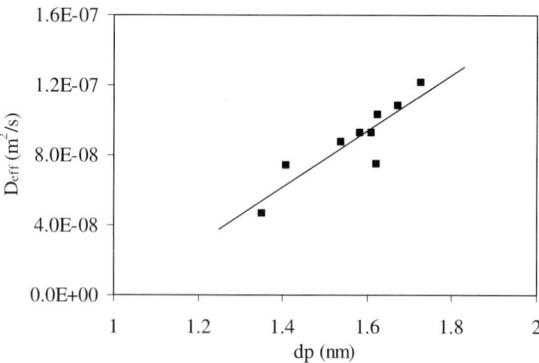

Figure 2. $D_{s,eff}$ estimated at 473 K versus the mean micropore diameter of the activated carbons.

$D_{s,eff}$ is calculated as an average value that takes into account the residence time of the molecules in the different pore sizes. More sophisticated models that take into account the pore size distribution are needed to more accurately calculate the mean pore diameter from transient experiments.

## 5. CONCLUSIONS

The TAP reactor allows the fast and simultaneous determination of adsorption and diffusion parameters in activated carbons for the low-pressure limit. The adsorption and diffusion parameters determined agree well with those values obtained by other authors for n-butane adsorption over different activated carbons. The determined adsorption enthalpies are very high and almost do not depend on carbons, which means that all the samples have micropores whose dimensions are close to the n-butane molecular size. A TAP pulse response experiment allows gas access into the micropores. The value of the diffusion coefficient decreases linear with decreasing pore size.

### ACKNOWLEDGEMENTS
This research was made possible in part by financial support from MCYT (project PPQ2002-04201-CO02), DURSI (2001SGR00323 and 2002AIRE) and ALFA Program (project ALFA II 0412 FA FI). V. Fierro acknowledges the MCYT and the Universitat Rovira i Virgili (URV) for the financial support of her 'Ramón y Cajal' research contract.

## REFERENCES

[1]　A.C. van Veen, D. Farrusseng, M. Rebeilleau, T. Decamp, A. Holzwarth, Y. Schuurman and C. Mirodatos, J. Catal., 216 (2003) 135.
[2]　V. Fierro, M.T. Izquierdo, Y. Schuurman, B. Rubio and C. Mirodatos, Stud. Surf. Sci. Catal., 144 (2002) 255.
[3]　V. Fierro, V. Torne, D. Montané and J. Salvadó, 'Activated Carbons Prepared from Kraft Lignin by Phosphoric Acid Impregnation', Paper presented in Carbon'03, 2003.
[4]　J.T. Gleaves, G. S. Yablonskii, P. Phanawadee and Y. Schuurman, Appl. Catal. A : General 160 (1997) 55.
[5]　J.P. Huinink, J.H.B.J. Hoebink and G.B. Marin Can. J. Chem. Eng., 74 (1996) 580.
[6]　Y. Schuurman, A. Pantazidis and C. Mirodatos, Chem. Eng. Sci., 54 (1999) 3619.
[7]　O.P. Keipert and M. Baerns, Chem. Eng. Sci., 53 (1998) 3623.
[8]　T.A. Nijhuis, L.J.P. van den Broeke, M.J.G. Linders, J.M. van de Graaf, F.Kapteijn, M.Makkee and J.A. Moulijn, Chem. Eng. Sci., 54 (1999) 4423.
[9]　D.W. Marquardt, J. Soc. Indust. Appl. Math., 11 (1963) 431.
[10] X. Hu, B. King and D.D. Do, Gas Sep. Purif., 8 (1994) 175.
[11] E.G. Derouane, J.M. André and A.A. Lucas, J. Catal. 110 (1988) 58.
[12] E.G. Derouane, J. Mol. Catal. A: Chem. 134 (1998) 29.
[13] D.D. Do and H.D. Do, Sep. Purif. Technol., 20 (2000) 49.
[14] J.L. Allen, J.L. Gatz and P.C. Eklund, Carbon, 37 (2001) 1485.
[15] Y. Liu, J.A. Ritter and B.K. Kaul, Sep. Purif. Technol., 20 (2000) 111.
[16] M.J.G. Linders, L.J.P. van den Broeke, T.A. Nijhuis, F. Kapteijn and J.A. Moulijn, Carbon, 39 (2001) 2113.
[17] I. Prasetyo, H.D. Do and D.D. Do, Chem. Eng. Sci., 57 (2002) 133.
[18] D.D. Do, X. Hu and P.L.J. Mayfield, Gas Sep. Purif., 5 (1991) 35.

Studies in Surface Science and Catalysis 160
*P.L. Llewellyn, F. Rodriquez-Reinoso, J. Rouqerol and N. Seaton (Editors)*

# Porous carbon deposits in controlled fusion reactor: adsorption properties and structural characterization

**C. Martin[a], M. Richou[a], C. Brosset[b], W. Sakaily[c], B. Pégourié[b] and P. Roubin[a]**

[a]PIIM-UMR 6633, Université de Provence, Centre Saint-Jérôme (service 242), F-13397 Marseille cedex 20, France

[b]Association EURATOM-CEA, CEA Cadarache, F-13108 Saint Paul Lez Durance, France

[c]CP2M, Université Paul Cezanne, Centre Saint-Jérôme (service 221), F-13397 Marseille cedex 20, France

## 1. INTRODUCTION

In the fusion devices (tokamaks) using carbonaceous plasma facing components (PFC), erosion as well as re-deposition of carbon atoms have been observed. In the current configuration of Tore Supra (TS), the tokamak located in Cadarache, France, the largest part of PFCs is the toroidal pumped limiter, which is in close interaction with the plasma. It is composed of actively cooled (120°C) carbon fiber composites (CFC) tiles. Analysis of long plasma discharges has pointed out an abnormal deuterium retention which can be due to re-deposited layers, as observed on several places on the PFCs [1]. These carbon layers (up to 800 micrometers thick) have been collected and their chemical analysis performed using nuclear reaction analysis shows a D/C deuterium concentration less than 10 % [2]. This low co-deposition rate makes difficult explanation of the particle retention and it is possible that porosity and structural properties of deposits play an important role in deuterium retention.

The first part of this paper focuses on the structural study of TS deposits using scanning and transmission electron microscopy in order to characterize the mesopore and macropore network, as well as the microstructure. In the second part, we present a comparative study of adsorption properties of TS deposits and some reference samples. We have measured volumetry adsorption isotherms of various probe molecules like $CH_4$, $N_2$, $C_6H_6$ and have determined parameters like energy adsorption, micropore volume and pore size distribution (PSD), using empirical methods and both Dubinin-Asthakov and Stoeckli models [3].

## 2. EXPERIMENTAL PART AND METHODS

### 2.1. Samples

The TS samples were scraped-off layers from reactor walls collected on the toroidal pumped limiter (LIM sample) and on the leading edge of the neutralizers of pumped limiter (NTR sample).

The reference samples were:

- the original carbon fiber composite (CFC sample) provided by SNECMA which is composed of PAN fibers and pyrolytic carbon matrix. This sample is mainly graphitic and non-microporous.
- a commercial microporous activated carbon (AC-C03 sample) provided by MeadWestvaco corporation which was a nuchar FAC activated carbon with a specific surface area between 600 and 1000 $m^2 g^{-1}$.

## 2.2. Electron microscopy

Transmission electron microscopy (TEM) has been carried out at CP2M laboratory (Marseille, France) ; a high resolution, field emission, transmission electron microscope, JEOL 2010F, was used with a structural resolution of 0.18 nm and a spot size of 0.2 nm. Focused Ion Beam (FIB) proved to be very suitable for localisation of a relevant zone and preparation of TEM thin foils. The FIB was a Philips FIB 200 TEM with a spatial resolution of 5 nm and with a platinum gas injector.

## 2.3. Adsorption isotherms

Volumetric point by point isotherms were measured on a home made experimental setup which was composed of a quartz adsorption cell connected to a stainless calibrated volume equipped with three pressure gauges (MKS Baratron, manometer capacitance) in the $10^{-4}$- 1000 mbar range. This cell was maintained under a vacuum better than $10^{-7}$ mbar, its temperature being regulated by a cryogenic bath (liquid nitrogen or CO2 sticks). Before each experiment, the samples were slowly outgassed during 6 hours to 600° C under $10^{-7}$ mbar. The probe molecules were nitrogen (Alphagaz, with a purety of 99.9990 %), methane (Air Liquide, 99.95 %), benzene (Carlo Erba Reagenti, RPE, ACS for analysis). We have checked the reproducibility of both isotherm shapes and heights.

We used the Brunauer-Emmett-Teller (BET) theory [4], as a standard procedure, to determine specific surface area of samples, $a_{BET}$ and net heat of adsorption, $\Delta Q$. For the various adsorbates $a_{BET}$ was calculated at relative pressure, $p/p^0$ (where $p$ and $p^0$ were the equilibrium and the saturation pressure, respectively), in the range 0.01-0.35, which was the validity domain of the linear BET plot.

In order to characterize the adsorbents micropore system we used the Dubinin theory of the volume filling of micropores. The isotherm for slit-shaped micropores of activated carbon was described by the Dubinin-Asthakov (DA) equation in the relative pressure range $10^{-4}$-0.2 [5]:

$$V(p) = V_0 \exp[ -(A/\beta E_0)^n ] \tag{1}$$

where $V$ was the adsorbed volume ; $A = RT \ln(p^0/p)$ ; $\beta$ the affinity coefficient ; $E_0$ the characteristic energy. For carbonaceous materials, the values of exponent $n$ lie usually in the 1-3 range, depending strongly on the adsorbent type. Values of $n$ close to three or higher than three, are the most frequently found for adsorbents with narrow micropores of small size range. The most simple slit-shaped microporosity method of characterization is the estimation of the average micropore size ($L_0$) defined by the following relation [6]:

$$L_0 (nm) = 10.4 / [ E_0 (kJ \; mol^{-1}) - 11.4 ] \tag{2}$$

Then, assuming that microporosity is composed of different pore sizes and that each type of adsorption can be described by the DA equation with exponent $n$ equal to 3, we used the Stoeckli method [7] to determine the gaussian PSD, $f(L)$:

$$V(p) = V_0 \{ a \, / \, [a + (A/\beta E_0 L_s)^3] \}^m \tag{3}$$

$$f(L) = 3 \; V_0 \; a^m \, L^{(3m - 1)} \exp(-a \, L^3) \, / \, \Gamma(m) \tag{4}$$

where $L_s$ was mean pore size in Eq. (3) and $a$, $m$ were parameters of pore size distribution.

## 3. RESULTS AND DISCUSSION:

### 3.1. Electron microscopy
Scanning electron microscopy (SEM) micrographs of NTR sample are displayed in Fig. 1. The deposited layer shows egg-like structures with self-similarity character. This self-similarity has been analyzed in a previous work [8] leading to a fractal dimension of 2.15.

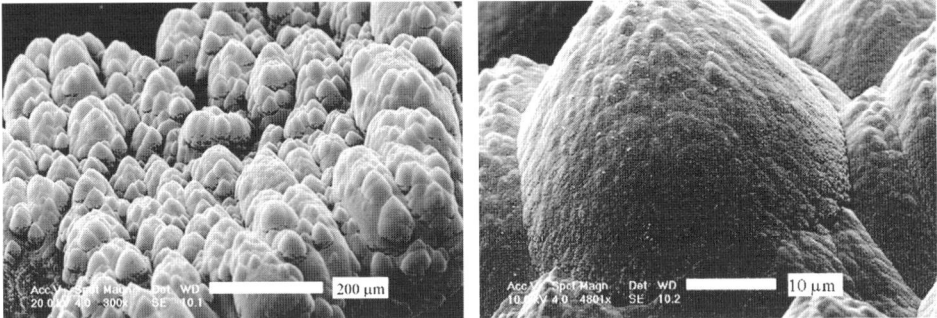

Fig. 1. Scanning electron microscopy micrographs of NTR sample.

Two FIB foils ($150 \times 15 \times 0.1$ μm) have been prepared from this layer: F1 and F2, parallel and perpendicular to the egg symmetry axis, respectively. TEM micrograph of F1 is displayed in Fig. 2(a), it clearly shows a macropore (width about 100 nm) and a network of parallel mesopores (mean width about 20 nm). The same experiment performed on F2 reveals that these pores are slit-shaped.

F1 is observed at the atomic scale in Fig. 2(b) using high resolution transmission electron microscopy. The observation of the parallel lines due to contrast between different intensities indicates that the NTR sample is mainly graphitic: these lines correspond to the graphene sheets whose c axis is parallel to the foil. The inter-planar distance, $d_{002}$, has been measured on several micrographs with several sheets to be of $3.5 \pm 0.2$ Å. The coherent length along c which gives information on the long range order has been estimated to be 7 nm (20 sheets). This local analysis is in agreement with mean values obtained for NTR using X-ray diffraction [9].

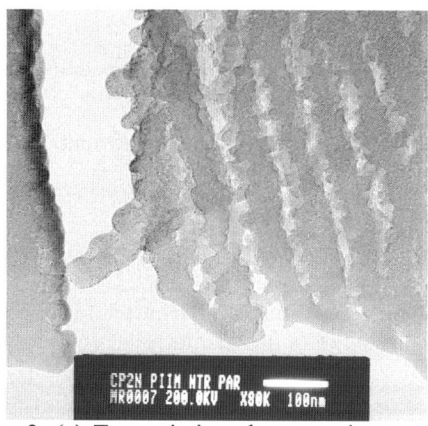 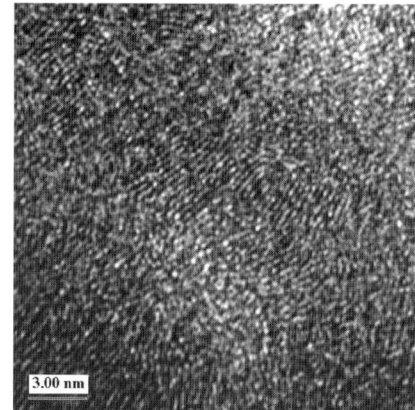

Fig. 2. (a) Transmission electron microscopy and (b) high resolution transmission electron microscopy micrographs of NTR sample thin foil obtained using FIB.

### 3.2. Adsorption isotherms

Methane isotherms measured at 77 K on TS and CFC samples are displayed in Fig. 3. CFC isotherm is a type II isotherm according to the IUPAC classification [10], while TS samples isotherms are type I (as evidenced by the logarithm plot of Fig. 5). A type IV isotherm should be measured for mesoporous adsorbents: the NTR isotherm is therefore not consistent with the previous TEM results, indicating that the mesopore network is not fully open. This result is consistent with the NTR low density ($\rho = 0.8$ g cm$^{-3}$) we have measured by helium picnometry.

BET results ($a_{BET}$ and $\Delta Q$) are gathered in Table 1. Specific surface area of TS samples are much higher than that of CFC, ~ 80 times for NTR and ~ 60 times for LIM [11]. The LIM sample presents a higher $a_{BET}$ value than that of CFC, but it remains lower than the NTR one. These results are consistent with the non-porous nature of CFC and with the relative roughness of NTR, LIM and CFC surfaces observed at micrometer scale by SEM (micrographs not shown here).

Fig. 4 shows the $\alpha$-plot obtained by plotting the adsorbed volume on NTR versus the adsorbed volume on CFC at the same $p/p^0$. The non-linearity observed for low adsorbed volume is due to microporosity: the first part of the curve corresponds to the adsorption in the micropores whereas the second part corresponds to the adsorption on the external surface. We deduce from this $\alpha$-plot a micropore volume of 34 cm$^3$ STP g$^{-1}$ for NTR.

Table 1

BET parameters for CH$_4$ adsorption on samples at 77 K: $a_{BET}$, specific surface area ; $\Delta Q$, net heat of adsorption. $a_{BET}$ / m$^2$ g$^{-1}$ is calculated using the CH$_4$ molecular area equal to 0.178 nm$^2$.

|  | $a_{BET}$ | | $\Delta Q$ |
|---|---|---|---|
|  | cm$^3$ STP g-1 | m$^2$ g-1 | kJ mol$^{-1}$ |
| CFC | 0.3 | 2 | 5.2 |
| LIM | 25.0 | 120 | 5.6 |
| NTR | 39.0 | 190 | 5.2 |
| AC-C03 | 213.0 | 800 | 5.5 |

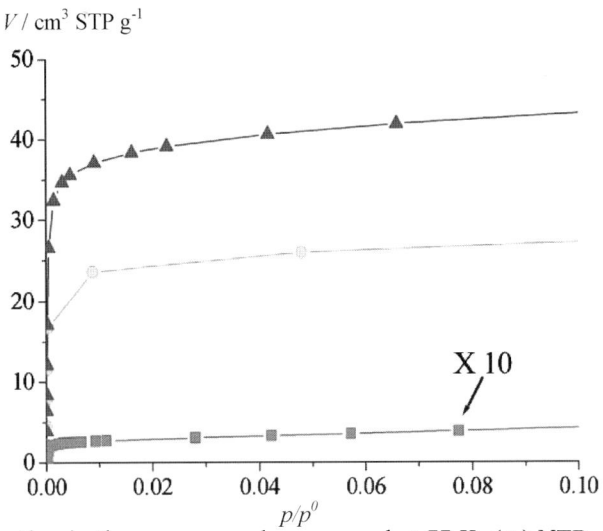

Fig. 3. CH$_4$ adsorption isotherms on samples measured at 77 K: (▲) NTR, (●) LIM and (■) CFC samples. For clarity the isotherm on CFC is multiplied by 10.

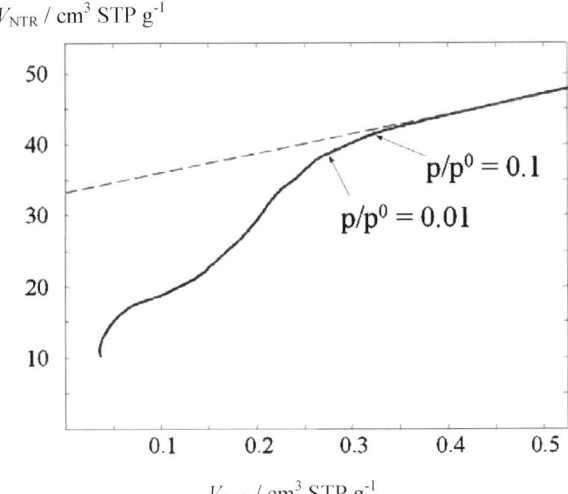

Fig. 4. α-plot using CH$_4$ adsorption isotherms on NTR and CFC measured at 77 K.

Fig. 5 shows the CH$_4$ isotherms measured at 77 K on NTR, LIM, CFC and AC-C03 samples and displayed in a logarithm scale in order to magnify the adsorption range for relative pressure below 0.1. Both isotherm step shape and position ($p/p^0$ at $V/V_0 = 0.5$) are different. BET and Dubinin theories show, that lower the position of the step, lower the adsorption energy of the adsorbent ($\Delta Q$ and $E$ respectively). Obviously, the position values are lower for microporous adsorbents (NTR, LIM, AC-C03) than that of CFC. Isotherm shapes of the microporous adsorbents are dominated by the absolute pore width and geometry [12].

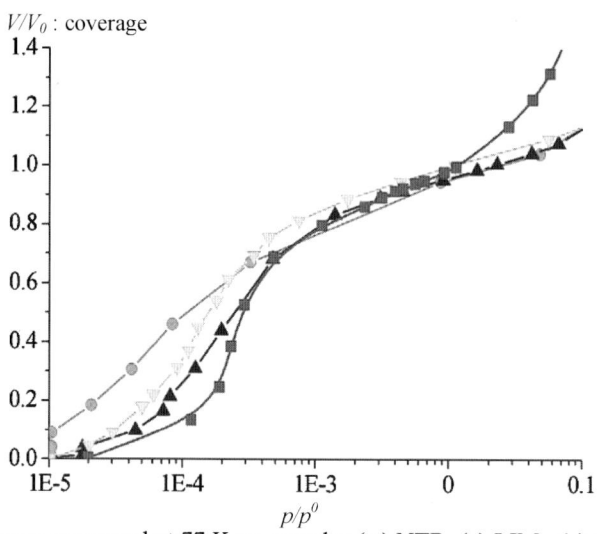

Fig. 5. $CH_4$ isotherms measured at 77 K on samples (▲) NTR, (•) LIM , (■) CFC and (▼) AC-C03.

We performed a comparative study of NTR and AC-C03 isotherm shapes, in order to quantitatively characterize micropores, using DA and Stoeckli equations. We have first applied the DA equation to $C_6H_6$ / AC-C03 isotherm measured at 293 K. The fit parameters were $E_0$, $n$ and $V_0$, and $\beta$ was fixed to 1, the commonly used value for $C_6H_6$ on activated microporous carbons. Thus, we have deduced $E_0 = 19$ kJ mol$^{-1}$ for the AC-C03 and the $\beta$ values for each gas, our results being consistent with published values [13-16].

The results of DA modeling of $N_2$ isotherms on NTR and AC-C03 at 77 K are shown in Fig. 6 and Table 2. NTR has a low $n$ and a high $E$ values compare to AC-C03, we have obtained similar results with the $C_5H_{12}$ isotherms at 195 K, nevertheless with a reduce gap between $E$ values and slightly different $n$. The lowest $n$ value corresponds to the broadest PSD whereas the highest energy corresponds to the lowest mean pore size. Fig. 7 shows PSD obtained from the Stoeckli model, a mean pore size of 1.8 and 1.9 nm are found for NTR and AC-C03 respectively: this is consistent with the previous Dubinin-Asthakov analysis (Table 2).

Table 2
Dubinin-Asthakov parameters for $N_2$ adsorption on samples at 77 K: $n$, exponent ; $E$, characteristic energy ; $L_0$ average micropore size ; $\beta$ affinity coefficient.

| $N_2$ (77 K) | $n$ | $E$ kJ mol$^{-1}$ | $\beta$ | $L_0$ nm |
|---|---|---|---|---|
| AC-C03 | 3.65 | 5.0 | 0.26 | 1.37 |
| NTR | 2.7 | 5.6 | 0.26 | 1.00 |

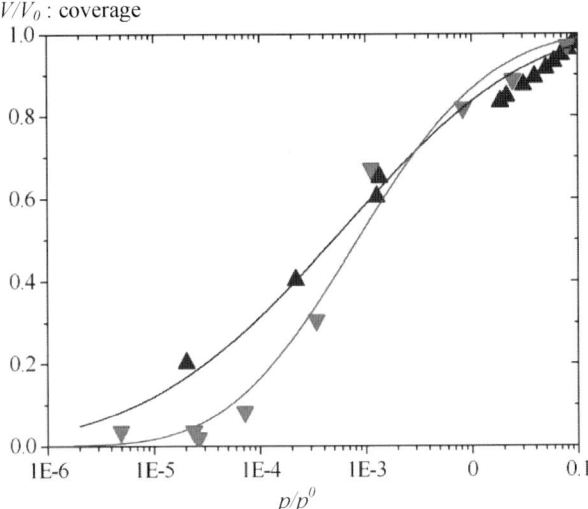

Fig. 6. Adsorption isotherms of $N_2$ on NTR and AC-C03 measured at 77 K: (▲) NTR, (▼) AC-C03 and (—) Dubinin-Asthakov models.

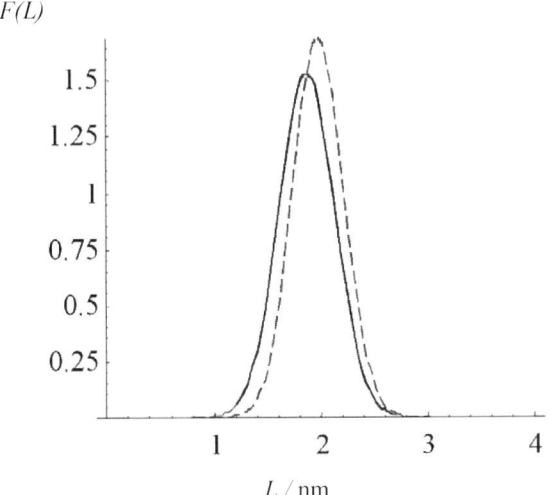

Fig. 7. Pore size distribution for $N_2$ adsorption on samples using Eq. (4): (—) NTR and (– –) AC-C03.

## 4. CONCLUSION

The local investigation of microstructure using high resolution transmission electronic microscopy have shown that NTR deposit is graphite-like with nanoparticles size about 7 nm. These results are in good agreement with previous X-ray diffraction study [9]. Transmission electronic microscopy investigation has shown that the deposited layers in Tore Supra exhibit an important porosity which is composed of slit-shaped pores in the mesopore as well as in the macropore range. No mesopores evidence on adsorption isotherms and low density of

NTR suggest that these mesopores are probably closed. In order to confort this hypothesis, it could be interesting to perform experimental methods such as liquid intrusion and thermoporosity [17]. Gas adsorption study has revealed a TS deposits microporosity and its quantitative characterization has been done using Dubinin-Asthakov and Stoeckli theory. To valid these results, we will extend the number of probe molecules and compare our experimental results with DFT a statistical method [18]. Finally, this observed multiscale porosity can play a role in diffusion and retention of hydrogen, studies are on progress to put in evidence this effect.

## REFERENCES

[1] B. Pégourié et al., Physica Scripta T111 (2004).23-28.
[2] C. Brosset, H. Khodja, Tore Supra team, Journal of Nuclear Materials 337–339 (2005) 664–668.
[3] F. Rouquerol, J. Rouquerol, K. Sing, Adsorption by Powders & Porous Solids, Academic Press, London, 1999.
[4] S. Brunauer, P. Emmet, E. Teller, J. Am. Chem. Soc. 60 (1938) 309.
[5] M. M. Dubinin, A.V Astakhov,. Izv. Akad.Nauk SSSR Ser. Khim, (1971) 125.
[6] F. Stoeckli, P. Rebstein, L. Ballerini, Carbon 28 (1990) 907-909.
[7] F. Stoeckli, Porosity in carbons – Characterization and applications, (ed.) J. Patrick, London: Arnold, (1995) 67-92.
[8] E. Delchambre et al., 30th EPS Conference on Controlled Fusion and Plasma Physics, ECA 27A (2003) P-3.169.
[9] P. Roubin, C. Martin, C. Arnas, Ph. Colomban, B. Pégourié, C. Brosset, Journal of Nuclear Materials 337-339 (2005) 990-994.
[10] K.S.W. Sing, D.H. Everett, R.A.W. Haul et al., Pure Appl. Chem., 57 (1985) 604.
[11] C. Martin et al., 30th EPS Conference on Controlled Fusion and Plasma Physics, ECA 27A (2003) P-1.158.
[12] K.S. Sing, Adv. Colloid Interface Sci. 76-77(1998)3-11.
[13] G.O Wood, Carbon, 39 (2001) 343-356
[14] D. Hugi-Cleary, S. Wermeille, F. Stoeckli, Chimia, 57 (2003) 611-615
[15] P.J.M Carrott, M.M.L Ribeiro Carrott, Carbon, 37 (1999) 647-656
[16] F. Stoeckli, A. Guillot, A.M. Slasli, D. Hugi-Cleary, Carbon 40 (2002) 383-388.
[17] R. Denoyel, P. Llewellyn, I. Beurroies, J. Rouquerol, F. Rouquerol and L. Luciani, Part. Part. Syst. Charact. 21 (2004) 128-137.
[18] PI Ravikovitch, A. Vishnyakov, R. Russo, AV Neimark, Langmuir 16 (2000) 2311-2320.

Studies in Surface Science and Catalysis 160
P.L. Llewellyn, F. Rodriquez-Reinoso, J. Rouqerol and N. Seaton (Editors)

257

# Characterization of the porosity of a microporous model carbon

**Q. Cai, A. Buts, N. A. Seaton and M. J. Biggs**

Institute for Materials and Processes, University of Edinburgh, Kenneth Denbigh Building, Mayfield Road, Edinburgh EH9 3JL, UK.

The intersecting capillaries model (ICM), combined with the Monte Carlo simulation approach, was applied to characterize a computer-generated microporous "model carbon" with known structure, in order to evaluate the realism of this characterization method. The "partial" PSDs for three species ($CH_4$, $CF_4$ and $SF_6$) were obtained by comparing the Monte Carlo simulated isotherms in the slit pores of the ICM with the isotherms generated from the model carbon. There is good agreement between model carbon-generated isotherms and the isotherms predicted based on the overall PSDs (by combining the partial PSDs). The overall PSD agree well with the real PSD of the model carbon in their dominant pore size range. These results support the validity and the realism of this characterization method for the characterization of porous carbons.

## 1. INTRODUCTION

To interpret the pore size distribution of porous carbons, an "intersecting capillaries model (ICM)" is often used to approximate the microstructure of nanoporous carbons. The pores are assumed to be a poly-disperse ensemble of slit pores with smooth graphite walls, which are connected in some way. Despite its simple geometry, this ICM has been used successfully to characterize carbon adsorbents and to predict adsorption in these materials and thus will likely remain the dominant approach to characterization by adsorption [1].

The PSD is obtained by solving the adsorption integral equation:

$$N(T,P)_i = \int_0^\infty \rho(w,T,P)_i f(w)dw \qquad i = 1 \cdots n \qquad (1)$$

where $N(T,P)_i$ is the experimentally determined adsorption at temperature $T$ and bulk fluid phase pressure $P$. $\rho(w,T,P)_i$ is the density of adsorptive in a model pore of width $w$, and $n$ is the number of adsorption measurements used in the analysis. The solution of eq.1 to give $f(w)$ is strongly dependent on both the model-pore isotherms and the experimental isotherms.

Several methods [2-9] have been in use to interpret the PSD of microporous carbons, by correlating adsorption in the slit-shaped model pores with the experimental adsorption on the real carbons. Recently, the grand canonical Monte Carlo (GCMC) simulation method has received more attention [9-13]. Based on a molecular description of adsorption, this method can reliably determine the PSD throughout the micro- and meso-pore size range. Although many researchers have demonstrated that this method gives self-consistent PSD, until now there has been no independent evidence that the structural information obtained is accurate. This work is intended to test this, by applying the GCMC simulation method to characterize a computer-generated microporous model carbon.

## 2. THE MODEL CARBON

The foundations of the solid model are based on the understanding that nanoporous carbons are built up from domains of two-dimensional short-range order that may be reasonably represented by small polyaromatic molecules or similar structures. These domains assemble in a roughly aligned manner with out-of-plane spacing somewhat greater than that of graphite to form nanoscale regions of *local molecular orientation* (LMO), which then combine to form mesoscopic structures. Several solid model parameters are used to define the structure of the model carbons: the LMO extents, the inter-LMO spacing, the intra-LMO spacing distribution, and the domain misorientation (the tilting of superdomains). Varying these parameters would yield complex pore structures that contain a wide variety of pore shapes, extents, and surfaces. The detailed description of the model carbons and the algorithm for building them can be found in the papers by Biggs *et al.* [14]. The model carbon used in this work, shown in Figure 1, is Model $1_{p(15,15)}$ listed in their papers. This model has no inter-LMO pores and a random $\pm 15^{\circ}$ tilting of superdomains imposed about both centroidal in-plane axes, which makes it a microporous carbon with tilting pores. This model carbon is able to produce realistic adsorption isotherms and heats of adsorption behaviour.

## 3. THE REAL PSD OF THE MODEL CARBON BY ETHING PROCEDURE

The pore space of the model microporous solid is divided into separate pores using algorithm similar to the one developed in [15]. In brief, this algorithm could be described as follows: A cubic lattice is built over pore domain and energy of fluid-solid interaction is calculated on each node of the lattice. The lattice cell size is chosen substantially less than the characteristic size of the interaction. Nodes where the interaction energy is positive and higher than some limit are marked as belonging to solid. All other nodes form pore volume. The pore volume then splits into separate pores through pore space "etching" procedure. This procedure consists in removal of consequent layers of pore space until convex separate areas of pore volume are formed. For lattice it means consequent marking (deletion) of nodes. The convex areas form pore's nuclei, which then expanded back, gaining layer after layer of adjoined pore nodes until the whole pore volume becomes occupied by pores. This process is similar to a crystal growth from a nucleus. The size of the pore, obtained in this way, is then defined as the size of maximum sphere, which can be fitted into this pore (See [15] for the details of the algorithm and its comparison with more mathematically rigorous algorithm of fitting a critical sphere.).

## 4. ANALYSIS OF THE MODEL CARBON IN TERMS OF THE INTERSECTING CAPPILLARIES MODEL

There are four stages to obtain the PSD of a real carbon or – in this case – a model carbon: (1) determine a molecular model for gas-gas and gas-pore interaction, (2) obtain the experimental isotherms for the carbon of interest, (3) generate a database of model-pore isotherms using GCMC simulation, and (4) invert the adsorption integral eq.1 to obtain $f(w)$. The model carbon-generated isotherms, as "experimental isotherms", are shown in Fig.2. The inversion of the adsorption integral eq.1 has been discussed in detail in the papers of Davies and coworkers [11].

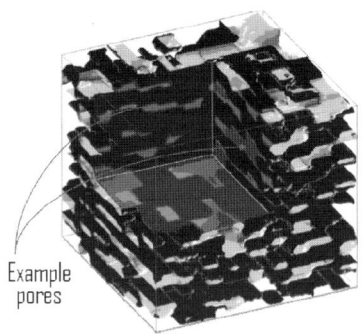

Figure 1. Isoenergy map of a microporous region of the model carbon. Dark and light indicate regions of inaccessible and accessible volume, respectively.

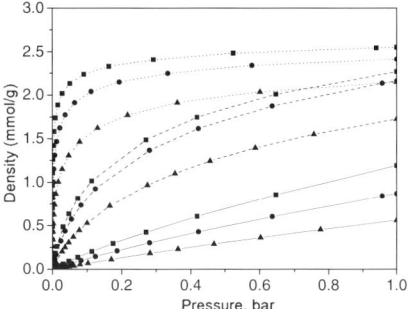

Figure 2. Adsorption isotherms generated by the model carbon. Lines indicate different adsorptives (solid-CH₄, dash-CF₄, dot-SF₆) and symbols indicate different temperatures (square-258K, cycle-275K, triangle-296K).

### 4.1. Molecular Models.

Three different gases ($CH_4$, $CF_4$ and $SF_6$) are used as adsorptives to probe the pore size distribution of the model carbon. As $CH_4$, $CF_4$ and $SF_6$ are all approximately spherical, we describe their interactions by the Lennard-Jones 12-6 potential:

$$U_{ff}(r) = 4\varepsilon_{ff}[(\frac{\sigma_{ff}}{r})^{12} - (\frac{\sigma_{ff}}{r})^{6}] \qquad (2)$$

where $\sigma$ and $\varepsilon$ are the molecular size and energy parameters respectively. The Lennard-Jones parameters of these three gases used in GCMC simulations are given in table 1. In the simulations, the potential is truncated at 15 Å, beyond which the fluid-fluid interactions are ignored.

Table 1. Lennard-Jones Parameters Used in GCMC Simulations

| Molecule | $\varepsilon/\kappa_B$, K | $\sigma$, Å |
|---|---|---|
| $CH_4$ | 149.92 | 3.7327 |
| $CF_4$ | 152.5 | 4.70 |
| $SF_6$ | 200.9 | 5.51 |
| C | 28.0 | 3.4 |

The interaction between a fluid molecule and the carbon is calculated by Steele's 10-4-3 potential [16]:

$$U_{sf}(z) = 2\pi\rho_s\varepsilon_{sf}\sigma_{sf}^{2}\Delta[\frac{2}{5}(\frac{\sigma_{sf}}{z})^{10} - (\frac{\sigma_{sf}}{z})^{4} - \frac{\sigma_{sf}^{4}}{3\Delta(0.61+z)^{3}}] \qquad (3)$$

Here, $\Delta$ (= 0.335 nm) is the distance between the two graphite layers, $\rho_{sf}$(= 114 nm⁻³) is the carbon number density in the graphite and $z$ is the perpendicular distance between the site in an adsorbate molecule and the adsorbent surface. $\sigma_{sf}$ and $\varepsilon_{sf}$ are the solid-fluid LJ parameters which are determined using the standard Lorentz-Berthelot combination rules.

**4.2 Adsorption in intersecting capillaries model.**

The Grand Canonical Monte Carlo (GCMC) simulation method is used to generate single-pore isotherms for the adsorptive species of interest ($CH_4$, $CF_4$ and $SF_6$) in a series of assumed slit-like pores (6 Å – 42 Å). The temperature, the volume and the chemical potentials of all species are kept constant, while the total number of molecules fluctuates during the course of the simulation, as for adsorption in real porous materials.

The initial configuration for the first point on the isotherm was generated by placing molecules at random in the simulation cell, checking that they did not overlap. For subsequent points on the isotherm, at progressively higher pressures, the final configuration generated at the previous pressure was used as the initial configuration. The Monte Carlo steps involved in the GCMC simulation are random creations, destructions and moves, obeying the appropriate acceptance criteria [17]. For each isotherm point, the system was allowed to equilibrate over 500,000 steps. After equilibration data were collected over a further 100,000 steps. The length of a Monte Carlo simulation is sufficiently long that the mean values of the quantities of interest, averaged over the simulation, are as accurate as necessary.

**5. RESULTS AND DISCUSSION**

Fig.3-5 show selected single-pore isotherms for $CH_4$, $CF_4$ and $SF_6$ respectively. Very little adsorption occurs in pores smaller than 7 Å for $CH_4$, and in pores smaller than 8 Å for $CF_4$ and $SF_6$, so they are not shown in these Figures. A decrease in adsorption as the temperature increases can be seen for all the three gases. Fig.3 shows that the adsorption of $CH_4$ decreases as the pore width increases. Since the adsorption of $CF_4$ is stronger than $CH_4$, its isotherms reflect a more rapid pore filling, as shown in Fig.4. The adsorption also decreases as the pore width increases. However, there are crossings for 10 Å and 12 Å, 14 Å isotherms at 258K, and another crossing for 10 Å and 12 Å isotherms at 275K. This crossing of isotherms is not seen in the simulation results at higher temperature, although it would be at sufficiently high pressures. $SF_6$ shows the strongest adsorption, compared with $CH_4$ and $CF_4$. The isotherms reach the highest plateau at 14 Å, from where the adsorption begins to decrease with increasing pore width. Again some of the isotherms cross. The complex variation between the isotherms for different adsorbates, including the fact that some cross others, is typical of adsorption in micropores. The crossings of some isotherms are due to that the packing effect in bigger pores outweighs the effect of the stronger energetic interaction of the smaller pores at some pressure points.

The PSDs obtained by inverting eq.1 are given in Fig.6. There are differences in the PSDs using various adsorptives at different temperatures. This may be mainly attributed to the difference in molecular sizes and the adsorption energy. An observation which is worth noting is that at lower temperatures (258K and 275K), $SF_6$ seems to detect smaller pores than $CF_4$ does, and the first peaks of $CF_4$ PSD and $SF_6$ PSD are very close to each other. This is surprising because from the theoretical point of view, the smaller molecular size of $CF_4$ should have facilitated it going into smaller pores. However, since the significant adsorption is from pores wider than 8 Å for both $CF_4$ and $SF_6$, as shown in Fig.4 and 5, the fitting of the PSD is insensitive to the volume of pores smaller than this size, so this apparent discrepancy is not significant. Actually, only parts of the PSDs shown in Fig.6 can give reliable pore structure information. Fig.3 shows that in the pressure range studied (up to 1 bar), the $CH_4$ single-pore isotherms become linear above about 10 Å at lower temperature (258K) and 9 Å at higher temperatures (275 and 296K). As the single-pore isotherms become more linear functions of pressure, the contributions of the various isotherms to the amount adsorbed become linearly dependent and these isotherms carry essentially no additional information about adsorption within the pores. As a result, the calculated PSD from eq.1 in this pore size range is not reliable and so should be discarded from the whole range of the pore size. The reliable pore size range defines the "window of reliability" which is bounded on the left by the smallest accessible pore size and on the right by the pore size at which adsorption becomes substantially linear. The reliability of the PSD is improved when a more strongly adsorbing adsorbate is used. For example, the $CF_4$ PSD at 258K is reliable up to 16 Å as shown in Fig.4, while the $SF_6$ PSD at the same temperature includes pores up to 36 Å, maybe even larger, as shown in Fig.5. The window of reliability can also be extended by measuring the isotherm at lower temperatures (As in this work, we can place a higher confidence in larger pores of the PSDs obtained at 258K.) or extending the measurements to include adsorption at higher pressures.

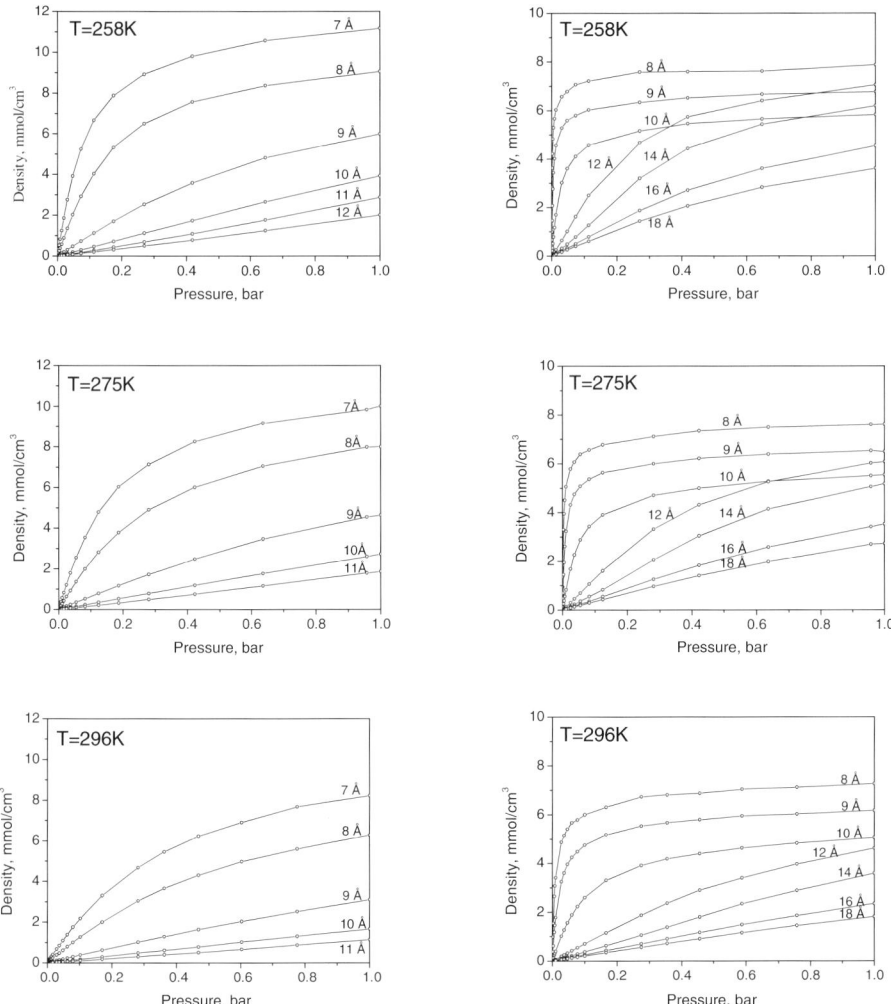

Figure 3. Adsorption isotherms for $CH_4$ in slit-shaped pores of various widths.

Figure 4. Adsorption isotherms for $CF_4$ in slit-shaped pores of various widths.

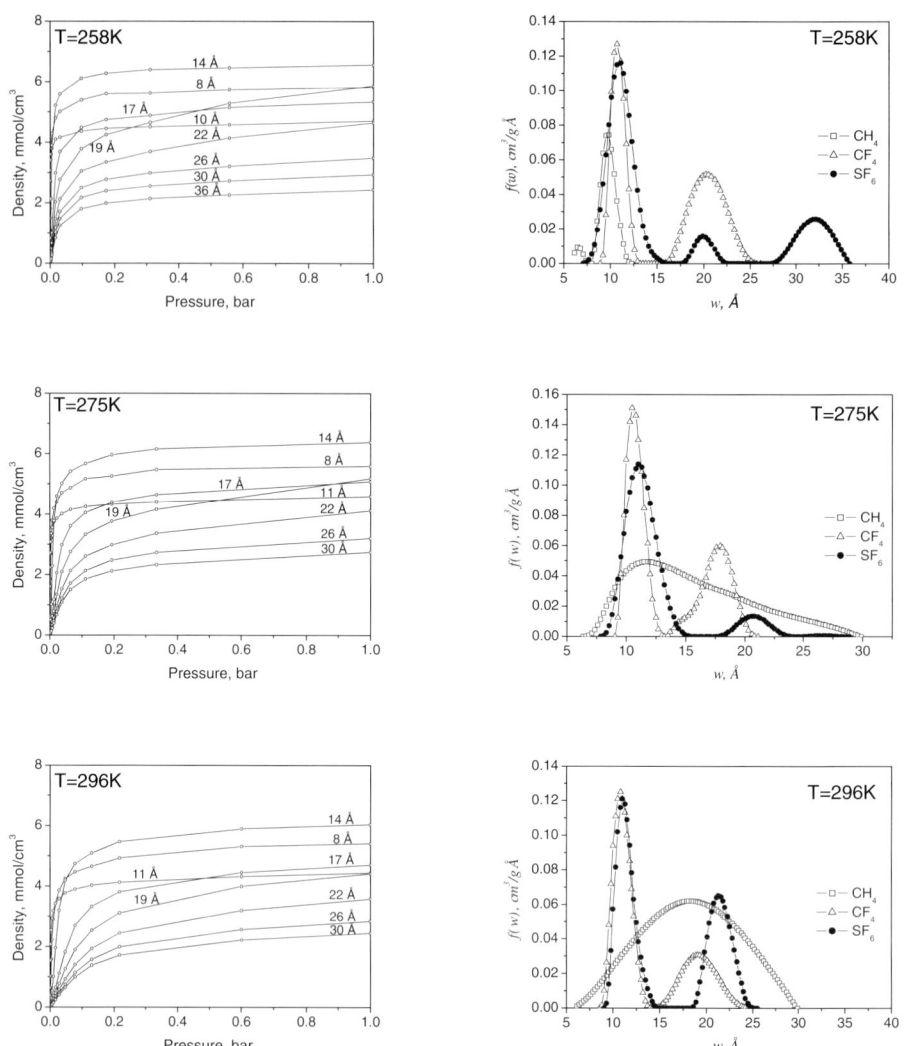

Figure 5. Adsorption isotherms for $SF_6$ in slit-shaped pores of various widths.

Figure 6. Pore size distributions obtained at three temperatures.

Fig.7 shows the overall PSDs at three temperatures. These PSD are obtained by choosing the $CH_4$ PSD for pores smaller than 10 Å and the $SF_6$ PSD for the remining part, while the $CF_4$ PSD is totally ignored. This is based on how of the accuracy of the predictions based on the overall PSDs (shown in Fig.9). It is clear that there is fairly good agreement between predicted and experimental isotherms at all the three temperatures. At 258K, the predicted $CH_4$ and $SF_6$ isotherms are in excellent agreement with the experimental isotherms, while the predicted $CF_4$ isotherm is slightly higher than the experimental one. The overall PSD obtained at 258K is then used to predict adsorption at higher temperatures. Again, the predicted isotherms are in good agreement with the experimental isotherms for $CH_4$ and $SF_6$, while $CF_4$

is slightly overpredicted. At 296K, the 258K PSD-based prediction overestimates the adsorption for all the three gases, while the 296K PSD-based prediction shows a good agreement with the experimental isotherms. As discussed previously, the stronger adsorption at lower temperature (258K in this case) gives a wider window of reliability for the PSD that fits the "experimental" isotherm at that temperature. That means the 258K PSD includes more pores than the PSDs that fit the "experimental" isotherms at higher temperatures, and so overpredicts adsorption at higher temperatures. The above observation provides strong evidence that the model carbon is behaving in the same way as the real carbon studied by López-Ramón *et al.* [13].

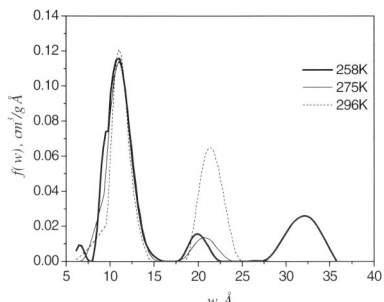

Figure 7. Overall PSDs of the model carbon.

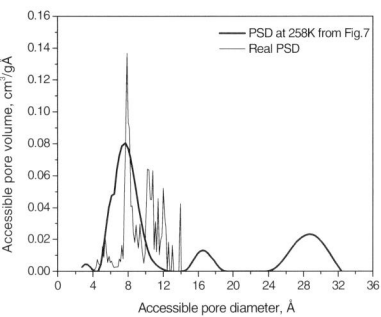

Figure 8. PSD comparison

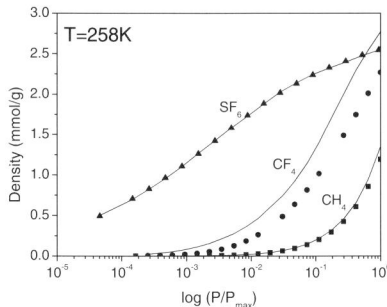

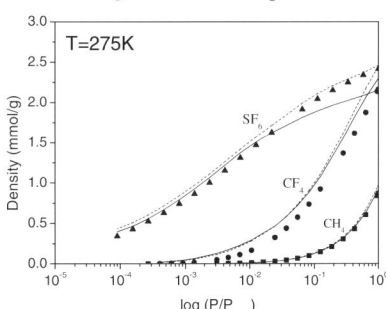

Figure 9. Fit of predicted isotherms (lines) to the experimental isotherms (symbols). Solid lines indicate prediction from the overall PSD obtained at the corresponding temperature; dotted lines at 275 and 296K indicate prediction using the overall PSD obtained at 258K. ($P_{max}$ is around 1 bar.)

We further compare the 258K overall PSD (ICM-PSD) with the real PSD of the model carbon, in terms of accessible volume, shown in Fig.8. At 7.8Å, both of these two PSDs show their highest peaks. In the pore size range of 0 – 14 Å, the ICM-PSD includes more smaller pores (smaller than 7.8Å) in, while the real PSD has more pores larger than 7.8 Å. It seems that the ICM-PSD has shifted to the left, compared with real PSD. This shifting may be due to the effect of the intra-corners formed by intersected pore walls. When the molecules come to the corners of big pores, the interaction is stronger than that in the middle of the pores, which makes adsorption in these big pores similar to adsorption in smaller, slit-shaped pores in the ICM. This effect has also been demonstrated by Davies and Seaton [11]. The ICM-PSD shows another two peaks in the range of 14 – 34 Å. This is unphysical at first sight but can be explained. The junctions between pores represent regions of low adsorption energy relative to the intra-pores. These junctions thus behave like wider "micropores" or even small "mesopores", within the ICM.

## 5. CONCLUSIONS

We have investigated the applicability of the GCMC approach, combined with an ICM, for the characterization of microporous carbons. A more complete picture of the PSD was obtained by combining the partial PSDs probed by three different gases. The predicted adsorptions from the overall PSDs are in good agreement with those generated from the model carbon, indicating the realism, and the predictive power, of these overall PSDs. This model carbon, as discussed, is behaving in the same way as the real carbon. The further comparison of the overall PSD with the real PSD of the model carbon shows that the pore structure obtained from the ICM is a realistic representation of the real structure. These results support that the use of the ICM to represent, in a tractable way, a much more complex and realistic structure, is a valid approach to the characterization of carbons.

**ACKNOWLEDGEMENT:** Q. Cai gratefully acknowledges an ORS award from Universities U.K. The U.K. Engineering and Physical Sciences Research Council are acknowledged for the financial support of this work.

**REFERENCES**
[1]  P. I. Ravikovitch, A. Vishnyakov; R. Russo and A. V. Neimark, Langmuir, 16 (2000) 2311
[2]  S. J. Gregg and K. S. W. Sing, Adsorption, surface area and porosity; 2nd ed.; Academic Press, London, 1982.
[3]  F. Rodríguez-Reinoso and A. Linares-Solano, Micro- porous Structure of Activated Carbons as Revealed by Adsorption Methods; Marcel Dekker, New York, 1988.
[4]  H. F. Stoeckli, J. Colloid Interface Sci., 59 (1977) 184.
[5]  M. M. Dubinin, Progress in Surface and Membrane Science; Academic Press, New York, 1975.
[6]  G. Horváth and K. Kawazoe, J. Chem. Eng. Japan, 16 (1983) 470.
[7]  N. A. Seaton, J. P. R. B. Walton and N. Quirke, Carbon, 27 (1989) 853.
[8]  P.N. Aukett, N.Quirke, S.Riddiford, S.R.Tennison, Carbon, 30 (1992) 913.
[9]  V. Y. Gusev, J. A. O'Brien and N. A. Seaton, Langmuir, 13 (1997) 2815.
[10] V. Y. Gusev and J. A. O'Brien, Langmuir, 13 (1997) 2822.
[11] G. M. Davies and N. A. Seaton, Carbon, 36 (1998) 1473.
[12] M. B. Sweatman and N.Quirke, Langmuir, 17 (2001) 5011.
[13] M. V. López-Ramón, J. J agiełło, T. J. Bandosz, and N. A. Seaton, Langmuir, 13 (1997) 4435.
[14] M. J. Biggs, A. Buts and D. Williamson, Langmuir, 20 (2004) 5786.
[15] J. F. Thovert, J. Salles and P. M. Adler, Journal of Microscopy, 65 (1992) 170.
[16] W. A. Steele, The interaction of gases with solid surfaces, Pergamon, Oxford, 1974.
[17] D. Frenkel and B. Smit, Understanding molecular simulation: from algorithms to applications, Academic Press, USA, 1996.

Studies in Surface Science and Catalysis 160
P.L. Llewellyn, F. Rodriquez-Reinoso, J. Rouqerol and N. Seaton (Editors)
© 2007 Elsevier B.V. All rights reserved

# Qualitative assessment of the purity of multi-walled carbon nanotube samples using krypton adsorption

C.J. Gommes[a], F. Noville[a], C. Bossuot[b], J.-P. Pirard[a]

[a]Laboratoire de Génie Chimique, Université de Liège B6a, B-4000 Liège, Belgium

[b]Nanocyl S.A., rue de l'Essor 4, B-5060 Sambreville, Belgium

*Abstract*

Krypton is a subcritical vapour at the nitrogen boiling temperature. As such, its adsorption on crystalline surfaces leads to condensation steps, typical of type VI isotherms according to IUPAC, while its adsorption on rough surfaces is BET-like. Based on this property of krypton adsorption at 77 K, a methodology is proposed to determine the purity of carbon nanotubes samples. The method is tested on model samples obtained by mixing mechanically purified multi-walled carbon nanotubes with various amounts of the same catalyst as used for their synthesis.

## 1. INTRODUCTION

Since their discovery by Ijima in the early nineties as a byproduct of fullerene synthesis [1], carbon nanotubes have received a growing interest. A huge number of synthesis routes have been proposed, ranging from laser ablation of carbon target, catalytic chemical vapour deposition (CCVD), liquid phase synthesis, plasma methods, and so forth [2]. Also, a large variety of application niches have been identified that render nanotubes a promising material [3].

In order to meet the expected industrial needs for such materials, a large scale CCVD production facility has been developed in Nanocyl S.A., with a production capacity of the order of one kg of multi-walled carbon nanotubes (MWCNTs) per hour. In a typical CCVD reactor [e.g. 4], the contact between a gaseous hydrocarbon and an appropriate solid catalyst at high temperature results in the deposition of carbon nanotubes on the catalyst. The raw product that exits the reactor is therefore expected to contain carbon nanotubes, a catalyst residue and possibly amorphous carbon. The latter two products are usually referred to as impurities. A general method to assess the amount of impurities in a MWNT sample is not available yet. The present work reports on the use of krypton adsorption to determine the amount of nanotubes actually present in as-synthesized sample.

Krypton at 77 K is a subcritical vapour, the adsorption of which on crystalline surfaces is known to give rise to stepped isotherms, classified as type VI by IUPAC [5]. For such vapours, when a given pressure is reached the intermolecular forces between adsorbed molecules overwhelms their thermal energy, by which a phase transition occurs between and adsorbed gas-like and adsorbed dense phases [6, 7]. This phenomenon leads to a riser in the isotherms that corresponds to the complete coverage, at a given pressure, of the surface by a 2D dense phase. Such isotherms have been reported for the adsorption of several subcritical vapours carbon nanotubes [8, 9]. As type VI isotherms are never observed for amorphous solids, even with so-called subcritical vapours [7], we propose to exploit this

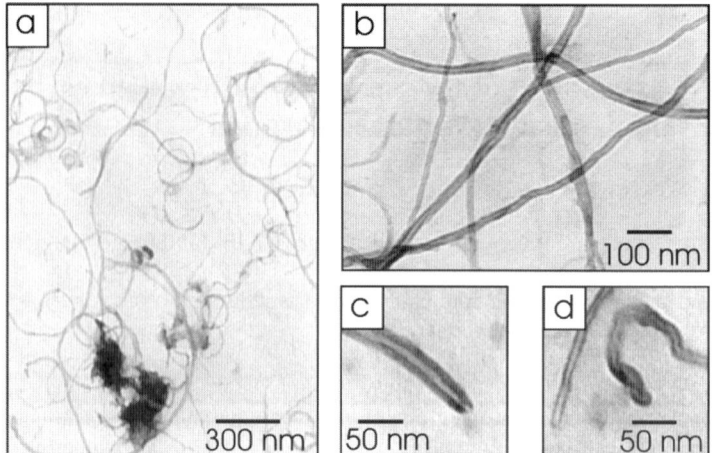

Figure 1. Example of transmission electron micrographs of the purified multi-walled nanotubes sample at various magnifications.

phenomenon to discriminate between the crystalline surface of the MWNTs and the amorphous surface of the impurities. The proposed methodology is tested on model samples obtained by mixing purified MWNTs with various amounts of the same catalyst as used for their synthesis.

## 2. EXPERIMENTAL SECTION

### 2.1. Preparation and characterization of the samples

The MWNTs are synthesized in a fixed bed CCVD reactor at 700°C. A quartz boat containing about 1 g of catalyst is placed in the centre of the reactor fed with a 1-2 l(STP)/min flow of a 50-50% mixture of nitrogen and ethylene. A typical reaction time is 20 min. The used catalyst is $Fe_x$-$Co_y$ supported on alumina, prepared by impregnation as described elsewhere [10]. The purification of the MWNTs proceeds in two steps: (i) the sample is leached with concentrated fluoric acid in order to dissolve the catalyst support and the metallic particles; (ii) an acidic $KMnO_4$ solution is used to selectively oxidize the amorphous carbon. After these treatments, the sample is filtered and washed with distilled water, and dried for 48 h in a vacuum oven heated at 120°C.

Figure 1 displays several typical transmission electron micrograph of the purified sample, taken at various magnifications. As is visible in Figure 1a, a small amount of unidentified impurities are still present in the sample after the oxidation. At larger magnification, the hollow cavities of the tubes are visible as a bright line lying in their middle (Figure 1b-d). The question of whether the inner cavity of tube is accessible for adsorption cannot be cogently answered from TEM observations, as some tubes seem to be open (Figure 1c) while others seem to be closed (Figure 1d).

A series of 15 micrographs was acquired at the same magnification as Figure 1b, in order to assess the inner and outer diameter of the tubes. The micrographs were analyzed automatically using a digital image analysis procedure described elsewhere [11]. The obtained statistical distributions are plotted in Figure 2. The average values of the outer and inner diameters, together with their standard deviations are $d_O = 20 \pm 8$ nm and $d_I = 6 \pm 3$ nm. This analysis is made in order to help analysing the adsorption data.

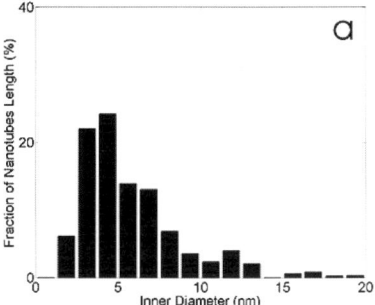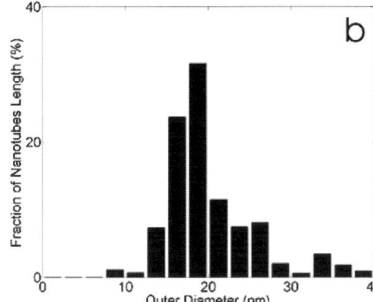

Figure 2. Statistical distributions of (a) the inner and (b) outer diameter of the nanotubes, estimated from digital image analysis of TEM micrographs.

Three mixtures were prepared by mixing mechanically the purified MWNTs materials with various amounts of catalyst. The nomenclature of the samples is given as $X(Y)$, where $Y$ is the weight percentage of nanotubes.

### 2.2 Krypton and nitrogen adsorption measurements

The krypton and nitrogen adsorption isotherms are determined at nitrogen boiling temperature by the classical volumetric method with a CE Instruments SORPTOMATIC 1990 series of THERMO ELECTRON. The device is equipped with an additional 10 torr pressure gauge and a turbomolecular pump. The used high purity krypton (99.997 %) and nitrogen (99.999 %) were purchased from Air Liquide.

The samples were first outgassed at $10^{-4}$ Pa during 16 h, then heated from 25° to 150°C at a rate of 1°C/min and kept at 150°C during 12 h. The weight loss of each sample during this treatment was determined and removed from the mass of the samples. Typically, the isotherms were determined on 150 mg of sample. Preliminary experiments allowed establishing the optimal experimental conditions to obtain reproducible isotherms starting at a relative pressure of $10^{-4}$.

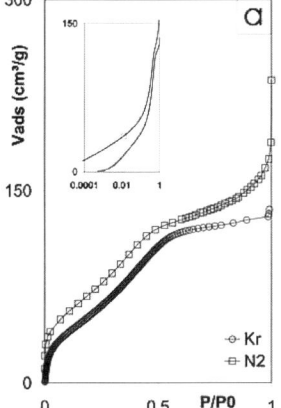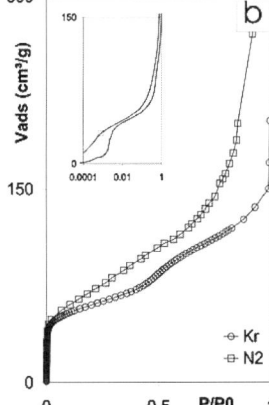

Figure 3. Krypton and nitrogen adsorption isotherms on the catalyst (a) and on the nanotubes (b). The insets magnify the low pressure adsorption with a logarithmic scale.

## 3. RESULTS AND DISCUSSION

### 3.1 Qualitative description of the isotherms

Adsorption of Kr and $N_2$ on the catalyst leads to isotherms with a very similar shape (Figure 3a). A type IV isotherm is obtained in both cases, with a strong adsorption near $P/P_0 \sim$ 0.4, characteristic of very small mesopores [5]. On the contrary, qualitatively different isotherms are obtained for Kr and $N_2$ adsorbed on the nanotubes (Figure 3b). The shape of the $N_2$ adsorption isotherm is reminiscent of a type II isotherm, typical of a nonporous or macroporous solid (The nitrogen adsorption-desorption isotherm shows a narrow hysteresis loop for $P/P_0$ larger than 0,8) while the adsorption of Kr leads to type VI stepped isotherms. The origin of the steps in type VI isotherms is the occurrence of a phase transition in the adsorbed layer, between two adsorbed gas-like and dense phases [7], which phenomenon can be qualitatively captured by a 2D van der Waals equation for the adsorbate [6]. The first step, that is visible in the inset of Figure 3b, corresponds to the condensation of the first Kr monolayer on the surface of the nanotubes, and the second step near $P/P_0 = 0.5$ corresponds to the condensation of a second monolayer on the top of the first one [7, 8].

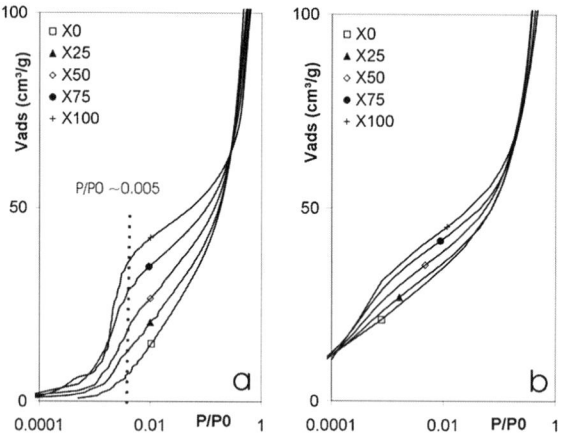

Figure 4. Krypton (a) and nitrogen (b) adsorption isotherms on the samples with increasing nanotube purity (X0 to X100).

Figure 4 displays the Kr and $N_2$ isotherms on the whole series of samples. Figure 4a clearly highlights the fact that the krypton condensation step is a characteristic of nanotubes. For nitrogen adsorption, only a slight hump is visible, whose amplitude increases when increasing the nanotubes content of the samples. A similar isotherm has been reported for long for the adsorption of $N_2$ on well graphitized surfaces and its shape is sometimes referred to as the Joyner-Emmet step [12]. It corresponds to the completion of the first adsorbed layer, which is progressive because adsorbed nitrogen is supercritical at 77 K. In the litterature devoted to the comparative adsorption of different gases, the vapors are classified according to if they are 2D-subcritical or 2D-supercritical at the considered temperature [7, 15].

This refers to the possibility of 2-dimensional phase transitions in the adsorbed layers. No phase transition is observed for $N_2$ at 77 K which is therefore referred to as supercritical. On the contrary, in the case of a 2D-subcritical vapors, such as Kr at 77 K, a transition between a gas-like adsorbed phase and a liquid-like adsorbed phase occurs at a pressure independent of the surface coverage. This phenomenon occurs only on uniform and almost perfectly crystalline surfaces. This leads to a riser in the isotherms which are classified a type VI.

The finite width of the adsorption steps of Kr contrasts with what is observed on planar graphite, in which a neat adsorption vertical riser is present [*e.g.* 7]. Also, as observed by other authors [8], the pressure at which the krypton monolayer condensation occurs is higher on carbon nanotubes than on graphite. This shift originates in the positive curvature of the adsorbent surface, which delays adsorption. Since the condensation of krypton on thick nanotubes should occur at a lower pressure than on thin nanotubes, the polydispersity of the samples can broaden the adsorption step. The width of the adsorption step of Kr on nanotubes could therefore be related the width of the diameter distribution of the tubes.

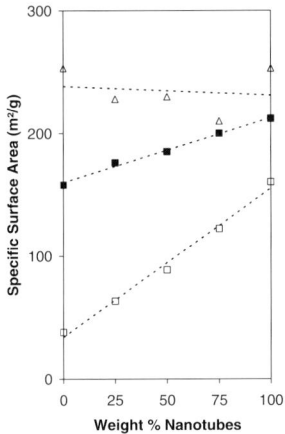

Figure 5. Specific surface areas estimated from the BET model applied to the krypton and nitrogen adsorption, $S_{Kr}$ (■) and $S_{N2}$ (Δ), and from the height of the krypton condensation step, $S_{step}$ (□).

## 3.2. Quantitative analysis of the isotherms

Although this model is not suited for analyzing stepped adsorption, the Kr and $N_2$ isotherms were fitted with the standard BET model [6]. The estimated specific surface areas with the two different vapours are referred to hereafter as $S_{Kr}$ and $S_{N2}$. They were estimated using the standard values for the surface covered by a single molecule 16.2 $\text{Å}^2$ for $N_2$ [5] and 15.7 $\text{Å}^2$ for Kr [7, 8]. These surfaces are plotted in Figure 5 as a function of the nanotubes content of the samples. There is an important scattering of the estimated surface areas (especially for $N_2$) that probably stems from the fact that the BET model could only be applied to the adsorption data over a very narrow pressure range. The very linearity of the plots in Figure 5 suggests that the specific surface area of any sample be the average of the specific surface areas of the catalyst and nanotubes, weighed by their relative mass fractions. However, as is visible in this figure, the specific surface area of the catalyst is very close to that of the nanotubes, so that the BET surface area cannot serve as a measure of the nanotubes content of the samples.

As already mentioned stepped isotherms are generally observed for the adsorption of subcritical vapours on crystalline surface [7]. Since the nanotubes are the only species with a crystalline surface in the studied samples, the height of the step should be associated with the presence of nanotubes, which is confirmed by Figure 4a. In order to make this statement more quantitative, the height of the krypton adsorption step is estimated as the amount adsorbed at $P/P_0 = 5 \; 10^{-3}$. The choice of the pressure limit of the step is arbitrary as it only results from the visual inspection of the isotherms. This approximation however proofs to be useful for an order of magnitude analysis of the adsorption process. The so estimated volume of the step is converted to the specific area $S_{step}$ of the surface on which stepped adsorption occurs, as described elsewhere [13]. This latter quantity is plotted in Figure 5.

$S_{step}$ is a linear function of the of the weight fraction, $y$, of purified nanotubes in the sample, with parameters

$$S_{step} \; (m^2 / g) = 34 + 122 \; y \tag{1}$$

The line does not pass through the origin because the adsorption on the pure catalyst is such that the amount adsorbed on it at $P/P_0 = 5 \; 10^{-3}$ account for ca. 34 m²/g. Moreover, the height of the step for the purified sample corresponds to 156 m²/g, while the total specific surface of the same sample is $S_{Kr} = 212$ m²/g. Since the BET surface $S_{Kr}$ accounts for the total specific surface area, and $S_{step}$ accounts only for the crystalline fraction of the same surface, the difference between these two numbers could mean that there is still a significant amount of impurities left in the sample. This would be confirmed by the TEM micrograph of Figure 1a. Another origin of this difference could be that the acidic treatment used to purify the sample significantly damaged the surface of the nanotubes. A previous study showed that this would result in a smoothing of the adsorption step [14].

Another possibility to explain the difference between $S_{Kr}$ and $S_{step}$ of the purified sample is that the nanotubes are so entangled in sample that only a fraction of their surface is accessible for layer by layer adsorption. To gain some insight into this issue, it is interesting to compare $S_{step}$ to the geometrical surface of the tube. The nanotubes can be modelled as ideal hollow cylinders with inner and outer diameters $d_I$ and $d_O$, and made of a material with density $\rho$. This results in the following surface over mass ratios [13]

$$\frac{S_T}{m} = 4 \frac{d_O + d_I}{\rho \, (d_O^2 + d_I^2)} \qquad \frac{S_O}{m} = 4 \frac{d_O}{\rho \, (d_O^2 + d_I^2)} \tag{2}$$

where $S_T$ and $S_O$ stand for total and outer surface. Using the diameters estimated from image analysis (section 2.1) and assuming that the nanotubes wall has the same density as graphite with $\rho \sim 2$ g/cm³, leads to $S_T/m = 143$ m²/g and $S_O/m = 110$ m²/g. Although these figures have to be considered as orders of magnitude, it can be noticed that the outer surface of the nanotubes alone could hardly account for the value of $S_{step}$. However, the total surface compares well with the height of the step. It seems therefore that the inner surface of the tubes be accessible for adsorption. Furthermore, it is likely that the difference between $S_{Kr}$ and $S_{step}$ stems from the presence of impurities, rather than from the entanglement of the tubes.


## 4.    CONCLUSION

Krypton adsorption can be used to qualitatively assess the purity of multi-walled carbon nanotubes samples. The exploited phenomenon is the krypton monolayer condensation that occurs exclusively on the crystalline surface of the nanotubes but not on the catalyst residue, nor on amorphous carbon. In principle, the same methodology could be used with the adsorption of any vapour below its 2D critical temperature. On the contrary, this kind of analysis is not feasible with 2D supercritical vapours, such as $N_2$ at 77 K, for which the step is replaced by a mere hump.

The proposed methodology was tested on a series of model samples obtained by mixing mechanically a purified nanotubes sample with known amounts of catalyst. It was shown that the height of the first condensation step in the Kr adsorption isotherm is indeed directly related to the amount of nanotubes in the sample. Comparing the height of the adsorption step with the geometric surface of the nanotubes further suggests that the purified nanotubes sample still contained a significant amount of impurities, probably amorphous carbon.

**Acknowledgements** C.J. Gommes is grateful to the National Funds for Scientific Research (FNRS, Belgium) for a Ph.D. research fellow position. The authors thank the Ministère de la Région Wallonne (DGTRE) and the Ministère de la Communauté Française de Belgique (Action de recherche concertée 00/05-265) for their financial support.

# REFERENCES

[1] S. Ijima, Nature 354 (1991) 56.
[2] C. Journet, P. Bernier, Appl. Phys. A 67 (1998) 1.
[3] Nanoscience and Nanotechnologies: Opportunities and Uncertainties. London: The Royal Society & The Royal Academy of Engineering, 2004.
[4] C. Gommes, S. Blacher, C. Bossuot, P. Marchot, J. B.Nagy, J.P. Pirard, Carbon 42 (2004) 1473.
[5] J. Rouquerol D. Avnir, C.W. Fairbridge, D.H. Everett, J.H. Haynes, N. Pernicone, J.D.F. Ramsay, K.S.W. Sing, K.K. Unger, Pure Appl. Chem. 66 (1994) 1739.
[6] D.D. Do, Adsorption Analysis: Equilibria and Kinetics, Series in Chemical Engineering, Volume 2, Imperial College Press, London, 1998.
[7] A. Thomy, X. Duval, J. Regnier, Surf. Sci. Rep. 1 (1981) 1.
[8] K. Masenelli-Varlot, E. Mc Rae and N. Dupont-Pavlovsky, Appl. Surf. Sci. 196 (2002) 209.
[9] A. Bougrine, N. Dupont-Pavlovsky, J. Ghanbaja, D. Billaud and F. Beguin, Surf. Sci. 506 (2002) 137.
[10] K. Hernandi, A. Fonseca, J. B.Nagy, D. Bernaerts, A. Fudala, A.A. Lucas, Zeolites 17 (1996) 416.
[11] C. Gommes, S. Blacher, K. Masenelli-Varlot, C. Bossuot, E. McRae, A. Fonseca, J. B.Nagy, J.-P. Pirard, Carbon 41 (2003) 2561.
[12] L. Joyner, P. Emmet, J. Am Chem Soc., 70 (1948) 2353.
[13] C. Gommes, S. Blacher, N. Dupont-Pavlovksy, C. Bossuot, D. Marguillier, A. Fonseca, J. B.Nagy, J.P. Pirard, Colloid Surf. Sci. A 241 (2004) 155.
[14] R. Babaa, E. McRae, C. Gommes, S. Delpeux, G. Medjahdi, S. Blacher, F. Beguin, American Institute of Physics Conference Proceedings 723 (2004), 133.
[15] A. Thomy, X. Duval, Surf. Sci. 299 (1994), 415.

Studies in Surface Science and Catalysis 160
P.L. Llewellyn, F. Rodriquez-Reinoso, J. Rouqerol and N. Seaton (Editors)

# Study of the Anomalous Behaviour of MFI Zeolites Towards Nitrogen Adsorption

**Elpiniki Panayi and Charis R. Theocharis**

Porous Solids Group, Department of Chemistry, University of Cyprus, P.O. Box 20537, 1678, Nicosia, Cyprus

In this presentation we discuss the effect of the presence of small amounts of template and of the nature of the exchangeable cation on the shape of the nitrogen adsorption isotherm of ZSM-5 zeolite samples. It was found that for some exchangeable cations a low-pressure hysteresis loop is shown for Si: Al ratio of 120 but not 80. An explanation of the conditions under which the loop appears is given.

## 2. INTRODUCTION

Zeolites have been of considerable interest in the past several years because of their diverse applications as catalysts, ion-exchangers, sorbents, and latterly because of their environmental applications. It has also been proposed, that zeolites may be used as model sorbent materials that could be used for calibration and comparison purposes. An EU-funded project proposed the use of zeolites of the so-called MFI structural type, such as ZSM-5. The reason for this lay in the fact that synthetic routes existed that could consistently produce well-characterised samples with crystallite size in the region of 0.1 to 0.2mm [1]. Such samples would enable the measurement of adsorption isotherms that would not exhibit secondary capillary adsorption characteristics, or effects of activated adsorption. However, Carrott and Sing [2], Unger *et al* as well as Rouquerol *et al* observed that the nitrogen adsorption isotherms of such Silicalite or H-ZSM-5 samples, exhibited low-pressure hysteresis loops at a $p/p^0$ value of 0.15 for the former, and close to 0.10 for the latter [1, 3, 4]. Clearly such a hysteresis loop could not be attributed to capillary condensation. Other ZSM-5 samples did not exhibit such behaviour towards nitrogen adsorption but did so towards other gases [5]. Silicalite I exhibited similar behaviour with adsorbates other than nitrogen. A theoretical simulation of the adsorption of argon in Silicalite failed to reproduce the above-mentioned step in the isotherm, but has nevertheless indicated that adsorption in the micropores would proceed in at least two steps, the first one being localised adsorption on the strongest sites, and the second corresponding to molecules clustering around other adsorbed molecules [6,7].

The appearance of low-pressure hysteresis was attributed to a solid-liquid transition or a phase transition of the adsorbed phase in the porous system of the MFI lattice, and neutron diffraction data appear to corroborate this. Some work with ZSM-5 samples has indicated a correlation between the pressure at which the low-pressure hysteresis loop appeared as well as its height with the presence of aluminium [4]. Sing has discussed in detail the occurrence of low-pressure in hysteresis in Silicalite in several articles [8, 9]. A discussion of low pressure hysteresis was presented in [10]. However, very little systematic work has been carried out on the nature of this unusual behaviour, especially of the factors influencing its position or shape. Furthermore, all data available thus far, were on large crystalline samples, while no data are available on commercially available ones. This paper presents work, as part of an ongoing investigation in this laboratory [11, 12] on commercially available ZSM-5 samples, which attempts to clarify the nature of this anomalous sorptive behaviour. In a previous publication, we presented the first results of a study of the parameters affecting the occurrence of low-pressure hysteresis in these samples [13]. In this paper we continue this investigation, examining the effect of heat treatment as well as of the nature of the exchangeable cation on the shape of the low-pressure part of the nitrogen adsorption isotherm for commercial samples of ZSM-5. A preliminary investigation of the ion-exchange conditions was also carried out.

## 3. EXPERIMENTAL

Nitrogen adsorption isotherms were carried out using a Micromeretics ASAP 2000 automated apparatus at 77K, from a starting pressure of 0.13Pa. Prior to adsorption isotherm determination, samples were outgassed at 0.13Pa overnight at various temperatures as specified in Table 1. ZSM-5 and Silicalite samples used in this work were donated by Degussa and were free of binder. We chose to use ZSM-5 samples with Si: Al=120. Some data were also obtained on samples with Si: Al=80. The pristine form of ZSM-5 had $Na^+$ as counterions. Other counter-ions, specifically $NH^+_4$, $Cs^+$, $Ca^{2+}$, $Cu^{2+}$, $K^+$, and $Ce^{4+}$ were introduced by repeated overnight suspensions of the samples in appropriate solutions at room temperature. Different lengths of contact times were employed in an effort to examine the effect of the extent of ion-exchange on the adsorptive properties of the samples. In certain experiments, the sample was calcined before outgassing. The calcination step was carried out prior or after ion-exchange. Temperatures and calcinations times are presented in Table 1. The sample with Si: Al=80 contained a large amount of templating agent, which made the calcination step prior to adsorption measurement or ion exchange obligatory. Thermogravimetric analysis was carried out in the region 300-1073K in flowing air using a Shimadzu TGA-50 apparatus. FTIR spectra were obtained using an 8500 Shimadzu Spectrometer, and DRIFTS spectra using a Spectratech attachment.

## 4. RESULTS AND DISCUSSION

Table 1 summarises the BET surface area, information on the presence of low-pressure hysteresis and heat treatment for all samples examined, with Si:Al=120. No sample with Si:Al=80 exhibited a low-pressure hysteresis loop, for any counter-ion.

Table 1
Samples Examined, $S_{BET}$ Values, and Outgassing Temperatures

| Sample | | $S_{BET}$ /m$^2$g$^{-1}$ | Low Pressure Hysteresis |
|---|---|---|---|
| Exchangeable Cation | Heat Treatment | | |
| Na$^+$ | Outgass 383K | 429 | Yes |
| Na$^+$ | Outgass 673K | 433 | Yes |
| Na$^+$ | Calcine 673K 4h, Outgass 383K | 445 | Yes |
| Na$^+$ | Calcine 673K 4h, Outgass 673K | 440 | Yes |
| Na$^+$ | Calcine 673K 8h, Outgass 383K | 428 | Yes |
| NH$_4^+$ | Outgass 383K | 429 | Yes |
| NH$_4^+$ | Calcine before exchange 673K 4h, Outgass 383K | 448 | Yes |
| NH$_4^+$ | Calcine before exchange 673K 8h, Outgass 383K | 439 | No |
| NH$_4^+$ | Calcine before and after exchange 673K 4h, Outgass 383K | 464 | Yes |
| Cs$^+$ | Outgass 383K | 421 | No |
| Ca$^{2+}$ | Outgass 383K | 420 | Yes |
| Ca$^{2+}$ | Calcine after exchange 673K, 8h, Outgass 383K | 421 | No |
| Cu$^{2+}$ | Calcine before exchange 673K 4h, Outgass 383K | 428 | Yes |
| Cu$^{2+}$ | Calcine before exchange 673K 8h, Outgass 383K | 418 | Yes |
| K$^+$ | Calcine before exchange 673K 4h, Outgass 383K | 444 | Yes |
| Ce$^{4+}$ | Calcine before exchange 673K 4h, Outgass 383K | 468 | Yes |

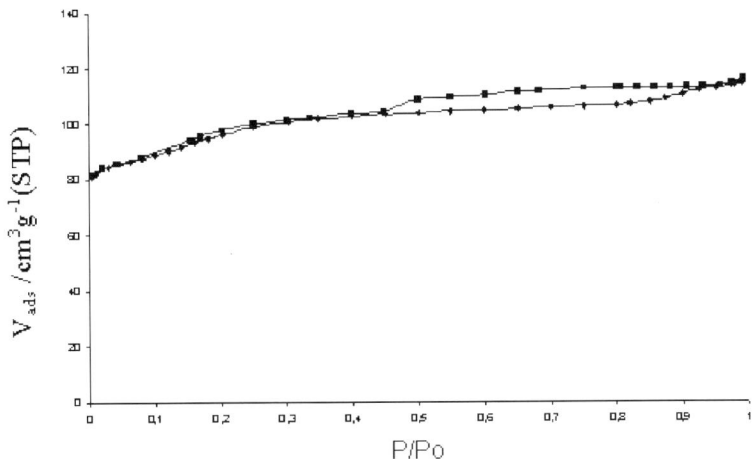

Figure 1 Nitrogen adsorption isotherm for Ca$^{2+}$-ZSM-5 (Si:Al=120)

It can be seen from Table 1 that for the $NH_4^+$ sample, calcination at 673K for 8h prior to ion exchange led to a loss of the low-pressure hysteresis loop, observed for that sample, without the calcination step. FTIR spectra and thermogravimetry have confirmed that under such conditions, there was no appreciable decomposition of the ammonium ion to $NH_3$ and $H^+$. This was also confirmed by surface acidity measurements. Under the stated conditions there was no change in the surface acidity of the solid, as measured by acid-base titrations. On the contrary, if calcinations was carried out for 24h, then there was a six-fold increase in the surface acidity of the sample, indication of the generation of $H^+$-ZSM-5. Examination by FTIR has indicated that the pristine ZSM-5 (Si:Al=120) contains small amounts of templating agent, which is removed by the calcinations step, prior to ion exchange. From Table 1, it can be seen that for $Cs^+$ and $Ca^{2+}$ (after calcination) as exchangeable cations there was no low-pressure hysteresis loop. The outgassing conditions did not appear to influence either the shape or occurrence of the loop.

This observation was contrary to what was previously observed for microcrystalline Silicalite [11,12,14] where the shape of the low-pressure hysteresis loop was dependent upon the calcination conditions. Figure 1 shows the nitrogen adsorption isotherm measured for $Ca^{2+}$-ZSM-5 after calcination at the conditions described in Table 1. There is no hysteresis loop, unlike the case that obtains for the uncalcined sample.

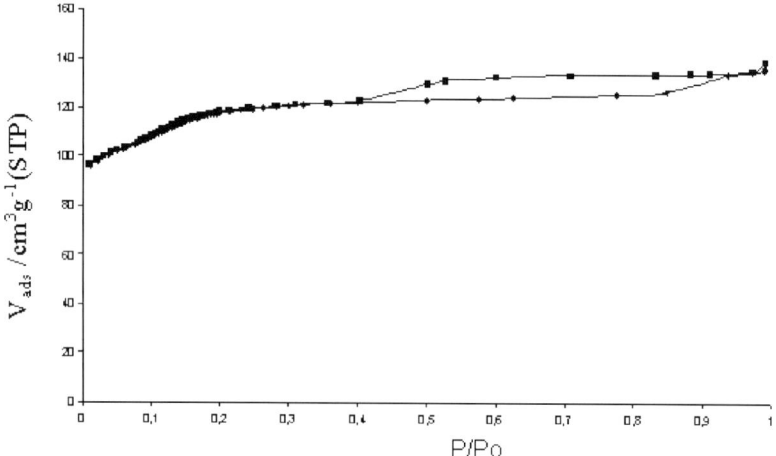

Figure 2 Nitrogen adsorption isotherm for $Cs^+$-ZSM-5 (Si:Al=120)

Figure 2 contains the nitrogen adsorption isotherm for $Cs^+$-ZSM-5 (Si:Al=120), Figure 3 the corresponding isotherm for $Na^+$-ZSM-5 (Si:Al=120) and Figure 4 the isotherm for sample $NH_4^+$-ZSM-5 (Si:Al=120). It can be seen from the three isotherms, in addition to the absence or presence of a hysteresis loop, that the low-pressure portion of the isotherm differs in shape from sample to sample. Classically, such changes would be interpreted as being due changes in pore diameter.

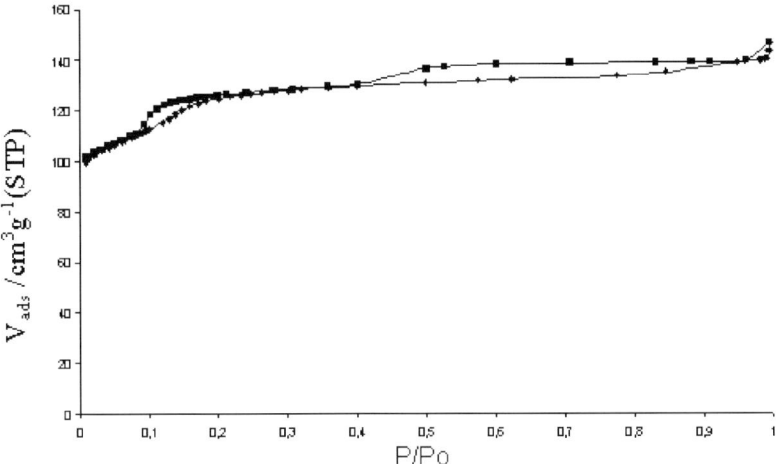

Figure 3 Nitrogen adsorption isotherm for Na$^+$-ZSM-5 (Si:Al=120)

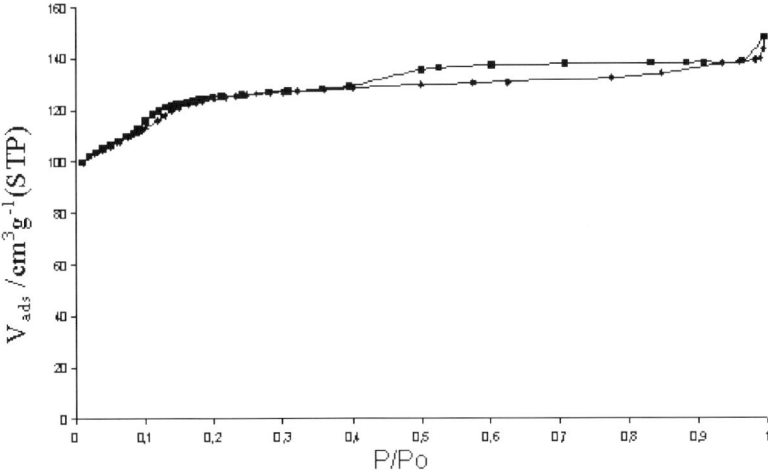

Figure 4 Nitrogen adsorption isotherm for NH$_4$$^+$-ZSM-5 (Si:Al=120)

However, the changes brought about by changing the exchangeable cation of the zeolite would not be extensive enough to explain completely such changes. X-ray diffractometry was employed to confirm that no loss of crystallinity or change in crystal structure occurred during the ion exchange.

It is evident that the Si:Al ratio is one factor affecting the occurrence of the low-pressure hysteresis loop. It is further evident that the nature of the exchangeable cation also affects the shape, including the occurrence of a hysteresis loop, of the isotherm at low partial pressure. It is now further shown that the presence of small amounts of impurities, such as remnants of the matrix, would also influence the shape of the early part of the isotherm. As previously suggested [13], the low-pressure hysteresis loop would be caused by a reorganisation of the adsorbed phase at the pore walls. The fact that the outgassing conditions, which would affect the amount of water present in the pores, does not appear to influence the occurrence of the loop, would indicate that the effect is not a function of pore size. Water molecules would, presumably, be coordinated at the exchangeable cations or framework charges and their number might therefore affect pore size or shape. The exchangeable cation, would presumably, have a similar effect. It is suggested that the presence or absence of the loop, as well as the overall shape of the early part of the isotherm, would be linked with the strength of interaction between the sorbed molecules and the pore walls. This would be influenced by the nature, and number, of exchangeable cations, as the charge density would be different. The influence of a small amount of matrix present, might be linked to the masking of charge sites on the walls.

## REFERENCES

[1] U. Muller, H. Reichert, E. Robens, K.K. Unger, Y. Grillet, F. Rouquerol, J. Rouquerol, D. Pan and A. Mersmann, Fresenius Z. Anal. Chem., 333, 433 (1989)

[2] P.J.M. Carrott and K.S.W. Sing, Chem. & Ind., 786, (1986)

[3] H. Reichert, U. Muller, K.K. Unger, Y. Grillet, F. Rouquerol, J. Rouquerol, J.P. Coulomb, Studies in Surface Science and Catalysis, (Eds F. Rodriguez-Reinoso, J. Rouquerol, K.S.W. Sing and K.K. Unger), Elsevier, Amsterdam, 62, 535, (1991)

[4] U. Muller and K.K. Unger, Studies in Surface Science and Catalysis, (Eds K.K. Unger, J, Rouquerol, K.S.W. Sing and H. Kral), Elsevier, Amsterdam, 39, 101, (1988)

[5] P.L. Lllewellyn, PhD Thesis, Brunel University, (1992)

[6] R J-M Pellenq and D. Nicholson, Studies in Surface Science and Catalysis, (Eds J. Rouquerol, F. Rodriguez-Reinoso, K.S.W. Sing and K.K. Unger), Elsevier, Amsterdam, 87, 21, (1994)

[7] D. Nicholson, R.W. Adams, R.F. Cracknel and G.K. Papadopoulos, Characterisation of Porous Solids IV, (Eds B. McEnaney, T.J. Mays, J. Rouquerol, F. Rodriguez-Reinoso, K.S.W. Sing and K.K. Unger), Royal Society of Chemistry, London, 57, (1997)

[8] K.S.W. Sing, Colloids and Surfaces, 38, 113, (1989)

[9] K.S.W. Sing, $3^{rd}$ Fundamentals of Adsorption (Ed. A.B. Mersmann and S.E. Scholl), Engineering Foundation, New York, p78, (1991)

[10] F. Rouquerol, J. Rouquerol and K.S.W. Sing, in "Adsorption by Powders and Porous Solids", Academic Press, New York, pp 389-366 (1999)

[11] M.-E. Eleftheriou, PhD Thesis, University of Cyprus (1995)

[12] M.-E. Eleftheriou and C.R. Theocharis, Characterisation of Porous Solids IV, (Eds B. McEnaney, T.J. Mays, J. Rouquerol, F. Rodriguez-Reinoso, K.S.W. Sing and K.K. Unger), Royal Society of Chemistry, London, 475, (1997)

[13] C.R. Theocharis and G. Kyriacou, "Studies in Surface Science and Catalysis", Elsevier Science Publishers, 144, 709-716 (2002)

[14] G. Kyriacou, PhD Thesis, University of Cyprus (2005)

Studies in Surface Science and Catalysis 160
P.L. Llewellyn, F. Rodriquez-Reinoso, J. Rouqerol and N. Seaton (Editors)

# Characterization of alkaline post-treated ZSM-5 zeolites by low temperature nitrogen adsorption

**Yousheng Tao[a*], Hirofumi Kanoh[a], Johan C. Groen[b] and Katsumi Kaneko[a]**

[a]Department of Chemistry, Faculty of Science, Chiba University, Chiba 263-8522, Japan
*email: tao@pchem2.s.chiba-u.ac.jp

[b]DelftChemTech, Delft University of Technology, 2628 BL Delft, The Netherlands

ZSM-5 zeolites having $SiO_2/Al_2O_3$ molar ratios of 23.8 and 200 were treated in 0.05 M NaOH solutions at 325 K. The pore structure changes were investigated by low temperature nitrogen and argon adsorption. The mesopores with broad mesopore size distributions were developed after alkaline treatment of ZSM-5 with a low $SiO_2/Al_2O_3$ molar ratio 23.8. Alkaline treatment of ZSM-5 with a high $SiO_2/Al_2O_3$ molar ratio 200 gave no mesopores from both nitrogen and argon adsorption measurements. Formation of defect-cluster associated pores in ZSM-5 zeolite upon alkaline post-treatment is discussed.

## 1. INTRODUCTION

Introduction of mesopores into conventional zeolites is of great interest due to the improvement of guest molecular transport in microporous zeolites [1-5]. The mesopore-added zeolites find a wide range of applications such as catalysis, adsorption and separation, particularly for heterogeneous catalysis [4-8].  They can be prepared via several routes. Although novel dual templating methods using secondary mesostructural carbon materials as template have recently been developed [9-13], different post-treatments on the synthesized zeolites with hydrothermal treatments and chemical treatments are the most frequently applied [14-21]. The dual templating methods with carbon sources such as multiwall carbon nanotubes (MWNTs), carbon nanofibers (CNFs), carbon aerogels (CAs), and resorcinol-formaldehyde aerogels (RFAs) as templates offer a high degree of control over the porosity of mesopores in zeolites [10-13]. The post-treatments mostly provide zeolites having inhomogeneous mesopores from defect domains [14-16].

Leaching post-treatment with NaOH solution was recently applied to create mesopores in MFI zeolite [22-25]. In contrast to acid treatment, which preferentially removes framework Al atoms, alkaline post-treatment was found to extract framework Si atoms selectively, leading to substantial formation of mesoporosity [26,27]. Ogura et al. concluded from TEM observation that mesopores were created by dissolving the portion of ZSM-5 with poor crystallinity and a part of a zeolite particle, weak against alkali-treatment, became a defect in the zeolite framework [28]. Groen et al. showed that tetrahedrally coordinated aluminum in ZSM-5 frameworks controls the mechanism of intracrystalline mesopore formation by desilication in alkaline medium [27], and the created mesopores do not have a uniform size, but a broad pore size distribution around 10 nm [24]. As a consequence of created mesoporosity the external surface area increases from 40 $m^2g^{-1}$ in the original samples up to 225 $m^2g^{-1}$, while alkaline-treatment decreases the micropore volume from 0.17 to 0.13 $cm^3g^{-1}$ [24].

In this contribution the changes in structural properties of alkaline-treated ZSM-5 zeolites of different $SiO_2/Al_2O_3$ molar ratios upon treatment under mild conditions at low concentration of NaOH solutions, which have not been investigated in previous studies, are discussed using low temperature nitrogen and argon adsorption, and X-ray diffraction measurements.

## 2. EXPERIMENTAL

### 2.1. Materials

ZSM-5 zeolites with high and low $SiO_2/Al_2O_3$ molar ratios of 200 (synthesized with the established procedures [29]) and 23.8 (HSZ-820NAA, Tosoh Corp.) in Na form have been investigated by treatment in 0.05 M NaOH aqueous solutions at 325 K for different periods of time. The resulting slurry was immediately filtered by suction. The filtered cake was then rinsed with distilled water at 325 K and filtered. It was repeated three times to eliminate silicate materials precipitating during the alkaline treatment. Finally the samples were dried in an oven at 383 K overnight. The samples are denoted using the $SiO_2/Al_2O_3$ molar ratios and the alkaline post-treatment periods, for example ZSM-5(H0.5) is the ZSM-5 zeolite sample prepared from ZSM-5 zeolite with high $SiO_2/Al_2O_3$ molar ratio of 200 in NaOH aqueous solutions for 0.5 h, and ZSM-5(L0.5) is the ZSM-5 zeolite sample from ZSM-5 zeolite with low $SiO_2/Al_2O_3$ molar ratio of 23.8 treated for 0.5 h.

### 2.2 Characterization

Nitrogen adsorption isotherms of all samples were measured at 77 K. Argon adsorption measurements at 87 K were performed to examine the anomalous nitrogen adsorption behavior of low-pressure hysteresis loops at a $P/P_0$ ~0.2 of ZSM-5 zeolites upon alkaline post-treatments. Argon was of > 99.998 % purity. Both nitrogen and argon adsorption data were obtained using a Quantachrome Autosorb-1 gas adsorption analyzer. The samples were outgassed at 623 K for 12 h under a vacuum of $10^{-4}$ Pa prior to the adsorption measurements. Mesopore size distributions (MPSDs) were calculated from the nitrogen adsorption isotherm with BJH method [30]. The powder X-ray diffraction (XRD) patterns were acquired on a Miniflex X-ray automatic diffracometer (Rigaku Corporation.) using a monochromatized X-ray beam from nickel-filtered Cu Kα ($\lambda$ = 0.15406 nm) radiation and operated at 30.0 kV and 15.0 mA.

(A)                                                    (B)

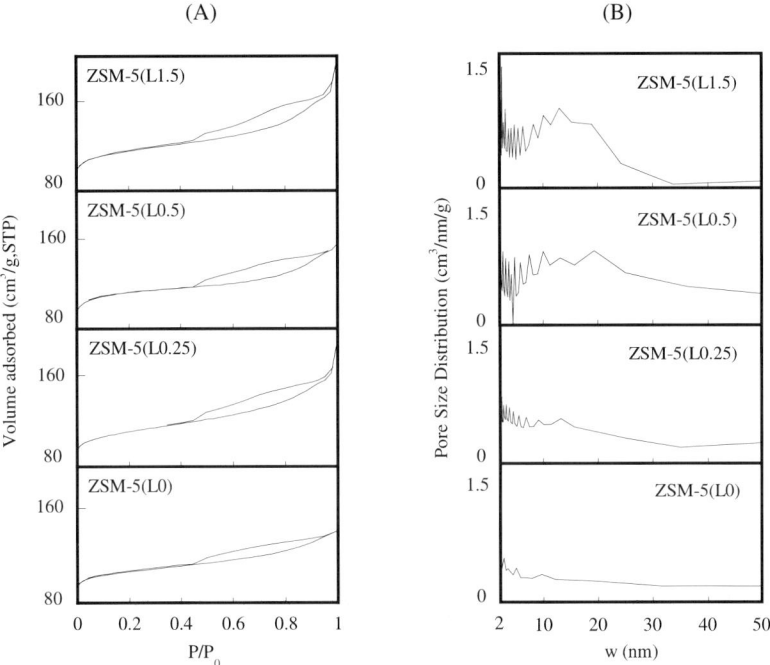

Fig. 1. (A) Nitrogen adsorption isotherms at 77 K for ZSM-5 zeolite with $SiO_2/Al_2O_3$ molar ratio 23.8 and alkaline-treated ZSM-5 zeolites. (B) The mesopore size distribution of alkaline-treated ZSM-5 zeolites.

## 3. RESULTS AND DISCUSSION

The $N_2$ adsorption/desorption isotherms of ZSM-5 zeolite with $SiO_2/Al_2O_3$ molar ratio 23.8 and the alkaline-treated samples are shown in Fig. 1 (A). A slight increase in uptake and change in the shapes of hysteresis loops at $P/P_0 = 0.4$ to ~1 for alkali-treated samples of ZSM-5(L0.25), ZSM-5(L0.5) and ZSM-5(L1.5) are observed compared to the untreated zeolite ZSM-5(L0), indicating small changes in the mesopore structures upon the alkaline-treatment. The MPSDs derived from the BJH analysis of the $N_2$ adsorption isotherms are shown in Fig. 1 (B). As expected from the proposed mechanism of mesopore formation due to selective dissolution of siliceous species from the framework of the zeolite (a lower amount of Al was also eluted) in previous studies [22-25,27], the mesopores with relatively broad size distributions are gradually developed with a prolonged alkaline-treatment period. ZSM-5(L0.25) has a fuzzy MPSD of 2-35 nm, ZSM-5(L0.5) has MPSD with a median size between 2 and 30 nm, and ZSM-5(L1.5) with a median size between 2 and 25 nm.

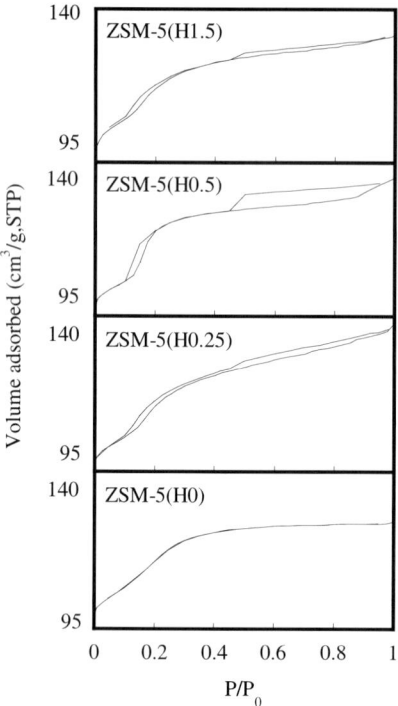

Fig. 2. Nitrogen adsorption isotherms at 77 K for ZSM-5 zeolite with $SiO_2/Al_2O_3$ molar ratio 200 and alkaline-treated ZSM-5 zeolites.

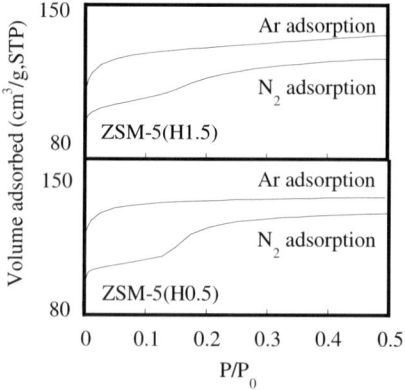

Fig. 3. Nitrogen and argon adsorption on alkaline-treated zeolite ZSM-5(H0.5) and ZSM-5(H1.5).

The situation is quite different for ZSM-5 zeolites having a high $SiO_2/Al_2O_3$ molar ratio of 200. The $N_2$ adsorption/desorption isotherms of ZSM-5 zeolite with $SiO_2/Al_2O_3$ molar ratio 200 and the alkaline-treated samples are shown in Fig. 2. The $N_2$ adsorption isotherm of untreated zeolite ZSM-5(H0) was basically of IUPAC Type I. Predominant adsorption ended below $P/P_0$ = 0.02, which is a characteristic of uniform microporous solids. After alkaline-treatment, the feature in the nitrogen isotherms showed a hysteresis loop at $P/P_0 \sim 0.2$ and a hysteresis loop at $P/P_0$ above 0.4. It is also interesting to note that the hysteresis loop shapes of these materials are different. Those of ZSM-5(H0.25) and ZSM-5(H1.5) are obvious and shoulder steps in the range of $0.1 < P/P_0 < 0.3$, while that of ZSM-5(H0.5) is a clear square in the range of $0.1 < P/P_0 < 0.2$. The higher Si/Al ratio framework provides less electrostatic interaction with the adsorbate and results in a sharp transition in the isotherm according to Llewellyn et al. [31]. This sharpness is broadened by the electrostatic effects induced by the aluminum-rich framework [31,32]. However, these results appear to be contradictory because the alkali treatment was found to extract framework Si atoms selectively. Although the hysteresis loops of nitrogen isotherms of some ZSM-5 samples at $P/P_0 \sim 0.2$ have been discussed since the first report on a unique feature of nitrogen isotherms on Silicalite-I by Carrott and Sing [33], recently some researchers claimed those to pore filling into supermicropores or small mesopores of prepared zeolitic materials [34-36]. We checked our samples with argon adsorption measurements. All these hysteresis loops "disappeared" (Fig. 3), confirming phase change in the adsorbed layer and absence of supermicropores or small mesopores [33,37,38]. Indeed, the transition feature for nitrogen in the vicinity of $P/P_0$ = 0.15-0.20 shifts to much lower $P/P_0$ by two orders of magnitude for argon, presuming that this is a consequence of weaker adsorbate-adsorbent interactions for argon than nitrogen [32]. Thus the obvious hysteresis loops in the nitrogen isotherms of ZSM-5(L0.25) and ZSM-5(L1.5) at $P/P_0 \sim 0.2$ should be caused by the formation of a less ordered nitrogen adsorbed phase, while the square one of ZSM-5(H0.5) could be ascribed to the formation of an ordered structure of nitrogen (orientation solid-like phase) due to a local change.

From the literature it is known that argon adsorption presents a purely structural characterization of the samples due to little specific interaction with the Al sites, while nitrogen shows larger differences between samples with a homogenous and heterogeneous Al distribution [39,40]. The low pressure hysteresis loops were observed for the alkaline-treated materials and was not for the untreated zeolite reference of ZSM-5(H0), suggesting that the defect-cluster associated pores appeared in ZSM-5 zeolites upon alkaline post-treatment, for which further studies are needed.

As could be derived from the minor changes in the isotherm at $P/P_0$ = 0.4 to ~1, the limited mesoporosity developed in the samples of alkaline-treated ZSM-5 zeolites from one of high $SiO_2/Al_2O_3$ molar ratio 200 (PSDs not shown here).

The structure and crystallinity of all samples were examined with X-ray diffraction measurements. The diffraction peaks assigned to MFI topology did not change at all in their position and intensities after alkali treatment (Fig. 4), confirming that the crystallinity was not

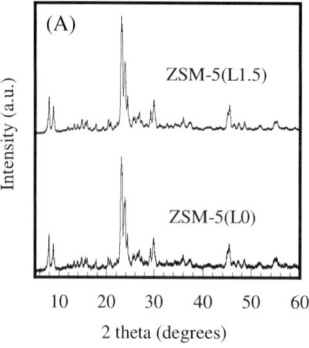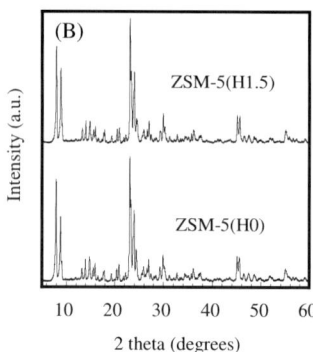

Fig. 4. X-ray diffraction patterns of ZSM-5 untreated and alkaline-treated. (A) ZSM-5(L0) and ZSM-5(L1.5), and (B) ZSM-5(H0) and ZSM-5(H1.5).

altered by the alkaline treatment. Hence, mesopores are produced by nonuniform aggregation of defect clusters around the vacancies coming from dissolution without change in the mother frame structure of ZSM-5.

## 4. CONCLUSIONS

Characterization of alkaline post-treated ZSM-5 zeolites by low temperature nitrogen and argon adsorption has shown that the changes in pore structure of the alkaline post-treatment materials depend on the $SiO_2/Al_2O_3$ molar ratios of zeolites and the alkaline post-treatment conditions. After alkaline-treatment of ZSM-5 with a low $SiO_2/Al_2O_3$ molar ratio of 23.8 in 0.05 M NaOH solutions at 325 K for different periods of time, mesopores with broad mesopore size distributions are developed. While upon alkaline-treatment of ZSM-5 with a high $SiO_2/Al_2O_3$ molar ratio of 200, no mesopores are evident from both nitrogen and argon adsorption examinations. Comparison of the nitrogen isotherms of ZSM-5 zeolites upon alkaline post-treatment to the untreated zeolite references suggests the nonuniform association of defect clusters for mesopores in ZSM-5 zeolite upon alkaline post-treatment. Since the XRD analysis confirmed that the crystallinity was not altered by the alkaline post-treatment under the experimental conditions, in accordance with previous work, numerous commercial applications could benefit from these ZSM-5 zeolites upon alkaline post-treatment [41].

## ACKNOWLEDGMENTS

This study has been supported by the Grant-in-Aid (Fundamentals) from the Japanese Government. The authors thank Dr. M. Nakano and Mr. S. Yoshida, Tosoh Corporation for kind supply of zeolite samples.

# REFERENCES

[1] A. Corma, Chem. Rev., 97 (1997) 2373.

[2] M. E. Davis, Nature, 417 (2002) 813.

[3] A. Corma, M. T. Navarro, Stud. Surf. Sci. Catal., 142 (2002) 487.

[4] J. Karger, D. M. Ruthven, Diffusion in Zeolites and Other Microporous Materials; Wiley: New York, 1992.

[5] S. van Donk, A. H. Janssen, J. H. Bitter, K. P. de Jong, Catal. Rew., 45 (2003) 297.

[6] C. H. Christensen, K. Johannsen, I. Schmidt, C. H. Christensen, J. Am. Chem. Soc., 125 (2003) 13370.

[7] S. van Donk, J. H. Bitter, A. Verberckmoes, M. Versluijs-Helder, A. Broersma, K. P. de Jong, Angew. Chem. Int. Ed., 44 (2005) 1360.

[8] M. Hartmann, Angew. Chem. Int. Ed., 43 (2004) 5880.

[9] C. J. H. Jacobsen, C. Madsen, J. Houzvicka, I. Schmidt, A. Carlsson, J. Am. Chem. Soc., 122 (2000) 7116.

[10] I. Schmidt, A. Boisen, E. Gustavsson, K, Stahl, S. Pehrson, S. Dahl, A. Carlsson, C. J. H. Jacobsen, Chem. Mater., 13 (2001) 4416.

[11] A. H. Janssen, I. Schmidt, C. J. H. Jacobsen, A. J. Koster, K. P. de Jong, Micropor. Mesopor. Mater., 65 (2003) 59.

[12] Y. Tao, H. Kanoh, K. Kaneko, J. Am. Chem. Soc., 125 (2003) 6044.

[13] Y. Tao, Y. Hattori, A. Matsumoto, H. Kanoh, K. Kaneko, J. Phys. Chem. B, 109 (2005) 194.

[14] J. Lynch, F. Raatz, P. Dufresne, Zeolites, 7 (1987) 333.

[15] C. Choi-Feng, J. B. Hall, B. J. Huggins, R. A. J. Begerlein, Catal., 140 (1993) 395-405.

[16] Y. Sasaki, T. Suzuki, Y. Takamura, A. Saji, H. J. Saka, J. Catal., 178 (1998) 94-100.

[17] R. Dutartre, L. C. D. Menorval, F. Di Renzo, D. McQueen, F. Fajula, P. Schulz, Microporous Mater., 6 (1996) 311.

[18] D. McQueen, B. H. Chiche, F. Fajula, A. Auroux, C. Guimon, F. Fitoussi, P. Schulz, J. Catal., 161 (1996) 578.

[19] G. R. Meima, CATTECH, 2 (1998) 5.

[20] R. M. Lago, W. O. Haag, R. J. Mikovsky, D. H. Olson, S. D. Hellring, K. D. Schmitt, G. T. Kerr, Stud. Surf. Sci. Catal., 28 (1986) 677.

[21] M. Rozwadowski, J. Kornatowski, J. Włoch, K. Erdmann, R. Gołembiewski, Appl. Surf. Sci., 191 (2002) 352.

[22] M. Ogura, S. Shinomiya, J. Tateno, Y. Nara, E. Kikuchi, M. Matsukata, Chem. Lett., (2000) 882.

[23] T. Suzuki, T. Okuhara, Microporous Mesoporous Mater., 43 (2001) 83-89.

[24] J. C. Groen, J. Pérez-Ramírez, L. A. A. Peffer, Chem. Lett., (2002) 94.

[25] J. C. Groen, L. A. A. Peffer, J. A. Moulijn, J. Pérez-Ramírez, Colloid Surf. A, 241 (2004) 53.

[26] R. M. Dessau, E. W. Valyocsik, N. H. Goeke, Zeolites, 12 (1992) 776.

[27] J. C. Groen, J. C. Jansen, J. A. Moulijn, J. Péerez-Ramírez, J. Phys. Chem. B, 108 (2004) 13062.

[28] M. Ogura, S. Shinomiya, J. Tateno, Y. Nara, M. Nomura, E. Kikuchi, M. Matsukata, Appl. Catal. A, 219 (2001) 33.

[29] Z. Gabelic, E. G. Derouane, ACS Symp. Ser., (1984) 248.

[30] P. Barret, L. G. Joyner, P. P. Halenda, J. Am. Chem. Soc., 73 (1951) 373.

[31] P. L. Llewellyn, M. P. Coulomb, Y. Grillet, J. Patarin, H. Lauter, H. Reichert, J. Rouquerol, Langmuir, 9 (1993) 1846.

[32] A. Saito, H. C. Foley, Microporous Mater., 3 (1995) 543.

[33] P. J. M. Carrott, K. S. W. Sing, Chem. Ind. (London), 17 (1986) 786.

[34] X. Chen, S. Kawi, Chem. Commun., (2001) 1354.

[35] C. Zhang, Q. Liu, Z. Xu, K. Wan, Microporous Mesoporous Mater., 62 (2003) 157.

[36] Z. Yang, Y. Xia, R. Mokaya, Adv. Mater., 16 (2004) 727.

[37] H. Reichert, U. Müller, K. K. Unger, Y. Grillet, F. Rouquerol, J. Rouquerol, J. P. Coulomb, Stud. Surf. Sci. Catal., 62 (1991) 535.

[38] Y. Tao, H. Kanoh, K. Kaneko, Adv. Mater., in press.

[39] P. L. Llewellyn, Y. Grillet, J. Rouquerol, Langmuir, 10 (1994) 570.

[40] Y. Grillet, P. L. Llewellyn, M. B. Kenny, F. Rouquerol, J. Rouquerol, Pure Appl. Chem., 65 (1993) 2157.

[41] Y. Tao, H. Kanoh, K. Kaneko, Chem. Rev., in press.

Studies in Surface Science and Catalysis 160
P.L. Llewellyn, F. Rodriquez-Reinoso, J. Rouqerol and N. Seaton (Editors)
© 2007 Elsevier B.V. All rights reserved

# Kureha activated carbon characterized by the adsorption of light hydrocarbons

W. Zhu, J.C. Groen, A. van Miltenburg, F. Kapteijn, and J.A. Moulijn

Reactor & Catalysis Engineering, DelftChemTech, Delft University of Technology, Julianalaan 136, 2628 BL, Delft, The Netherlands

The light hydrocarbons ethane, ethene, propane, propene, $n$-butane, and isobutane were used as probes to characterize Kureha activated carbon, which is a purely microporous material. The Tóth model gives a good description of the adsorption isotherms of all the adsorptives investigated. For butane isomers, the adsorbed amount for $n$-butane is slightly higher than that for isobutane over the whole experimental range investigated. This is attributed to the fact that the linear $n$-butane molecules can adsorb in the smaller micropores. For the $C_2$ and $C_3$ hydrocarbons, the saturation capacity for the alkene extracted by the Tóth model is higher than for the corresponding alkane due to the higher packing efficiency of the alkene molecules in the micropore space. An interesting reversal in alkane/alkene adsorption selectivity with pressure is observed: at low pressures the selectivity towards the alkanes is driven by energetic effects while at high pressures the selectivity is towards the alkenes due to entropic effects.

## 1. INTRODUCTION

The detailed knowledge of the dimensions of adsorptive molecules is crucial to understand molecular exclusion as well as shape and size selectivity on microporous materials. In general, kinetic diameters or Lennard-Jones potential constants are employed to determine the accessibility of adsorbing molecules to channels and/or apertures on microporous materials. These kinetic or collision diameters are the intermolecular distances of the closest approach for two molecules colliding as the potential is equal to zero. In this approach, the calculated kinetic diameter of an alkane molecule is slightly smaller than that of the corresponding alkene molecule with the same carbon number [1]. This, however, contradicts some experimental observations. For example, propene molecules can adsorb on the all-silica DD3R, entering via eight-membered rings of the adsorbent whereas propane molecules cannot [2]. This is ascribed to the fact that the cross section of a methyl group is circular while that of a methylene group is more elliptical. In addition, a double bond can decrease the curvature of the molecule. Therefore, it is logical to expect that the alkene molecule has a smaller critical diameter, compared to that of the corresponding alkane [2].

The molecular shape and dimensions also affect a packing efficiency on microporous materials. For instance, linear alkane molecules have a higher packing efficiency inside silicalite-1 structure, compared to the corresponding branched alkane molecules, resulting in selectivity for the linear molecules [3-6]. Recently, Pascual et al. [7] found that in alkane/alkene mixtures, the configurational entropy effects favour alkenes because such molecules "pack" more efficiently within silicalite-1 pores, observed from grand canonical Monte Carlo simulations. Experimentally, however, only limited work has been carried out on a comparative and detailed investigation on the packing efficiency of the adsorption of both alkanes and alkenes on microporous materials, although some single-component and binary isotherm data have been measured and these data are accumulated in a handbook by Valenzuela and Myers [8].

This paper presents experimental results for the equilibrium adsorption of the shorter unbranched hydrocarbons, ethane, ethene, propane, and propene and of the linear and branched $C_4$ alkanes n-butane and isobutane on Kureha activated carbon, a purely microporous material. The aim of the present study is to investigate comparative packing efficiencies of these light alkanes and alkenes and of the linear and branched $C_4$ alkanes inside the adsorbent pores. An interpretation of the difference in the adsorption behaviour for these six adsorptives is given. In addition, thermodynamic properties like isosteric heat associated with adsorption are presented to characterize interactions between adsorptive and adsorbent and an outlook on mixture adsorption is discussed for this carbon.

## 2. EXPERIMENTAL

The commercial sample, spherical bead activated carbon, was supplied by Kureha Chemical Industry. This activated carbon is referred to as Kureha carbon, which has a total micropore volume of 0.56 $cm^3 g^{-1}$ and a BET surface area of 1300 $m^2 g^{-1}$. The detailed textural properties of Kureha carbon are reported elsewhere [9]. The pore size distribution was evaluated in terms of the simulation of the density functional theory (DFT) using the isotherm data of nitrogen adsorption at 77 K and relative pressures up to 0.2. Only micropores contribute to the total pore volume and surface area. This was further confirmed by mercury intrusion porosimetry, no significantly additional porosity was observed in the pore size range from 2 nm to 100 μm. So, the investigated adsorbent is a purely microporous material and its pore size distribution covers the range from 0.4 to 1.9 nm [9].

Table 1. Summary of physical and molecular properties of the adsorptives investigated[a]

| Adsorptive | $MW$ | $\alpha$ | $\sigma_k^b$ | $\sigma_c^b$ | $T_b$ | $T_c$ | $p_c$ | $\Delta H_v$ | $MVL^c$ |
|---|---|---|---|---|---|---|---|---|---|
| | g mol$^{-1}$ | Å$^3$ | nm | | K | K | MPa | kJ mol$^{-1}$ | cm$^3$ mol$^{-1}$ |
| ethane | 30.07 | 4.47 | 0.38 | 0.37 | 184.5 | 305.3 | 4.872 | 14.69 | 55.21 |
| ethene | 28.05 | 4.25 | 0.39 | 0.34 | 169.5 | 282.3 | 5.041 | 13.53 | 49.40 |
| propane | 44.11 | 6.37 | 0.43 | 0.45 | 231.1 | 369.8 | 4.248 | 19.04 | 75.50 |
| propene | 42.08 | 6.26 | 0.45 | 0.43 | 225.5 | 364.9 | 4.600 | 18.42 | 68.94 |
| n-butane | 58.12 | 8.20 | 0.43 | | 272.7 | 425.1 | 3.796 | 22.44 | 96.64 |
| isobutane | 58.12 | 8.14 | 0.50 | | 261.4 | 407.8 | 3.640 | 21.30 | 97.94 |

[a] Ref. [10]. All data from this source unless otherwise specified. $\alpha$ is the molecular polarizability, $T_b$ is the boiling point at 101.325 kPa, $T_c$ and $p_c$ are the critical temperature and pressure, respectively, and $\Delta H_v$ is the molar enthalpy of vaporization at the component boiling point.

[b] Kinetic diameter ($\sigma_k$), data from ref. [1] and critical diameter ($\sigma_c$), data from ref. [2].

[c] Molar volume, calculated from the molecular weight ($MW$) and the liquid density at the boiling point.

A Micromeritics ASAP 2010 gas adsorption analyser (stainless steel version) was used to measure the adsorption isotherms of ethane, ethene, propane, propene, n-butane, and isobutane on Kureha carbon in the pressure range from 0.002 to 120 kPa. The instrument was equipped with a turbomolecular vacuum pump and three different pressure transducers (0.13, 1.33, and 133 kPa, respectively) to enhance the sensitivity in different pressure ranges. The static-volumetric technique was used to determine the volume of the gas adsorbed at different partial pressures: upon adsorption a pressure decrease was observed in the gas phase, which is a direct measure for the amount adsorbed.

The sample cell was loaded with 155.7 mg of Kureha carbon particles. Prior to the adsorption measurements the adsorbent particles were outgassed *in situ* in vacuum at 623 K for 16 h to remove any adsorbed impurities. The obtained dry sample weight was used in the calculation of isotherm data. Adsorption measurements were subsequently done at different temperatures from 194 to 338 K for ethane and ethene, from 273 to 358 K for propane and propene, and from 298 to 393 K for butane isomers. Five different temperatures for the adsorption of each adsorptive were used to reduce the uncertainty in the derived adsorption parameters.

The gaseous adsorptives such as ethane, ethene, propane, propene, *n*-butane, and isobutane were 3.5 grade (>99.95%). The physical and molecular properties of the adsorptives investigated are listed in Table 1.

## 3. RESULTS AND DISSCUSSION

### 3.1 Isotherms

The isotherms of ethane, ethene, propane, propene, *n*-butane, and isobutane on Kureha carbon have been measured with the volumetric technique. Fig. 1 shows an example for propene on Kureha carbon. Figs. 2 and 3 present a comparison of the isotherm data of the alkanes with those of the corresponding alkenes and of the linear *n*-butane with those of the branched isobutane. For ethane and ethene at 194 K, the adsorbed amount for ethane is slightly higher than for ethene over a lower pressure range while the adsorption becomes favourable for ethene at higher pressures. For the adsorption of propane and propene at 273 K, the difference in loading between two adsorptives is very small at pressures up to 20 kPa, although

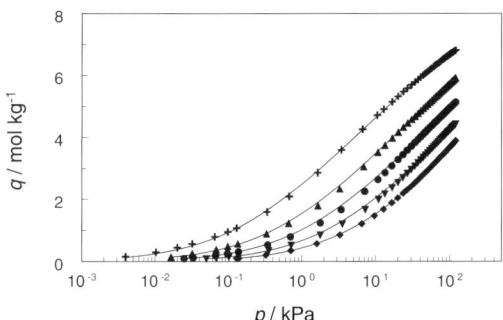

Fig. 1. Isotherms of propene on Kureha carbon. Lines are the Tóth model correlations. (:) 273 K; (▲) 298 K; (●) 318 K; (▼) 338 K; (■) 358 K.

adsorption slightly favours propane. However, at higher pressures the adsorbed amount of propene is higher than that of propane, similarly as for the ethane-ethene system. For butane isomers, the adsorbed amount for *n*-butane is higher than that for isobutane over the whole pressure range investigated.

Although Kureha carbon is a purely microporous material, the pore size distribution is wide. Therefore, Kureha carbon is considered as a heterogeneous adsorbent. For the adsorption on heterogeneous adsorbents such as activated carbon, the Tóth model is often used to correlate isotherm data [8,11],

$$q = q^{sat} \frac{Kp}{\left[1 + (Kp)^t\right]^{\frac{1}{t}}} \tag{1}$$

where $q$ is the amount adsorbed, $q^{sat}$ is the saturation adsorption capacity, $K$ is the equilibrium adsorption constant, $p$ is the pressure, and $t$ is the parameter that characterizes the system heterogeneity [11]. In the Tóth isotherm, the parameters $K$ and $t$ are temperature dependent, with the parameter $K$ taking the usual van't Hoff relation for the adsorption affinity that can be written as:

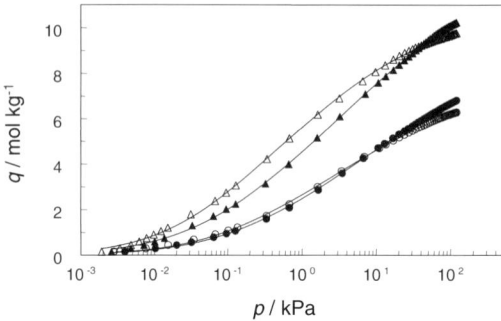

Fig. 2. Comparison of the isotherms of the alkanes with those of the corresponding alkenes. Lines are the Tóth model correlations. ($\triangle$) Ethane and ($\blacktriangle$) ethene at 194 K; ($\bigcirc$) propane and ($\bullet$) propene at 273 K.

$$K = K_0 \exp\left[\frac{Q_0^{st}}{R_g T_0}\left(\frac{T_0}{T}-1\right)\right] \qquad (2)$$

where $K_0$ is the affinity at reference temperature $T_0$, $R_g$ is the universal gas constant, and $Q_0^{st}$ is the isosteric heat of adsorption at zero coverage. The parameter $t$ and the maximum adsorption capacity $q^{sat}$ can take the following functional forms of temperature dependence [11],

$$t = t_0 + \beta\left(1 - \frac{T_0}{T}\right) \qquad (3)$$

$$q^{sat} = q_0^{sat} \exp\left[\gamma\left(1 - \frac{T}{T_0}\right)\right] \qquad (4)$$

The temperature dependence of the parameter $t$ does not have a theoretical basis. The saturation capacity, in principle, slightly decreases with temperature due to a thermal expansion of the adsorbed phase [5]. However, the value of $q^{sat}$ should remain almost

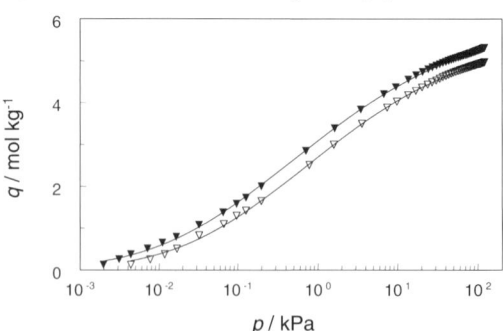

Fig. 3. Comparison of the isotherm of $n$-butane ($\blacktriangledown$) with that of isobutane ($\triangledown$) at 298 K. Lines are the Tóth model correlations.

constant, because the thermal expansion of the adsorbed phase is usually negligible in the temperature range investigated. It means that the parameter $\gamma$ in Eq. (4) should be close to zero. The lowest temperature investigated is used as $T_0$, since the measured isotherm at this temperature covers the widest range of loading.

The nonlinear parameter estimation is carried out in two steps. First, the three parameters in Eq. (1) are extracted from the isotherm data at $T_0$, referred to as $K_0$, $t_0$, and $q_0^{sat}$. Second, the parameters $Q_0^{st}$, $\beta$, and $\gamma$ are estimated from the isotherm data at all temperatures by Eqs. (1)-(4); called the combined fitting.

The isotherm data can be appropriately described by the combined fitting, as shown in Figs. 1-3 by the drawn lines. The estimated parameter values are listed in Table 2. The values of $\gamma$ in all the cases are statistically equal to zero. This means that the estimated saturation capacities are constant at the five temperatures investigated. In addition, the extracted saturation capacities for the alkenes are higher than for the corresponding alkanes and *n*-butane has a slightly higher saturation capacity than isobutane. The saturation capacity, $q^{sat}$, reflects the number of moles of adsorptive that can be adsorbed in the adsorbent pore space. When combined with the molar volume of the adsorptive in its liquid state, *MVL* presented in Table 1, the $q^{sat}$ can be converted to the accessible pore volume or the pore filling volume $V_p$. This assumes that the molecules of the adsorbed phase pack as effectively in the porous solid as they do in the liquid. As shown in Table 2, the calculated values of $V_p$ for the alkenes are larger than those for the corresponding alkanes. So, the alkene molecules have a higher packing efficiency inside the Kureha carbon pores and/or penetrate smaller pores. For butane isomers, the molar volumes (*MVL*) are comparable while the kinetic diameter of *n*-butane is smaller than that of isobutane (Table 2). In Kureha carbon, part of the micropores are smaller than 0.5 nm [9] and *n*-butane molecules can access these micropores while isobutane cannot. This probably gives the explanation for the difference in the saturation capacity between butane isomers.

Table 2. Estimated parameter values for the combined fitting of the adsorption data by the Tóth model

| Adsorptive | $T_0$ | $t_0$ | $K_0$ | $q_0^{sat}$ | $Q_0^{st}$ | $\beta$ | $\gamma$ | $V_p^{a}$ |
|---|---|---|---|---|---|---|---|---|
| | K | | kPa$^{-1}$ | mol kg$^{-1}$ | kJ mol$^{-1}$ | | | cm$^3$ g$^{-1}$ |
| ethane | 194 | 0.364 | 31.4 | 11.2 | 28.2 | 0.0722 | 0 | 0.618 |
| ethene | 194 | 0.287 | 26.9 | 14.3 | 29.5 | 0.0146 | 0 | 0.706 |
| propane | 273 | 0.326 | 11.4 | 8.37 | 41.5 | 0.145 | 0 | 0.632 |
| propene | 273 | 0.293 | 9.73 | 10.3 | 38.3 | 0.109 | 0 | 0.710 |
| *n*-butane | 298 | 0.333 | 56.6 | 6.19 | 44.0 | 0 | 0 | 0.598 |
| isobutane | 298 | 0.362 | 23.8 | 5.78 | 42.5 | 0 | 0 | 0.566 |

$^{a}$ Pore volume calculated from the saturation capacity and the molar volume of the adsorptive (*MVL*).

The parameter *t* in the Tóth model reflects a degree of the system heterogeneity [11]. The larger the deviation from unity, the system is said to be more heterogeneous. The extracted values of *t* for all the adsorptives are much smaller than unity in the temperature range investigated. This implies a strong degree of heterogeneity for the adsorption of all the adsorptives on Kureha carbon. In addition, for the alkenes the adsorbent appears to be slightly more heterogeneous than for the corresponding alkanes in terms of the *t* values. This may be due to the size of the adsorptive molecule, i.e. the smaller alkene molecules would "see" a wider pore distribution. The same holds for the butane isomers on Kureha carbon.

Other isotherm models, such as the Unilan and Sips models [11] and those based on the theory of micropore volume filling (Dubinin [12]), have also been considered, but the Tóth model describes the present case much better over the full range. For engineering purposes the Tóth isotherm is quite attractive to be applied due to its simplicity. In addition, because of its correct behaviour at low and high pressures, the Tóth model is usually recommended as isotherm expression for correlating adsorption data [11]. In the further analysis, the Tóth isotherm is used for the estimation of thermodynamic adsorption properties.

### 3.2 Adsorption thermodynamics

The extracted values of $Q_0^{st}$ for all the adsorptives (Table 2) are approximately two times larger than their corresponding enthalpies of vaporization shown in Table 1. It is noted that the derived value of $Q_0^{st}$ for ethene is slightly larger than that for ethane. If one assumes that in the adsorption the non-specific interactions are involved, the adsorption potential is almost entirely the product of dispersion forces, where the interaction between adsorptive and adsorbent is proportional to the polarizability $\alpha$ of the adsorptive [13,14]. The polarizability of the alkane is larger than that of the corresponding alkene (Table 1), and therefore the derived value of $Q_0^{st}$ should be larger for the alkane, like is the case for propane. Thus, the larger value of $Q_0^{st}$ for ethene may be due to some specific interactions with this alkene molecule. Although, in general, activated carbon is hydrophobic in nature, some carbon-oxygen surface groups can be formed during manufacturing [15]. These carbon-oxygen groups can create some specific interactions with the double bond in ethene molecule, resulting in a larger value of $Q_0^{st}$. But the effect of the double bond in an alkene molecule on overall interactions between adsorbent and adsorptive is expected to become smaller as the number of carbon atoms increases in the molecule. Indeed, the extracted value of $Q_0^{st}$ for propene is slightly smaller than that for propane, indicating that the contribution from these specific interactions, if present, is already reduced. For butane isomers, the derived value of $Q_0^{st}$ for $n$-butane is larger than that for isobutane, following the order of the polarizability $\alpha$ for these two adsorptives.

From the derived Tóth isotherms, the isosteric heat of adsorption as a function of coverage can be calculated by the following equation [11]:

$$Q^{st} = -R_g \left[ \frac{\partial \ln p}{\partial (1/T)} \right]_\theta = Q_0^{st} - \frac{1}{t}(\beta R_g T_0) \left\{ \ln \left[ \frac{\theta}{\left(1 - \theta'\right)^{1/t}} \right] - \frac{\ln \theta}{\left(1 - \theta'\right)} \right\} \quad (5)$$

where $\theta$ is the adsorption coverage. Fig. 4 shows an example of the isosteric heat of adsorption as a function of coverage for the $C_2$ and $C_3$ hydrocarbons at 273 K. The dependence of $Q^{st}$ on $\theta$ is generally used to characterize the energetic heterogeneity of an adsorption system. The isosteric heat of adsorption for propane and propene monotonically decreases with increasing loading. It implies that the adsorbent is energetically heterogeneous. Kureha carbon consists of micropores with different widths relative to the molecular diameter of the adsorptive. Physically, molecules prefer to adsorb onto sites of high energy and hence as adsorption progresses molecules then adsorb onto sites of decreasing interaction energy. Consequently, $Q^{st}$ decreases with increasing loading.

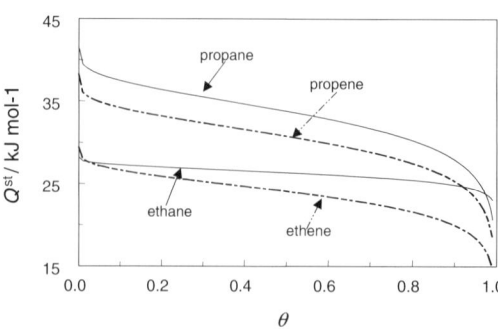

Fig. 4. Isosteric heat of adsorption as a function of coverage for ethane, propane (solid line), ethene and propene (dashed line) on Kureha carbon at 273 K.

As shown in Fig. 4, only at extremely low coverage, $Q^{st}$ for ethene is higher than for ethane due to some specific interactions between ethene and carbon-oxygen groups on the solid surface, suggesting that the number of the carbon-oxygen surface groups is limited. Apart from these limited adsorption sites having the specific interactions with ethene molecules, the adsorption potential is proportional to the polarizability of the adsorptives, resulting in the higher isosteric heat of adsorption for ethane. For propane and propene, $Q^{st}$ is higher for propane in the whole range of coverage.

At low coverage, the strength of the interactions between adsorptive and adsorbent dominates the adsorption of single components and also will determine the ideal selectivity based on two single-component isotherms. A reversal in the ideal selectivity with pressure is observed for propane and propene in Fig. 2, reflecting a change in the mechanism governing the selectivity. Based on the analysis of the isosteric heat of adsorption as a function of coverage, propane molecules have stronger interactions with the adsorbent than propene molecules. Under these conditions, the selectivity strongly depends on attractive adsorptive interactions, and thus adsorption on Kureha carbon is favourable for propane.

On the other hand, at high coverage or pressures, the size and shape of adsorptive molecules are by far the most important parameters; usually the smaller molecules adsorb favourably [16]. It is well known that extensive entropic contribution to adsorption is the greatest at high loadings [17]. Thus the ability to pack better becomes crucial to minimize the free energy, resulting in selectivity for propene at high pressures. This explains the experimentally observed reversal of propane/propene selectivity with pressure. The same holds for ethane/ethene selectivity as a function of pressure. It also implies that selectivity will be favourable for the alkenes in the alkane/alkene mixtures at elevated pressures.

The ideal adsorbed solution (IAS) theory is the thermodynamically consistent and most useful model that enables to predict multi-component isotherms by using the single-component isotherms as input [18].

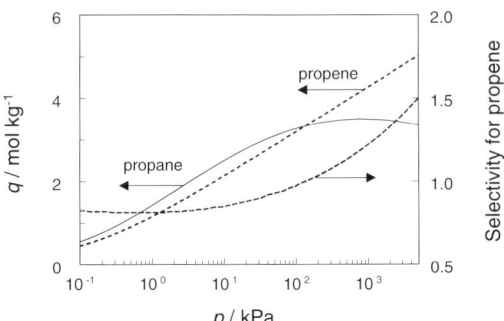

Fig. 5. IAS simulation for propane and propene (50:50) mixture adsorption on Kureha carbon as a function of total pressure at 273 K.

Fig. 5 gives an example of the binary (50:50) adsorption as a function of total pressure, simulated by the IAS theory. At low pressures, the loadings of both propane and propene increase with increasing pressure and the adsorption is slightly favourable for propane. At these conditions the adsorption is determined by the most favourable energy, resulting in the mixture selectivity for propane. As the pressure is increased beyond 1000 kPa, the thermodynamic tendency to attain a higher loading is achieved by "replacing" the adsorbed propane by propene. Thus the adsorptive species that can pack more efficient begin to dominate the pore space [16], resulting in an entropy-controlled selectivity for propene. The same holds for ethane/ethene mixture adsorption as a function of total pressure.

Like predicted for linear/branched mixture adsorption on silicalite-1 [4], for butane isomers over the whole pressure range the adsorption is favourable for the linear *n*-butane. This is attributed to the fact that energy and entropy as well as molecular sieving control the mixture adsorption of *n*-butane and isobutane on the activated carbon.

## 4. CONCLUSIONS

The single-component adsorption isotherms of ethane, ethene, propane, propene, *n*-butane, and isobutane on Kureha carbon have been accurately measured at pressures up to 120 kPa and over a wide temperature range. The Tóth model gives a good description of the adsorption isotherms of all the adsorptives investigated. The difference in the limiting adsorption capacity between alkane and alkene can be explained by the fact that these molecules have different molecular sizes and shapes. The alkene molecules have a higher packing efficiency inside the micropores of the adsorbent and may probe smaller pores. A reversal in alkane/alkene selectivity with pressure has been observed. At low pressures, the selectivity is dominated by the energetics of the system and adsorption favours the alkanes; at high pressures, the entropic effects begin to dominate the adsorption process and the selectivity is favourable for the alkenes. For butane isomers, the adsorbed amount for *n*-butane is higher than that for isobutane over the whole range investigated due to the fact that the linear *n*-butane molecules can adsorb in the smaller micropores while isobutane cannot. The present study indicates that the activated carbon can be well characterized by the probe molecules having different molecular sizes and shapes.

## REFERENCES

[1]    D.W. Breck (ed.), Zeolite Molecular Sieves-Structure, Chemistry, and Use, Wiley, New York, 1974.
[2]    W. Zhu, F. Kapteijn, J.A. Moulijn, M.C. den Exter and J.C. Jansen, Langmuir, 16 (2000) 3322.
[3]    T.J.H. Vlugt, W. Zhu, F. Kapteijn, J.A. Moulijn, B. Smit and R. Krishna, J. Am. Chem. Soc., 120 (1998) 5599.
[4]    T.J.H. Vlugt, R. Krishna and B. Smit, J. Phys. Chem. B, 103 (1999) 1102.
[5]    W. Zhu, F. Kapteijn and J.A. Moulijn, Phys. Chem. Chem. Phys., 2 (2000) 1989.
[6]    W. Zhu, F. Kapteijn, B. van der Linden and J.A. Moulijn, Phys. Chem. Chem. Phys., 3 (2001) 1755.
[7]    P. Pascual, P. Ungerer, B. Tavitian and A. Boutin, J. Phys. Chem. B, 108 (2004) 393.
[8]    D.P. Valenzuela and A.L. Myers (eds.), Adsorption Equilibrium Data Handbook, Prentice-Hall, New Jersey, 1989.
[9]    W.D. Zhu, J.C. Groen, F. Kapteijn and J.A. Moulijn, Langmuir, 20 (2004) 5277.
[10]   D.R. Lide (ed.), CRC Handbook of Chemistry and Physics, CRC Press, Boca Raton, 2001.
[11]   D.D. Do, Adsorption Analysis: Equilibrium and Kinetics, Imperial College Press, London, 1998.
[12]   M.M. Dubinin, Chem. Rev., 60 (1960) 235.
[13]   W. Zhu, J.C. Groen, A. van Miltenburg, F. Kapteijn and J.A. Moulijn, Carbon, 43 (2005) 1416.
[14]   R.E. Richards and L.V.C. Rees, Langmuir, 3 (1987) 335.
[15]   M.J.G. Linders, L.J.P. van den Broeke, F. Kapteijn, J.A. Moulijn and J.J.G.M. van Bokhoven, AIChE J., 47 (2001) 1885.
[16]   S. Mohanty and A.V. McCormick, Chem. Eng. J., 74 (1999) 1.
[17]   R. Krishna, B. Smit and T.J.H. Vlugt, J. Phys. Chem. A, 102 (1998) 7727.
[18]   A.L. Myers and J.M. Prausnitz, AIChE J., 11 (1965) 121.

Studies in Surface Science and Catalysis 160
P.L. Llewellyn, F. Rodriquez-Reinoso, J. Rouqerol and N. Seaton (Editors)

# Water adsorption/desorption isotherms for characterization of microporosity in sandstone and carbonate rocks

**H. Fischer[a], N. R. Morrow[a], and G. Mason[b]**

[a]Department of Chemical and Petroleum Engineering, University of Wyoming,
Dept. 3295, 1000 E. University Ave., Laramie, WY, 82071, USA

[b]Chemical Engineering Department, Loughborough University,
Leicestershire LE11 3TU, UK

Porous rocks from hydrocarbon bearing formations have microporosity, which contributes to the retention of connate water and can have significant effect on reservoir productivity and in formation evaluation such as in interpretation of electric logs for shaly sands. In the present work microporosity of a selection of sandstone and limestone outcrop rocks has been investigated through measurement of water adsorption/desorption isotherms using glycerol/water solutions to control the relative humidity, $H_r$. Duplicate plugs gave closely reproducible results. Three cycles of adsorption/desorption were obtained for initially dry and initially water saturated rock samples. For desorption of initially water saturated rocks, water retention was consistently higher than secondary desorption down to $H_r$ of about 40 %. The second and third adsorption/desorption cycles were closely reproducible. Water retention at 60 % $H_r$ correlated with BET surface areas using nitrogen.

## 1. INTRODUCTION

Measurement of very small pore sizes using mercury penetration techniques is somewhat limited because of the possible effects of high injection pressures on fragile pore structure, definition of total pore space, and inability to restore a tested sample back to its original state. Furthermore, in many locations, environmental concerns have lead to a pressing need to adopt mercury-free laboratory procedures. Measurement of sorption isotherms is a commonly used approach to investigation of microporosity. Measurement of water sorption isotherms through use of glycerol [1] has many advantages over other commonly used methods of humidity control such as saturated salt solutions or variation in sulfuric acid concentration [2–4]. Glycerol is a trihydric aliphatic alcohol with a density of 1.261 g/cm³ (as supplied) at 20 °C and at atmospheric pressure. It is non-toxic and non-corrosive. In the anhydrous state glycerol is very hygroscopic. It is soluble in water in all proportions and is absent from the vapor phase at ambient conditions [5-7].

Microporosity in reservoir rocks is important to many aspects of hydrocarbon reservoir evaluation and production. These include retention of connate water, formation damage, ion exchange capacity, electrical resistivity and spontaneous potential, log interpretation for shaly sands, and the production potential of low permeability gas sands.

Measurement of adsorption/desorption isotherms provides an alternative approach to investigation of microporosity. Such sorption isotherms constitute an extension of capillary pressure data to very high pressures. Microporosity is a widely used generic term. Pore sizes have been subclassified by IUPAC (Everett [8]) as micropores (pore radius < 2.1 nm), mesopores (pore radius between 2.1 and 53 nm and macropores (pore radius > 53 nm). A $H_r$ of 60 %, which corresponds to a water/air capillary pressure of 69.36 MPa (10,060 psi), constitutes the dividing line between micropores and mesopores. A $H_r$ of 98 % (equivalent to a capillary pressure of around 2.76 MPa (400 psi)) designates the dividing line between mesopores and macropores. The format of the sorption plots was basically adopted from [3]. In this study sorption isotherms were measured for a series of carbonate and sandstone rocks.

## 2. EXPERIMENTAL

### 2.1. Cores
Cores cut from outcrops that are quarried for building stone are commonly used in studies of oil recovery. For this study a total of 61 core plugs, each with a nominal length of 0.5 in and a nominal diameter of 1.5 in, were cut from 19 different rocks. Some of them are currently in use for laboratory studies of oil recovery. 18 of the core plugs were carbonates (limestones) and 43 were sandstones. The cores were washed, dried at ambient temperature for one day and then oven dried at 110 °C for two days. The permeability to nitrogen, $k_g$, was measured in a Hassler-type core holder at a confining pressure of 2.07 MPa (300 psi). Porosity was calculated from the increase in mass after saturation under vacuum with water. Permeability, porosity, BET surface area and cation exchange capacity (CEC) of each rock sample are listed in Table 1.

Table 1
Rock properties

| | Sandstones | | | | | Carbonates | | | |
|---|---|---|---|---|---|---|---|---|---|
| | $k_g$ (md) | $\Phi$ (%) | BET (m²/g) | CEC (meq/g) | | $k_g$ (md) | $\Phi$ (%) | BET (m²/g) | CEC (meq/g) |
| Berea EV1 | 50 | 16.7 | 1.570 | 0.00183 | Edwards (GC), (EGC) | 12 | 19.6 | 0.195 | 0.00026 |
| Berea EV2 | 47 | 17.6 | 1.441 | 0.00199 | Whitestone, (WUZ) | 5 | 20.8 | 0.899 | 0.00029 |
| Berea EV3 | 49 | 17.1 | 1.138 | 0.00206 | Fort Riley Limestone, (FRL) | 2 | 18.8 | 1.639 | 0.00076 |
| Berea EV4 | 133 | 18.9 | 0.896 | 0.00182 | Lueders Limestone, (LL) | 2 | 16.7 | 0.790 | 0.00024 |
| Berea EV5 | 62 | 18.3 | 1.304 | 0.00196 | Gambier, (GA) | 3138 | 54.9 | 0.867 | 0.00065 |
| Berea EV6 | 122 | 18.5 | 1.154 | 0.00203 | Whitestone, (WLZ) | 1 | 19.7 | 0.466 | 0.00036 |
| Berea EV7 | 57 | 17.0 | 1.515 | 0.00195 | | | | | |
| Berea EV8 | 68 | 16.9 | 1.516 | 0.00179 | | | | | |
| Berea PH2 | 707 | 22.1 | 0.801 | 0.00145 | | | | | |
| Berea PH5 | 182 | 18.7 | 0.881 | 0.00129 | | | | | |
| Pink Berea PH3 | 3082 | 21.9 | | | | | | | |
| Berea C3 | 70 | 17.0 | | | | | | | |
| Berea C4 | 72 | 16.1 | 1.111 | 0.00112 | | | | | |

Sandstone permeabilities varied from 47 to 182 md, except for PH2 and PH3, (707 and 3082 md, respectively). Sandstone porosity ranged from 16.1 % to 18.9 %, except for the samples PH2 and PH3 (22.1 and 21.9 %, respectively). BET surface area and CEC ranged from 0.801 to 1.570 m²/g and 0.00112 to 0.00206 meq/g, respectively. Carbonate permeabilities were

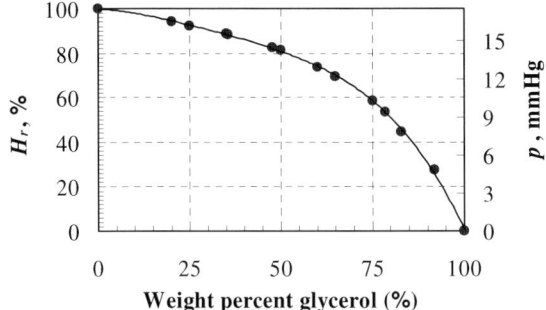

Fig. 1. Relative humidity and vapor pressure as a function of weight fraction glycerol

generally lower and varied between 1 and 12 md, except for the highly permeable Gambier reef boundstone (3138 md). Porosities for the tested carbonate samples were comparable to the sandstone porosities with values between 16.7 and 20.8 %, except for the Gambier limestone (54.9 %). BET surface area for carbonates exhibited a wider spread (0.195 to 1.639 $m^2/g$) than the sandstones. However, as might be expected, CEC were significantly lower than compared to sandstones and ranged from only 0.00024 to 0.00076 meq/g.

## 2.2 Procedure

Fig. 1 displays data plotted as $H_r$ and vapor pressure, $p$, versus weight percent glycerol in the aqueous solution [1]. High glycerol concentrations are needed to attain a significant reduction in $H_r$.

Aqueous glycerol solutions contained in small open jars were prepared to give nominal $H_r$ of 10, 20, 30, 40, 50, 60, 70, 75, 80, 85, 90, 93, 96 and 98 % based on Fig. 1. Each jar was then placed in a sealable container (humidity chambers). The basic technique was to place pre-weighed cores into a particular humidity chamber on a grid that separates the cores from the glycerol solution and periodically monitor the core weight until equilibrium was reached. In general, the time until the core reached equilibrium ranged from a few days for $H_r$ up to about 80 % to as much as a few weeks for $H_r$ higher than 80 %. However, many samples could be accommodated in each container, the technique only involves periodic weighing and for most practical purposes much fewer data points would suffice. Water sorption on cores was measured relative to a standardized dry weight measured at 10 % $H_r$.

Accurate humidity control was ensured by regular checks on the respective glycerol solution density with a PAAR Densitymeter DMA 48. Any slight departure from the original solution density was corrected through addition of glycerol or distilled water. Measurements were performed at a closely controlled ambient temperature of 20 °C ± 0.5. To avoid the consequences of possible temperature gradients, the humidity containers were packed in styrofoam insulation inside well insulated ice chests.

Sorption isotherms were measured for two initial core conditions: initially water saturated core plugs and initially dry core plugs. In the case of initially water saturated core plugs, the samples were first placed into the container of highest humidity (98 % $H_r$). The core plugs were weighed every 2-3 days until constant weight was attained. Subsequently, the core plugs were moved to the container with the next lower humidity (96 % $H_r$). This process was repeated down to 10 % $H_r$ to give a desorption isotherm described by 14 data points starting at 100 % initial water saturation. Adsorption isotherms were obtained by reversing this procedure starting at 10 % $H_r$. Cores were transferred sequentially into the containers of next higher humidity after constant weight was obtained for up to 98 % $H_r$. For initially water saturated cores, after measurement of the initial desorption isotherm, two more complete sorption cycles (2 adsorption and 2 desorption branches) were measured for $H_r$ ranging from 10 to 98 %, to give a total of 2 ½ sorption cycles, as a test of reproducibility and to investigate how subsequent sorption cycles compare with the initial cycle. For initially dry plugs 3 sorption cycles were measured.

## 3. RESULTS AND DISCUSSION

### 3.1. Sorption Isotherms

Sorption isotherms are plotted as $H_r$ vs. water saturation of pore space, $S_w$. The corresponding capillary pressure as given by the Kelvin equation (1), which relates the relative vapor pressure $p/p^0$ to the capillary pressure $P_c$, and the equivalent pore size and/or crack thickness is shown on a separate axis in Fig. 2a. In the Kelvin equation, $P_c$ is expressed in terms of surface tension, $\sigma$, and mean radius of curvature, $r_m$; $v$ is the liquid molar volume, $R$ the universal gas constant and $T$ the temperature.

$$RT \ln\left(\frac{p}{p^0}\right) = -\frac{2\sigma}{r_m} v = -P_c v \qquad (1)$$

$p$ is the reduced vapor pressure over a curved meniscus and $p^0$ is the saturated vapor pressure of the bulk liquid. The main mechanisms that determine the shape of sorption isotherms are physical adsorption (desorption) on rock surfaces at low $H_r$ and capillary condensation and evaporation in pore spaces at higher $H_r$. The concept of capillary condensation is based on the classical Kelvin equation. A $H_r$ of about 60 % corresponds to the transition from adsorbed water at mineral surfaces to additional adsorption by capillary condensation. $H_r$ is related to the relative vapor pressure by

$$H_r = \left(\frac{p}{p^0}\right) 100\% \qquad (2)$$

### 3.2. Reproducibility of Sorption Isotherms

Sorption isotherms for selected core plugs, plotted as capillary pressure (and the corresponding $H_r$) versus $S_w$ are shown in Figs. 2 and 3. In Fig. 2a the corresponding pore size is also displayed. The initial desorption isotherm from 100 % initial water saturation as well as sorption isotherms obtained from the subsequent 2nd and 3rd cycle are presented for the initially water saturated carbonate, Fort Riley limestone FRL (Fig. 2a), and two sandstones, Berea sandstone C3 and Berea sandstone PH2 (Fig. 2b). For all three samples, the initial desorption isotherm, starting at 100 % $H_r$, is located to the right of the respective subsequent desorption isotherms started at 98 % $H_r$. The results obtained from the 2nd and 3rd cycle for the initially dry carbonate Edwards (GC) are shown in Fig. 2a. The 2nd and 3rd cycle for all four core plugs were closely reproducible apart from only minor deviations. Hysteresis is observed for $H_r$ in the range of around 60 to 98 %. Comparison of the two carbonates shows that adsorption for the Edwards (GC) sample was very low with a maximum pore space saturation of less then 0.5 %, compared to about 4 % for the Fort Riley limestone.

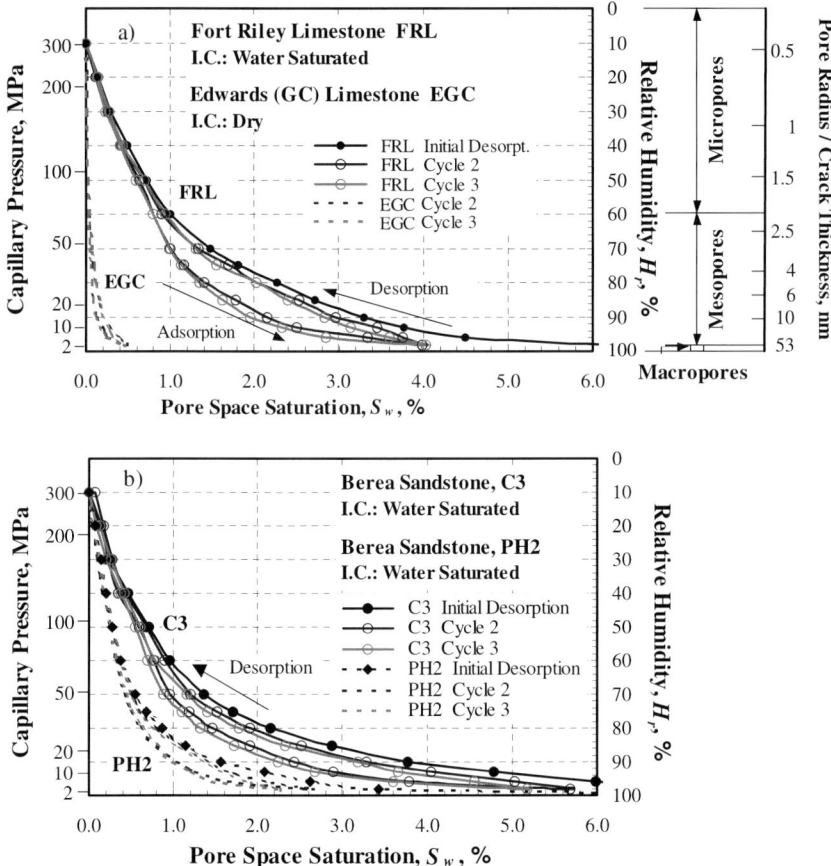

Fig. 2: Initial desorption from 100 % saturation and reproducibility of sorption isotherms for (a) limestones and (b) sandstones with distinct differences in fractions of microporosity.

Hence, the Edwards (GC) rock consists mainly of macropores. In contrast, the Fort Riley carbonate is characterized by about an eight times higher fraction of micro- and mesopores than the Edwards (GC). Berea sandstone C3 exhibits the largest fraction of micro- and mesopores (over 5 %), which is about twice as much as for the Berea sandstone PH2. In Berea sandstone, chert, lithic fragments, partially dissolved feldspars, sheet pores at crystal boundaries and clays contribute to microporosity.

Desorption isotherms for initially water saturated cores exhibited higher pore space saturations than the subsequent desorption isotherms. This phenomenon is due to the inaccessibility of certain pores during the course of initial desorption. Certain pores might be only accessible through pores with a smaller pore diameter. Consequently, during initial desorption these larger pores cannot be emptied until the surrounding smaller pores are emptied. For subsequent adsorption, the smaller pores fill first. Results for initially water saturated carbonates and sandstones are shown in Fig. 4 as the average of the absolute differences in water saturation, $\Delta S_w$, between the initial and the subsequently measured $2^{nd}$ desorption isotherm vs $H_r$.

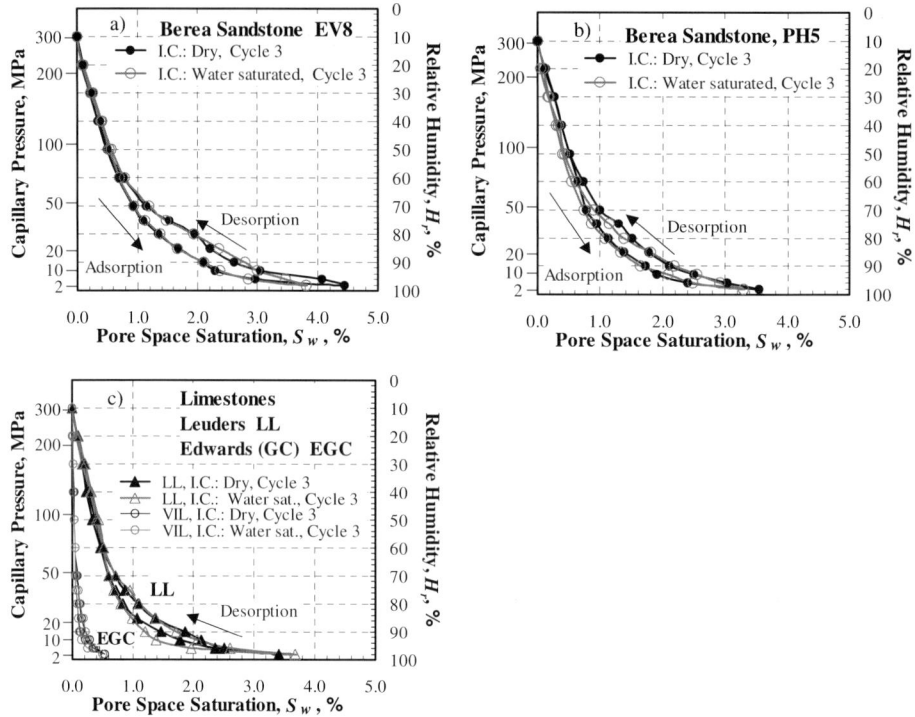

Fig. 3: Comparison of 3$^{rd}$ cycle sorption isotherms for initially water saturated and initially dry rocks: (a) Berea Sandstone EV8, (b) Berea Sandstone PH5, (c) Leuders LL and Edwards (GC) EGC.

For both rock types, the difference in saturation increased steadily with $H_r$. For saturations up to $H_r$= 85 % the difference was small (less than 0.2 %). At $H_r$ above 85 %, however, there is a distinct departure of the initial desorption isotherms from the subsequently measured desorption isotherms. A comparison of the initial and the subsequent desorption isotherms for carbonates and sandstones shows that the absolute difference in water saturation between the first and subsequent desorption isotherms is comparable for the tested rock types. Fig. 2 also shows that sorption isotherm hysteresis occurred for all four samples, but it is far less pronounced for the Edwards (GC) limestone because of the lack of mesopores.

### 3.3. Reproducibility of sorption isotherm cycles for different initial conditions
Fig. 3 is a compilation of sorption isotherms for the two Berea sandstones EV8 (Fig. 3a) and PH5 (Fig. 3b), and the two carbonates Edwards (GC) EGC and Leuders Limestone LL (Fig. 3c). Each of the four plots compares the sorption isotherm determined from the 3$^{rd}$ sorption cycle, but for different initial conditions (initially dry and initially water saturated). Apart from minor deviations, the 3$^{rd}$ cycle sorption isotherms were generally closely reproducible in terms of similarity of shape, the extent of the hysteresis loop, and the degree of water adsorbed at any given $H_r$.

### 3.4. Correlation of $S_w$ and BET

At 60 % $H_r$ (the transition from surface adsorption to capillary condensation), pore space saturations were taken from the respective desorption isotherm and averaged for each rock type (sandstones and grainstone carbonates). These saturations, which depend mainly on surface adsorption, gave a linear correlation with BET surface areas (Fig. 5). The linear trend between BET surface area and the amount of adsorbed water held for other values of $H_r$.

### 3.5. Fine Structure (Microporosity)

Average values of micropores as well as the sum of micro- and mesopores as a fraction of the total pore space are presented in Fig. 6 for both carbonates and sandstones. (Individual values for a certain rock type were within a narrow range.) The datasets for individual rock types are superimposed and hence all bars commence at zero % of total pore space. In general, the carbonates exhibit a significantly lower fraction of micro- and mesopores than the lower permeability sandstones (~ 0.25 vs. ~ 2 % of the total pore space, respectively).

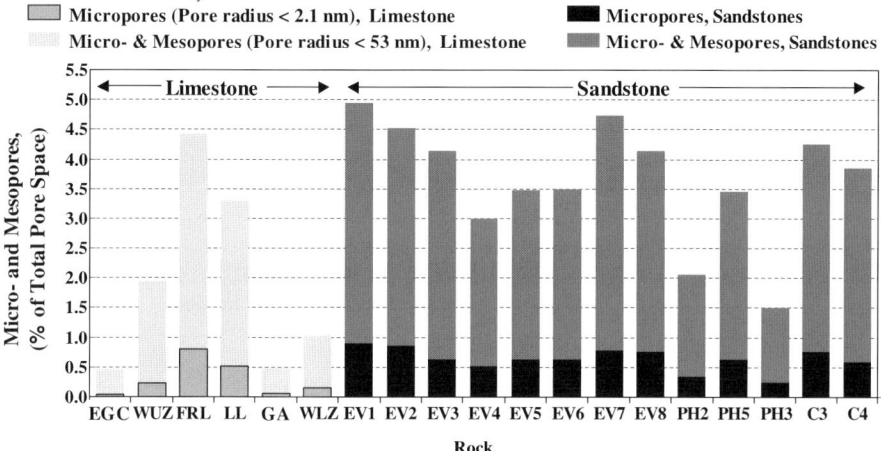

Fig. 6: Fraction of micropores (10 % < $H_r$ < 60 %) and the sum of micro- and mesopores (10 % < $H_r$ < 98 %) for each rock type from sorption isotherms for water.

Fort Riley Limestone FRL and Leuders Limestone LL possess micro- and mesopores in amounts comparable to Berea sandstones with permeabilities in the range of 47 – 182 md. However the limestone permeabilities are only about 2 md. Overall, the Edwards (GC) EGC serves as a useful model rock for study of oil recovery because it has moderate permeability and porosity and complications related to the presence of microporosity are minimal.

### 4. CONCLUSION

1. Use of glycerol for humidity control has many advantages over sulfuric acid or saturated salt solutions.
2. Chambers containing aqueous solutions of glycerol provide a low cost and straightforward approach to measurement of water sorption isotherms for porous media such as sedimentary rocks.
3. Sample weight at 10 % $H_r$ provides a convenient operational definition of dry rock weight and 98 % $H_r$ is a practical upper limit for study of adsorption/desorption cycles and the

relative fractions of micro- (pore radius < 2.1 nm) and meso- (pore radius between 2.1 and 53 nm) pores in the total pore space.

4. The initial desorption isotherm for initially water saturated rocks showed consistently higher retention of water than subsequent desorption (from 98 % $H_r$) because a larger fraction of water filled pores are only accessible after smaller pores empty.

5. Adsorption isotherms from 10 % $H_r$ were independent of whether the rock was initially dry or 100 % saturated or the number of subsequently measured hysteresis cycles.

6. For all rock samples, adsorption/desorption isotherms exhibited reproducible hysteresis loops in the range of about 60 % to the adopted upper value of $H_r$ of 98 %.

7. A linear trend was obtained between the nitrogen BET surface area and the pore space saturation at 60 % $H_r$ that persisted at higher values of $H_r$ up to 98 % with only modest increase in scatter.

8. The Edwards (GC) limestone has the lowest fraction of microporosity and is especially suited as a model limestone for oil recovery studies.

## ACKNOWLEDGEMENTS

This work is supported by U.S. DOE under the contract No. DE-PS26-01NT41048

## REFERENCES

[1] D.W. Grover and J.M. Nicol, J. Chemical & Industry, Transactions, 59 (1940) 175.

[2] N.R. Morrow, M.E. Cather and J.S. Buckley, paper SPE 21880 presented at the Rocky Mountain Regional Meeting and Low-Permeability Reservoir Symposium, Denver, CO, Apr.15-17 1991.

[3] J.C. Melrose, paper SPE 16286 presented at the 1987 SPE Int. Symposium on Oilfield Chemistry, San Antonio, TX, Feb. 4-6

[4] J.S. Ward and N.R. Morrow, SPE Formation Evaluation, (1987) 345-56.

[5] CRC Handbook of Chemistry and Physics, 1976-1977, CRC Press, 57th Edition, Cleveland, OH.

[6] L.J. Root and B. J. Berne, J. Chem. Phys., 107 (11), (1997) 4350.

[7] J.A. Monick, Alcohols: their chemistry, properties and manufacture, Reinhold Book Corp., New York (1968).

[8] D.H. Everett, Pure and Appl. Chem. (1972) 578.

Studies in Surface Science and Catalysis 160
P.L. Llewellyn, F. Rodriquez-Reinoso, J. Rouqerol and N. Seaton (Editors)

303

# Determination of pore-size distributions of highly-connected networks with assisted-filling characteristics

**Fernando Rojas, Carlos Felipe, Isaac Kornhauser, and Salomón Cordero**

Departamento de Química, Universidad Autónoma Metropolitana-Iztapalapa,
P.O. Box 55-534, México D.F. 09340, México

Hg porosimetry and $N_2$ sorption processes have been replicated in simulated 3-D porous networks constructed by Monte Carlo procedures. These two characterization techniques render complementary information about the pore structural parameters of highly-connected porous networks. Through this study, it has been possible to depict a phenomenon labeled as *delayed adsorption*. This phenomenon consists in that condensation is not taking place inside a cavity unless the pore throats that surround this void have been already filled with liquid. This phenomenon arises when pore necks are comparable in size to the cavities to which they are connected. If condensation occurs this way, the pore-size distributions calculated from $N_2$ adsorption are biased toward overvalued pore sizes. Under this circumstance, Hg porosimetry analysis can still be suitable for realizing and assessing the latter problem since a complementary cavity-size distribution can be calculated from the Hg retraction curve.

## 1. INTRODUCTION

The calculation of the pore-size distribution (PSD) from either $N_2$ sorption or Hg porosimetry is frequently based on the assumption that the porous sample is formed by a collection of independent capillaries of various sizes [1, 2]. This conceptualization implies that $N_2$ condensation in a single pore occurs disregarding of what can be happening in neighboring pores. Likewise, the intrusion or retraction of Hg inside this kind of porous networks, is thought to occur irrespectively of the state (Hg-filled or not) of nearby pores. Nonetheless, a realistic picture about the structure of an interconnected porous network can still lie far beyond this simple schematization of independent pores. Consequently, the modeling of credible sorption and intrusion-retraction phenomena occurring in simulated porous media can be seriously affected by any structure misconception.

Hg intrusion is frequently a percolating process that is highly influenced by the connectivity of the porous network and by the spatial distribution of pore sizes [3]. In turn, Hg retraction is highly dependent on this spatial distribution as well as on the occurrence of phenomena such as Hg snap-off or the entrapment of non-wetting liquid in the form of disconnected islands [4, 5] throughout the porous structure.

Several phenomena complicate the understanding of the mechanisms that control sorption processes. Two of them are the variations in density of the adsorbed phase along the pore axis [6] and in the changes in the potential field that emanates from the pore walls towards the adsorbed molecules. Also, the conditions at which condensation occurs in an isolated pore can change drastically when this same void is connected to other pores. It is important to note that connected pores share adsorbed-interfaces, something that represents a prime factor for promoting condensation. Nevertheless, only few authors have examined the problem of how condensation in a given pore can effectively alter the conditions at which neighboring pores can fill with condensate [7, 8]. In a recent work, it has been suggested that highly connected porous networks are likely to undergo a phenomenon branded as *delayed adsorption* [9], which basically consists in the influence that the states (liquid-filled or not) of surrounding pore elements can exert on the filling of a central cavity. In this work, we show how the combination of $N_2$ sorption and Hg porosimetry experiments can be used conjointly to assess more accurately the PSD of porous networks that can undergo delayed adsorption. This study is based on a Monte Carlo modeling of mesoporous networks, which is followed by simulations of $N_2$ sorption and Hg porosimetry processes inside these substrates.

## 1.1 Delayed adsorption

For the sake of simplicity, the porous materials to be studied in this work have been idealized as 3-D cubic networks made of spherical cavities (sites) interconnected through cylindrical tubes (bonds). Besides, $N_2$ sorption and Hg intrusion-extrusion processes have been simulated in these networks. During an adsorption process, $N_2$ molecules form an adsorbed layer of growing thickness on the surfaces of sites and bonds as the vapor pressure is increased. Capillary condensation can occur spontaneously in a single bond if an adsorbed film of a critical thickness is formed on the pore walls. In the case of capillary condensation to take place in a site, it is not only required that this pore has the right size but also that the liquid-vapor interface can fully circumscribe the cavity in question. Possible filling sequences of a site in a porous network can be visualized in Figs. 1a-b. A liquid-vapor interface that collapses towards the center of the cavity is formed until all delimiting bonds are occupied by condensate (Fig. 1a). Moreover, the site can be overtaken by condensate when menisci of neighboring bonds merge together to form and advancing interface (see Fig. 1b). In this last case, a remaining empty bond can be invaded by condensate (i.e. *advanced adsorption* of bonds).

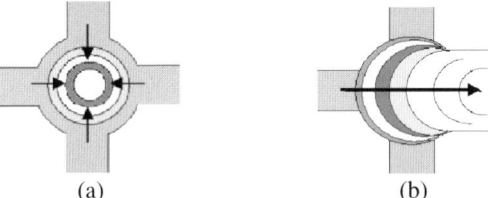

(a)                                                         (b)

**Fig. 1.** Alternative schemes of delayed condensation in sites: (a) condensation only occurs if all delimiting bonds are already occupied by condensate then forming an encircling interface that moves towards the center of the site; (b) an advancing meniscus is produced when most delimiting bonds are filled with condensate; the merge of neighboring interfaces creates the advancing meniscus, which not only overtakes the site completely but its movement continues until a neighboring empty bond is surpassed by the condensate front.

Two extreme situations arise during capillary condensation in porous networks as a consequence of pore interconnection. When the site sizes of are considerably bigger than their surrounding neck sizes, these latter throats are then filled with condensate at pressures significantly lower than those required for the filling of cavities. Consequently, condensation develops as in a bundle of isolated pores: first, at low pressures, bonds experiment condensation, and afterward, at higher pressures, condensation takes place in sites. Conversely, if sites and bonds are of comparable sizes, condensation in some of the latter pores should still occur first. An encircling interface that will ensure condensation in sites is formed progressively as their linked bonds are being filled with condensate. At a certain pressure, interconnected pore elements (cavities and necks) will be filled with condensate almost conjointly because of their similarities in size. Consequently, adsorption in a correlated porous network happens differently that in a collection of independent pores of the same sizes.

The possibility that sites in a porous network can be filled at pressures larger than those corresponding to identical isolated voids has been labeled as delayed adsorption [9], and this phenomenon is more likely to occur in highly connected porous materials having pores of comparable sizes. Delayed adsorption yields PSD curves (when calculated from the adsorption boundary curve) biased toward bigger pore sizes if the bonds linking cavities are just passages or windows of negligible volumes.

### 1.2 Detection of delayed adsorption
Delayed adsorption can be made evident by comparing the PSD curves calculated from the adsorption boundary curve (ABC) and from a primary ascending scanning curve (PASC) [9] of a sorption isotherm. While the PSD-ABC outcome can yield a unimodal PSD curve, the PSD calculated from a PASC can be subdivided into two regions: one lying in a zone of relatively large pore sizes and another one resting in a region of fairly small void sizes (Fig. 2b). These PSD characteristics arise from the two following facts: (i) the irruption of the delayed adsorption phenomenon during the development of the ABC, (ii) the absence of this same phenomenon in the zone close to the inversion pressure point of the PASC. This latter effect is partly due to the fact that the onset of a PASC occurs with some necks already filled with condensate; contrastingly, the ABC course begins with no pores filled whatsoever. This last condition makes delayed adsorption to manifest more virulently during the development of the ABC rather than throughout the occurrence of a PASC.

Fig. 2 exemplifies the effect of delayed adsorption over the position and shape of $N_2$ sorption isotherms and their corresponding PSD functions. Fig. 2a shows several isotherms, one (○) corresponding to a porous network with cavities joined to bonds of comparable sizes, then experimenting delayed adsorption; another one (□) corresponding to the very same cavities of the former network, but this time the pores are isolated (not connected to any bond). As can be seen from Fig. 2a, the ABC of the connected network is biased toward larger pressure values if compared to the ABC of the same isolated cavities. Fig. 2b depicts the corresponding PSD functions of the isotherms shown in Fig. 2a. As a consequence of the position of the isotherms, the PSD function of the connected cavities is located to the right of the PSD function of the isolated pores.

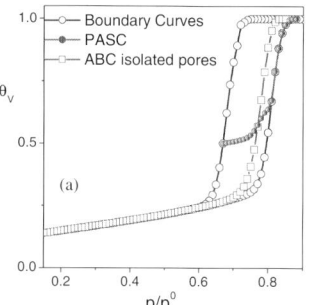 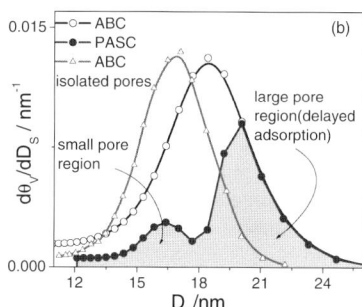

**Fig. 2.** Detection of delayed adsorption through $N_2$ sorption. (a) Sorption isotherm showing a PASC. (b) PSD curves determined from the ABC and PASC isotherms depicted in (a). $\theta_v$ is the adsorbed volume fraction (with respect to the total pore volume) and $D_s$ represents the cavity (site) sizes.

## 2. SIMULATION OF 3-D POROUS NETWORKS

Complete details about the method of construction of 3-D porous networks through Monte Carlo simulation can be found elsewhere [10]; similarly, the precise algorithm employed to replicate sorption processes in porous networks has been reported somewhere else [11]. Here, we will only mention several key aspects regarding these porous network and sorption simulations. First, the critical conditions required for cavities (hollow spheres) and necks (hollow cylinders open at both ends) to be fully occupied by either condensate or vapor have been calculated by means of the Broekhoff-de Boer (BdB) equation [12], while the thickness of the adsorbed film has been approximated via the Harkins-Jura equation [13]. Some other important assumptions that are made in this work are: (i) the pore volume is exclusively due to sites; (ii) bonds are considered as volumeless windows that communicate neighboring sites; (iii) bonds can merge into a site without suffering of any geometrical interference with adjacent throats.

The algorithm used to simulate the Hg intrusion and extrusion processes was derived from [14]. It is important to note that this algorithm takes into account the *canthotaxis* effect during the intrusion process, while the snap-off effect was assumed to occur during the course of the extrusion process [5]. Canthotaxis occurs at the joint formed between a bond and a site and consists in the anchoring and curvature increase of the liquid-vapor meniscus until the pressure of the liquid phase is enough for this interface to overcome this entrance while establishing its equilibrium contact angle with the new wall. The snap-off phenomenon is simulated by assuming that can only take place in a fraction of the smallest bonds. Additionally, it is required for a bond to undergo snap-off that the two extremes of this throat are connected through continuous liquid paths to the external Hg supply.

Three types of porous networks were chosen to perform this $N_2$ sorption-Hg porosimetry PSD comparison work. The corresponding statistical parameters are shown in Table 1.

Table 1. Structural parameters of porous networks

| Network | $\bar{R}_B$ /nm | $\bar{R}_S$ /nm | $\sigma_S, \sigma_B$ /nm | C | $\Omega$ |
|---------|---------|---------|---------|---|----|
| N-I | 13.0 | 25.0 | 1.5 | 6 | 0.0 |
| N-II | 22.5 | 25.0 | 1.5 | 6 | 0.6 |
| N-III | 24.2 | 25.0 | 1.5 | 6 | 0.9 |

$R_S \equiv$ mean radius of sites; $R_B \equiv$ mean radius of bonds; $\sigma_S, \sigma_B \equiv$ standard deviations of sites and bonds; $C \equiv$ connectivity of the network; $\Omega \equiv$ overlap between the site and bond size distributions

The above values of the structural parameters were chosen with the purpose of simulating highly connected ($C=6$) porous networks with pore sizes not far from the upper limit of the mesopore region; this was made to avoid strong potential interactions between the pore walls and the absorbed phase, and also to make more representative the Hg porosimetry simulations. The three porous networks have the same site-size distribution but differ in the bond-size-distributions. N-I is the network with the smallest neck sizes; N-III has the largest neck sizes; and N-II is a substrate of characteristics in-between those of N-I and N-III. Since delayed adsorption is a consequence of the size correlation existing between cavities and necks, the extent of this phenomenon in each type of network can be predicted. At this respect, N-I should be the substrate that less suffers of the delayed adsorption effect, as it has the smallest necks connected to cavities; conversely, N-III should display the most virulent delayed adsorption effect, as it has cavities and necks of comparable sizes; finally, N-II should represent an intermediate situation between the cases of N-I and N-III.

The topological properties of Monte Carlo simulated porous networks have been previously described in detail [10]. Particular characteristics are as follows. The similarity in the sizes of cavities and throats in the case of N-III is the cause of a strong size correlation between the two kinds of void elements and the network structuralizes in a patch-wise fashion according to pore sizes. In turn, the N-I substrate consists of spherical hollows delimited by narrow windows, the sizes of both kinds of pore elements can be randomly allocated throughout the space since every cavity can accommodate any neck size. Finally, the N-II substrate structuralizes in a way amid the two other networks.

## 3. N$_2$ SORPTION RESULTS

Fig. 3 shows the N$_2$ sorption isotherms at 77.4 K of substrates N-I, N-II and N-III; in every one of these isotherms a PASC is also depicted. The sorption curves differ substantially with respect to the position of the desorption knee of their corresponding descending boundary curves (DBC), whilst not so strongly in the location of their respective ABC's. The different positions adopted by the desorption knee are dictated by the bond sizes since it is a well-known fact that desorption is usually a percolating process that is controlled by pore throats [15]. The network with the smallest neck sizes (N-I) presents a desorption knee situated at a small pressure value, while the network with the largest neck sizes (N-III) has a desorption knee positioned at a high pressure value. Although the three types of networks involve the same cavity size distribution, the trajectories followed by their respective ABC's are somewhat different. This fact can be satisfactorily explained under the framework of the delayed adsorption effect. As it was said before, the correlation in sizes among cavities and necks is the prime cause for the occurrence of delayed adsorption in sites. The more alike the sizes of connected cavities and necks are, the more intense the delayed adsorption effect is. Consequently, the network having the smallest necks (N-I) depicts an ABC (Fig. 3) that starts rising from the smallest pressure values; this means that sites are being filled with condensate according to their sizes and no delayed adsorption is happening. Contrastingly, the network with the largest necks (N-III) follows an ABC trajectory that is located within the highest pressure values; this indicates that delayed adsorption is taking place to some extent. Finally, the N-II substrate follows an ABC trajectory that is located in-between the ABC curves of the N-I and N-III materials.

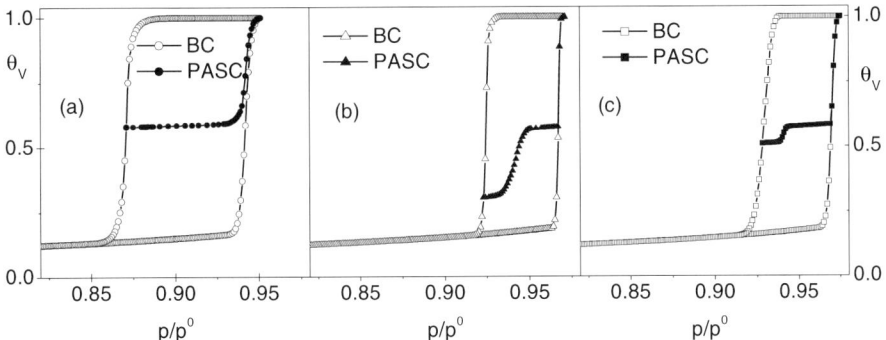

**Fig. 3.**   Simulated N$_2$ sorption isotherms at 77.4K on: (a) N-I, (b) N-II, and (c) N-III. BC refers to the boundary curves of the sorption isotherm.

## 3.1 PSD curves calculated from N$_2$ adsorption

Fig. 4 presents the PSD results obtained from the ABC and PASC paths of each type of porous network. From this figure, it can be appreciated that the PSD curves obtained from the analyses of the ABC and PASC isotherms of the N-I material fit reasonably well the real PSD curve (note that the PASC-PSD outcome occupies only a portion of the real PSD since, along this path, the porous network is being refilled with liquid from the point of inversion up to saturation). Contrastingly, the PSD outcomes arising from the ABC isotherms of samples N-II and N-III are seriously overestimating the real PSD. In the case of the PASC paths of N-II and N-III samples, an interesting situation emerges since the PSD outcome is subdivided in two zones: one located at large pore sizes and another one situated at small pore sizes. This would mean that some sites are filled independently according to their sizes whilst some other cavities suffer of delayed adsorption. The extent of delayed adsorption can be quantified from the PASC-PSD outcome in terms of the areas of the two regions, thus confirming that the strongest manifestation of this effect occurs in the N-III network. In summary, the PSD analysis of PASC isotherms indicates that: (i) in the N-I network all bonds are filled with condensate in advance to sites; the cavities are then filled with condensate according to their sizes; (ii) in the N-II and N-III networks there exist sites that are refilled in agreement with their sizes, but there are some others that are subject to delayed adsorption. The greater the overlap between the sizes of sites and bonds is, the greater the intensity of the delayed adsorption phenomenon becomes.

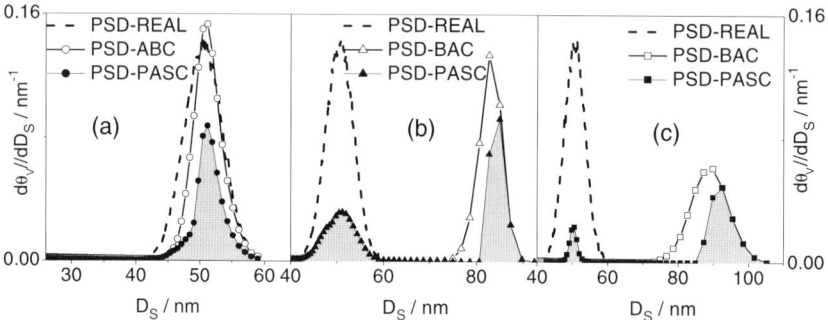

**Fig. 4**. Pore Size Distributions calculated from the isotherms of Fig. 3. (a) N-I. (b) N-II. (c) N-III.

## 4. MERCURY POROSIMETRY RESULTS

Fig. 5 exhibits the Hg porosimetry curves simulated for the diverse types of porous networks. Crudely, Hg intrusion is the simile of $N_2$ desorption; both are capillary processes in which percolating, pore blocking or entrapping effects can arise. These effects are highly influenced by the connectivity of the porous network. The shapes of the Hg intrusion curves are a function of the bond size distribution. Comparatively, the Hg intrusion curve of N-I has an intrusion zone that is located at relatively high pressures, whilst N-III depicts an intrusion process at fairly low pressures. Finally, the Hg intrusion course corresponding to the N-II substrate occurs at intermediate pressures.

Likewise, Hg retraction can be considered as the equivalent of $N_2$ adsorption; both processes are controlled a great deal by the sizes of sites when a piston-like retraction is assumed. Consequently, all Hg retraction curves that are shown in Fig. 5 are located within a range of similar pressure values. Nevertheless, the three Hg retraction curves differ in the slopes of the boundary intrusion-extrusion curves and in the amount of Hg trapped.

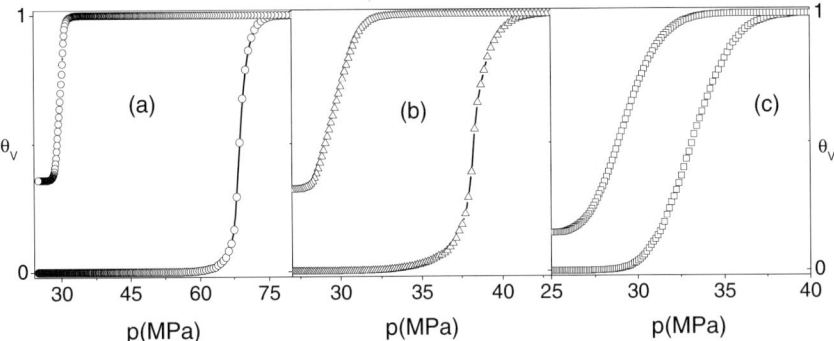

**Fig. 5**. Hg porosimetry curves. (a) N-I. (b) N-II. (c) N-III.

Even though Hg withdrawal is supposed to occur in every pore according to the Washburn equation [16], the phenomenon of snap-off makes retraction from pores to be dependent on the states of neighboring elements. Since snap-off is supposed to occur at the smallest bonds during Hg retraction, Hg threads in the N-I network suffers disconnection at multiple centers. These latter points are uniformly disseminated throughout the porous structure then favoring the sudden (percolative) disconnection of the liquid phase from the external source and, consequently, significant amounts of Hg are trapped in the form of isolated islands. On the other hand, as N-III is structuralized in domains made of pores of similar sizes, snap-off in this structure only occurs in those regions that comprise pores of the smallest sizes, thus leading to a low Hg trapping. The rest of pore domains are drained gradually of Hg, each domain being totally dislodged of condensate at the right pressure. As expected the N-II substrate displays Hg retracting characteristics that are intermediate between those of the N-I and N-III networks.

### 4.1 PSD functions calculated form Hg retraction curves
In view of the apparent similarities that are supposed to exist between Hg retraction and $N_2$ desorption processes, the PSD curves of the different types of substrates were re-evaluated from the Hg extrusion curves. Fig. 6 shows that the PSD curves calculated from Hg retraction curves coincide according to different degrees with the real PSD of every network.

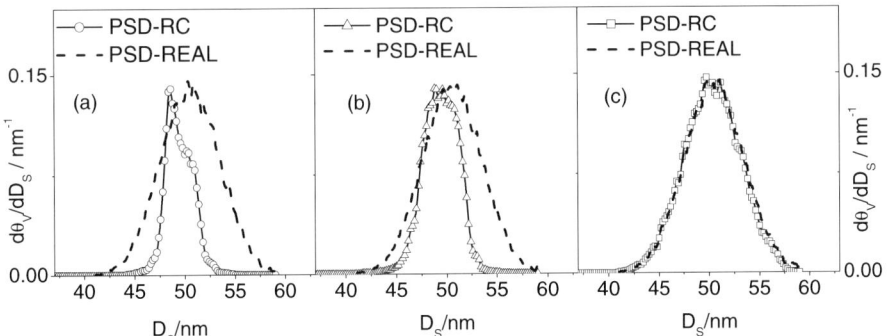

**Fig. 6**. PSD curves calculated from the Hg retraction curves presented in Fig. 5. (a) N-I. (b) N-II. (c) N-III. The acronym RC refers to retraction curves.

The best fitting is found for N-III and the poorest one for N-I, while the N-II outcome fits the real PSD with an accuracy in-between the previous ones. The PSD calculated for N-III coincides practically altogether with the true PSD. These results turn out to be complementary to those found from $N_2$ adsorption. In this way, while the PSD calculated from the ABC isotherm of N-I provides an excellent estimate of the actual PSD, the result obtained from the corresponding Hg retraction curve is biased toward smaller sizes; besides, Hg is not completely removed from this porous network because of the snap-off effect. On the contrary, whereas the PSD calculated from the ABC of the N-III network renders a somewhat deceptive outcome, the PSD calculated from the Hg retraction curve gives an excellent agreement with the real curve. The PSD results calculated from $N_2$ sorption and Hg extrusion taking place in the N-II substrate render intermediate results between those proceeding from N-I and N-III.

The above results mean that Hg withdrawal develops according to the size of the pores when a piston-like mechanism is adopted, so that the PSD calculated from the retraction curve gives useful information about the sizes of cavities. Nevertheless, the snap-off mechanism can still bias the calculated PSD from the real value. N-III provides the best PSD estimation simply because this porous network has cavities of sizes similar to the sizes of its connected necks. On the other hand, N-I renders the poorest PSD estimation due to the fact that this network has the smallest necks of all substrates, thus having as a consequence a lot of disconnected islands filled with Hg.

## 5. CONCLUSIONS

Delayed adsorption biases the site PSD toward overvalued values when the adsorption boundary curve is employed to perform this task. Nevertheless, primary adsorption scanning curves as well as Hg porosimetry can advise about this pore size overestimation. The accuracy of a PSD determination depends both on the mechanistic aspects that are involved during the development of the capillary process selected for the analysis, and on the topological characteristics of the porous network. This latter issue involves as an important aspect the relative sizes existing between cavities and throats.

## REFERENCES

[1]   E. P. Barrett, L. G. Joyner and P. H: Halenda, J. Am. Chem. Soc., 73 (1951) 373.
[2]   H. L. Ritter and L. C. Drake, Ind. Eng. Chem. Analyt. Ed., 17 (1945) 782. C. A. León y León, Adv. Colloid Interf. Sci., 76-77 (1998) 341.
[3]   C. D. Tsakiroglou and A. C. Payatakes, Adv. Colloid Interf. Sci., 75 (1998) 215.
[4]   C. D. Tsakiroglou and A. C. Payatakes, J. Colloid Interf. Sci., 137, (1990) 315.
[5]   A. V. Neimark, Langmuir, 11 (1995) 4183.
[6]   V. Mayagoitia, J. Chem. Soc., Faraday Trans. 1, 81 (1985) 2931.
[7]   G. P. Androutsopoulos and C. E. Salmas, Ind. Eng. Chem. Res., 39 (2000) 3747.
[8]   S. Cordero, F. Rojas, I. Kornhauser, M. Esparza and G. Zgrablich, Adsorption, 11 (2005) 91.
[9]   S. Cordero, F. Rojas and J.L. Riccardo, Colloids Surf. A 187-188 (2001) 425.
[10]  F. Rojas, I. Kornhauser, C. Felipe, J. M. Esparza, S. Cordero, A. Domínguez and J. L. Riccardo, Phys. Chem. Chem. Phys., 4 (2002) 2346.
[11]  J.C.P. Broekhoff and J. H. De Boer J. Catalysis, 9 (1967) 8.
[12]  W.D. Harkins and G. Jura, J. Amer. Chem. Soc., 66 (1944) 1362.
[13]  K. Kaneko, H. Kanoh and Y. Hanzawa, (eds.) Fundamentals of Adsorption 7, International Adsorption Society-IK International, Nagasaki, 2002, pp. 1030 – 1037.
[14]  S. J. Gregg and K. S. W. Sing, Adsorption, Surface Area and Porosity, Academic Press, London, 1982.
[15]  E. W. Washburn, Proc. Nat. Acad. Sci. USA, 7(1921) 115.

Studies in Surface Science and Catalysis 160
*P.L. Llewellyn, F. Rodriquez-Reinoso, J. Rouqerol and N. Seaton (Editors)*
© 2007 Elsevier B.V. All rights reserved

# Large-scale simulations of poly(propylene oxide)amine/Na+-montmorillonite and poly(propylene oxide) ammonium/Na+-montmorillonite using a molecular dynamics approach

P. Boulet[a,*], H.C. Greenwell[b], B. Chen[c], A.A. Bowden[d], I. Beurroies[a], F. Salles[e], P.V. Coveney[b], J.R.G. Evans[c] and A. Whiting[d]

[a]MADIREL, Université de Provence Aix-Marseille 1, Site de Saint-Jérôme, 13395 Marseille Cedex, France

[b]Centre for Computational Science, Department of Chemistry, University College London, 20 Gordon Street, WC1H 0AJ London, United Kingdom

[c]Department of Materials, Queen Mary, University of London, Mile End Road, E1 4NS, London, United Kingdom

[d]Department of Chemistry, University of Durham, South Road, DH1 3LE Durham, United Kingdom

[e]LMTE, CEA Cadarache, 13108 Saint-Paul les Durance, France

We present a study of novel clay-polymer nanocomposite materials using a combination of both atomistic simulation and experiment. We show how computational simulation can bring new insight to our understanding of the structure and dynamics of these materials. Experiments of the intercalation of low molecular weight amine functionalized (poly propylene) oxide oligomers into Na+-montmorillonite clay show that we can control the interlayer separation of the organo-clay system to obtain non-exfoliated nanocomposites. Infrared spectroscopy experiments indicate that an increase in hydrogen-bonding occurs when amine and protonated amine monomers are intercalated into the clay galleries. Molecular dynamics simulations offers new information that corroborates these finds by giving a detailed description of the structure of the nanocomposites.

## 1. INTRODUCTION

In recent years, both academic and industrial research on new nanocomposites based on clays fillers within a polymer matrix have rapidly developed [1-3]. Advances in this field have shown that these materials may exhibit enhanced properties (for example, increased heat and mechanical resistance, electrical conductivity and gas barrier properties) compared to the corresponding unfilled polymer matrix. Despite recent progress in the design of these nanocomposites, many structure-properties relationships are still to be elucidated.

---

[*] Author to whom correspondance should be sent. Pascal Boulet: Pascal.Boulet@up.univ-mrs.fr

In this article we present recent results obtained for clay-oligomer nanocomposites where the clay is sodium-montmorillonite and the organic species are either poly(propylene oxide) bis(2-aminopropyl ether) or its protonated ammonium form. Both experimental and modelling approaches are presented. The work described here is part of a wider project in which, a large range of functionalized poly(ethylene oxide) and poly(propylene oxide) compounds are used as oligomers for the preparation of clay-polymer nanocomposites through a diverse-discovery approach (see http://www.exclaim.org.uk). This wider project aims to study the design, characterization and materials properties of new high clay-fraction nanocomposite materials.

The article is divided into the following sections. Section 2 briefly reviews the theoretical methods and the experimental approaches used to model, synthesize, and characterize the composites. In Sections 3 and 4 we present some selected results and discuss our interpretation of the data, respectively. Finally, we summarize our findings and offer some concluding remarks in Section 5.

## 2. METHODOLOGY

As stated in the introduction, this study is part of a more general project that deals with the synthesis, characterization, materials properties investigation, and molecular modelling of clay-polymer nanocomposites. This section presents the experimental and theoretical approaches used in this work.

### 2.1. Experimental details

Poly(propylene oxide) bis(2-aminopropyl ether) (PPO-NH$_2$), hereafter called functionalized oligomers or organic oligomers, with a low molecular weight (230 g mol$^{-1}$) was purchased from Aldrich and used without further purification. PPO-NH$_2$ was protonated by addition of 10% HCl solution to obtain PPO-NH$_3$Cl. Montmorillonite clay was purchase from the Southern Clay Repository at Purdue University. The procedure to obtain intercalated composites is now described. The clay (0.4 g) was stirred for an hour in water and sonicated for further 30 minutes to obtain a suspension. The organic oligomers were added to the suspension which was subsequently sonicated for 30 minutes. Finally, water was removed by evaporation at 80°C.

The resulting nanocomposites were characterized by a variety of methods (see Ref. [4] for a full description). The intercalation of organic compounds within the clay platelets was ascertained by X-ray diffraction (XRD) measurements using a Siemens d5000 diffractometer. Fourier transform infrared (FTIR) measurements were carried out on a Perkin Elmer 1600 series instrument. The amount of both the organic material and water intercalated within the clay were measured by thermogravimetric analysis (TGA) experiments using a Perkin Elmer Pyris 1, scanning from 25 to 600°C at a rate of 10°C min$^{-1}$.

### 2.2. Theoretical approach

A poly(propylene oxide) amine (PPO-NH$_2$)/Na$^+$-montmorillonite model system was built. The structure was minimized with respect to the energy to remove close contacts and excessive repulsion interaction using the Discover simulation engine within the Cerius$^2$ program package [5]. Inter- and intra-molecular forces were defined by Teppen's force field [6]. This force field has been specifically designed, using the CFF91 force field as a basis [7], to correctly reproduce the behaviour of both the clay and the functionalized oligomers. For sake of consistency with experimental data, a mixture of water and monomers were intercalated within the clay sheets and the amount of both species corresponded as closely as possible to the experimental observations obtained from TGA measurements.

For poly(propylene oxide) ammonium (PPO-NH$_3^+$)/Na$^+$-montmorillonite models, some Na$^+$ cations were removed from the amine-based model, and some amine functional groups were protonated so as to maintain the electroneutrality of the whole system.

Subsequently, the minimized structures were enlarged up to about 20,000 atoms and simulated by molecular dynamics (MD) using LAMMPS (Large-scale Atomic/Molecular Massively Parallel Simulator) [8,9] and the Teppen force field. LAMMPS uses specific algorithms that allow efficient use of parallel computers to simulate very large models. The details of these algorithms have been presented elsewhere [9,10]. A constant isobaric-isothermal (NPT) ensemble was used, the temperature and pressure being fixed at 300 K and 101325 Pa, respectively. The simulations were run for 1 ns. Data were collected for the last 0.9 ns, after a period of 0.1 ns during which the system had reached equilibrium as adjudged from monitoring the thermodynamics quantities. Data processing was done using an in-house program package.

## 3. RESULTS

The intercalation of PPO-NH$_2$ and PPO-NH$_3^+$ oligomers was probed by XRD experiment. For the PPO-NH$_2$ species, upon intercalation the $d$ spacing increased from about 9.5 Å for the pristine Na$^+$-montmorillonite clay to between 13.5 Å and 14.3 Å, independently of the amount of organic added (Table 1). Lin and coworkers [11], who carried out experiments on the intercalation of amine-based polymers within clays, obtained a basal spacing of 15 Å. Similar results were observed for the intercalation of the protonated PPO-NH$_3^+$ oligomers (14 Å), in agreement with results published by others (15 Å) [12]. This correspond to a monolayer arrangement of intercalated organic compounds within the clay gallery. From MD simulations it was obtained that the corresponding $d$ spacing amounts to 13.98 Å ± 0.002 Å for the PPO-NH$_2$-based composites and 14.39 Å ± 0.002 Å for the PPO-NH$_3^+$ ones. In both cases, allowing for differences between the model structure and the actual clay and the assumption that the TGA data all pertains to intercalated material, the agreement is satisfactory with experiment.

Table 1
Monomer loading in Na$^+$-montmorillonite and corresponding $d$ spacing

| monomers | wt% (based on total) | $d$ spacing / Å |
|---|---|---|
| PPO-NH$_2$ | 62 | 13.5 |
| PPO-NH$_2$ | 30 | 14.3 |
| PPO-NH$_2$ | 6 | 13.8 |

The structure of the nanocomposite has been investigated by FTIR measurements, the results for which are shown in Table 2. Only the NH$_2$ bending modes are presented as the NH stretching modes (3000-3600 cm$^{-1}$) fall in the range of the hydroxyl (-OH) stretching modes of the clay and, hence, they are too obscure for interpretation to be carried out. The main band of the bending modes of the pure PPO-NH$_2$ (without clay) appear at 1593 cm$^{-1}$ with a shoulder at 1571 cm$^{-1}$. For the PPO-NH$_3^+$ oligomers these bands appear at 1608 cm$^{-1}$ and 1506 cm$^{-1}$.

When intercalated, the lowest peak of the PPO-NH$_2$ bending modes is shifted to higher frequencies by 47 cm$^{-1}$ and the highest peak of the PPO-NH$_3^+$ bending mode is shifted to lower frequencies by 15 cm$^{-1}$. In contrast, the higher frequency band for the PPO-NH$_2$-clay and the lowest bands for the PPO-NH$_3^+$-clay composite remained unchanged. The shifts of the PPO-NH$_2$ and PPO-NH$_3^+$ bending modes are indicative of both the interaction of the organic compounds within the clay sheets and of the formation of a hydrogen bonding network within the clay interlayer space. Additionally, these results indicate that PPO-NH$_2$ can intercalate as such, that is without being protonated [13].

Table 2
FTIR measurements of the PPO-NH$_2$ and PPO-NH$_3^+$-based nanocomposites

| monomers | modes | frequencies / cm$^{-1}$ |
|---|---|---|
| PPO-NH$_2$ | NH$_2$ bending | 1640, 1576 |
| PPO-NH$_3^+$ | NH$_2$ bending | 1623, 1507 |

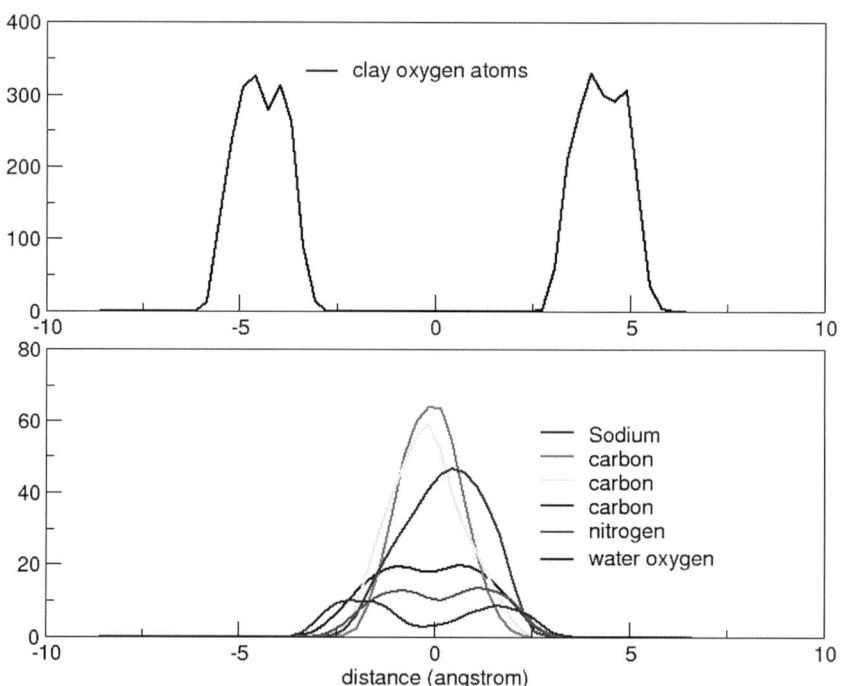

Fig. 1. One-dimensional atom density map for the PPO-NH$_2$ composites. These maps depict the density along the c axis of some atoms pertaining to the organo-clay. The y axis is in arbitrary units.

The structure of PPO-NH$_2$ and PPO-NH$_3$$^+$-clay composites obtained from MD simulations are depicted in Fig. 1 and Fig. 2. For the PPO-NH$_2$-based composites (Fig. 1), it can be seen that the Na$^+$ cations are located on either side of the interlayer mid-plane, close to the nitrogen atoms of the organic oligomers. The carbon backbone is situated in the middle of the clay interlayer. Water oxygen atoms arrange near the clay sheets oriented in the direction of the mid-plane, therefore reinforcing the hydrogen bonding network.

The structure of the PPO-NH$_3$$^+$-clay composite is substantially different. In Fig. 2 it can be seen that, the ammonium protons are mainly located near the clay sheets, therefore playing the same role as the one usually played by inorganic Na$^+$ cations. These ammonium protons form strong hydrogen bonds with the tetrahedral Si oxygen atoms of the clay surfaces. Consequently, and in contrast to PPO-NH$_2$, the carbon backbone of PPO-NH$_3$$^+$ is slighly offset from the middle of the interlayer. Finally, again in contrast with the PPO-NH$_2$-clay composites, water molecules are located between the ammonium protons and the nitrogen atoms of the amine functional groups, again reinforcing the hydrogen bonding network.

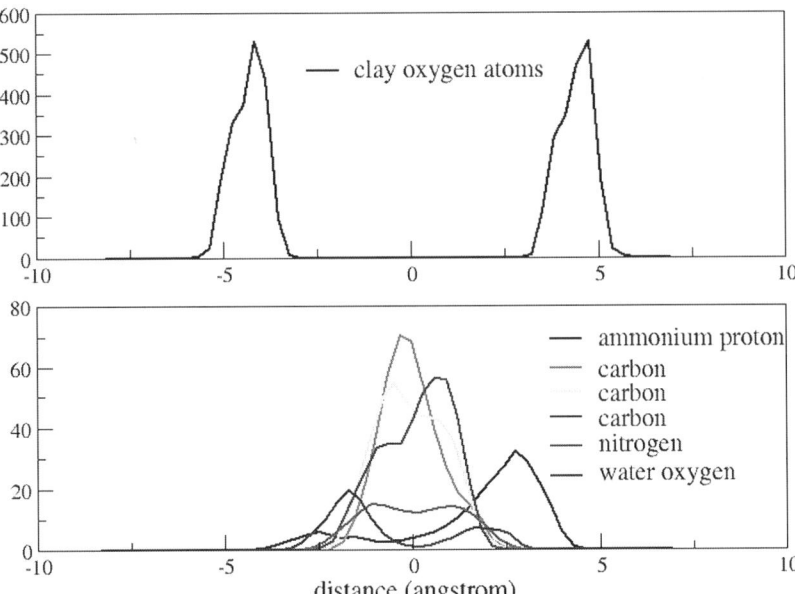

Fig. 2. One-dimensional atom density map for the PPO-NH$_3$$^+$ composites. These maps depict the density along the c axis of some atoms pertaining to the organo-clay. The y axis is in arbitrary units.

## 4. DISCUSSION

We have seen that the interlayer of nanocomposites based on PPO-NH$_2$ and PPO-NH$_3^+$ exhibit very different structures. This may have important implications for the bulk properties of these materials. From our results we can infer, from the increase hydrogen bonding observed, that the PPO-NH$_2$ and PPO-NH$_3^+$ composites exhibit enhanced resistance to mechanical stress compared to the bare polymer. This is indeed what we have observed from simple manual tests: PPO-NH$_2$-clay composites are more resistant to mechanical crushing than any other nanocomposites we have studied so far [4,14]. This behaviour can be explained by the fact that, first, these organic compounds tend to form only monolayers when intercalated. As a consequence, they are more tightly bound to the clay sheets. This is even more justified for the PPO-NH$_3^+$-clay nanocomposites as we have seen that for these materials the ammonium head groups behave like alkali cations as they are mainly located at the vicinity of the clay sheet surfaces. Secondly, we have evidenced, from FTIR experiment and computer modelling, that, upon intercalation, organic amine/ammonium compounds form a hydrogen bond network between the clay platelets which is probably strengthening the nanocomposites. Such a behaviour could be acertained by measurement of the elastic constants. This work has already been undertaken for other nanocomposites [15].

A key feature of our simulations is the use of a model that is as similar as possible to the experimental sample. We have seen that a way to do so is to intercalate between the clay sheets the same amount of water and monomers molecules as observed by TGA measurements. As to the clay itself, we are using an idealised model, that is a 3D periodic model system that cannot account for isolated particules or grain junctions. A way to compare this model, idealised clay to a real sample is then to calculate the volume accessible by the oligomers in both systems.

Recently, we have shown [16] that thermoporometry measurements can reveal useful information about the thermodynamics and the structure of porous materials. Using a fully hydrated Na$^+$-montmorillonite clay sample it has been possible to determine both the pore size distribution and the porous volume. The pore size distribution is presented in Fig. 3.

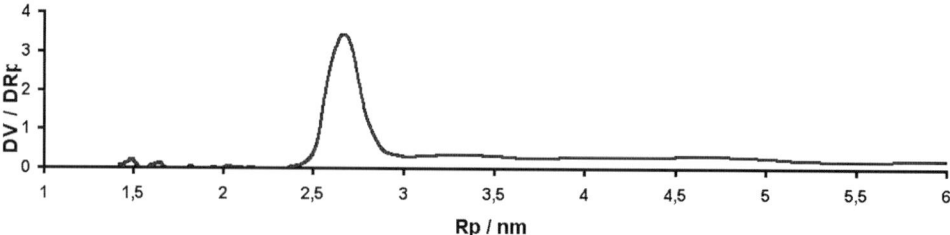

Fig.3. Pores size distribution of a fully hydrated Na$^+$-montmorillonite sample clay

We can first note that the distribution in pore size of the clay is essentially monodisperse with a pore radius of 2.6 nm. Our model is by nature in agreement with this observation. By integrating the area under the main peak we obtain the interlayer specific volume of the clay, that is 0.85 cm$^3$ g$^{-1}$. So, the specific surface amounts to about 327 m$^2$ g$^{-1}$, assuming that the pores are rectangular. If we consider that the specific surface remains constant with changes of the $d$ spacing, the corresponding interlayer specific volume amounts to 0.15 cm$^3$ g$^{-1}$ for a interlayer space of 4.7 Å, which corresponds to the interlayer space of our model system. Using a simple theoretical approach, it was possible to determine the volume accessible by a probe molecule of radius 1.6 Å. For the model we used, with stoichiometry $[(Al_{56}Mg_8)(Si_{124}Al_4O_{320})(OH)_{64}]Na_{12}$ and molecular weight 11892 g mol$^{-1}$, we estimate the volume at about 3943 Å$^3$, that is, about 0.2 cm$^3$ g$^{-1}$ of clay. This estimate is therefore slightly larger than the experimental measurements. However, considering the fact that we use an idealized model and that the order of magnitude for the interlayer pore volume is recovered, this result is rather satisfactory and we can assume that our model is an adequate representation of a real sample.

## 5. CONCLUSION

In this article we have illustrated the simulation of new, high clay-fraction polymer-clay nanocomposites containing low molecular weight poly(propylene oxide) based species terminated by either amine or ammonium functional groups. We have shown that, whatever the starting concentrations of organo-amine/ammonium species, a monolayer of intercalated organics is formed. The organic material tends to form a hydrogen bond network within the clay interlayer. These results, based on FTIR, XRD and TGA measurements, have been confirmed by large-scale molecular dynamics simulations. Additionally, there is clear evidence from the simulations that the structure of the PPO-NH$_3^+$-clay nanocomposite is very different from that of the PPO-NH$_2$-clay one. The particular arrangement of the organo-amine/ammonium, water and cationic species within the clay interlayer space may lead to increased reinforcement of the nanocomposite compared with the polymer matrix. Finally, we have shown, by comparing with thermoporometry measurements, that our model is a good representation of the real system. By contrast, in term of size, these models are still poorly representative of real samples. We are now involved in the process of simulating larger model structures that contain from 350,000 to 400,000 atoms [17] and beyond. We hope that these computer simulations will offer better insight to the chemistry of these materials.

**ACKNOWLEDGEMENTS**
The authors are thankful to the EPSRC for funding this work under the grant number GR/30907 (including access to the CSAR supercomputing service at the University of Manchester, UK). The author wish to thank the HEFCE for funding the SGI Onyx2 located at University College London, the Solide-State NMR and Thermal Analysis Services at the University of Durham and the Centre Informatique National de l'Enseignement Supérieur (CINES) of Montpellier (France) for allowing us to access their computing resources (project number pmc2430).

**REFERENCES**

[1] T.J. Pinnavaia, G.W. Beall (eds.), Polymer-clay Nanocomposites, Chicester, John Wiley & Sons Ltd (2000).
[2] E.P. Giannelis, Adv. Mater., 8 (1996) 29.
[3] B.K.G. Theng (eds.), The Chemistry of Clay-Organic Reactions, John Wiley & Sons, New York, 1974.
[4] P. Boulet, A.A. Bowden, B.Q. Chen, P.V. Coveney and A. Whiting, J. Chem. Mater., 13 (2003) 2540.
[5] Discover, Cerius$^2$, version 4.2, San Diego, Accelrys, USA, (2001).
[6] B.J. Teppen, private communication (2002).
[7] J.R. Maple, M.J. Wang, T.P. Stoskfisch, U. Dinur, M. Waldman, C.S. Ewig and A.T. Hagler, J. Comput. Chem., 15 (1994) 162.

[8] S. Plimpton, Large-Scale Atomic/Molecular Massively Parallel Simulator 2001, Sandia National Laboratories, Albuquerque, (2001).

[9] S. Plimpton, J. Comput. Chem., 117 (1995) 1.

[10] P. Boulet, P.V. Coveney and S. Stackhouse, Chem. Phys. Letters, 389 (2004) 261.

[11] J.-J Lin and Y.-M. Chen, Langmuir, 20 (2004) 4261.

[12] C.-C. Chou, F.-S. Shieu and J.-J. Lin, Macromolecules, 36 (2003) 2187.

[13] S. Yariv and L. Heller, Isr. J. Chem., 8 (1970) 391.

[14] P. Boulet, A.A. Bowden, B.Q. Chen, P.V. Evans, J.R.G. Evans, H.C. Greenwell and A. Whiting, Intercalation and *in situ* polymerisation of poly(akylene oxide) derivatives within $M^+$-montmorillonite towards high clay fraction clay-polymer nanocomposites (M=Li, Na, K), in preparation, 2005.

[15] B.Q. Chen, A.A. Bowden, H.C. Greenwell, P. Boulet, P.V. Coveney, A. Whiting and J.R.G. Evans, J. Polym. Sci. B, in press.

[16] I. Beurroies, R. Denoyel, P. Llewellyn and J. Rouquerol, Thermochimica Acta, 421 (2004) 11.

[17] H.C. Greenwell, M.J. Harvey, P. Boulet, A.A. Bowden, P.V. Coveney and A. Whiting, Macromolecules, *in press*.

Studies in Surface Science and Catalysis 160
P.L. Llewellyn, F. Rodriquez-Reinoso, J. Rouqerol and N. Seaton (Editors)

# A comparison of characterization methods based on $N_2$ and $CO_2$ adsorption for the assessment of the pore size distribution of carbons

C.O. Ania, J.B. Parra*, F. Rubiera, A. Arenillas and J.J. Pis

Instituto Nacional del Carbón, CSIC, Apartado 73, 33080 Oviedo, Spain

The determination of an accurate pore size distribution of activated carbons is still a complex issue and several methods and adsorbates are currently used to this purpose. In this work, different methods have been applied to characterize the microporosity of activated carbons with different burn-off degrees. A deep analysis of the $N_2$ and $CO_2$ adsorption data of the samples has been done. The results of the different methods applied to the gas adsorption isotherms have been compared and their predictions discussed, in order to throw some light on the characterization of microporous materials.

## 1. INTRODUCTION

The preparation of nanoporous templated materials and other micro-patterning technologies, along with the possibility of a rational design of porous materials for different applications [1, 2] are attracting the attention of the scientific community. With increasing environmental concerns worldwide, porous materials have become more important and useful for the separation of polluting species, recovery of useful products, sensors and catalysis. Consequently, textural characterization of porous solids becomes essential in the design of porous solids for a given application. The development of reliable methods for characterization of porosity in porous carbons has been the focus of numerous research efforts for almost 50 years [3-6]. Recently, some studies have focused on the use of different adsorbates for the characterization of the porous structure of carbonaceous materials [7-10]. However, the issue of the determination of an accurate pore size distribution (PSD) of porous materials by a standard procedure is still under discussion.

Commonly, textural characterization of porous solids is carried out by physical adsorption of gases, which can be analyzed using several theories, to provide detailed information about the carbon micropore structure. A number of attempts have been made to establish standard procedures for the interpretation of the adsorption data in the characterization of porous solids [11-13]. However, there is still a lack of agreement on the assessment and interpretation of the adsorption data [14], and the results found in the literature depend upon the theory used to interpret the isotherms [15-18]. Usually, the PSD of porous solids is evaluated from $N_2$ adsorption at 77 K, and the structural heterogeneity of the microporosity is determined from the Dubinin-Radushkevich method (DR) and its modifications (Dubinin-Asthakov, DA,

Dubinin-Stoeckli, DS). These methods are based on Dubinin's theory of the volume filling of micropores (TVFM), the density functional theory (DFT) and the Horvath-Kawazoe method. However, $CO_2$ provides a complement to $N_2$ adsorption for the assessment of the narrow microporosity [19]. A frequently observed disagreement between the PSD obtained from adsorption isotherms of different gases is mostly attributed to molecular sieving and networking effects [20], and to specific adsorbate-carbon interactions [9, 21]. Although these factors are important, possible inconsistencies in the PSD may also be caused by the choice of parameters for intermolecular interactions [9].

In previous studies [22, 23] the structural characterization of a series of carbonaceous materials obtained from pyrolysis and subsequent activation of PET waste was conducted by means of $N_2$ and $CO_2$ adsorption isotherms. The gas adsorption data were interpreted by the BET, Dubinin-Asthakov, t-plot and Horvath-Kawazoe methods. For the t-plot method, the standard data given by Sellés-Pérez [24] for a nonporous active carbon treated at 2073 K, was used. Significant discrepancies between the results from the t-plot method and those obtained from the DR and/or DA methods, applied to both $N_2$ and $CO_2$ adsorption isotherms, were found.

In this work, the $N_2$ and $CO_2$ adsorption data obtained for a series of activated carbons with increasing activation degrees, were analyzed by several methods. The usefulness of those methods to achieve a consistent and meaningful pore size distribution for the samples studied, was evaluated. The t-plot method was applied by using nonporous carbon-coated silica (Sooty Silica) proposed by Carrot et al. [25] as a reference material. In addition, the DR and DA methods were also applied to the adsorption data of the activated carbons. The DFT model [26] was applied to the $N_2$ adsorption isotherms and it was taken in this work as the method of reference as suggested by other authors [8, 27]. In order to compare the PSD of the different samples, the accumulated pore volume curve was divided in the pore size ranges: narrow microporosity (pore width, w < 0.7 nm), medium-sized microporosity ( 0.7 < w < 2 nm) and mesoporosity (2 < w < 50 nm). This subdivision, although not in strict accordance with that given by the IUPAC, is widely accepted.

## 2. EXPERIMENTAL

The carbon materials were obtained from pyrolysis and $CO_2$ activation of PET waste. A complete explanation about the pyrolysis experimental procedure is given in a previous work [22]. Batches of around 6 g of the carbon material (0.5-1.0 mm) were activated in 10 $cm^3$ $min^{-1}$ of $CO_2$ at 1198 K. Samples with different activation degrees (12, 35, 58 and 76% burn-off) were obtained and denoted as PC12, PC35, PC58 and PC76, respectively. Adsorption isotherms of $N_2$ at 77K and $CO_2$ at 273 K were carried out in an ASAP 2010 M and Gemini 2375 from Micromeritics, respectively.

## 3. RESULTS AND DISCUSSION

The $N_2$ adsorption isotherms used in this work to compare the predicted pore-size distribution functions are shown in Fig. 1. Briefly, all $N_2$ isotherms are of type I but the shape progressively changes with burn-off. There is a clear enhancement of the amount of $N_2$ adsorbed with burn-off, along with an opening of the knee of the isotherms with a more gradual approximation to the plateau. This change is usually associated with a widening of microporosity. The desorption isotherms were reversible except for PC76 sample, which presented a small hysteresis loop, being indicative of a slight development of mesoporosity.

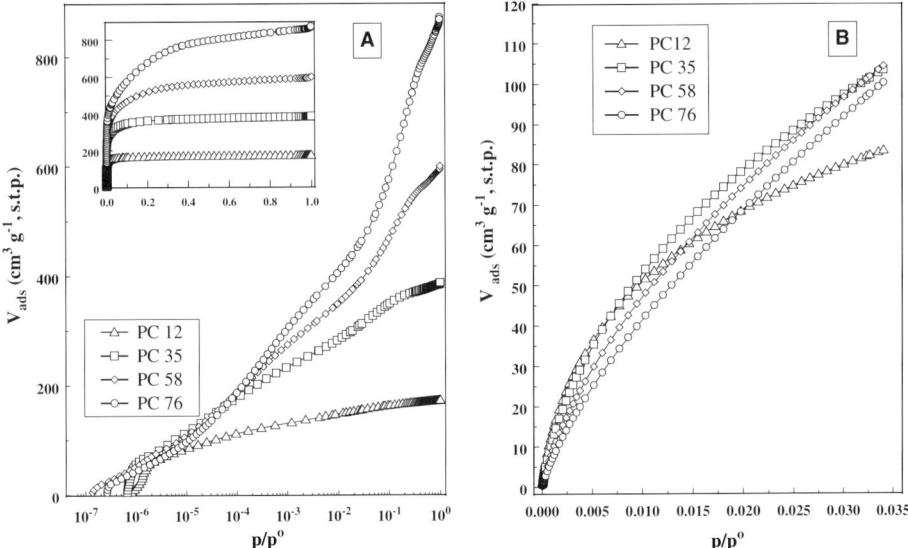

Fig. 1. Nitrogen adsorption isotherms at 77 K (A) and $CO_2$ adsorption isotherms at 273 K (B) of the carbons studied.

The use of $N_2$ at 77 K has been complemented with the adsorption of $CO_2$ at 273 K (Fig.1). It has been established that in order to follow and analyze adsorption in micropores it is more convenient and beneficial to use $CO_2$ at 273 K rather than $N_2$ at 77 K, as suggested by other authors [8,10,19]. The critical dimensions of both molecules are alike, but the higher temperature of the $CO_2$ experiments allows to avoid the kinetic restrictions for the accessibility of the $N_2$ probe in the narrow micropores. Besides, because of the much lower relative pressure range covered in the $CO_2$ adsorption isotherms (up to $p/p^0 < 0.035$), the data correspond to the domain of the narrow microporosity [19]. As a result, the combination of both techniques provides complementary information [10,19] and including $CO_2$ data allows extending the range of the PSD analysis to the narrow microporosity.

## 3.1 DFT and t-plot methods

The PSD obtained by the DFT model for the different samples are shown in Fig. 2. The macropore volume has been determined from mercury porosimetry [22]. It can be seen in this figure a significant development of medium-sized micro and mesoporosity with an increase in the burn-off degree. It should be noted that the mesoporosity was negligible for the sample activated at 12% but the mesopores volume reaches a value of 0.500 $cm^3g^{-1}$ for 76 % burn-off. It can also be seen that the volume associated to narrow microporosity increased up to an activation degree of 35 %, slightly decreasing for higher burn-off values.

In Fig. 2 the PSD obtained by application of the t-plot method, using the Sooty Silica [25] and the nonporous carbon [24] as reference materials, are also shown. When the standard proposed by Sellés-Pérez [24] was used, a gradual increase of medium-sized micropores and mesopores was observed, while the volume of narrow micropores promptly decreased with the burn-off degree. As inferred from the application of this method, the PC76 sample did not presented narrow microporosity. This prediction is highly in disagreement with the PSD obtained by the DFT method. It seems that the Sellés-Pérez reference material fails to predict the pore structure of microporous activated carbons, especially in the assessment of microporosity.

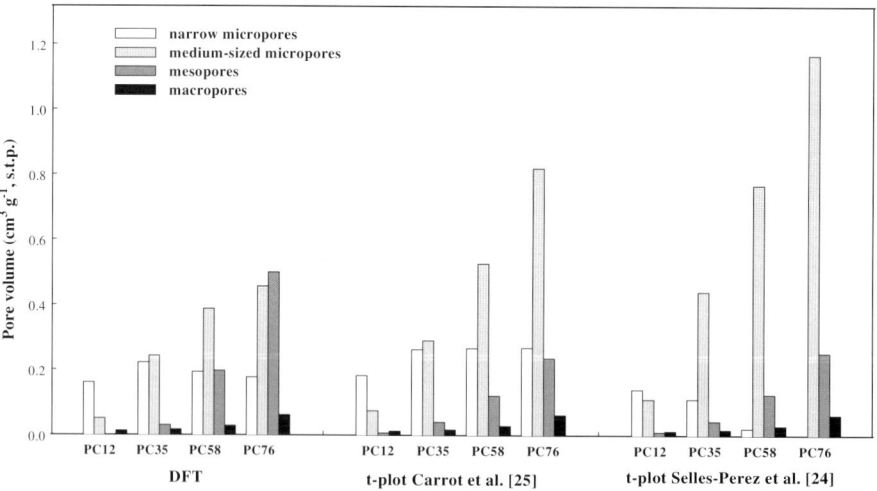

Fig. 2. PSD obtained from $N_2$ adsorption data applying DFT model and t-plot with different reference materials.

When the Sooty Silica is used as reference, the narrow micropore volume increased with the burn-off degree, as opposed to the gradual decrease predicted by the use of the Sellés-Pérez reference material. The evolution of medium-sized micropores and mesopores follows, in both cases, a similar trend although the increase in microporosity is lower for the Sooty Silica case. Taking into account the PSD obtained by application of the DFT method, the results shown in Fig. 2 indicate that the t-plot method obtained by using Sooty Silica is more reliable for predicting the evolution of narrow and medium-size microporosity than the t-plot obtained with the nonporous carbon of Sellés-Pérez. The total pore volumes assessed by the DFT method are noteworthy lower than those obtained by the t-plot method. These differences are attributed to the assumptions of each method. The preceding results indicate that for highly microporous carbons the t-plot might be considered as an adequate method for microporosity characterization if the appropriate reference material is chosen.

## 3.2 Application of the DA, DR and DS methods

When the DA equation was applied to the $N_2$ adsorption isotherms, an increase in the micropore volume ($W_o$) with burn-off was found [22], as it can be seen in Fig. 3. However, suitable information about the development of narrow micropores cannot be attained from $N_2$ adsorption. It is known that $CO_2$ provides an alternative to $N_2$ adsorption for the assessment of the narrow microporosity [19]. Usually, the structural heterogeneity of the microporosity is also determined from Dubinin's theory of the volume filling of micropore, by applying the Dubinin-Radushkevich equation (DR) or any of its modifications (Dubinin-Asthakov DA, or Dubinin-Stoeckli, DS). Although it possesses some limitations, the large amount of literature available on its applicability on the assessment of micropores, suggest that the DA equation is one of the most universally applied equations in the adsorption field [3, 4, 6, 8].

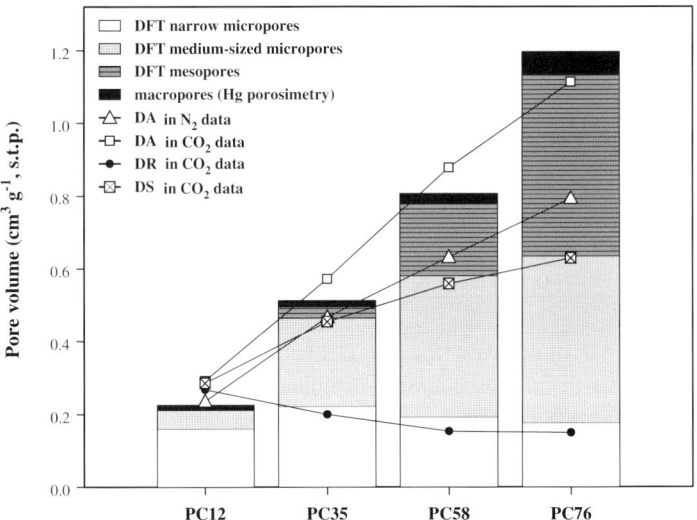

Fig. 3. PSD of the studied carbons and micropore volumes evaluated from $N_2$ and $CO_2$ adsorption isotherms.

Many works have discussed the difficulty in the choice of the interval of experimental data of adsorption to be fitted to the DR and DA equations, and their influence on the values of both $E_o$ and $n$ parameter [28, 29]. The $CO_2$ isotherms of the studied samples have been previously reported [22]. The corresponding fitting curves to the DA equation presented linearity in the whole range of relative pressures, indicating the goodness of the fitting. However, the micropore volumes ($W_o$) obtained from the DA equation applied to the $CO_2$

adsorption isotherms were larger than those achieved from the application of the same equation to the $N_2$ adsorption data (cf. Fig. 3). Furthermore, the values were also larger than the total micropore volumes obtained by either the DFT or the t-plot methods.

In order to throw some light on these contradictory results and on the choice of the method to evaluate the microporosity of activated carbons, the $CO_2$ adsorption isotherms were fitted to the DR equation. Only in the case of PC12 the experimental data showed a linear fit to the DR equation over the entire range of relative pressures. The exponent of PC12 was n=1.95, close to the ideal value (n=2) of the DR equation. In the rest of the series, there are two linear ranges, at either low or high relative pressures.

In this work the data measured in the low relative pressures range ($p/p_o < 0.003$) were used to fit the DR equation to the $CO_2$ experimental data of adsorption. The micropore volumes obtained are compiled in Table 1. When the micropore volumes obtained by the DR fitting of the $CO_2$ adsorption isotherms, over the aforementioned range of relative pressures, were compared with the narrow microporosity from the $N_2$ adsorption data, a better agreement was obtained than in the case of the application of the DA equation to the $CO_2$ adsorption data (see Fig. 3).

Table 1
Structural parameters calculated from DR applied to the $CO_2$ adsorption data

| Sample | Wo $(cm^3 g^{-1})$ | Eo $(kJ\ mol^{-1})$ | L $(nm)$ | Smi $(m^2\ g^{-1})$ |
|---|---|---|---|---|
| PC 12 | 0.268 | 28.5 | 0.63 | 848 |
| PC 35 | 0.200 | 28.8 | 0.62 | 646 |
| PC 58 | 0.146 | 28.8 | 0.62 | 470 |
| PC 76 | 0.150 | 27.6 | 0.67 | 448 |

Contrary to what have been deduced from the DA equation applied to the $CO_2$ isotherms, it can be seen in Fig. 3 that the narrow micropore volume reached a maximum. These results indicated that for low burn-off degrees (i.e. 12%) the activation process developed narrow microporosity. The micropore volume of PC12 given by the DR equation is higher than the total microporosity evaluated by application of the DFT method. This fact is corroborated by the microporous nature of a material activated at a low burn-off degree (12%), presenting pores mainly of small sizes. Thus, this porosity remains most likely non accessible to nitrogen at 77 K, while it is detected by the $CO_2$ molecule. For higher activation degrees, the narrow microporosity decreases at expenses of an increase of the total micropore volume (as calculated from nitrogen adsorption data). This fact corroborates that the microporosity was developed by a widening of narrow micropores. These results are in good agreement with those reported by Linares et al. [19]. Consequently, it seemed that the values obtained with the DA equation were somewhat unrealistic and not consistent with nitrogen adsorption data. In contrast, DR equation fits perfectly if the range of data to be fitted is carefully chosen.

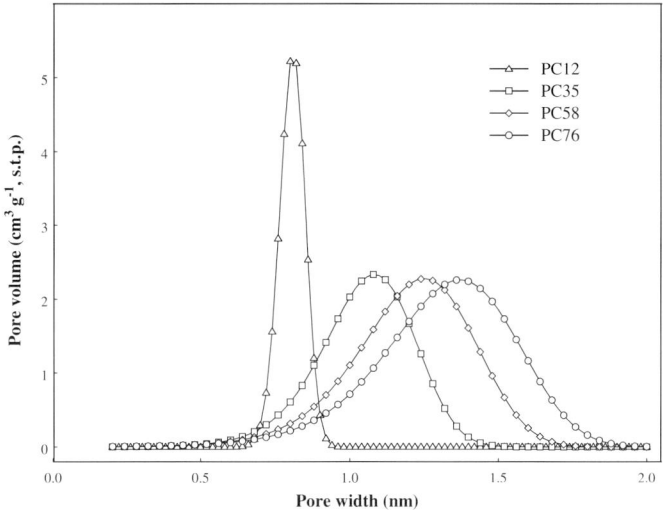

Fig. 4. PSD obtained by application of DS equation to $CO_2$ adsorption isotherms.

The $CO_2$ adsorption data can also be analysed upon the Dubinin-Stoeckli (DS) theory [30, 31]. In this work, the DS equation was applied to the $CO_2$ adsorption isotherms in the range of relative pressures from $3.4 \times 10^{-5}$ to $3.4 \times 10^{-2}$. The application of the DS method allowed to obtain a distribution of the microporosity, and it can be seen in Fig. 4. The micropore size widened with the increase in the burn-off degree, along with a broadening in the mean pore width, which is in good agreement with the $N_2$ adsorption data.

## 4. CONCLUSIONS

In this work several methods were applied in order to predict the pore size distributions of a series of activated carbons with different activation degrees. The results obtained by application of the DFT and the t-plot methods to the $N_2$ adsorption data indicate that the t-plot method provides a means for a reliable characterization of porous materials, when a suitable reference material is employed.

By using $CO_2$ adsorption, both the micropore volume and micropore size distribution of samples with very narrow micropores can be obtained, giving complementary information to $N_2$ adsorption. However, the interpretation of the results obtained from the application of the different available theories is not always straightforward. When applied to the $CO_2$ adsorption isotherms, the predictions obtained with the DA method were unrealistic and no consistent with the nitrogen adsorption data; this method overestimates the micropore volumes. The results obtained by application of the DS equation to the $CO_2$ adsorption isotherms were in agreement with the results predicted by application of the DFT method to the $N_2$ adsorption data.

**REFERENCES**

[1]     S. Yang, P. Mirau, J.N. Sun, D. Gidley, Rad. Chem. Phys., 68 (2003) 351.

[2]     C. Boissiere, A. Larbot, A. van der Lee, P. J. Kooyman, E. Prouzet, Chem. Mater., 12 (2000) 2902.

[3]     S.J. Gregg, K.S.W. Sing, Adsorption, Surface Area and Porosity; Academic Press: New York, 1982.

[4]     M. Jaroniec, R. Madey, in Physical Adsorption on Heterogeneous Solids; Elsevier, Amsterdam 1998.

[5]     V.B. Fenelonov, Porous Carbon, Institute of Catalysis: Novosibirsk, 1995.

[6]     Y.F. Yin, B. McEnaney, T.J. Mays, Carbon, 36 (1998) 1425.

[7]     D.L. Valladares, F. Rodríguez-Reinoso, G. Zgrablich, Carbon, 36 (1998) 1491.

[8]     P.A. Gauden, A.P. Terzyk, G. Rychlicki, P. Kowalczyk, M.S. Cwiertnia, J.K. Garbacz, J. Colloid Interf. Sci., 273 (2004) 39.

[9]     P.I. Ravikovitch, A. Vishnyakov, R. Russo, A.V. Neimark, Langmuir, 16 (2000) 2311.

[10]    J. Jagiello, M. Thommes, Carbon, 42 (2004) 1227.

[11]    F. Rodríguez-Reinoso, A. Linares-Solano. (P.A. Thrower, Ed.), Chemistry and Physics of Carbon, vol. 21, New York: Marcel Dekker, 1988, p. 1-146.

[12]    J.W. Patrick, in Porosity in Carbons, Ed. Edward Arnold, London 1995.

[13]    K. Kaneko, T. Ohba, Y. Hattori, M. Sunaga, H. Tanaka, H. Kanoh, Stud. Surf. Sci and Catal. 144 (2002) 11.

[14]    K.S.W. Sing, D.H. Everett, R.A.W. Haul, L. Moscou, R.S. Pierotti, J. Rouquerol, T. Siemieniewska, Pure Appl. Chem., 57 (1985) 603.

[15]    P. Lodewyckx, L. Verhoeven, Stud. Surf. Sci. Catal., 144 (2002) 731.

[16]    P.J.M. Carrot, M.M.L. Ribeiro-Carrot, Carbon, 37 (1999) 647.

[17]    D. Lozano-Castelló, D. Cazorla-Amorós, A. Linares-Solano, D.F. Quinn, J. Phys Chem. B, 106 (2002) 9372.

[18]    J. Rouquerol, P. Llewellyn, R. Navarrete, F. Rouquerol, R. Denoyel, Stud. Surf. Sci. Catal., 144 (2002) 171.

[19]    D. Lozano-Castelló, D. Cazorla-Amorós, A. Linares-Solano, Carbon, 42 (2004) 1233.

[20]    M.V. Lopez-Ramon, J. Jagiello, T.J. Bandosz, N.A. Seaton, Langmuir, 13 (1997) 4435.

[21]    N. Quirke, S.R.R. Tennison, Carbon, 34 (1996) 1281.

[22]    J.B. Parra, C.O. Ania, A. Arenillas, J.J. Pis. Stud. Suf. Sci. Catal., 144, (2002) 537.

[23]    J.B. Parra, C.O. Ania, A. Arenillas, F. Rubiera, J.M. Palacios, J.J. Pis, SECV, 43 (2002) 547.

[24]    M.J. Sellés-Pérez, J.M. Martín-Martínez. J. Chem. Soc. Faraday Trans., 87 (1991) 1237.

[25]    P.J.M. Carrott, R.A. Roberts, K.S.W. Sing, Carbon, 25 (1987) 769.

[26]    J.P. Olivier, Carbon, 36 (1998) 1469.

[27]    P.A. Gauden, P. Kowalczyk, A.P. Terzyk, Langmuir, 19 (2003) 4253.

[28]    J. Garrido, A. Linares-Solano, J.M. Martín-Martínez, M. Molina-Sabio, F. Rodríguez-Reinoso, R. Torregrosa, Langmuir, 3 (1987) 76.

[29]    P.J.M. Carrot, R.A. Roberts, K.S.W. Sing, Carbon, 25 (1987) 59.

[30]    R.C. Bansal, J.B. Donnet, F. Stoeckli. Active Carbon, Marcel Dekker, Inc. New York, 1988

[31]    M.M. Dubinin, H.F. Stoeckli. J. Colloid Interf. Sci., 75 (1980) 34.

Studies in Surface Science and Catalysis 160
P.L. Llewellyn, F. Rodriquez-Reinoso, J. Rouqerol and N. Seaton (Editors)

# Adsorption of nitrogen, hydrogen and carbon dioxide on alumina-pillared clays

A. Gil[a], R. Trujillano[b], M.A. Vicente[b] and S.A. Korili[a]

[a]Departamento de Química Aplicada, Edificio Los Acebos, Universidad Pública de Navarra, Campus de Arrosadía, s/n, E-31006 Pamplona, Spain

[b]Departamento de Química Inorgánica, Facultad de Ciencias Químicas, Universidad de Salamanca, Plaza de la Merced, s/n, E-37008 Salamanca, Spain

A comparative study of the structure developed by two alumina-pillared clays using the results of adsorption from nitrogen, hydrogen and carbon dioxide is reported. The samples considered come from a montmorillonite and a saponite that have been treated with solutions of hydrolysed aluminium. The solids resulting of the intercalation have been calcined at 473, 623 and 773 K for 4 h. The textural properties of the samples derived from the results of nitrogen adsorption at 77 K, show a decrease of the specific surface area and the micropore volume as the temperature of calcination increases. Results from hydrogen adsorption indicate the presence of micropores of widths smaller than the diameter of the molecule of nitrogen, while carbon dioxide adsorption reveals the presence of specific sites of chemisorption.

## 1. INTRODUCTION

The analysis of the pore structure of materials continues to attract attention of researchers interested in fuel storage, heterogeneous catalysis, removal of trace impurities and separation processes. Gas adsorption is a fast and convenient characterization technique. These measurements depend on the internal physical and chemical structure of the material and on the nature of the adsorbate molecule. Thus the results of these measurements, mainly at low pressures, are sources of valuable information about adsorbate-adsorbent interaction, and structural and energetical properties of the surface materials [1].

The development of inorganic pillared interlayered clays, in short PILCs, an important category of microporous materials, has created remarkable new opportunities in the field of the synthesis and applications of clay-based solids [2]. These materials are prepared by exchanging the charge-compensating cations present in the interlamellar space of swelling clays by hydroxy-metal polycations. On calcining, the inserted polycations yield rigid, thermally stable oxide species, named *pillars*, which prop apart the clay layers and prevent their collapse. This process results in an interesting two-dimensional porous structure of molecular dimensions.

In the last two decades there has been an increasing interest in the development of reversible systems for hydrogen storage. Systems in which hydrogen is concentrated by physical adsorption above 70 K show high energy efficiency, which is critical to the large-scale application of hydrogen fuel cells, in particular for mobile applications. Microporous materials such as activated carbons [3], carbon nanofibers [4] and carbon nanotubes [5], are potential adsorbents for such high-capacity systems.

The aim of this work is to contribute to the design and development of adsorbents for a hydrogen storage system suitable for mobile applications. To this end, emphasis is put on the exploitation of the results of nitrogen, hydrogen and carbon dioxide adsorption at very low relative pressures in order to evaluate the microporous structure of two alumina-pillared clays treated at several temperatures.

## 2. EXPERIMENTAL

### 2.1. Preparation of the intercalated and pillared samples

Two natural clay minerals were used as raw materials, namely a saponite from Ballarat (California, USA) and a montmorillonite from Gador (Almería, Spain). The as-received material was purified by careful aqueous dispersion and decantation and the fractions with particle size less than 2 μm were separated and subjected to intercalation experiments with aluminium oligomers.

The parent clays were intercalated with $[Al_{13}O_4(OH)_{24}(H_2O)_{12}]^{7+}$ polycation following a standard procedure [6]. The Al polycation solution was prepared by slow titration of a solution of $AlCl_3 \cdot 6H_2O$ (Panreac, p.a.) with a solution of NaOH under vigorous stirring, using an $OH^-/Al^{3+}$ mole ratio equal to 2.2 (pH = 4.1) [7, 8]. The hydrolysed solution was allowed to age for 24 h at room temperature under constant agitation. The interlayered clays were obtained by addition of 8 g of each clay to an aqueous solution of hydroxy-aluminium, using an $Al^{3+}$/clay ratio of 5 mmol/g. The slurries were stirred for 24 h at room temperature and then centrifugated and washed by dialysis with distilled water until no chloride was present in the filter wash waters.

The resulting intercalated clays were dried in air at 323 K for 16 h and then calcined at various temperatures for 4 h (heating rate of 5 K/min) in order to obtain the alumina-pillared clays. The solids are designated by the following nomenclature: BAsap and GAmont refer to the Ballarat saponite and the Gador montmorillonite used in the intercalation process; Al indicates that aluminium polycations were used for intercalation, and it is followed by the temperature, in K, at which the samples were calcined; e.g., (GAmont-Al)473 or (Basap-Al)623.

## 2.2. Characterization techniques

X-ray powder diffraction (XRD) patterns were obtained by using a Siemens D-5000 diffractometer employing nickel filtered Cu Kα radiation and operating at 40 kV and 30 mA.

Elemental analyses of the solids were carried out by Activation Laboratories Ltd., Ancaster, Ontario, Canada, using Inductively Coupled Plasma Spectroscopy (ICPS).

Textural analyses were carried out from the corresponding nitrogen (Air Liquide, 99.999 %), carbon dioxide (Air Liquide, 99.998 %) and hydrogen (Praxair, 99.999 %) adsorption at 77 K, 273 K and 77 K, respectively, with a static volumetric apparatus (Micromeritics ASAP 2010 adsorption analyser). The nitrogen adsorption data were collected in the relative pressure range of $10^{-5} \leq p/p^o \leq 0.99$. To obtain an adequate characterization of the microporous region, sufficient data points at low pressures are needed, and this requires the addition of constant small nitrogen volumes. The nitrogen adsorption data were obtained using about 0.2 g of sample and successive nitrogen doses of 4 $cm^3$/g until $p/p^o = 0.01$ was reached. Each point of the adsorption isotherm in this range was equilibrated for at least 2 h in order to characterize correctly the smallest micropores. Further nitrogen was added and the volumes required to achieve a fixed set of $p/p^o$ were measured. The carbon dioxide and hydrogen adsorption data were also obtained using 0.2 g of sample and, as in the case of nitrogen adsorption, each point of adsorption was equilibrated for at least two hours. Before analysis, samples were degassed at 473 K for 24 h ($p < 10^{-3}$ mmHg).

## 3. RESULTS AND DISCUSSION

The basal spacing (d(001)) of the intercalated and pillared samples determined from the (001) reflection in the XRD powder patterns ranged from 18.6 to 20.3 Å for the (BAsap-Al) series, and from 18.5 to 19.9 Å for the (GAmont-Al) one. These values fit well into the ranges generally reported in the literature for alumina-pillared clays and indicate that the intercalating and pillaring processes have been successfully accomplished in every case.

The nitrogen adsorption at 77 K of the samples is shown in Figure 1. The adsorption isotherms are of type I+II in the Brunauer, Deming, Deming and Teller (BDDT) classification

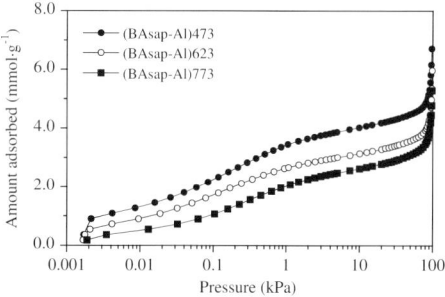 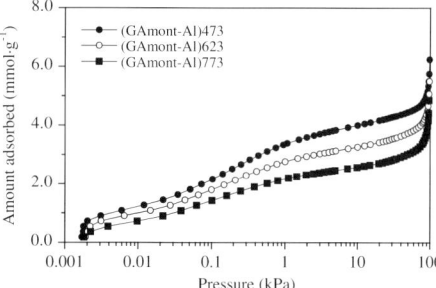

Fig. 1. Nitrogen adsorption at 77 K

[9]. The nitrogen adsorption experiments at low pressures ($p/p^o < 0.1$) show the greatest differences between the isotherms of the samples. More information in this pressure range can be obtained from the adsorption of hydrogen and carbon dioxide.

The textural properties of the samples are more explicitly given in Table 1. From these results, for a given series differences among the solids are found depending on the temperature of calcination. Intercalated and pillared samples showed a decrease of specific surface area and micropore volume on calcining. The specific microporous volumes, calculated according to the Horvath-Kawazoe model [10] and applying the method proposed by Gil and Grange [11], as well as the maxima of the micropore-size distributions, are also summarized in Table 1. Specific surface areas from 391 to 249 $m^2/g$ and specific microporous volumes from 0.145 to 0.093 $cm^3/g$ have been obtained, depending on the temperature of calcination. These results also indicate that the loss of specific micropore volume is comparable with that of the specific surface area.

The hydrogen adsorption at 77 K of the samples is shown in Figure 2. Comparing these results with the ones presented in Figure 1, there is a reasonable possibility that micropores with lower widths than the nitrogen molecule diameter exist in the structure of the pillared clays. From hydrogen adsorption, it is deduced that (BAsap-Al)473 and (BAsap-Al)623 show the same microstructure, and also (BAsap-Al)773, although with significantly lower adsorption. However, (GAmont-Al)473 clearly shows micropores with lower widths than the presented in the structure of (GAmont-Al)623, while a decrease in the adsorption capacity similar to that observed in the saponite series is observed when heating the solid at 773 K.

Table 1
Textural properties derived from the nitrogen adsorption at 77 K

| Sample | $A_{Lang}$[a] $(m^2 g^{-1})$ | $A_{ext}$[b] $(m^2 g^{-1})$ | $V_p$[c] $(cm^3 g^{-1})$ | $V_{\mu p(HK)}$[d] $(cm^3 g^{-1})$ | $dp_{HK}$[e] (Å) | $\Sigma Vp$[f] $(cm^3 g^{-1})$ |
|---|---|---|---|---|---|---|
| (BAsap-Al)473 | 391 ($C^g$=518) | 30 | 0.232 | 0.145 | 5.3;6.6 | 0.071 |
| (BAsap-Al)623 | 304 (C=486) | 32 | 0.207 | 0.114 | 5.2;6.5 | 0.074 |
| (BAsap-Al)773 | 256 (C=308) | 33 | 0.183 | 0.096 | 5.4;7. | 0.072 |
| (GAmont-Al)473 | 388 (C=481) | 29 | 0.216 | 0.144 | 5.2;6.5 | 0.057 |
| (GAmont-Al)623 | 316 (C=501) | 31 | 0.190 | 0.118 | 5.3;6.5 | 0.060 |
| (GAmont-Al)773 | 249 (C=487) | 33 | 0.168 | 0.093 | 5.3;6.7 | 0.063 |

[a] Specific surface area from the Langmuir method ($0.01 \leq p/p^o \leq 0.05$, interval of relative pressure).
[b] Specific external surface areas obtained from the t-method.
[c] Specific total pore volumes al $p/p^o = 0.99$.
[d] Specific micropore volume derived from the Horvath-Kawazoe method.
[e] Maxima of the Horvath-Kawazoe micropore size distributions.
[f] Cumulative pore volume from the BJH method (for pores in the range 17-500 Å).
[g] Langmuir C-value, characteristic of the intensity of the adsorbate-adsorbent interactions.

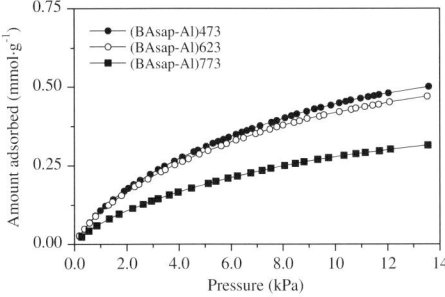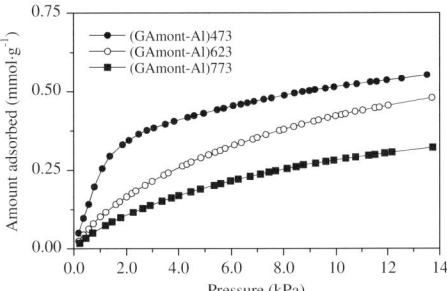

Fig. 2. Hydrogen adsorption at 77 K

The carbon dioxide adsorption at 273 K of the samples is shown in Figure 3. In the case of (BAsap-Al) series, the treatment at 623 K produces a notable decrease of the carbon dioxide adsorption with respect to the adsorption capacity of the sample treated at 473 K. An additional, but lower, decrease of the adsorption capacity is also observed after treatment at 773 K. In the case of (GAmont-Al) series, the evolution of the carbon dioxide adsorption capacity is not the same with the previously described for (BAsap-Al). In this case, the treatment at 623 K produces an increase of the carbon dioxide adsorption that decreases after a new treatment at 773 K. Thus, comparing the two series, the most important difference between them is the adsorption capacity of the samples treated at 473 K.

The complementary use of nitrogen and carbon dioxide adsorption to characterize microporous materials as activated carbons has been recommended by various authors [12,13]. In the case of pillared clays, little work has been done using carbon dioxide as adsorbate. In a previous work [14], we found that the adsorption of carbon dioxide and that of nitrogen are not sensible to the same type of microporosity. It is also necessary to take into account the possibility of chemisorption when carbon dioxide is used as adsorbate. In this way and from the results presented in this work, there is an effect between the temperature of calcination of the samples (GAmont-Al) and the adsorption of carbon dioxide that cannot be explained from a modification of the texture of the pillared clays; the nature of the surface sites created at 473 K would merit further research.

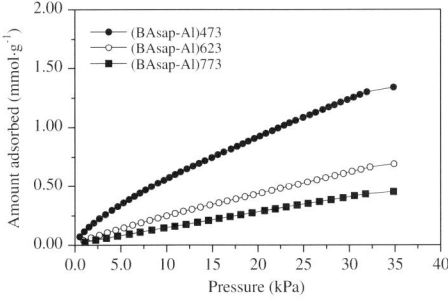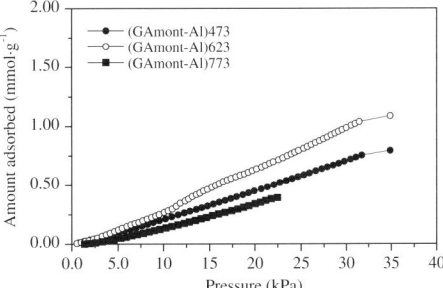

Fig. 3. Carbon dioxide adsorption at 273 K

Several isotherm equations have been proposed to describe the experimental data from adsorption of gases adsorption on microporous materials [15]. Isotherms based on the Langmuir and Gibbs approaches and the potential theory have been considered for this proposal. In this work, Langmuir [15] and Virial [16,17] equations have been applied to the experimental adsorption results.

Langmuir equation:

$$V = (K \cdot P)/(1 + B \cdot P) \tag{1}$$

Virial equation:

$$\ln(P/V) = A_0 + A_1 \cdot V + A_2 \cdot V^2 \tag{2}$$

and from this last equation:

$$K = \exp(-A_0) \tag{3}$$

where $V$ is the amount adsorbed per unit weight of the sorbent and $K$ is the Henry's constant. Langmuir and Virial equations reduce Henry's Law in the low-pressure region. The fitted Langmuir and Virial parameters are given in Table 2.

## 4. SUMMARY AND CONCLUSIONS

This work reports the adsorption of nitrogen, hydrogen and carbon dioxide on two alumina-pillared clays treated at various temperatures. The textural properties of the samples estimated from nitrogen adsorption at 77 K indicate that intercalated and pillared samples exhibit a decrease of specific surface area, from 391 and 249 $m^2/g$, and micropore volume, from 0.145 and 0.093 $cm^3/g$, as the temperature of calcination increases.

The comparison among nitrogen, hydrogen and carbon dioxide at 77 K and 273 K indicates that the adsorption of these molecules is not sensitive to the same type of microporosity and surface. The temperature of adsorption, the diameter of these molecules and the surface nature can explain this behaviour.

The experimental adsorption results from hydrogen and carbon dioxide have been fitted with Langmuir and Virial equations in order to estimate the Henry's constant.

## 5. ACNOWLEDGEMENTS

Financial support by the Spanish Ministry of Education and Science and FEDER funds (MAT2003-01255) and the Navarre Government is gratefully acknowledged. S.A.K. acknowledges financial support by the Ministry of Education and Science through the Ramon-y-Cajal program.

Table 2
Langmuir and Virial parameters for the hydrogen and carbon dioxide adsorption at 77 and 273 K respectively

| | (BAsap-Al)473 | (BAsap-Al)623 | (BAsap-Al)773 | (GAmont-Al)473 | (GAmont-Al)623 | (GAmont-Al)773 |
|---|---|---|---|---|---|---|
| **Hydrogen** | | | | | | |
| Langmuir | | | | | | |
| $K$ (*mmol·g$^{-1}$·atm$^{-1}$*) | 11.10 | 10.73 | 6.62 | 36.62 | 10.47 | 6.63 |
| $B$ (*atm$^{-1}$*) | 14.82 | 15.58 | 13.75 | 61.62 | 14.74 | 13.38 |
| Virial | | | | | | |
| $A_0$ | 0.837 | 0.805 | 1.283 | 0.262 | 0.902 | 1.400 |
| $A_1$ | 0.129 | 0.157 | 0.237 | -0.079 | 0.134 | 0.179 |
| $A_2$ | -0.0010 | -0.0022 | -0.0068 | 0.019 | -0.0011 | -0.0013 |
| $K$ (*mmol·g$^{-1}$·atm$^{-1}$*) | 14.70 | 15.18 | 9.41 | 26.11 | 13.77 | 8.39 |
| **Carbon dioxide** | | | | | | |
| Langmuir | | | | | | |
| $K$ (*mmol·g$^{-1}$·atm$^{-1}$*) | 7.02 | 2.74 | 1.60 | 2.10 | 2.89 | 1.11 |
| $B$ (*atm$^{-1}$*) | 2.38 | 1.08 | 0.55 | -0.39 | -0.44 | -1.79 |
| Virial | | | | | | |
| $A_0$ | 0.946 | 2.139 | 2.754 | 2.918 | 4.323 | 2.888 |
| $A_1$ | 0.085 | 0.101 | 0.121 | -0.035 | -0.460 | -0.070 |
| $A_2$ | -0.0016 | -0.0039 | -0.0079 | 0.0012 | 0.037 | 0.0021 |
| $K$ (*mmol·g$^{-1}$·atm$^{-1}$*) | 13.18 | 4.00 | 2.16 | 1.83 | 0.45 | 1.89 |

## REFERENCES

[1]  M. Jaroniec, T.J. Pinnavaia and M.F. Thorpe (eds.), Access in Nanoporous Materials, Plenum Press, New York, 1995, p. 259.

[2]  A. Gil, L.M. Gandía and M.A. Vicente, Catal. Rev.-Sci. Eng., 42 (2000) 145.

[3]  A.C. Dillon and M.J. Heben, Applied Physics A, 72 (2001) 133.

[4]  T.V. Hughes and C.R. Chambers, U.S. Patent No 405,480 (1889).

[5]  S. Iijima, Nature, 354 (1991) 56.

[6]  N. Lahav, U. Shani and J. Shabtai, Clays Clay Miner., 26 (1978) 107.

[7]  J.Y. Bottero, J.M. Cases, F. Fiessinger and J.E. Poirier, J. Phys. Chem., 84 (1980) 2933.

[8]  S.M. Bradley, R.A. Kydd and R. Yamdagni, J. Chem. Soc., Dalton Trans., (1990) 2653.

[9]  S.J. Gregg and K.S.W. Sing, Adsorption, Surface Area and Porosity, Academic Press, London, 1991.

[10] G. Horvath and K. Kawazoe, J. Chem. Eng. Jpn., 16 (1983) 470.

[11] A. Gil and P. Grange, Langmuir, 13 (1997) 4483.

[12] F. Rodríguez-Reinoso and A. Linares-Solano, Chemistry and Physics of Carbon, Marcel Dekker, Inc., New York, 1989.

[13] M.B. Sweatman and N. Quirke, Langmuir, 17 (2001) 5011.

[14] A. Gil and L.M. Gandía, Chem. Eng. Sci., 58 (2003) 3059.

[15] R.T. Yang, Gas Separation by Adsorption Processes, Imperial College Press, London, 1997, pp. 26-48.

[16] C.R. Reid, I.P. O'koye and K.M. Thomas, Langmuir, 14 (1998) 2415.

[17] M.S. Sun, D.B. Shah, H.H. Xu and O. Talu, J. Phys. Chem. B, 102 (1998) 1466.

Studies in Surface Science and Catalysis 160
P.L. Llewellyn, F. Rodriquez-Reinoso, J. Rouqerol and N. Seaton (Editors)

# CH$_4$ adsorption in Faujasite systems: Microcalorimetry and Grand Canonical Monte Carlo simulations

**G. Maurin$^a$, P.L. Llewellyn$^b$ and R.G. Bell$^c$**

$^a$ Laboratoire LPMC, UMR CNRS 5617, Université Montpellier II, Place E. Bataillon, 34095 Montpellier cedex 05, France.
$^b$ Laboratoire MADIREL, UMR CNRS 6121, Université de Provence, Centre St Jérôme, Av. Escadrille Normandie Niemen, 13397 Marseille cedex 20, France.
$^c$ The Davy Faraday Research Laboratory, Royal Institution of Great Britain, London W1S 4BS, United Kingdom.

Microcalorimetry measurements are combined with Grand Canonical Monte Carlo simulations in order to understand more deeply the interactions between methane and two types of faujasite systems. The modelling study, based on newly derived force fields for describing the adsorbate/adsorbate and adsorbate/adsorbent interactions, provide isotherms and evolutions of the differential enthalpy of adsorption as a function of coverage for DAY and NaX which are in very good accordance with those obtained experimentally. The influence of the location of the extra-framework cations within the supercages on these thermodynamics properties is also pointed out. Furthermore, the microscopic mechanisms of CH$_4$ adsorption is then carefully analysed in each faujasite system which are consistent with the trend observed for the differential enthalpies of adsorption.

## 1. INTRODUCTION

Methane has attracted much attention over the last past decade as this undesirable greenhouse gas which contributes to 23% of the total greenhouse forcing [1], leads to global warming and health problems. It is also the main constituent of the natural gas and its conversion to other useful products is of great economical interest. One of the major processes for the utilisation of CH$_4$ resources consists in its conversion to synthesis gas (H$_2$, CO) via reforming reactions [2]. These syngas can be either transformed to liquid fuels by the Fischer-Tropsch process [3] and to various chemicals especially methanol and gasoline via the methanol to gasoline process [4]. Due to their high selectivity and their stability at high temperature, zeolites are very promising materials as catalysts for such application [5,6]. They have been found to be strongly active for methane conversion to produce more valuable hydrocarbons [7]. Furthermore, a deeper understanding of the adsorption and separation processes of the hydrocarbons in zeolite materials is of great interest for the petrochemical field [8].

For such ambitious applications, it is first necessary to understand more deeply the interactions between methane and the microporous adsorbent surface. The enthalpy of CH$_4$ adsorption has been evaluated in various zeolite systems by using isosteric methods via the Clapeyron equation or performed using microcalorimetry measurements which allow direct access to this thermodynamic data [9-12]. This experimental technique has been extensively used to characterise the adsorbent surfaces and various adsorption phenomena occuring in nanoporous materials [13]. Here, the selected faujasite system which is an aluminosilicate zeolite has a wide range of industrial applications partly because of its large pore size and void volume [14]. The adsorption and catalytic properties of this material depend strongly on its chemical nature of the extra-framework cations [15] and it presents a high ability to change the degree of energetic heterogeneity of its surface by modifying the Si/Al ratio. The adsorption properties of the purely siliceous form named DAY and the cation-containing NaX were investigated by microcalorimetry over a wide range of pressure (0-35 bars) which is quite uncommon although it is of crucial interest for gas storage technology. This experimental approach has been combined with Grand Canonical Monte Carlo simulations which are most appropriate to establish a correlation between the microscopic behaviour of the zeolite/adsorbate and the macroscopic properties which are measured experimentally such as isotherms and enthalpies of adsorption [16]. They rely on accurate interatomic potentials needed to reproduce as closely as possible, the interactions between the adsorbate and the zeolite framework and between the adsorbate themselves [16]. Much effort has been expended on the development of new reliable interatomic potentials for different types of adsorbate/zeolite pairs by using quantum mechanical methods, which may then be transferable to any zeolite structure [16]. In the case of hydrocarbon - zeolite interactions, several studies mainly consisted on deriving transferable interatomic potentials via a fitting procedure from existing experimental data in order to further predict the adsorption properties of various branched and linear alkanes and alkenes in these microporous materials [17, 18]. They have proposed united atom (UA) or anistropic united atom (AUA) potentials where $CH_4$ was represented by a neutral center of force. Other investigations reported adsorbate/zeolite force fields where each atom of $CH_4$ was represented by a separate center of force (AA model) [19-23]. The GCMC simulations of $CH_4$ in NaY were thus investigated by several authors. Woods et al [19] and Maddox et al [20] used force fields based on the Lennard Jones potentials calculated by Kiselev and Pham Quang Du [24] where as Pellenq et al [21] introduced a new adsorbate-zeolite potential function. In these later studies no coulombic interactions were taken into account. Yashonath et al [22] and more recently Macedonia et al [23], using commercial force field (CVFF), used an interatomic potential where partial charges were assigned to each atom of $CH_4$. The first step of this work is then to derive a realistic though simple enough forcefield combining Lennard Jones and Coulombic contributions for describing the interactions between methane-methane and methane-zeolite via ab initio cluster calculations [25]. The validation of this force field is then obtained by a direct comparison between simulated isotherms and differential enthalpies of adsorption with experimental ones recorded up to 35 bars. For NaX, it is thus pointed out that depending on the exact locations of the extra-framework cations within the supercages, the calculated thermodynamics properties are significantly different. From these successful simulations, the next step consists in proposing the microcospic mechanisms for $CH_4$ adsorption in each of the faujasite systems which are consistent with the evolution of the differential enthalpy of adsorption as a function of the coverage.

## 2. EXPERIMENTAL

### 2.1. Samples and characterisation

The structure of Faujasite used in this study is characterised by a three dimensional pore network of large cavities of roughly spherical geometry (supercages with diameter around 12.5 Å) connected via windows to four others in a tetrahedral arrangement. The structure also contains sodalite cage units linked together by double six rings (Fig. 1). The monovalent extra-framework cations preferentially occupy different crystallographic sites named I, I', II and III [26].

The structural and textural characteristics of the two investigated samples DAY and NaX kindly supplied by Air Liquide (France) are reported elsewhere [15]. The dealuminated Y zeolite (DAY), corresponds to the highly siliceous form of Faujasite. The sample was obtained by dealumination treatment via a steaming process. The chemical analysis gave the following composition: $Na_{1.9}Al_{1.9}Si_{190.1}O_{384}$ (Si/Al ratio = 100) which corresponds to the presence of some residual extra-framework cations. Furthermore it has been previously shown that the DAY surface presents some textural defects created by the dealumination process [15]. NaX which is a cation-containing faujasite form, is characterised by a Si/Al ratio equal to 1 corresponding to the chemical formula $Na_{96}Al_{96}Si_{96}O_{384}$.

Methane used in the present study was obtained from Air-Liquide (France) with a minimum purity of 99.995 %.

## 2.2. Microcalorimetry measurements

Prior to each adsorption experiment, the sample was outgassed using Sample Controlled Thermal Analysis (SCTA) [27] which consisted of heating the sample under a constant residual vacuum pressure up to a final temperature of 450°C with specific conditions previously reported in detail [13].

The pure gas adsorption properties of the adsorbents with respect to methane, were investigated at ambient temperature (300 K) up to 35 bars. This investigation was performed by coupling a Tian-Calvet type isothermal microcalorimeter and a manometric device built in house [17]. The absolute isotherms of adsorption were obtained from the correction of the primary excess values using an appropriate expression for gas non ideality in this range of pressure. In this way, the GPR equation of state was taken into account [28]. A point by point introduction adsorptive procedure was used to evaluate a pseudo-differential enthalpy of adsorption noted $\Delta_{ads}\dot{h}$ via the measured exothermic thermal effect associated with each dose. These calculations have been already detailed in our previous publications [13]. For each sample, the values of $\Delta_{ads}\dot{h}$ were obtained with a maximum bare error of 0.6 % in the whole range of pressure.

## 3. COMPUTATIONAL METHODOLOGY

The crystal structure of the zeolite systems was modelled as follows:

(a) Siliceous faujasite $Si_{192}O_{384}$ with a cubic unit cell and lattice parameter of 24.8 Å [29] was considered to represent the DAY zeolite. This assumption is a reasonable first approximation because the DAY sample has a Si/Al ratio of 100 which corresponds only to 1.9 residual $Na^+$ per unit cell and no experimental data are available in the literature about the location of these extra-framework cations.

(b) The crystal of NaX sample was modelled by using two refinement structures reported by Vitale *et al* [30] and Zhu *et al.* [31] corresponding to the stoechiometry $Na_{96}Al_{96}Si_{96}O_{384}$ and $Na_{92}Al_{92}Si_{100}O_{384}$ respectively. These structures mainly differ by the cristallographic positions of the extra-framework cations in the supercage which are found closer to the zeolite surface in those proposed by Zhu *et al*. In this way, the extra-framework cations are distributed as follows: 32 $Na^+$ in sites I' located in the sodalite cage in front of the 6-ring window connected to the hexagonal prism, 32 $Na^+$ in sites II 4-ring winfows of the supercages. For Vitale's structure the additional 32 $Na^+$ are located in sites III, 6-ring windows, whereas for Zhu's structure the remaining 28 $Na^+$ partially occupy 12-ring positions named sites III' and III'' depending on the neighbouring O-Al-O or O-Si-O sequences.

The adsorbate-adsorbent and adsorbate-adsorbate interactions were modelled by an interatomic potential consisting of a Lennard Jones (LJ) dispersion-repulsion term and a Coulombic contribution. Faujasite was assumed to be semi-ionic with atoms carrying the following partial charges (in electron unit): Si (+2.4), Al (+1.4), O (-1.2) and Na (+1) as previously defined [32]. Methane was represented by an atomic point charge model with the following partial charges carried by each atom (in electron unit): C (-0.48) and O (+0.12). The LJ parameters for modelling both $CH_4$/zeolite and $CH_4$/$CH_4$ interactions were extracted from *ab initio* cluster calculations [25]. The potential parameters for each interacting pairs as well as the details of the computing procedure can be found elsewhere [15, 25,33].

Absolute adsorption isotherms were computed using a Grand Canonical Monte Carlo calculation algorithm, as implemented in the *Sorption* module of the Cerius2 software suite [34]. These simulations consisted of evaluating the average number of adsorbate molecules whose chemical potential equals those of the bulk phase for given pressure and temperature. All these simulations were performed at 300 K using one unit cell of faujasite with typically $3.10^6$ Monte Carlo steps. The Ewald summation was used for calculating electrostatic interactions and the short range interactions were calculated with a cutoff distance of 12 Å. The zeolite structure was assumed to be rigid during the sorption process. Dummy atoms with appropriate van der Walls radius were introduced in the sodalite cages in order to avoid any introduction of adsorbates in this space, thus leading to only accessibility for methane in the supercages as previously mentioned [15,33]. The evolution of the absolute differential enthalpy of adsorption as a function of the loading was then calculated at 300 K through the fluctuations of the number of particles in the system and from fluctuations of the internal energy [15,33]. Furthermore, from the ensemble average, the radial distribution functions between both adsorbate-adsorbate and adsorbate-adsorbent were evaluated in order to provide information on the location of the methane molecules within the supercages.

## 4. RESULTS AND DISCUSSIONS

Fig. 1. reports the absolute isotherms for methane adsorption on DAY and NaX obtained both experimentally and theoretically at 300 K. It has to be mentioned that this adsorbate deviates from ideal gas behaviour in the whole range of pressure and consequently, the experimental and simulated data were corrected to take into account this non-ideal state. We observe that the simulated absolute isotherm reproduces well the experiments for DAY. For NaX, the isotherm calculated from model 1 is in very good agreement with the experimental data where as an under-estimation of the loading is observed in the low and intermediate pressures by considering the model 2. The higher average loading observed for model 1 can be explained by the most favourable adsorbate-adsorbent interactions due to the higher accessibility for $CH_4$ in model 2. At high pressure, the two models containing almost the same number of extra-framework cations within the supercages, lead to similar saturation loading (~65 $CH_4$ molecules/u.c.) which is controlled by the space available to the adsorbed molecules in NaX. This maximum calculated amount is close to the values reported in the literature [19,21]. Furthermore, the CH$_4$ affinity, which can be estimated from the slope of the isotherms in the initial low domain of pressure, is much pronounced for NaX than for DAY due to additional interactions between the polarisable CH$_4$ molecules and the electric field generated by the sodium ions.

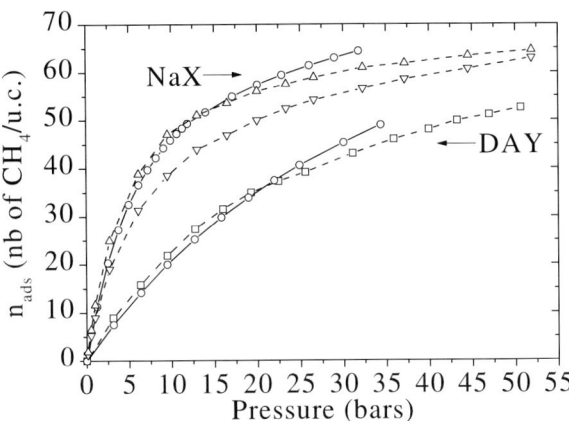

Fig. 1. Absolute isotherms for methane adsorption on DAY (($\Box$) simulation; ($\bigcirc$) experiment) and NaX (simulation : Vitale's structure-model 1 ($\triangle$), Zhu's structure- model 2 ($\nabla$); experiment ($\bigcirc$)) at 300 K in the range of pressure 0-35 bars.

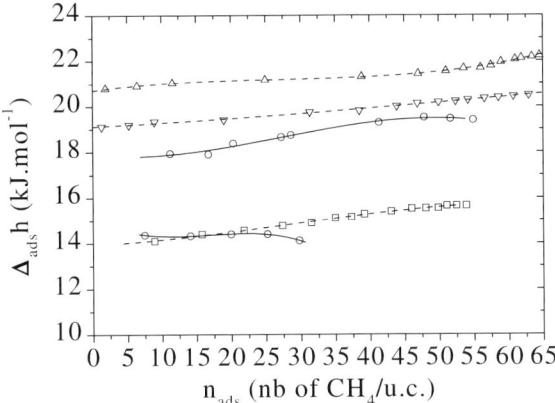

Fig. 2. Evolution of the differential enthalpies of adsorption ($\Delta_{ads}h$) as a function of the coverage for DAY (($\Box$) simulation; ($\bigcirc$) experiment) and NaX (simulation : Vitale's structure-model 1 ($\triangle$), Zhu's structure- model 2 ($\nabla$); experiment ($\bigcirc$)) at 300 K.

The evolutions of the differential enthalpies of adsorption as .a function of the coverage for the two different Faujasite forms are reported in Fig.2. We observe that the adsorption of CH$_4$ on DAY gives within the experimental error, almost constant differential enthalpy values before slightly decreasing. This type of evolution suggests a relatively heterogeneous adsorbate/adsorbent interaction. The heterogeneity of the DAY sample comes from the residual 1.9 Na$^+$ extra-framework cations or from a low concentration of textural defect sites formed during the dealumination process as was previously revealed [33], thus leading to some preferential sites for the adsorbate. Our simulation performed on a purely siliceous faujasite without any structural defects reproduces well the experimental differential enthalpy of adsorption at low coverage. These values about 13.6 and 14.4 kJ.mol$^{-1}$ for experiment and simulation respectively are slightly higher than those previously reported by Siperstein *et al.*[12]. However, we observe an increase of the simulated differential enthalpy with increasing coverage as it was already pointed out for other purely siliceous zeolite system such as silicalite [35]. The discrepancy thus observed between experiment and simulation is due to the characteristics of the DAY sample (textural defects and presence of residual cations) not included in our computational model. We further observed that whatever the loading, the methane molecules are homogeneously distributed within the supercage with several preferential adsorption sites close to sites II and III in the 4- and 6-ring widows and to the center of 12-membered window. This observation is quite similar with a previous energy minimisation investigation which locate the most energetically adsorption site between the 12-ring window and SII [36]. The radial distribution functions (rdfs) for DAY/CH$_4$ system are reported Fig.3. The absence of pronounced peaks in the rdf between oxygen of the framework and hydrogen of methane suggests that the methane molecules are not confined in the previous mentioned adsorption sites but are rather homogeneously delocalized within the whole supercage. This observation is similar to those previously reported by a Molecular Dynamics investigation in NaY [22] which stated a more fluidlike behaviour of CH$_4$ at room temperature. Furthermore, Fig.3. shows that the average distance between the carbon atoms of the adsorbate (d(C-C)) becomes significantly shorter when the loading increases. This later observation means that the methane molecules are closer to each other at higher loading, leading to an increase of the adsorbate-adsorbate interaction energy. Indeed, methane probes DAY as a homogeneous energetic surface with an almost constant CH$_4$/adsorbent interaction energy. As the CH$_4$/CH$_4$ contribution increases with the loading, an increase of the differential enthalpy of adsorption with coverage results, as reported in Fig.2.

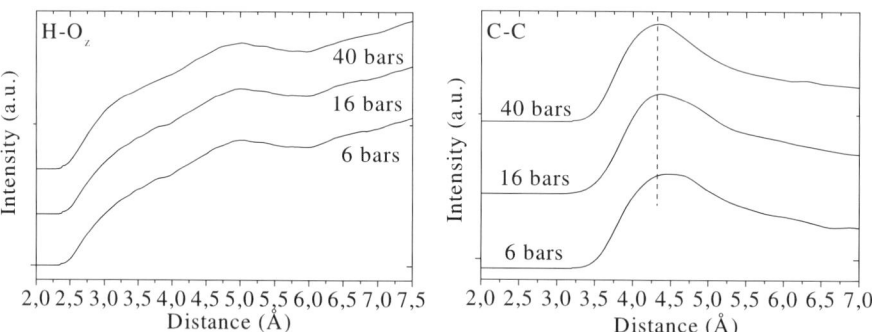

Fig. 3. Radial distribution functions between adsorbate - zeolite (H-O$_z$) and adsorbate - adsorbate (C-C) calculated at different loading for methane adsorbed in DAY.

The experimental and simulated differential enthalpy of adsorption for NaX (Fig. 2) increases with the loading. Similar enthalpy profile has been previously reported for both NaX and NaY [11, 19-21] with differential enthalpies of adsorption at low coverage ranging from 15.5 kJ.mol$^{-1}$ to 18.9 kJ.mol$^{-1}$ [9-12, 19-22] in accordance with the experimental extrapolated value of 17.8 kJ.mol$^{-1}$ and with those of 19.1 kJ.mol$^{-1}$ and 20.5 kJ.mol$^{-1}$ simulated by means of the model 1 and 2 respectively. The difference in enthalpy of about 1.4 kJ.mol$^{-1}$ between the two models can not be explained by the slight difference of the number of cations between the two models (96 for model 1 vs 92 for model 2), but rather comes from the enhancement of the adsorbate/adsorbent interactions due to the higher accessibility of sites III in the model 1. It has been observed that the methane molecules preferentially interact with the extra-framework cations in sites III (III' and III'') as previously suggested by Woods *et al* [19]. The radial distribution functions reported in Fig.4. arbitrary calculated for the model 2 show that the average distance between extra-framework cations and hydrogen of the methane remains almost unchanged when the loading increases whereas those between the carbon atoms of the adsorbate (d(C-C)) becomes significantly shorter. This later observation means that the methane molecules surrounded the extra-framework cations in sites III (III' and III'') are closer to each other at higher loading, leading to an increase of the adsorbate-adsorbate interaction energy. The combination of these two contributions leads to the increase of the differential enthalpy of adsorption with the loading.

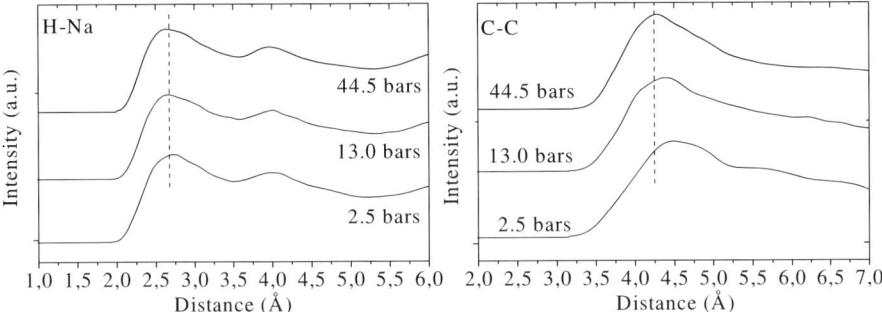

Fig. 4. Radial distribution functions between adsorbate - zeolite (H-Na$^+$) and adsorbate - adsorbate (C-C) calculated at different loading for methane adsorbed in NaX calculated for model 2.

## 5. CONCLUSIONS

This work clearly shows that the force field derived from *ab initio* calculations for representing the interactions between methane and zeolite framework was revealed to be very well transferable. It allowed to reproduce accurately the microcalorimetry data across a wide range of pressure, for two different faujasite forms, DAY and NaX via our Grand Canonical Monte Carlo simulations. Our simulations describe the nature of the CH$_4$/zeolite interaction depending on the chemical composition of the faujasite. It has been shown that the positions of the extra-framework cations within the supercage significantly modify the adsorption properties of the zeolite material. This work is of high interest for predicting the performance of different types of zeolite materials with respect to CH$_4$ and thus for defining the main characteristics of the adsorbent materials able to store or separate this gas for environmental or petrochemical applications.

## REFERENCES

[1] K. Zhang, U. Kogleschatz and B. Eliasson, Energy & Fuels 15 (2001) 395.
[2] B.S. Liu, L.Z. Gao and C.T. Au, Applied Catalysis A: General 235 (2002) 193.
[3] J.R.H. Ross, A.N.J. Van Keulen, M.E.S. Hegarty, K. Seshan, Catal. Today 30 (1996) 193.
[4] C.D. Chang, Stud. Surf. Sci. Catal. 36 (1988) 127.
[5] K. Zhang, B. Eliasson and U. Kogelschatz, Ind. Eng. Chem. Res. 41 (2002) 1462.
[6] S.M. Gheno, S. Damyanova, B.A. Riguetto, C.M.P. Marques, C.A.P. Leite, J.M.C. Bueno, Journal of Molecular Catalysis A: Chemical 198 (2003) 263.
[7] N.D. Parkyns, C.I. Warburton, J.D. Wilson, Catal. Today 18 (1993) 385.
[8] R. W. Neuzil, US Pat. 3558732, 1971.
[9] R.M. Barrer and J.W. Sutherland, Proc. R. Soc. London A, 237 (1956) 439.
[10] S.Y. Zhang, O. Talu and D.T. Hayhurst, J. Phys. Chem. 95 (1991) 1722.
[11] J.A. Dunne, M. Rao, S. Sircar, R.J. Gorte and A.L. Myers, Langmuir 12 (1996) 5896.
[12] F. Siperstein, S. Savitz, R.J. Gorte and A.L. Myers, in Fundamentals of Adsorption, Proc. 6$^{th}$ Int. Conf. Fundamentals of Adsorption; F. Meunier (Ed.), Elsevier, Paris (1998) 111.
[13] P.L. Llewellyn and G. Maurin, Cr. Acad. Sci. Sc. II. 8 (2005) 283.
[14] M.W. Ackley, S.U. Rege and H. Saxena, Microporous and Mesoporous Materials 61 (2003) 25.
[15] G. Maurin, P.L. Llewellyn, Th. Poyet and B. Kuchta, J. Phys. Chem. B., 109 (2005) 125.
[16] A.H. Fuchs and A.K. Cheetham, J. Phys. Chem. B 105 (2001) 7375.
[17] P. Pascual, P. Ungerer, B. Tavitian, P. Pernot and A. Boutin, Phys. Chem. Chem. Phys. 5 (2003) 3684.
[18] T.J.H. Vlught, R. Krishna and B. Smit, J. Phys. Chem. B 103 (1999) 1102.
[19] G.B. Woods and J.S. Rowlinson, J. Chem. Soc. Faraday Trans. 2 85 (1989) 765.
[20] M.W. Maddox and J.S. Rowlinson, J. Chem. Soc. Faraday Trans. 89 (1993) 3619.
[21] R.J.M. Pellenq, B. Tavitian, D. Espinat and A.H. Fuchs, Langmuir 12 (1996) 4768.
[22] S. Yashonath, P. Demontis and M.L. Klein, J. Phys. Chem. 95 (1991) 5881.
[23] M.D. Macedonia, D.M. Moore, E.J. Maginn and M.M. Olken, Langmuir, 16 (2000) 3823.
[24] A. Kiselev and P. Quang Du, J. Chem. Soc. Faraday Trans. 2 77 (1981) 1.
[25] R.G. Bell et al, manuscript in preparation.
[26] Mortier, W.J. "Compilation of Extraframework Sites in Zeolites", Butterworth, Guildford, 1982.
[27] J. Rouquerol, Thermochim. Acta 144 (1989) 209.
[28] K. A. M. Gasem, W. Gao, Z. Pan, R. L. Robinson Jr., J. Phase Equilibria, 181 (2001) 113.
[29] A.N. Fitch, H. Jobic and A. Renouprez, J. Phys. Chem. B 90 (1986) 1311.
[30] G. Vitale, C.F. Mellot, L.M. Bull and A.K. Cheetham, J. Phys. Chem. B 101 (1997) 4559.
[31] L. Zhu and K. Seff, J. Phys. Chem. B 103 (1999) 9512.
[32] G. Maurin, R. Bell, S. Devautour, J.C. Giuntini, J. Phys. Chem. B, 108 (2004) 3739.
[33] G. Maurin, R. Bell, B. Kuchta, P.L. Llewellyn and Th. Poyet, Adsorption 11 (2005) 331.
[34] Cerius$^2$. v. 4.0, Accelrys Inc., San Diego (1999).
[35] J.A. Dunne, R. Mariwala, M. Rao, S. Sircar, R.J. Gorte and A.L. Myers, Langmuir 12 (1996) 5888.
[36] J.O. Titiloye, S.C. Parker, F.S. Stone and C.R.A. Catlow, J. Phys. Chem. 95 (1991) 4038.

Studies in Surface Science and Catalysis 160
*P.L. Llewellyn, F. Rodriquez-Reinoso, J. Rouqerol and N. Seaton (Editors)*

# Amino-functionalized low density silica xerogels seen by different characterization methods

**J. Mrowiec-Białoń**

Institute of Chemical Engineering, Polish Academy of Science, Bałtycka 5, 44-100 Gliwice, Poland, e-mail: j.bialon@iich.gliwice.pl

Amino-functionalized low density xerogels with attractive textural properties were prepared using ethyl silicate 40 as a main silica precursor. Their pore sizes and overall pore volumes were larger than in the corresponding gels prepared using TEOS. The properties of xerogels appeared to depend not only on the silica precursor but also on the type of amine group introduced. Samples with aminopropyl groups had larger specific surface area, $S_{BET}$ ca. 600-700 $m^2g^{-1}$ and smaller pores with diameter, ca. 6 nm, than those with propyl ethylendiamine, the pores of which were 13-17 nm in diameter. The latter had more robust structure, shown by SAXS.

## 1. INTRODUCTION

Amino-functionalized porous silicas are widely used in many important applications, e.g. as stationary phases in chromatography, in immobilization of enzymes and antibodies, preconcentration of metals, and many others [1]. They can conveniently be obtained in the liquid phase using the sol-gel one- or two-step preparative route. Despite the exiting discovery of ordered mesoporous materials many applications still pose a considerable demand for materials with very open structure, featured by low-density xerogels. More recently Alie *et al.* [2, 3] observed that low-density silica xerogels can be readily prepared by incorporating an aminofunctionalsilanes, e.g. aminopropylalkoxysilane, to the reaction mixture containing the main silica precursors: tetraethyl- or tetramethyl orthosilicate (TEOS, TMOS). Also more recently our group reported [4-6] that the use of prepolymerized silica precursors, including the low cost commercial ethyl silicate 40 (ES), can result in gels with attractive textural properties, notably larger pore sizes and overall pore volumes, than in the corresponding gels prepared using TEOS or TMOS alone.

The aim of this work was to explore the potentials of the application of ES together with aminofunctionalsilanes to obtain functionalized low-density xerogels from a cost-effective preparative route, and to compare their properties with those obtained from the conventional procedure.

## 2. EXPERIMENTAL

### 2.1. Sample preparation

Alcogels were prepared in a direct procedure using as a silica precursor: commercial ethyl silicate 40 (Unisil-Tarnow, Poland) or, for comparison, tetraethoxysilane (TEOS,

ABCR, Germany). The additive aminofunctionalsilanes were 3-(aminopropyl)-triethoxysilane (APT) or N-[3-trimethoxysilyl)propyl ethylendiamine (ATM) (both from ABCR, Germany). The synthesis of samples was carried out at room temperature and the molar ratio of compounds in all reaction mixtures involving ES was $Si:EtOH:H_2O:NH_3 = 1:8:3:x$, where x took the values of 0; 0.008; 0.016 and 0.032, whereas in those involving TEOS the molar ratio was 1:8:4:x. After gelation all samples were aged for seven days and then dried at 50°C for 100 h to give xerogels. Then they were additionally dried at 100°C for 5 h. The prepared sample abbreviation was: silica precursor-amine precursor-molar percent of amine precursor (Table 1).

Table 1.
Characteristic parameters of silica xerogels

| Sample | $NH_3 : Si$ ratio | $S_{BET}$ $m^2g^{-1}$ | $V_{pN2}$ $cm^3g^{-1}$ | $d_{BJH}$ nm | $\rho_b$ $gcm^{-3}$ | $V_t$ $cm^3g^{-1}$ |
|--------|------|------|------|------|------|------|
| TEOS | 0.008 | 1004 | 0.60 | 2.9 | 0.870 | 0.69 |
| ES | 0.032 | 898 | 1.63 | 7.4 | 0.400 | 2.04 |
| ES-APT-5 | 0 | 708 | 1.27 | 5.9 | 0.281 | 3.10 |
| ES-APT-10 | 0 | 648 | 1.13 | 5.5 | 0.338 | 2.50 |
| ES-APT-20 | 0 | 592 | 1.16 | 6.1 | 0.330 | 2.57 |
| ES-ATM-5 | 0 | 498 | 2.02 | 17 | 0.237 | 3.77 |
| ES-ATM-10 | 0 | 441 | 1.70 | 14.0 | 0.250 | 4.00 |
| ES-ATM-20 | 0 | 373 | 1.47 | 13.5 | 0.293 | 2.96 |
| TEOS-ATM-2 | 0.008 | 274 | 1.11 | 14.7 | 0.468 | 1.68 |
| TEOS-ATM-5 | 0.008 | 166 | 0.40 | 13.4 | 0.206 | 4.41 |
| ES-ATM-2 | 0.008 | 632 | 1.98 | 14.4 | 0.325 | 2.62 |
| ES-ATM-5a | 0.016 | 524 | 1.82 | 13.2 | 0.237 | 3.77 |
| ES-ATM-5b | 0.032 | 531 | 1.85 | 15.0 | 0.337 | 2.40 |

## 2.2. Characterization

Nitrogen adsorption-desorption isotherms measured at 77 K with a Micromeritics ASAP 2000 instrument were used to obtain values of the specific surface area, $S_{BET}$, estimated from a linear section of adsorption isotherm, taking five points from 0.05-0.2 $p/p_o$ range, and pore volume $V_{pN2}$, determined from the amount adsorbed at $p/p_o$ of about 0.98. Bulk density, $\rho_b$, was determined from mercury porosimetry data afforded by Micromeritics Auto Pore 9220. The later quantity was used to calculate the total pore volume = $V_{pN2}$ + $V_{MACROPORES}$ using the expression $V_t = 1/\rho_b - 1/\rho_s$ [7], where $\rho_s$ is skeletal density, taken equal to 2.2 g cm$^{-1}$, i.e. density of amorphous silica obtained by the sol-gel method [8].

SAXS measurements were performed using a Kratky camera with line-collimated primary beam system and CuKα radiation. The samples investigated were granules loosely packed between two thin foils.

The TEM images of selected samples were obtained on a JOEL 2000SX instrument operating at 160 kV. Powder sample, ca. 10 mg was sonicated 1 min with 5 ml of ethanol and a drop of this solution was put on 400 mesh Cu grids covered with a carbon holey film.

Infrared spectra (IR) were recorded at room temperature, after heat treatment, on a Carl-Zeiss Jena SPECORD M80 spectrometer using the KBr pellet technique.

Thermogravimetric measurements were made with Mettler Toledo thermobalance (TGA/SDTA831$^e$, LF/1100°C) using platinum crucibles. The samples (about 10 mg) were heated in flowing air (60 ml min$^{-1}$) from 25 to 700°C at 5 K min$^{-1}$.

## 3. RESULTS AND DISCUSSION

The results obtained have shown the pronounced influence of the aminofunctionalsilane additives on textural properties of xerogels. Compared to pure TEOS or ES sample they have larger pores and larger total pore volumes (Table 1). This more open structure is also well seen in the TEM images presented in Fig. 1.

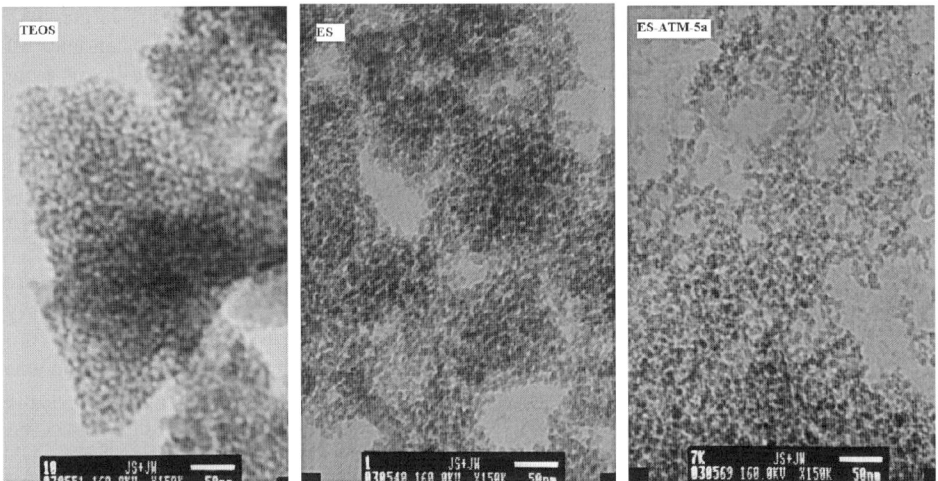

Fig.1. TEM images of xerogels.

Moreover, the surface properties of the samples functionalized using APT or ATM and prepared using ES as the main silica precursor are more attractive than those of the corresponding samples synthesized using TEOS. On the whole, in addition to the larger pore volume, these samples have larger specific surface area (Table 1), the value of which depends not only on the silica precursor but also on the additive and its concentration. Considerable differences between the properties of ES-APT and ES-ATM samples with 5, 10 and 20 mol.% amine content were observed. This is evidenced by notably different shapes of nitrogen adsorotion/desorption isotherms (cf. Fig. 2).

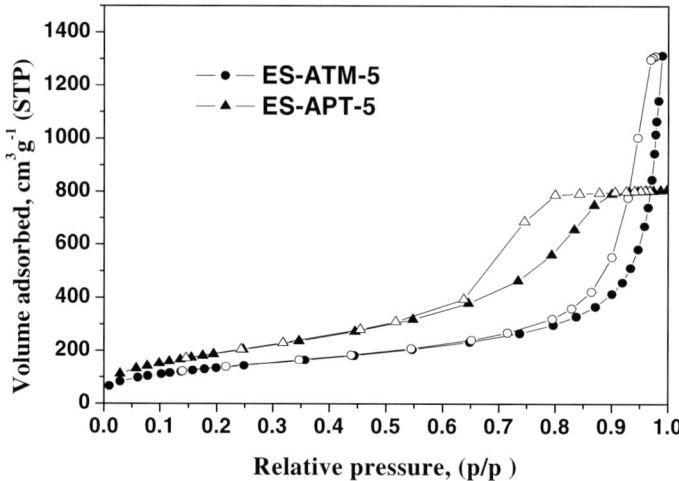

Fig.2. Nitrogen adsorption/desorption isotherms for xerogels with aminopropyl and propyl ethylendiamine groups.

Compared to ES-APT series, the ES-ATM family of xerogels exhibits larger pore volumes $V_{pN2}$ and also larger pore sizes, as can be inferred from the larger amounts of nitrogen adsorbed at notably higher relative pressures (>0.9). Moreover, the isotherms from ES-ATM xerogels do not level off at relative pressure close to the saturation vapor pressure, similarly as observed earlier for aerogels prepared by a two step process [10]; a strong indicative of the presence of very large mesopores and macropores seen in TEM images (Fig.1). The type H2 hysteresis loop was observed for ES-APT samples indicating the presence of pores with narrow mouths (ink-bottle pores), while H3 type was characteristic for ES-ATM series.

The shapes of isotherms from the ES-ATM series implied the broad pore size distributions, with pores in the range of 5 - 50 nm, whereas those from the ES-APT family were characterized by more narrow pore size distributions, with maximum located at 7 nm (Fig. 3). Both types of materials contained 1.3 to 1.8 $cm^3g^{-1}$ macropores. The larger value was for smaller amine concentration (5 mol%) and it appeared not to depend on a type of amine precursor. Contrary to pure silica xerogels [6] ammonia concentration had little influence on the porosity of the amino-functionalized xerogels.

Clearly, the differences between ES-APT and ES-ATM samples are caused by different effects of the amine precursors. As shown in [2, 9], the additives containing more reactive methoxy group (ATM) react first to form primary particles, to which the principal precursor, containing less reactive ethoxy groups condenses to form silica particles. This nucleation mechanism is absent in case of the APT doped systems. The TEOS-ATM xerogels appeared to have larger total pore volumes and larger pores than a pure TEOS sample but their values were still lower than in the ES-ATM xerogels (cf. Table 1). Moreover, the latter had notably larger specific surface area.

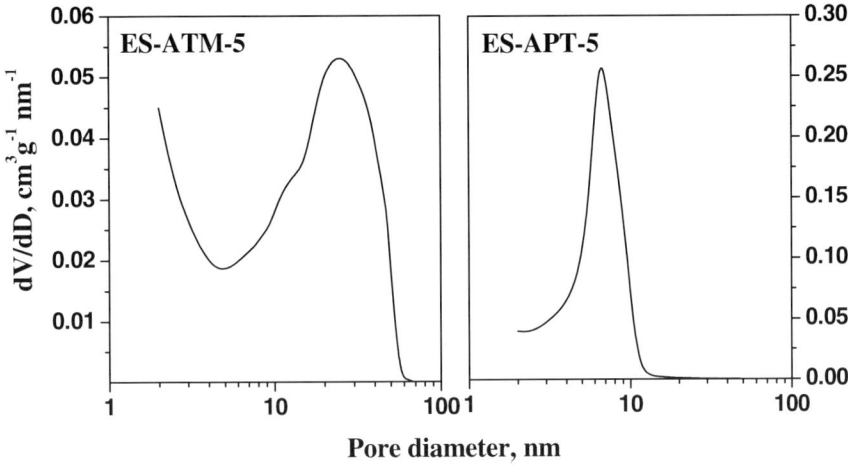

Fig. 3. Pore size distributions in xerogels with different amines.

Apparently, these differences should be reflected by differences in the structure of silica skeleton networks examined by the small-angle X-ray scattering. Indeed, the analysis of scattering curves (SAXS) is often used to obtain averaged values of colloidal particles making up the skeleton and also the degree of materials homogeneity in very short scales. The lack of a characteristic maximum in $Iq^3$ vs. q plots given in Fig. 4 (I is smeared scattering intensity, q is scattering vector) indicates that the particles could not be clearly distinguished, similarly as before in the pure ES low-density xerogels [6].

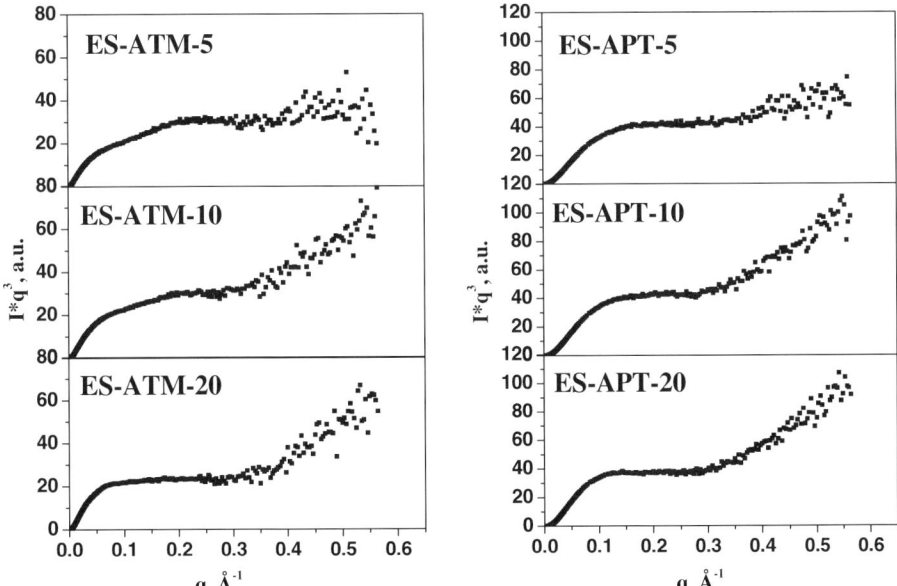

Fig. 4. Scattering spectra from xerogels with different amine concentrations.

However, the calculation of chord lengths, $l_m$ [11] in the solid phase gave the values in the range of 23-30 Å and 38-55 Å, for ES-ATP and ES-AMP series, respectively. The larger values of the latter, imply the more robust structure of ATM xerogels, less susceptible to shrinkage in drying, in good agreement with the larger pore volumes observed in this series. In very short scales the scattering curves gradually evolved from a rapid slope, consistently seen in xerogels with large amine content, toward a plateau at their low concentration, an indicative of larger microstructural homogeneity of xerogels with lower amine content.

Infrared spectra shown in Fig. 5 confirmed the incorporation of the aminopropyl and propyl ethylendiamine groups into the silica skeleton. After incorporation of amine groups into silica vibrations from strong bands N-H, N-H$_2$ and C-H$_2$ can be seen in the IR spectra [12]. The asymmetric stretching vibration $v_{as}(CH_2)$ at 2934 cm$^{-1}$, symmetric stretching vibration $v_s(CH_2)$ at 2862 cm$^{-1}$ and the bending vibration $\delta$ (NH$_2$) at 1596 cm$^{-1}$ are observed in both samples. In sample ES-ATM-20 additional asymmetric $v_{as}(N-H)$ and symmetric vibrations $v_s(N-H)$ at 3375 cm$^{-1}$ and at 3305 cm$^{-1}$, respectively were present. This is in agreement with the data obtained for aminoalkoxysilanes deposited using a gas-phase technique [13].

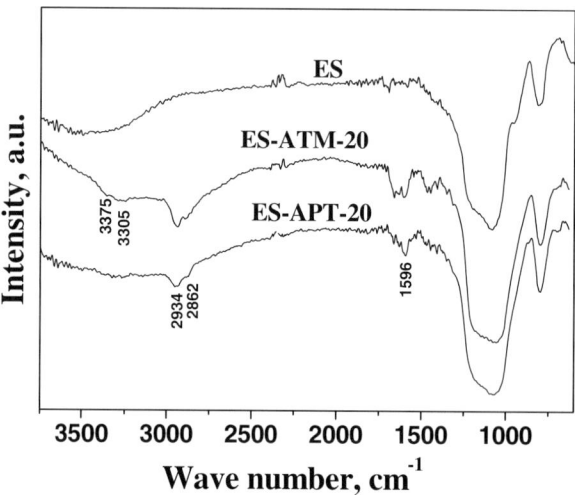

Fig. 5. Infrared spectra of amine functionalized xerogels.

The thermal stability of amino-functionalized silica was examined using thermogravimetry. Samples functionalized with aminopropyl and propyl ethylendiamine groups are stable up to 250°C and 200°C, respectively. Fig. 7 shows TG, DTG and SDTA data obtained during thermal analysis of ES-APT-5 (left) and ES-ATM-5 (right) samples. The mass lost in the range of 25-250°C for ES-APT-5 and 25-200°C for ES-ATM-5 is due to physically adsorbed and chemically bound water. The decomposition of aminopropyl groups starts at about 250°C and is accompanied by well seen exothermic effect on SDTS curve. The propyl ethylendiamine groups are less stable and its decomposition starts about 200°C.

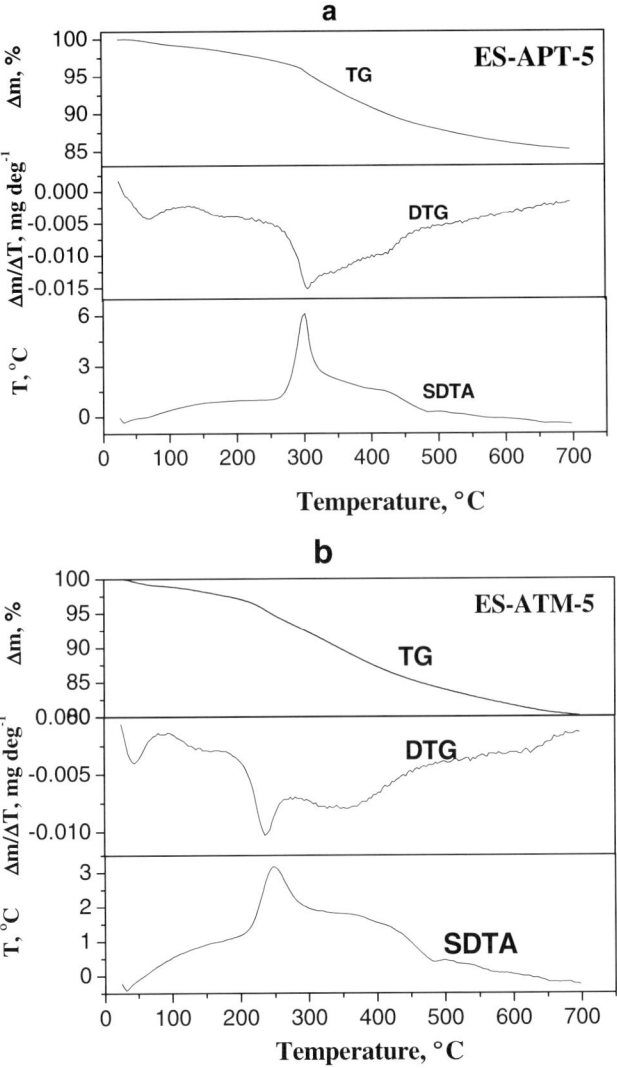

Fig. 6. TG, DTG and SDTA curves for ES-APT-5 (a) and ES-ATM-5(b) xerogels.

## 4. CONCLUSIONS

Low density xerogels functionalized with amines and exhibiting very attractive textural properties can be obtained by one step procedures using low cost commercial ethyl silicate 40 (ES) instead of TEOS. These materials have notably larger pores, overall pore volume and specific surface area, than the corresponding gels prepared using TEOS or TMOS alone. Structure properties of xerogels depend on the type of aminofunctionalsilane additive. The use of compound containing more reactive methoxy group (ATM), acting as nucleation agent, resulted in xerogels with larger mesopores and more robust structure, similarly as observed before. The ES-APT and ES-ATM xerogels were thermally stable up to 250 and 200°C, respectively.

**ACKNOWLEDGMENT.** This work was partially financed by grant No 4 T09C 023 23 from Polish Ministry of Scientific Research and Information Technology. The author wishes to thank Prof. Andrzej Jarzębski for helpful discussion.

## REFERENCES

[1] E.F.Vansant, P. Van Der Voort, K.C. Vrancken, Characterization and Chemical Modification of the Silica Surfaces, Elsevier, Amsterdam, 1995, Chapters 8 and 9.
[2] C. Alié, R. Pirard, A.J. Lecloux, J.P. Pirard, J. Non-Cryst. Solids, 246 (1999) 216.
[3] C. Alié, R. Pirard, A.J. Lecloux, A.J.; Pirard, J. Non-Cryst. Solids 285 (2001) 135.
[4] J. Mrowiec-Białoń, A.B. Jarzębski, Langmuir 17 (2001) 626.
[5] O.A. Kholdeeva, N.N. Trukhan, M.P. Vanina, V.N. Rommannikov, V.N. Parmon, J. Mrowiec-Białoń, A.B. Jarzębski, Catal. Today, 75 (2002) 203.
[6] J. Mrowiec-Białoń, A. B. Jarzębski, L. Pająk, Z. Olejniczak, M. Gibas, Langmuir, 20 (2004) 10389.
[7] G.W. Scherer, J. Non-Cryst. Solids, 225 (1998) 192.
[8] H.E. Bergna, The Colloidal Chemistry of Silica, Eds.: American Chemical Society, Washington, DC, 1994.
[9] C. Alié, R. Pirard, J.P. Pirard, J. Non-Cryst. Solids, 311 (2002) 304.
[10] C.J. Brinker, G.W. Scherer, Sol-Gel Science. The Physics and Chemistry of Sol-Gel Processing, Academic Press: San Diego, CA, 1990.
[11] A.B. Jarzębski, J. Lorenc, L. Pająk, Langmuir, 13 (1997) 1280.
[12] C.-H Chiang, H. Ishida, J.L. Koenig, J. Colloid Interf. Sci.,74, 2 (1980) 396
[13] S. Ek, E.I. Iiskola, L. Niinistö, Langmuir, 19 (2003) 3461.

Studies in Surface Science and Catalysis 160
P.L. Llewellyn, F. Rodriquez-Reinoso, J. Rouqerol and N. Seaton (Editors)

# $CO_2$ adsorption in synthetic hard carbons

G. Reichenauer[a,b]

[a]Physikalisches Institut, Universität Würzburg, Am Hubland, 97074 Würzburg, Germany

[b]Bavarian Center for Applied Energy Research, Am Hubland, 97074 Würzburg, Germany

## 1. INTRODUCTION

Carbon aerogels, sol-gel derived, open porous hard carbons, show a two step adsorption kinetics for $CO_2$ at 273K (and 295K) that is not detected for $N_2$ at 273 K. A similar behavior, however, with different adsorption rates has also been reported for other types of hard carbons [1-4]. A detailed investigation of the two adsorption steps for carbon aerogels as a function of relative pressure reveals that the slower step actually corresponds to a larger micropore size; this suggests two well defined types of micropores to be present in the same sample, an easily accessible one and a wider, however, bottleneck type set of micropores [5]. The comparison of samples with different macropore sizes but similar microporosity shows that the adsorption rate for the bottleneck type micropores is not affected by the morphology of the samples on the macropore scale nor the sample dimensions.
The investigation presented here focuses on the adsorption kinetics of the fast adsorption step of $CO_2$ in these synthetic hard carbons.

## 2. EXPERIMENTAL

### 2.1. Samples

The carbon aerogels investigated were monolithic solids prepared from sol-gel derived organic precursors via pyrolysis at 1050 °C [6,7]. The starting solution used for the synthesis of the organic gels consisted of resorcinol, formaldehyde (37,5 wt.% in water), water as the solvent and $Na_2CO_3$ as the catalyst. The relative volume of the solid phase in the organic precursors was adjusted via the mass ratio

$$R_M = \frac{M_{resorcinol} + M_{formaldehyde}}{M_{resorcinol} + M_{formaldehyde} + M_{water} + M_{catalyst}} , \qquad (1)$$

with $M_x$ the mass of the respective component; for the samples presented here the mass ratio $R_M$ was set to 30 % resulting in a carbon aerogel density of about 0.3 $g/cm^3$. For all samples the molar ratio of resorcinol to formaldehyde was kept constant (R/F= 0.5). At fixed mass ratio the meso- and macrostructure of the organic gels was varied via the catalyst concentration and by changing the time that the gelling sol was kept at room temperature (see Table 1) prior to further aging the samples for 1 day at 90 °C. The so called R/C ratio, i.e. the molar ratio of resorcinol to sodium carbonate catalyst, was set to 1000 and 1500, respectively. The time at room temperature was 0 and 4 days, respectively. After aging, the water in the pores of the aquagels was replaced by acetone prior to drying the samples at 50 °C and ambient pressure.

Table 1
Synthesis parameters for the three different samples prepared.

| sample | $R_M$ | R/C | R/F | aging /gelling time at room temperature |
|--------|-------|------|-----|------------------------------------------|
| #1 | 0.3 | 1500 | 0.5 | 4 |
| #2 | 0.3 | 1500 | 0.5 | 0 |
| #3 | 0.3 | 1000 | 0.5 | 0 |

### 2.2. Morphological characterization of the samples

The morphology of the porous carbons was characterized by small angle X-ray scattering (SAXS) as well as nitrogen sorption at 77K and sorption of carbon dioxide at 273 K.

SAXS was performed at the beam-line JUSIFA at the synchrotron source HASYLAB, Hamburg. To determine the scattering cross section for the sample investigated on an absolute scale, the scattering of the sample was calibrated using a glassy carbon standard as a reference.

The sorption isotherms were taken with a commercially available volumetric instrument (ASAP 2000, software 2010, Micromeritics).

### 2.3. Measurements of adsorption kinetics

The adsorption kinetics was recorded using cylindrical carbon aerogel monoliths (diameter about 1.6 cm, height about 1cm) in combination with a volumetric sorption instrument (designed at Wuerzburg University) that is equipped with a standard and a fast pressure transducer [8]. Upon dosing of carbon dioxide the maximum pressure increase in the sample chamber (volume about 125 ccm) of the instrument was set to about 5 mbar. Due to the restrictions of the instrument the experiments were performed at a temperature of 295 K rather than 273 K.

## 3. RESULTS

### 3.1. Morphological characteristics

Fig. 1 shows the small angle X-ray scattering for the three samples investigated. The curves qualitatively show that the micropore characteristics are almost identical for all aerogels investigated while the scattering related to the envelope surface of the particle forming the open porous skeleton of the hard carbons clearly varies.

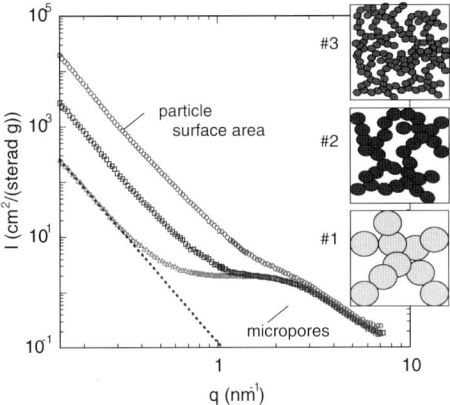

Fig.1. SAXS data for the three samples investigated; note that the data are given on an absolute scale. The cartoons indicate the morphology of the different carbon aerogels consisting of a network of microporous particles with different average diameters.

The data were quantitatively evaluated with respect to the total micropore volume and the external surface area of the carbon skeleton. Details of the procedure applied can be found in ref. [5]. The characteristic particle and pore sizes as well as the micropore volume of the samples are given in Table 2.

Table 2
Morphological parameters derived from SAXS and sorption data, with $\rho$ the sample density, $V_{micro}/m$ the specific micropore volume, $S/m_{particle}$ the external specific surface area of the microporous particles and $d_{mesopore}$ and $d_{particle}$ the average diameter of the mesopores and the particles, respectively.

| sample | density ($g/cm^3$) | SAXS | | | | $N_2$ sorption (77K) | $CO_2$ sorption (273 K) |
| | | $V_{micro}/m$ ($cm^3/g$) | $(S/m)_{particle}$ ($m^2/g$) | $d_{pore}$ (nm) | $d_{particle}$ (nm) | $V_{micro}/m$ ($cm^3/g$) | $V_{micro}/m$ ($cm^3/g$) |
|---|---|---|---|---|---|---|---|
| #1 | 0.33±0.02 | 0.27±0.02 | 1.7±0.06 | 5400±280 | 2700±160 | not detectable | 0.24±0.01 |
| #2 | 0.31±0.02 | 0.29±0.02 | 17.9±0.6 | 550±29 | 261±15 | 0.24±0.01 | 0.30±0.01 |
| #3 | 0.34±0.02 | 0.36±0.02 | 155±5 | 55±3 | 33±2 | 0.20±0.01 | 0.26±0.01 |

## 3.2. Adsorption kinetics

The adsorption kinetics for carbon dioxide at 295 K is shown versus square root of time in Fig.2, both, for sample #2 at different relative pressures as well as for all samples at a relative pressure of 0.0017. The plots clearly show two distinct adsorption steps except for sample with the smallest macropores (#3).

Fitting of the curves with a superposition of two adsorption steps yields for the slow adsorption step an almost sample and pressure independent diffusion coefficient.

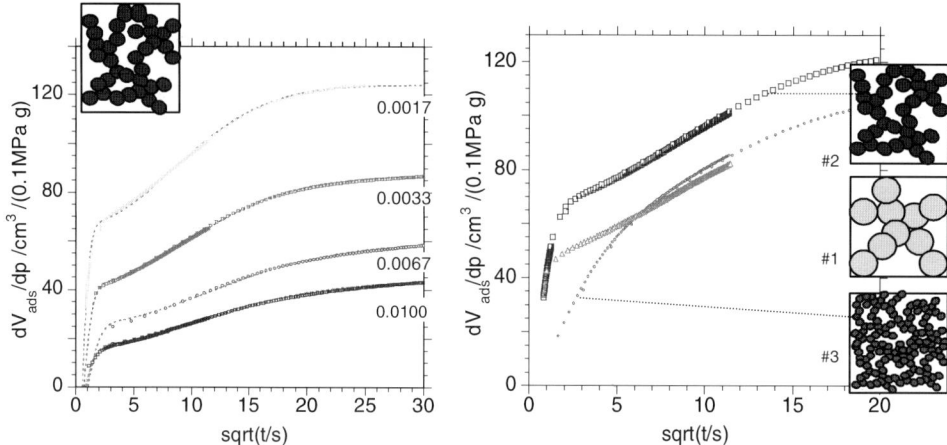

Fig.2. Left: Adsorption kinetics of carbon dioxide at 295 K for sample #2; the dashed lines represent the corresponding fit curves (see text). The parameters given are the relative pressures of each run. Right: Adsorption kinetics of carbon dioxide at 295 K for all three samples at a rel. pressure of 0.0017

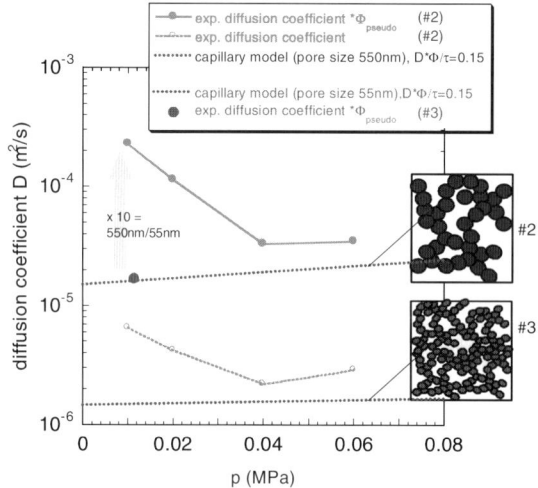

Fig.3. Diffusion coefficients $D$ as derived from a fit of the experimental data in Fig.2 with the solution of the diffusion equation for cylindrical sample (open circles); in addition, values corrected for the amount of gas to be transported ($D*\Phi$pseudo, full circles) are depicted. For comparison the theoretical diffusion coefficients for gas phase diffusion in cylindrical pores are also included (dashed lines); hereby the value of the macroporosity (50%) and a tortuosity factor of 3 are taken into account. The macroporosity was calculated from the bulk density of the sample and the micropore volume (macroporosity=total porosity – microporosity= 86 % - 36%).

The fast adsorption provides the diffusion coefficients plotted in Fig.3. To account for the different amounts of gas to be transported to the sorption site, the experimental diffusion coefficients have been corrected by the so called "Pseudoporosity"

$$\Phi_{Pseudo}(p/p_0) = \Phi + \frac{V_{ads}(p/p_0)}{V_{pore,total}} \qquad (2)$$

that includes both, the actual porosity $\Phi$ of sample and the apparent porosity that corresponds to the amount removed from the gas phase via adsorption. $V_{ads}$ is the volume adsorbed (in STP) and $V_{pore,\,total}$ is the total pore volume of the sample. For comparison, the diffusion coefficient (molecular diffusion plus viscous flow) for capillaries of different sizes are included.

## 4. DISCUSSION

The analysis of the adsorption kinetics of the fast adsorption step (sample #2) yields a pressure dependence that is characteristic for surface diffusion: the diffusion coefficient varies as the derivative of the corresponding sorption isotherm which is governed by micropore adsorption [9].

The "surface" diffusion is therefore related to transport along the micropores rather than the external surface of the particles that form the skeleton of the aerogels. In addition, gas phase transport takes place that can be described by a superposition of molecular diffusion and viscous flow in capillaries with an average diameter of about the pore size in the respective sample and a tortuosity of about 3. The value of 3 was chosen to match the experimental data; it is slightly higher than the value of about 1.8 detected for other carbon aerogels ([10]: 1.8 =1.35$^2$).

Comparing aerogels with different pore macropore sizes (samples #2 and #3), however, a similar microporosity yields diffusion coefficients that differ by about a factor of 10, i.e. the ratio of the respective macropore sizes. I.e. despite the same microporosity and an even larger external surface the surface transport component decreases with the pore size (see also ref. [10]).

## 5. SUMMARY AND CONCLUSIONS

$CO_2$ adsorption at 295K in carbon aerogels pyrolized at 1050 °C takes place via two sorption steps:

The step with the slow adsorption kinetics is essentially independent of the morphology of the samples on the meso- and macropore scale indicating restricted diffusion at the mouth of the micropores.

In contrast, the fast adsorption step reflects the different macropore sizes of the samples under investigation. In particular the diffusion coefficients determined from the experimental data reveal a superposition of "surface" and gas phase diffusion, with the latter corresponding to the theoretically expected value for a bundle of capillary that have about the size of the pores present in the samples investigated.

The "surface" contribution clearly correlates with the isotherm resulting from the adsorption in the micropores. Comparing samples with different macropore sizes, the absolute value of the (total, transient) diffusion coefficients is found to be proportional to the macropore size of the carbons; this indicates a transport that is dominated by molecular diffusion coupled with surface transport at low relative pressures.

## REFERENCES

[1] J. Alcaniz-Monge, C. Blanco, A. Linares-Solano, R. Brydson, B. Rand, *Carbon* **40**, *541-550* (2002).

[2] J. Przepiorski, B. Tryba, A. W. Morawski, *Applied Surface Science* **196**, *296-300 (2002).*

[3] M. Nakashima, S. Shimada, M. Inagaki, A. Centeno, *Carbon* **33**, *1301-1306* (1995).

[4] R.T. Yang, *Series on Chemical Engineering, Vol.1.* : *"Gas Separation by Adsorption Processes"*, *Imperial College Press* (1997), *5 and 269 ff*

[5] G. Reichenauer, to be published in *Adsorption* (2005).

[6] R. Saliger, G. Reichenauer, J. Fricke, Characterization of Porous Solids V, Unger K.K. et al. (Eds.), 381, 2000.

[7] J. Fricke, R. Petricevic, Carbon Aerogels, Handbook of Porous Solids, Eds. F. Schüth, K. Sing, J. Weitkamp, Wiley-VCH, Weinheim, p. 2037–2062, 2002.

[8] G. Reichenauer, H.J. Fella, J. Fricke, *Monitoring Fast Pressure Changes in Gas Transport and Sorption Analysis,* Proceedings of COPS IV, Studies in Surface and Catalysis 144, F. Rodriguez-Reinoso, B. McEnaney, J. Rouquerol and K. Unger (Eds.), Elsevier Science B.V. (2002), p. 443.

[9] K.J. Sladek, E.R. Gilliland, R.F. Baddour, Ind. Eng. Chem. Fundam. 13 (1974) 100.

[10] G. Reichenauer, J. Fricke, Proceed. MRS Symposium, Boston, ed. by J. Drake (1996) 345.

Studies in Surface Science and Catalysis 160
P.L. Llewellyn, F. Rodriquez-Reinoso, J. Rouqerol and N. Seaton (Editors)

# Characterisation of Nanoporous Aluminosilicate Monoliths Derivatised with Metal Cations for Selective Propene-Propane Adsorption

**M. Kargol [a], J. Zajac [b], D. J. Jones [b], J. Rozière [b] and A. B. Jarzębski [a,c]**

[a] Institute of Chemical Engineering, Polish Academy of Sciences, Baltycka 5, 44-100 Gliwice, Poland.

[b] Laboratoire des Agrégats Moléculaires et Matériaux Inorganiques, UMR 5072, Université Montpellier 2, Place E. Bataillon, 34095 Montpellier Cedex 5, France.

[c] Department of Chemical Engineering, Silesian University of Technology, M. Strzody 7, 44-100 Gliwice, Poland.

Monoliths $SiO_2$-$Al_2O_3$ with a desired shape and size have been synthesized via direct liquid crystal templating pathway (DTLC) combined with sol-gel method and subsequently functionalised with metal cations. Effects of metal loading and metal addition method on structural properties of functionalised materials were tested and discussed. The techniques of copper and silver addition used in the present work included the conventional incipient wetness impregnation and ion exchange, as well as the direct incorporation during the synthesis. Additionally, for the functionalised adsorbents, the metal cation effect, $Cu^+$, $Cu^{2+}$ and $Ag^+$, on the propene adsorption capacity was investigated.

## 1. INTRODUCTION

Cryogenic distillation has been used for over 60 years for the recovery and refining of ethylene and propylene from olefin plants, refinery gas streams, and other sources [1]. In addition to this highly energy intensive distillation process, a number of alternative separation technologies, e.g. extractive distillation, physical or chemical adsorption or absorption, and membrane separation have been investigated [2, 3-5]. Especially, adsorption has been explored in the last years to carry out olefin/paraffin separations because the adsorbent can increase the separation factor several times and also diminish cost demands. Yet only a few reports are devoted to separation by selective adsorption onto solid substrates, such as carbon black [6], zeolite 13X [7], activated carbon, silica gel and zeolite A [8, 9] or ordered mesoporous silica [10, 11-12]. The most promising technological approach is separation based on the mechanism of π-complexation between the alkene double bond and transition metal cations, mostly Cu or Ag. CuCl in the form of crystals has been considered for olefin/paraffin separations as early as 1939 [13, 14]. Preferential sorption of the alkene via the formation of π-complexes, resulting from charge donor-acceptor interactions between the appropriate atomic and molecular orbital, was first explained by Dewar [15]. The bonds formed between the metal cations and the olefins are stronger than those formed by Van der Waals forces alone and thus, selective olefin adsorption is expected. Therefore, it is possible to achieve high selectivity and capacity for the component to be boned. Nevertheless, in the vast majority of cases, the forces that bind adsorbates to adsorbents are weaker than those of covalent forces, and this fact allows the adsorption to be reversed by simple engineering means such as rising the temperature of the adsorbent or reducing the concentration or partial pressure of the adsorbate. The early attempts to separate alkene/alkane mixtures based on $\pi$-complexation employed liquid solutions containing silver ($Ag^+$) or copper ($Cu^+$) ions had limited success [2, 4]. Several other studies dealt with gas/solid systems [16, 17-20]. More recently, new promising sorbents based on $\pi$-complexation have been prepared for selective olefin adsorption. These include $Ag^+$-exchanged resins [21], monolayer $CuCl/\gamma\text{-}Al_2O_3$ [16], monolayer CuCl on pillared clays [22, 23], monolayer $AgNO_3/SiO_2$ and $AgNO_3/MCM41$ [19, 20] as well as ordered mesoporous materials SBA-15 and silver-ion deposited SBA-15 [10, 11-12]. The solid adsorbents are attractive in comparison with the other because gas/solid operations can be simpler as well as more efficient, especially with pressure/vacuum swing adsorption processes [23].

The main objective of the present study was to characterize the monolithic Cu and Ag-supported adsorbents having high capacity and selectivity for alkene-alkane adsorption in the form suitable for industrial use. Using direct liquid crystal templating method (DLCT) and commercially available, cheap and non-toxic, non-ionic surfactant as template the monolithic material with a regular pore network, required properties and without defects were synthesized. The monolithic aluminosilicate supports were modified by three different techniques: ion-exchange, incipient wetness impregnation and direct incorporation during synthesis with copper and silver to induce a $\pi$ – complexation effect.

## 2. EXPERIMENTAL

### 2.1 Chemicals for Synthesis

The polyoxyethylenated surfactant, Brij 30, and tetraethyl orthosilicate, TEOS, were obtained from Aldrich, aluminium nitrate, $Al(NO_3)_3 \cdot 9H_2O$, and copper nitrate, $Cu(NO_3)_2 \cdot 3H_2O$, were purchased from Merck. Silver nitrate, $AgNO_3$, and a 0.1 M nitric acid solution, used to adjust the pH of the aqueous phase, were Fisher products. Gaseous ammonia, nitrogen, helium, propane and propene of high purity (99.9999 %) were supplied by Air Liquide.

### 2.2 Synthesis

Typical synthesis route used to prepare the $SiO_2\text{-}Al_2O_3$, designated as SiAl20 via the direct liquid crystal templating (DTLC) pathway in high-concentration surfactant solution and acidic medium was detailed in [24]. The mass ratio of 0.1 M $HNO_3$:Brij 30:TEOS was 1:1:3. Aluminium nitrate was added as the last reagent to a homogenous mixture of Brij30, dilute acid and TEOS. In the case of functionalisation by ion exchange and impregnation methods the calcined crack-free cylinders of *ca.* 2 mm in diameter and 3-5 mm in length with Si:Al molar ratio of 20 were used as supports for varying amounts of copper and silver.

*Direct Incorporation Method* (DI)

The incorporation of Cu or Ag was achieved in the initial synthesis stage through the stirring of the desired amounts of $Cu(NO_3)_2 \cdot 3H_2O$ or $AgNO_3$ salts with TEOS for 5 minutes, and then with dilute $HNO_3$ for about 30 minutes to obtain a transparent solution. Next, Brij30 was added to a homogeneous mixture of TEOS-acid-salt and stirred for 10 minutes, prior to introduction of $Al(NO_3)_3 \cdot 9H_2O$. Further synthesis route leading to monoliths in the form of cylinders was described in previous work [24]. The metal content in a sample was varied between 2 to 20 wt %. Thus, following samples were obtained: $Cu2SiAl_{DI}$, $Cu5SiAl_{DI}$, $Cu10SiAl_{DI}$, $Cu20SiAl_{DI}$, $Ag2SiAl_{DI}$, $Ag5SiAl_{DI}$, $Ag10SiAl_{DI}$ and $Ag20SiAl_{DI}$.

*Ion exchange* (IE)

The dry sample was first stored above ammonium hydroxide for 2 h and then put into contact with 10 mL of 0.5 mol $L^{-1}$ $Cu(NO_3)_2$ or 1 mol $L^{-1}$ $AgNO_3$ aqueous solutions. The ion exchange was carried out at room temperature for 24 h under constant shaking, and then the sample was filtered, rinsed with water at least five times and dried at 373 K. The following samples were prepared with Cu and Ag contents determined by atomic absorption spectroscopy (AAS) 7.5 and 19.6 wt %, respectively, designated as $Cu7SiAl_{IE}$ and $Ag20SiAl_{IE}$.

*Impregnation* (IM)

A solid sample of 0.5 g was first dried overnight at 373 K, and then placed in glass vials, immersed in the previously prepared $Cu(NO_3)_2$ or $AgNO_3$ aqueous solution and sealed tightly. The impregnation stage was carried out at room temperature for 2 days. Afterwards, the impregnated sample was dried at 373 K. The metal content in a sample was varied between 2 to 20 wt %. The following samples were obtained: $Cu2SiAl_{IM}$, $Cu5SiAl_{IM}$, $Cu10SiAl_{IM}$, $Cu20SiAl_{IM}$, $Ag2SiAl_{IM}$, $Ag5SiAl_{IM}$, $Ag10SiAl_{IM}$ and $Ag20SiAl_{IM}$.

### 2.3 Characterization Methods

A PYE UNICAM SP9 spectrometer equipped with a hollow cathode lamp at a wavelength of 324.8 nm for copper and 328.1 nm for silver was used to determine Cu and Ag concentrations. Solid state $^{27}Al$ NMR spectra were recorded at 104.3 MHz using a Bruker ASX400 spectrometer. Cu(I)-functionalised adsorbent was obtained by a partial reduction of a Cu(II)-exchanged precursor, $Cu7SiAl_{IE}$ sample. Obtained monolithic sample was referred to as $Cu7SiAl_{IE+RED}$. The surface reduction of Cu(II) into Cu(I) and then into metallic Cu was investigated by temperature-programmed reduction (TPR) and carried out in the temperature range from 223 to 1073 K (heating rate 2 K $min^{-1}$) under a flow of $H_2$ using a Micromeritics AutoChem 2910.

### 2.4 Adsorption Measurements

The surface area and pore structure parameters, $S_{BET}$, $S_{mes}$, $V_{mes}$ and $d_p$, of monolithic samples were evaluated from nitrogen adsorption-desorption measurements carried out using an automated volumetric Analsorb 9011 apparatus. Adsorption/desorption isotherms were measured at 77 K on samples previously outgassed overnight at 493 K. The number of surface acid sites in modified samples was evaluated by the two-cycle adsorption of ammonia at 373 K using a Micromeritics ASAP 2010 apparatus. The equilibrium pressure was measured after each adsorption step and the amount of $NH_3$ adsorbed was calculated. At the end of the first adsorption cycle, the sample was outgassed under vacuum at 373 K for 30 min and a second adsorption cycle was performed at the same temperature. Prior to adsorption experiments, the samples were outgassed overnight at 423 K. The individual adsorption of hydrocarbons from gas phase at 298 K was studied using the same apparatus and procedures as for ammonia adsorption. Prior to adsorption measurements, a solid sample of *ca.* 150 mg was outgassed overnight at 423 K.

### 3. RESULTS AND DISCUSSIONS

The $SiO_2$-$Al_2O_3$ monoliths were functionalised either by framework substitution, i.e. incorporation of heteroelements during synthesis, called direct incorporation (DI) or by post-synthesis surface modification: incipient wetness impregnation (IM) or ion exchange (IE). The specific surface area, $S_{BET}$, and the pore structure parameters for the derivatised samples are given in Table 1 and Table 2 for the Cu and Ag, respectively. The surface parameters of the initial material SiAl20 without metal were included for comparison. Generally, the derivatised materials have larger pores and lower specific surface areas than the initial SiAl20 sample, most pronounced for ion-exchanged samples. We observed that the incorporation of Cu or Ag by ion-exchange produced uncontrolled changes in the pore structure.

The changes in the porosity are well evidenced in Fig. 1 by the isotherms measured on the ion-exchanged samples. The isotherm for initial aluminosilicate sample SiAl20 is also present for comparison purpose. As is shown, both adsorption and desorption curves for sample without metal have two characteristic features of a type I isotherm, a steep initial portion and a well-defined horizontal plateau at larger relative pressures. In the case of ion-exchanged samples a hysteresis loop can be seen at moderate relative pressures. The appearance of this adsorption phenomenon is seen in the region of greater pore diameters.

In the case of cylindrically shaped aluminosilicate monoliths derivatised with Cu(II) and Ag(I) by impregnation and direct incorporation methods, all isotherms (not shown here) of nitrogen gas adsorption at 77 K are characteristic of materials with small mesopores. On the whole, functionalisation by impregnation reduces the values of $S_{BET}$, $S_{mes}$ and $V_{mes}$ parameters and increases their $d_p$ values; the effects being more pronounced in the case of copper.

Table 1
Specific surface area, $S_{BET}$, mesopore surface area, $S_{mes}$, mesopore volume, $V_{mes}$, mean pore diameter, $d_p$, amount of metal incorporated per unit mass of the derivatised sample and number of acid sites, $n_a$, per unit mass and per unit surface, for the Cu-containing monoliths.

| Sample | $S_{BET}$ $(m^2\,g^{-1})$ | $S_{mes}$ $(m^2\,g^{-1})$ | $V_{mes}$ $(m^2\,g^{-1})$ | $d_p$ (nm) | Cu content $(mmol\,g^{-1})$ | $n_a$ $(mmol\,g^{-1})$ | $n_a$ $(\mu mol\,m^{-2})$ |
|---|---|---|---|---|---|---|---|
| SiAl20 | 944 | 647 | 0.41 | 2.5 | - | 0.44 | 0.48 |
| $Cu7SiAl_{IE}$ | 384 | 347 | 0.25 | 2.9 | 1.1 | 0.65 | 1.69 |
| $Cu2SiAl_{IM}$ | 848 | 595 | 0.38 | 2.6 | 0.31 | 0.42 | 0.49 |
| $Cu5SiAl_{IM}$ | 871 | 564 | 0.41 | 2.9 | 0.79 | 0.81 | 0.93 |
| $Cu10SiAl_{IM}$ | 489 | 277 | 0.21 | 3.0 | 1.57 | 1.18 | 2.41 |
| $Cu20SiAl_{IM}$ | 454 | 260 | 0.18 | 2.7 | 3.15 | 1.31 | 2.89 |
| $Cu2SiAl_{DI}$ | 784 | 536 | 0.30 | 2.2 | 0.31 | 0.45 | 0.57 |
| $Cu5SiAl_{DI}$ | 813 | 547 | 0.31 | 2.3 | 0.79 | 0.59 | 0.73 |
| $Cu10SiAl_{DI}$ | 960 | 819 | 0.44 | 2.2 | 1.57 | 0.63 | 0.66 |
| $Cu20SiAl_{DI}$ | 875 | 670 | 0.46 | 2.7 | 3.15 | 0.63 | 0.72 |

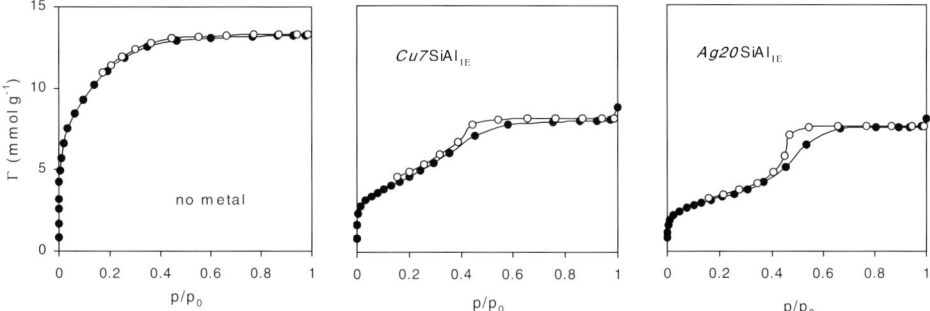

Fig. 1. Adsorption (solid circles)-desorption (open circles) isotherms of nitrogen at 77 K onto calcined and functionalised with copper and silver aluminosilicate monoliths SiAl20.

When the transition metal content does not exceed 5 wt %, small alterations are observed. This may be ascribed to some hydrolytically induced changes in the pore structure. Moreover, no decrease in the pore volume is observed, thus pore blockage can be excluded.

Table 2
Specific surface area, $S_{BET}$, mesopore surface area, $S_{mes}$, mesopore volume, $V_{mes}$, mean pore diameter, $d_p$, amount of metal incorporated per unit mass of the derivatised sample and number of acid sites, $n_a$, per unit mass and per unit surface, for the Ag-containing monolits.

| Sample | $S_{BET}$ $(m^2 g^{-1})$ | $S_{mes}$ $(m^2 g^{-1})$ | $V_{mes}$ $(m^2 g^{-1})$ | $d_p$ (nm) | Cu content (mmol $g^{-1}$) | $n_a$ (mmol $g^{-1}$) | $n_a$ ($\mu$mol $m^{-2}$) |
|---|---|---|---|---|---|---|---|
| SiAl20 | 944 | 647 | 0.41 | 2.5 | - | 0.44 | 0.48 |
| $Ag20$SiAl$_{IE}$ | 258 | 252 | 0.25 | 3.9 | 1.85 | 0.51 | 1.98 |
| $Ag2$SiAl$_{IM}$ | 917 | 678 | 0.41 | 2.4 | 0.18 | 0.34 | 0.37 |
| $Ag5$SiAl$_{IM}$ | 927 | 670 | 0.46 | 2.7 | 0.46 | 0.53 | 0.57 |
| $Ag10$SiAl$_{IM}$ | 603 | 404 | 0.24 | 2.4 | 0.93 | 1.09 | 1.81 |
| $Ag20$SiAl$_{IM}$ | 577 | 355 | 0.23 | 2.6 | 1.85 | 1.24 | 2.15 |
| $Ag2$SiAl$_{DI}$ | 909 | 633 | 0.34 | 2.1 | 0.18 | 0.50 | 0.55 |
| $Ag5$SiAl$_{DI}$ | 890 | 632 | 0.34 | 2.2 | 0.46 | 0.51 | 0.57 |
| $Ag10$SiAl$_{DI}$ | 841 | 584 | 0.32 | 2.2 | 0.93 | 0.52 | 0.62 |
| $Ag20$SiAl$_{DI}$ | 585 | 424 | 0.25 | 2.4 | 1.85 | 0.74 | 1.26 |

Nevertheless, the changes in the porosity are still less pronounced than those induced by ion exchange, especially in the case of silver. For ion-exchanged samples, a significant decrease in the specific adsorbent area and increase in the pore diameters compared either to the initial SiAl20 sample or to the impregnated one and modified by direct incorporation is observed. This is probably due to the blockage of smaller pores, and thus only a part of metal ions are accessible to alkene molecules. Direct incorporation of Ag(I) or Cu(II) nitrate is accompanied by a small increase in the mean pore diameter. This procedure better preserves the porosity of the cylindrically shaped samples compared to ion exchange and incipient wetness impregnation. For samples containing silver, the $S_{BET}$ value decreases with increasing metal content. The specific surface area of $Ag20SiAl_{DI}$ is almost half that of the other samples, whereas the mean pore diameter is only a little greater. This adsorbent also has the smallest pore volume. It seems probable that large part of active silver sites are within the walls, where they form bigger metal clusters. In consequence, pore blockage cannot really be deduced from the observed changes.

For the materials derivatised during the synthesis, the effect of copper or silver addition is illustrated by [27]Al MAS NMR spectra in Fig. 2. The NMR spectra of the calcined samples with largest metal content of 20 wt % shows an intense line, especially for Cu-containing sample, at $\delta \approx 52$ ppm from tetrahedrally coordinated Al. The appearance of a second signal at $\delta \approx 0$ ppm indicates that 'extra-framework' aluminium attributed to octahedrally coordinated Al is only present in sample without metal, SiAl20. Tetrahedrally coordinated Al was of special interest and highly desirable, because it is incorporated into the framework and therefore is responsible for the acid sites formation, and thus enhance the propene adsorption selectivity and capacity [16]. The octahedral aluminium species are occluded in the pores or exist as an amorphous by-product.

The number of surface acid sites was determined with the use of the same procedure for all samples and was detailed elsewhere [24]. Fig. 3 shows typical ammonia adsorption isotherms obtained in two cycles for aluminosilicate sample, SiAl20. The total number of acid sites in the samples, $n_a$, is given above (Table 1 and Table 2). The linear segments of each adsorption isotherm at higher pressures were suitably extrapolated to zero pressure. The difference between the adsorption values corresponding to the two intersection points was taken as the estimate of the total number of acidic sites in the sample [25]. We clearly observe that for both series of samples, impregnated and modified by direct incorporation, the strength of acidic sites increases with increasing metal content. Nevertheless, this dependence appears to be more pronounced for impregnated samples.

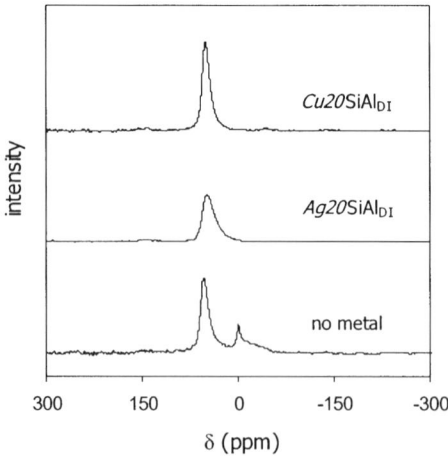

Fig. 2. [27]Al MAS NMR spectra of cylindrically shaped aluminosilicates prepared with a Si:Al ratio of 20 and functionalised with copper and silver by direct incorporation method.

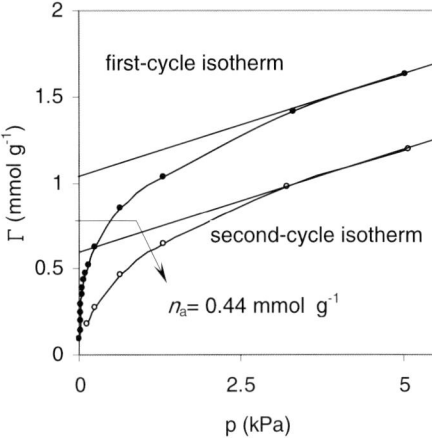

Fig. 3. Two-cycle adsorption of gaseous ammonia onto calcined SiAl20 at 373 K.

The number of surface acidic sites for monoliths functionalised by impregnation method, especially with higher copper or silver amounts (10 and 20 wt %) is almost double compared to the samples modified by direct incorporation method. These results confirm the presence of the active Cu and Ag sites (introduced during the synthesis, DI method) within the walls of resulted materials. Both, ion-exchanged and impregnated samples show a significant rise in the number of surface acid sites. We can thus conclude that the incorporation method of Cu and Ag is of greater importance than the metal content itself and this considerably affects the properties of material. The introduction of Cu or Ag was necessary for selective propene-propane adsorption under equilibrium conditions. In the previous work [24] the individual adsorption isotherms of $C_3H_8$ and $C_3H_6$ with equilibrium selectivity of propene towards propane were presented. The propene adsorption capacities on different adsorbents shown in Fig. 4 are twice as large as those of propane.

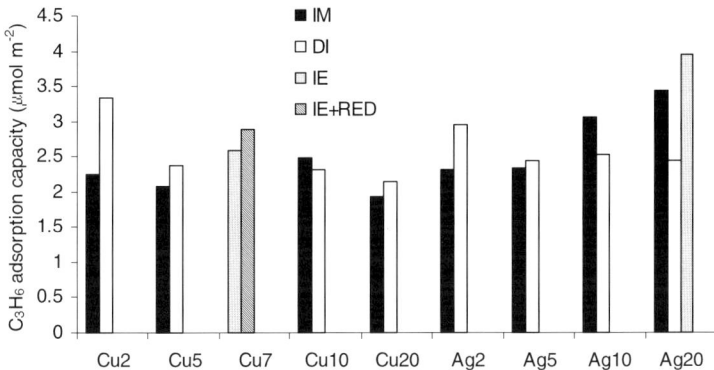

Fig. 4. Propene adsorption capacities at 101.3 kPa and 298 K for samples functionalised by three different techniques.

In the series of Cu-functionalised samples, $Cu7SiAl_{IE+RED}$ has a quite high adsorption capacity of propene comparing to two other methods. For Ag-functionalised materials, the functionalisation by ion-exchange method allowed to attain the highest $C_3H_6$ adsorption capacity. Thus, we can claim that monovalent cationic sites, $Cu^+$ and $Ag^+$, are more appropriate for the selective adsorption of alkene. Nevertheless, direct incorporation during synthesis seems to represent a more suitable technique to obtain selective Cu-containing sorbents for propene.

New effective sorbents were synthesized by dispersing $Cu(NO_3)_2$ and $AgNO_3$ salts over $SiO_2$-$Al_2O_3$ substrates using three different techniques. The attained high capacities make them suitable for propane/propene separation It is clear that functionalisation method play an important role in selective olefin adsorption via $\pi$-complexation.

**Acknowledgments** This work was partially financed by grant Nr 3 T09C 034 26 from Polish Ministry of Scientific Research and Information Technology.

## REFERENCES

[1] G. E. Keller, A. E. Marcinkowsky, S. K. Verma, K. D. Williamson in Separation and Purification Technology, N. N. Li, J. M. Calo, Eds.: Marcel Dekker: New York, 1992.
[2] R. B. Eldridge, Ind. Eng. Chem. Res., 32 (1993) 2208.
[3] T. K. Ghosh, H.-D. Lin, A. L. Hines, Ind. Eng. Chem. Res., 32 (1993) 2390.
[4] D. J. Safarik, R. B. Eldridge, Ind. Eng. Chem. Res., 37 (1998) 2571.
[5] M. Teramoto, N. Takeuchi, T. Maki, H. Matsuyama, Sep. Pur. Tech., 28 (2002) 117.
[6] P. Glanz, B. Körner, G. H. Findenegg, Adsorpt. Sci. Technol., 1 (1984) 41.
[7] C. M. Shue, S. Kulvaranon, M. E. Findlay, A. I. Liapis, Sep. Technol., 1 (1990) 18.
[8] H. Järvelin, J. R. Fair, Ind. Eng. Chem. Res., 32 (1993) 2201.
[9] M. E. Patiño-Iglesias, G. Aguilar-Armenta, A. Jiménez-López, E. Rodríguez-Castellón, Colloids and Surfaces A: Physicochem. Eng. Aspects, 237 (2004) 73.
[10] B. L. Newalkar, N. V Choudary, P. Kumar, S. Komarneni, S. G. T. Bhat, Chem. Mater., 14 (2002) 304.
[11] B. L. Newalkar, N. V. Choudary, U. T. Turaga, R. P. Vijayalakshimi, P. Kumar, S. Komarneni, S. G. T. Bhat, Chem. Mater., 15 (2003) 1474.
[12] C. A. Grande, J. D. P. Araujo, S. Cavanti, N. Firpo, E. Basaldella, A. E. Rodrigues, Langmuir, 20 (2004) 5291.
[13] E. R. Gilliland, J. E. Seebold, J. R. FitzHugh, P. S. Morgan, J. Am. Chem. Soc., 61 (1939) 1960.
[14] E. R. Gilliland, H. L. Bliss, C. E. Kip, J .Am. Chem. Soc., 63 (1941) 2088.
[15] M. J. S. Dewar, Bull. Soc. Chim. Fr., 18 (1951) C71.
[16] R. T. Yang, E. S. Kikkinides, AIChE Journal, 41 (1995) 509.
[17] J. Padin, R. T. Yang, Ind. Eng. Chem. Res., 36 (1997) 4224.
[18] J. Padin, R. T. Yang, C. L. Munson, Ind. Eng. Chem. Res., 38 (1999) 3614.
[19] J. Padin, R. T. Yang, Chem. Eng. Sci., 55 (2000) 2607.
[20] S. U. Rege, J. Padin, R. T . Yang, AIChE Journal, 44 (1998) 799.
[21] Z. Wu, S.-S. Han, S.-H. Cho, J.-N. Kim, K.-T. Chue, R. T. Yang, Ind. Eng. Chem. Res., 36 (1997) 2749.
[22] L. S. Cheng, R. T. Yang, Adsorption, 1 (1995) 61.
[23] N. V. Choudary, P. Kumar, S. G. T. Bhat, S. H. Cho, S. Han, J. N. Kim, Ind. Eng. Chem. Res., 41 (2002) 2728.
[24] M. Kargol, J. Zajac, D. J. Jones, Th. Steriotis, J. Rozière, P. Vitse, Chem. Mater., 16 (2004) 3911.
[25] J. Zajac, R. Dutartre, D. J. Jones, J. Rozière, Thermochim. Acta 379 (2001) 123.

Studies in Surface Science and Catalysis 160
P.L. Llewellyn, F. Rodriquez-Reinoso, J. Rouqerol and N. Seaton (Editors)
© 2007 Elsevier B.V. All rights reserved

# Comparison of nitrogen and carbon dioxide as molecular probes of low surface area carbonaceous materials

**M. Schneemilch**[a], **L. Bellarosa**[a], **K. Nakai**[b] and **N. Quirke**[a],[♦]

[a]Department of Chemistry, Imperial College of Science, Technology and Medicine, South Kensington SW7 2AY, United Kingdom

[b] BEL Japan, Inc. Research & Development
11-27, 2-CHOME, SHIN-KITANO, YODOGAWA-KU, OSAKA 532-0025 Japan

We investigate the pore size distributions obtained from the inversion of the adsorption integrals of nitrogen and carbon dioxide on low surface area ($\sim$20 m$^2$.g$^{-1}$) and medium surface area ($\sim$70 m$^2$.g$^{-1}$) carbonaceous materials. Simple single-site (SS) Lennard-Jones models and multiple-site (MS) models incorporating partial charges have been compared for each probe gas. The multiple-site model produced an improved fit for the nitrogen isotherm compared to the single-site model, while the carbon dioxide isotherm was equally well matched by both models. The pore size distribution obtained from the carbon dioxide isotherm was able to successfully predict the nitrogen adsorption isotherm up to moderate pressures for both materials.

## 1. INTRODUCTION

The ability to easily estimate the pore size distribution of porous materials is important for practical applications such as gas separation, purification and storage. Carbonaceous porous materials constitute a widely used class of adsorbents. The measurement of an adsorption isotherm for carbons using a probe gas such as nitrogen or carbon dioxide is a routine task. For many applications but especially materials selection and design, it is important to be able to convert a measured isotherm into an accurate estimate of the PSD i.e. to invert the adsorption integral

$$V(P) = \int f(w)v(w,P)dw$$

where $V(P)$ is the experimentally determined excess volume of adsorbate at STP per gram of adsorbent, $f(w)$ is the PSD to be determined and $v(w,P)$ is the excess density of adsorbate in a pore of width $w$ at pressure $P$. Since most porous materials contain nonideal pores of a range of sizes it is not possible to determine $v(w,P)$ experimentally.

Traditionally, adsorption onto surfaces is described by 'isotherm' equations that represent the isotherm with a few fitted parameters. These approaches although widely used have provided limited insight or predictive power. By contrast molecular based methods including molecular simulation and density functional methods offer both insight and accurate prediction, albeit for idealised models[1].. Initially, density functional theory (DFT) simulations were widely used [2,3] due to their speed. The trade off was that the adsorbate model was effectively limited to a simple single-site model. Grand canonical Monte Carlo (GCMC) simulations, which allow more sophisticated models to be employed, have become increasingly popular as computer technology has become cheaper and more powerful. Many studies have established quantitative agreement between NLDFT and MC simulations [4].

In what follows our aim is to compare the merits of single-site and multiple-site molecular models for the calculation of pore size distributions for low surface area carbons using DFT and MC methods. Nitrogen and carbon dioxide are widely used as probe gasses and have been represented in the literature [1] as both single-site and multiple-site models. The former is commonly used at 77K so that the bulk coexistence pressure is close to atmospheric pressure. This temperature is very close to the triple point and means that nitrogen is susceptible to freezing, particularly at high pressure and or pore junctions, which may in turn give rise to pore blocking [5]. Carbon dioxide is commonly used at ambient temperatures which is very close to the critical temperature so that freezing is not an issue. Note that while previous studies have used only a limited range of pressures, our experimental measurements and theoretical calculations include the full range up to the bulk coexistence pressure.

| Parameter | nitrogen | carbon dioxide |
|---|---|---|
| $\sigma_{ff}$ (nm) | 0.3572 | 0.3411 |
| $\varepsilon_{ff}/k_B$ (K) | 93.98 | 220.63 |

Table 1. DFT parameters.

| parameter | nitrogen | carbon dioxide |
|---|---|---|
| $\sigma_{ff}$ (nm) | 0.33 | C: 0.275 |
|  |  | O: 0.304 |
| $\varepsilon_{ff}/k_B$ (K) | 36 | C: 28.3 |
|  |  | O: 84.2 |
| $l_x$ (nm) | ±0.0547 | C: 0 |
|  |  | O: ±0.1149 |
| $l_q$ (nm) | ±0.847 | 0 |
|  | ±0.1044 | ±0.1149 |
| $q$ (e) | 0.373 | 0.6512 |
|  | -0.373 | -0.3256 |

Table 2. MC parameters.

## 2. METHODOLOGY

In order to be able to perform effective characterization it is crucial to use appropriate interaction potentials. Parameters for fluid-fluid interaction potentials have been studied extensively and in the present work were taken from the literature. The solid-fluid interaction potential parameters were calibrated separately for each material considered. The calibration procedure consists of the following steps [8]:
1. Simulate the adsorption of gas in graphitic pores to generate a database consisting of the excess adsorbed amount as a function of pressure and pore width using an initial estimate for the fluid-surface interaction.

2. Use this database to find a best fit PSD based on the experimental adsorption isotherm for a given material.
3. Compare the surface area of the fitted PSD to the measured BET surface area of the reference material and adjust the surface fluid interaction.

These steps are repeated until the surface area of the best fit PSD matches the experimental value. The surface area was used as the main criterion since it is widely used as a standard.

Two gases have been modeled - nitrogen at 77 K and carbon dioxide at 298 K- using both GCMC and DFT. All DFT simulations use a single Lennard-Jones (LJ) site model and therefore require only two parameters, the interaction energy $(\varepsilon)$ and length $(\sigma)$. These parameters are determined from the equation of state using experimental values of the coexistence fluid densities and pressure and are listed in **Table 1**. Intermolecular potentials for single-site (DFT) models were cut and shifted at the cutoff distance (1.87 nm). MC simulations used a six site model for nitrogen (2 LJ sites and 4 coulomb sites) and a three site model for carbon dioxide (all dual LJ-coulomb sites). The Lennard-Jones parameters for the nitrogen interaction were taken from the thesis of Delhomelle [6] and the partial charges are those of Kuchta and Etters [7]. The parameters for the multiple-site carbon dioxide were taken from Sweatman and Quirke [8]. All parameters are tabulated in **Table 2**. The molecular potentials for multiple-site models were ramped so that the energy linearly decreased to zero over the last 10 percent of the distance to the cutoff [1]. The coexistence densities and pressure for the multiple-site molecular models were determined using Gibbs ensemble calculations not reported here. The bulk density-pressure dependence for the nitrogen models are shown in Erreur ! Source du renvoi introuvable. with the experimental curve at 77 K. Both the single-site and multiple-site models tend to overestimate the coexistence pressure and density but are otherwise very accurate. Similarly, the bulk pressure density isotherm for carbon dioxide at 298 K is shown in Figure 1. The bulk density-pressure dependence for nitrogen. Solid line: experimental data, squares: DFT equation of state, diamonds: GCMC simulations.
**Figure 2**. The agreement is good at low densities but both models tend to overestimate the pressure as the density increases. The single-site model accurately reproduces the density at coexistence but overestimates the pressure while the multiple-site model accurately represents the pressure but underestimated the density.

The graphitic surfaces were modeled using the Steele potential in the slit pore geometry,

$$u_{sf}(z) = 2\pi\rho_s\Delta\varepsilon_{sf}\sigma_{sf}^2\left[\frac{2}{5}\left(\frac{\sigma_{sf}}{z}\right)^{10} - \left(\frac{\sigma_{sf}}{z}\right)^4 - \frac{\sigma_{sf}^4}{3\Delta(z+0.61\Delta)^3}\right]$$

where $\rho_s = 114$ nm$^{-3}$, $\Delta = 0.335$ nm and $z$ is the distance of the LJ site from the plane of the carbon atoms in the first layer of the surface. The value of $\varepsilon_{sf}$ and $\sigma_{sf}$ were calculated using the Lorentz-Bertholet mixing rules. Databases are constructed over a range of 60 pressures up to the coexistence pressure and over a range of 40 pore widths. The width of the pore is measured between the centres of carbon atoms in the first layer of opposing walls of the pore.

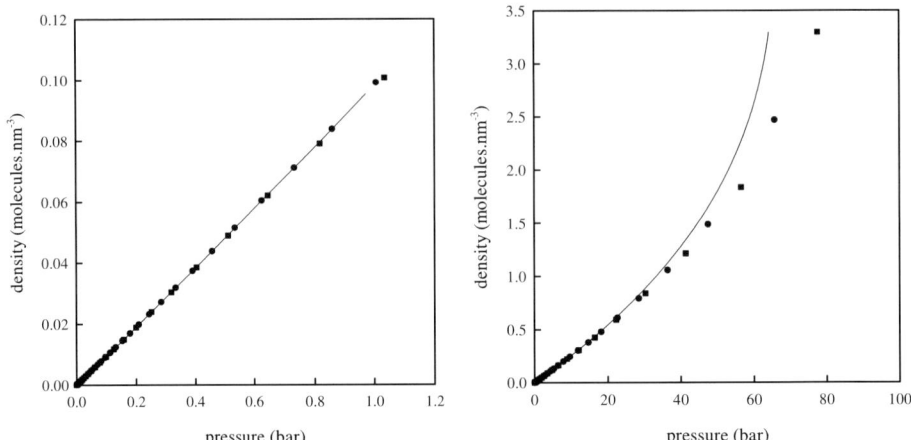

Figure 1. The bulk density-pressure dependence for nitrogen. Solid line: experimental data, squares: DFT equation of state, diamonds: GCMC simulations.
Figure 2. The bulk density-pressure dependence for carbon dioxide. Solid line: experimental data, squares: DFT equation of state, diamonds: GCMC simulations.

Once a database has been constructed it is used to invert the measured isotherm for each of our materials. The procedure is similar to that used in previous work [8,9,11] where $f(w)$ is assumed to be a sum of log-normal functions

$$f(w) = \sum \alpha_i /(\gamma_i w(2\pi)^{1/2}) \exp(-(\ln w - \beta_i)^2 /(2\gamma_i))$$

The free parameters $\alpha_i$, $\beta_i$, $\gamma_i$ were varied using a downhill simplex routine until a satisfactory fit to the experimental isotherm via equation 1 is achieved. Five modes were used and the PSD was evaluated at 100 points distributed exponentially in order to favour the smaller pore width. The objective function was the rmsd of the calculated isotherm from the experimentally measured isotherm

$$F = \sum e^{-P^*W} \sqrt{\left(V(P^*) - A \int f(w)v(w, P^*)dw\right)^2}$$

where the sum is over the range of experimental pressures. The exponential term weighted the objective function to favor the lower pressures. The degree of weighting was determined by the value of the weighting factor, $W$; large values of $W$ resulted in greater weighting on lower pressures. In most cases $W$ was set to zero and the resulting objective function was not weighted.

In previous work [9,11] it has been  common practice to average over a large range of trial PSDs, as the problem was underconstrained and there existed a large number of local minima, so that the final PSD was dependent on the starting point. We have databases with a higher resolution and as a result the objective function appears to be much smoother and the global minimum can be reached from almost any starting point. However, the fitting procedure requires approximately 10 minutes on a 1 GHz processor.

Nitrogen and Argon adsorption isotherm at 77K and 87K were measured by volumetric instrument (BELSORP-18plus, BEL Japan,Inc.) High pressure gas adsorption isotherm up to 50 bar at 298K was measured by gravimetric instrument (MSB-AD-H, BEL Japan, Inc.) which was used Magnetic suspension balance from Rubotherm GMBH. The ambient saturation pressure was continuously recorded during the measurement of the isotherms and used to calculate the experimental relative pressure. The relative pressures of the databases were calculated using the Gibbs ensemble saturation pressure for the multiple-site models and the equation of state saturation pressure for the single site models

Table 3. Calibrated values of the surface interaction strength, $\varepsilon_{ss}$ and the micropore volume

| Adsorbate | $\varepsilon_{SS}$, K | | Volume, cm$^3$.g$^{-1}$ | |
|---|---|---|---|---|
| | Vulcan | CB#51 | Vulcan | CB#51 |
| SS nitrogen | 30 | 22 | 0.15 | 0.044 |
| MS nitrogen | 28.5 | 24 | 0.12 | 0.042 |
| SS carbon dioxide | 25 | 18 | 0.21 | 0.047 |
| MS carbon dioxide | 23 | 14 | 0.15 | 0.032 |

## 3. RESULTS

The results of the calibration procedure are tabulated in Table 3. Note the surface-surface interaction strengths are listed rather than the surface-fluid strengths to allow comparison between single-site and multiple site models.

*Nitrogen*

The nitrogen databases were calibrated on Vulcan by changing the surface fluid interaction to yield a best fit PSD with a surface area of approximately 70 m$^2$.g$^{-1}$. Note that it was assumed that the Vulcan was porous consistent with our previous work [9,11]. For the single-site model it was necessary to weight the fit in favour of the low pressure data as it proved impossible to achieve a best fit PSD with the correct surface area otherwise. The optimal $\varepsilon_{ss}$ value was 30 K at weighting W=4 while for the multiple-site model the unweighted optimal $\varepsilon_{ss}$ was 28.5 K. The fits for both models are shown with the experimental adsorption isotherm in **Figure 3**. The RMS error was 9.9 for the single-site model and 0.9 for the multiple-site model. The PSDs for both molecular models are shown in Figure 4. The single-site PSD comprises a main peak dominated at 4.5 nm and a minor peak at 0.7 nm. The main peak in the multiple-site PSD is shifted to 4 nm with smaller peaks at 3 and 0.8 nm.

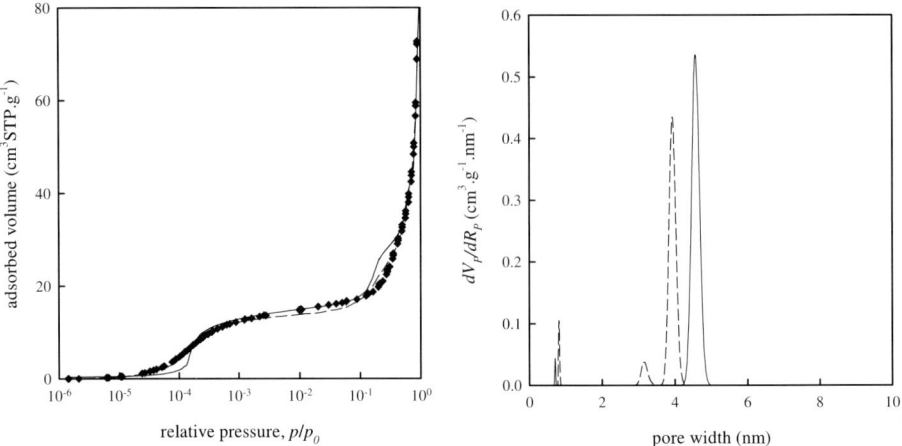

Figure 3. The fits to the experimental isotherm for Vulcan using the single-site nitrogen model (solid line) and the multiple-site nitrogen model (dashed line).
Figure 4. The PSDs obtained for Vulcan using nitrogen databases. Full line: single-site model, dashed line: multiple-site model.

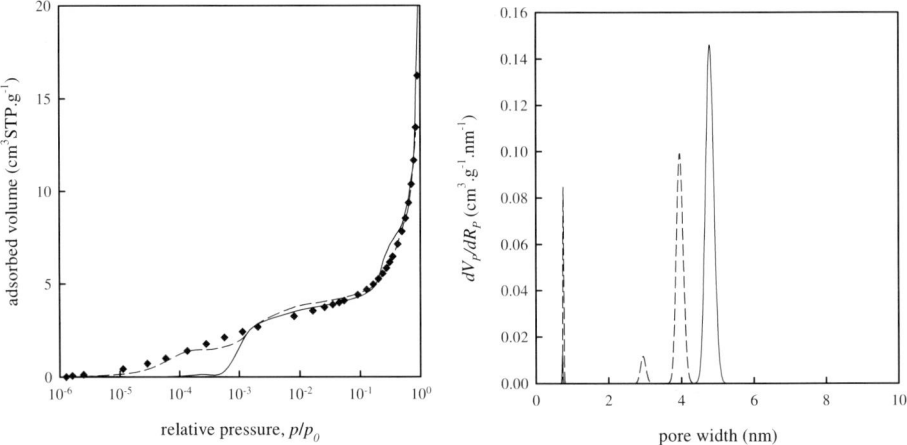

Figure 5. The fits to the experimental isotherm for Carbon Black #51 using the single-site nitrogen model (solid line) and the multiple-site nitrogen model (dashed line).
Figure 6. The PSDs obtained for Carbon Black #51 using nitrogen databases. Full line: single-site model, dashed line: multiple-site model.

The BET surface area of Carbon Black #51 was measured at around 20 $m^2.g^{-1}$. The optimal $\varepsilon_{ss}$ value was 22 K for the single site model while for the multiple-site model the optimal $\varepsilon_{ss}$ was 24 K. No weighting was necessary to yield the appropriate surface area. The fits for both models are shown with the experimental adsorption isotherm in **Figure 5**. The RMS error was 0.7 for the single-site model and 0.2 for the multiple-site model. The PSDs for both molecular models are shown in Figure 6. The single-site PSD comprises a single peak at 5 nm. The multiple-site PSD comprises three peaks at 4 nm, 3 nm and 0.7 nm, respectively.

*Carbon Dioxide*

When carbon dioxide databases were calibrated on Vulcan the surface fluid interaction was optimised to yield a best fit PSD with a surface area of 70 $m^2.g^{-1}$. For single-site model the optimal $\varepsilon_{ss}$ value was 25 K while for the multiple-site model the optimal $\varepsilon_{ss}$ was 19 K. Both databases were able to accurately fit the adsorption isotherms with an RMS error of only 0.12 and 0.016 for the single-site and multiple-site models, respectively. These fits are shown in **Figure 7**. The PSDs for both molecular models are shown in **Figure 8**. The single-site PSD comprises four peaks dominated by a large peak at 8 nm and smaller peaks a 3, 2 and 0.8 nm. The main peak in the multiple-site PSD is shifted to 7.7 nm with smaller peaks around 2 nm and 1 nm.

The optimal values of the surface interaction for Carbon Black #51 were significantly lower than for Vulcan with $\varepsilon_{ss} = 18$ K for the single-site model and only 14 K for the multiple site model. Surprisingly, as shown in **Figure 9**, the single site model actually achieved a slightly better fit to the experimental isotherm than the multiple-site model with errors of 0.08 and 0.1, respectively. The PSD corresponding to these fits are shown in **Figure 10**. Once again the single-site PSD is dominated by a main peak at 8 nm with smaller peaks at 2.3 and 0.8 nm. The main peak in the multiple-site is shifted to 6.5 nm with smaller peaks at 0.7 nm, 1 nm and 1.5 nm.

*Adsorption Prediction*

We took the PSD generated using the single-site carbon dioxide database optimised for Vulcan and attempted to predict the adsorption isotherm for nitrogen on Vulcan using the single-site database we had previously optimised for Vulcan. Even though the PSD was very different from that obtained from the optimisation process, the predicted adsorption isotherm was almost as good a fit as the optimised isotherm. In other words we were able to successfully predict the adsorption isotherm for nitrogen on Vulcan using the PSD obtained from inverting the carbon dioxide adsorption isotherm. We arbitrarily define a prediction as successful if the rmsd between the generated isotherm and the fitted experimental isotherm is less than twice the rmsd for the best fit isotherm and the experimental isotherm at the same weight. Note that the prediction was successful at W=4 but not at W=0.

Similar results were achieved for the Carbon Black isotherms; even though a best fit PSD with the appropriate surface area was identified at W=0 the predicted isotherm matched the experiment only when weighted with W=4. When the process was repeated using the multiple-site databases the normalised error was 5 times the error for the optimised isotherm. *The carbon dioxide PSD appears to be able to predict the nitrogen adsorption isotherm only for single site models.* The reverse process, using the nitrogen PSD to predict the carbon dioxide adsorption isotherm was wholly unsuccessful.

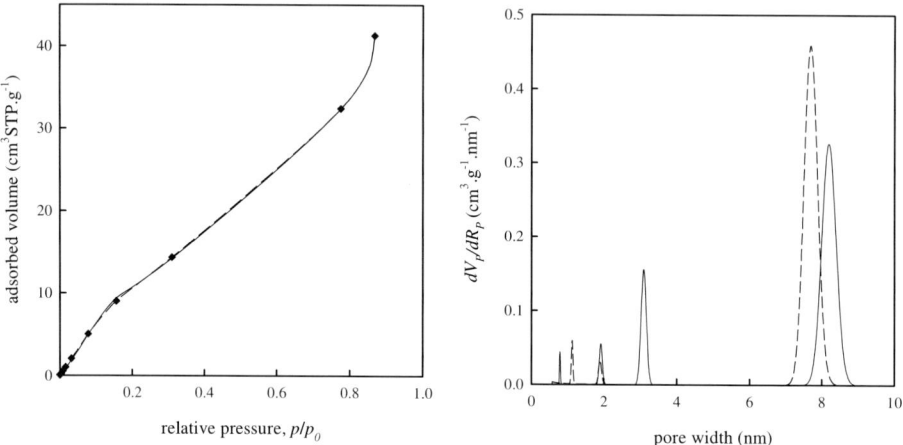

Figure 7. The fits to the experimental isotherm for Vulcan using the single-site carbon dioxide model (solid line) and the multiple-site carbon dioxide model (dashed line).
Figure 8. The PSDs obtained for Vulcan using carbon dioxide databases. Full line: single-site model, dashed line: multiple-site model.

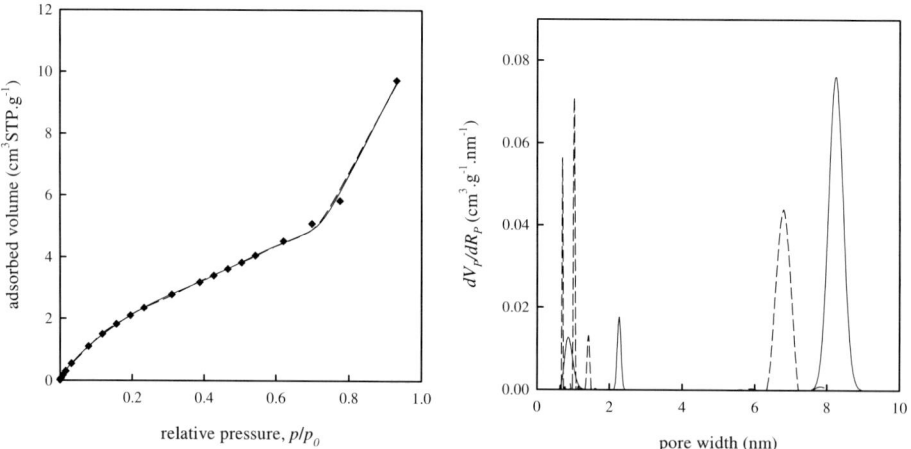

Figure 9. The fits to the experimental isotherm for Carbon Black #51 using the single-site carbon dioxide model (solid line) and the multiple-site carbon dioxide model (dashed line).
Figure 10. The PSDs obtained for Carbon Black #51 using carbon dioxide databases. Full line: single-site model, dashed line: multiple-site model.

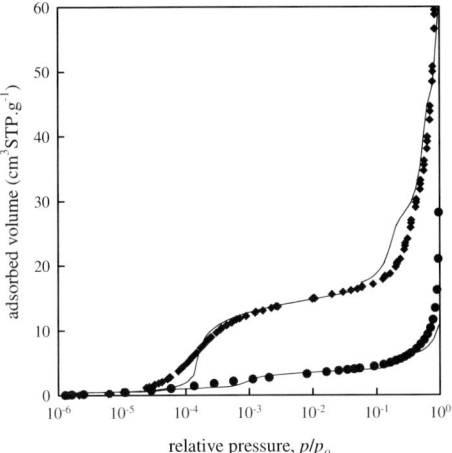

Figure 11. The predicted isotherms (full lines) and experimental isotherms (symbols) of nitrogen on Vulcan (diamonds) and Carbon Black (circles).

## 4. DISCUSSION

We find that the single site model of carbon dioxide does a very good job of reproducing the experimental isotherms on both Vulcan and Carbon Black. Ravikovitch et al[10] also reported good agreement between the NLDFT single site model and the multiple-site GCMC models of carbon dioxide. On the other hand the multiple-site model of nitrogen has a clear advantage over the single-site version on both materials. In particular, the single site model was not able to match the Vulcan surface area without weighting against the high pressure data.

A major difference between the single-site and multiple-site models is that the effective diameter of the single site model is larger than the effective diameter of the multiple-site model when viewed along the axis connecting the centres (see **Table 1** and **Table 2**). As a result, the multiple-site model will adsorb in a smaller pore than the single site model, for a given epsilon value. It is therefore expected that the calibrated epsilon values for the surface-fluid interaction should be greater for the single-site model compared to the multiple site model. The effect is more pronounced for carbon dioxide which has a greater difference between cross section than nitrogen.

The clear shift in the main PSD peak when moving from a single-site to a multiple-site model requires explanation. A similar shift was observed by Sweatman and Quirke[11] in their comparison of PSDs for AX21 generated from a carbon dioxide isotherm measured at 293K using multiple-site GCMC databases and single-site DFT databases. In their case the experimental data were truncated at 1 bar and the PSD was limited to 2 modes; consequently the PSD showed only one peak in each case; centred on 0.71nm for the single-site DFT database and centred on 0.56 nm for the multiple-site model.

It is apparent that the optimised surface interaction is stronger for nitrogen than carbon dioxide for both molecular models and both adsorbents. This may be due to the fact that the coulomb interaction has not been included in the surface fluid potential. The potential used by Zhao and Johnson[12] which includes the multipole terms is inversely dependent on temperature so will have a larger effect on the nitrogen than carbon dioxide databases. Inclusion of the coulomb interaction will undoubtedly reduce the epsilon value for the Steele potential and is therefore likely to close the gap between the nitrogen and carbon dioxide values. This remains to be demonstrated by further work.

The ability of the single site carbon dioxide optimised PSD to predict the low pressure nitrogen adsorption isotherm, but not the reverse, recommends carbon dioxide as a probe gas.

## ACKNOWLEDGEMENT
We thank Martin Sweatman for useful conversations on many aspects of this work

## REFERENCES

* Author to whom correspondence should be addressed.

[1] M. Sweatman and N. Quirke, (2005), 'Modelling gas adsorption in amorphous nanoporous materials', chapter in *Handbook of Theoretical and Computational Nanotechnology'*, Michael Rieth, Wolfram Schommers (Editors), American Scientific Publishers.
[2] N. A. Seaton, J. P. R. B. Walton and N. Quirke, Carbon, 27 (1989) 853.
[3] C. Lastoskie, K. E. Gubbins and N. Quirke, J. Phys. Chem., 97 (1993) 4786.
[4] P. I. Ravikovitch, A. Vishnyakov and A. V. Neimark, Physical Review E, 64 (2001) 011602.
[5] M.W. Maddox, C.M. Lastoskie, K.E. Gubbins and N. Quirke, Simulation Studies of Pore Blocking Hysteresis in Model Porous Carbon Networks, Fundamentals of Adsorption: Proceedings of the Fifth International Conference on Fundamentals of Adsorption, ed. M.D. LeVan, Kluwer Academic Publ., Boston, pp. 571-578 (1996).
[6] J. Delhomelle, Ph.D. Thesis, Université de Paris-Sud U. F. R. Scientifique d'Orsay, 2000.
[7] B. Kuchta and R. D. Etters, Phys. Rev. B, 36 (1987) 3400.
[8] M. B. Sweatman and N. Quirke, J. Phys. Chem., in press.
[9] M. B. Sweatman and N. Quirke, Langmuir, 17 (2001) 5011.
[10] P. I. Ravikovitch, A. Vishnyakov, R. Russo and A. V. Neimark, Langmuir, 16 (2000) 2311.
[11] M. B. Sweatman and N. Quirke, J. Phys. Chem. B, 105 (2001) 1403.
[12] X. Zhao and J. K. Johnson, Mol. Sim., 31 (2005) 1.

Studies in Surface Science and Catalysis 160
P.L. Llewellyn, F. Rodriquez-Reinoso, J. Rouqerol and N. Seaton (Editors)

375

# Water sorption in hydrophobic porous materials: isotherm shapes and their meanings for the mesoporous MCM-41 and the microporous AlPO₄-5.

**N. Floquet[a], J.P. Coulomb[a], G. André[b] and R. Kahn[b]**

[a]Centre de Recherche en Matière Condensée et Nanosciences, CNRS, Campus de Luminy, Case 901, 13288 Marseille Cedex 9 - France.

[b]Laboratoire Léon - Brillouin, CEA - Saclay, 91191 Gif - sur - Yvette, Saclay - France.

We report on extensive neutron diffraction and incoherent quasi-elastic neutron scattering analyses for the water sorption in two hydrophobic porous materials: the mesoporous material MCM-41 and the microporous zeolite AlPO₄-5. Water sorption isotherms have, in the both porous materials, the characteristics of type V isotherms: vertical step at $p/p_0 > 0.3$ and H1 hysteresis loop. Whatever the pore diameter (either mesoporous 20 Å < Ø < 40 Å or microporous Ø =7.3 Å), whatever the pore wall structure (either amorphous $SiO_2$, or crystalline $AlPO_4$), water sorption phenomenon looks like the so-called capillary condensation phase transition. Our neutron scattering results clearly validate such an expected behaviour in the mesoporous confinement range (20 Å < $Ø_{MCM-41}$ < 40 Å). Concerning water confinement in the microporous range ($Ø_{AlPO4-5}$ = 7.3 Å), our results are more surprising. Type V sorption isotherm is the signature of a crystallization phenomenon at room temperature (T = 300 K). The confined water crystallizes in two helices that are commensurate with the AlPO₄-5 micropore structure. The confined ice has a density of 1.2 g.cm$^{-3}$.

## 1. INTRODUCTION

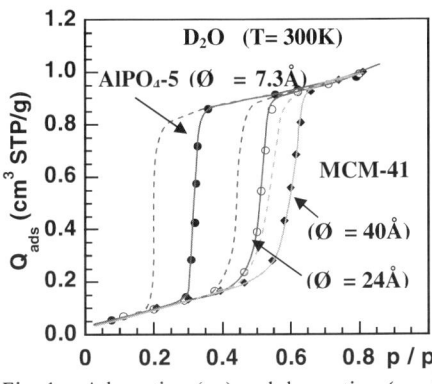

Fig. 1. : Adsorption (—) and desorption (- - -) isotherms of water at T= 300K on MCM-41 (Ø = 40 Å (◆) and 24 Å(O)) and AlPO₄-5 (●). respectively.

Confined water is actively involved in a lot of natural and chemical process. For many years, the sorption of water has been studied in a large variety of hydrophobic porous materials. The more recent studies concern water confined phase in mesoporous materials such as vycor glass, MCM-41 and SBA-15 [1-9] and in microporous materials such as microporous carbons (activated carbon fibers ACFs, superhigh surface area carbons HSACs) [10-12], carbon nanotubes [13-17] and zeolites (VPI-5, AlPO₄-5, wairakite, bikitaite...) [18-27]. The water sorption behaviour due to hydrophobic interaction is generally characterized by a sorption isotherm of type V: sigmoid form, vertical step at $p/p_0 > 0.3$ and H1 hysteresis loop (Fig. 1.). This particular adsorption behaviour is generally known as a "capillary condensation" phenomenon, although the mechanisms of water adsorption and desorption are most often not well characterized and still open for discussion.

The main goal of the present paper is to outline how could be different the structural and dynamic properties of water phases confined in two model hydrophobic porous materials, the mesoporous MCM-41 material and the microporous $AlPO_4$-5 zeolite, although their sorption behaviours are characterized by the same type V of adsorption isotherm.

MCM-41 material is well suited for sorption phenomenon studies, due to its model porosity, composed of an hexagonal structure of cylindrical pores, having diameter in the mesoporous confinement range (20 Å < $Ø_{MCM-41}$ < 100 Å) (Fig. 2.). In addition, our MCM-41 samples (20 Å < $Ø_{MCM-41}$ < 40 Å), have been extensively characterized by microcalorimetric and neutron diffraction measurements of many other adsorbates such as Ar, $N_2$, $O_2$, $C_2$, $H_2$ and $CO_2$ [4, 6, 28-31].

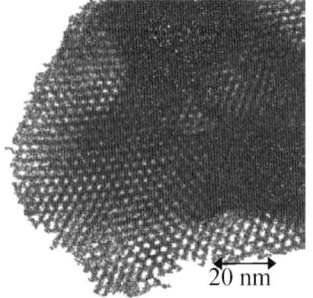 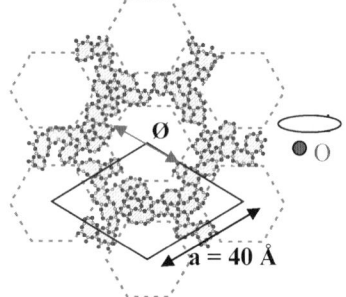

Fig. 2. Transmission electron micrograph of our MCM-41 samples and illustration of the hexagonal network of the MCM-41 mesopores ($a$ = 40 Å) showing the quasi cylindrical mesopores $Ø$ ~ 25 Å and the rough amorphous silica walls

$AlPO_4$-5 zeolite is largely used as a model host material in the microporous range due to its hexagonal network of quasi one dimensional channel ($Ø_{AlPO4-5}$ = 7.3 Å) (Fig. 3.). As for our MCM-41 samples, we have extensively characterized our $AlPO_4$-5 samples [29-31].

 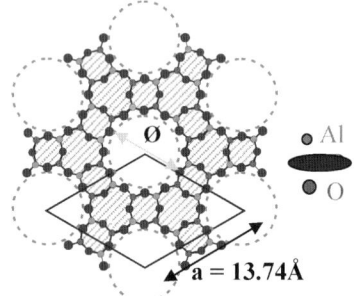

Fig. 3. Scanning electron micrograph of our $AlPO_4$-5 crystals and view along the $c$ micropore direction of the $AlPO_4$-5 hexagonal structure ($a$ = 13.74 Å and $c$= 8.48 Å) showing the quasi one dimensional channels $Ø$= 7.3 Å and the crystalline $AlPO_4$ walls.

MCM-41 material as well $AlPO_4$-5 zeolite have induced a lot of studies in the field of confinement analysis. Only few of them concern water confinement [3-6]. Studies on MCM-41 material are rather focused on the measurement of the depression of the freezing point that is expected proportional to the reciprocal of the pore diameter. Concerning $AlPO_4$-5 crystals, several papers [21-25] have pointed out the unexpected water sorption isotherm, the exceptionally large quantity of confined water compared to other sorbates such as hydrogen (Fig. 4.).

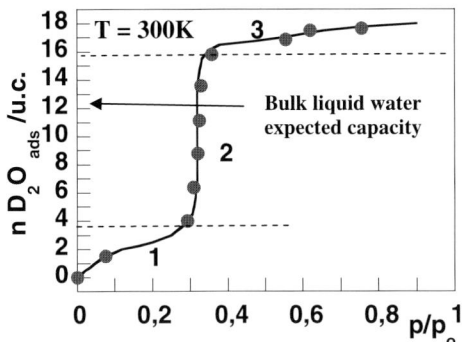

Fig. 4. Calibration sorption isotherm measured at T = 300 K during the neutron diffraction experiment concerning the $D_2O$ / $AlPO_4$-5 system. Full circle (●) represents the water loading ($Q_{ads.}$) corresponding to the recorded diffraction patterns (Fig. 12). The three adsorption stages are noted 1, 2 and 3. Around 11.7 $D_2O$ / u.c. of bulk liquid water are expected to full fill the $AlPO_4$-5 main channels (Ø = 7.3 Å)

To better understand such particular features of the confined water in porous hydrophobic materials, we performed neutron scattering measurements during the adsorption process. Our results point out definitive differences of the properties between confined water in MCM-41, ultra-confined water in $AlPO_4$-5 and bulk water.

## 2. EXPERIMENTAL SECTION

The $AlPO_4$-5 zeolite sample and the MCM-41 (Ø = 19 Å and 24 Å) sample used in the present experiments were synthesized at the Laboratoire des Matériaux Minéraux (Mulhouse – France). Concerning the MCM-41 (Ø = 25 Å) and (Ø = 40 Å) samples, they were prepared at the Laboratory of Inorganic Chemistry (Mainz – RFA). Prior to any experiments, the samples are outgassed under vacuum (P ≤ $10^{-6}$ Torr) over a night at T > 250 °C. Neutron scattering experiments were performed at the Laboratoire Léon – Brillouin (C.E.N. – Saclay). For neutron diffraction (ND) and incoherent quasi-elastic neutron scattering (IQNS) measurements, the two axis diffractometer G4-1 and the time of flight MIBEMOL were used respectively. Both adsorption and desorption isotherms with $D_2O$ (neutron diffraction experiments) or with $H_2O$ (incoherent quasi-elastic neutron scattering measurements) were performed in the neutron scattering cryostat.

## 3. RESULTS

### 3.1. Water in MCM-41 (Ø = 19 Å to 40 Å)
Adsorption stages of the confined water in MCM-41 have been structurally characterized by "in situ" neutron diffraction. As shown in Fig. 5., the hexagonal network of MCM-41 produces diffraction peaks at low diffraction angle (Q < 0.5 $Å^{-1}$). One intense peak (100) and two weak peaks (110) and (200) are generally well observed. The MCM-41 walls made of the amorphous silica give a very large bump near Q = 1.6 $Å^{-1}$. And the confined water diffracts in this range between Q = 1 to 3 $Å^{-1}$. Intensity modifications of the MCM-41 diffraction peaks tell about the location of the confined water within the MCM-41 mesopores [28]. It allows to identify the two adsorption stages observed in the isotherms. Intensity modifications are well observed for the first strong (100) peak (Fig. 6.). Its intensity increases during the first stage of adsorption ($Q_{ads.}$ < 15% for MCM-41 (Ø = 24 Å)), when some water molecules physisorb on the rough MCM-41 walls. Then, the intensity is decreasing during the so called capillary stage, when the water fills the entire mesopore.

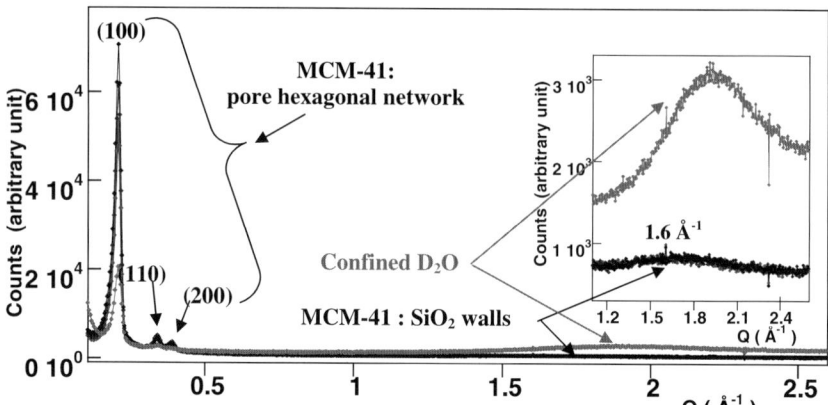

Fig. 5 Neutron diffractograms of MCM-41 (Ø=24 Å) measured at T=273K : empty (black plot) and for a D$_2$O loading of 95% (red plot).

Diffraction of the capillary water phase appears at T = 300 K as a broad peak (Fig. 7.). Its position is around Q = 1.8 Å$^{-1}$. In the case of MCM-41 (Ø = 24 Å), its width corresponds to a coherence length of around L$_{coher.}$ = 8 Å. It means that the capillary water phase is characterized by a short range order as expected for a liquid phase. Decreasing the temperature above T = 300K, the bump maximum remains at the same position Q = 1.8 Å$^{-1}$ but its width presents a sudden decrease between T = 230K and T = 220K ( for MCM-41 (Ø = 24 Å)). This step is well evidenced by plotting the coherence length as a function of temperature (Fig. 8). Thus, for MCM-41 (Ø = 40 Å), in the higher temperature range, the coherence length is around L$_{coher.}$ = 8 Å. At lower temperatures, the coherence length increases steeply to L$_{coher.}$ = 28 Å. This sharp increase at T$_{3t}$ = 234K corresponds to the solidification of the confined liquid water in MCM-41 (Ø = 40 Å). This transition was also observed by calorimetry measurements and by quasi elastic neutron scattering experiments [4, 6].

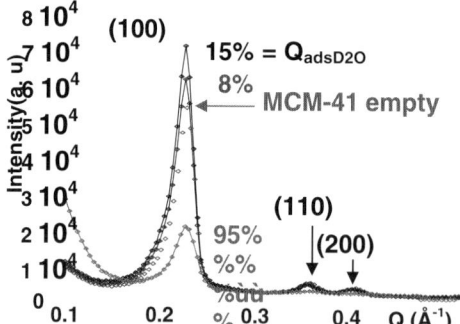

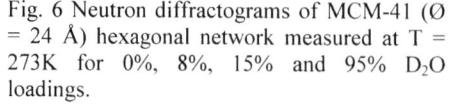

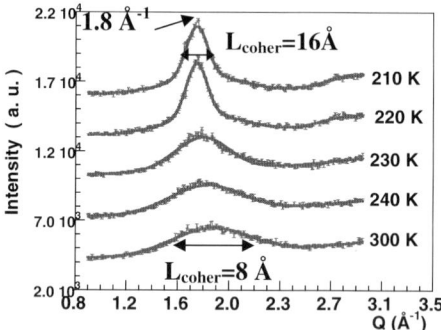

Fig. 6 Neutron diffractograms of MCM-41 (Ø = 24 Å) hexagonal network measured at T = 273K for 0%, 8%, 15% and 95% D$_2$O loadings.

Fig. 7 Neutron diffractograms of confined D$_2$O (95% loading) in MCM-41 (Ø=24 Å) measured at increasing T.

In our IQNS experiments, the spectra were measured at a water loading of Q$_{ads.}$ = 95 % and for increasing diffraction angles. A broadening of these peaks is well observed. The translational mobility calculated from these broadenings is around D$_t$ = 5 10$^{-6}$ cm$^2$s$^{-1}$ at T =

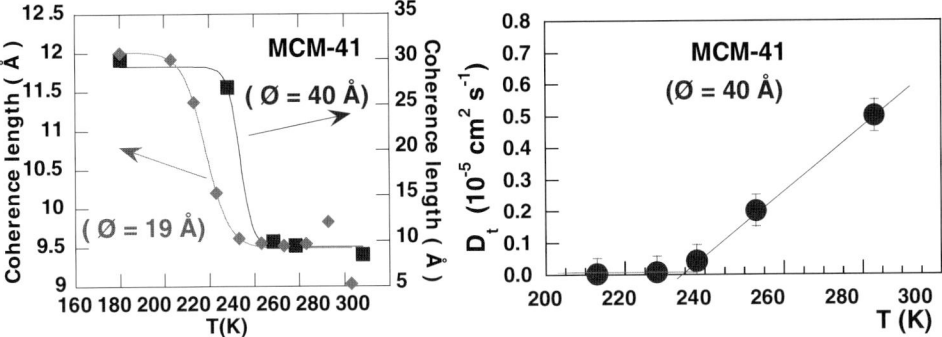

Fig. 8 : Coherence length of confined $D_2O$ (95% loading) in MCM-41 (Ø = 19 Å and 40 Å) as a function of T.

Fig. 9 : Translational mobility of confined $H_2O$ (95% loading) in MCM-41 (Ø = 40 Å) as a function of T.

283 K, then it decreases linearly down to the melting/freezing temperature of T = 234 K, in agreement with the diffraction results (Fig. 9.). These IQNS data confirm the strong triple point depression of the confined water in MCM-41.

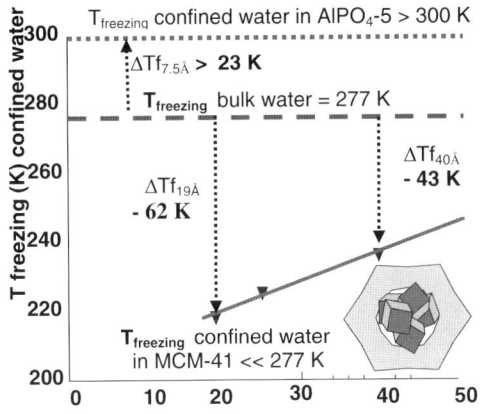

Fig. 10. Freezing temperature $T_{3t}$ of $D_2O$ (95% loading) confined in MCM-41 (Ø = 19 Å to 40 Å) and in AlPO$_4$-5 (Ø = 7.5 Å). Illustration of the nanocrystallites of cubic ice structure grown in MCM-41 below $T_{3t}$.

The water triple point position shifts with the pore size (Fig. 10). It is of $\Delta Tf_{40Å}$ = - 43 K for 40 Å size and it increases up to $\Delta Tf_{19Å}$ = - 62K for the 19 Å size. In addition, this capillary "solid" water phase is found to be tiny cubic single crystals. Note that external bulk water would crystallize in its hexagonal form. This result comes from the analysis of the diffraction pattern obtained at T = 173K. In the diffraction range Q = 1 to 3 Å$^{-1}$, there are two distinct diffraction peaks, whose positions characterizes the (111) and (220) peaks of the cubic ice structure. Their wide shape indicates a lack of crystallinity and a nanometric size. Mixtures of (3d) hexagonal and (3d) cubic ices have been already observed in Vycor (40 – 70 Å) [1].

### 3.2. Water in AlPO₄-5 (Ø = 7.3 Å)

From our IQNS experiments realized with $H_2O$ (Fig. 11), no molecular translational mobility is observed even at T = 300K [32]. There is no quasi elastic broadening of the signal at the different water loadings. The IQNS signal is a pure elastic line similar to the instrumental resolution. Thus the melting temperature of the confined water in AlPO₄-5 nanopore is displaced towards the high temperature side, at the opposite side of the water confined in MCM-41 mesopore (Fig. 10). Even when water is confined in the nanoporous carbon materials [10-12], the rule of the depression of the freezing point of water in a small space was confirmed. Such outstanding phenomenon is expected to be due to a

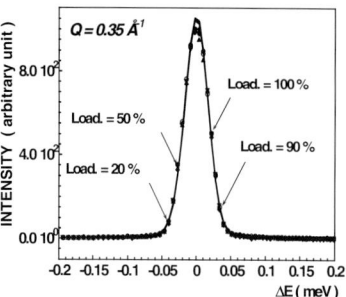

Fig. 11. Measured incoherent quasi-elastic neutron scattering spectra of $H_2O$ confined phase in $AlPO_4$-5 zeolite at T = 300K. The different water loading (20 %, 50 %, 90 % and 100 %) are deduced from the calibration isotherm. Intensity of the different spectra are normalised to an arbitrary value $10^3$.

commensurability effect between water and $AlPO_4$-5 structure, as it was previously observed in the case of methane confined in $AlPO_4$-5 [30].

Adsorption stages of the nanoconfined water in $AlPO_4$-5 have been structurally characterized from neutron diffraction experiments realized with heavy water (Fig.12). The hexagonal network of $AlPO_4$-5 crystals gives diffraction peaks at well known positions. Adsorption of water produces intensity modifications well seen for the first strong (100) peak. But the fine analysis shows that intensity of each peak is modified. In Fig. 13, we have plotted the relative intensity for the three peaks (100), (210) and (220) as a function of the water loading. It clearly appears that the three slope changes observed are correlated to the three parts of the adsorption isotherm (Fig. 4.). Each slope change indicates a change of the $AlPO_4$-5 filling mode by water molecules. To characterize the water location within the $AlPO_4$-5 nanopores at these three stages, we have proceeded to a Rietveld analysis of each recorded diffraction pattern [32]. Numerous experimental and theoretical structures of bulk and confined ice were tested. Finally, only one acceptable fit of each diffraction pattern was obtained for water positions initially chosen from a comparison between the hexagonal ice structure and the $AlPO_4$-5 framework.

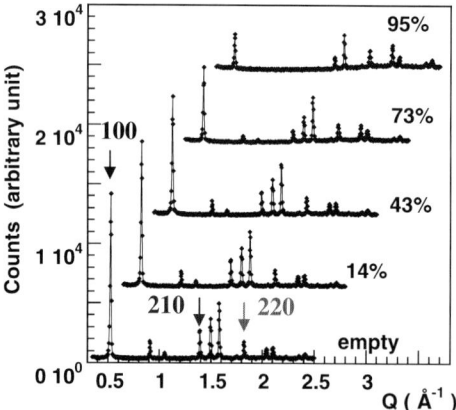

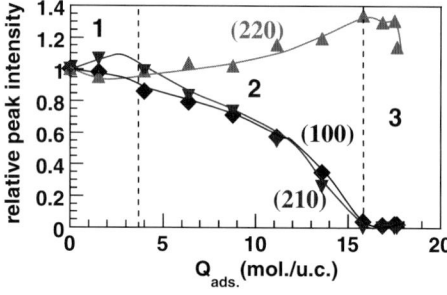

Fig. 12. Neutron diffractograms of $AlPO_4$-5 (Ø = 7.3 Å) at T = 300 K for increasing $D_2O$ loading : 0%, 14%, 43%, 73% and 95%. Diffraction peaks (100), 210) and (220) are shown.

Fig. 13. Relative intensity of $AlPO_4$-5 diffraction peaks (100), (210) and (220) as a function of $D_2O$ loading (T = 300 K ). The slope changes 1, 2 and 3 correspond to the three steps of the adsorption isotherm (see Fig. 4).

These three adsorption stages are illustrated in the Fig. 14. The unexpected uptake step observed in the sorption isotherm corresponds to the growth of two helices of ice. The high density (d =1.2 g.cm$^{-3}$) and the stability of the confined ice is driven by structural commensurability with the AlPO$_4$-5 hexagonal channel structure. Thus the crystallization phenomenon occuring at T = 300K would explain the exceptionally large quantity of water adsorbed in AlPO$_4$-5.

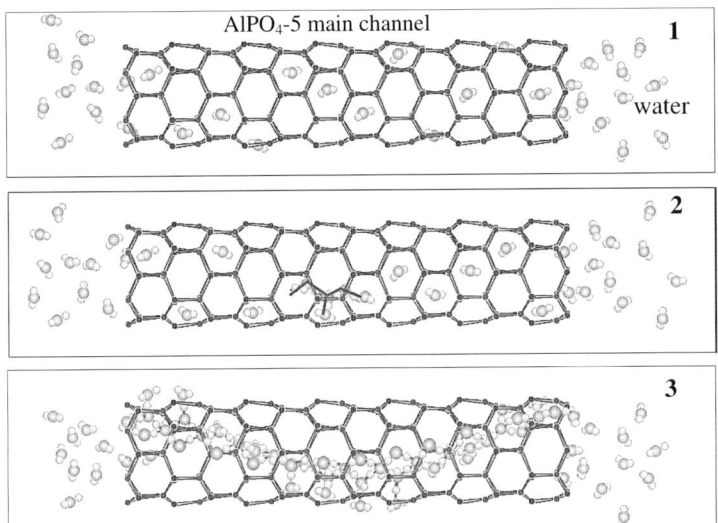

Fig. 12. Illustration of the three adsorption stages of water in AlPO$_4$-5 main channel : (1) At low relative pressure, up to 4 molecules per unit cell, water adsorbs on the AlPO$_4$-5 wall defects.(2) at the relative pressure of the step, water full fills AlPO$_4$-5 by forming two helices of ice that are commensurate with the AlPO$_4$-5 structure. (3) On the plateau, the confined ice reaches a density of 1.2. Only one helice is drawn in the illustration.

## 4. CONCLUSIONS

Our neutron experiments realized during the adsorption is proved to be a very successful way in the knowledge of the adsorption phenomena, and especially of the water adsorption phenomenon. Our choice for the two model porous materials MCM-41 and AlPO$_4$-5 appears to have improved our knowledge on the confinement effects.

In the mesoporous range for MCM-41, each steps of the water adsorption mechanism are well identified by neutron diffraction: the physisorption of the first molecules on the rough MCM-41 wall, then the whole filling of the MCM-41 mesopore by water molecules. The usual molecular water organisation is measured (confined liquid phase, confined nanocrystallites of cubic ice). The confinement effect, the displacement of the whole water phase diagram towards the low temperature side is quantified :decreasing the MCM-41 pore size increases the temperature displacement.

In the microporous range, the water adsorption mechanism is host material highly dependent: in addition to the usual confinement effect, a large influence of the host material walls will act. Each microporous host material has its own specificity. The water adsorption mechanism in carbon slits [12] or nanotubes [17] is different from the one observed in AlPO$_4$-5. Our finding is a new molecular water organisation characterized by a high density (around 1.2 g.cm$^{-3}$) and a double helix structure which is stable at room temperature T = 300 K. Our neutron diffraction results proved the crucial function of the AlPO$_4$-5 wall surface in the adsorption of water. Its structure and chemical nature help the confined water molecules to stabilize in a organized molecular arrangement. Such commensurability effect was first observed in the case of methane [30]. The determination of its temperature range stability is in progress. A simulation study is underway to confirm and refine our findings.

## REFERENCES

[1] J. D. Dore,. B. Weber, M. Hartl, P. Behrens and T. Hansen, Physica A 314 (2002) 501.
[2] P.L. Llewellyn, F. Schüth, Y. Grillet, F. Rouquerol, J. Rouquerol, and K.K. Unger, Langmuir 11 (1995) 574.
[3] K. Morishige and K. J. Nobuoka, Chem. Phys. 107 (1997) 6965
[4] N. Floquet, J.P. Coulomb, C. Martin, Y. Grillet, P.L. Llewellyn and G. André, Proceedings of the 12th Int. Zeolite onference, Ed. M.J. Treacy et al., Material Research Society (1999) 659
[5] K.Morishige and K. J. Kawano, Chem. Phys., 110 (1999) 4867
[6] J.P. Coulomb, N. Floquet, Y. Grillet, P.L. Llewellyn, R. Kahn and G. André, Studies in Surface Science and Catalysis, 128 (2000) 235.
[7] K. Morishige and H. Iwasaki, Langmuir, 19 7 (2003) 2809
[8] S. Takahara, M. Nakano, S. Kittaka, Y. Kuroda, T. Mori, H. Hamano and T. Yamaguchi, J. Phys. Chem. B, 103 (1999) 5814
[9] J. S. Oh, W. G.Shim, J. W. Lee Kim, H. Moon and G. Seo, Journal of Chemical and Engineering Data, J. Chem. Eng. Data, 48 (2003) 1458.
[10] T. Iiyama, K Nishikawa, T Suzuki and K Kaneko, Chem. Phys. Lett. 274 (1997) 152
[11] J. Pires, M L Pinto, A. Carvalho and M. B. de Carvalho, Adsorption 9 (2003) 303
[12] T. Kimura, H. Kanoh, T. Kanda, T. Ohkubo, Y Hattori, Y. Higaonna, R. Denoyel and K Kaneko, J. Phys. Chem. B 108 (2004) 14043
[13] G. Hummer, J. C.Rasaiah and J. P. Noworyta, Nature, 414 (2001) 188
[14] K. Koga, G. T. Gao, H Tanaka and X. C. Zeng, Physica A, 314 (2002) 462
[15] Y. Maniwa, H. Kataura, M. Abe, S. Suzuki, Y. Achiba, H. Kira and K. J. Matsuda, Phys. Soc. Jpn., 71 12 (2002) 2863
[16] A. P. Kolesnokov, J. M. Zanotti, C-K. Loong and P. Thiyagarajan, Phys. Rev. Lett. 93 3 (2004)35503
[17] Y. Maniwa, H. Kataura, M. Abe, A Udaka, S. Suzuki, Y. Achiba, H. Kira, K. Matsuda, H. Kadowaki and Y. Okabe, Chem. Phys. Lett. 401 (2005) 534
[18] L. B. McCusker, Ch. Baerlocher, E. Jahn and M. Bülow, Zeolites, 11 (1991) 308
[19] E. Fois, A. Gamba and A. Tilocca, J. Phys. Chem. B, 106 (2002) 4806
[20] P. P. Knops-Gerrits, H. Toufar, X.Y. Li, P. Grobet, R. A.Schoonheydt, P. A. Jacobs and W. A. Goddard, J. Phys. Chem. A, l04 (2000) 2410
[21] M. E.Davis, C.Montes, P. E. Hathaway, J. P Arhancet,. D. L Hasta,. and J. M Garces, J. Am. Chem. Soc., 111 (1989) 3919
[22] S. G.Izmailova, E. A. Vasiljeva, I. V. Karetina, N. N.Feoktistova, and S. S. J. Khvoshchev, Colloid Interface Sci., 179 (1996) 374
[23] P. B. Malla, and S. Komarneni, Zeolite, 15 (1995) 324
[24] K. Tsutsumi, K. Mizoe, and K Chubachi,. Colloid Polym. Sci., 277 (1999) 83
[25] B. L. Newalkar, R. V. Jasra, V. Kamath, and S. G. T. Bhat, Micropor. Mesopor. Mater., 20 (1998) 129
[26] C. M. B. Line and G. J. J. Kearley, Chem. Phys., 112 20 (2000) 9058
[27] E. Fois, A. Gamba, G. Tabacchi, S. Quartieri and G. J. Vezzalini, Phys. Chem. B, 105 (2001) 3012
[28] N. Floquet, J.P. Coulomb and G. Andre, Microp. Mesopor. Mater. 72 (2004) 143
[29] C. Martin, J.P. Coulomb and M. Ferrand, Europhys. Lett. 36 7 (1996) 503
[30] C. Martin, N. Tosi-Pellenq, J. Patarin and J.P. Coulomb, Langmuir 14 (1998) 1774
[31] J.P. Coulomb, N. Floquet, C. Martin, R. Kahn, Eur. Phys. J. E 12 (2003) 25
[32] N. Floquet, J.P. Coulomb, N. Dufau and G. Andre, J. Phys. Chem. B 108 (2004) 13107

Studies in Surface Science and Catalysis 160
P.L. Llewellyn, F. Rodriquez-Reinoso, J. Rouqerol and N. Seaton (Editors)

# Uncertainty in $\alpha_S$ analyses and pore volumes propagated from uncertainty in gas adsorption data

A. Badalyan, P. Pendleton

Center for Molecular and Materials Sciences
University of South Australia, Mawson Lakes, South Australia, 5095, Australia

The effect of experimental uncertainties in the amount adsorbed on uncertainties in $\alpha_S$-analysis and pore volumes was studied for microporous FM1/250 activated carbon cloth (ACC) during low pressure adsorption experiments at 77 K. Non-porous materials with different surface chemistry, such as carbon black and Aerosil 200, were used as standards for the development of $\alpha_S$-plots. A weighted mean least squares method was applied to $\alpha_S$-data for the evaluation of primary and total micropore volumes and their uncertainties. Primary micropore volumes for FM1/250 ACC calculated using $\alpha_S$-data via these non-porous materials only agree within 14.12% with relative combined standard uncertainties (RCSU) equal to 4.15 and 7.46%, respectively. In contrast, total micropore volumes agree within 0.39%, falling within RCSU of 1.68 and 2.29% respectively. The surface chemistry of the non-porous material does not affect the calculated total micropore volume values. The primary micropore volume values significantly depend on the similarity of surface chemistry of the standard materials and the material under investigation.

## 1. INTRODUCTION

Manometric gas adsorption is the most frequently applied method to determine BET specific surface areas (BET-SSA), specific pore volumes and pore size distributions of adsorbents. The analysis of uncertainty in the measured data and their propagation to the combined standard uncertainty in BET-SSA is well established [1]. An evaluated uncertainty in BET-SSA and pore volume has important contributions to testing adsorption theory models and appropriate industrial process economics. The role of pore volume becomes important when investigating gas adsorption at high-pressures and at supercritical conditions, since this parameter is required to determine the absolute amount adsorbed compared with the surface excess amount adsorbed.

In 1968, Sing introduced the $\alpha_S$-analysis comparison plot methodology for specific surface area determination [2]; the method found a wide application to identify the presence of porosity and evaluate (micro)pore volumes in test adsorbents. No detailed uncertainty analyses exist for pore volumes in porous materials; no internationally recognized standard porous materials exist. The comparison of the amount adsorbed by standard and test adsorbents leads to a complex interplay of the combined standard uncertainty ( $u_c$ ) in the amount adsorbed and in the clamped cubic spline functions employed to interpolate common relative pressures and their dependent amounts adsorbed.

Both Sing [3, 4] and Kaneko [5, 6] have published data which are accepted as standard reference data, however their CSU are unpublished. Uncertainty in micropore and/or pore volumes derived from comparison plots analyses are developed from CSU in the primary standard data and from the CSU in the test adsorption data. A second important consideration for an $\alpha_S$ -analysis is an appropriate selection of a reference standard whose surface chemistry simulates that of the test adsorbent. Thus, a comparison plot uncertainty analysis needs to differentiate between the contributions of experimental uncertainty and adsorbent chemistry differences.

To establish the criteria for total pore and/or micropore volume uncertainty evaluation, we develop expressions for uncertainty propagation and demonstrate the influence of standard surface chemistry on the pore volumes identified in an activated carbon cloth.

## 2. EXPERIMENTAL

Single sheet, plain-weaved activated carbon cloth FM1/250 (ex. Calgon Carbon, Pittsburgh, PA, USA), non-graphitized carbon black (ex. Micromeritics, Athens, GA, USA), and Aerosil 200 (ex. Degussa AG, Düsseldorf, GERMANY) were used for nitrogen gas adsorption at 77 K using an automated manometric gas adsorption apparatus described elsewhere [1, 7]. Physically adsorbed moisture and other gases' removal occurred by exposing these samples to 200 °C heat treatment and a background vacuum of 0.1 mPa for 8 hours. Thermal transpiration corrections were applied to adsorption data at pressures below 266 Pa ($\approx$ 2 mmHg) [8]. Adsorbent samples were contained in a glass holder, thus no thermal gradient corrections were necessary [9].

## 3. UNCERTAINTY IN $\alpha_S$-DATA AND PORE VOLUMES

Fig. 1 shows the nitrogen adsorption isotherms for FM1/250 and the two reference adsorbents. A weighted, mean least-squares analysis of the data in the ranges $0.04P^0 \leq P \leq 0.25\ P^0$ rendered BET-SSA; combined standard uncertainty, $u_c(area)$ are summarized in Table 1. Although the FM1/250-nitrogen isotherm is classically Type I, and thus (principally) contains micropores, we evaluated the BET-SSA for comparison purposes. Relative combined standard uncertainty (RCSU) for each data point in the pressure range $8 \times 10^{-7}P^0 \leq P \leq 0.99P^0$, in Table 1, reflect the cumulative nature of the experimental uncertainties in the amount of nitrogen adsorbed. The $u_c(V_{ads})$ for each material, which contribute to the $u_c(\alpha_S)$ and $u_c(micropore\ volume)$, are shown in Fig. 1.

Table 1
Summary of adsorbent physical properties and their uncertainty

| Adsorbent | BET-SSA, m²/g | $u_c$(area), m²/g | % RCSU ($V_{ads}^{STP}$) |
|---|---|---|---|
| Carbon black | 109.42 | 0.72 | 0.247 – 1.006[*] |
| Aerosil 200 | 207.80 | 2.27 | 0.484 – 1.794[†] |
| FM1/250 | 1033.9 | 10.9 | 0.371 – 1.809[•] |

[*] over the pressure range: $8 \times 10^{-6}P^0 \leq P \leq 0.97P^0$
[†] over the pressure range: $1 \times 10^{-5}P^0 \leq P \leq 0.99P^0$
[•] over the pressure range: $8 \times 10^{-7}P^0 \leq P \leq 0.96P^0$

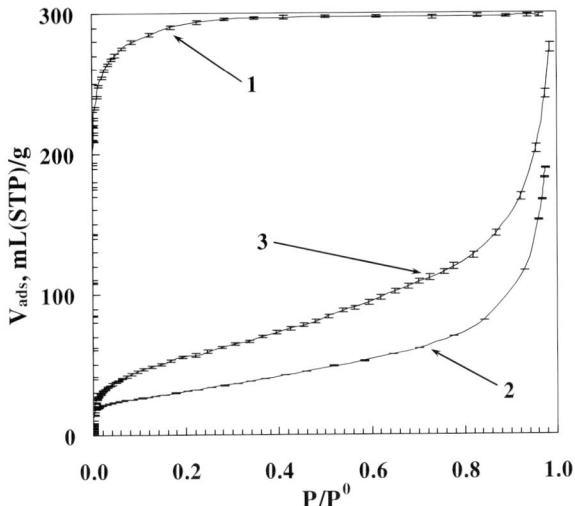

Fig. 1. Nitrogen adsorption isotherms for FM1/250 (1), carbon black (2) and Aerosil 200 (3).

### 3.1. Uncertainties in α<sub>S</sub>-data for non-porous standard materials

The $\alpha_s$ -data for non-porous carbon black and Aerosil 200 are determined in the usual manner [2]. The amount of nitrogen adsorbed at $0.4P^0$ is not always available directly from the experimental data and thus should be evaluated via a clamped-boundary, spline curve polynomial fit of $V_{ads} = f\left(P/P^0\right)_T$ over the pressure range $0.30P^0$-to-$0.50P^0$. The expression should render $V_{ads}$ -values with an error less than the RCSU in the experimentally determined $V_{ads}$ . In the present work, where $x = P/P^0$, we determined

$$V_{ads}^{calc.} = a + \frac{b \ln x}{x} + ce^{-x} \tag{1}$$

Eq. (1) gave errors in $V_{ads}$ from 0.035-to-0.212 % for carbon black and from 0.038-to-0.140 % for Aerosil 200. Over the same pressure range, the RCSU ($V_{ads}$) are 0.870-to-0.949 % and 1.652-to-1.692 %. Since the errors are less than the RCSU, the calculated $V_{ads}$ value at $0.4P^0$, $V_{ads}^{0.4P^0}$, is accepted; for uncertainty propagation calculations, we selected the RCSU ($V_{ads}$) for the datum closest to the $V_{ads}^{0.4P^0}$ -value, *viz.* 0.943 % (for carbon black) and 1.663 % (for Aerosil 200). Eq. (2) combines these data with $V_{ads}$ and $u_c\left(V_{ads}\right)$ resulting in $u_c\left(\alpha_S\right)$. The $u_c\left(\alpha_S\right)$- values exhibit a cumulative nature, as shown in Fig. 2.

$$u_c\left(\alpha_S\right) = \sqrt{\left\{\left(\frac{\partial \alpha_S}{\partial V_{ads}}\right)_{V_{ads}^{0.4P^0}} dV_{ads}\right\}^2 + \left\{\left(\frac{\partial \alpha_S}{\partial V_{ads}^{0.4P^0}}\right)_{V_{ads}} dV_{ads}^{0.4P^0}\right\}^2} \tag{2}$$

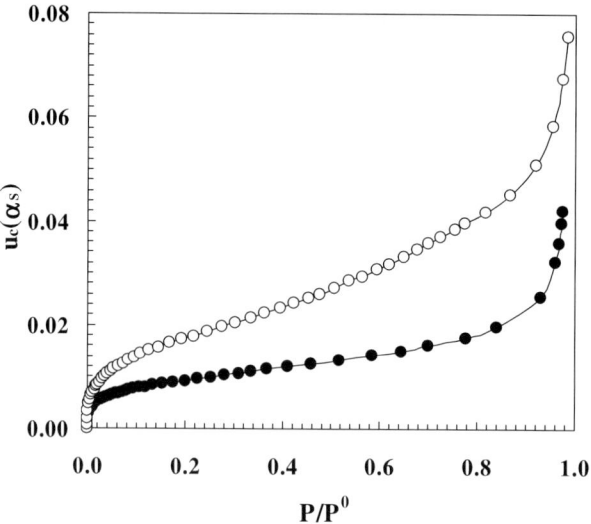

Fig. 2. $u_c(\alpha_S)$ vs $P/P^0$ (● carbon black, ○ Aerosil 200).

We used both non-porous materials as standards for FM1/250 to explore the influence of surface chemistry on $\alpha_S$-analysis and pore volume evaluation and its uncertainty. Normal practice is to establish a relationship between $\alpha_S$ and $P/P^0$; we used a quartic polynomial over three separate sections of the data set. Each polynomial gave $\alpha_S$-values agreeing with those from the experimental data within average errors of 0.95 % for carbon black and 1.29 % for Aerosil 200. Since Eq. (2) gives corresponding average RCSU ($\alpha_S$) values of 1.061 % and 2.197 %, for subsequent calculations of the $\alpha_S$-values of the test sample we used the quartic polynomial curve-fit. In contrast, for $u_c(\alpha_S)$ calculations, we use Eq. (2). The form of the polynomial depends upon the adsorption data, which are dependent upon the experimental conditions. Thus, it is essential that these conditions be accurately reproduced and reported.

### 3.2. Uncertainties in $\alpha_S$-data and micropore volume for FM1/250

In making an $\alpha_S$-analysis one correlates the relative pressures of the test adsorption data with appropriate $\alpha_S$-data of the standard. To propagate the uncertainty in the $\alpha_S$-data to the corresponding pressure of the test adsorption data, an expression is obtained by fitting the data in Fig. 2 as $u_c(\alpha_S) = f(P/P^0)$. The uncertainty in the amount adsorbed by the test sample is determined from the combination of the uncertainty in the volume adsorbed (at STP) and in the conversion factors to generate the uncertainty in the liquid amount adsorbed [1]. RCSU ($V_{ads}^{mL.(liq)}$) for FM1/250 vary from 0.371 %-to-1.809 % over the range $8 \times 10^{-7}P^0 \leq P \leq 0.96P^0$, comparable to those for the amount adsorbed at STP.

The $\alpha_S$-analyses of FM1/250 using carbon black and Aerosil 200 as the standards are shown in Fig. 3. The graphs reveal an independence of standard adsorbent surface chemistry on the evaluation of the total micropore volume. The choice of standard influences the shape of the curves for $\alpha_s$-values $< 0.5$, the range affecting the calculated primary micropore volume, and hence the secondary micropore volume. The inclusion of $u_c\left(V_{ads}^{ml.(liq)}\right)$ and $u_c(\alpha_s)$ limits for each data point restricts the flexibility of any weighted mean least squares (WMLS) analyses for pore volumes evaluation. We used a WMLS analysis to calculate intercept (and slope) values, yielding total and primary micropore volumes and their uncertainties. The total pore volume via the carbon black standard is evaluated over the range $1.1467 \le \alpha_s \le 3.7154$, and via Aerosil 200 $0.9277 \le \alpha_s \le 2.0695$. The pore volumes are summarized in Table 2.

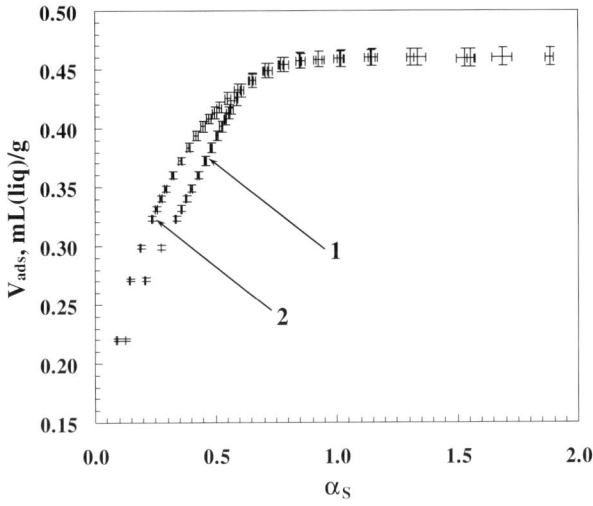

Fig. 3. $\alpha_s$-plots for FM1/250 ACC from carbon black-data (1) and Aerosil 200-data (2)

Table 2
Pore volumes of FM1/250 evaluated via carbon black and Aerosil 200 standards

| Standard | Total pore volume, mL(liq)/g | Primary pore volume, mL(liq)/g | Secondary pore volume, mL(liq)/g |
|---|---|---|---|
| Carbon black | $0.4595 \pm 0.0077$ | $0.1927 \pm 0.0080$ | $0.2668 \pm 0.0111$ |
| Aerosil 200 | $0.4577 \pm 0.0105$ | $0.2199 \pm 0.0164$ | $0.2378 \pm 0.0195$ |

The total micropore volumes values agree within 0.39 %. The combined standard uncertainty in both total pore volumes implies that this property (of our particular sample) is constant via either standard adsorbent, which may be a consequence of the lack of mesopores in the material, or a demonstration that this volume is insensitive to the adsorbent standard's surface chemistry. We suggest that since the external specific surface area is considerably less than the monolayer equivalent area (or volume) of the micropores, any surface chemistry differences between the standard and test adsorbents are nullified during adsorption processes up to gas pressures equivalent to $\alpha_s$-values of 1.

The primary micropore volume and/or filling swing, via the carbon black standard, can be evaluated by recognizing linearity in the data in the range $0.2 \leq \alpha_S \leq 0.5$. Both filling and cooperative swings are evident in the data supported by the carbon black standard, simplifying the determination of the primary micropore volume. In contrast, the Aerosil 200 standard delivers a continuous curvature resulting in a lack of distinction between these processes, usually interpreted as the presence of a steadily increasing micropore size distribution. This lack of filling and cooperative swings implies that the difference in standard and test adsorbents' surface chemistry becomes an important parameter in a comparison plot analysis. To define the linear range leading to primary micropores, the WMLS analysis was confined to a minimum of 5 $\alpha_S$-values leading to a regression coefficient of at least 0.999. The resulting primary micropore volumes are given in Table 2. The difference between the two volumes is statistically significant, 14.12 %, with a minimum difference of 1.40 %.

Secondary micropore volumes are determined from the difference between the total and the primary micropore volumes. Consequently, the CSU in this volume increases over those of the other volumes.

## 4. CONCLUSION

By accounting for the source and propagation in uncertainty in the amount adsorbed by adsorbent materials of different surface chemistry that may be regarded as standards, and in the amount adsorbed by a porous activated carbon cloth, we conclude: standard adsorbent surface chemistry has a minor influence on the evaluation of the total micropore volume of a wholly microporous adsorbent; standard adsorbent surface chemistry greatly influences the identification of filling and cooperative swing processes in the micropore range; $u_c(V_{ads})$ and $u_c(\alpha_S)$ strongly influence the identification of linear ranges in pore volume evaluation.

## ACKNOWLEDGEMENT

We thank the Center for Molecular and Materials Sciences and the University of South Australia Research Office for their generous support of this work.

## REFERENCES

[1] A. Badalyan and P. Pendleton, Langmuir, 19 (2003) 7919.
[2] K.S.W. Sing, Chem. Ind., 44 (1968) 1520.
[3] M.R. Bhambhani, P.A. Cutting, K.S.W. Sing and D.H. Turk, J. Colloid Interface Sci., 38 (1972) 109.
[4] P.J.M. Carrott, R.A. Roberts and K.S.W. Sing, Carbon, 25 (1987) 769.
[5] K. Kaneko, J. Membr. Sci., 96 (1994) 59.
[6] K. Kaneko, C. Ishii, H. Kanoh, Y. Hanzawa, N. Setoyama and T. Suzuki, Adv. Colloid Interface Sci., 76-77 (1998) 295.
[7] A. Badalyan, P. Pendleton and H. Wu, Rev. Sci. Instrum., 72 (2001) 3038.
[8] T. Takaishi and Y. Sensui, Trans. Faraday Society, 59 (1963) 2503.
[9] A. Badalyan and P. Pendleton, J. Colloid Interface Sci., 283 (2005) 605.

Studies in Surface Science and Catalysis 160
P.L. Llewellyn, F. Rodriquez-Reinoso, J. Rouqerol and N. Seaton (Editors)

# Uncertainty in amount adsorbed and surface excess from uncertainty in high-pressure gas adsorption data

**P. Pendleton**

Center for Molecular and Materials Sciences
University of South Australia, Mawson Lakes, South Australia, 5095, Australia

The method of analysis of the uncertainty in the amount of fluid or density in the continuous phase in manometric adsorption measurements, calculated via Peng-Robinson and Bender equations of state is presented. It is applied for the evaluation of the specific surface excess amount and its uncertainty during high-pressure nitrogen adsorption by a microporous activated carbon cloth at pressures up to 17 MPa at 252.40 K. Adsorption data were analysed via the use of bilinear-interpolated data from a $P$-$\rho$-$T$ matrix developed from the NIST Chemistry WebBook fluid physical properties database. Deviations of calculated specific surface excess amounts from those calculated using NIST density data approach ± 0.2 %, considerably superior to either Peng-Robinson or Bender EoS, ranging from 6.4 to 3.0 %.

## 1. INTRODUCTION

How to define the uncertainty in any general expression is well described by Taylor [1], and specifically for low-pressure gas adsorption by Badalyan and Pendleton [2]. The general form for combined standard uncertainty (CSU) analysis for a function $y = f(x_{1...n})$, in which the variables might be independent of one another, is given as:

$$u_c(y) = \sqrt{\left[\left(\frac{\partial y}{\partial x_1}\right)_{2...n} .dx_1\right]^2 + \left[\left(\frac{\partial y}{\partial x_2}\right)_{1,3...n} .dx_2\right]^2 + .... + \left[\left(\frac{\partial y}{\partial x_n}\right)_{1...n-1} .dx_n\right]^2} \qquad (1)$$

The derivatives $dx_{1...n}$ are known standard uncertainty (SU) values in the parameters $x_{1...n}$, defined as $u(x_{1...n})$, and the partial derivatives have their usual meaning. The calculation of the CSU in the amount of fluid in the continuous phase above the adsorbent, $u_c(n)$, is an application of Eq. (1) to the expressions for the various equations of state (EoS) applied to high-pressure adsorption isotherm analysis. The relative combined standard uncertainty (RCSU) is quoted in the usual way [1].

Analyses of high-pressure manometric adsorption measurements demands the need to determine which EoS is best suited to describe the amount of fluid in the continuous phase material balance calculations, and thus to determine which of these EoS provides the lowest uncertainty. The least RCSU, often referred to as the error, needs to be calculated against a known reference value. These are determined by comparing the calculated amount of fluid and/or continuous phase density with those obtained via the NIST "Chemistry WebBook" fluid-phase reference tables [3].

We present a discussion of the uncertainty contributions to the amount of fluid in the continuous phase or the density of the same from the SU in sample and dosing volumes and associated pressure data, and the manifold and adsorption system temperature. Each of the coefficients in the EoS also has its inherent uncertainty, which also needs to be considered. We use nitrogen adsorption data (relative to helium dead-space measurements) for a microporous activated carbon cloth (ACC) to demonstrate uncertainty in the various EoS evaluation and its propagation to the combined standard uncertainty in the (Gibbs) specific surface excess amount, $n_{GSE}$, equivalent to the traditionally known "amount adsorbed" (in low pressure measurement), shown in Eq. (2)

$$n_{GSE} = n_{abs} - \rho V_a \qquad (2)$$

where, $n_{abs}$ is the absolute amount adsorbed, $\rho$ is the density of an adsorbate in the bulk phase, and $V_a$ is the volume of the adsorbate layer.

Although a calculation of the uncertainty in the absolute amount adsorbed is best suited for molecular simulation [4], such a calculation requires a detailed knowledge of the pore volume for porous materials or specific surface area for non-porous materials [5]. Despite the volume calculation omission, we believe that it is equally important to make an analysis of the uncertainty in the selected EoS as their selection impacts initially on the uncertainty in the specific surface excess amount and then on the uncertainty in the absolute specific amount adsorbed, due to their linear combination, Eq. (2).

## 2. EXPERIMENTAL

### 2.1. Materials

A traceable standard carbon black supplied by Micromeritics Corporation (Sydney, Australia) with a recommended multiple-point BET-SSA of $113\pm5$ m$^2$/g was used to calibrate and commission a newly designed, all stainless steel, high-pressure manometric adsorption apparatus.

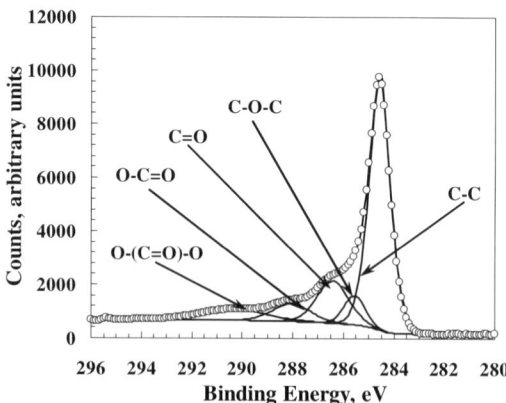

Fig. 1. Analysis of $C_{1_S}$ peak XPS spectrum for FM1/250 showing types of C-O surface groups.

The porous adsorbent used for the present discussion was a low oxygen content ACC, FM1/250 (ex Calgon Carbon Corp., Pittsburgh, PA); the low oxygen content, consists of ether, carbonyl, acid, and anhydride oxygen structures (see Figure 1).

Ultra-high purity, 99.999 %, nitrogen and helium gases were used as supplied by CIG, Adelaide, Australia.

## 2.2. Experimental apparatus and measurement method

The dosing volume and physical volume of the sample tube were determined as previously reported [2, 6]. Since the sample tube is stainless steel, a temperature correction was applied to the fluid phase in the volume up to 15 cm above the level of the ethanol constant temperature bath [7], resulting in a value of 110.48 $m^2/g$, within 0.15 % of our previously reported average value of five independent measurements [2].

Prior to use, we washed FM1/250 with de-ionized water until the electrical resistivity of the effluent matched that of the influent, 18 MOhm.m. The sample was dried in an ambient-pressure oven at 378 K for 24 hours. A typical sample mass of 0.2 g of this washed and dried sample was used for low- and high-pressure adsorption measurements. Typical out-gassing conditions consisted of heating at a rate of 2 K/min to 473 K, soaking for 8 hours then cooling at natural conduction and convection rates to room temperature. This procedure rendered a background pressure of $< 1 \times 10^{-4}$ Pa.

Dead-space volumes were measured via helium expansion at 298 K. For high-pressure adsorption measurements, the sample and its stainless steel container (sampling tube) were immersed in a cooled ethanol bath maintained at 252.40 K, controlled by an N-type thermocouple in the temperature control-loop with a standard deviation of $\pm$ 0.03 K. The part of the stainless steel sampling tube up to 15 cm above the level of ethanol was subjected to a temperature gradient (from 252.40 K to room temperatures), which was accounted for by introduction of a thermal correction [7]. A platinum resistance temperature detector (RTD) ($R_0$=25.0231 Ohm), was connected to a high-accuracy Keithley multimeter as part of a 4-wire connection arrangement. The previously calibrated RTD, with $u(T) \approx \pm$ 0.05 K, monitored and recorded the temperature for each dosing and equilibrium measurement. The manifold containing the dosing volume was maintained at slightly above room temperature, 300.57 $\pm$ 0.06 K. Low-pressure measurements up to 0.103 MPa were made using an MKS-698 Series Baratron capacitance differential transducer manometer with a calibrated uncertainty of 0.05 % of reading. High-pressure readings up to 20 MPa were made using a Keller pressure gauge, also with uncertainty of 0.05 % of reading.

## 3. RESULTS AND DISCUSSIONS

The propagation of uncertainties requires knowledge of the sources and values of uncertainty in variables, which when combined, contribute to the uncertainty in the (final) dependent variable. In following the path, one often finds interdependence of variables and sometimes the reduction of uncertainty magnitudes [1]. In manometric adsorption uncertainty analyses, the dependent variable is the specific surface excess amount, derived from EoS applications. In EoS uncertainty analysis, the dependent variable is the amount of fluid in the continuous phase above the sample, regardless of whether it is calculated directly via a cubic EoS or indirectly via a multiple parameters EoS. The SU-values used in this work were either measured or identified in manufacturer's information and taken to be 3-$\sigma$ standard deviations. These values are summarized in Table 1. Since the standard deviation of the sample temperature is less than the uncertainty in the calibrated temperature detector, for uncertainty calculations the detector uncertainty is used.

Each calculation of uncertainty in the various sections of the adsorption apparatus derives from (the uncertainty in) the amount of fluid in the continuous phase via the applied EoS. These uncertainties then contribute to uncertainties in the amount of fluid for subsequent pressure analyses. Ultimately, these uncertainties are combined (in a manner similar to the material balance analysis) to give the $u_c(n)$. This uncertainty is then combined with the $u\left(w_{sample\ weight}\right)$ to give $u_c\left(n_{GSE}\right)$, the CSU in the (Gibbs) specific surface excess amount.

Table 1
Uncertainty in parameters used in propagation calculations.

| Parameter | SU | Units |
|---|---|---|
| $P_{MKS-1}$ | 0.005 x scale reading | MPa |
| $P_{Keller-2}$ | 0.005 x scale reading | MPa |
| $V_{dosing-1}$ | $6.133 \times 10^{-8}$ | $m^3$ |
| $V_{dosing-2}$ | $2.111 \times 10^{-8}$ | $m^3$ |
| $V_{sample\ tube\ at\ ambient\ temperature}$ | $3.153 \times 10^{-8}$ | $m^3$ |
| $V_{sample\ at\ experiment\ temperature}$ | $9.876 \times 10^{-8}$ | $m^3$ |
| $R$ | $10^{-9}$ | $MPa.m^3.mole^{-1}.K^{-1}$ |
| $T_{sample\ temperature}$ | 0.05 | K |
| $T_{ambient}$ | 0.06 | K |
| $P_c$ | $5 \times 10^{-3}$ | MPa |
| $T_c$ | 0.04 | K |
| $w_{sample\ weight}$ | $5.0 \times 10^{-3}$ | g |
| $\rho_{NIST\ data}$ | $2.0 \times 10^{-4}$ | $mole.L^{-1}$ |

In the following discussion we explain how to account for the uncertainty introduced via solutions to EoS and how to combine these with the uncertainty in the data and variables within each EoS, leading to the CSU in the calculated amount of fluid. As a final analysis we compare these CSU with the uncertainty in the data obtained via NIST [3], which we will regard as "standard" data with a known uncertainty, and calculate the RCSU in the surface excess amount adsorbed.

### 3.1. Peng-Robinson cubic EoS uncertainty evaluation and analysis

In the solution of the Peng-Robinson (P-R) cubic EoS, the coefficients contain mixed terms (see Table 2) and have their own uncertainty contributing to $u_c(n)$. The general procedure for the solution of the cubic EoS is to rearrange the EoS to generate an expression in the form:

$$\alpha n^3 + \beta n^2 + \gamma n + \varepsilon = 0 \qquad (3)$$

For the P-R EoS, the coefficients $\alpha$, $\beta$, $\gamma$, and $\varepsilon$ in Table 2 are expressions in terms of $P$, $V$, $T$, $R$, $a(T)$ and $b$. To define the CSU, as $u_c(n)$, the solution of Eq. (3) for $n$ requires us to define the uncertainty in the quotients $\beta/\alpha$, $\gamma/\alpha$, and $\varepsilon/\alpha$. Values of $R$, $T_c$, and $P_c$ are regarded constant, with their uncertainty listed in Table 1. The expressions for the $\partial n/\partial(R, T_c, \text{or } P_c)$ have varying degrees of complexity and numbers of terms contributing to the final CSU-value.

Table 2
Summary of coefficients of number of moles, $n$, in the P-R EoS.

| EoS | $\alpha$ | $\beta$ | $\gamma$ | $\varepsilon$ |
|---|---|---|---|---|
| P-R | $b^3P + b^2RT - a(T)b$ | $V(a(T) - 2bRT - 3b^2P)$ | $V^2(bP - RT)$ | $PV^3$ |

Accuracy is defined as how close the measured value is to the actual value, if one exists [8]. Often an actual value is one accepted as a standard; NIST provide isothermal and isobaric density data electronically [3] with an uncertainty of $\pm$ 0.02 %. We refer to all calculations associated with data thus obtained as NIST-derived. Comparison of the specific surface excess amount via the P-R EoS analysis with that from NIST-derived density data is shown in Fig. 2. The P-R EoS delivers values of the surface excess amount within 6.4 % of the NIST-derived over the range $\approx$ 0.25 < P < 17 MPa. To demonstrate which of the parameters listed in Table 1 has the greatest impact on the surface excess amount and its RCSU, we calculate the RCSU by varying individually each of the SU in Table 1 over the range – 100 % < SU < + 1000 %, to give the $u_c(n)$. For the P-R EoS-derived data, $u(T_{amb})$ has the greatest impact on the CSU decreasing in the order: $u(T_{amb}) > u(P_c) > u(V_{sample\ tube\ at\ ambient\ temperature})$ $> u(V_{dosing}) > u(P_{Keller}) > u(V_{sample\ tube\ at\ experiment\ temperature}) > u(R) > u(T_c) >$ $u(w_{sample\ weight}) > u(T_{sample\ temperature})$. Fig. 3 shows results of this impact in the case of $u(P_{Keller})$.

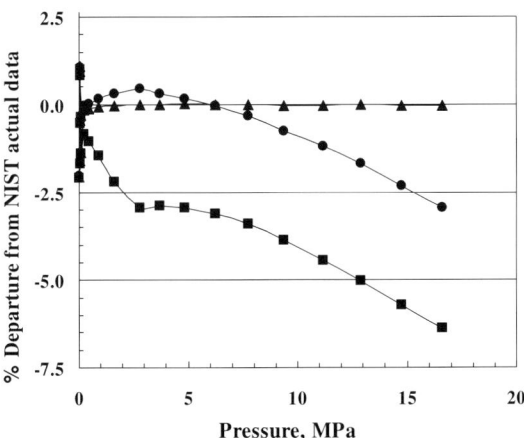

Fig. 2. Departure of specific surface excess amounts from NIST amounts for NIST interpolated data (▲), P-R EoS (■), and Bender EoS (●).

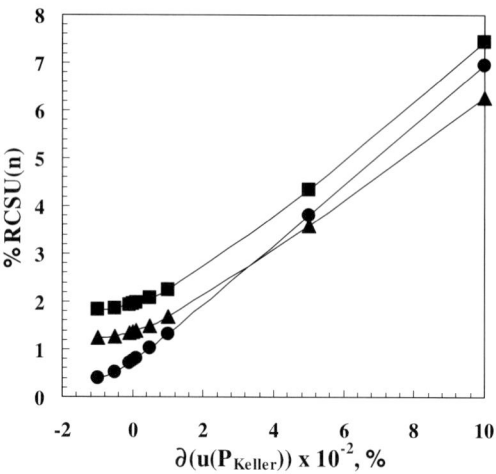

Fig. 3. Sensitivity of NIST interpolated data (▲), P-R EoS (■), and Bender EoS (●) to changes in $u(P_{Keller})$.

### 3.2. Bender multiparameter EoS uncertainty evaluation and analysis

The polynomial format of the Bender multiparameter EoS yields a simple solution via a Newton-Raphson iteration procedure [9]. We found three iterations provided fluid-phase density values with an associated RCSU of $< 2 \times 10^{-4}$ % between the third and fourth iteration. We took the value of the $6^{th}$-iteration, which gave an RCSU of 0.0% between the $4^{th}$- and $6^{th}$-iteration. The form of the Bender EoS requires implicit differential expressions [10] to evaluate and propagate the uncertainty to generate the CSU in the amount of fluid change from the material balance, leading to $u_c(n)$. Implicit differentials were developed for $(\partial \rho/\partial P)_{T,R}$, $(\partial \rho/\partial T)_{P,R}$, and $(\partial \rho/\partial R)_{P,T}$. The SU-values used in these calculations are listed in Table 1. The Bender EoS is a polynomial curve-fit to the nitrogen $P$-$\rho$-$T$-data, where the coefficients are allowed to vary. The Bender EoS-derived data compare very favorably with the NIST-derived data in Fig. 2, reaching $\approx$ 3.0 %-departure at $\approx$ 17 MPa, and never more than the NIST-data $\approx \pm$ 0.9 %-departure up to 10 MPa, which is normally accepted for process design specification [11]. To determine which of the variables listed in Table 1 has the greatest influence on the CSU, we applied the same independent variable analyses made for the P-R EoS. $u(T_{amb})$ has the greatest impact on the CSU decreasing in the order:

$$u(T_{amb}) \; > \; u(P_{Keller}) \; > \; u(R) \; > \; u(T_c) \; > \; u(w_{sample\;weight}) \; > \; u(V_{dosing}) \; \approx$$

$$u(V_{sample\;tube\;at\;ambient\;temperature}) \; \approx \; u(V_{sample\;tube\;at\;experiment\;temperature}) \; > \; u(P_c) \; =$$

$$u(T_{sample\;temperature}).$$

### 3.3. Comparison between NIST-interpolated and NIST-derived data

NIST suggests that the $u(\rho)$ in their nitrogen density data is $\pm$ 0.02 % for temperatures from the triple point up to 523 K and up to pressures of 12 MPa and from 240 to 523 K with pressures < 30 MPa [12]. We created our own $P$-$\rho$-$T$ matrix from the NIST-data over the temperature range 150–400 K and over the pressure range 0.001–25 MPa. We then determine the density at the measured pressure and temperature via a bilinear interpolation method [13]. $u_c(n)$ from the NIST-derived data are defined via the $u(\rho_{NIST})$, the SU in the system volumes and in the sample weight. The interpolated data contain these uncertainties and the uncertainty due to each interpolation. Values of the specific surface excess amount via interpolation of NIST-data and their CSUs are compared with those from the NIST-derived data in Fig. 2. Differences are attributed to CSU accumulation and to the effects of interpolation, i.e. to the combination of the uncertainty in the four separate temperatures and pressures used in each interpolation.

The NIST-interpolated data compare very well with those defined as NIST-derived. The departure of the majority of the data varies from -0.04 to + 0.02 % with the maximum value of ≈ 0.16 %. Factors such as the size of the interpolation interval and those ascribed to CSU differences determine the "polynomial nature" of the departure. We used a temperature interval of 1.25 K and a series of pressure intervals: 0.001, 0.005, 0.1, and 0.25 MPa. Reducing the intervals generates values closer to the NIST-derived data however, their CSUs remain similar in magnitude to those in Fig. 2 due to the interpolation methodology. Replacing the bilinear interpolation with a cubic interpolation reproduces the exact NIST-values from the same $P$-$\rho$-$T$ matrix, however, since this method employs a series of differential expressions over three levels of data, greater uncertainty results. The interpolation method of analysis for density determination and amount of fluid in the material balance provides a preferred method over the cubic and multiple parameters EoS methods, and, from a computing speed perspective we recommend the bilinear method.

## 4. CONCLUSIONS

P-R and Bender EoS render density data and specific surface excess amount adsorbed of similar magnitudes when applied to high-pressure and/or supercritical manometric adsorption. Comparison of specific surface excess calculated using these EoS with those from NIST data shows that these data agree within 6.4 % of the NIST-derived data for the conditions examined.

If possible, one should interpolate the NIST data from a $P$-$\rho$-$T$ matrix since this method is quick, seamless and provides density and the specific surface excess amount within ± 0.2 % of the NIST-derived data over the pressure range examined. The uncertainty in these data is similar in magnitude to those of the Bender EoS, but always overlaps the value and CSU of the NIST-derived specific surface excess amount. Consequently, interpolation is the superior method of choice for high-pressure and supercritical manometric adsorption analyses. Cubic or bilinear interpolation methods can be used with the former rendering exact NIST values but with greater CSU than the bilinear method, which gives a maximum %-departure of ≈ − 0.2 %. Use of the latter method is recommended.

## ACKNOWLEDGEMENT

We thank the Centre for Molecular and Materials Sciences of University of South Australia and the Australian Academy of Sciences for their financial assistance with this work.

# REFERENCES

[1] J.R. Taylor, An introduction to error analysis. The study of uncertainties in physical measurements, University Science Books, New York, 1982.
[2] A. Badalyan and P. Pendleton, Langmuir, 19 (2003) 7919.
[3] NIST, Chemistry WebBook, Department of Commerce, Gaithersburg, MD, 2004, http://webbook.nist.gov.
[4] D. Nicholson and N.G. Parsonage, Computer simulation and the statistical mechanics of Adsorption, Academic Press, London, 1982.
[5] A.L. Myers and P.A. Monson, Langmuir, 18 (2002) 10261.
[6] A. Badalyan, P. Pendleton and H. Wu, Rev. Sci. Instrum., 72 (2001) 3038.
[7] A. Badalyan and P. Pendleton, J. Colloid Interface Sci., 283 (2005) 605.
[8] J.R. Taylor and C.E. Kuyatt, Guidelines for evaluating and expressing the uncertainty of NIST measurement results, National Institute of Standards and Technology, 1997.
[9] R.L. Burden and J.D. Faires, Numerical analysis, McGraw-Hill, New York, 1997.
[10] M.L. Boas, Mathematical methods in the physical sciences, Wiley, New York, 1983.
[11] S.I. Sandler, Chemical Engineering Thermodynamics, Wiley, New York, 1999.
[12] R. Span, E.W. Lemmon, R.T. Jacobsen, W. Wagner and A. Yokozeki, J. Phys. Chem. Ref. Data, 29 (2000) 1361.
[13] W.H. Press, S.A. Teukolsky, W.T. Vetterling and B.P. Flannery, Numerical recipes in Fortran 90. The art of parallel scientific computing, Cambridge University Press, 1996.

Studies in Surface Science and Catalysis 160
P.L. Llewellyn, F. Rodriquez-Reinoso, J. Rouqerol and N. Seaton (Editors)

# A new methodology to characterize the porosity of Y zeolites by liquid chromatography

**L. Teyssier, E. Guillon and M. Thomas**

Institut Français du Pétrole, Catalysis and Separation Division, B.P. 3, 69390 Vernaison, France.

Liquid chromatography breakthrough curves have been used to characterize the porosity of a dealuminated Y zeolite, and more precisely the mesopores in cavities and the cylindrical mesopores. The methodology presented in this paper shows that the use of several probe molecules with different molecular size and adsorption strength can give an estimation of the zeolite porosity.

## 1. INTRODUCTION

Zeolites are highly crystalline microporous aluminosilicates. Because of their strong acidity and shape-selectivity properties, they are widely used in catalytic processes in oil refining, petrochemistry and fine chemistry. However, these materials are not suitable to convert bulky molecules which can not access the microporosity. Dealumination processes are then carried out to tune the acidity and porosity properties of the zeolite [1,2]. In particular, steaming (high-temperature treatment usually above 773 K) or acid leaching (with strong inorganic acids) can be used to generate a second pore network composed of mesopores (pore size between 2 and 50 nm). These mesopores may prevent diffusional limitations and therefore enhance catalytic activity as already reported in literature [3]. For dealuminated faujasite zeolites, 3D TEM shows [4] that two kinds of mesopores can be highlighted : (i) mesopores in cavities, only accessible *via* the micropores and (ii) cylindrical mesopores, connected to the external surface of the crystal, and so directly accessible *via* the macropores.

Molecular probes have already been used to characterize nanoporous materials, in term of pore size determination (gas adsorption) or surface chemistry investigation (use of polar probe molecules) [5]. In this paper, we describe a new methodology based on liquid chromatography in order to characterize the two kinds of mesopores discussed above. The results obtained so far on a highly dealuminated (Y, Zeolyst CBV780) and non-dealuminated (Y, Zeolyst CBV300) zeolites are reported.

## 2. METHODOLOGY

### 2.1 Theoretical considerations

Liquid chromatography is a technique widely used for the determination of adsorption and diffusion parameters in microporous materials such as zeolites [6], in meso- and macroporous materiels such as aluminas [7], as well as in biporous pelletized materials [8]. This technique enables also the development of suitable adsorbent for a given separation, for example the separation of isomers of xylenes on X or Y zeolites [9,10].

The mass transfer in the adsorption column can be described by a set of differential partial equations, but the analytical solution is often difficult to establish. The adsorption parameters can also directly be obtained by the determination of the moments of the beakthrough curve. The fisrt moment, $\mu$, is only related to equilibrium parameters, while the second moment, $\sigma^2$, is related to mass transfer parameters [11], and they can be directly calculated by integration of the response curve.

The expression for the 1$^{st}$ moment, for a step response, is :

$$\mu = \int_0^\infty \left(1 - \frac{C}{C_0}\right) dt \tag{1}$$

where $C_0$ is the inlet concentration (feed) and $C$ the outlet concentration.

This technique can also be used to characterize porous solid materials. As the mean retention time, $\mu$, is related to the accessible volume to a given molecule, the use of several probe molecules, different in size as well as in chemical structure, can be helpful to determine the different accessible volumes within the adsorbent.

## 2.2. Methodology

First of all, we have to define rigorously the different volumes (porosities) which are going to be characterized. The column volume ($V_C$) is composed of the interparticle volume ($V_I$) and the particle volume ($V_P$). The particle volume consists in the solid volume (volume of the atoms, $V_S$), the macroporous volume ($V_M$), the microporous volume ($V_\mu$) and the mesoporous volume ($V_m$). The latter can be divided in two volumes : the mesoporous cavities volume, called "internal mesoporous volume" ($V_{im}$), and the cylindrical mesoporous volume, called "external mesoporous volume" ($V_{em}$) :

$$V_C = V_I + V_P = V_I + \left(V_M + V_{em} + V_{im} + V_\mu + V_S\right) \tag{2}$$

A simple mass balance leads to the following equation :

$$FC_0\mu = C_0 V_I + C_M V_M + C_{em} V_{em} + C_{im} V_{im} + C_\mu V_\mu \tag{3}$$

in which $F$ represents the flow rate of the feed (cm$^3$.min$^{-1}$), $C_0$ the feed concentration in the liquid phase (mmol.cm$^{-3}$), $\mu$ the first moment (retention time) of the breakthrough curve (min), and $C_M$, $C_{em}$, $C_{im}$, $C_\mu$ the compound concentration (mmol.cm$^{-3}$) in the macropores, the "external mesopores", the "internal mesopores" and the micropores, respectively. In view of the size of the macropores (pore size larger than 50 nm) and the "external mesopores" (pore size that can reach 20-30 nm, see ref. [4]), we can make the reasonable hypothesis that the compounds concentration in these volumes equals the feed concentration $C_0$. This leads to the general equation, giving the first moment related to the different volumes :

$$FC_0\mu = C_0\left(V_I + V_M + V_{em}\right) + C_{im} V_{im} + C_\mu V_\mu \tag{4}$$

The principle of the methodology is to use several feeds and solvents with components of different molecular size and adsorption selectivity properties. This will allow to make some hypotheses on the concentration of the compounds in the different volumes, and therefore to simplify the general equation (4). The methodology presented in this paper will consist in three determinations : (i) the total pore volume $(V_l + V_M + V_m + V_\mu)$, (ii) the « non-selective » volume $(V_l + V_M + V_{em})$ and (iii) the volumes $V_{im}$ and $V_\mu$. The couples solvent/feed will be called « systems » in the following.

*2.2.1 Determination of $(V_l + V_M + V_m + V_\mu)$*
The first moment of the molecular probe has to depend on all the different porosities. Therefore, the molecular probe has to fulfill the following conditions : (i) its size should allow to enter the micropores (no steric exclusion) and (ii) it must be able to easily replace the molecules of the solvent on the adsorption sites. For this latter condition, the molecular probe must have, at least, the same adsorption properties of the solvent. We may choose for the feed and the solvent two molecules of the same chemical family, that is to say two paraffines or two aromatics. A better choice would be a molecular probe which adsorbs more strongly on the adsorption sites than the molecules of the solvent. It would be the case with an aromatic as the feed and a paraffine as the solvent : the aromatic adsorbs preferentially because of the stronger interaction between the electronic cloud of the aromatic and the adsorption sites (the cations and the –OH groups) of the zeolite. The "Gurvich rule" [12] says that densities in the adsorbed phase and the liquid phase are close to that of the bulk liquid adsorptive. If we make this hypothesis $(C_\mu = C_{im} = C_0)$, the equation (4) becomes :

$$F\mu = V_l + V_M + V_{em} + V_{im} + V_\mu = V_l + V_M + V_m + V_\mu \qquad (5)$$

Note that the Gurvich rule is probably more valid with flexible molecules such as *n*-alkanes, which can fit the spherical supercages, than with rigid ones like aromatics.

*2.2.2 Determination of $(V_l + V_M + V_{em})$*
For this determination, the molecular probe must not penetrate into the zeolite micropores (and so not in the mesoporous cavities). Therefore, we must choose a bulky probe molecule whose size is larger than the zeolite micropore dimensions. In these conditions, the concentrations of the probe molecule in the micropores and in the « internal mesopores » equal zero, so the equation (4) becomes :

$$F\mu = V_l + V_M + V_{em} \qquad (6)$$

Knowing $V_l$ and $V_M$ from mercury porosimetry data, $V_{em}$ is accessible. In our case, this volume includes the binder mesopores. By difference with the previous determination $(V_l + V_M + V_m + V_\mu)$, the volume $(V_{im} + V_\mu)$ can be estimated :

$$V_{im} + V_\mu = V_m + V_\mu - V_{em} \qquad (7)$$

The next determination will allow to measure $V_{im}$ and $V_\mu$ separatly.

*2.2.3 Determination of $V_{im}$ and $V_\mu$*
The previous determination allowed to determine $V_{em}$. The determination of the other mesoporous volume, $V_{im}$, is much more complicated, since this volume is related to the microporous volume $V_\mu$ (the mesoporous cavities are only accessible via the micropores). To obtain $V_{im}$ and $V_\mu$ separately, we have to employ two molecular probes at the same time : these two molecules A and B must have a similar structure to prevent an adsorption selectivity in

$V_{im}$, but their structure must be sufficiently different to induce a selectivity in $V_\mu$. The solvent molecules have also to be replaced by the two molecular probes. Making the hypothesis that the concentration of A and B in the mesopores are the same than in the feed, we can write :

$$\begin{cases} V_{im}C_L^A + V_\mu C_\mu^A = FC_L^A \mu_A - \left(V_l + V_M + V_{em}\right)C_L^A \\ V_{im}C_L^B + V_\mu C_\mu^B = FC_L^B \mu_B - \left(V_l + V_M + V_{em}\right)C_L^B \end{cases} \qquad (8)$$

in which $C_L^i$ stands for the concentration of the molecular probe $i$ ($i$ = A or B) in the liquid phase and $C_\mu^i$ the concentration of the molecular probe $i$ in the micropores.

If we choose two molecular probes similar in size, we can make the assumption that the total number of molecules per supercage in the micropores equals a constant value $C_s$ :

$$C_\mu^A + C_\mu^B = C_s \qquad (9)$$

Moreover, from the previous determination :

$$V_{im} + V_\mu = V_m + V_\mu - V_{em} = V_0 \qquad (10)$$

with $V_0$ a numerical value. The equations (8), (9) and (10) then constitute a system of 4 equations with 4 unknowns ($V_{im}$, $V_\mu$, $C_\mu^A$ and $C_\mu^B$). The solutions of this system exist only if the concentrations of the probe molecules A and B in $V_\mu$ are different. This means that an adsorption selectivity is necessary.

## 3. EXPERIMENTAL

### 3.1 Materials

The two zeolite samples (Y CBV300 and Y CBV780, with a Si/Al molar ratio of 2,5 and 40, respectively) were provided by Zeolyst International. Zeolite Y CBV300 (that will be called HY in this paper) is a Na-Y zeolite which has been ammonium-exchanged. Zeolite Y CBV780 (that will be called HDaY) is prepared by two steamings of zeolite Y CBV300, followed by a mineral acid leaching [13]. The samples were used in their H form. To ensure a good mechanical resistance of the samples, the zeolites have been pelletized with a binder (30% weight of boehmite alumina). The pellets have been calcined 4 hours at 823 K under an air flow (2 dm³.g⁻¹.h⁻¹) to remove the nitric acid used for the binder peptization.

The solvents and feeds used (see Table 1) were n-heptane, i-octane, m-xylene, p-xylene, toluene, mesitylene, 1,3,5-triisopropylbenzene and 1,3,5-triisopropylcyclohexane, and were of analytical grade (at least 99 %), except for 1,3,5-triisopropylbenzene and 1,3,5-triisopropylcyclohexane (pure at 95 %). A molecular sieve (zeolite 3A) was introduced in the bottles of solvents and feeds to remove moisture before injection in the column.

Table 1
Characteristics of the solvents and feeds used.

| Compound | Chemical formula | Molecular dimensions (Å) [a] | | |
|---|---|---|---|---|
| | | x | y | z |
| *n*-heptane (nC7) | $CH_3(CH_2)_5CH_3$ | 11.6 | 4.5 | 4.0 |
| *i*-octane (iC8) | $(CH_3)_3CCH_2CH(CH_3)_2$ | 6.5 | 6.2 | 6.6 |
| toluene (tol) | $C_6H_5CH_3$ | 8.3 | 6.7 | 4.2 |
| *m*-xylene (mXyl) | $C_6H_4(CH_3)_2$ | 9.2 | 7.4 | 4.2 |
| *p*-xylene (pXyl) | $C_6H_4(CH_3)_2$ | 9.2 | 6.7 | 4.2 |
| mesitylene (mes) | $C_6H_3(CH_3)_3$ | 8.4 | 8.9 | 4.2 |
| 1,3,5-tri-*i*-propylbenzene (iPr3-bzn) | $C_6H_3((CH(CH_3)_2)_3$ | 10.7 | 9.8 | 6.7 |
| 1,3,5-tri-*i*-propylcyclohexane (iPr3-CyC6) | $C_6H_9((CH(CH_3)_2)_3$ | - | - | - |

a : from ref. [14] and/or calculated with Cerius[2] software marketed by Accelrys.

## 3.2 Experimental setup

The experimental setup is shown in Fig. 1. The dimensions of the stainless-steel column used are 20 cm in length and 1 cm of internal diameter. The column was filled with the zeolite sample (about 10 g), which was then dried overnight at 723 K under a nitrogen flow in a tubular furnace. The column was then placed in a HP 5890 - GC oven. The solvent and the feed were injected in the column by two independent pumps (flow rate : 0.5 cm$^3$.min$^{-1}$). A 4-way valve allowed switching between the solvent and feed injections.

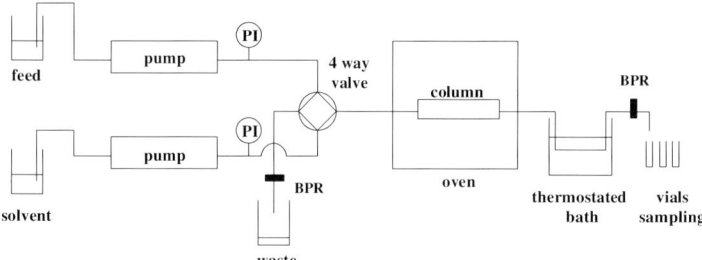

Fig. 1. Experimental setup. BPR : backpressure regulator ; PI : pressure indicator.

The experiments were carried out at 423 K. A backpressure regulator allowed to ensure that the hydrocarbons were in their liquid form throughout the experiments. A pressure of 12 bar was generally applied. The effluents were collected every minute (collect time : 1 minute) in 80 sampling vials. The vials were analysed on a Agilent 6890N-GC equipped with a FFAP or HP5 column depending on the components to be separated.

## 3.3 Zeolites characterization

### 3.3.1 Nitrogen adsorption

The nitrogen adsorption and desorption isotherms of the zeolites are given in Fig. 2. Isotherms of the non-pelletized zeolites are shown for comparison. The BET surface areas, the microporous volumes (*t*-plot method) and the mesoporous volumes (adsorbed nitrogen volume at $P/P_0 = 0.96$ minus the microporous volume) are reported in Table 2. The BJH method has been applied to determine the pore size distribution of the samples (Fig. 3).

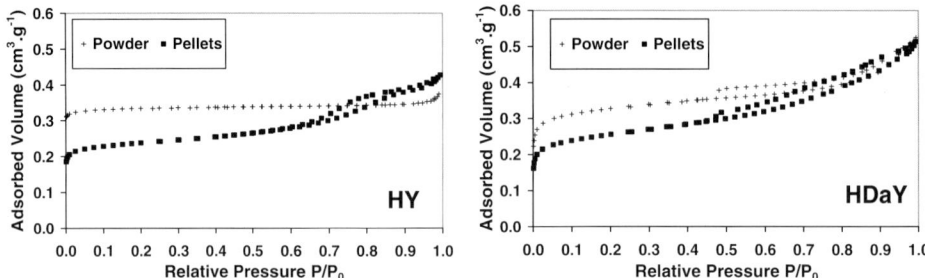

Fig. 2 . Nitrogen adsorption and desorption isotherms of zeolites HY (left) and HDaY (right).

Table 2
Physical properties of the zeolites HY (CBV300) and HDaY (CBV780)

| Samples | BET Surface Area $(m^2.g^{-1})$ | Microporous Volume from $t$-plot $(cm^3.g^{-1})$ | Mesoporous Volume (V at $P/P_0$=0.96 - Microp. Vol.) $(cm^3.g^{-1})$ | Macroporous Volume from mercury porosimetry $(cm^3.g^{-1})$ |
|---|---|---|---|---|
| HY (powder) | 783 | 0.33 | 0.02 | - |
| HY (pellets) | 555 | 0.18 | 0.22 | 0.20 |
| HDaY (powder) | 764 | 0.29 | 0.20 | - |
| HDaY (pellets) | 595 | 0.17 | 0.31 | 0.09 |

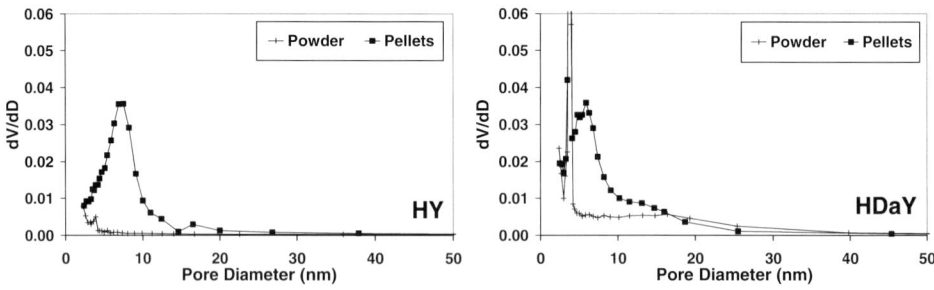

Fig. 3. Pore size distribution of the zeolite HY (left) and HDaY (right) using the BJH method (desorption branch).

### 3.3.2 Mercury porosimetry

The macroporous volume of the samples have been determined by mercury intrusion (see Table 2). The macroporous and interparticle volumes $V_M$ and $V_I$, respectively, can be calculated according to :

$$V_M + V_I = V_C \left[ \varepsilon_I + \left( 1 - \varepsilon_I \right) \varepsilon_M \right]$$

(11)

in which $\varepsilon_M$ and $\varepsilon_I$ are the macroporous and interparticle porosities, defined as :

$$\varepsilon_M = \frac{V_M}{V_G} = \frac{V_P - (V_m + V_\mu + V_S)}{V_P} = 1 - \frac{V_m + V_\mu + V_S}{V_P} = 1 - \frac{\rho_P}{\rho_S^{Hg}} \qquad (12)$$

$$\varepsilon_I = \frac{V_I}{V_C} = \frac{V_C - V_P}{V_C} = 1 - \frac{V_P}{V_C} = 1 - \frac{\rho_B}{\rho_P} \qquad (13)$$

with $\rho_P$ the particle density (g.cm$^{-3}$) obtained by Hg low pressure, $\rho_S^{Hg}$ the strutural density (g.cm$^{-3}$) obtained by Hg medium pressure and $\rho_B$ the bulk density (g.cm$^{-3}$).

## 4. RESULTS AND DISCUSSION

All the pelletized samples show an hysteresis loop (type IV isotherm, see Fig. 2), indicating the presence of mesopores. For the HY zeolite, a non-dealuminated material, this is clearly the consequence of the binding with alumina (see the type I isotherm of the non-pelletized HY zeolite). Zeolite HDaY (powder) has a much larger mesoporous volume (0.20 cm$^3$.g$^{-1}$) than zeolite HY (0.02 cm$^3$.g$^{-1}$), because of the dealumination treatment carried out on this zeolite. The binder alumina has a drastic effect on the samples : it decreases the BET surface area and the microporous volume, and increases the mesoporous volume (see Table 2). It is still not clear why there is such a difference between the macroporous volume of zeolite HY (0.20 cm$^3$.g$^{-1}$) and zeolite HDaY (0.09 cm$^3$.g$^{-1}$), although the binder amount (30 % weight) is the same in the two zeolites.

The pore size distributions determined by the BJH method show that, as expected, the non-pelletized zeolite HY has no mesopore. The non-pelletized zeolite HDaY has mesopores in the range 5-30 nm. As far as the pelletized samples are concerned, it appears that the binder has a drastic effect since it adds a large amount of mesopores centered at about 7 nm. Note that the peaks at about 4 nm are an artefact of the BJH method [4].

The expected volumes and the experimental volumes found with the methodology presented above are gathered in Tables 3 and 4. The expected volumes have been calculated based on the nitrogen adsorption-desorption data ($V_\mu$ and $V_m$, see section *3.3.1*) and mercury porosimetry data ($V_M$ and $V_I$, see section *3.3.2*).

Table 3
Comparison between the expected volumes and the eluted volumes for the different systems (S : solvent ; F : feed) on zeolite HY.

| Systems | Measured Volumes | Expected Volumes (cm$^3$.g$^{-1}$) | Eluted Volumes (cm$^3$.g$^{-1}$) |
|---|---|---|---|
| S: mXyl<br>F: pXyl | $V_I + V_M + V_m + V_\mu$ | 1.25 | $V_{pXyl} = 1.15$ |
| S: nC7<br>F: toluene | $V_I + V_M + V_m + V_\mu$ | 1.25 | $V_{tol} = 1.16$ |
| S: mXyl<br>F: 10 % iPr3-bzn / 90 % mXyl | $V_I + V_M + V_m$ | 1.07 | $V_{iPr3-bzn} = 1.10$ |

The determination of the total porous volume ($V_I + V_M + V_m + V_\mu$) of zeolite HY has been carried out with the systems mXyl/pXyl and nC7/tol (see Fig. 5).

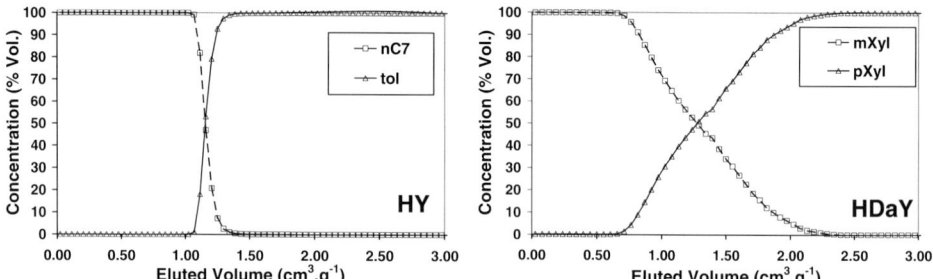

Fig. 5. Breakthrough curves allowing the determination of the total porous volume on zeolite HY (left) and zeolite HDaY (right) ; flow rate : 0.5 cm³.min⁻¹, temperature : 423 K, total pressure : 12 bar.

The volumes eluted with these systems are close to but smaller than the expected volumes (see Table 3). Maybe *p*-xylene and toluene are too rigid molecules and do not fit well the supercages, so these latters can not be totally filled up with that molecule. Therefore, a better molecular probe for the determination of the total porous volume would be an alkane, which could better fit the supercages because of its higher flexibility compared to aromatic molecules (experiments to be carried out).

The bulky molecule used for the determination of the « non-selective » volume ($V_I + V_M + V_{em}$) of zeolite HY was 1,3,5-triisopropylbenzene, diluted in *m*-xylene (see Table 3). The eluted volumes found (1.10 cm³.g⁻¹) is close to the expected volume (1.07 cm³.g⁻¹), but slightly larger. This result may stem from the adsorption of this molecule on the surface of the zeolite crystal. A second explanation would be that 1,3,5-triisopropylbenzene can penetrate into the faujasite micropores (pore opening : 7.4 Å), which has already been reported in the literature [15].

Table 4
Comparison between the expected volumes and the eluted volumes for the different systems (S : solvent ; F : feed) on zeolite HDaY.

| Systems | Measured Volumes | Expected Volumes (cm³.g⁻¹) | Eluted Volumes (cm³.g⁻¹) |
|---|---|---|---|
| S: mXyl<br>F: pXyl | $V_I + V_M + V_m + V_\mu$ | 1.39 | $V_{pXyl} = 1.34$ |
| S: mXyl<br>F: 5 % iPr3-bzn / 95 % mXyl | $V_I + V_M + V_{em}$ | 1.22 | $V_{iPr3\text{-}bzn} = 1.27$ |
| S: mXyl<br>F: 5 % iPr3-cC6 / 95 % mXyl | $V_I + V_M + V_{em}$ | 1.22 | $V_{iPr3\text{-}cC6} = 1.25$ |
| S: mXyl<br>F: 55 % tol / 45 % mes | $V_I + V_M + V_m + V_\mu$ | 1.39 | $V_{tol} = 1.42$<br>$V_{mes} = 1.49$ |
| S: nC7<br>F: 80 % iC8 / 20 % tol | $V_I + V_M + V_m + V_\mu$ | 1.39 | $V_{iC8} = 1.23$<br>$V_{tol} = 1.52$ |

Fig. 5 shows the determination of the total porous volume ($V_I + V_M + V_m + V_\mu$) of zeolite HDaY with the system mXyl/pXyl. For the same reasons as in the case of zeolite HY, the eluted volume is smaller than the expected one. If we look at the breakthrough curves of Figure 5, we see that the front of the system nC7/tol on zeolite HY is much sharper than the front of the system mXyl/pXyl on zeolite HDaY. This is the consequence of the stronger adsorption of toluene on the adsorption sites of the zeolite compared to *n*-heptane, as discussed in section *2.2.1*. Note that the system mXyl/pXyl has also been carried out on zeolite HY (breakthrough curves not shown here), and similar curves than on zeolite HDaY, that is, more spread out curves than the ones obtained with the system nC7/tol, have been obtained.

The determination of the « non-selective » volume ($V_I + V_M + V_{em}$) of zeolite HDaY has been carried out with 1,3,5-triisopropylbenzene, diluted in *m*-xylene (see Table 4). Once again, the eluted volume found (1.27 cm$^3$.g$^{-1}$) is slightly larger than the expected volume (1.22 cm$^3$.g$^{-1}$). The two explanations given for zeolite HY are still valid here. A third explanation could be that the supercage structure is partially destroyed by the dealumination treatment, leading to a greater pore opening and to an easier accessibility to this molecule. Another test has been carried out on zeolite HDaY with 1,3,5-triisopropylcyclohexane. This molecule has no aromatic ring and, therefore, has fewer probabilities than 1,3,5-triisopropylbenzene to adsorb at the crystal surface. This is confirmed by the eluted volume of 1,3,5-triisopropylcyclohexane (1.25 cm$^3$.g$^{-1}$), which is an intermediate volume between the expected one and the eluted one obtained with 1,3,5-triisopropylbenzene.

As discussed in section *2.2.3*, the determination of $V_{im}$ and $V_\mu$ is possible if an adsorption selectivity is induced in the micropores but not in the « internal mesopores ». The probe molecules toluene and mesitylene have first been chosen because of their different molecular size. As it can be seen on Fig. 6 and Table 4, the eluted volumes of toluene and mesitylene are nearly the same, which means that there is virtually no adsorption selectivity between these two molecules in the zeolite micropores. Therefore other probe molecules have to be chosen.

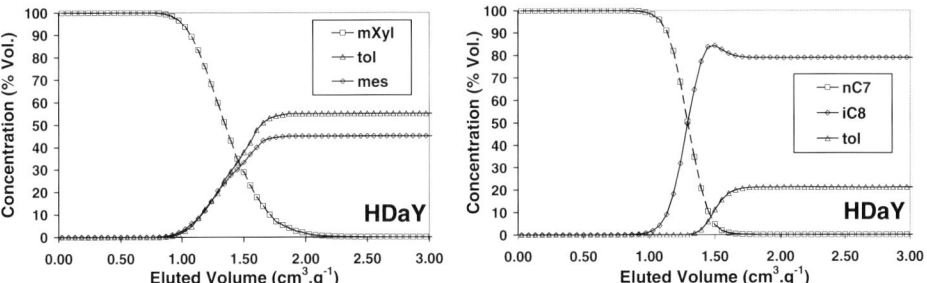

Fig. 6. Breakthrough curves of the system mXyl/(55 % tol + 45 % mes) and nC7/(80 % iC8 + 20 % tol) on zeolite HDaY ; flow rate : 0.5 cm$^3$.min$^{-1}$, temperature : 423 K, total pressure : 12 bar.

Toluene and *i*-octane have totally different adsorption strengths : toluene adsorbs much stronger on the adsorption sites because of its aromatic ring. Fig. 6 shows the breakthrough curves for the system nC7/(80 % iC8 + 20 % tol). The eluted volume of toluene (1.52 cm$^3$.g$^{-1}$) is much larger than the eluted volume of *i*-octane (1.23 cm$^3$.g$^{-1}$), although the *i*-octane concentration in the feed is much higher than the toluene concentration. That means that there is a strong adsorption selectivity for toluene in the zeolite micropores. However, as discussed above, it is likely that toluene molecules do not fit well the zeolite supercages. In addition, it is not excluded that an adsorption selectivity occurs in the "internal mesopores" (whereas there should not be for the sake of the methodology) because of the strong adsorption capacity of toluene compared to *i*-octane. This would lead to misleading conclusions on $V_{im}$ and $V_\mu$. Accordingly, another system has to be found for the characterization of $V_{im}$ and $V_\mu$.

## 5. CONCLUSIONS

The methodology presented herein is, to our knowledge, the first attempt to characterize the different porosities of dealuminated Y zeolites by breakthrough curves. Even if the results are not fully satisfactory, these first sets of experiments carried out on two pelletized Y zeolites can lead to the following conclusions.

The presence of the binder alumina leads, not surprisingly, to an «extra» mesoporous volume, difficult to characterize by itself. Experiments have to be carried out on agglomerated zeolite cristals without binder, even if high pressure mercury characterization (macroporosity) will be difficult to perform accuratly due to a weak mechanical resistance of the samples.

The molecular probes have to be carefully chosen. They should have the following characteristics : (i) probe molecules similar in size have   preferably to be retained ; (ii) aromatic molecules should be avoided as they don't fit well the supercage of zeolites. Due to their flexibility, normal and iso-paraffinic molecules should be prefered when possible in order to characterize the microporosity ; (iii) olefinic molecules could be potentially used together with paraffinic ones to lead to an adsorption selectivity in micropores, but their reactivity with cationic zeolites could be problematic.

Finally, only the first moment $\mu$ has been exploited in the methodology presented in this paper. The second moment of the breakthrough curves ($\sigma^2$), related to the diffusion of the molecules within the particle and crystal, could also give useful information about the pore accessibility.

## REFERENCES

[1] A. Corma, V. Fornés, and F. Rey, Appl. Catal., 59 (1990) 267.
[2] J. Lynch, F. Raatz, and P. Dufresne, Zeolites, 7 (1987) 33.
[3] A. Corma, Stud. Surf. Sci. Catal., 49 A (1989) 49.
[4] A.H. Janssen, A.J. Koster, and K.P. de Jong, J. Phys. Chem. B, 106 (2002) 11905.
[5] K.S.W. Sing; R.T. Williams, Part. Part. Syst. Charact., 21 (2004) 71.
[6] Y.H. Ma and Y.S. Lin, AIChE Symp. Series 259, 83 (1987) 1.
[7] Y.H. Ma, Y.S. Lin and H.L. Flemming, AIChE Symp. Series 264, 84 (1988) 1.
[8] C.H. Ho, C.B. Ching and D.M. Ruthven, Ind. Eng. Chem. Res., 26 (1987) 1407.
[9] E. Santacesaria, M. Morbidelli, A. Servida, G. Storti and S. Carrà, Ind. Eng. Chem. Process Dev., 21 (1982) 446.
[10] S. Carrà, E. Santacesaria, M. Morbidelli, G. Storti and D. Gelosa, Ind. Eng. Chem. Process Dev., 21 (1982) 451.
[11] D.M. Ruthven, Principles of Adsorption and Adsorption Processes, Wiley, New York (1984).
[12] F. Rouquerol, J. Rouquerol and K. Sing, Adsorption by powders and porous solids, Academic Press (1998).
[13] M. J. Remy, D. Stanica, G. Poncelet, E. J. P. Feijen, P. J. Grobet, J. A. Martens and P. A. Jacobs J. Phys. Chem., 100 (1996) 12440.
[14] E.W. Webster, R.S. Drago and M.C. Zerner, J. Am. Chem. Soc., 120 (1998) 5509.
[15] S.F. Zaman, K.F. Loughlin and S.S. Al-Khattaf, Ind. Eng. Chem. Res., 44 (2005) 2027.

Studies in Surface Science and Catalysis 160
P.L. Llewellyn, F. Rodriquez-Reinoso, J. Rouqerol and N. Seaton (Editors)

# Study of the microporous texture of active carbons by Small Angle Neutron Scattering

**N. Cohaut[a], J.M. Guet[a], O. Manfroi[b], A. Albiniak[c] and G. Furdin[b]**

[a] Centre de Recherche sur la Matière Divisée-UMR 6619, 1[b] rue de la Férollerie, 45071, Orléans Cédex 2, France.

[b]Laboratoire de Chimie du Solide Minéral-UMR 7555, BP 239, Université Henri Poincaré, 54506 Vandoeuvre les Nancy Cédex, France.

[c]Institute of Chemistry and Technology of Petroleum and Coal, Wroclaw University of Technology, ul Gdanska 7/9, 50-344 Wroclaw, Poland.

## INTRODUCTION

The behaviour of porous carbons is important for use as gas storage media, and the role played by porosity development during activation and accessibility of fluids to this porosity is fundamental in the selection and production of these materials.

As presented in previous works [1-2], SAXS provides scattering data from both "open" and "closed" porosity, which cannot be decoupled. In this study we have used neutron scattering, presenting, compared to X-ray scattering, the ability to apply contrast matching [3-4] in trying to perform SANS in filling accessible pores with a liquid having a neutron scattering cross section close to that of the carbonaceous matrix.

## 1. PREPARATION AND ADSORPTIOMETRY.

### 1.1. Raw materials

Three different activated carbons (AC) were studied. The first one, named CSa, is derived from saccharose pyrolysed at 800°C under nitrogen flow, and next activated 4 hours with steam at the same temperature. The second AC is an experimental material, called Wa, elaborated from wood through a physical activation process. The third named Maxsorb 30 is manufactured by KANSAY company.

CSa and Wa contain a great amount of oxygen due to the conditions of preparation of the materials (5.7 wt% for CSa sample and 14 wt% for Wa sample).

The oxygen content of carbons is related to the amount of surface functions. Such high amounts are favorable for substituting the hydrogen of the samples with deuterium.

### 1.2. Modification of the surface chemistry of the activated carbons.

The nature and the concentration of surface groups may be modified by suitable chemical treatments. Oxidation in the liquid phase can be used to increase the concentration of surface groups. In order to replace hydrogen of surface functions with deuterium, samples were treated at room temperature with $DNO_3$ - 15% in $D_2O$. In the meantime, reference materials were prepared by immersion of AC with $D_2O$.

The experimental procedure is the following: 30 $cm^3$ of 15% acid solution was added to 1 g of activated carbon. The medium containing the AC particles was filtrated after 15 minutes of stirring. Finally the particles were washed two times with 25 $cm^3$ of $D_2O$. Three conditions of drying were compared : drying at 100°C or 120°C during 1 hour or degassing (5 $10^{-4}$ mbar, 100°C).

The reference treatment consists on the mixing of 1g AC with 25 ml of $D_2O$ during 15 minutes and filtration.

## 2. EXPERIMENTS

### 2.1. Adsorptiometry measurements

The textural parameters of each AC were characterized by adsorption studies of carbon dioxide and benzene molecules at 298 K, using a gravimetric technique. Porosity parameters were evaluated using Dubinin-Radushkevitch micropore filling theory and capillary condensation theory as well as BET theory. BET surface area develop by M30, WA and CSa are 3000 $m^2/g$, 579 $m^2/g$ and 1275 $m^2/g$ respectively. pore volume distributions are given on Fig.1. Porous textures of these AC are rather different. CSa and Wa possess almost identical total pore volumes (0.590 $cm^3/g$) but CSa is microporous at 85 %, whereas 34 % of the pore volume of Wa is attributable to microporosity. M30 is highly microporous (1.63 $cm^3/g$)

It may be noticed that raw coke obtained by carbonization of saccharose (CS) develops a very low surface area, less that 30 $m^2/g$.

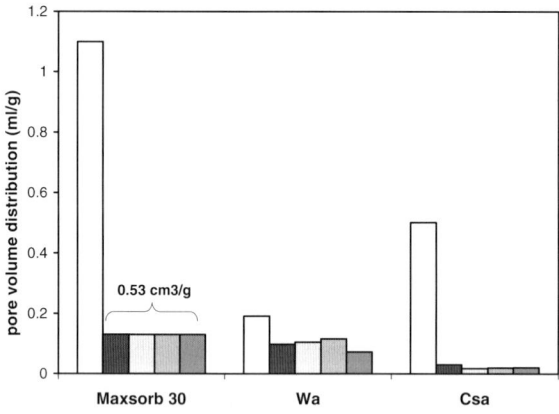

Fig.1: Pore volume distribution obtained from $CO_2$ and $C_6H_6$ adsorptiometry
☐ 0.4-2nm ■ 2-3nm ☐ 3-5nm ☐ 5-10nm ■ 10-50nm

### 2.2. Small Angle Neutrons Scattering (SANS)

Small Angle Scattering measurement of both X-rays and neutrons provide structural details of porous materials on a scale covering a range from 10Å to $10^3$ Å, and the theory of SAS has been covered extensively in several comprehensive reviews and books.

A recent and important development of SAS in the characterization of porous solids is the application on the contrast variation technique. Classically we consider a two phase system where the scattering $d\sigma/d\Omega(q)$, expressed as the length of the scattering vector, is proportional to the contrast, i.e. the square of the scattering length densities difference between the two phases 1 and 2:

$$d\sigma/d\Omega(q) = K[\rho(1) - \rho(2)]^2 = K(\Delta\rho)^2$$

For an evacuated porous solid, where $\rho(1) = \rho$ (solid) the situation is simple since $\rho(2) = 0$. However if pores are filled with an adsorbed fluid, the scattering arising from the contrast pores/matter may be matched if pores are filled with a condensed liquid with a scattering length density, $\rho(2)$ which is the same as the solid, viz ($\Delta\rho)^2 = 0$. This feature may be used to distinguish open and closed porosity or to prove the accessibility of some pores sizes to the adsorbate.

SANS experimentation has been performed at the LLB laboratory (CEA, Saclay) on the PAXY line of the Orphee reactor. The objective was to study the pore volume distribution of studied AC in comparison with results given by gas adsorption and also verify the enhancement of an acid treatment on the penetration of water. Because the penetration of deuterated solutions permit to match the SANS contrast the nitric treatment were done with $DNO_3$ and the washing with deuterated water ($D_2O$) was followed by different air drying conditions.

## 3. DISCUSSION

On SANS plots of AC given on Fig. 2, the flat curve of the CS sample (saccharose pyrolysed at 800°C) shows a weak scattering, meaning a low development of porosity and confirmed by the value of specific surface area lower than $30 m^2/g$.

The curve of CSa shows that the activation during 4h develops essentially a microporosity in the range $q > 0.6$ $nm^{-1}$. For M30 the shape of the SANS curve shows the same features: assuming that the amount of micropores is also predominant on mesopores. The $q^{-4}$ dependence observed for lower q-values may arise from the scattering by the surface of mesopores or by grains with micrometric size.

On contrast with Wa, SANS curve is intense in the small q-range, so that the profile gives evidence for a large contribution of mesopores. These qualitative observations are in agreement with pore volume distributions obtained by $CO_2$ and $C_6H_6$ adsorptiometry since the specific pores volume in mesopores for Wa is 4 times higher than those of CSa sample.

Near the origin, intensity decays first with $q^{-2.6}$ up to 0.4 $nm^{-1}$. In the fractal approach, a decay of SAS intensity with a slope $\alpha$ and $2 < \alpha < 3$ is ascribed to a mass fractal behaviour characterized by the mass fractal dimension $Dm = \alpha$. In the present case, the extension of the q-range where this law is observed covers only one decade so that the mass fractal behaviour may not be reasonably substantiated. We prefer to explain this power law regime with non entire exponent as the combination in the same q-range of features arising from a large polydispersity associated with a lamellar shape of scattering domains, as already observed on carbon materials. [5].

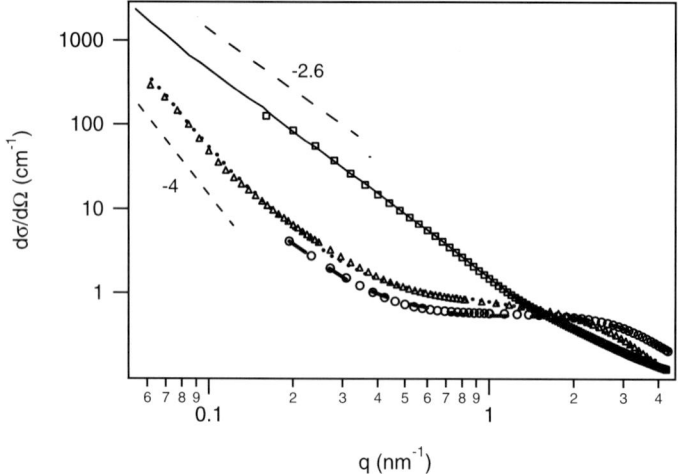

Fig.2. Experimental SANS curves and PVD fits

— CSa   o   CSa fit PVD ····· M30   ▴   M30 fit PVD —— Wa   ▫   Wa fit PVD

### 3.1. Pore volume distribution (PVD)

We have modeled SANS curves of these active carbons, using Indra2 routines from Argonne National Laboratory, running on IGOR software. The fit assumes that the SANS scattering is attributed to the sum of N pore volume distributions. At each distribution is associated the value of the neutronic contrast, the volume fraction of scatterers, the shape of the scattering objects and the type of distribution. Results of best fits are plotted with symbols on Fig. 2. They were obtained assuming log-normal distributions of disk-like pores characterized by a mean diameter and an aspect ratio $R<1$ so that the product ($R*$mean diameter) corresponds to the height of disk-like pores. The fits parameters concerning the microporosity are printed on Table 1.

Only the size of micropores may be studied because pores larger than the limit of the experiments are present but obviously their size may not be accurately deduced from the limited q-range of experiments. Nevertheless the fits (Fig.2.) and the pore volume distribution plotted (Fig. 3) show that the fit of the whole curve needs to take into account of the presence of micro and mesopores even if it is clear that the proposed distribution is truncated at larger size.

The model, sensitive to mean diameter and the anisometry of pores displays that the largest diameters of pores are noticed on M30 which also possesses the lowest anisometric ratio, showing that for this activated carbon the term " slit shaped pores" is the more convenient. Wa and CSa have approximatively the same shape of pores, 1.1-1.5 nm in mean diameter and thus 0.5-0.55nm in height but the distribution is narrower for Wa.

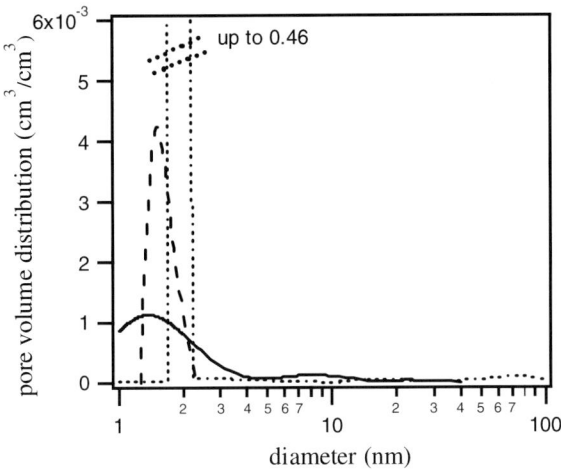

Fig.3. Pores volume distribution corresponding to fits of experimental SANS curves
– – CSa —— Wa·····  M30

It must be pointed out that the PVD obtained by SANS and adsorption are not strictly identical because the first gives the pore distribution in diameter and the second the pore distribution in height. Nevertheless the pores range given by both techniques are in agreement even if SANS provides in our study a better accuracy on the dimension and a measure of the anisometry of micropores.

Table 1
Parameters deduced from SANS fits.(FWHM is the Full Width at Half Maximum).

| Samples | Aspect ratio | Mean diameter (nm) | FWHM (nm) | Scattering volume |
|---|---|---|---|---|
| Maxsorb 30 | 0.18 | 2.25 | 0.14 | 0.86 |
| CSa | 0.35 | 1.5 | 0.4 | 0.58 |
| Wa | 0.5 | 1.1 | 0.78 | 0.25 |

### 3.2. Effect of the acidic post treatment

Fig. 4a, b and c are devoted to the effect of the acidic treatment on the different AC.

Whatever the AC and the choice of an acidic treatment or not, the degassing of the AC permit to remove all solvents used during treatments.

For Wa (Fig.4a), we show that without pre-treatment (15%$DNO_3$/$D_2O$), the deuterated water is removed from micropores even after a simply air drying at 120°C. But if Wa is pretreated with the acidic solution and then, air dried, some $D_2O$ is retained inside the porous network, decreasing significantly the contrast between the two phases. The air drying at 120°C is more efficient than at 100°C since at this last temperature $D_2O$ is always in some pores.

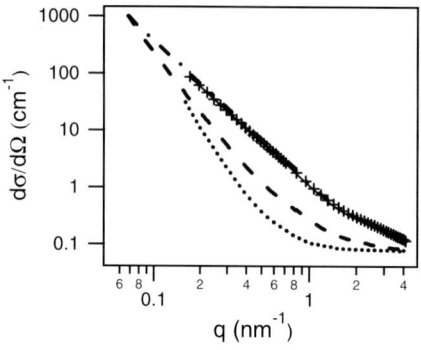 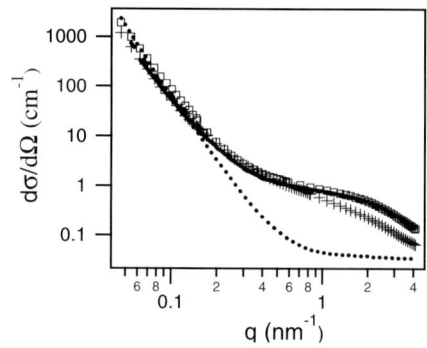

Fig. 4a. ——Wa;
.-.-.-treated 15%DNO$_3$, washing with D$_2$0 and degassing;
+++ washing with D$_2$0 and drying (1h, 120°C);
----- treated 15% DNO$_3$, D$_2$0 washing and drying (1h, 120°C);
....treated 15% DNO$_3$, D$_2$0 washing and drying (1h, 100°C).

Fig. 4b ——M30;
-----treated 15%DNO$_3$, washing with D$_2$0 and degassing;
+++ washing with D$_2$0 and drying (1h, 120°C);
....treated 15% DNO$_3$, D$_2$0 washing and drying (1h, 120°C).

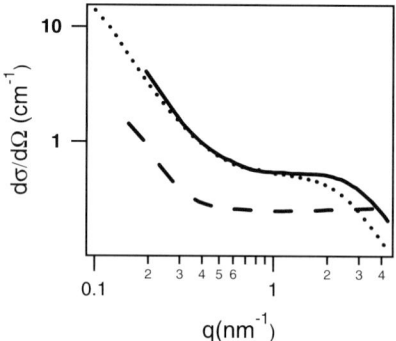

Fig. 4c. ----- Saccharose based coke; —— CSa; ·······CSa treated 15% DNO$_3$, D$_2$0 washing and drying (1h, 100°C).

By contrast with Wa, if the sample M30 (Fig.4b) is treated by the acidic solution a complete retention of D$_2$0 is observed even after drying at 120°C. The lower amount of mesopores is evidently responsible for the remaining of water in micropores.

In the case of CSa (Fig.4c), if it is reasonably assumed that main pores are opened, only the very weak fraction of mesopores may explain the limited penetration of aqueous solution inside micropores so that the treatment is not efficient.

To summarize, a preliminary introduction of acidic groups inside pores tends to make easier the penetration and the holding of deuterated liquid in the porous network. Obviously this is enhanced if the amount of mesopores is not negligible.

## CONCLUSION

Completing the panel of SAS and non-intrusive technique, we have in this paper tested SANS to characterize activated carbon with much efficiency.

In modeling the SANS curves by appropriate pore volume distributions, we showed that the mean size and the width of the distribution in micropores differs from one AC to another and fits are useful to distinguish the degree of anisometry of pores.

By using contrast matching technique we confirm, using a nitric acid attack, the role played by the surface chemistry.

## REFERENCES

[1] J.M. Guet, N. Cohaut, I. Gérard, D. Bégin, G. Furdin and A. Albiniak, (Ed. B.Q.Li and Z.Y. Liu) Prospects for Coal Science in the 21[th] Century, Shanxi Science & Technology Press, Taiyuan, China P.R., 1999, 913.
[2] J.M. Guet, A. Perrin, N. Cohaut, A. Celzard, T. Siemienieska and G. Furdin, Proceedings of Jumelage : « Matériaux Carbonés et Catalytiques pour l'Environnement », Poland, 2002, 71.
[3] J.M. Calo, P.J. Hall, M. Antxustegi, Colloids and Surfaces A: Physicochem. Eng. Aspects 187-188 (2001) 219-232.
[4] N. Cohaut, J.M. Guet, O. Manfroi, J.F. Marêché, A. Celzard, A. Albiniak and G. Furdin, Proceedings of Jumelage « Matériaux Carbonés et Catalytiques pour l'Environnement », Poland, 2004, 43.
[5] N. Cohaut, C. Blanche, C. Dumas, J.M Guet and J.N Rouzaud, Carbon, 38(2000) 1391.

Studies in Surface Science and Catalysis 160
P.L. Llewellyn, F. Rodriquez-Reinoso, J. Rouqerol and N. Seaton (Editors)

# Characterisation of nanostructured materials by combination of neutron scattering and 3D stochastic reconstruction techniques

E.S. Kikkinides[a, b], K.L. Stefanopoulos[c], Th.A. Steriotis[c], A.Ch. Mitropoulos[d] and N.K. Kanellopoulos[c]

[a]Laboratory of Inorganic Materials, Chemical Process Engineering Research Institute, P.O. Box 361, 57001, Thessaloniki, Greece

[b]Department of Engineering and Management of Energy Resources, University of Western Macedonia, Kastorias and Fleming St., 50100, Kozani, Greece

[c]NCSR "Demokritos", Institute of Physical Chemistry, 15310, Agia Paraskevi Attikis, Greece

[d]Department of Petroleum Technology, Kavala Institute of Technology, 65404, Ag. Lucas, Kavala, Greece

Ceramic nanostructured materials have recently gained scientific and industrial interest due to their unique properties. A series of such nanoporous structures were characterized by Small Angle Neutron Scattering (SANS) and by Ultra Small Angle Neutron Scattering (USANS) techniques. The scattering data have been evaluated by using a generalised form of the indirect Fourier transformation method suitable for concentrated systems and compared with those obtained from nitrogen and mercury porosimetry measurements. In addition, by Fourier inversion, the autocorrelation functions were evaluated from the scattering data. A stochastic reconstruction model was employed to generate 3D images of the porous material with the same porosity and autocorrelation function as the original ceramic materials. A comparison between the experimental autocorrelation functions and the reconstructed media autocorrelation functions is presented too. Finally, simulation results of permeation on the reconstructed images provide good agreement with experimental data.

## 1. INTRODUCTION

The characterisation of ceramic materials is, nowadays, a challenging issue worldwide motivated by the wide range of applications (membranes, sorbents, sensors, etc.) involving these solids. SANS is a non-destructive and non-intrusive technique for providing structural details about inhomogeneities of scattering objects, such as pores, in the size range from 10 Å to about 200 nm. In addition, during last decade there is a gradually increasing interest for

USANS facilities. USANS technique is a useful tool for the investigation of larger inhomogeneities (such as macropores) having sizes extending up to few micrometers in real space [1]. On the other hand, recent developments in reconstructing 3D images of actual porous media have prompted research on simulations of diffusion in such structures [2-4]. These studies deal with the development of stochastic methods that generate reconstructed structures based on information from two-dimensional (2D) images of the actual porous media obtained through serial tomography and/or SEM-TEM techniques. Alternatively this information can be obtained indirectly by Small Angle Scattering (SAS) data. The basic underlying principle is that both the real and the model structures should have identical statistical properties, such as the average porosity and the autocorrelation function which are used as input for the creation of the simulated structures under the assumption of statistical homogeneity [5]. Recent studies in reconstructing 3D images of Vycor porous glass based on structural information obtained by SAS have opened up the possibility of employing more sophisticated adsorption and diffusion simulations in such structures [6-9].

## 2. EXPERIMENTAL

In the present work two membranes representing different classes of porous materials were studied: (a) a $\gamma$-Al$_2$O$_3$ mesoporous pellet, prepared by symmetrical compaction of Degussa (type-C) aluminium oxide powder and (b) a commercial $\alpha$-Al$_2$O$_3$ macroporous membrane in fiber form. The membranes were characterised by nitrogen ($\gamma$-Al$_2$O$_3$) and mercury ($\alpha$-Al$_2$O$_3$) porosimetry. The measured porosities are reported in Table 1.

The SANS measurements of the $\gamma$-Al$_2$O$_3$ membrane were carried out on V4 instrument, HMI, Berlin at a $Q$ range 0.1-2.4 nm$^{-1}$. The $\alpha$-Al$_2$O$_3$ was also measured at V4 ($0.04 < Q < 1$ nm$^{-1}$). In both cases the wavelength was 0.457 nm. The raw data were further corrected for background and empty cell scattering by using the BerSANS software developed at BENSC [10]. In order to cover a wider range of sizes, additional USANS measurements were performed for macroporous alumina on the double crystal diffractometer V12a, HMI, Berlin. The wavelength was 0.4763 nm and the covered $Q$ range varied between 0.002 to 0.07 nm$^{-1}$. The scattering curves (empty cuvette, cuvette + sample) were corrected for the instrumental background and for the attenuation of neutrons passing into the analyser crystal [11]. Fig. 1a and 1b (insets) show the scattering curves from the two samples.

The $\gamma$-Al$_2$O$_3$ membrane was attached to a permeability rig, described in [12], outgassed under high vacuum at 200 °C and the differential steady state permeability of N$_2$ and He was measured at mean pressures ranging from 1 to 60 bar. The measurements were carried out by keeping both sides of the membrane under pressure while maintaining a constant small pressure head (1 bar) on the high-pressure side. The permeability was calculated after monitoring the pressure change on the low-pressure side. Following Barrer [13], proper corrections were performed in order to subtract the Knudsen and slip flow contributions and thus extract only the viscous component of the overall permeability measured. The validity of corrections was confirmed, since the value calculated for both gases were identical. The permeability of the macroporous $\alpha$-Al$_2$O$_3$ membrane was measured after attaching the membrane on an air-gas cylinder equipped with a pressure reducer. Air (1-10 bar) was admitted to the high pressure side of the membrane, the volume flow was measured by a soap flowmeter attached on the low pressure side, while the pressure head was monitored by means of a differential manometer. The aforementioned corrections for Knudsen and slip flow contribution were also applied.

## 3. EVALUATION OF THE SCATTERING DATA

In the general case, the scattering from a homogeneous and isotropic dispersion of particles (pores) is proportional to the product of the single particle (pore) form factor $P(Q)$ and the structure factor $S(Q)$ arising from interference effects in the scattering from particles (pores) being in close separation. The scattering data were further evaluated by using a generalized form of the indirect Fourier transformation method for concentrated systems that allows the determination of both the form factor and the structure factor simultaneously [14-15]. This is possible due to the different analytical behaviour of these functions, which leads in most cases to the existence of a global minimum in the parameter surface. The averaged structure factor, containing the interference effects, is based on an analytical solution of the Percus-Yevick approximation, which is a model for hard spheres without any charge. As a result, the pore radius distribution is calculated for a polydisperse pore system [15]. The method takes also into account the slit-smearing effect. As a result the USANS data were desmeared using the slit length profile of the V12a instrument. The resulting pore size distributions are in good agreement with those deduced from the porosimetry data (Fig. 1a, 1b).

## 4. REPRESENTATION OF THE POROUS STRUCTURE

The spatial distribution of matter in a porous medium can be typically represented by the phase function $Z(\mathbf{x})$, defined as follows:

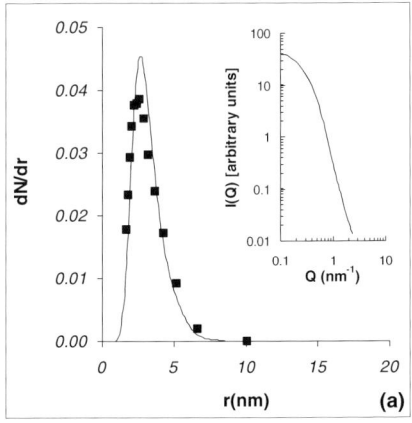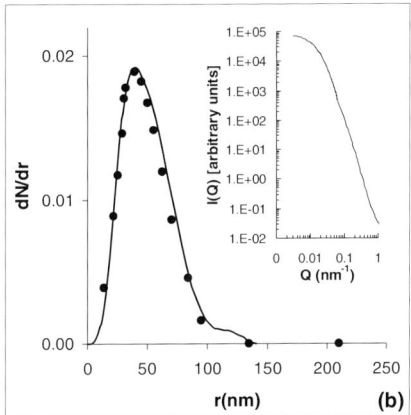

Fig. 1. Pore size distributions deduced from porosimetry data (symbols), evaluated from scattering data (lines) and scattering curves (insets) for (a) $\gamma$-Al$_2$O$_3$ and (b) for $\alpha$-Al$_2$O$_3$.

$$Z(\mathbf{x}) = \begin{cases} 1 & \text{if } \mathbf{x} \text{ belongs to the pore space} \\ 0 & \text{otherwise} \end{cases} \tag{1}$$

where $\mathbf{x}$ is the position vector from an arbitrary origin.

Due to the disordered nature of porous media, $Z(\mathbf{x})$ can be considered as a stochastic process, characterized by its statistical properties. The porosity, $\varepsilon$, and the autocorrelation function $R_z(\mathbf{r})$ can be defined by the statistical averages [5]:

$$\varepsilon = \langle Z(\mathbf{x}) \rangle \tag{2a}$$

$$R_z(\mathbf{r}) = \frac{\langle (Z(\mathbf{x}) - \varepsilon) \cdot (Z(\mathbf{x} + \mathbf{r}) - \varepsilon) \rangle}{\varepsilon - \varepsilon^2} \tag{2b}$$

Note that $<\cdot>$ indicates spatial average. For an isotropic medium, $R_z(\mathbf{r})$ becomes one dimensional as it is only a function of $r = |\mathbf{r}|$ [5].

## 5. RELATION BETWEEN SANS SPECTRUM AND AUTOCORRELATION FUNCTION

According to SAS theory, for an isotropic scatterer, the spherically averaged intensities $I(Q)$ may be represented by the integral [4, 6]:

$$I(Q) = 4\pi\varepsilon(1-\varepsilon)\rho^2 V \int_0^\infty r^2 \gamma(r) \frac{\sin Qr}{Qr} dr \tag{3}$$

where $V$ is the volume of the sample, $\rho$ is the neutron scattering length density, $\gamma(r)$ is the density fluctuation autocorrelation function at point $r$, which is essentially the same as the autocorrelation function of the porous material, $\varepsilon$ is the porosity of the sample and $Q$ is the scattering vector ($Q = 4\pi\sin\theta/\lambda$, where $\lambda$ is the wavelength and $2\theta$ is the scattering angle). Since isotropic media have been assumed, vectors have been substituted by scalars. Eq. (3) shows how the observable intensity can be calculated as a function of the scattering vector, provided the correlation function is known for all distances. Conversely, using the Fourier-inversion of Eq. (3), $\gamma(r)$ can be calculated as a function of the distance, provided the intensity is known as a function of the scattering vector.

## 6. STOCHASTIC RECONSTRUCTION

The purpose of the stochastic reconstruction procedure applied in the present work is the generation of a digitised 3-dimensional snapshot of $Z(\mathbf{x})$ with a specified statistical behaviour assumed to be described by the first two moments of $Z(\mathbf{x})$, namely the porosity and the two point correlation function. The algorithm used for the reconstruction was first proposed by Joshi [16] and was extended in three dimensions by Quiblier [17] and Adler *et al.* [2]. In brief, the space is discretised in $N^3$ cubic elements, the position of which is characterized by the vector $\mathbf{x}' = (i, j, k)$ where $i, j, k$ integers with values 1, 2,..., $N$ and a random value $X(\mathbf{x}')$ is assigned to any element. The values $X(\mathbf{x}')$ are uncorrelated and normally distributed with a mean equal to 0 and a variance equal to 1. A correlated field $Y$ with a correlation function $R_y(\mathbf{r})$ can be deduced from the $X$ field by the inverse Fourier transform:

$$Y(\mathbf{x'}) = N^{3/2} \sum_{\mathbf{m}} \left( \hat{R}_{y\mathbf{m}} \right)^{1/2} \cdot \hat{X}_{\mathbf{m}} \cdot e^{-2i\pi \mathbf{k}_\mathbf{m} \mathbf{x'}} \qquad (4)$$

where $\hat{R}_{y\mathbf{m}}$ and $\hat{X}_{\mathbf{m}}$ are the coefficients of the discrete Fourier transform of $R_y$ and $X$ respectively. The values $Y(\mathbf{x'})$ are real and normally distributed with zero mean and unit variance, hence the distribution function $P(y)$ is given by:

$$P(y) = (2\pi)^{1/2} \int_{-\infty}^{y} e^{-t^2/2} dt \qquad (5)$$

The extraction of the binary phase function $Z(\mathbf{x'})$ from the real array $Y$ can be accomplished by the condition:

$$Z(\mathbf{x}) = \begin{cases} 1 & \text{if } P[Y(\mathbf{x'})] \le \varepsilon \\ 0 & \text{otherwise} \end{cases} \qquad (6)$$

The most difficult step of the overall technique is the determination of the correlation function $R_y(r)$ from the experimentally observed $R_z(r)$. After a series of manipulations [5], $R_z(r)$ can be expressed as a series of $R_y(r)$, as follows:

$$R_z(r) = \sum_{m=0}^{\infty} C_m^2 \cdot R_y^m(r) \qquad (7)$$

The coefficients $C_m$ are given by

$$C_m = (2\pi m!)^{-1/2} \int_{-\infty}^{+\infty} c(y) e^{-y^2/2} H_m(y) dy \qquad (8)$$

where

$$c(y) = \begin{cases} \dfrac{\varepsilon - 1}{\left[ \varepsilon(1-\varepsilon) \right]^{1/2}} & \text{if } P(y) \le \varepsilon \\[3mm] \dfrac{\varepsilon}{\left[ \varepsilon(1-\varepsilon) \right]^{1/2}} & \text{if } P(y) > \varepsilon \end{cases} \qquad (9)$$

and $H_m(y)$ is the Hermite polynomial of m[th] order:

$$H_m(y) = (-1)^m e^{y^2/2} \frac{d^m}{dy^m} e^{-y^2/2} \qquad (10)$$

$\gamma\text{-Al}_2\text{O}_3$ $\qquad\qquad\qquad$ $\alpha\text{-Al}_2\text{O}_3$

Fig. 2. Typical cross sections for the reconstructed samples (white: solid, black: pores): $\gamma\text{-Al}_2\text{O}_3$ ($\varepsilon$=0.42, sample length=110 nm), $\alpha\text{-Al}_2\text{O}_3$ ($\varepsilon$=0.355, sample length=1000 nm).

Ideally, a representative reconstruction of a porous medium in three dimensions should have the same correlation properties as those measured on a single two-dimensional section, expressed properly by the various moments of the phase function. In practice, matching of the first-two moments, that is, porosity and autocorrelation function, has been customarily pursued since the first application of the method [2].

The analysis is based on averages of four realizations of reconstructions of two different porous media with $\varepsilon=0.42$ and $\varepsilon=0.355$, respectively. Typical cross sections for each medium are shown in Fig. 2. A comparison between the experimental autocorrelation functions (SANS data) and the autocorrelation functions measured on the reconstructed media is presented in Fig. 3a and 3b. It is evident that the stochastic reconstructions exhibit nearly identical autocorrelation functions with the experimentally observed ones. This indicates that the reconstructed materials respect the basic statistical content of the actual porous materials.

## 7. CALCULATION OF PERMEABILITY

The creeping flow of an incompressible fluid is described by the Stokes equation coupled with the continuity equation:

$$\nabla p = \mu \nabla^2 \boldsymbol{v} \tag{11a}$$

$$\nabla \boldsymbol{v} = 0 \tag{11b}$$

where $\boldsymbol{v}$, $p$, $\mu$ are the local velocity, the pressure and the viscosity of the fluid respectively. The boundary conditions for $\boldsymbol{v}$ are spatial periodicity and no-slip at the surface of the solid unit elements. The above set of equations applies locally at each point of the interstitial space. In addition, a macroscopic pressure gradient $\nabla p$ is specified. The seepage velocity, $<\boldsymbol{v}>$, is the superficial velocity averaged over a cross-section of the medium. This quantity is related to the macroscopic pressure gradient by the permeability tensor $\boldsymbol{K}$, as follows:

$$\langle \boldsymbol{v} \rangle = -(\boldsymbol{K} / \mu)\nabla p \tag{12}$$

$\boldsymbol{K}$ is a symmetric tensor that depends only on the geometry of the system. For isotropic media, $\boldsymbol{K}=K\boldsymbol{I}$, where $\boldsymbol{I}$ is the unit tensor.

A finite difference scheme using a staggered marker-and-cell mesh is used, with the pressure defined at the centre of the cell, and the velocity components defined along the corresponding surface boundaries of the rectangular cell [3]. In Table 1 we compare the computed predictions of permeability on the reconstructed porous media with the experimentally measured values on the original samples. It is evident that in both cases the prediction is remarkably good considering the fact that reconstruction of these materials has been based solely on the information obtained from the SANS spectra and the porosity of each material.

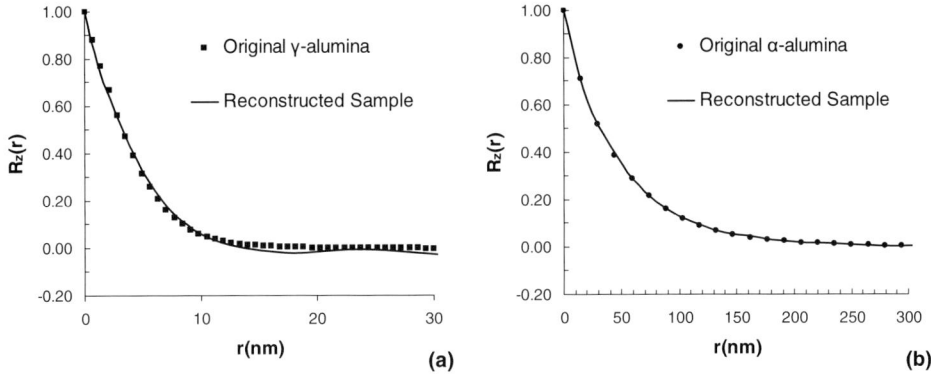

Fig. 3. Autocorrelation functions evaluated from the scattering data (symbols), measured on the reconstructed media (lines) for (a) $\gamma$-Al$_2$O$_3$ and (b) for $\alpha$-Al$_2$O$_3$.

Table 1
Comparison between experimental permeability $P_{exp}$ and calculated permeability ($P_{sim}$) of the reconstructed materials.

| Material | $\varepsilon$ | $P_{exp}$ (m$^2$) | $P_{sim}$ (m$^2$) |
|---|---|---|---|
| $\gamma$-Al$_2$O$_3$ | 0.42 | $4.0 \times 10^{-19}$ | $3.1 \times 10^{-19}$ |
| $\alpha$-Al$_2$O$_3$ | 0.355 | $2.5 \times 10^{-17}$ | $2.0 \times 10^{-17}$ |

## CONCLUSIONS

Two alumina membranes with different classes of pores were characterised by nitrogen adsorption, mercury porosimetry, permeability and SANS (USANS). The pore size distributions obtained from porosimetry data are in good agreement with those evaluated from the scattering data by using the indirect Fourier transformation method. The experimental autocorrelation functions obtained from the scattering data and the autocorrelation functions measured on the reconstructed media based on the stochastic reconstruction model are in reasonable agreement too. Finally, the computed predictions of permeability on the reconstructed porous media with the experimentally measured values on the original samples are remarkably good. Differences between simulation and experimental values can be attributed to (a) slip effects in the flow of gases through the porous materials during the experiment that are not accounted for in the simulation, (b) inability to represent accurately the pore structure of the materials by matching only the first two moments of the phase function and (c) heterogeneity and/or anisotropy in the porous structure that could not be accounted for when transforming the SANS spectra to autocorrelation functions.

### Acknowledgements
We gratefully acknowledge the support by BENSC, HMI, Berlin, Germany, by the EC (contract HPRI-CT-1999-00020) and by the Archimides Research Program.

## REFERENCES

[1]     P.K. Makri, K.L. Stefanopoulos, A.Ch. Mitropoulos, N.K. Kanellopoulos and W. Treimer, Physica B, 276-278 (2000) 479.

[2]     P.M. Adler, C.J. Jacquin and J.A. Quiblier, Int. J. Multiphase Flow, 16 (1990) 691.

[3]     E.S. Kikkinides and V.N. Burganos, Phys. Rev. E, 59 (1999) 7185.

[4]     P. Levitz and D. Tchoubar, J. Phys. I , 2 (1992) 771.

[5]     P.M. Adler, Porous Media, Geometry and Transports, Butterworth, London, 1992.

[6]     P. Debye, H.R. Anderson and H. Brumberger, J. Appl. Phys., 28 (1957) 679.

[7]     O. Glatter, Appl. Cryst. 12 (1979) 166.

[8]     E.S. Kikkinides, M.E Kainourgiakis, K.L. Stefanopoulos, A.Ch. Mitropoulos, A.K. Stubos and N.K Kanellopoulos. J. Chem. Phys., 112 (2000) 9881.

[9]     J.D.F. Ramsay and E. Hoinkis, Physica B, 248 (1998) 322.

[10]    U. Keiderling, Physica B, 234-236 (1997) 1111.

[11]    P.J. McMahon and W. Treimer, Cryst. Res. Technol., 33 (1998) 625.

[12]    T.A. Steriotis, F. Katsaros, A.K. Stubos, A.Ch. Mitropoulos, P. Galiatsatou and N.K. Kanellopoulos, Rev. Scient. Instrum., 67 (1996) 2545.

[13]    R.M. Barrer, The solid-gas interface, E.A. Flood (ed.), Marcel Dekker Inc, NY, vol. 2, 1967.

[14]    J. Brunner-Popela and O. Glatter, J. Appl. Cryst., 30 (1997) 431.

[15]    B. Weyerich, J. Brunner-Popela and O. Glatter, J. Appl. Cryst., 32 (1999) 197.

[16]    M.Y. Joshi, A Class of Stochastic Models for Porous Media, PhD thesis, Univ. of Kansas, 1974.

[17]    J.A. Quiblier, J. Colloid Interface Sci. 98, 84 (1986).

Studies in Surface Science and Catalysis 160
P.L. Llewellyn, F. Rodriquez-Reinoso, J. Rouqerol and N. Seaton (Editors)

# Hydrogen storage in nanoporous carbons

**O.Y. Odunsi, Y. He and T.J. Mays**[*]

Department of Chemical Engineering, University of Bath, Bath BA2 7AY, United Kingdom

Hydrogen is of much current interest as an energy store and/or carrier in sustainable energy conversion systems. The attractions are that $H_2$ generates ~143 MJ kg$^{-1}$ when it reacts with $O_2$, with only $H_2O$ as the product. This compares to fossil fuels, which have lower specific energy densities (in the range 10–50 MJ kg$^{-1}$), and which generate environmentally–damaging $CO_2$. However, there are many challenges to be met before energy systems based on $H_2$ are technically feasible or socially/economically acceptable. One of the main challenges is to store $H_2$ safely, effectively (principally at low volume) and at low cost.

This paper presents the results of an experimental study of supercritical $H_2$ adsorption in nanoporous carbons plus some supporting data from molecular simulations of adsorption. This is part of a broad examination of the potential of these materials as storage media for $H_2$ in sustainable energy systems, in comparison with other proposed storage media such as carbon nanofibres and metal hydrides, and with other storage systems such cryogenic liquid and high–pressure gas. Note that the new term nanoporous refers to pores in the size range 0.1 to 100 nm, as proposed in another Mays paper in these Proceedings.

Recent results of this work appear to show high $H_2$ uptakes in certain nanoporous carbons, but that these uptakes only occur for very pure hydrogen. We propose a pore blocking mechanism to account for this. While at present high uptakes only appear possible at liquid nitrogen temperatures, the pressures are 'reasonable' (below 2000 kPa).

## 1. INTRODUCTION

Sustainable energy supply is crucial in helping to tackle issues related to fossil fuels such depleting energy resources, decreasing energy security, reducing air quality and global warming. Hydrogen is a suitable and likely candidate as an energy vector to help in combating these issues because the products of its combustion are environmentally benign (water vapour). However, one challenge on the road to a 'hydrogen economy' is that as the lightest element hydrogen is difficult to store in small volumes. This is unsuitable if hydrogen is to be used for mobile applications where on-board space is a premium. For example, 5 kg of $H_2$ could provide enough energy for a standard saloon car to cruise for 500 km. However, in ambient conditions this would occupy a spherical vessel of about 5 m internal diameter, which is impractical. By contrast 5 kg of liquid $H_2$ would occupy a spherical vessel of only 0.5 m internal diameter, which is potentially more manageable.

Currently, hydrogen can be stored as a liquid (its most energy dense form) and as a compressed gas at very high pressures and in solid-state systems such as metal hydrides.

[*] Corresponding author: Tel: +44 (0)1225 386528, Fax: +44 (0)1225 385713,
E-mail: t.j.mays@bath.ac.uk

There are issues surrounding these methods such as the energy intensive nature of the hydrogen liquefaction process and problems due to boil-off, and the associated high pressures for gas storage, which can pose safety issues on board a vehicle. Metal hydride storage systems require high desorption temperatures and may only store hydrogen irreversibly.

Nanoporous carbons have been identified as potentially good hydrogen storage media as they are widely available, are low cost, they have high surface areas and pore volumes and may be easily regenerated for recycling purposes. Hydrogen storage in these materials is the main focus of this paper. We have measured adsorption isotherms on these materials using industrial or low-grade and research or high-grade grade hydrogen to determine whether the concentration of impurities in the gas has an impact on the storage capacity of these materials. We have also carried out grand canonical Monte Carlo simulations to investigate these effects.

## 2. EXPERIMENTAL DETAILS

The adsorbents used in this work are AC-1 (supplied by Calgon Carbon Corporation) and AC-2 (supplied by Norit Activated carbon). AC-1 is a bituminous coal based granular material activated at high temperature in steam. AC-2 is a pelletised chemically activated carbon, produced from a renewable raw material source via a version of the phosphoric acid process. It is a high activity, medium density grade carbon. These materials have surface areas of 1,000 and 1,262 $m^2$ $g^{-1}$ respectively obtained from BET analysis of nitrogen adsorption measurements at 77 K below relative pressures of 0.4 using a Micromeritics ASAP 2010 volumetric system. The hydrogen gas used was of two grades: 99.995 % (low grade, LG-$H_2$) and 99.9995 % (high grade, HG-$H_2$) purity supplied by BOC Special Gases Division. The lower grade hydrogen has no more than 50 ppm of impurities while the high-grade hydrogen has about 5 ppm impurities, see Table 1.

Table 1
Hydrogen gas specification (supplied by BOC Special Gases)

| Grade | Purity (% min) | $O_2$ ppm | $N_2$ ppm | THC* ppm | $CO_2$ ppm | $H_2O$ ppm | CO ppm |
|-------|----------------|-----------|-----------|----------|------------|------------|--------|
| LG-$H_2$ | 99.995 | | Total impurities 50 ppm maximum | | | | |
| HG-$H_2$ | 99.9995 | 1 | 2 | 0.5 | 0.5 | 1 | - |

* Total hydrocarbons

Equilibrium hydrogen gas adsorption isotherms were obtained gravimetrically using an Intelligent Gravimetric Analyser (IGA) (Hiden Analytical Ltd.). This instrument combines computer-control and measurement of weight change, pressure and temperature to enable determination of gas adsorption-desorption isotherms in a wide range of operating conditions. After equilibrium is established at a particular pressure, the pressure is increased to the next set value and the subsequent uptake is measured over time until equilibrium is re-established. We have carried out experiments at 77 K, 195 K, 273 K and 303 K over a pressure range of 0 – 2000 kPa.

Prior to the measurement of each isotherm samples of AC-1 (~150 mg) were degassed to a constant weight at 423 K and $10^{-6}$ bar for 16 hours. For the isotherms measured with high-grade hydrogen on AC-2, the sample was prepared by heating to 523 K and $10^{-6}$ bar for 4 hours; the higher temperature having being determined to achieve similar degassing weight losses but over a shorter time. The pressure was monitored using three pressure transducers with ranges of 0 – 1 bar, 0 – 10 bar and 0 – 20 bar and maintained at the relevant set-point by active computer control of the feed and exhaust valves throughout the duration of the experiments to an accuracy of +/- 0.02 % of the range used. The isotherms generated are reproducible when the experimental set-up is kept the same as described above. The hydrogen adsorption amount (% mass uptake) is calculated as follows

$$\% \text{ mass uptake} = ( w_E / w_S ) \times 100 \qquad (1)$$

where $w_E$ is the excess uptake (weight of adsorbate less the weight of adsorptive that would occupy the same volume as the adsorptive at the same pressure and temperature) and $w_S$ is the weight of the sample in vacuum after degassing. The excess uptake is calculated as follows

$$w_E = w - w_S + \rho_B V_S \qquad (2)$$

where $w$ is the detected weight of the sample, $\rho_B$ is the bulk mass density of the adsorptive and $V_S$ is the inaccessible volume of the sample (which displaces the bulk gas). The bulk mass density of the adsorptive is calculated using the Peng-Robinson equation of state, while $V_S$ is obtained in ambient conditions using a He pycnometer (AccuPyc 1330, Micromeritics) assuming no He uptake.

## 3. MONTE CARLO SIMULATION

Grand canonical Monte Carlo (GCMC) simulations, in which the temperature, the volume of the simulation cell and the chemical potential of the adsorbate-adsorptive system are kept constant, have been carried out to probe the effect of impurities on the storage of hydrogen on nanoporous carbons. Details of the GCMC simulations can be found in the work of Frenkel and Smit [1]. The impurities are taken as nitrogen, since it is the dominant balance gas and oxygen and carbon dioxide have similar effects as nitrogen. Nitrogen molecules are modelled also as two-site Lennard-Jones (LJ) spheres, which interact via the 12-6 LJ potential. A two-site LJ spheres model is also applied to hydrogen. The Lennard-Jones parameters (collision diameter, $\sigma$, energy parameter, $\varepsilon / k_B$, where $k_B$ is Boltzmann's constant, and bond length) for the adsorptive species are given in Table 2.

Table 2
Lennard-Jones parameters for the adsorptive species in the molecular simulations

| Adsorptive | $\sigma$ / nm | ( $\varepsilon / k_B$ ) / K | Bond length / nm |
|---|---|---|---|
| Hydrogen [2] | 0.259 | 12.5 | 0.074 |
| Nitrogen [3] | 0.331 | 37.3 | 0.1098 |

The pores in the activated carbon adsorbent are modelled as structureless slits, and each pore wall consists of an infinite number of structureless graphitic layers composed of Lennard-Jones sites. The interaction between a site on an adsorptive molecule and a single semi-infinite slab of graphite is given by Steele's 10-4-3 potential [4]:

$$u_{sf}^{*}(z_{i}) = 2\pi\varepsilon_{sf}\rho_{s}\sigma_{sf}^{2}\Delta\left[\frac{2}{5}\left(\frac{\sigma_{sf}}{z_{i}}\right)^{10} - \left(\frac{\sigma_{sf}}{z_{i}}\right)^{4} - \frac{\sigma_{sf}^{4}}{3\Delta(z_{i}+0.61\Delta)^{3}}\right] \qquad (3)$$

Where $\rho_s$ is the number of carbon atoms per unit volume in the graphitic layer ($114\ nm^{-3}$), $\Delta$ is the separation distance between layers of graphitic carbon (0.3354 nm), and $z_i$ is the distance between the site and the surface. The subscript sf refers to LJ parameters for solid-fluid interactions obtained using Lorentz-Berthelot combination rules.  All the simulations were carried out in a rectangular simulation cell, which is bounded in the z direction by the pore walls and replicated in the x and y directions. The cell length in the x and y directions is 4.44 nm and periodic boundary conditions are applied in these directions.  The cut-off distance, beyond which the potential is neglected, is set to be 1.524 nm. The system is equilibrated with $6\times10^6$ Monte Carlo steps, then the variables of interest are averaged over another $1\times10^7$ Monte Carlo steps  Output of the simulations is the total number of molecules of each gas species in a pore at a specified pressure and temperature.  This can easily be converted to excess uptake as the pore volume in known. Further details of the GCMC simulations can be found in the work of Frenkel and Smit [1].

## 4. RESULTS AND DISCUSSION

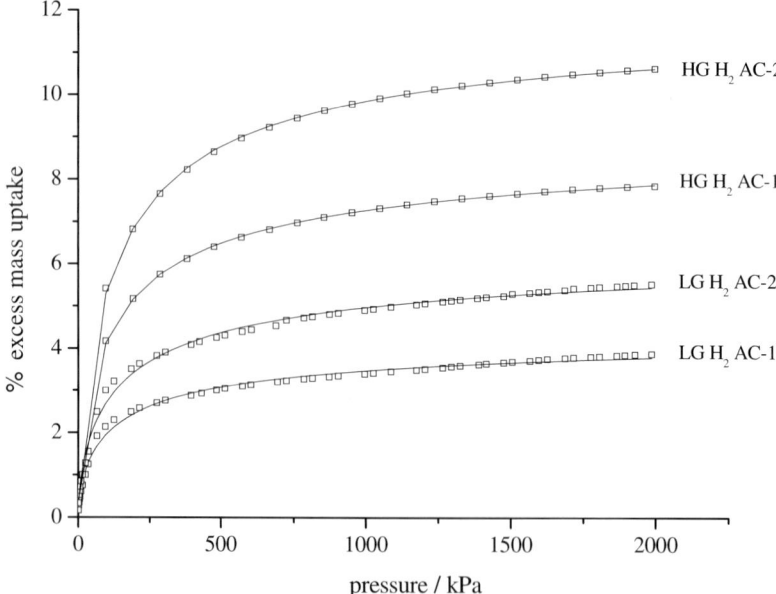

Fig. 1. Experimental adsorption isotherms at 77 K for the two activated carbons (AC-1 and AC-2) and for low and high grade (LG and HG) hydrogen.  The experimental data have been fitted to the Langmuir-Freundlich (or Sips) equation used commonly for representing supercritical Type I isotherms.

Fig. 1. shows the excess adsorption isotherms measured gravimetrically at 77 K on the two activated carbon samples for the two grades of hydrogen. All isotherms are of Type I in the IUPAC classification - this is typical of supercritical adsorption (where there is no condensation in pores) at pressures where the density of the bulk gas is very much lower then the adsorbate. For both samples the % mass uptake increases when high-grade hydrogen is used, with AC-2 always showing a higher uptake than AC-1. The latter can be linked to the higher BET surface area of AC-2 (1,262 $m^2$ $g^{-1}$) compared to AC-1 (1,000 $m^2$ $g^{-1}$). Isotherms at higher temperatures were similar, but reached much lower maximum uptakes (< 2 %).

The main observation from Fig. 1 is the very high maximum uptakes for both carbons (~10 % for AC-2 and ~7.5% for AC-1) in the HG-$H_2$ compared with capacities about half these values for the LG-$H_2$, which are commonly observed for activated carbons. This implies that the storage of hydrogen in porous carbons may be affected by the purity of the gas used and that these carbons may be sensitive to ppm amounts of impurities such as nitrogen, which therefore may compromise their storage capacity. This is in agreement with the work of Amankwah and Schwarz [5], where the amount of hydrogen adsorbed by an activated carbon appeared to be reduced in the presence of nitrogen impurities. The difference between uptakes for the two adsorptive grades may be due to differences between the molecular sizes and interactions energies of the hydrogen and impurities (see Table 1) and how these affect uptake in pores of different sizes. Results from the simulations clarify this.

Fig. 2 shows the GCMC simulation results of the adsorption of binary $H_2$/$N_2$ mixtures (including HG-$H_2$, 5ppm $N_2$, and LG-$H_2$, 50 ppm $N_2$, see Table 1) in slit-shaped pores of size 0.74 nm and 2.664 nm. Those simulated isotherms are presented as plots of mass density for the component as a function of pressure. The simulated pure hydrogen adsorption isotherms are also shown in Fig. 2 as a comparison. In the 0.74 nm pore, the existence of trace amounts of $N_2$ greatly reduces the amount of $H_2$ adsorbed in the pore, and the amount of $H_2$ adsorbed is less than 10% in comparison to pure $H_2$ adsorption at 2000 kPa. The adsorbed phase still contains $N_2$ even when using HG-$H_2$ where the amount of $N_2$ in the gas phase is only 5 ppm, though the amount is less than for 50 ppm. This very high weight uptake for $N_2$ at such a low partial pressure (about $10^{-4}$ kPa) is caused by the strong overlap between the interactions of $N_2$ with the two parallel slit pore walls; such interactions are much smaller for $H_2$-wall interactions. For LG-$H_2$, the amount adsorbed for hydrogen is small even at high pressures. In the 2.664 nm pore, the existence of $N_2$ seems not to influence the adsorption of $H_2$, which decreases only marginally as the content of $N_2$ increases from 5 ppm to 50 ppm, and is quite similar to pure hydrogen adsorption. The amount of $N_2$ adsorbed is considerably reduced as the $N_2$ content in the gas phase decreases from 50 ppm to 5 ppm. The adsorbed phase is mainly $H_2$, especially at high pressures and with HG-$H_2$. The overlap effect disappears in these larger slit pores since gas-wall interactions are short-ranged.

Fig. 3 shows the adsorbed-phase density for hydrogen adsorption in slit-shaped pores at 77 K by GCMC simulations. Small pores with pore sizes of less than 1 nm are filled with hydrogen at quite low pressure while the adsorbed-phase density approaches liquid hydrogen density at high pressure in bigger pores. The maximum density at 2000 kPa is obtained in a pore with a pore size of about 0.9 nm, which is sonly just less than the density of liquid hydrogen (0.071 g $cm^{-3}$). Fig. 4 shows a relationship between the adsorbed-phase density and the accessible pore volume of an adsorbent sample for fixed weight uptake. The dotted line parallel to the *y*-axis represents the liquid hydrogen density. Clearly, for a 'reasonable' high hydrogen weight uptake, a large accessible pore volume or a high adsorbed phase density of hydrogen or both is needed. For example, a combination of 1.0 $cm^3$ $g^{-1}$ accessible pore volume and 0.063 g cm$^{-3}$ of adsorbed hydrogen density will lead to 6-wt% of hydrogen in a carbon sample.

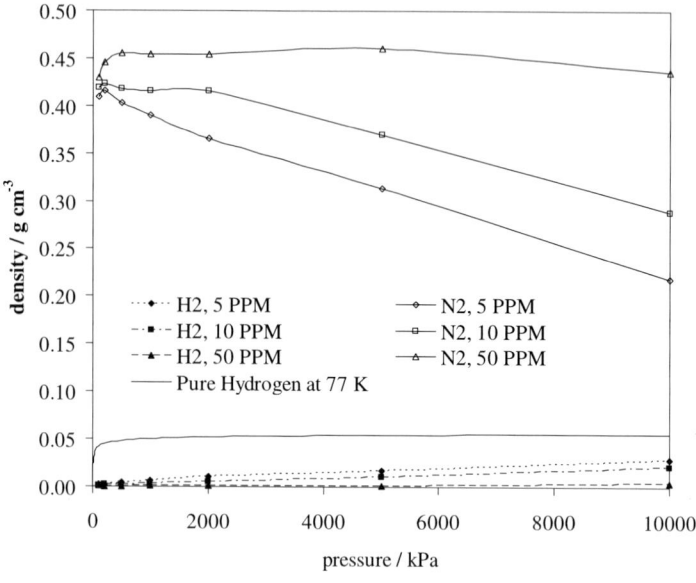

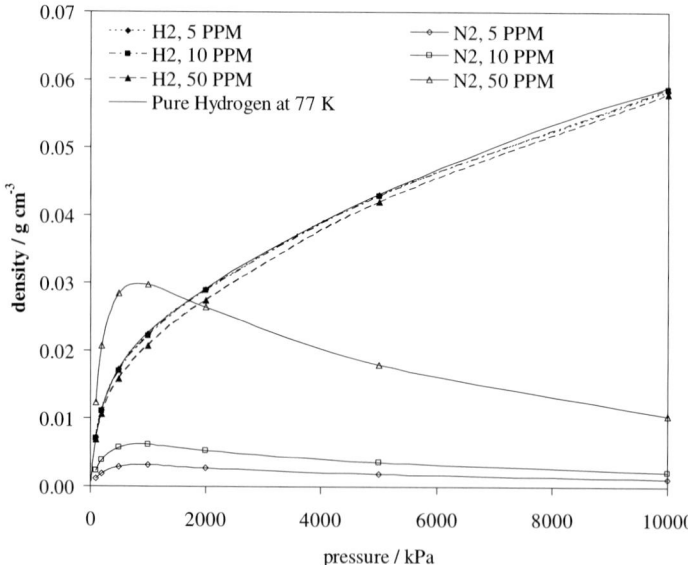

Fig. 2 GCMC simulation of the adsorption of H$_2$/N$_2$ mixtures in slit-shaped carbon pores at 77 K. Top: 0.74 nm pore width; bottom: 2.664 nm pore width.

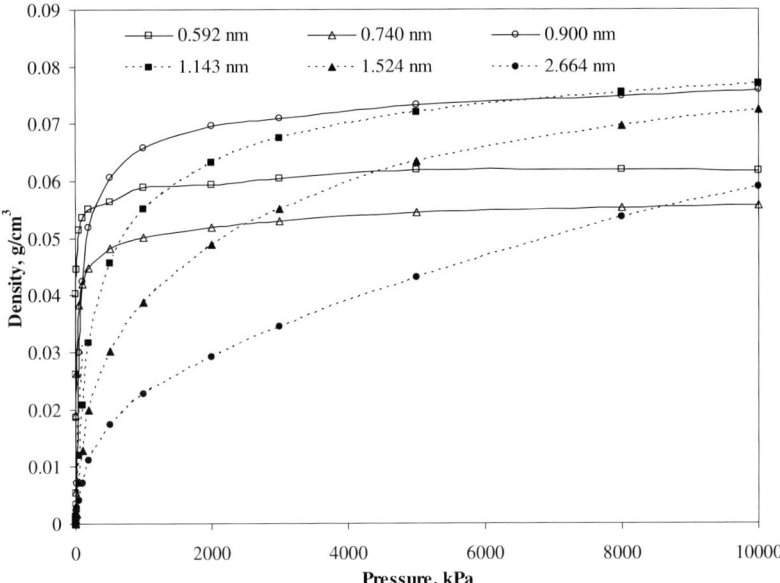

Fig. 3. Adsorbed-phase density of hydrogen in slit pores at 77 K from GCMC simulations.

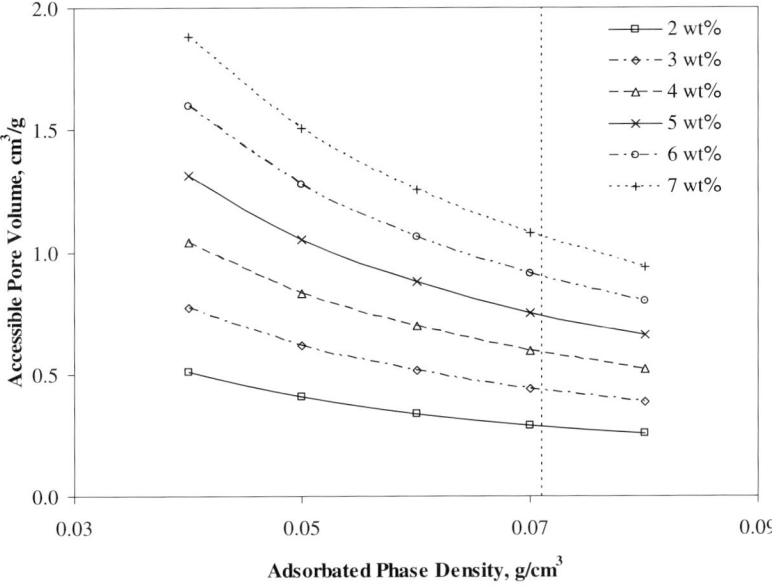

Fig. 4. Theoretical relationships between adsorbed-phase density for pure hydrogen and accessible pore volume for different mass uptakes. The dotted vertical line is the density for liquid hydrogen.

However, the pore volume for the AC-1 carbon in this work is 0.66 cm$^3$ g$^{-1}$, estimated by N$_2$ adsorption at 77 K. It would therefore seem impossible for this carbon to have an uptake of about 7 wt%, since the adsorbed-phase density is equal or less than the density of liquid hydrogen (Fig. 4). Therefore, it is reasonable to suspect that there are some pores not accessible to N$_2$ at 77 K in the carbons. But these pores may still be accessible to pure H$_2$ since the smallest slit pore accessible to N$_2$ is 0.58 nm while H$_2$ can access a slit pore with a pore size of 0.52 nm. Accordingly, we propose a pore blocking mechanism as follows.

Simulations suggest that in low concentrations ($<$ 50 ppm) nitrogen will be preferentially adsorbed in certain small pores compared to hydrogen due to higher interaction energies with pore walls. Nitrogen will not be important in larger pores, where affects due to differences in interaction energies are reduced (Fig. 2), nor in inaccessible smaller pores. These filled pores may restrict access to the rest of the pore network accessible to hydrogen, and hence reduce capacity, even though they do not contribute much to total uptake. Similar pore blocking arguments have been presented before [6]. The extent of pore blocking will be less for lower impurity concentrations (Fig. 2), and here it seems that reducing impurities from 50 to 5 ppm increases capacity by about a factor of about two. We are currently testing these ideas with further experimental measurements and simulations of hydrogen adsorption.

## 5. CONCLUDING REMARKS

We have observed that low cost, readily available commercial nanoporous activated carbons are capable of storing large amounts of hydrogen (up to 10 wt.%) at cryogenic temperatures (77 K), at pressures below 2000 kPa and in very pure gas. Molecular simulations of adsorption suggest that the reduced capacity of the carbons in lower purity gas is due to restriction of the pore network by the blocking of some small pores by impurities. While the simple materials, high capacities and low pressures are certainly attractive for hydrogen storage, the requirements of cryogenic temperatures and high purity gas would add to the cost.

### ACKNOWLEDGEMENTS

We thank the EPSRC for funding and members of the United Kingdom Sustainable Hydrogen Energy Consortium (UK-SHEC, see http://www.uk-shec.org/) for their support.

### REFERENCES

[1]    D. Frenkel and B. Smit, Understanding Molecular Simulation, Academic Press, San Diego, 1996.
[2]    R.F. Cracknell, Phys. Chem. Chem. Phys., 3 (2001) 2091.
[3]    A.R. Turner and N. Quirke, Carbon, 36 (1998) 1439.
[4]    W.A. Steele, Interaction of Gases with Solid Surfaces, Pergamon Press, Oxford, 1974.
[5]    K.A.G. Amankwah and J.A. Schwarz, Int. J. Hydrogen Energy, 16 (1991) 339.
[6]    M.W. Maddox, N. Quirke and K.E. Gubbins, Molecular Simulation, 19 (1997) 267.

Studies in Surface Science and Catalysis 160
P.L. Llewellyn, F. Rodriquez-Reinoso, J. Rouqerol and N. Seaton (Editors)
431

# Migration of siloxane polymer in ordered mesoporous MCM-41 silica channels

**J. Goworek [a], A. Borówka [a] and R. Zaleski [b]**

[a] Faculty of Chemistry, Maria Curie-Skłodowska University,
M. Curie-Skłodowska Sq. 3, 20- 031 Lublin, Poland

[b] Institute of Physics, Maria Curie-Skłodowska University,
M. Curie-Skłodowska Sq. 1, 20-031 Lublin, Poland

Penetration of mesopore structure of MCM-41 silica by the siloxane polymer 50%methyl-50%phenyl polysiloxane (OV-17) has been investigated by positronium annihilation lifetime spectroscopy (PALS) using a series of silica samples with different coating levels. The changes of parameters characterizing porosity of investigated materials with increase of the coating was also monitored using low temperature nitrogen adsorption experiments (LN). It is shown that the gradual changes in the specific surface area and the pore volume are observed in the studied coating range. This work shows the usefulness of the PALS to monitor the changes in the pore structure of coated MCM-41 samples. Positron Annihilation Lifetime Spectroscopy provides the information on free volume size and their concentration in porous solids, independently if they are open or closed - inaccessible for adsorptives.

## 1. INTRODUCTION

Silicate mesoporous materials called MCM-41 are promising as host materials due to their high specific surface area, large pore volume and high adsorption capacity. One of the outstanding problems concerning the applicability of these mesoporous silicas is preparation of composite materials with organic or inorganic guest substances in the pores. Potential applications for such composite materials range from size selective catalysts [1] and chemical sensors to nanoscale electronics [2,3]. However, MCM-41 pore dimensions are of several nanometers, therefore modification of their internal surface by external treatment is very difficult, particularly for macromolecules. In preparation of periodic mesoporous organo-silicas knowledge of thermodynamic driving force for pore filling and the kinetics barriers which restrict that process is needed.

After external treatment i.e. adsorption of macromolecules from the liquid phase it is important to know the degree of guest molecules incorporation in pores of the host. In the case of semiconducting polymers it is achieved using optical techniques [3-5]. Some information can be obtained from the results of adsorption method e.g. low temperature adsorption of nitrogen. The major problem is determining the distribution of no conducting polymers in silica matrix.

The adsorption/absorption of polymers causes significant decrease of the specific surface area and total pore volume of silica [6]. However, on the basis of adsorption data it is difficult to distinguish if macromolecules are adsorbed on internal surface of channels or block only their openings. Adsorption method leads to satisfactory results in the case when whole pore interior is accessible for adsorptive while PALS method gives information also about closed pores. Thus, in the case of the silica materials with deposited macromolecules this last method seems to be mostly attractive. In the present study the evolution of the porous properties and external morphologies of MCM-41 during a siloxane impregnation is investigated using positronium annihilation lifetime spectroscopy (PALS) and the nitrogen adsorption methods.

## 2. EXPERIMENTAL

### 2.1. Materials

Mesoporous silica MCM-41 was prepared following the synthesis procedure reported previously [7]. MCM-41 material was synthesized using tetraethyl orthosilicate (TEOS) as silica source, octadecyltrimethylammonium bromide (C18TMAB) as a template and ammonia as a catalyst. The precipitated surfactant-silica mesophase was filtered, washed with distilled water and dried at 323K. These as-synthesized samples were next calcined at 823K in air atmosphere. A residual carbon left in the silica sample after pyrolysis is one of the most important factors influencing the penetration of the channels by guest molecules as well as distorting the positron lifetime spectra. Thus, after calcination the silica material was additionally thermally treated in oxygen to remove the carbon deposits located inside the channels. Because the coating of the internal surface of MCM-41 channels with hydroxyl groups is not complete and silica surface is partially hydrophobic (only 30% of the Si atoms bear OH group [8]), for experiments polysiloxane OV-17 (50%methyl-50%phenyl polysiloxane) of intermediate polarity according to McReynolds scale was chosen.

Three portions of silica MCM-41 were coated with 13 %, 20 % and 33 % by weight of OV-17 polymer (Applied Science Inc., USA) usually used as stationary phase in gas chromatography. The modified silicas were obtained by dissolving a known amount of polymer in dichloroethane and adding appriopriate amount of MCM-41 silica. The silica-polymer mixture was equilibriated for 24 h. Next the solvent was removed in vacuum at 333K. Samples were dried in 333 K during 2 hours.

### 2.2. Methods

The nitrogen adsorption/desorption isotherms were measured at 77 K using ASAP 2010 volumetric adsorption analyzer (Micromeritics Inc., Norcross, GA). The pore diameters were obtained by means of the Barret-joyner-Halenda method [9]. The pore size distributions (PSD) were calculated from the desorption branch of the isotherm. The mesopore radii ($R_p^{meso}$) given in Table 1 correspond to the first maximum on the pore size distribution. The specific surface areas of the investigated samples were evaluated using the standard BET method at the relative pressure $p/p_0$ range of 0.04 to 0.25. The total pore volume was estimated from the amount adsorbed at the relative pressure of about 0.99. The volumes of the primary pores (hexagonally arranged) were estimated for $p/p_0 \approx 0.5$ corresponding to the end of the first step on the adsorption isotherms.

The PALS spectra were collected using a standard "fast-slow" lifetime spectrometer. The air was evacuated from the measurements chamber to $p \approx 0.3$ Pa in order to avoid oxygen influence on o-Ps lifetime. Positronium lifetime measurements (PALS method) are based on the relation between ortho-positronium (o-Ps) lifetime and the size of free volume, in which o-Ps is trapped. PALS spectra were processed as described in Ref. [10].

Additionally, the surface structure of silica samples was examined with atomic force microscope (AFM) NanoScope III (Digital Instruments, USA).

## 3. RESULTS AND DISCUSSION

### 3.1. Nitrogen adsorption

The low temperature adsorption/desorption isotherms of nitrogen of pure MCM-41 silica and MCM-41 coated with OV-17 polymer 13 %, 20 % and 33 % by weight are shown in Fig.1. All these isotherms exhibit capillary condensation at a relative pressure $p/p_0 \approx 0.4$, what is usually observed for MCM-41 materials. Moreover, one can observe that the increase of the siloxane phase causes flattening of the first step on the isotherm. In comparison to the isotherm on pure siliceous MCM-41, these condensation steps in the relative pressure range $p/p_0 = 0.35 - 0.45$ for samples with polysiloxane coating are less pronounced. This is a result of penetration of the primary pores by the siloxane molecules and consequent the decrease of the pore diameter and the pore volume. All adsorption/desorption isotherms in Fig. 1 exhibit the presence additional hysteresis loop above $p/p_0 = 0.9$, which may be ascribed to interparticle space.

The BJH pore size distribution curves of the studied samples are shown in Fig. 2. The volumes of larger pores (the radii about 15 nm) after polymer treatment are similar. It suggests that the greater part of OV-17 phase is adsorbed inside the mesopores.

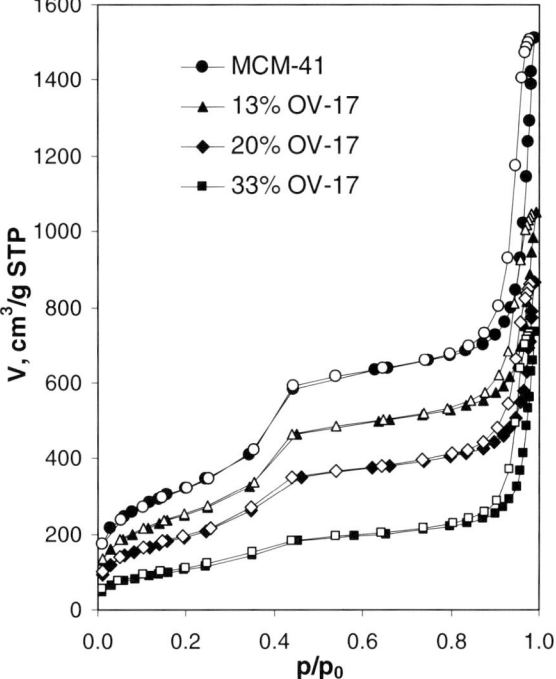

Fig.1. Nitrogen adsorption/desorption isotherms for samples: pure MCM-41 and MCM-41 covered with 13%, 20% and 33% w/w of OV-17 phase

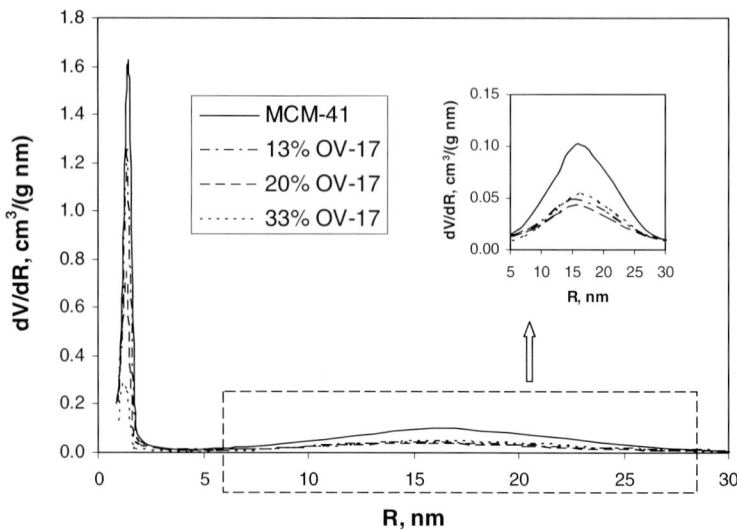

Fig.2. Pore size distributions obtained from desorption branch of nitrogen adsorption/ desorption isotherm for pure MCM-41 and MCM-41 covered by 13%, 20% and 33% w/w of OV-17 phase

The BET surface areas, total pore volumes, mesopore volumes (recalculated per 1 g of pure silica) and primary mesopore sizes corresponding to first maximum on the pore size distribution and obtained using PALS method are collected in Table 1.

Table 1
Structural parameters of the parent MCM-41 and modified MCM-41 obtained from nitrogen desorption data and PAlS experiment

| Sample | Nitrogen desorption | | | | PALS |
|---|---|---|---|---|---|
| | $S_{BET}$ $m^2\,g^{-1}$ | $V_{total}^{BJH}$ $cm^3\,g^{-1}$ | $V_{meso}^{BJH}$ $cm^3 g^{-1}$ | $R_p^{meso}$ nm | $R_p$ nm |
| MCM-41 | 1155 | 2.25 | 0.91 | 1.45 | 1.13 |
| 13% OV-17 | 1037 | 1.52 | 0.83 | 1.44 | 1.10 |
| 20% OV-17 | 866 | 1.26 | 0.68 | 1.42 | 1.08 |
| 33% OV-17 | 570 | 1.08 | 0.42 | 1.28 | 0.63 |

After absorption/adsorption of polysiloxane a large decrease of the specific surface area is observed. Simultaneously the total pore volume of the primary pores also decreases. The mesopore radii become little smaller as amount of the coating polymer increases. However, estimating the parameters characterizing the porosity from liquid nitrogen experiment may be not reliable since part of pores may be blocked by the polymer molecules located on the external surface of silica grains (see Fig.3). Thus, part of pores is eliminated for adsorption experiments, even if cores of pores are empty and walls of these pores are coated with polymer. The PALS method can supply additional information about mechanism of polymer adsorption.

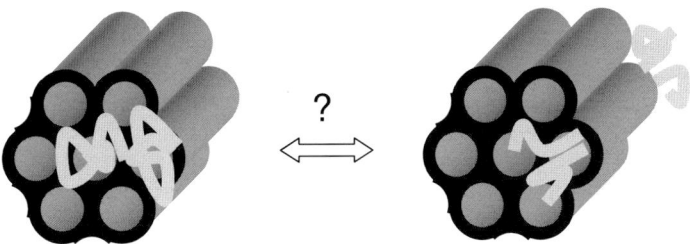

Fig.3. Schematic distribution of siloxane polymer in pores of MCM-41 material

## 3.2. Atomic force microscopy

The surface microstructures of silica samples were additionally investigated by atomic force microscopy (AFM). Fig. 4 shows AFM images of pure MCM-41 silica (part a), MCM-41 coated with 33 % OV-17 (part b) and MCM-41 coated with 33 % highly polar dicyanoallylpolysiloxane (OV-275) (part c).

Two samples i.e. MCM-41 and MCM-41 (OV-17) exhibit similar morphology. The image of third sample shows the presence of polymer macromolecules on microcrystalites. It suggests poor polymer penetration into pores. Moreover, one can assume that whole amount of OV-17 phase is present within the pores and its surface is more homogeneous.

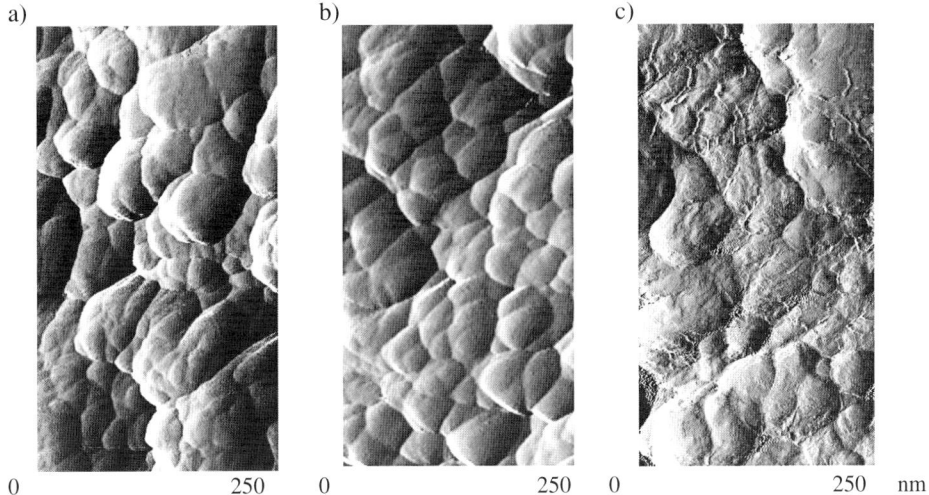

a)  b)  c)

0          250    0          250    0          250    nm

Fig.4. Atomic force microscopy images of: a) pure MCM-41 silica; b) MCM-41 silica coated with 33% OV-17 phase; c) MCM-41 silica coated with 33% OV-275 phase

## 3.2. Positronium lifetime spectroscopy

The positron annihilation lifetime spectroscopy provides the information on free volume size and their concentration in porous solids independently if they are open and closed - inaccesible for odsorptives. Ortho-positronium (o-Ps) forms in free volumes and its pick-off annihilation probability depends on free volume size. o-Ps lifetimes are related to free volume size in the way described by the extended Tao-Eldrup model (ETE) [11]. The intensities of respective spectrum components depend on free volume concentration.

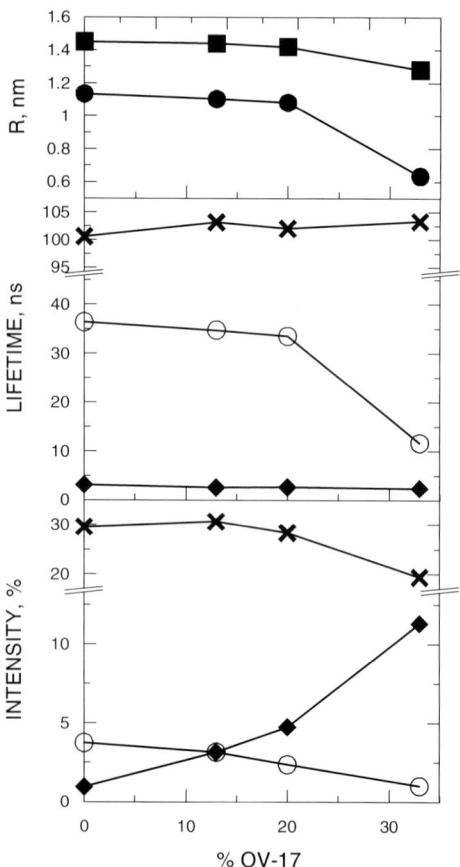

Fig.5. Lifetimes and intensities of ortho-positronium components (diamonds – annihilation in the polymer structure, circles – inside the pores, triangles – outside the grains) and pore diameters (squares – nitrogen desorption, full circles – PALS) for various concentrations of OV-17 phase

For investigated MCM-41 samples with polymer coating there can be found three ortho-positronium components, which can be identified as positronium annihilation: in the polymer structure (lifetimes ~2.5 ns, short spectrum component, SSC), inside the pores (20-40 ns, medium spectrum component, MSC) and outside the grains (100-110 ns, long spectrum component, LSC).

Appearance of low intensity of the shorter component (3 ns) related to ortho-Ps annihilation in MCM-41 without the polymer can be explained as the result of presence of organic template remnants in closed pores after calcination. Organic deposits may be the source of permanent small distortion of the short spectrum component. The rise of OV-17 concentration causes almost linear increase of SSC intensity and similar decrease of MSC one due to gradual pore filling by the polymer molecules (Fig.5).

Almost constant MSC lifetime, except the case of 33% OV-17 concentration, indicates that pore size does not change. According to the ETE model the radii of infinite cylindrical pores for lower polymer coating are about 1.1 nm. For 33% OV-17 sample the medium lifetime is three times shorter in comparison to initial sample and corresponds to smaller pores. They may be approximated by infinite cylinders with radii of 0.6 nm, or more probably by closed ones with the radii of 1.1 nm and 1 nm length. Intensity of this spectrum component is very low. Moreover, the long component intensity is lower for this sample indicating that there is less positronium escaping open pores or some parts of intergranular space is filled with the polymer.

The change of $R_p$ vs. % OV-17 (Fig. 5) shows similar tendency in the case of both techniques. The decrease of pore radius is most pronounced for the sample containing 33% of OV-17 phase and larger in the case of the PALS method. It confirms that at higher coating levels a part of pores is inaccessible for nitrogen molecules, while the PALS method gives information also about the dimensions of closed pores.

## 4. CONCLUSIONS

The PALS spectra provide valuable information about evolution of micro and mesopores in MCM-41 silica coated with different amounts of polysiloxane. Depression of pore dimensions derived from nitrogen adsorption data is less sensitive on polymer concentration in comparison to those from PALS experiment. Moreover from PALS data one can obtain information about distribution of siloxane within silica material.

## REFERENCES

[1] F. Marquez, H. Garcia, E. Palomares, L. Fernandez and A. Corma, J. Am. Chem. Soc., 122 (2000) 6520.
[2] Y. Wei, J.M. Yeh, D. Jin, X. Jia and J. Wang, Chem. Mater., 7 (1995) 969.
[3] T.-Q. Nguyen, J. Wu, V. Doan, B.J. Schwartz and S.H. Tolbert, Science, 288 (2000) 652.
[4] J. Wu, A.F. Gross and S.H. Tolbert, J. Phys. Chem. B, 103 (1999) 2374.
[5] N.K. Mal, M. Fujiwara and Y. Tanaka, Nature, 421 (2003) 350.
[6] A. Bose, R.K. Gilpin and M. Jaroniec, J. Coll. Interf. Sci., 240 (2001) 224.
[7] M. Grün, K.K. Unger, A. Matsumoto and K. Tsutsumi, in: Characterization of Porous Solids IV, The Royal Society of Chemistry, B. McEnaney, J.T. Mays, J. Rouquerol, F. Rodriguez-Reinoso, K.S.W. Sing and K.K. Unger (eds), 1997, p. 81.
[8] H. Landmesser, H. Kosslick, W. Storek and R. Fricke, Solid State Ionics, 101-103 (1997) 271.
[9] E.P. Barrett, L.G. Joyner and P.P. Halenda, J. Am. Chem. Soc., 73 (1951) 373.
[10] R. Zaleski, J. Wawryszczuk, A. Borówka, J. Goworek and T. Goworek, Micropor. Mesopor. Mater., 62 (2003) 47.
[11] T. Goworek, K.Ciesielski, B. Jasińska and J. Wawryszczuk, Chem.Phys., 230 (1998) 305.

Studies in Surface Science and Catalysis 160
P.L. Llewellyn, F. Rodriguez-Reinoso, J. Rouqerol and N. Seaton (Editors)

# The sorption dynamics of propane, *i*-butane and neopentane in carbon nanotubes

**J. Valyon,[a] Zs. Ötvös,[a] Gy. Onyestyák,[a] and L. V. C. Rees[b]**

[a]Institute of Surface Chemistry and Catalysis, Chemical Research Center, Hungarian Academy of Sciences, P.O. Box 17, Budapest, Hungary 1525.

[b]School of Chemistry, University of Edinburgh, West Mains Road, Edinburgh EH9 3JJ, Scotland, UK

Carbon nanotube (CNT) was prepared by the catalytic vapor deposition (CVD) method. Various procedures were applied to get nanotubes in pure and well-defined form. The sorption kinetics of $C_3$-$C_5$ methyl methanes was studied in the CNT samples taking advantage of the unique potentials of the frequency response (FR) technique. Applying a batch-type FR system rate spectra were recorded in the 0.001-10 Hz frequency range. The temperature and the mean pressure of the measurements were varied in the −20 to 100 °C and the 60 to 600 Pa range, respectively. Over a CNT sample similar FR spectra were recorded for the sorption of the alkanes, different in molecular size and shape. Results suggested two parallel sorption processes, characterized by different sorption capacities and time constants. The processes were assigned to sorption on the convex and concave surfaces of the carbon tubes. On tube surfaces, having large number of structural defects and oxygen-containing functional groups, the sorption rate was found to be governed by the rate of diffusion.

## 1. INTRODUCTION

The relationship between the structure of microporous adsorbents, such as zeolites, or activated carbons, and the sorption equilibrium and dynamics of light alkanes are quite well known. Concerning the sorption properties both geometric factors, i.e., the size and shape of the adsorptive molecule and the pores, and energetic factors has significance. The roles, played by these factors, have been studied by both experimental and theoretical methods. Since many possible applications of the nanoporous CNT involve gas diffusion and sorption, there is a strong interest in learning more about the equilibrium and dynamic adsorption properties of these materials [1-9]. Sorption on faultless tubes can be easily described by molecular dynamics or Monte Carlo simulations [10-14]. However, the value of the reported results can be judged when they can be confronted with results of measurements. Relatively few experimental study concerns the sorption mass transport in CNTs.

CNTs represent a unique group of the mesoporous materials where the inner and outer diameters of the tubes is in the range of about one to ten nanometers, while the tubes can be as long as a few micrometers. The CNTs are adsorbents, where fluids can be adsorbed both within the hollow pores of the open tubes and on the external tube surfaces. In bundles of tubes the outer tube surfaces surround intertubular spaces that are also special environments for adsorption.

In the past decade the frequency response (FR) method has been demonstrated to be useful for studying the mass transfer kinetics of gases in various adsorbents, including activated carbons [15]. A recent FR study showed the relationships between the structure of activated carbon and CNT samples and their equilibrium, as well as, dynamic sorption behavior [16]. Lately, results suggested that the surface functional groups of the CNT, generated by oxidative treatment, have significant influence on the adsorption properties of the carbon tubes [17, 18]. The FR technique is a macroscopic transient method. At the transient methods the equilibrium is perturbed and the re-equilibration process is monitored. In the FR method, unlike to other transient macroscopic methods, the perturbation of the adsorption equilibrium is periodic and very small. The frequency of perturbation is additional degree of freedom, which provides the method with the potential of distinguishing parallel processes, having different time constants [19, 20]. Due to the formal resemblance of the FR and optical spectra, the FR method is often referred to as rate spectroscopy [19]. The method gives information also about the rate-controlling mechanism of the transport [19, 21].

The aim of this FR study was to learn more about the adsorption and diffusion of alkanes in different CNT preparations.

## 2. EXPERIMENTAL SECTION

*Materials:* The catalytic vapor deposition (CVD) method was used to prepare the carbon nanotubes. Single-walled carbon nanotubes (SWCNT) were obtained by methane carbonization over Co/MgO catalyst. The "as-synthesized" material was designated as SWCNT-4, referring to the 4-wt % carbon content of the sample. The purified SWCNT-59 sample, containing 59 wt % carbon, was obtained by dissolving a large fraction of the Co/MgO catalyst in concentrated HCl solution. Multi-walled carbon nanotubes (MWCNT-47) were produced by the catalytic decomposition of acetylene over a $Fe,Co/Al_2O_3$ catalyst. From the MWCNT-containing material the $Fe,Co/Al_2O_3$ was dissolved using first NaOH and then HCl solution. The obtained product is designated as MWCNT-95. The morphology of the MWCNT was modified by ball milling the tubes in air. According to results of electron microscopic examinations tubes having lengths of 300-400-nm with narrow length distribution and amorphous carbon debris were obtained (MWCNT-95B). In order to get the fragmented tubes in pure form the amorphous carbon was oxidized to $CO_2$ with $KMnO_4/H_2SO_4$ solution. The oxidized sample was designated as MWCNT-81. Some of this material was heated up to 900 °C in argon to remove the O-containing functional groups from the surface (MWCNT-81H).

*FR measurements:* In order to avoid effects of bed formation and particle size heterogeneity samples of narrow CNT sieve fractions were distributed evenly over a glass-wool plug, placed in the sample holder of the batch-type FR system. A detailed description of the FR system is given in ref. [22]. Samples were pre-treated *in situ* in high vacuum at 300 °C for 1 hour. Then, sorption equilibrium was established between the adsorbent and the sorptive alkane. The temperature and pressure of the FR measurements were in the -20 to 100 °C and 60-600 Pa range, respectively. A periodic, ±1% square-wave modulation was applied on the volume of the FR chamber. The modulation frequency was changed in the range of 0.001 and 10 Hz. A response pressure wave ($P_Z$) was recorded at each frequency. Reference, blank wave functions ($P_B$) were obtained from measurements without adsorbent in the sample holder. From the ratio, $P_Z/P_B$, and the phase difference $(\Phi_{Z-B})$

response wave functions were generated for each perturbation frequency. The FR spectrum was obtained by plotting the in-phase (real) and out-of-phase (imaginary) components of the response functions against the modulation frequency. The experimental FR spectra were fitted by theoretical FR functions, derived by Yasuda [19] or Jordi and Do [21]. These functions were obtained by solving the general mathematical model of a periodically perturbed sorption system containing gas and isotropic adsorbent particles of uniform size and shape. The model includes coupled mass-transfer resistances, such as transport through an external film or surface barrier and through macro and micropores. The dynamic parameters of the transport processes were obtained as parameters of the best-fit theoretical FR functions.

## 3. RESULTS AND DISCUSSION

Fig. 1 shows FR spectra for the neopentane diffusion in sieve fractions of MWCNT-81 sample. When the characteristic time of periodic modulation approaches the time constant of the transport process, resonance occurs, which is indicated by the appearance of an inflexion and a peak in the in-phase and out-of-phase component curves, respectively. In Fig. 1 the component curves of the FR spectrum meet asymptotically at high frequencies. The theoretical analysis of the FR measurement suggested the appearance of this relative curve position is characteristic for rate-controlling diffusion process. However, the rate-controlling transport resistance can not be assigned on the basis of the FR results only to the transport in the macro, meso or micropores of the sample.

It was found that the originally loose and fluffy CNT sample becomes more and more compact upon chemical treatments. The change results in the development of a felt-like texture containing mesopores among the tubes with sizes smaller than about 50 nm [23]. The chemical treatments affect the volume and the size distribution of this mesopores, but hardly change the size distribution of the smaller, intratubular nanopores.

The diffusion time constant ($r^2/D$) depends on the diffusivity (D) and the length of the diffusion path (r). Fig. 1 shows that the FR signals appear at higher frequencies for the smaller particle size fractions of the ball-milled and oxidized MWCNT sample, i.e., the diffusion time constant depends on the particle size. The CNT tubes must be of about the same length in the different fractions. Thus, in the largest particles, the rate-controlling diffusion proceeds most probably in the macro or mesopores among the felted carbon nanotubes. The FR spectrum of the 150-600 μm particles (Fig. 1) could be fitted with two peaks suggesting that sample contains particles characterized by distinctly different diffusion time constants. The low and high-frequency resonance signals are assigned to the intertubular diffusion process in the particles of nanotube aggregates and to the diffusion within the nanotubes, respectively. The single high-frequency resonance signal obtained for the sieve fraction, containing particles smaller than 150 μm suggests that, for this sample, only the rate-governing intratubular transport could been detected. The effect of particle size on the FR spectrum was similar for all the alkane sorptives and in the whole temperature range, examined. An earlier study shows similar relationship between the diffusion time constant and the size of particles for zeolite samples, comprising aggregates of microporous zeolite crystallites [15].

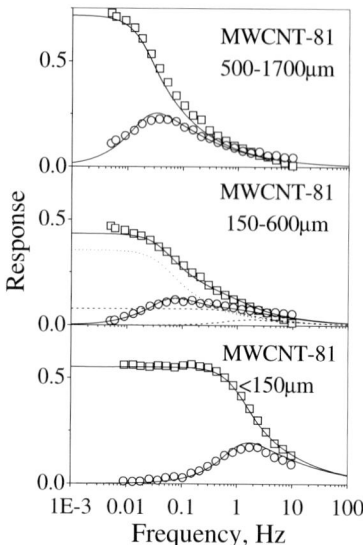

Fig. 1. FR spectra of neopentane sorption on MWCNT-81 at 50 °C and 133 Pa. Each measurement was made on 200 mg CNT. Symbols correspond to the in-phase (□) and out-of-phase (○) components of the experimentally determined response functions. Full lines are the best-fit characteristic functions. FR rate spectra of parallel processes, represented by dotted lines, were summarized to get the characteristic FR function of the sample, comprising 150-600-μm particles.

In Fig. 2 the FR spectra of the sorption interaction with various CNT samples are shown for alkanes, having about the same length ($C_3$), but different cross-sectional dimensions. The different size and shape of the adsorptive molecules are with minor affect on the FR rate spectra. The geometrical constraints to molecular transport, often posed by zeolites and other microporous materials, is absent here since the kinetic diameter of the alkane molecules is approximately ten times smaller than the diameter of the carbon tubes. A single FR resonance signal, defined by intersecting in-phase and out-of-phase component curves appear in the spectra of the MWCNT-95 and 95B samples. Similar spectra were generated theoretically on the bases of two different models. One of the models assumes that the rate of the sorption process is governed by the rate the adsorption-desorption step. Another possibility is that the pore openings represent the largest resistance to the transport towards the sorption sites. On the FR spectrum of the sample MWCNT-95B, obtained by ball-milling sample MWCNT-95, the intersection point of the component curves appeared at slightly higher frequency than on the spectrum of MWCNT-95 (Fig. 2, cf. spectra MWCNT-95 and MWCNT-95B). One result of the ball milling is that the longer tubes are fragmented and, thereby, the inner surface of the shorter tubes becomes more easily accessible through two open pore mouths. The shorter nanotubes of the MWCNT-95B sample form more compact aggregates than the longer tubes of the MWCNT-95 sample. It was also found that, upon ball milling in air, oxygen-containing functional groups were generated at locations where the tubes were fractured. The treatment of MWCNT-95B sample in oxidizing solution further increased the surface concentration of the oxygen-containing functional groups, terminating the tubes and linking to imperfections on the tube walls. The presence of polar functional groups may hinder the molecular motion and contribute to the increased diffusion resistance of the intratubular nanopores of the

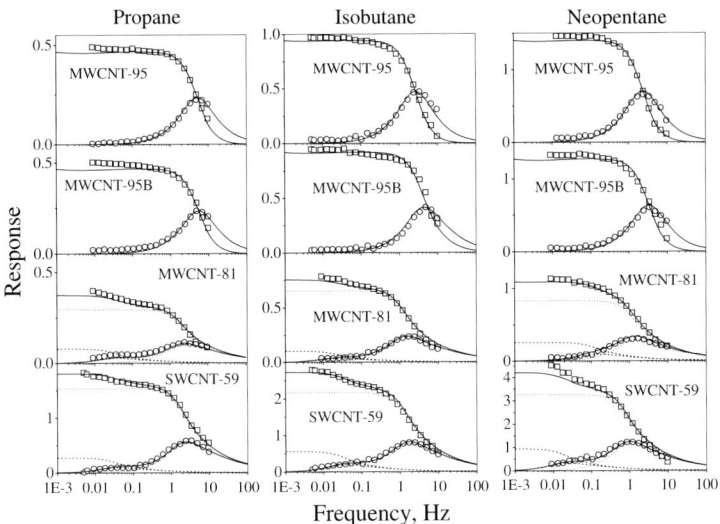

Fig. 2. FR rate-spectra for the sorption of propane, isobutane and neopentane in the carbon nanotube samples recorded at 30 °C and 133 Pa. 200 mg of <150-μm size particles were used. For some of the samples the characteristic FR spectrum (full lines) was obtained by summarizing FR spectra of parallel processes (dotted lines).

MWCNT-81 sample (Fig. 2). Due to the increased diffusion resistance of the nanopores diffusion became the rate-controlling mechanism of the sorption process of alkanes.

Regarding the rate-governing mechanism of alkane sorption the nanotube samples SWCNT-59 and MWNCT-81 are very similar. It is well known that the extremely thin walls of the single-wall CNT are susceptible to the formation of imperfections. The defect sites in the tube may hinder the molecular motion the same way as the O-containing functional groups of the MWCNT-81 sample.

The intensity of the FR signals is related to the change of sorption capacity induced by the pressure perturbation. The FR resonance signal of the SWCNT-59 sample are much more intense than the corresponding signal of the MWCNT-81 sample (Fig. 2). This is in accordance with the difference in the surface area of the samples.

From each spectrum of the SWCNT-59 and MWCNT-81 samples resonance signals of two parallel sorption processes could be resolved (Fig. 2, dotted lines). The resonance signals are suggested to come from sorption processes on the convex and concave surfaces of the carbon tubes. The process with larger time constant is, most probably, related to the intertubular diffusion to sorption sites on the convex outer surface of the nanotubes. The process with a smaller time constant is related with transport through the intratubular nanopores as it was discussed in relation with Fig. 1. Regarding the rate and the mechanism of the mass transport the FR spectra of the different alkanes are quite similar on a given sample.

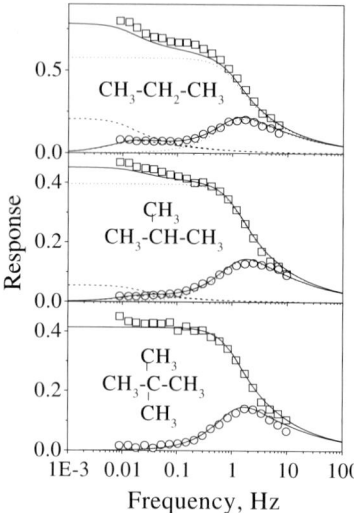

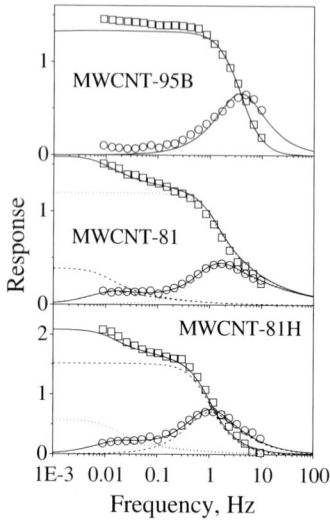

Fig. 3. FR rate spectra for the MWCNT-81 sample. Spectra were recorded using 200 mg of <150-μm size particles and 133 Pa. The temperature was selected to get similar adsorption coverage for the different gases: propane, 0 °C; isobutane, 50 °C; neopentane, 100 °C.

Fig. 4. FR sorption rate spectra for the sorption of propane, recorded at -20 °C and 133 Pa using 200 mg of <150-μm size particles. Sample MWCNT-81 was obtained from MWCNT-95B by oxidative treatment. The MWCNT-81H was prepared from the MWCNT-81 by thermal treatment at 900 °C in argon.

The diffusivity, deduced from the FR resonance, was found to be invariant of the alkane pressure in the range of 60-600 Pa (not shown), suggesting that Knudsen type diffusion prevails. However, the ratio of diffusivities does not correspond to the ratio of the molecular weights, suggesting that the adsorbate-carbon interaction is different for the different alkanes.

It is to be noticed that under the identical measurement conditions, the intensity of the resonance shows correlation with equilibrium coverage of the sorption sites that increases from $C_3$ to $C_5$. It is also known that the time constant of the sorption process also depends on the coverage [19]. Obviously, the high-frequency signal, assigned to intratubular diffusion, appears at lower frequencies for the heavier alkanes (Fig. 2). The FR spectra, shown in Fig. 3 relates to the sorption of methyl methanes in the sample MWCNT-81. The pressure of the measurements was the same, but the temperatures were adjusted to get about the same adsorption coverage for the different alkanes. The FR spectra are quite similar, but the intensity ratio of the resonance signals assigned to intra- and intertubular sorption processes paralleled the change of the molecular weight of the alkane. This change seems to be in accordance with the extent of the possible interaction between the adsorbate alkane and the convex or concave surfaces of the carbon tubes. The branched alkanes may find a preferential environment on the convex inner walls of the carbon nanotubes relative to the concave outer walls of the tubes, while the sites on the inner and outer tube surfaces compete with near to equal chance for the adsorption of linear alkanes.

The FR signals for the sorption of the different alkanes appeared practically at the same frequencies. This observation suggests the same diffusivity for the different alkanes, i.e, that the molecular weight differences of the alkanes were compensated by the temperature differences of the measurements.

In previous studies various techniques has been applied to study O-containing surface groups of carbon (e.g. IR, XPS, TPD, titration, etc.). The present FR study suggested that the mechanism of the sorption transport is sensitive to the presence of surface functional groups. Different mechanism of alkane transport prevails in the carbon tubes prior to and after oxidative treatment of the tubes. The adsorption-desorption step was the rate-controlling step of sorption on purified and ball-milled MWCNT, while the diffusion controlled the rate of transport following an oxidative treatment (see Fig. 1 and 4). The low and high-frequency resonance signals are assigned to the intertubular diffusion process in the particles of nanotube aggregates and to the diffusion within the nanotubes, respectively. The single high-frequency resonance signal obtained for the sieve fraction, containing particles smaller than 150 μm suggests that, for this sample, only the rate-governing intratubular transport could been detected. The effect of particle size on the FR spectrum was similar for all the alkane sorptives and in the whole temperature range, examined. An earlier study shows similar relationship between the diffusion time constant and the size of particles for zeolite samples, comprising aggregates of microporous zeolite crystallites [15].

The higher population of the oxygen-containing functional groups on the surface posed higher resistance to diffusion mass transport. As a result the diffusion became the rate-governing step of the sorption mass transport. The effect of carbon functionalization on the mechanism of sorption transport is demonstrated by the FR spectra of propane, shown in Fig. 4. In inert atmosphere at elevated temperature surface groups decompose, while water, carbon monoxide and carbon dioxide is produced [24]. In this process the diffusion barrier, posed by the functional groups, is also removed. This is shown by the FR spectrum of the heat-treated MWCNT-81H sample. The in-phase and out-of-phase components of FR response functions gave an intersection, suggesting that the rate controlling diffusion resistance was removed by the high-temperature treatment and adsorption-desorption became again the rate-governing process step. However, it is to be noted, that the resonance signal appeared at lower frequency than in the spectrum of the parent MWCNT-95B sample, indicating that the sample is not the same as before the treatments. The removal of O-containing groups from the carbon nanotube surface leaves back defects, which represent new sites for sorption.

## 4. CONCLUSION

In mesopores the sorption mass transport is less influenced by the structure and size of the alkane adsorptives than in micropores where the transport can be controlled also by geometric factors. The dynamic sorption properties of carbon nanotubes were shown to be strongly affected by the presence or absence of surface functional groups.

During ball milling in air and/or the procedure of chemical purification oxygen-containing groups are generated on the carbon surface. These groups can be removed by high-temperature treatment in inert atmosphere. Different rate-controlling mechanisms were obtained for the samples, functionalized to distinctly different extents.

The FR rate "spectroscopy" proved to be an effective method to characterize the sorption dynamics of alkanes in various carbon nanotube preparations.

## ACKNOWLEDGEMENTS

The financial support provided to the collaboration by Royal Society, London and the Hungarian Research Fund (OTKA No. T 037 681) is gratefully acknowledged.

## REFERENCES

[1]  R. Ströbel, L. Jörissen, T. Schliermann, V. Trapp, W. Schütz, K. Bohmhammel, G. Wolf and J. Garche, Journal of Power Sources, 84 (1999) 221.
[2]  F. Lamari Darkmin, P. Malbrunot and G.P. Tartaglia, International Journal of Hydrogen Energy, 27 (2002) 193.
[3]  M. Eswaramoorthy, R. Sen and C.N.R. Rao, Chemical Physics Letters, 304 (1999) 207.
[4]  C. Gommes, S. Blacher, N. Dupont-Pavlovsky, C. Bossuot, M. Lamy, A. Brasseur, D. Marguillier, A. Fonseca, E. McRae, J.B. Nagy and J.-P. Pirard, Colloid and Surfaces A, 241 (2004) 155.
[5]  L. Zhou, Y. Zhou, and Y. Sun, International Journal of Hydrogen Energy, 29 (2004) 475.
[6]  Q. Yang, P. Hou, S. Bai, M. Wang and H.Cheng, Chemical Physics Letters, 345 (2001) 18.
[7]  S. Inoue, N. Ichikuni, T. Suzuki, T. Uermatsu and K. Kaneko, Journal of Physical Chemistry B, 102 (1998) 4689.
[8]  M. Bienfait, B. Asmussen, M. Johnson and P. Zeppenfeld, Surf. Sci., 460 (2000) 243.
[9]  H. Cheng, Q. Yang and C. Liu, Carbon, 39 (2001) 1447.
[10] F. Zhang, J. Chem. Phys., 111 (1999) 9082.
[11] Z. Mao and S.B. Sinnott, J. Phys. Chem. B, 104 (2000) 4618.
[12] Z. Mao and S.B. Sinnott, J. Phys. Chem. B, 105 (2001) 6916.
[13] K.H. Lee and S.B. Sinnott, J. Phys. Chem. B, 108 (2004) 9861.
[14] D.J. Shu and X.G. Gong, J. Chem. Phys., 114 (2001) 10922.
[15] Gy. Onyestyák and L.V.C. Rees, J. Phys. Chem., B 103 (1999) 7469.
[16] Gy. Onyestyák, J. Valyon, K. Hernádi, I. Kiricsi and L.V.C. Rees, Carbon, 41 (2003) 1241.
[17] Zs. Ötvös, Gy. Onyestyák, J. Valyon, I. Kiricsi, Z. Kónya and L.V.C. Rees, Appl. Surf. Sci., 238 (2004) 73.
[18] Gy. Onyestyák, Zs. Ötvös, J. Valyon, I. Kiricsi and L.V.C. Rees, Helv. Chim. Acta, 87 (2004) 1508.
[19] Y. Yasuda, Heterog. Chem., 1 (1994) 103.
[20] S.C. Reyes and E. Iglesia, Catalysis, 11 (1994) 51.
[21] R.G. Jordi and D.D. Do, Chem. Eng. Sci., 48 (1993) 1103.
[22] L.V.C. Rees and D. Shen, Gas Sep. Purif., 7 (1993) 83.
[23] Zs. Ötvös, Gy. Onyestyák, J. Valyon, I. Kiricsi and L.V.C. Rees, Stud. Surf. Sci. Catal., 156 (2005) 617.
[24] N.V. Beck, S. Meech, P.R. Norman and L.A. Pears, Carbon, 40 (2002) 531.

Studies in Surface Science and Catalysis 160
P.L. Llewellyn, F. Rodriquez-Reinoso, J. Rouqerol and N. Seaton (Editors)

# Adsorption and diffusion kinetics of alkanes (C₃ & C₅) on different CaA adsorbents

F. Benaliouche, M. Fodil Cherif, S. Belkhiri and Y. Boucheffa[*]

EMP, UER de Chimie Appliquée, BP 17 Bordj El-Bahri 16111, Algiers, Algeria.

Adsorption and diffusion kinetics of propane and pentane on CaA molecular sieves having different degrees of calcium exchange (0, 33, 44, 57 and 75% of $Ca^{2+}$) have been studied using thermobalance under constant pressure of 27kPa and 373 and 523K. The results obtained show clearly the effect of calcium exchange on parameters such as the adsorbed quantities, the initial rates of adsorption and diffusion coefficients. Despite the smaller size of propane molecule, the pentane is more adsorbed and its initial rate of adsorption is more important. This is related to the stronger interaction molecule/wall of zeolite α cavity. Using the model resulting from second Fick's law, a good concordance appears between theoretical and experimental curves of diffusion.

## 1. INTRODUCTION

5A zeolite is used in various adsorption processes notably for separation of n/iso paraffin mixture within the framework of gasoline reformulation. The exchanging of monovalent cation $Na^+$ by divalent cation $Ca^{2+}$ can provide a pore size that admit normal paraffins and exclude other hydrocarbons [1,2]. When in zeolite NaA one exchanges $2Na^+$ by $Ca^{2+}$ in stages n-paraffins begin to be sorbed freely when about 30% of $Na^+$ is replaced [3]. In spite of the extensive studies of 5A zeolites and the calcium exchanging, numerous interesting phenomena related to the adsorption of paraffins can be observed. The objective of this study, carried out by thermogravimetric measurements, is to elucidate the behaviour of propane ($\varnothing_{kinetic}$=4,3Å) and pentane ($\varnothing_{kinetic}$=4,9Å) adsorption and diffusion in various CaA adsorbents (calcium exchange: 0, 33, 44, 57, 75%). The effect of calcium exchange degrees on the adsorption capacity, the rate of adsorption and diffusion coefficient will be studied.

## 2. EXPERIMENTAL

The adsorption isotherms of propane and n-pentane (purity > 99%, from Fluka) were obtained at constant pressure of 27kPa and at 373 and 523K in a Setaram MTB $10^{-8}$ thermobalance. The CaA adsorbents ($A_0$, $A_{33}$, $A_{44}$, and $A_{57}$) were supplied by IFP and the 75% exchanged adsorbent ($A_{75}$) was from Rhone-Poulenc. All these samples contain 80% wt.-% of zeolite and 20wt.-% of clay binder.

---

* Corresponding author

The adsorption capacity of nitrogen at 77K carried out in thermobalance were: 0, 0.02, 0.165, 0.216 and 0.222 respectively for $A_0$, $A_{33}$, $A_{44}$, $A_{57}$ and $A_{75}$. Therefore, in our case, the opening appears actually from 44% [4].

The propane or pentane was introduced in thermobalance system after the pretreatment of the sample (0.05g) in vacuum ($10^{-3}$Pa) at 693K for 5 hours.

## 3. RESULTS AND DISCUSSION

### 3.1. Adsorption isotherms of propane and pentane

The increase in weight of CaA adsorbents during the adsorption of propane and pentane was monitored at 373 and 523K., Fig. 1.

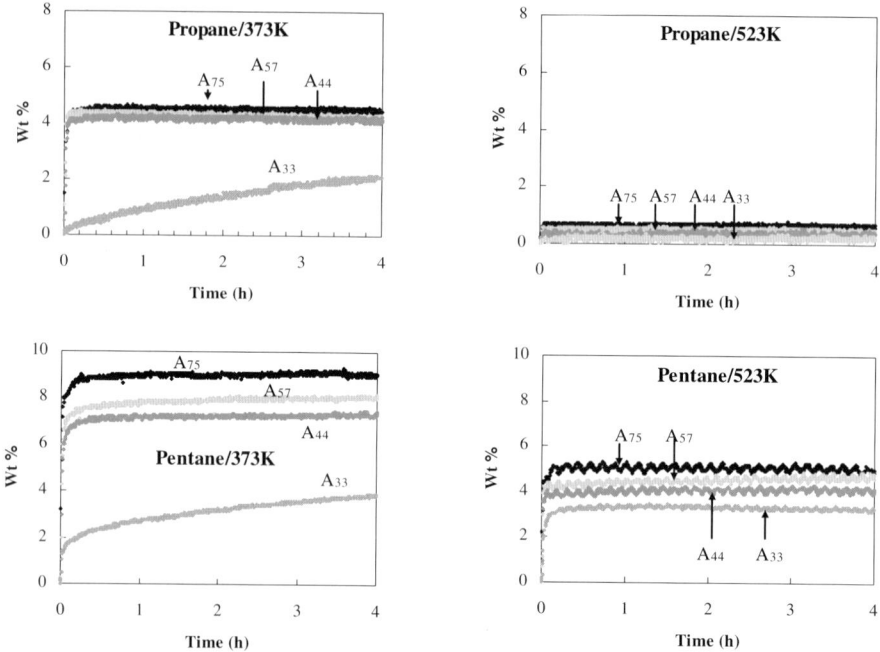

Fig. 1. Adsorption isotherms of propane and pentane on CaA adsorbents at 373 and 523K (P=27kPa).

From these results, we observe that pentane is more easily adsorbed. The greater ease of pentane adsorption is explained by the strong interaction between this molecule and the wall of the spherical CaA cavity [5]. Moreover, although the weak value of nitrogen adsorption capacity measured at 77K of $A_{33}$ adsorbent (0.02 cm³/g), the pentane (less for propane) was sufficiently adsorbed at 373K (4.11% pentane and 2.11% for propane). With value equal to 0.02 cm³/g as a porous volume (calculated from the liquid density), it was expected that adsorption of pentane can not exceed 1.25%. No adsorption was operated on $A_0$ adsorbent (zero calcium exchange).

At 523K, the adsorption of propane is less than 1%, while the pentane is considerably adsorbed and the order of adsorption is preserved versus calcium exchange. In this case, a particular disruption of isotherms is observed. This can be related to the diffusion difficulty and also to the heavy compound formation at 523K [6].

Considering that only 80%-wt. of each sample can adsorb the guest molecule, the number of molecules versus the degree of calcium exchange on CaA adsorbents is given in Fig. 1.

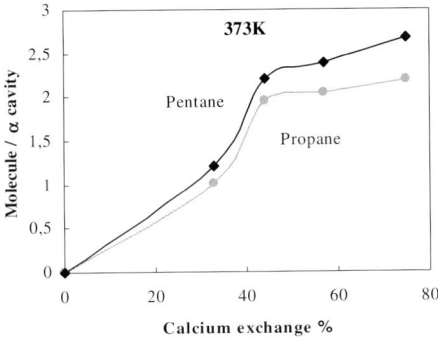 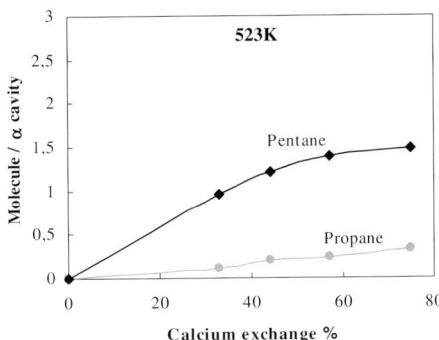

Fig. 2. Number of propane and pentane molecules per α cavity of CaA adsorbents versus the degree of calcium exchange.

At 373K, approximately 1 propane molecule can be located in α cavity at low degree of calcium exchange and the number exceeds 2 molecules at high degree. With the pentane, the number closes to 3 molecules. It should be remarked that the volume occupied by 3 pentane molecules ($\approx 580\text{Å}^3$) is smaller than the volume of the α cavity ($770\text{Å}^3$) [7]. It means the cavities are not completely filled. At 523K, the effect of the calcium exchange is weakly pronounced on the adsorption of the two paraffins (1 to 1.5 molecules for pentane and less than ½ molecule for propane).

The initial rate of adsorption estimated from origins of adsorption isotherms was given in Fig. 3. We notice that whatever the temperature, the initial rate increases with calcium exchange degrees. The evolution of these curves is similar to the growth of molecule number per α cavity. Therefore, a simple determination of initial rate can give a good prediction of calcium exchange effect on adsorption of this type of paraffins.

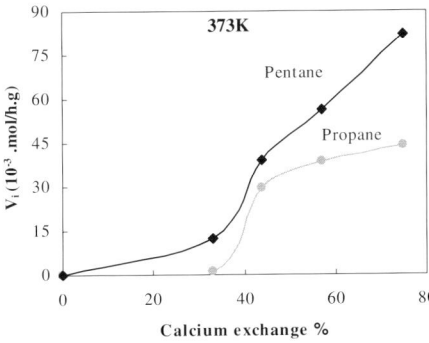 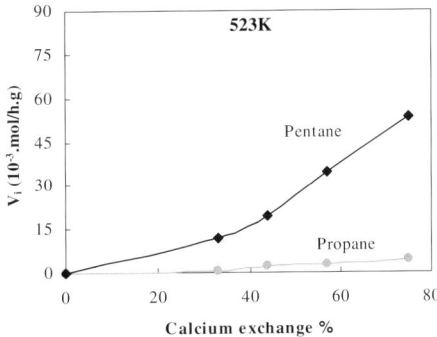

Fig. 3. Initial rate of propane and pentane adsorption versus the degree of calcium exchange.

### 3.2. Diffusion of propane and pentane

The thermogravimetric data are exploited to estimate the diffusion coefficient by using the second Fick's law [7]:

$$\frac{m_t}{m_\infty} = 1 - \frac{6}{\pi^2} \sum_{n=1}^{\infty} \frac{1}{n^2} \cdot exp\left(-\frac{D \cdot n^2 \cdot \pi^2 \cdot t}{r_0^2}\right) \quad (1)$$

For small time, this equation converges very slowly and can be presented (for $m_t/m_\infty < 0,3$) by:

$$\frac{m_t}{m_\infty} = \frac{6}{r_0}\left(\frac{D_i \cdot t}{\pi}\right)^{0,5} \quad (2)$$

where $m_t$ and $m_\infty$ are respectively the adsorbed amounts at time t and equilibrium, $D_i$ is diffusion coefficient, $r_0$ the radius of the zeolite crystallite and t the sorption time.

As first step, we have compared the theoretical Eq. (1) and experimental Eq. (2) curves. An example of this comparison is given at 373K for propane (Fig. 4), and pentane (Fig. 5).

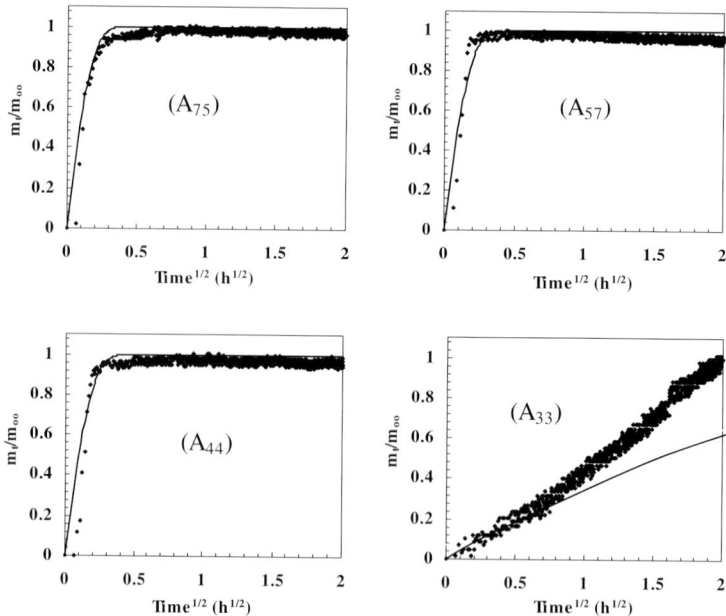

Fig. 4. Comparison of theoretical (—) and experimental (⋯) diffusion curves using thermogravimetric data obtained for propane at 373K.

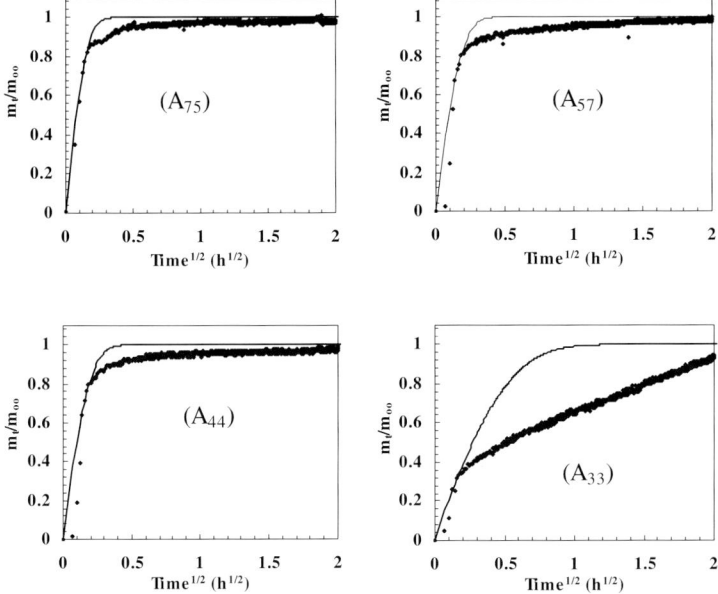

Fig. 5. Comparison of theoretical ($-$) and experimental ($\cdots$) diffusion curves using thermogravimetric data obtained for pentane at 373K.

Except for $A_{33}$ adsorbent, the experimental and theoretical curves are in good agreement. However the diffusion curves of propane are even better. Moreover, at 523K, the theoretical and experimental curves are in better conformity with pentane, whatever the calcium exchange. The values of initial diffusion coefficients given as $D_i/r_0^2$ versus calcium exchange degree are recapitulated in Table 1.

Table 1
Initial diffusion coefficient of propane and pentane in CaA adsorbents.

| %Ca in CaA adsorbent | $10^{-3} \cdot D_i/r_0^2$ (s$^{-1}$) (Propane) | | $10^{-3} \cdot D_i/r_0^2$ (s$^{-1}$) (Pentane) | |
|---|---|---|---|---|
| | 373K | 523K | 373K | 523K |
| 75 | 2.25 | 0.69 | 1.66 | 2 |
| 57 | 2.11 | 0.66 | 1.22 | 1.66 |
| 44 | 1.83 | 0.55 | 1.02 | 1.44 |
| 33 | 0.007 | 0.14 | 0.3 | 0.72 |

On the other hand, as presented in the example of Fig. 6, an inflection was observed in the beginning of all curves $m_t/m_\infty = f(t^{1/2})$. This idle period (noted $t_m$) corresponds to the time preceding the effective diffusion.

The comparison of this time for the both molecules indicates the existence of a relationship with calcium exchange degrees. An example of idle times estimated at 373K for propane and pentane is given in Fig. 7. This result showing the reduction of this time with the increase of the calcium exchange proves that the intercrystalline resistance is not the unique phenomenon existing during the diffusion as reported in literature [8]. In the case of pentane, a particular precaution was taken to avoid the considerable elapsed duration before the establishing the equilibrium vapor pressure in the thermobalance system. Therefore, the effect of calcium exchange on the idle period was evident in these experiences.

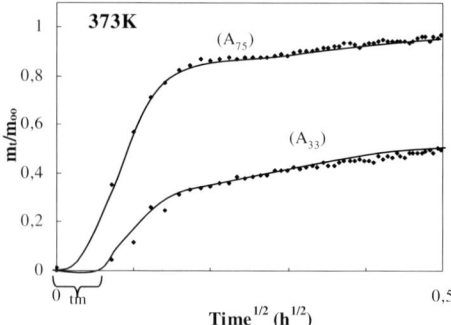

Fig. 6. Example of idle period preceding the effective diffusion obtained with pentane at 373K.

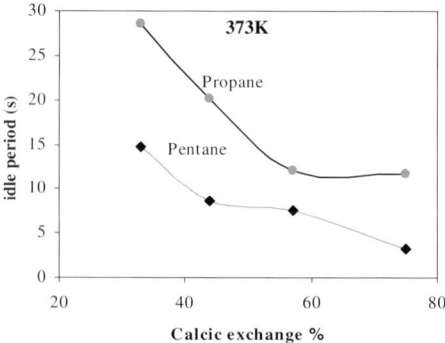

Fig. 7. Example of evolution of the idle period preceding the effective diffusion obtained at 373K versus calcium exchange degrees.

## CONCLUSION

According to the results presenting the initial rate of adsorption and/or the initial diffusion coefficient versus the calcium exchange degrees, it appears that the propane molecule seems be a very interesting model of adsorbat in highlighting the effect of calcium exchanging in A adsorbents. Actually, the parameters mentioned above, are affected identically and their behavior versus calcium exchange were similar to the evolution of nitrogen capacity adsorption. Moreover, the concordance of experimental and theoretical curves allows us to consider the propane molecule as an excellent probe molecule.

The idle period preceding the effective diffusion, using the approximate model ensuing from the second Fick's law, was evidently affected by the calcium exchange. This result excludes the hypothesis of the existence of the intercrystalline resistance as the unique phenomena during the diffusion.

## REFERENCES

[1] Johnson J. A. and Oroska A. R., Studies in Surface Science and Catalysis, Zeolites as Catalysis, Sorbents and Detergent Builders; Karge, H. G., Weitkamp, J., Eds.; Elsevier: Amsterdam, 46 (1989) 451.
[2] Jullian S., Mank L. and Minkkiner A., Fr. Patent 2.679.245, 1993; U.S. Patent 5,233, 1993, 120.
[3] Barrer R. M., Zeolites and Clay Minerals as Sorbents and Molecular Sieves, Academic Press, London, 1978.
[4] Benaliouche F. and Boucheffa Y., Chemical Engineering Transactions, 3 (2003) 1427.
[5] Shaozhou C., Xiolong Z., and Chao T., Acta. Petrolei Sinica (Petroleum processing section), 3 (1995) 40.
[6] Magnoux P. , Boucheffa Y., Guisnet M., Joly G. and Jullian S., Fundamentals of Adsorption VI, Meunier F. Eds., (1998) 135.
[7] Breck D. W., Zeolite Molecular Sieves, John Wiley and Sons Eds., New York, 1974, 29.
[8] Ruthven D. M., Principles of adsorption and adsorption Process, John Wiley and Sons, New York, 1984.

Studies in Surface Science and Catalysis 160
P.L. Llewellyn, F. Rodriquez-Reinoso, J. Rouqerol and N. Seaton (Editors)

# Transport properties of catalyst supports derived from a catalytic test reaction

**D. Enke, F. Friedel, T. Hahn and F. Janowski**

Institute of Technical Chemistry and Macromolecular Chemistry, University of Halle, Schloßberg 2, D-06108 Halle/Saale, Germany

The hydrogenation of benzene over $Ni/SiO_2$ model catalysts was used to investigate the transport characteristics of the mesopores inside the primary particles of silica supports. Glass beads with controlled mesoporosity in the range between 2 and 20 nm were used as model catalyst supports. The effective mesopore diffusivities were obtained from the Arrhenius plots. Furthermore, the pore diffusion coefficients were determined. The tortuosity factors were obtained from measurements of the permeability of nitrogen of porous glass membranes with comparable texture properties. A systematic correlation between the texture properties (mean pore size) and transport characteristics (pore diffusion coefficient) was found.

## 1. INTRODUCTION

Porous solids with bimodal pore structures are widely used as catalyst supports. Examples are formed bodies, such as tablets, beads, extrudates, and monoliths of molecular sieves, silica and alumina. The formed bodies are usually prepared by compacting fine micro- or mesoporous particles [1]. As a result, macropores are formed between the agglomerated particles. Recently, catalyst supports on the basis of silica and silica-alumina with hierarchical bimodal pore structures have been prepared by phase separation in sol-gel process [2, 3]. Here, an interconnected macroporous morphology is formed when transitional structures of spinodal decomposition of an organic polymer are frozen-in by the sol-gel transition of silica followed by a thermal treatment. The mesopores are located in the silica walls separating the macropores. Zhang et al. [4] prepared catalyst supports with bimodal pore structures by introducing silica, zirconia or alumina sol into the macropores of silica gel pellets.

In order to select optimal support material for a certain application, information regarding the transport characteristics of the micro/meso- and macropore system of the available bimodal catalyst supports is required. The transport parameters of the macropore system of a formed body can be determined using a Wicke-Kallenbach diffusion cell. Except for the well examined molecular sieves, there is only little information regarding the transport properties of the micro- or mesoporous primary pore system of catalyst supports [5].

Kotter et al. [6] proposed a procedure for the measurement of effective diffusivities of supported catalysts under reaction conditions. Here, the effective diffusivity is determined by reaction rate measurements. Additional information regarding the porosity and tortuosity factor of the catalyst support is necessary to obtain the pore diffusion coefficient of a reactant under reaction conditions. The uniform distribution of the active component is another requirement.

A suitable model system is necessary to study the correlation between the texture properties (mean pore size) and transport characteristics (pore diffusion coefficient) of the primary pore system of catalyst supports. The model catalyst support should be characterized by (i) a monomodal pore structure with pore sizes in the range between 2 and 20 nm (typical for the primary pore system of catalyst supports), (ii) a narrow pore size distribution and (iii) a flexible geometric form (beads and membranes with *comparable* texture properties).

Porous glasses as leaching products of phase-separated alkali borosilicate glasses meet all these requirements in a nearly ideal manner. Porous glasses are characterized by a tailorable pore size between 0.3 and 1000 nm, large specific surface areas and high thermal stability [7]. Additionally, they can be prepared in various geometric forms, i.e., as beads, rods and plates. The uniform pore structure represents another favourable property. Therefore, porous glasses were used extensively in the 1950s and 1960s as model systems in heterogeneous catalysis [8].

In this work, porous glasses in the shape of beads and ultrathin membranes with controlled mesoporosity in the range between 2 and 20 nm were used as model catalyst supports. The hydrogenation of benzene over nickel catalysts on the basis of the porous glass beads was used to investigate the transport characteristics of the mesopores inside the primary particles of silica supports. The tortuosity factors of the supports were obtained from measurements of the permeability of membranes with comparable texture properties.

## 2. EXPERIMENTAL

The preparation of the porous glass beads (S - X, S = support, X = number) and ultrathin membranes (M - X, M = membrane, X = number) was described in previous papers [9, 10]. An initial glass of the composition 70 wt.-% $SiO_2$, 23 wt.-% $B_2O_3$ and 7 wt.-% $Na_2O$ was used in all cases. The beads are 0.2 mm in diameter. A typical membrane geometry is 20 x 20 x 0.2 $mm^3$.

The nickel supported catalysts were prepared by equilibrium adsorption in $Ni(NO_3)_2$ aqueous solution. First, the catalyst supports were activated at 210°C for 24 h. After that, the samples were covered with $Ni(NO_3)_2$ aqueous solution (2 mole $Ni(NO_3)_2 \cdot 6H_2O$ dissolved in 1000 ml water; 3 g support in 10 ml solution), kept in vacuum for 20 h and washed with deionised water. The products were dried overnight at 120°C and then carefully calcined in air at 450°C for 5 h (heating rate 3.25 K $min^{-1}$). The resulting samples were reduced in a hydrogen flow of 50 ml $min^{-1}$ at 450°C for 5 h (heating rate 7 K $min^{-1}$) to obtain the nickel supported catalysts (S - X - Ni) and finally cooled down to the reaction temperature in a hydrogen flow. The metal loading of the nickel supported catalysts varied between 0.3 and 2 wt.-%.

Nitrogen sorption measurements were performed by using a Sorptomatic 1990 apparatus by ThermoFinnigan. All samples were degassed at 393 K before measurement for at least 24 hours at $10^{-5}$ mbar. Adsorption and desorption isotherms were measured over a range of relative pressures ($p/p^0$) from 0 to 1.0. Surface areas were determined from the linear part of the Brunauer-Emmett-Teller (BET) equation in a relative pressure range ($p/p^0$) of the adsorption isotherms between 0.05 and 0.25 [11]. A value of 0.162 $nm^2$ was used for the cross-sectional area per nitrogen molecule. The total pore volume was estimated from the amount of gas adsorbed at the relative pressure $p/p^0 = 0.99$ assuming that pores were filled subsequently with condensed adsorptive in the normal liquid state. The mercury intrusion measurements were carried out on a Pascal 440 apparatus by ThermoFinnigan. A contact angle of 141.3° for Hg was used. The cumulative pore volume at a given pressure represents the total volume of mercury taken up by the sample at that pressure. The mean pore diameter $d_P$ of samples with pore sizes below 5 nm was calculated from the total pore volume $V_P$ (nitrogen adsorption) and the BET surface area $S$ according to $d_p = 4V_P/S$ (cylindrical pore model). Mercury intrusion was used for pore sizes > 5 nm. Here, the mean pore diameter was calculated by applying the Washburn equation and a cylindrical pore model.

The ultrathin porous glass membranes were examined for the permeability of nitrogen at 25°C and 1.1 bar. The measurements were performed as follows: The membrane was stuck on a brass plate containing a bore with 2 and 5 mm diameter, respectively. After that, the plate was fixed with a special dome on a turbo molecular pump (TMU 261, Pfeiffer). Now, the dome was evacuated. The gas flow was determined by measuring the pressure below the membrane. Finally, the integral permeability was estimated using the pumping speed, the measured pressures above and below the membrane and the membrane thickness.

The benzene hydrogenation on the Ni catalysts based on the porous glass beads was carried out in a continuous-flow reactor at atmospheric pressure. A gaseous mixture of benzene and hydrogen flowed through the catalyst bed (0.1 or 0.2 g catalyst) at temperatures between 80 and 140°C. The partial pressure of benzene was 0.141 atm. Only cyclohexane was detected as a product under the present conditions.

## 3. RESULTS AND DISCUSSION

Light micrographs of the porous glass beads and an ultrathin porous glass membrane are shown in Fig. 1. The porous glass beads (A) used as catalyst supports are characterized by an ideal spherical shape, a narrow particle size distribution and a mean particle diameter of 0.2 mm. The porous glass membranes (B) are completely porous, free of cracks or defects and show a homogeneous microstructure.

The texture properties of the catalyst supports and the membranes prepared in our laboratory were initially characterized by the equilibrium based methods nitrogen gas adsorption and mercury intrusion. The texture characteristics are summarized in Table 1. The BET surface areas of the model catalyst supports are in the range between 80 and 200 $m^2 g^{-1}$. The pore volumes vary between 0.1 and 0.5 $cm^3 g^{-1}$. The mean pore diameter can be controlled in the range between 2 and 20 nm, which is typical for the primary pore structure of silica based catalyst supports. The ultrathin membranes are characterized by texture properties comparable to the support beads. The nitrogen sorption isotherms of a porous glass support and an ultrathin porous glass membrane are shown in Fig. 2. A typical type IV isotherm shape with a hysteresis of type H1 (IUPAC classification) is obtained in both cases. This indicates the presence of fairly uniform mesopores. Generally, all porous glass supports and ultrathin membranes are characterized by a relatively narrow pore size distribution.

The transport characteristics of the ultrathin porous glass membranes are given in Table 2 and Fig. 3. The values of the porosity $\varepsilon$ are determined form the total pore volume assuming a frame density of the porous glass of 2.2 $g cm^{-3}$.

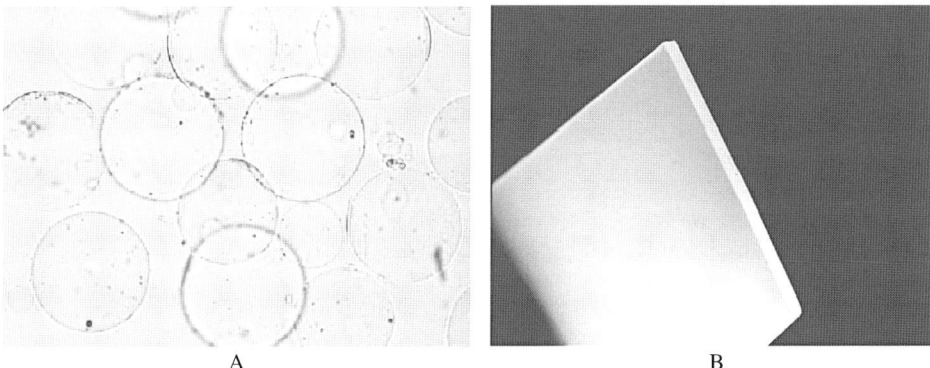

A                    B

Fig. 1 Porous glass beads (A) and an ultrathin porous glass membrane (B)

Table 1
Texture properties of the porous glass beads and the ultrathin porous glass membranes

| Sample | Spec. surface area [m² g⁻¹] | Spec. pore volume [cm³ g⁻¹] | Mean pore diameter [nm] |
|---|---|---|---|
| S - 1 | 201[1] | 0.125[1] | 2.5[1] |
| S - 2 | 210[1] | 0.150[1] | 4.5[1,2] |
| S - 3 | 110[1] | 0.269[1] | - |
|  | 116[3] | 0.260[3] | 8.4[3] |
| S - 4 | 109[1] | 0.465[1] | - |
|  | 151[3] | 0.414[3] | 11.2[3] |
| S - 5 | 105[1] | 0.375[1] | - |
|  | 138[3] | 0.470[3] | 14.0[3] |
| S - 6 | 79[1] | 0.213[1] | - |
|  | 111[3] | 0.434[3] | 16.2[3] |
| M - 1 | 250[1] | 0.127[1] | 2.0[1] |
| M - 2 | 163[1] | 0.139[1] | 4.0[1,2] |
| M - 3 | 97[1] | 0.258[1] | - |
|  | 93[3] | 0.203[3] | 8.6[3] |
| M - 4 | 220[1] | 0.676[1] | - |
|  | 175[3] | 0.442[3] | 10.2[3] |
| M - 5 | 181[1] | 0.719[1] | - |
|  | 179[3] | 0.564[3] | 12.0[3] |
| M - 6 | 107[1] | 0.269[1] | - |
|  | 195[3] | 0.751[3] | 18.0[3] |

[1] $N_2$ sorption at 77 K, [2] BARRETT-JOYNER-HALENDA, adsorption branch, [3] Mercury intrusion

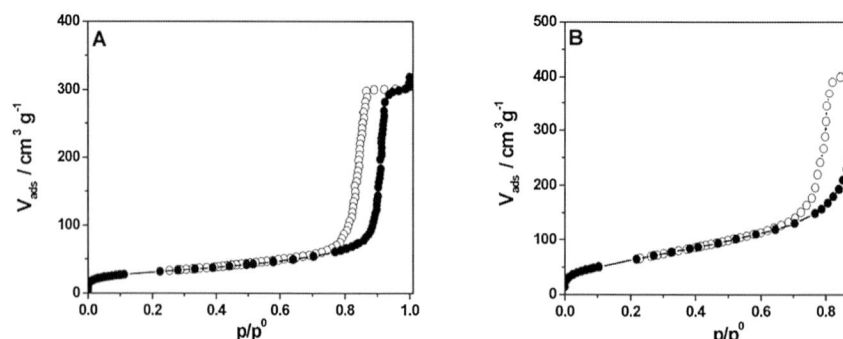

Fig. 2 Nitrogen sorption isotherms at 77 K of porous glass support (A) and an ultrathin porous glass membrane (B)

Table 2
Transport characteristics of the porous glass membranes ($N_2$, 25°C, 1.1 bar)

| membrane | $d_{p,m}$ [nm] | $\varepsilon$ | $D_{eff(Mem)}$ [cm$^2$ s$^{-1}$] | $D_{G,K}$ [cm$^2$ s$^{-1}$] | $\tau$ |
|---|---|---|---|---|---|
| M - 1 | 2.0 | 0.21 | $7.6 \cdot 10^{-5}$ | $3.2 \cdot 10^{-3}$ | 40 |
| M - 2 | 4.0 | 0.23 | $2.2 \cdot 10^{-3}$ | $6.3 \cdot 10^{-3}$ | 2.9 |
| M - 3 | 8.6 | 0.31 | $5.8 \cdot 10^{-3}$ | $1.4 \cdot 10^{-2}$ | 2.4 |
| M - 4 | 10.2 | 0.60 | $1.0 \cdot 10^{-2}$ | $1.6 \cdot 10^{-2}$ | 1.6 |
| M - 5 | 12.0 | 0.55 | $1.3 \cdot 10^{-2}$ | $1.9 \cdot 10^{-2}$ | 1.5 |
| M - 6 | 18.0 | 0.62 | $1.8 \cdot 10^{-2}$ | $2.8 \cdot 10^{-2}$ | 1.6 |

The effective diffusivities $D_{eff\,(Mem)}$ are calculated from the permeability $P$ of nitrogen through the membrane

$$D_{eff}(Mem) = \frac{P}{\varepsilon} \qquad (1).$$

The tortuosity factors of the membranes are given by

$$\tau = \frac{D_{G,K}}{D_{eff(Mem)}} \qquad (2)$$

where $D_{G,K}$ is the Knudsen diffusion coefficient of nitrogen at 25°C in a straight cylindrical pore with the corresponding pore diameter. The permeability of all membranes is independent of the pressure in the investigated range (1 - 1.5 bar). Except for the smallest pore size (2.0 nm), the mass transfer through all membranes under the experimental conditions is in the Knudsen region. The tortuosity factors vary between 1.5 and 40 depending on the pore diameter and the porosity of the membranes. An extremely high tortuosity factor of 40 is obtained for the 2 nm membranes. This indicates that configurational diffusion is involved. Here, the effect of the porosity can be excluded. (Samples M - 1 and M - 2 are characterized by comparable porosities but very different tortuosity factors.)

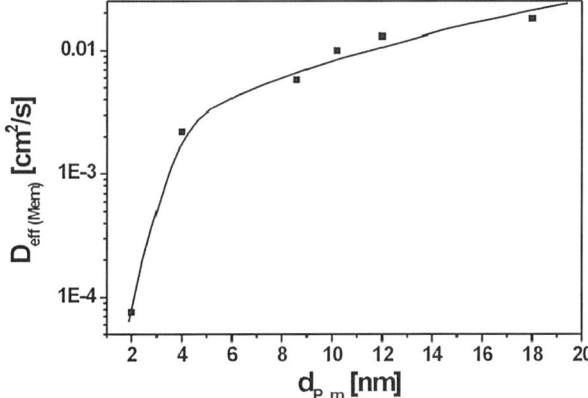

Fig. 3 Effective diffusivities of nitrogen at 25°C vs. mean pore size of the porous glass membranes

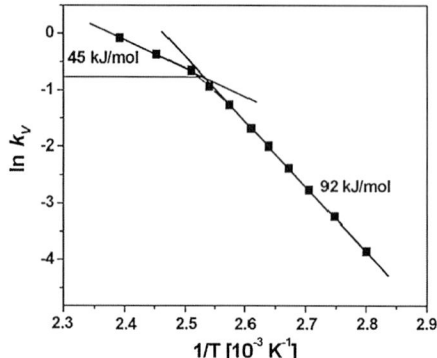

Fig. 4 Arrhenius plot of a nickel supported catalyst based on porous glass beads with a mean pore diameter of 16.2 nm

The effective diffusivities of benzene under reaction conditions are determined from the point of intersection of the lines separating the kinetic from the mass transfer controlled region in the Arrhenius plots according to [6]. This is shown in Fig. 4 for a nickel supported catalyst with a mean pore diameter of 16.2 nm. The intrinsic rate constant per unit of catalyst particle volume $k_V$ was estimated using plug flow behaviour and first order kinetics for the observed benzene conversion. The effective diffusivity is determined by the expression

$$D_{eff(benzene)} = \frac{R_P^2 \cdot k_{V(PI)}}{\phi^2} \tag{3}$$

where $k_{V(PI)}$ is the intrinsic rate constant at point of intersection, $R_P$ is the particle radius and $\phi$ is the Thiele diffusion modulus. The porous glass beads used as model catalyst supports are characterized by an ideal spherical shape and a narrow particle size distribution. Therefore, a particle radius of 0.01 cm and a Thiele diffusion modulus of 3 at the point of intersection [6] reflect the "reality" with high accuracy.

The pore diffusivities of benzene under reaction conditions are estimated by

$$D_{P(benzene)} = \frac{\tau_{(membrane)} \cdot D_{eff(benzene)}}{\varepsilon_{(beads)}} \tag{4}$$

where $\tau_{(membrane)}$ is the tortuosity factor of a membrane with comparable texture characteristics. The pore diffusivities from the reaction kinetics obtained for the nickel supported catalysts are given in Table 3.

Table 3
Pore diffusivities from the reaction kinetics obtained for the nickel supported catalysts

| support | $d_{p,\,m}$ [nm] | $D_{eff(benzene)}$ [cm$^2$ s$^{-1}$] | $\varepsilon$ | $\tau_{(membrane)}$ | $D_{P(benzene)}$ [cm$^2$ s$^{-1}$] |
|---|---|---|---|---|---|
| S - 1 - Ni | 2.5 | $5.4 \cdot 10^{-9}$ | 0.18 | 40 | $1.2 \cdot 10^{-6}$ |
| S - 2 - Ni | 4.5 | $6.4 \cdot 10^{-7}$ | 0.21 | 2.9 | $8.8 \cdot 10^{-6}$ |
| S - 3 - Ni | 8.4 | $1.0 \cdot 10^{-6}$ | 0.37 | 2.4 | $6.5 \cdot 10^{-6}$ |
| S - 4 - Ni | 11.2 | $3.0 \cdot 10^{-6}$ | 0.51 | 1.6 | $9.4 \cdot 10^{-6}$ |
| S - 5 - Ni | 14.0 | $3.6 \cdot 10^{-6}$ | 0.51 | 1.6 | $1.1 \cdot 10^{-5}$ |
| S - 6 - Ni | 16.2 | $5.1 \cdot 10^{-6}$ | 0.49 | 1.6 | $1.7 \cdot 10^{-5}$ |

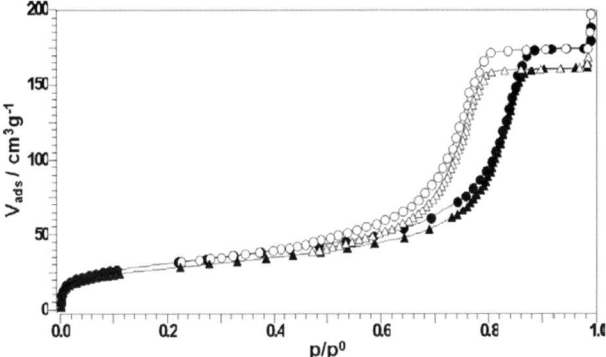

Fig. 5 Nitrogen sorption isotherms of a porous glass support before (circles) and after (triangles) the modification procedure

The preparation procedure of the nickel supported catalysts (equilibrium adsorption in $Ni(NO_3)_2$ aqueous solution, optimised washing step, activation conditions) in combination with a strong interaction between the nickel and the support [12] lead to a uniform distribution of the active component. One important precondition for the determination of effective diffusivities from reaction rate measurements is fulfilled.

The preservation of the characteristic texture and transport properties of the supports during the modification procedure (i.e. no pore blocking by the active component) represents another important precondition. A systematic investigation of all supports and membranes used in this study showed that the characteristic texture and transport properties were *retained* during the modification procedure. As an example, Fig. 5 shows the nitrogen sorption isotherms of a porous glass support before and after the modification procedure.

The pore diffusivities obtained are in the range between $1 \cdot 10^{-6}$ and $2 \cdot 10^{-5}$ cm$^2$ s$^{-1}$. There is a linear correlation between the pore diffusivity of benzene under reaction conditions and the mean pore diameter of the support in the investigated pore size range (Fig. 6). The low absolute values of the pore diffusivities compared to theoretical calculations (Knudsen diffusivity of benzene at reaction temperature in a straight cylindrical pore with the corresponding pore diameter) may arise from strong interactions of the diffusing reactant with the surface hydroxyl groups of the support [13]. Pore roughness effects are another possible explanation. Further investigations are in progress.

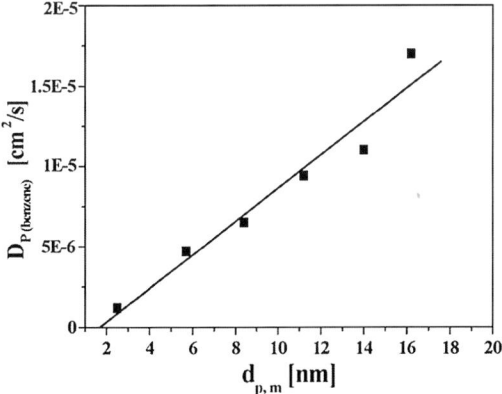

Fig. 6 Pore diffusivities of benzene vs. mean pore diameter of the porous glass support

## 4. CONCLUSIONS

Porous glasses in the shape of beads and ultrathin membranes with comparable texture properties are an ideal model system to investigate the transport characteristics of the mesopores inside the primary particles of silica supports. The hydrogenation of benzene over a nickel catalyst based on the porous glass beads is a suitable test reaction to estimate the effective diffusivities. The tortuosity factors can be obtained from measurements of the permeability of membranes with comparable texture properties. Now, the pore diffusivity of benzene under reaction conditions can be calculated. The low absolute values of the pore diffusivities obtained in this study indicate that interactions between the diffusing reactant and the surface of the support or pore roughness effects have to be considered.

## REFERENCES

[1]  U. Hammon, M. Kotter, Chem.-Ing.-Tech., 56 (1984) 455.
[2]  N. Nakamura, R. Takahashi, S. Sato, T. Sodesawa, S. Yoshida,
     Phys. Chem. Chem. Phys., 2 (2000) 4983.
[3]  R. Takahashi, S. Sato, T. Sodesawa, K. Arai, M. Yabuki, J. Catal., 229 (2005) 24.
[4]  Y. Zhang, Y. Yoneyama, K. Fujimoto, N. Tsubaki, J. Chem. Eng. Jap., 36 (2003) 874.
[5]  T. Dogu, Ind. Eng. Chem. Res., 37 (1998) 2158.
[6]  M. Kotter, P. Lovera, L. Riekert, Ber. Buns. physk. Chemie, 80 (1976) 61.
[7]  F. Janowski, D. Enke, in: Handbook of Porous Solids, F. Schüth, K.S.W. Sing,
     J. Weitkamp (eds.), Volume 3, Wiley-VCH, Weinheim, 2002, 1432 - 1542.
[8]  C.N. Satterfield, Mass Transfer in Heterogeneous Catalysis,
     Robert E. Krieger Publishing Company, Malabar, 1970.
[9]  D. Enke, F. Janowski, W. Schwieger, Microporous and Mesoporous Materials, 60 (2003) 19.
[10] D. Enke, F. Friedel, F. Janowski, T. Hahn, W. Gille, R. Müller, H. Kaden, in:
     Characterization of Porous Solids VI, F. Rodriguez-Reinoso, B. McEnaney,
     J. Rouquerol, K. Unger (eds.), Studies in Surface Science and Catalysis, Volume 144,
     Elsevier, Amsterdam, 2002, 347 - 354.
[11] S. Brunauer, P.H. Emmett, E. Teller, J. Am. Chem. Soc., 60 (1938) 309.
[12] M.P. Gonzalez-Marcos, J.I. Gutieerrez-Ortiz, C. Conzalez-Ortiz de Elguea,
     J.A. Delgado, J.R. Gonzalez-Velasco, Applied Catalysis A: General, 162 (1997) 269.
[13] J.A. Cusumano, M.J.D. Low, J. Phys. Chem., 74 (1970) 792.

Studies in Surface Science and Catalysis 160
P.L. Llewellyn, F. Rodriquez-Reinoso, J. Rouqerol and N. Seaton (Editors)

# Ellipsometric study of porosity distribution in hybrid silica-based sol-gel films

**L. Więcław-Solny [a, b], A. Kudła [c], J. Mrowiec-Białoń [d], A.B. Jarzębski [a, d],\***

[a] Department of. Chemical Engineering, Silesian University of Technology, M. Strzody 7, 44-100 Gliwice, Poland, e-mail: Andrzej.Jarzebski@polsl.pl

[b] Institute of Chemical Processing of Coal, Zamkowa 1, 41-800 Zabrze, Poland

[c] Institute of Electron Technology, al. Lotników 32/46, 02-668 Warsaw, Poland

[d] Institute of Chemical Engineering, Polish Academy of Science, Bałtycka 5, 44-100 Gliwice, Poland

## 1. INTRODUCTION

Films and coatings obtained using the sol-gel method attract considerable attention and the development of methods suitable for their characterization is of importance. Porosity is often seen as their critical property; it may be a desirable or detrimental factor, depending on applications. Sometimes its level has to be precisely controlled. Conventionally the mean value of porosity, averaged over the normal to the sample, is determined using the photometric methods. Yet very little is known on the porosity distribution over the film cross-section and its possible variation with a rise in distance from the substrate surface. The variable angle spectroscopic ellipsometry (VASE) is the method sensitive to the gradients in material properties vs. depth in the film, e.g. optical anisotropy. Thus it makes possible to determine the refractive index profiles, and hence, indirectly the changes in the volume of voids. The aim of this work was to explore these potentials in a study of porosity of defect-free hybrid sol-gel films: silica-Nafion (SN) and silica-phosphotungstic acid (HPA) obtained using a dip coating method and deposited on silicon wafers and glass slides.

## 2. EXPERIMENTAL

### 2.1 Fabrication of films

As the fabrication of films was described in much detail elsewhere [1] for this reason it is given here only in brief. Coating liquids were prepared to obtain a nominal content of active compound (AC): 30wt% (HPA) or 50wt% (Nafion) in a dry composite (AC+$SiO_2$). Typically the coating liquid of the molar composition Si:EtOH:$H_2O$:HCl = 1:4:8:0.008 was prepared by mixing two solutions, one containing silica precursor, the other one the appropriate amount of AC dissolved in EtOH. Then the mixture was homogenized to afford a clear sol which was used after a certain period of aging (3 to 50 days) as a dip-coating liquid. After a conventional treatment [1] the substrates were dipped in the coating liquid, held in the bath for 5 min and pulled up at a constant speed varying in the range of 0.5-25 cm/min. After each coating the substrate/film was dried at 50°C for 30 min and then at 110°C for 10 min and stored. In the case of multiple coatings the procedure was repeated once the previous step was completed. The samples were labelled: withdrawal speed (cm/min)-AC-sol age(days).

## 2.2. Ellipsometric measurements and porosity determination

Ellipsometry measures the ratio of two complex reflection coefficients for light polarized in the p-(plane of incidence) and s-(perpendicular to the plane of incidence) direction. Spectroscopic ellipsometry (SE) measures this ratio as a function of wavelength, and Variable Angle Spectroscopic Ellipsometry (VASE) performs these measurements as a function of both wavelength and angle of incidence [2,3]. The method is highly accurate and reproducible and no reference material is necessary. Moreover, because of the measurement of phase difference, $\Psi$ and amplitude ratio, $\Delta$, the method is also very sensitive, even in very short scales, what makes it suitable for characterisation of thin films. To extract useful information about the sample, it is necessary to perform a model dependent analysis of the ellipsometric $\Psi$ and $\Delta$ data [4]. The measured and computed characteristics of $\Psi$ and $\Delta$ are compared and the values of model parameters giving the best fit of experimental data are determined. Conventionally the Mean Square Error is used as a fitting criterion. Optical models of the film structure, considered to be either single- or multi-layered, contain indexes for each layer and their thickness [4]. In the study we used the Bruggeman model of effective medium approximation (EMA) [5], regarded as one of the most reliable. The film was considered either to be uniform or it was divided by five nodes to four layers, each of which was further divided into five equal sub-layers. Then refraction index depth profile was determined.

SE measurements were carried out using a Woollam VASE apparatus with a rotating analyser at 21°C. The values of $\Psi$ and $\Delta$ were measured over the spectral range of $250 - 1000$ nm at two angles 65 and 75 deg. The samples on glass surfaces were roughened from the backside to avoid backside reflection. Light spot diameter was about 2 mm (at $0°$ angle). Analyses were performed using WVASE32 software, and optical constants for silicon and silica were taken from its library. Surface roughness had little effect on the values of refraction index.

## 3. RESULTS AND DISCUSSION

Analysis of a few dozens of films obtained from a single or a double coating process was performed. The refraction index of SN voids free material was determined to be $n = 1.62$, and for HPA compound $n = 1.56$ at wavelength 632.8 nm. These values and the Bruggeman formula [5] was used to calculate the values of refraction index and the corresponding values of voids given in Fig 1.

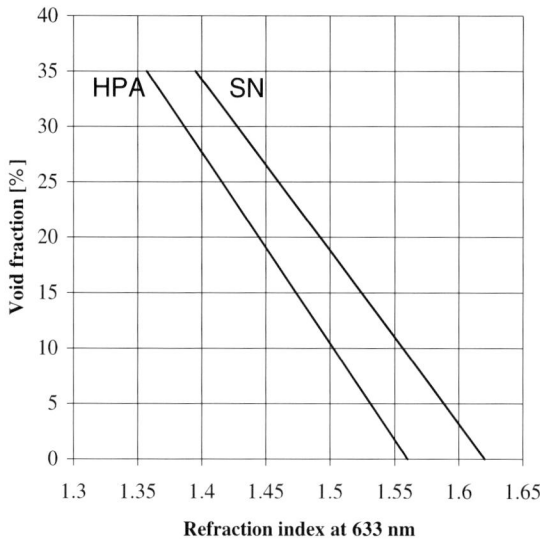

Fig. 1. Void fraction vs. refraction index values for HPA and SN films.

In case of some films obtained from a single coating process a single layer model gave the value of MSE less than 3, i.e. the value recognized as adequate to give a fair portrayal of experimental characteristics of $\Psi$ and $\Delta$. This is shown in Fig. 2 for 9.3-SN-3 deposited on silicon wafer (mean porosity ca. 22%). In most cases, however, we failed to obtain a fair fit of the experimental characteristics using a single layer model and application of the graded approach appeared necessary to obtain a satisfying value of MSE, and hence an agreement of modeling predictions and experiment (Fig. 3). From refractive index profiles we can draw conclusions regarding the variation of porosity as a function of distance from the substrate surface.

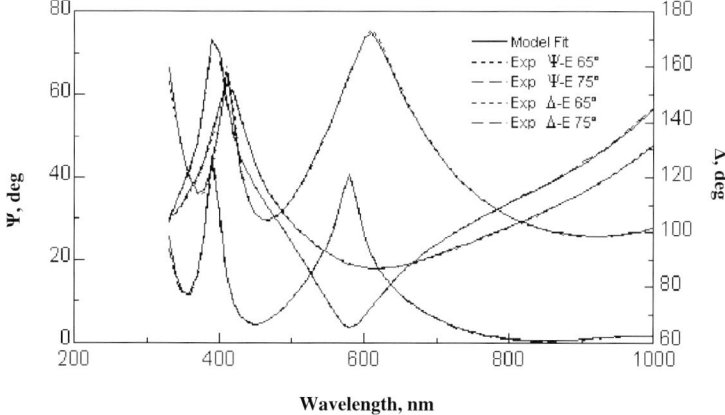

Fig. 2. $\Psi$ and $\Delta$ characteristics for 9.3-SN-3 sample.

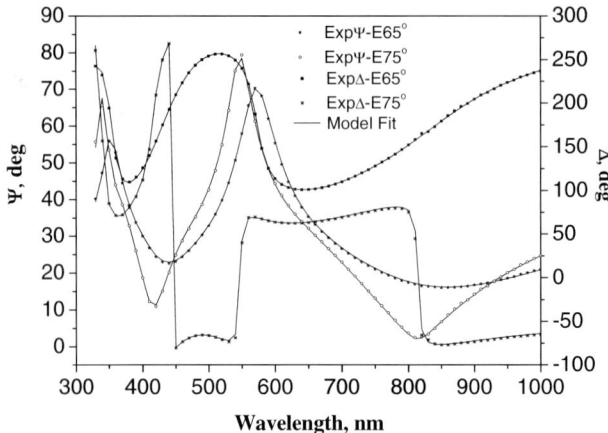

Fig. 3. Ψ and Δ characteristics for 19.5-SN-5 sample.

To check to what extent this concept is reliable we first made a number of measurements in different places of the same sample (single 19.5-SN-25 film on glass substrate). The profiles of refractive index given in Fig. 4 are quite similar in shape but show some difference is the value of film thickness. The latter appeared to be slightly larger in the lower section of the substrate than in the upper, but it is in accord with expectations. Interestingly, in both cases the highest value of refractive index, corresponding to that of very low porosity (cf. Fig.1), was in the region of substrate surface. This is a quantitative evidence of good contact between the film and the substrate, due to hydrophilisation of glass slide surface prior to coating [1]. We recorded a notable similarity in the shape of refractive index profiles from films deposited on glass slides using younger (3-5 days of age) sols (Fig. 5).

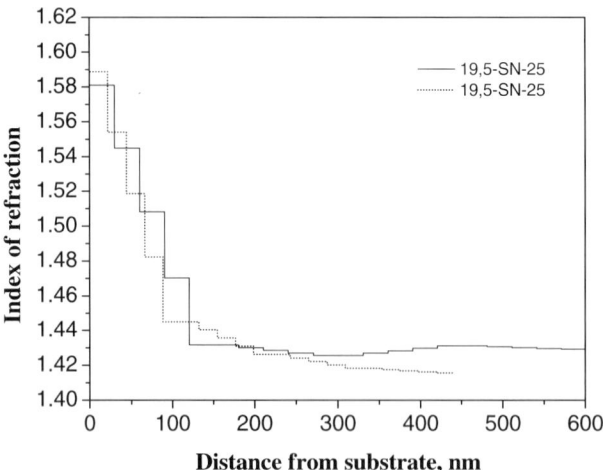

Fig. 4. Index of refraction vs. distance from glass substrate surface.

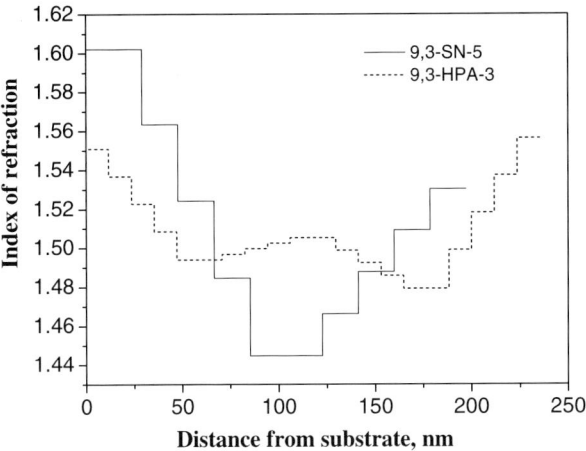

Fig. 5. Index of refraction vs. distance from glass substrate surface.

Regardless the type of an active compound, the largest value of refractive index was at the substrate surface, than it fell down to rise in the region close the film free surface. Apparently, the porosity changes trend was quite opposite. This demonstrates that in those samples a notable porosity of about 25% was in the internal portion of layers and the structure was much more compact at the free surface, with porosity ca. 15%. In the case of HPA sample the structure was as compact at the free surface as in the region close to the glass slide. In films obtained using more condensed (older) sols this surface compaction was less pronounced or it was not observed at all (Fig. 6). When sol was applied not on a pristine substrate but on a previous film this impregnated that layer leading to a decrease of its porosity from about 30% to some 20-25 %, evidenced by a rise of refractive index (cf. Fig. 6).

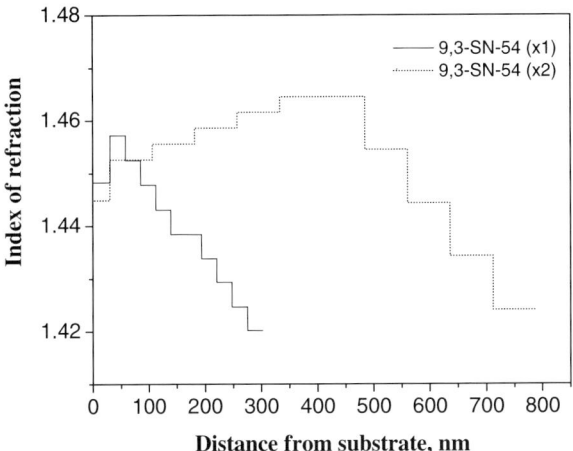

Fig. 6. Index of refraction vs. distance from glass substrate surface.

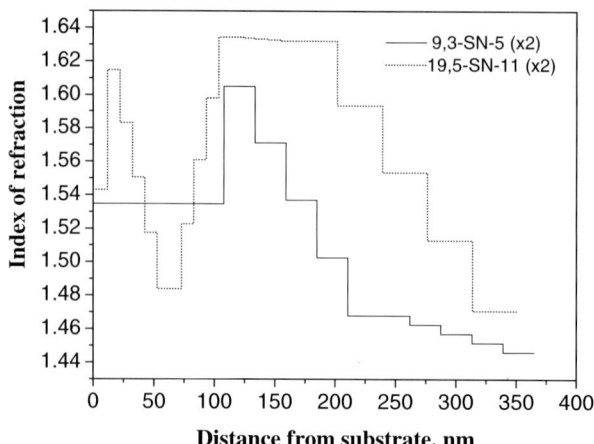

Fig. 7. Index of refraction vs. distance from silicon substrate surface.

The surface of silicon wafers was not hydrophilisized, unlike that of glass slides, and this appeared to affect the film structure, especially in the near substrate region (cf. Fig. 5 and Fig. 7). The values of refractive index from films, especially the older ones, supported on silicon were somewhat lower, an indicative of larger porosity (30-35 %). They fluctuated somewhat with a rise in distance from the substrate (cf. Fig. 8) and even attained a maximum value at some distance from its surface.

On the whole the mean value of porosity in the investigated coatings was estimated to be in the range of 15-30%. Yet these studies indicated that the local porosity value may strongly vary with distance from the substrate surface, sometimes even from less than 1% at the substrate surface up to about 25% in the internal part or free surface.

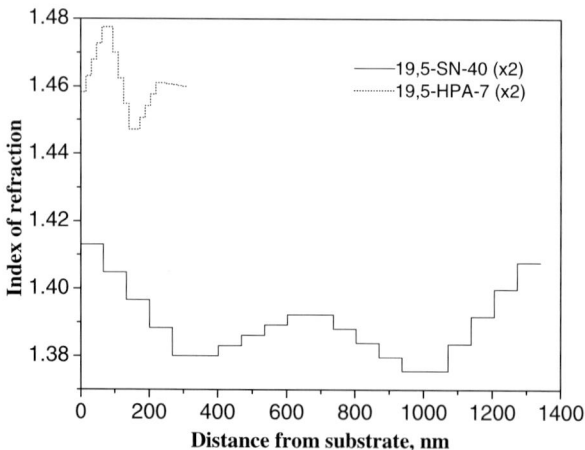

Fig. 8. Index of refraction vs. distance from silicon substrate surface.

## 4. CONCLUSIONS

The analysis of hybrid thin films obtained by a dip-coating sol-gel method shows that porosity within films markedly changes as a function of distance from the substrate surface.

As can be inferred from the refractive index profiles, the porosity profiles in films may be quite complex, depending on the coating sol structure and its condensation as well the substrate surface properties and its pretreatment. If highly condensed sols are used the profiles of porosity increase from the minimum values at the substrate surface to level out with a rise in distance from the substrate surface.

## REFERENCES

[1] L. Więcław-Solny, A.B. Jarzębski, J. Mrowiec-Białoń, W. Turek, Z. Ujma, M. Gibas, A. Kudła, Appl. Catal. A: General, (2005) 285, 79.
[2] H.G. Tompkins, E.G. Irene, Handbook of Ellipsometry, William Andrew Publishing, New York, 2005.
[3] H.G. Tompkins and W.A. McGahan, Spectroscopic Ellipsometry and Reflectometry, Wiley, New York, 1999.
[4] H. Arwin, Sens. Actuators A, (2001) 91, 43.
[5] Von D.A.G. Bruggeman, Annalen der Physik, (1935) 24, 636.

Studies in Surface Science and Catalysis 160
P.L. Llewellyn, F. Rodriquez-Reinoso, J. Rouqerol and N. Seaton (Editors)

# Positronium annihilation study of as-synthesized MCM-41 silica under pressure

**R. Zaleski [a], J. Goworek [b] and A. Borówka [b]**

[a] Institute of Physics, Maria Curie-Skłodowska University,
M. Curie-Skłodowska Sq. 1, 20-031 Lublin, Poland

[b] Faculty of Chemistry, Maria Curie-Skłodowska University,
M. Curie-Skłodowska Sq. 3, 20-031 Lublin, Poland

The lifetime spectra of positron annihilation in the MCM-41 ordered silica before template removal were measured as a function of pressure. The samples were compressed with various hydrostatic pressures by argon. Moreover, the same samples were compressed mechanically. Several lifetime components, well pronounced in the spectra, indicate the presence of various types of defects, voids or pores in the template as well as silica network. When the hydrostatic pressure increases, one can observe evolution of defects in the micellar interior and high stability of micropores in the silica walls of cylindrical pores. In the case of higher mechanical pressure, the samples exhibit strong degradation of silica framework. Simultaneously structural changes of template are less pronounced.

## 1. INTRODUCTION

Since 1992 the synthesis of ordered mesoporous materials using surfactants as pore directing agents has attracted increasing attention [1]. These materials are usually prepared using tetraethoxysilicate as the silica source and alkyltrimethylammonium halide as the template. The final stage of synthesis is the removal of the organic template from the as-synthesized organic/inorganic composite material. Many methods are used to eliminate the surfactant molecules, e.g. thermal methods (heating up to 550°C in air [1], nitrogen, ozone [2], argon [3]), washing-extraction with various solvents, thermal treatment in vacuum [4]. During calcination at 550°C in air final silica material contains up to several percent of carbon and traces of nitrogen species. To improve the chemical quality of MCM-41 further burn-off of the residual coke or other organic species can be performed by long-time treatment in strongly oxidative atmosphere, e.g. oxygen [2].

Mechanical stability is an important factor of MCM-41 for most practical applications. The ordered structure of MCM-41 with the evacuated template can be drastically changed by applying an external pressure of about 130 MPa and practically destroyed at pressures 250-500 MPa [5]. It should be noted that mechanical stability of MCM-41 depends strongly on the pore wall thickness, which may be controlled by adjusting the alkalinity of synthesis media and amount of surfactant in the synthesis mixture, and changed between 0.8 and 1.7 nm [6]. Over 1000 MPa complete amorphization of MCM-41 is observed [7].

Majority of the previous studies was devoted to surfactant degradation and structural properties of the silica product during and after calcination. The properties of as-synthesized material with micelle filling the silica network are rarely explored.

Recently, positronium annihilation lifetime spectroscopy (PALS) was applied for monitoring of micelle transformations during the thermal treatment. This technique can supply information about imperfections in the template structure; in particular, about the changes occurring at crossing the threshold of template degradation, e.g. the restructurization begins by splitting the cylindrical micelle into smaller parts. This way the free volumes among short cylinders are created. Thus, one can say that the micellar template exhibits some liquid-like properties.

The present paper reports the results of investigation of raw MCM-41 silica under high pressure. The main aim of these studies was more detailed description of structural changes in organic template. The PALS experiments were performed for MCM-41 prepared with octadecyltrimethylammonium bromide as a template. To the best of our knowledge, there are no data on the effect of compression under inert gas on properties and structural changes of MCM-41 with the organic template present in the pores of as-synthesized sample.

## 2. EXPERIMENTAL

### 2.1. Material
Mesoporous silica MCM-41 was prepared following the synthesis procedure reported previously [8]. MCM-41 material was synthesized, using tetraethyl orthosilicate (TEOS) as a silica source, octadecyltrimethylammonium bromide (C18TMAB) as a template and ammonia as the catalyst. The precipitated surfactant-silica mesophase was filtered, washed with distilled water and dried at 323K. Part of as-synthesized samples was next calcined at 823K in air atmosphere and additionally thermally treated in oxygen to remove the carbon deposits located inside the channels.

### 2.2. Methods
The nitrogen adsorption/desorption isotherms were measured at 77 K using the ASAP 2010 volumetric adsorption analyzer (Micromeritics Inc., Norcross, GA). The specific surface areas of the investigated samples were determined using the standard BET method at the relative pressure $p/p_0$ in the range 0.04 to 0.25. The total pore volume was estimated from single point adsorption at the relative pressure of about 0.99.

The samples for PALS experiments were prepared in the "sandwich" configuration (sample-positron source-sample). During the preparation it was formed into pastille (mechanically pressed to about 50 MPa) and placed in the argon pressure chamber. The other part of the same sample was mechanically compressed. The MCM-41 sample was put into a small tube, closed by two movable pistons sealed by O-rings. The air from the container was evacuated to the pressure of about 0.5 Pa. Such a set was placed in the argon pressure chamber, so that mechanical pressure exerted by the pistons pushed by argon could be applied to the sample. The PALS spectra were collected in both cases in the pressure range 0.1-490 MPa. The positronium annihilation method for characterization of porosity of solids is based on the relation between ortho-positronium (o-Ps) lifetime and the size of free volume, in which o-Ps is trapped. The PALS spectra were processed as described in Ref. [9].

## 3. RESULTS AND DISCUSSION

For investigation of pores in solids, the positron annihilation lifetime spectroscopy (PALS) uses a very small probe – the positronium atom, which is a bound state of positron and electron. The lifetime of the positronium triplet state (ortho-positronium, o-Ps) in the matter is shorter than in vacuum (142 ns) due to pick-off annihilation. The relation between the shortened lifetime and the free volume size is given by the Tao-Eldrup model (see Fig.1). Pore sizes from 0.1 nm to about 20 nm can be determined using its extended version [10]. Moreover, the pore concentration can be estimated from the intensities of lifetime spectrum components.

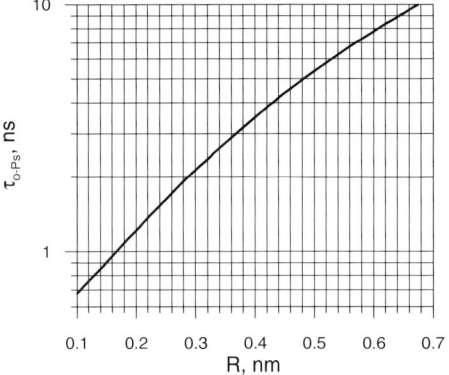

Fig. 1. Positronium lifetime dependence on the spherical free volume radius given by the Tao-Eldrup model.

### 3.1. Temperature experiment

The temperature dependences of lifetimes and intensities of positronium annihilating in MCM-41 filled with micelle are shown in Fig.2. The temperature rising over 400 K in vacuum (p $\approx$ 0.3 Pa) causes micelle degradation [9].

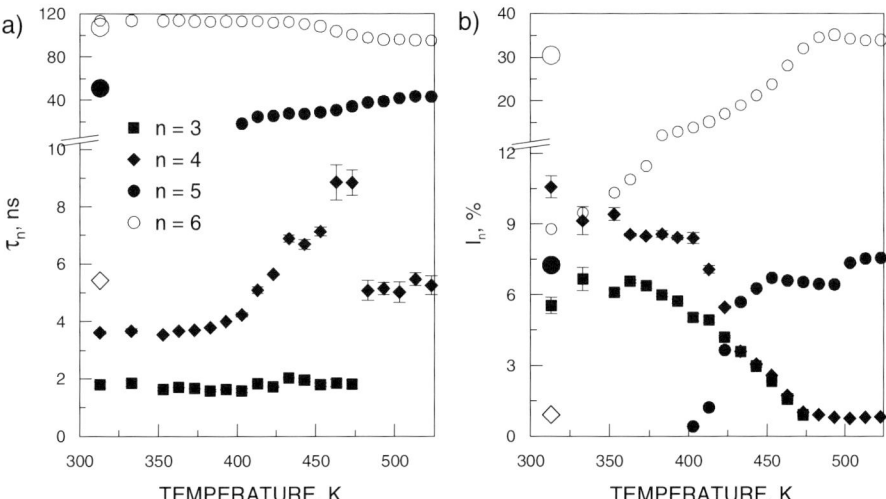

Fig. 2. Ortho-positronium lifetimes (a) and intensities (b) for as synthesized MCM-41 as a function of increasing temperature in vacuum. Large size symbols correspond to the parameters obtained after cooling the sample. *n* denotes the lifetime spectrum component.

It is confirmed by appearance of a spectrum component with the lifetime of 20 - 45 ns which corresponds to annihilation in emptied pores. The intensity of the intergranular component $I_6$ rises due to the increase of the total sample surface (both external and internal) combined with positronium escape from the pores. In vacuum its lifetime remains over 95 ns. Below 400 K the parameters of micellar components remain unchanged during heating. When micelle starts cracking ($\tau_5 \sim 20$ ns) rapid decrease of $I_4$ is visible indicating positronium escape from free volumes in the micelle to the micelle cracks. When cracks concentration establishes ($I_5$ becomes constant) the decrease in both $I_3$ and $I_4$ values becomes more slowly. In this stage of calcination, the main reason for changes is loss of the surfactant material due to its thermal decomposition. At a temperature below 523 K it consists mainly in detachment of $N(CH_3)_3$ group from the alkyl chain. If this fact is correlated to the rise of the lifetime $\tau_4$, the conclusion is that free volumes causing the existence of the longer micellar component are placed between the micelle and the silica wall. No change of $\tau_3$ means that temperature change does not affect the surrounding of appropriate free volumes. Probably changes of micelle structure are due to its free volume destruction. Over 475 K intensities the $I_3$ and $I_4$ drop below 1 % and these two components can not be distinguished any more. The remaining component is no longer connected with micelle and its presence is caused by the effect of positronium thermalization in large free volumes. After heating in vacuum changes of the sample structure are permanent (see Fig.2 – large size empty symbols).

### 3.2. Hydrostatic pressure experiments

Ortho-positronium lifetimes and intensities for MCM-41 silica with filled pores against pressure are shown in Fig.3. Three ortho-positronium components are present in the obtained PALS spectra at the beginning of experiment when the sample is under normal external pressure. Short lifetimes ($\tau_3 = 2.2$ ns and $\tau_4 = 3.5$ ns) correspond to the spherical empty spaces of the radii 0.31 nm and 0.40 nm. Existence of such free volumes is possible inside or in the neighbourhood of the micella.

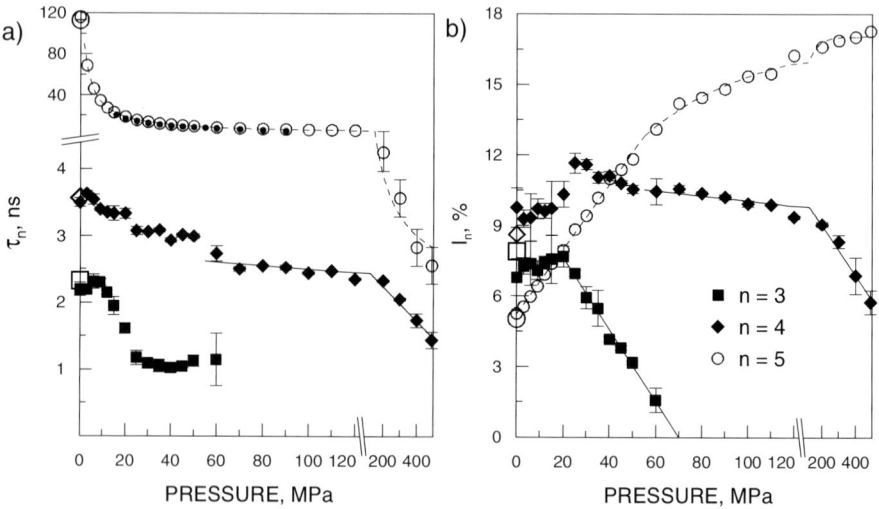

Fig. 3. Ortho-positronium lifetimes (a) and intensities (b) for as synthesized MCM-41 as a function of the increasing hydrostatic pressure of argon. Large size symbols correspond to the parameters obtained after the pressure release.

The third component ($\tau_5 = 115$ ns) is o-Ps annihilation outside the MCM-41 grains. These empty spaces have a mean radius about 8 nm. At lower pressures (p < 15 MPa) there are no significant changes in both spectrum components ascribed to the positron annihilation in the MCM-41 matrix. It leads to a conclusion that argon is not able either to change the micelle structure or to penetrate its interior. On the contrary, intergranular spaces are filled with argon which is confirmed by the characteristic change of the longest lifetime $\tau_5 \sim 1/p$ (dashed line in Fig.3).

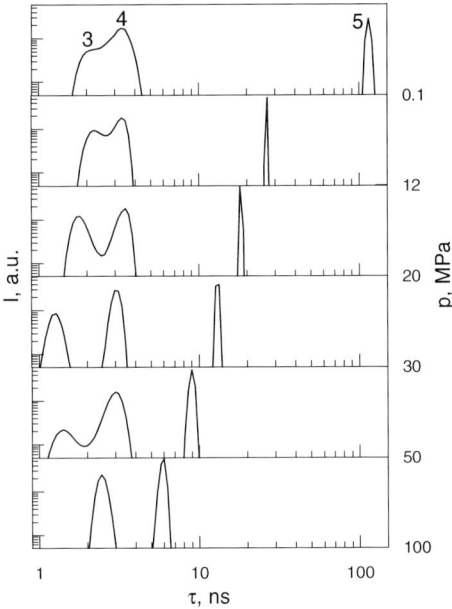

Fig. 4. Lifetime distribution histograms obtained by deconvolution of PALS spectra for the spectrum components: $\tau_3$, $\tau_4$ and $\tau_5$ collected at various pressures

The lifetime values are almost the same as those measured for argon in the absence of silica sample (small, filled circles in Fig.3a). The increase of this component intensity indicates increasing probability of positronium formation and annihilation in argon of increasing density.

In the pressure range 15 - 20 MPa the shortest o-Ps component lifetime $\tau_3$ decreases by about a half of its initial value. The reason for such $\tau_3$ shortening is the change of template structure. The radius of smaller free volumes in the micelle decreases quite rapidly to 0.17 nm (probably due to argon penetration of the micelle interior). The radius establishes but the intensity $I_3$ correlated to the free volume concentration shows linear decrease with the increasing pressure. Finally the shorter micellar component disappears at 70 MPa. The evolution of both short components is difficult to reconstruct. The problem is connected with a small lifetime difference and a rapid change of the intense longest component.

The deconvolution of the spectra performed using the MELT program [11] permits a closer look at the influence of pressure on the structure of template and silica. The lifetime distribution histograms obtained by deconvolution of PALS spectra are shown in Fig.4. At lower pressure two well separated peaks (lifetime distributions) corresponding to two different types of voids are observed. When argon pressure above the sample increases, $\tau_5$ becomes shorter. Moreover, the shorter spectrum component of the first bimodal peak initially becomes more separated and next at higher pressure disappears. The results presented in Fig. 4 illustrate disappearance of small free volumes in the micellar interior and stability of pores at high pressures in the silica framework (walls of pores). Generally, one can say that the results presented in Fig. 4 correlate reasonably well with those in Fig. 3a.

After the step at 20 - 25 MPa in the intensity of the longer component, $I_4$ decreases when the pressure increases similarly to $I_3$. However, the change is much slower and followed by lifetime $\tau_4$ shortening by about 2.6 ps/MPa. Even at 490 MPa this component is present with $\tau_4 = 1.44$ ns and $I_4 = 5.7$ %. Its resistance to argon hydrostatic pressure can be explained if we assume that the free volumes originating from this component are placed in the neighbourhood of the rigid silica skeleton, i.e. between the rough silica wall and the micelle surface. The hydrostatic argon pressure does not cause sample damage. If pressure is released, lifetimes and intensities come back to their initial values (large size empty symbols, Fig.3). It means that the pressure applied in argon experiments does not change MCM-41 powder morphology.

### 3.3. Mechanical pressure experiment

In order to separate argon related effects from the pressure affected changes, mechanical pressure was applied for compression of the silica sample. Without argon penetration, pressure dependence of the lifetimes and intensities of the longest lifetime $\tau_5$ refers to the sizes of intergranular spaces and the intensity of this spectrum component − to their concentration. The increasing pressure results in reduction of their amount and mean radius (open circles in Fig.5a). It means that under high mechanical pressure, the particles of MCM-41 are getting closer. At 140 MPa free spaces outside the grains disappear and all positronium annihilates in the proximity of the micelle.

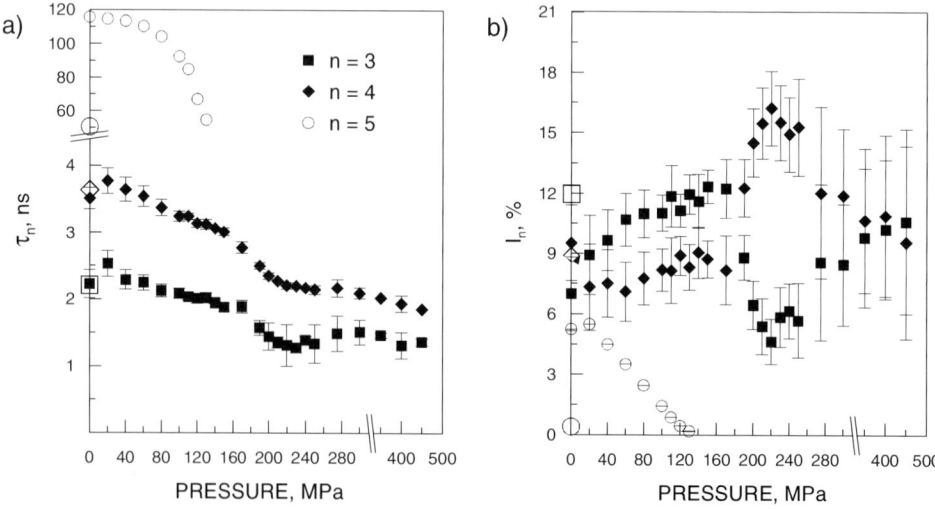

Fig. 5. Ortho-positronium lifetimes (a) and intensities (b) in the micelle filled MCM-41 as the function of increasing mechanical pressure.

Without argon penetration of the micelle, there are no rapid changes in the third spectrum component. Both short lifetimes $\tau_3$ and $\tau_4$ decrease slowly, while their intensities increase. At about 160 MPa, the linear dependence of the parameters vs. pressure is distorted, probably due to silica walls cracking. This effect correlates reasonably well with the experiments presented elsewhere [5,7]. Over 200 MPa the short lifetime values become almost independent of the pressure. Intensity changes occurring at higher pressures are difficult to reproduce due to large errors caused by a small difference between both lifetimes. After the pressure release, lifetimes of positronium annihilating in micelle recover. That indicates that the micelle structure was not destroyed by pressure. The longest component reappears, but its intensity is very small. It shows that the grains remain compressed.

### 3.4. Nitrogen adsorption
The effect of compression on the structure of MCM-41 with the template within the pores was examined by $N_2$ adsorption. Fig. 6 shows the $N_2$ adsorption/desorption isotherms for the as-synthesized MCM-41 before and after compression. The nitrogen adsorption/desorption isotherm for as-synthesized sample and samples after hydrostatic pressure experiment exhibit over the relative pressure $p/p_0 = 0.9$ hysteresis loop which is characteristic of the capillary condensation between silica particles. Some parameters characterizing porosity of the investigated samples are collected in Table 1. Two samples with the template do not exhibit porosity of MCM-41 silica. Only for MCM-41 after compression in argon the specific surface area and pore volume are little higher in comparison to the as-synthesized sample.

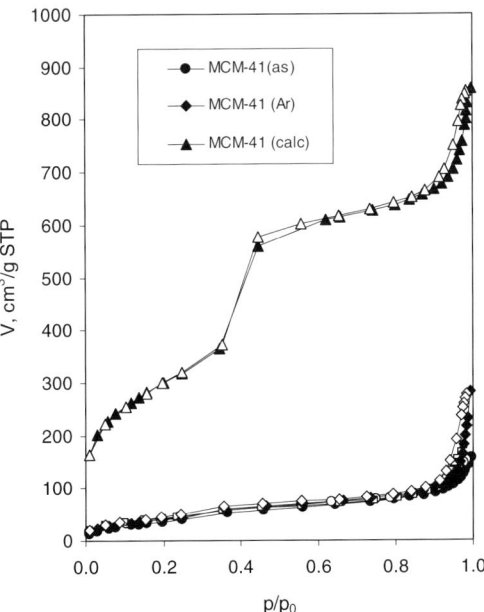

Fig. 6. Nitrogen adsorption/desorption isotherms for the samples: as-synthesized MCM-41, MCM-41 compressed in argon and MCM-41 calcined

Table 1
Parameters characterizing porosity of the investigated samples.

| Sample | $S_{BET}$ $m^2 g^{-1}$ | $V_p^{BJH}$ $cm^3 g^{-1}$ | $R_p^{meso}$ nm |
|---|---|---|---|
| MCM-41 (as) | 134 | 0.24 | - |
| MCM-41 (Ar) | 146 | 0.44 | - |
| MCM-41 (calc) | 1069 | 1.41 | 1.50 |

The nitrogen adsorption/desorption isotherms and the specific surface areas or the pore volumes derived from the $N_2$ desorption data show that during compression in argon the mesopore structure is not destroyed. Nitrogen adsorption for the calcined sample after pressure treatment in argon (not shown in Fig.6) and for the initial sample is identical and exhibits porosity characteristic of well ordered MCM-41 silica.

## 4. CONCLUSIONS

The positronium annihilation lifetime spectroscopy was successfully applied to the study of pore structure of the as-synthesized sample of MCM-41. The PALS technique can supply information about imperfections in the template structure as well as silica walls of MCM-41. The interior of cylindrical micelles encaged in the silica skeleton exhibits the presence of some kind of defects which disappear when pressure of argon increases. However, small voids present in the walls of silica network are resistant to compression and their dimensions are independent of pressure. Under mechanical pressure one can observe total destruction of interparticle pores at about 140 MPa. However, small voids in the sample are present up to 450 MPa. Temperature treatment leads to quite different effects than those observed for pressure experiment.

## REFERENCES

[1] J.S. Beck, J.C. Vartuli, W.J. Roth, M.E. Leonowicz, C.T. Kresge, K.D. Schmitt, C.T.W. Chu, D.H. Olson, E.W. Scheppard, S.B. McCullen, S.B. Higgins and J.L. Schlenker, J. Am. Chem. Soc., 114 (1992) 10834.
[2] M.T.J. Keene, R. Denoyel and P.L. Llewellyn, Chem. Commun., (1998) 2203.
[3] J. Goworek, A. Borówka, R. Zaleski and R. Kusak, J. Thermal Anal. and Calorimetry, 79 (2005) 555.
[4] R. Zaleski, A. Borówka, J. Wawryszczuk, J. Goworek and T. Goworek, Chem. Phys. Lett., 372 (2003) 794.
[5] M.A. Springuel-Huet, J.L. Bonardet, A. Gédéon, Y. Yue, V.N. Romannikov and J. Fraissard, Micropor. Mesopor. Mater., 44 (2001) 775.
[6] A. Galerneau, D. Desplantier-Giscard, F. Di Renzo and F. Fajula, Catal. Today, 68 (2001) 191.
[7] V.Y. Gusiev, X. Feng, Z. Bu,G.L. Haller and J.A. O'Brien, J. Phys. Chem., 100 (1996) 1985.
[8] M. Grün, K.K. Unger, A. Matsumoto and K. Tsutsumi, in: Characterization of Porous Solids IV, The Royal Society of Chemistry, B. McEnaney, J.T. Mays, J. Rouquerol, F. Rodriguez-Reinoso, K.S.W. Sing and K.K. Unger (eds), 1997, p. 81.
[9] R. Zaleski, J. Wawryszczuk, J. Goworek, A. Borówka, T. Goworek, Micropor. Mesopor. Mater., 62 (2003) 47.
[10] T. Goworek, K.Ciesielski, B. Jasińska and J. Wawryszczuk, Chem.Phys., 230 (1998) 305.
[11] A. Shukla, M. Peter, L. Hoffman, Nucl. Instr. and Meth., A 335 (1993) 310.

Studies in Surface Science and Catalysis 160
P.L. Llewellyn, F. Rodriquez-Reinoso, J. Rouqerol and N. Seaton (Editors)

# Structure-adsorptive characteristics of template-based mesoporous silicas containing residues of some phosphorus acids derivatives in their surface layer

O.A. Dudarko[a], I.V. Melnyk[a], Yu.L. Zub[a], and A. Dąbrowski[b]

[a]Institute of Surface Chemistry, NAS of Ukraine, 17 General Naumov Str., Kyiv 03164 Ukraine

[b]Faculty of Chemistry, Maria Curie-Sklodowska University, pl. M.Curie-Sklodowskiej 2, 20-031 Lublin, Poland

Mesoporous silicas (that contain such functional groups as $\equiv Si(CH_2)_3NHP(O)(OC_2H_5)_2$ and $\equiv Si(CH_2)_3NHP(S)(OC_2H_5)_2$ in surface layer) were synthesized by template method with 1-dodecylamine as a neutral template, which later has been removed from obtained mesophases by boiling methanol. It was shown, when the ratios of tetraethoxysilane/trifunctional silane were of 10:1, the samples with a high specific surface area ($>600$ $m^2/g$) were forming. In these samples the content of functional groups was 1.1 to 1.2 mmol/g. The resulting materials were thoroughly characterized by X-ray diffraction, IR spectroscopy, and thermogravimetry. It has been assumed, that these materials had disordered structure with short-range hexagonal symmetry.

## 1. INTRODUCTION

Recently a considerable attention in the field of application of sorbtion technologies is drawn to functionalized mesoporous silicas (FMS) obtained by template methods [1,2]. These sorbents have not only a high specific surface area and pores of similar size and form, but also surface ligand groups, that are capable to take up metal ions and organic substances. Now, if FMS with oxigen-, nitrogen- and sulfurcontaining functional groups in a surface layer have been researched rather good [3,4], the mesoporous silicas that contain residues of phosphoric acids are insufficiently known [5,6]. Nevertheless these materials are perspective as sorbents for extraction and separation of ions of rare earths metals and actinoids [7,8]. That is why the goal of this research was to develop techniques for template synthesis of FMS with such groups as $\equiv Si(CH_2)_3NHP(O)(OC_2H_5)_2$ and $\equiv Si(CH_2)_3NHP(S)(OC_2H_5)_2$, to define the factors, that influence on the structure-adsorptive characteristics of obtained sorbents, and the structure of these sorbents. 1-Dodecylamine (DDA) was used as a surfactant forming micelles.

## 2. MATERIALS AND METHODS

Initial materials used for synthesis of FMS were as follows: tetraethoxysilane, $Si(OC_2H_5)_4$ (TEOS, 98%, Aldrich), 1-dodecylamine, $CH_3(CH_2)_{11}NH_2$ (DDA, Aldrich), $NH_4F$ (98%,

Fluka), anhydrous methanol and ethanol (the purity of 99%). Diethyl phosphoric acid (3-triethoxysilyl)propyl amide, $(C_2H_5O)_3Si(CH_2)_3NHP(O)(OC_2H_5)_2$ (DPPA) and diethyl thiophosphoric acid (3-triethoxysilyl)propyl amide, $(C_2H_5O)_3Si(CH_2)_3NHP(S)(OC_2H_5)_2$ (DTPPA) were synthesized due to reaction of 3-aminopropyltriethoxysilane, $(C_2H_5O)_3Si(CH_2)_3NH_2$ (APTES, 99 %, Aldrich) with diethyl chlorophosphate or chlorothiophosphate, $(C_2H_5O)_2P(O$ or $S)Cl$ (97%, Aldrich) in the presence of triethylamine.

## 2.1. Synthesis of samples

Sample I (TEOS/DPPA = 5:1). 1.68 cm$^3$ (0.005 mol) of DPPA was dissolved in 5 cm$^3$ of ethanol and hydrolyzed by solution of 0.0111 g (0.0003 mol) of NH$_4$F in 1.04 cm$^3$ of H$_2$O during 5 min. After that the solution of 1.3902 g (0.0075 mol ) of DDA in 10 cm$^3$ of ethanol was added to obtained transparent mixture at stirring with magnetic stirrer, then 5.58 cm$^3$ (0.025 mol) of TEOS was added to the solution. 1 min later 12.5 cm$^3$ of H$_2$O was added drop-by-drop to the solution. A substantial amount of a white precipitate was formed. This precipitate was left to stay for 48 hrs at room temperature, then filtered off and dried in air for 48 hrs. The template was extracted trice by boiling methanol for 3 hrs (30 ml of MeOH per 1 g sample). The white precipitate was dried in vacuo for 4 hrs at 110°C. Yield was 1.98 g.

Sample II (TEOS/DPPA = 7.5:1). This FMS was obtained by the method similar to that for sample I with the following changes: first 1.68 cm$^3$ (0.005 mol) of DPPA was hydrolyzed by 0.0157 g (0.000425 mol) of NH$_4$F in 1.49 cm$^3$ of H$_2$O; 2.7804 g (0.015 mol) of DDA was dissolved in the mixture of 15 cm$^3$ of ethanol and 12.5 cm$^3$ of H$_2$O; then 8.36 cm$^3$ (0.0375 mol) of TEOS was added. The gelation was observed, and result precipitate was developed by analogous procedure for the sample I. Yield was 2.54 g.

Sample III (TEOS/DPPA = 10:1). This FMS was obtained as the sample I except the following changes: first 1.68 cm$^3$ (0.005 mol) of DPPA was hydrolyzed with 0.0204 g (0.00055 mol) of NH$_4$F in 1.94 cm$^3$ of H$_2$O; 1.3902 g (0.0075 mol) of DDA was dissolved in 6 cm$^3$ of ethanol and 6 cm$^3$ of water; 11.15 cm$^3$ (0.05 mol) of TEOS was added. The gelation was observed, and result precipitate was developed by analogous procedure for the sample I. The yield was 3.56 g.

Sample IV (TEOS/DTPPA = 5:1 ). This FMS was obtained analogously to the sample I with the following changes: first 2.48 cm$^3$ (0.005 mol) of DTPPA was hydrolyzed with 0.0111 g (0.0003 mol) of NH$_4$F in 1.04 cm$^3$ of water. Yield was 1.98 g.

Sample V (TEOS/DTPPA= 7.5:1). This FMS was obtained as the sample II except the following changes: 2.48 cm$^3$ (0.005 mol) of DTPPA was preliminary hydrolyzed with 0.0157 g (0.000425 mol) of NH$_4$F in 1.49 cm$^3$ of water; 1.3902 g (0.0075 mol) of DDA was dissolved in the mixture of 10 cm$^3$ of ethanol and 12.5 cm$^3$ water; 8.36 cm$^3$ (0.0375 mol) TEOS was used. The gelation was observed, and result precipitate was developed by analogous procedure for the sample I. Yield was 3.14 g.

Sample VI (TEOS/DTPPA = 10:1). This FMS was obtained as the sample III with the following changes: first 0.35 cm$^3$ (0.0007 mol) of DTPPA was hydrolyzed with 0.0029 g (0.000077 mol) of NH$_4$F in 0.27 cm$^3$ of water; 0.1946 g (0.00105 mol) of DDA was dissolved in 2 cm$^3$ of ethanol and 2.5 cm$^3$ of water; 1.56 cm$^3$ (0.007 mol) of TEOS was used. The gelation was observed, and result precipitate was developed by analogous procedure for the sample I.Yield was 0.50 g.

## 2.2. Methods

From the elemental analysis it is possible to calculate the contents of functional groups in the resulting materials (see Table 1).

Powder X-ray diffraction patterns were recorded using a DRON 4-07 diffractometer with Cu$_{K\alpha}$ radiation. DRIFT spectra were recorded on a Thermo Nicolet NEXUS FTIR (4000-400 см$^{-1}$) spectrometer at 4 cm$^{-1}$ resolution using the Spectra Tech collector diffuse reflectance accessory at room temperature. The samples were mixed with KBr (1:30) and were used to fill the DRIFT sample cup before measurements. The thin films between the KRS plates were used for registration of IR absorbtion spectra of DPPA and DTPPA. Thermal analysis was performed in the range of 20-800°C with a heating rate of 5°C min$^{-1}$ (in air stream; a Q-1500D E. Paulik, J. Paulik, L. Erdey System). Nitrogen adsorption isotherms for all the samples were measured at –196°C by the help of a "Kelvin-1042" adsorption analyzer (Costech Microanalytical). Before adsorption measurements, the samples were degassed at 110°C. The BET specific surface area [9] was calculated in the relative pressure range between 0.05 and 0.30 in order to exclude the points in the range of capillary condensation. The total pore volume was determined from the amount adsorbed at the relative pressure of 0.99. The pore size distributions were determined using the Barrett, Joyner and Halenda (BJH) algorithm [10].

## 3. RESULTS AND DISCUSSION

The synthesis of FMS were carried out at the room temperature, and applying bicomponent systems, with the use of neutral template agent – DDA, to them. The molar ratios of reacting components are: 0.1 TEOS : [0.01; 0.0133 or 0.02 DPPA (or DTPPA)] : (0.015-0.04) DDA : (0.9-3.0) H$_2$O. Therefore, the quantity of trifunctional silane relatively to TEOS in the initial solution was not higher than 20% (mol). It was stipulated by the fact, that if the content of trifunctional silane in the systems was higher, after template extraction the mesoporous structure of samples is not preserved [11-15]. Another interesting fact is, that the gel is observed for samples **II**, **III**, **V** and **VI** (in the chosen conditions of the synthesis), but after a while of staying the gel is forming the powder-like mass.

Table 1
Contain of surface functional groups and structure-adsorptive parameters of obtained FMS

| Sample | Molar ratio of TEOS/ DPPA (or DTPPA) | P/C | P/Si | $C_{f.gr.}$ mmol g$^{-1}$ | $d_{100}$, nm (BE/AE)* | $S_{sp.}$ m$^2$g$^{-1}$ | $V_s$ cm$^3$g$^{-1}$ | d nm |
|--------|------|------|------|------|------|------|------|------|
| I | 5:1 | 1/6 | 1/6 | 1.76 | 3.5/3.4 | 265 | 0.20 | 3.6 |
| II | 7.5:1 | 1/5.5 | 1/8 | 1.47 | 3.7/3.8 | 365 | 0.21 | 2.9 |
| III | 10:1 | 1/8 | 1/10 | 1.16 | 3.4/4.3 | 650 | 0.59 | 3.7; 6.5 |
| IV | 5:1 | 1/4.8 | 1/5.9 | 1.90 | 3.4/3.4 | 80 | 0.40 | 2.5 |
| V | 7.5:1 | 1/7.7 | 1/10 | 1.27 | 4.1/3.7 | 178 | 0.13 | 3.7 |
| VI | 10:1 | 1/10 | 1/11 | 1.06 | 3.4/3.7 | 640 | 0.61 | 3.6; 6.6 |

*BE – before extraction of template, AE – after extraction

Elemental analysis data for C, N, P and S was used to calculate the concentration of P-containing functional groups (Table 1). As it can be seen from this Table, the reported one-step synthesis give the possibility to get the porous materials with a high content of functional groups. Elemental analysis data indicate, that the P/Si ratio in some prepared samples was smaller than that was set by the alkoxysilanes in the initial mixture. It is obvious, that a small part of functional groups was washed off from the surface of the samples during template extraction by boiling methanol. As one can see from Table 1, the ratio P/C in obtained samples is close to one set by initial alkoxysilanes. At the same time, there is the evident tendency of diminishing of this ratio for some samples when the ratio of "TEOS/trifunctional silane" (Table 1) increases. It is possible, that the reason for this tendency is the presence of alcohol and/or residues of template. The content of functional groups in synthesized samples is in the range 1.06 - 1.90 mmol/g and primarily is defined by the ratio of initial alkoxysilanes.

The endothermic effects are observed for all the samples on the thermoanalytic curves in 90-110°C range, and can be explained by the loss of water and traces of solvent (2-6%). As an example the thermoanalytic curves of **III** and **VI** samples are presented on Fig.1. The exoeffects, that appeares at higher temperatures, confirms a destruction of surface layer and goes via the three stages (270-300°C, 320-330°C6, and 500-650°C). General weight loss of the samples (15-19%) correlates with the results of their elemental analysis.

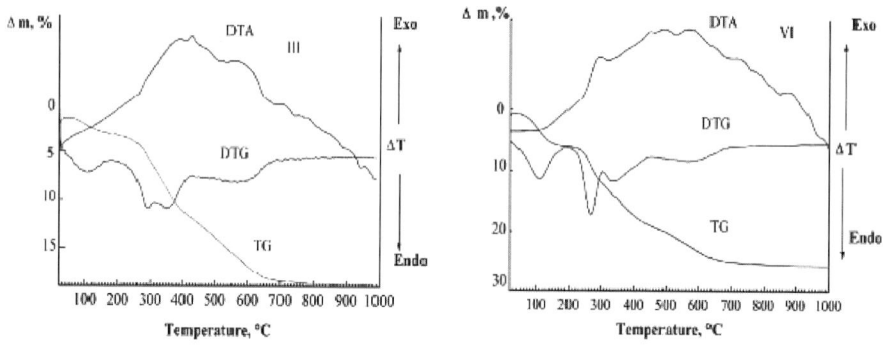

Fig.1. Thermoanalytic curves obtained for samples **III** and **VI**

A characteristic feature of the IR spectra for all synthesized samples (Fig.2) is the presence of an intensive absorption band in the 1070-1090 cm$^{-1}$ region. The appearense of this band is related to the stretching vibrations of Si-O-Si bonds in three-dimensional siloxane skeleton, that contains Si-C$_n$-R' groups [16]. As a rule, this absorbtion band on the high-frequency side of spectrum (at ~1120-1600 cm$^{-1}$) has a shoulder [16]. Indeed, this shoulder can be seen in IR spectra of all the samples (Fig. 2), however, in the case of **I** - **III** samples it transforms in a separate broaded absorption band with the maximum at ~1210 cm$^{-1}$. Obviously, this absorption band is corresponded to $\nu$(P=O) stretching vibration, which is identified in the IR spectrum of individual DPPA at 1237 cm$^{-1}$ (see Fig. 2). The movement of this band on ~20 cm$^{-1}$ in the low-frequency region can be the evidence of the participation of P=O groups in the forming of hydrogen bond. For the samples **IV-VI** this shoulder is situated at ~1200 cm$^{-1}$ and is attributed to stretching vibrations of O-Si-C. Also, in IR spectra of all these samples, there is a weak absorption band in the region of 556-642 cm$^{-1}$, which can be related to vibration of $\nu$(P=S) (in the IR spectrum of individual DTPPA it is indicated at 645 cm$^{-1}$) (Fig. 2). The absorption bands, that are related to the presence of water, are in the region of 1625-1647 cm$^{-1}$ and higher ~3100 cm$^{-1}$ ($\delta$(H$_2$O) and $\nu$(OH) accordingly), what is confirmed by thermogravimetry data. In addition, the IR spectra exhibit a set of absorption bands with a weak intensity at 2851 - 2991 cm$^{-1}$ related to the stretching vibrations of the C-H bond (Fig.2). The position and intensities ratio of these absorption bands differ from those in the IR spectra of mesophases of the same samples. Thus, the conclusion can be made, that there is no DDA in **I – VI** samples. This statement is also supported by the absence of absorption bands at 1300-1500 cm$^{-1}$, indicated for DDA. In conclusion, it needs to be mentioned, that the weak absorption band at ~2850 cm$^{-1}$ may be attributed to vibrations $\nu(CH)$ of a methoxy groups. These groups can be chemisorbed ($\equiv Si\text{-}OCH_3$) and/or belong to molecules of methanol. However, in the last case these groups need to be eliminated due to vacuum drying. The presence of methoxy groups in the composition of FMS obtained by applying of DDA and the same methods of synthesis, was shown earlie by use of $^{13}C$ CP/MAS NMR spectroscopy [17]. The fact of the presence of methoxy groups is confirmed with the data of elemental analysis. Therefore, the conclusion, made on the IR spectroscopy data, is, that the surface layer of samples **I – VI** is complicated in character and contains functional groups, that were incorporated during synthesis via trifunctional silanes, methoxygroups, and sorbed water.

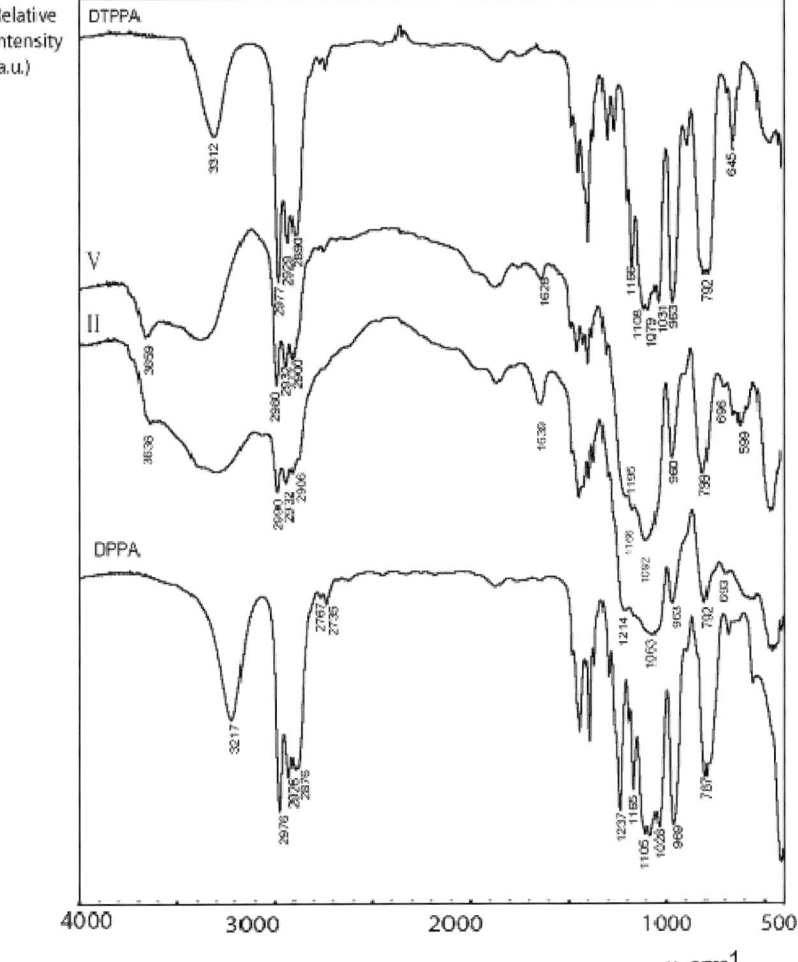

Fig. 2. IR spectra of initial trifunctional silanes and DRIFT spectra of **II** and **V** samples

Fig. 3 demonstrates the XRD patterns that were obtained for some synthesized mesoporous silicas. All the samples exhibit a single peak. It is notisible, that disordered mesoporous structures are formed using of DDA as template in absence of hydrothermal processing of mesophases. In most cases, these structures have a single reflection [18,19]. At the same time we can note, that in case of samples **III** and **VI** (containing the least amount of trifunctional silane in an initial solution) these reflections are less broadened and more intensive, than those for samples **I** and **IV** (containing the greatest amount of this alkoxysilane). It is possible to assume, that the structures of samples **III** and **VI** are more well-ordered, than those ones of the samples **I** and **IV**. Evidently, samples **III** and **VI** have a disordered structure with local hexagonal symmetry as it was shown in [19,20].

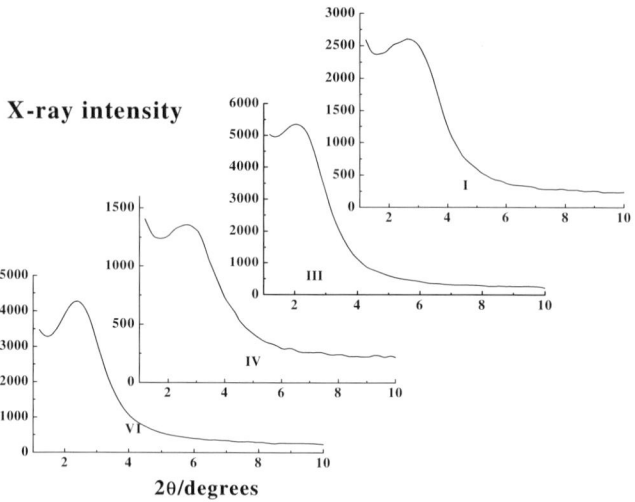

Fig. 3. Powder X-ray diffraction patterns for **I**, **III**, **IV** and **VI** samples (after extraction of template)

This assumption is inderectly supported by the structure-adsorbtive characteristics of synthesized samples (Table 1). The samples **I**, **II**, **IV** and **V** appeared to have moderate values of surface area and their isotherms attributed to the type I of IUPAC classification [21] (see Fig. 4). Meanwhile the samples **III** and **VI** have developed pore structures and their isotherms are attributed to IV type of IUPAC classification. Obviously, mesoporous structure is realized at ratio "TEOS/trifunctional silane" equal to 10:1, that is confirmed with their XRD patterns. It is possible, that with less containing of TEOS in such systems (samples **I**, **II**, **IV** and **V**) the appearanse of the inmurity of amorphous silica should be expected (that was shown for analogous systems with TEM [22]). Besides it can be noted, that $N_2$ sorption isotherms of samples **III** and **VI** clearly display two distinct capillary condensation steps [21] (Fig. 4). BJH model analysis provides two pore diameter maxima (Table I). The suggestion is therefore, that two pore systems are present and that the desorption hysteresis suggests a three-dimensional pore structure. It is interesting, that in case of mesoporous silicas, which were obtained with the use of DDA, but with smaller functional groups (amino-, thiol- and others [17]), the pore size is smaller than in the samples **III** and **VI** (Table I).

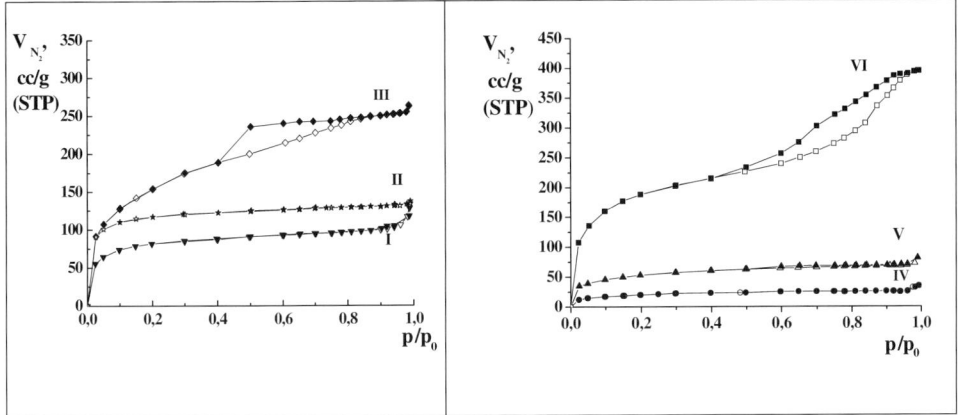

Fig. 4. Isotherms of nitrogen adsorbtion-desorbtion for the samples **I-VI**.

## 4. CONCLUSIONS

The FMS containing in surface layer such functional groups as $\equiv Si(CH_2)_3NHP(O)(OC_2H_5)_2$ and $\equiv Si(CH_2)_3NHP(S)(OC_2H_5)_2$ can be easily synthesized by use of template method (1-dodecylamine as a neutral template removed from obtained mesophases by washing in boiling methanol). It was shown, that the samples with a high developed surface ($S_{sp} > 600$ $m^2/g$) were formed when the ratio TEOS/DPPA (or DTPPA) was 10:1 (the content of functional groups in these samples was 1.1-1.2 mmol/g). The increase of this content (to 7.5:1 or 5:1) causes essential decrease of in specific surface area of the samples. This increase also causes the widening of the reflex and the decrease in its intensive on XRD patterns of the samples..

By means of IR spectroscopy, it has been found, that phosphorus-containing functional groups, that are in the surface layer of these samples, take part in the formation of hydrogen bonds and the surface layer itself contains methoxy groups and sorbed molecules of water.

The possibility of tailoring the mesoporous structure and the surface functionality make these materials very promising for various applications including environmental cleanup and separations.

## ACKNOWLEDGMENT

The support of this research by the NATO "Science for Peace" Programme through Grant 978006 is gratefully acknowledged.

## REFERENCES

[1] K. Moller and T. Bein, Chem. Mater., 10 (1998) 2950.
[2] G. A. Ozin, Chem. Commun., (2000) 419.
[3] J. Y. Ying, C .P. Mehnert, M. S. Wong, Angew. Chem. Int. Ed. Engl., 38 (1999) 56.
[4] A. Stein, B. J. Melde, R. C. Schroden, Adv. Mater., 12 (2000) 1403.
[5] R. J. P. Corriu, L. Datas, Y. Guari, A. Mehdi, C. Reye, C. Thieuleux, Chem. Commun., (2001) 763.

[6] R. J. P. Corriu, A. Mehdi, C. Reye, C. Thieuleux, Chem. Commun., (2004) 1440.

[7] G. E. Fryxell, H. Wu, Y. Lin, W. J. Shaw, J.C. Birnbaum, J. C. Linehan, Z. Nie, K. Kemmer, and S. Kelly, J. Mater.Chem., 14 (2004) 3356.

[8] S. Dai, M.C. Burleigh, Y. Shin, Molecular recognition imprint coatings for selective functionalized mesoporous sorbents for separation processes and sensors, US Patent No 6 251 280 (2001).

[9] S. Brunauer, P.H. Emmett, E. Teller, J. Am. Chem. Soc., 60 (1938) 309.

[10] E. P. Barrett, L. G. Joyner, P. P. Halenda, J. Am. Chem. Soc., 73 (1951) 373.

[11] S. L. Burkett, S. D. Sims, S. Mann, Chem. Commun., (1996) 1367.

[12] D. J. Macquarrie, Chem. Commun., (1996) 1961.

[13] C. E. Fowler, S.L.Burkett, S.Mann, Chem. Commun., (1997) 1769.

[14] S. R. Hall, C. E. Fowler, B. Lebeau, S. Mann, Chem. Commun., (1999) 201.

[15] S. Sadasivan, D. Khushalani, S. Mann, J. Mater. Chem., 13 (2003) 1023.

[16] L. P. Finn, I. B. Slinyakova, Colloid. J., 37 (1975) 723 (Russ.).

[17] I. V. Mel'nyk (Seredyuk), Yu. L. Zub, A. A. Chuiko, M. Jaroniec, and S. Mann, Stud. Surf. Sci. Catal., 141 (2002) 205.

[18] P. T. Tanev and T. J. Pinnavaia, Science, 267 (1995) 865.

[19] P. T. Tanev and T. J. Pinnavaia, Chem. Mater., 8 (1996) 2068.

[20] P. T. Tanev, M. Chibwe, T. J. Pinnavaia, Nature, 368 (1994) 321.

[21] F.Rouquerol, J. Rouquerol, K. Sing, Adsorption by Powders and Porous Solids. Principles, Methodology, and Application, Academic Press, San Diego, 1999.

[22] Yu. L. Zub, I. V. Melnyk, G. R. Yurchenko, O. K. Matkovskii, A. A. Chuiko, Chemistry, Physics and Technology of Surfaces (ISC of NAS of Ukraine), 10 (2004) 69.

Studies in Surface Science and Catalysis 160
*P.L. Llewellyn, F. Rodriquez-Reinoso, J. Rouqerol and N. Seaton (Editors)*

# Melting of atomic layers in carbon nanotubes.

**Lucyna Firlej[a] and Bogdan Kuchta[b]**

[a] Laboratoire des Colloïdes, Verres, Nanomatériaux (LCVN), Université Montpellier II,
   34095 Montpellier, France

[b] Laboratoire des Matériaux Divises, Revêtement, Electrocéramiques (MADIREL),
   Université de Provence, Centre de Saint-Jérôme , 13397 Marseille, France

## 1. INTRODUCTION

Studies of melting and freezing in confined geometry are interesting from a fundamental point of view as well as due to potential practical applications [1]. The mechanism of transitions in pores appears to be fundamentally different from those observed in bulk. The experiments performed on some porous materials show a melting point elevation in narrow pores [2,3,4], in contrast to the almost universally observed depression of transition temperature in larger pores. This observation raises a question of the role of the adsorbent-adsorbate interaction in the phase transition mechanism in nanometric confinements. Additionally, in nanometric pores, the adsorbed phase is expected to be heterogeneous and may have different properties in different layers, depending on the distance from the pore wall. This problem has been discussed in several theoretical studies [5-11].

In this paper we analyze the mechanism of melting in partially filled pores. Such particular situation should enhance our understanding of mechanism of melting in heterogeneous situations. As a modeled system, we have chosen Kr atoms adsorbed in hypothetical (with diameter d = 4 nm) carbon nanotubes with corrugated graphite-like wall surface. In this system, two krypton layers are adsorbed before the capillary condensation takes place. Therefore, the system allows us to study the mechanism of melting in single adsorbed layer as well as in two layers adsorbed in the corrugated carbon tube. The choice of krypton adsorbed in 4 nm carbon nanotube makes it possible to discuss the results in the context of similar calculation performed for a model of 4 nm MCM-41 pore [12]. The important differences between these two systems are (i) ordered walls in the carbon nanotube versus amorphous in MCM-41, (ii) stronger atom-wall interactions in the nanotubes than in silica pores, and, as a consequence, (iii) maximally two layers adsorbed in nanotubes before the capillary condensation and only one in MCM-41. It is the purpose of this study to enhance an understanding of the mechanism of melting in adsorbed layers before the capillary condensation occurs. In particular, we want to understand how the adsorption of subsequent layer modifies the melting properties of the layers that are already adsorbed.

The conventional grand canonical MC ensemble was applied. The simulation box, containing one carbon nanotube of 4 nm in diameter (with periodic boundary conditions along the tube axis) was assumed to be in equilibrium with the bulk gas, which obeyed the ideal gas law. This allowed us to use the external gas pressure as the thermodynamic parameter instead of the chemical potential. The adsorption is only allowed in the interior of the tube. Trial moves included the translations of atoms, insertion of new atoms and removal of existing ones. The system typically contained from 600 to 1300 adsorbed atoms in the box. Typical runs contained the minimum number of $10^6$ MC steps (per atom). The main results were extracted from the previously equilibrated runs.

The Kr-Kr interaction was modeled by Lennard-Jones (LJ) potential with the standard interaction parameters: $\varepsilon_{Kr-Kr}/k$ =171.0 K, $\sigma_{Kr-Kr}$ = 0.360 nm [13]. The Kr-nanotube wall interaction was computed by a pair-wise summation of LJ potential (with the LJ parameters for Kr-C interaction obtained from Lorentz-Berthelot mixing rules taking $\varepsilon_{C-C}/k$ =28.0 K, $\sigma_{C-C}$ = 0.34 nm [14]). The explicit atomic structure of the carbon nanotube is preserved in the calculations. This makes our nanotube walls corrugated, with energy minima at the centers of the hexagons and maxima at the positions of atoms. Although the amplitude energy of this corrugation is small compared to total atom-wall interaction, its influence on the structures of the first layers is non-negligible, as will be shown below.

Structural properties of planar atomic layer are usually represented by parameters which are sensitive to the deformation of the ideal triangular plane structure. The calculation of such parameter in the case of cylindrical surface requires unfolding of cylindrical layers into a plane geometry [6,7]. Then, one can apply the order parameter $\Phi_6$ defined by Mermin [15]:

$$\Phi_{6j} = \left| \frac{1}{N_b} \sum_{k=1}^{N_b} \exp(i6\theta_k) \right| \tag{1}$$

$\Phi_{6j}$ measures the average bond order within a plane triangular atomic j layer. Each nearest neighbor bond has a particular orientation in the plane that can be described by the polar coordinate $\theta_k$. The index k runs over the total number of nearest neighbor bonds $N_b$ in the adsorbed layer. One expects that $\Phi_6=1$ in the ideal solid 2D hexagonal layer (solid) and that $\Phi_6=0$ when the state of adsorbed layer corresponds to a two-dimensional fluid. Of course, such an ideal ordering is practically nearly impossible to be realized in confined geometries as the crystal phase always has some defects and the liquid phase exhibits some short range orientational order.

## 2. MELTING ALONG ISOTHERMS

Fig. 1 presents four isotherms of krypton adsorption in the nanopores, at T = 77, 90, 110 and 130 K. The logarithmic scale of pressure is used to visualize all four isotherms at the same time. Although the overall character of adsorption changes with temperature and becomes more continuous at higher temperature, some characteristic features are common. The

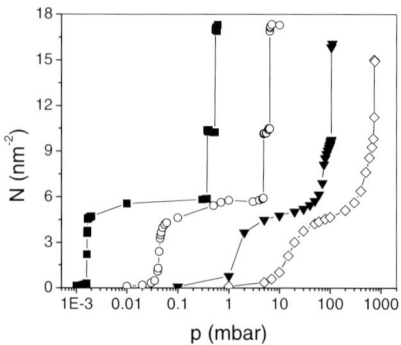

Fig. 1. The isotherms of adsorption of Kr in carbon nanotubes of diameter d = 4 nm.

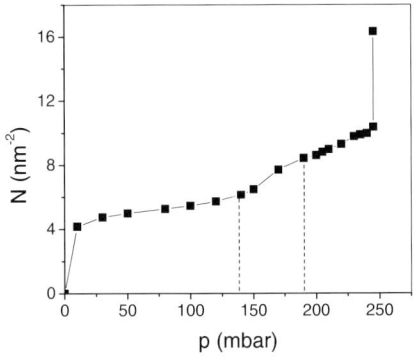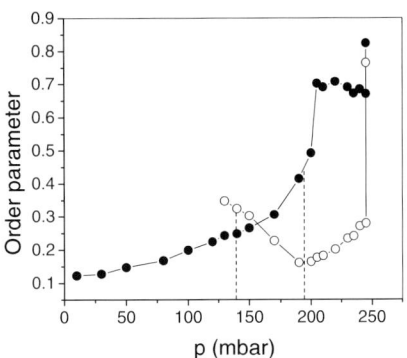

Fig. 2. The isotherm (T = 118 K) and order parameters (eq. 1) calculated for the first (solid circles) and the second (open circles) layer. The vertical dashed lines indicate the range of pressure where intermediate states, between first and second layer, are formed.

adsorption isotherms remain of step-wise type up to 130 K. There is always a well pronounced signature of first and second layer formation, before the pore is filled in the process of capillary condensation. This behavior indicates that the strong interaction of Kr with the nanotube wall affects the mechanism of adsorption up to the second layer formation. When the third layer starts to be formed, the direct influence of the already distant wall is negligible with respect to the Kr-Kr interaction and the capillary condensation takes place.

At T = 77 K the adsorbed layers are solid and at T = 130 K they are melted [16]. However, between these two temperatures, the melting observed along the isotherms shows some peculiar features. As an example, we present T = 118 K isotherm (Fig. 2) where the first layer is formed in a liquid state, exhibiting short range order ($\Phi_6$= 0.12). However, there is a continuous transition along the isotherm (in the first layer), when pressure increases, towards the solid state.

The adsorption of the second layer seems to play a crucial role affecting the mechanism of freezing of the first one. Fig. 2 shows that increasing density in the second layer accelerates the solidification of the first layer. At the same time, the second layer is adsorbed as a liquid. At the intermediate states of adsorption, between first and second layer, the isolated liquid patches are formed on the top of the first layer. Locally, they are more ordered than the underlying layer. They progressively merge to form a uniform second layer when the pressure increases and more atoms are adsorbed. The decreasing values of its order parameter reflect very large density fluctuations when the uniform second layer is formed. It does not solidify until the capillary condensation, even when the first layer freezes. The highest pressure points presented in the Fig. 2 show the effect of capillary condensation. The large value of the order parameter suggest that it is rather capillary 'freezing' of both layers.

## 3. THERMODYNAMIC PATHS OF MELTING

The analysis of the melting along isotherms does not characterize the evolution of the state of layers as a function of temperature. Therefore, to understand better the melting of partially filled nanotubes, we followed evolution of the state of the adsorbate along the thermodynamic paths corresponding to monolayer and two-layer states. Fig. 3 shows the phase diagram where the phase lines of first layer, second layer formations and capillary condensation are presented.

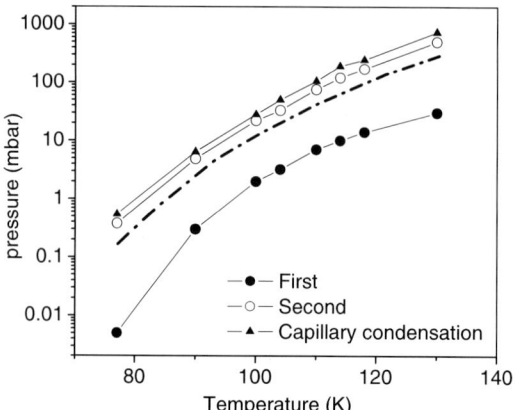

Fig. 3. Phase diagram of the krypton adsorbed in 4 nm carbon nanotube. There is 2D gas below the line of 'First' layer formation. The pore is completely filled above the line of 'Capillary condensation'. The dashed line shows the thermodynamic path chosen to study melting in the first layer. Similar line chosen to follow the two layer structure (not showed) goes in the middle of the narrow strip between 'First' and 'Second' lines, corresponding to the (p,T) range of existence of two layer structure of adsorbate.

On the base of this diagram, we have chosen two lines $p_i(T)$ (i = 1,2, see Fig.3) and calculated the in-layer order parameters following the one (i =1) and two-layer (i=2) structures, adsorbed before the capillary condensation occurs.

Fig. 4. shows the order parameters calculated for adsorbed system in one layer and two layers regions. When only one layer is adsorbed, it melts at $T_1 \approx 105$ K. Remembering that the triple point temperature for 3D krypton is $T_{3D} = 115$ K, one observes a decreasing of the melting temperature. However, when two layer are adsorbed, the situation changes. The melting temperature of the contact layer increases with respect to the one layer situation. Now it melts at $T \approx 125$ K, ten degrees above the bulk melting temperature. At the same time, the melting of the second layer is observed at $T < 100$ K, the temperature few Kelvins lower than the one found for the one layer system. Additionally, at low temperature one observes an

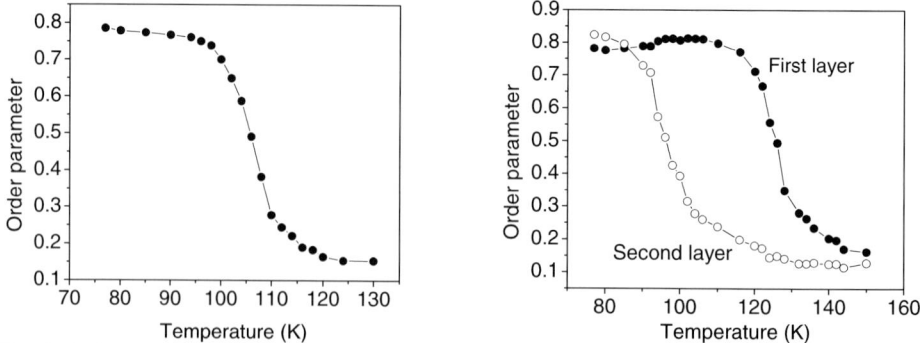

Fig. 4. Order parameters calculated for krypton adsorbed in the carbon nanotube, as a function of temperature. In the left panel $\Phi_6$ is shown for the first layer (in one layer regime). In the right panel, the contact and second layer parameters are reported, calculated in the two layer regime.

inversion of the order in two layers, that is, the second layer seems to be slightly better ordered than the first one. It seems, that the presence of the ordered second layer, forces the contact layer to deform. This deformation is released when the second layer starts to melt (see Fig. 4).

The lowering of the first layer melting temperature (in the monolayer phase) can be explained by the second layer promotion and partial desorption of atoms. Such mechanism of melting has been observed before in the case of Kr atoms adsorbed in a model of MCM-41 pores [12]. As a consequence, the in-layer density is lower; it facilitates the melting at lower temperature. However, when the second layer is present, the promotion of the first layer atoms into the second one starts to be energetically much more difficult. As a result, the first layer structure is stabilized by the interaction with the second one and the melting temperature of the first layer increases.

## 4. INFLUENCE OF THE WALL CORRUGATION ON THE MECHANISM OF MELTING

The influence of the adsorbate-adsorbent interactions is not only limited to the melting temperature. It is important to remember that the corrugation of the nanotube wall is a factor that may influence the structure of the layer and the mechanism of the transition. Such an effect has been observed in simulations of atomic (rare gases) and molecular ($N_2$, CO) layers adsorbed on graphite and carbon nanotubes [17-20]. Having this in mind, in this chapter we analyze the structures of krypton contact layer adsorbed in structured (corrugated) nanotube and compare these results with the calculations carried out in a typical approximation of structureless (smooth) cylindrical tube.

In the structureless cylindrical pore model one observes, even in the ordered solid phase, that the adsorbed system may move with respect to the nanotube, as a whole, because such displacement does not require any energy. Obviously, all directions of such movements are equally probable. However, in the corrugated nanotube, it is not the case. The competition between atom-atom and atom-wall interactions stabilizes the first layer in configurations that

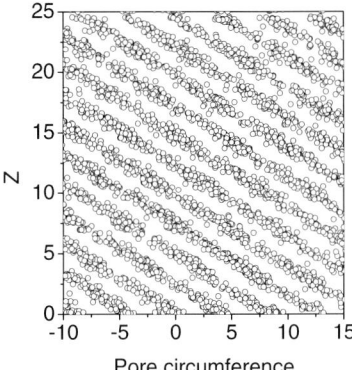
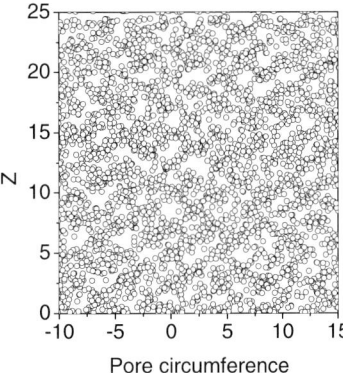

Fig. 5. Restrained fluctuations in the first layer, at T = 90 K (left) and at T = 100 K (right). The figures present 1000 instantaneous configurations taken from Monte Carlo runs of $10^6$ steps. Only one layer is adsorbed. The units of distance are angstroms.

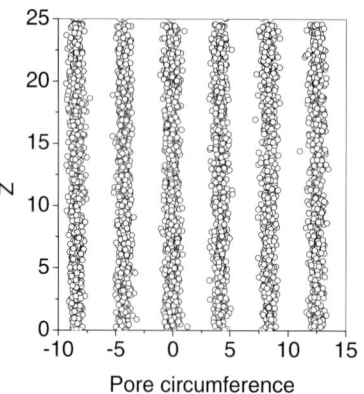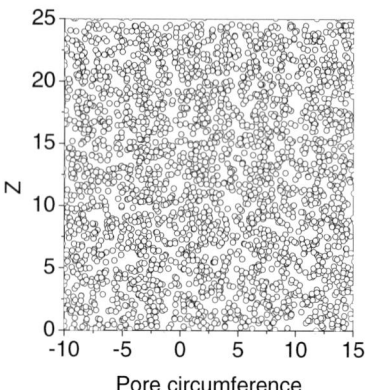

Fig. 6. Restrained fluctuations in the contact (left) and the second (right) layers, at T = 90 K. The figures present 1000 instantaneous configurations taken from Monte Carlo runs of $10^6$ steps. Only two layers are adsorbed in the pore. The units of distance are angstroms.

are more or less commensurate with the underlying carbon structure, depending on the equilibrium distance between the adsorbed particles. For krypton atoms, the structure is only slightly incommensurate. However, at very low temperature the amplitude of the corrugation is big enough to pin the adsorbed layer and the positions of atom are well localized in vicinity (but not exactly) of the centers of hexagons of the carbon wall. When temperature increases, we observed that the layer started to move in a very correlated manner (Fig.5). The direction of the movement is determined by chirality of the substrate. This problem will be discussed in a separate paper. Here, we use the 'armchair' type of nanotube [21] of chirality indexes (30,30). In this case, the epitaxy orientation of the layer with respect to the wall structure prefers one of three alternative displacements (as defined by the hexagonal graphite symmetry). Melting starts as a one directional displacements of atoms. Further increase of temperature provokes a crossover to a two-dimensional melting. Fig. 5 present two examples of quasi 'one-dimensional' and 'two dimensional' fluctuations of atom positions, observed at T = 90 K and T = 100 K, respectively, in the one-layer system.

The observed fluctuations have different characteristics, when two layers are adsorbed. In such a case the presence of the second layer seems to induce another choice of the preferred one-dimensional fluctuations at low temperatures, along the direction parallel to the pore axis (Fig. 6). At the same time, the second layer seems totally uncorrelated with the first one. It moves much more freely, seemingly not only unaffected by the pore wall corrugation but also not restrained by the first layer atoms.

## 5. CONCLUSIONS

We studied melting in the layers adsorbed in hypothetical single wall armchair carbon nanotube of diameter d = 4 nm. This study was limited to partially filled pores, where only one or two layers were adsorbed. Although the adsorbed system was krypton, most of the qualitative observation is valid in many other similar systems.

In general, melting temperature depends on a thermodynamic path chosen in the (p,T) space. We observed that the melting temperature of a monolayer adsorbed in the pore is lower than the temperature of the 3D triple point. However, the adsorption of the second layer pushes the temperature of melting of the contact layer up. It is worth noticing that the two-layer system behaves in a very heterogeneous way, that is, the melting of the second layer is observed at totally different temperature than the first one. In the case of krypton in carbon nanotubes, this temperature is even lower than the one reported for the monolayer system.

In general, a melting temperature of a given phase depends on the thermodynamic path chosen in the (p,T) space. In our system, there are different phase lines of melting points $T_m(p)$ in the monolayer as well as in the two-layer region. In heterogeneous systems, where one needs to characterize different layers by different melting lines, the detailed phase diagram is very rich. Further study are required to construct and understand such phase diagrams of multilayered systems in pores.

The mechanism of melting of the first layer may be strongly affected by the corrugation of the pore walls. In the case of carbon nanotubes, where the substrate possesses very regular and ordered structure, the mechanism of melting is biased by the initial incommensurability between the adsorbed system and the wall structure. To estimate quantitatively this effect we carried out additional simulations for pores with smooth walls. The temperature of melting of the first layer in smooth pore was shifted towards lower values by ~3 Kelvins. The influence of the atom-wall interaction on the melting of the second layer obviously should be weaker. This aspect will be the subject of further studies.

The relation between structures of different layers requires some comments. As it has been presented in the Fig.4, in two layer system we observed a crossover between the values of the order parameters corresponding to the first and the second layers. Similar, although not identical, situation was reported in the paper on $CCl_4$ melting in carbon nanotubes [9] where the order parameter of the second layer was larger than the one corresponding to the first layer. The explanation presented in the $CCl_4$ adsorbate was based on purely geometrical arguments, that is, that the pore circumference is incommensurate with the adsorbate for the contact layer, so that it can be less ordered than the others. This geometrical factor is undoubtedly very important and may also play a role in our system. However, this factor alone does not explain why the value of the first layer order parameter increases when the second layer starts to melts (Fig.4). Therefore, we claim that apart from the purely geometrical reasons, the interaction between the layers affects also the structures within the adsorbed layers. In the case of Kr we observe that the solid second layer forces the structure of the first layer to be less ordered.

The results presented in this paper provoke many questions concerning, in general, mechanism of structural transformations in pores with nanometric sizes. Among them is the problem of the melting/freezing hysteresis. In the monolayer system, the melting was continuous and no hysteresis was observed. It means that the melting mechanism in systems of lower dimension is different from the 3D situation where one expects a first order type transitions because of the important symmetry difference between crystal and liquid states. In a layered system, in the partially filled pore, the presence of the free surface affects the mechanism of the melting. The free interface allows the system to promote the particle to the higher layer. This mechanism facilitates a continuous melting. Apparently, such situation is not possible in filled pores. In such cases the melting should show the first order characteristics, as observed in recent paper on $CCl_4$ melting in carbon nanotubes [9]. In general, when the subsequent layers are adsorbed, a crossover into a first order type transition must appear, depending on the number of layers and the size of the pore. The mechanism of such 'cross-over' will also depend on correlations between the layers. These aspects of phase transitions in confined geometry will be pursued in the further studies.

We have also analyzed the influence of finite size effect on the results presented here. We have repeated the simulations of melting of the one-layer system using larger Monte Carlo boxes (two times and four times larger). There was a small size effect which made the transition more localized in temperature (as compared to the Fig.4) and slightly shifted towards lower temperature (about 2-3 Kelvins). However, the continuous character of melting seemed to be preserved. This aspect needs more studies, as its influence on melting temperature seems to be of the same weight as the corrugation of the pore wall, at least for the system sizes analyzed and in temperature range under consideration.

Metastability in simulations of freezing in confined geometries may trap the studied model in non-equilibrium configuration because in complex systems there may exist a number of local energy minima states. Simulations in generalized ensembles can overcome this difficulty [22]. Some of these techniques, such as parallel tempering techniques [23] and free energy calculations [1] have been already applied to simulations of nanopore systems. However, one may believe that the problem of trapping in local minima is less important when partially filled pores are modeled. In a natural way, the free volume of the pore facilitates the appearance of fluctuations (e.g., via the promotion of atoms into a higher layer) that help the system to avoid metastable states and allow it to find the real equilibrium configurations.

## REFERENCES

[1] L.D.Gelb, K.E.Gubbins, R.Radhakrishnan and M.Sliwinska-Bartkowiak, Rep.Prog.Phys. 62 (1999) 1573.
[2] J. Klein and E.Kumacheva, Science 269 (1995) 816
[3] K. Kaneko, A.Watanabe, T.Liyama, R.Radhakrishnan, and K.E.Gubbins, J.Phys.Chem. B103 (1999) 7061
[4] R. Radhakrishnan, K.E. Gubbins, M. Sliwinska-Bartkowiak, J.Chem.Phys.116 (2002) 1147
[5] R.Radhakrishnan and K.E.Gubbins, Mol.Phys. 96 (1999) 1247
[6] M.Miyahara and K.E.Gubbins, J.Chem.Phys. 106 (1997) 2865
[7] M.Maddox and K.E.Gubbins, J.Chem.Phys. 107 (1997) 9659
[8] J.Hoffman and P.Nielaba, Phys. Rev. E67 (2003) 036115
[9] F.R.Hung, B.Coasne, E.E.Santiso, K.E.Gubbins, F.R.Siperstein and M.Sliwinska-Bartkowiak, J.Chem.Phys. 122 (2005) 144706
[10] B.Coasne, J.Czwartos, K.E.Gubbins, F.R.Hung and M.Sliwinska-Bartkowiak, Adsorption 11 (2005) 301
[11] R. Evans, J.Phys.Condens.Matter, 2 (1990) 8989
[12] B.Kuchta and L.Firlej, J.Low Temp.Phys. (in press, 2005).
[13] R.O. Watts and I.J. McGee, Liquid State Chemical Physics, Willey, New York, 1976.
[14] W.A. Steele, Surf.Sci. 36 (1973) 317.
[15] N.D. Mermin, Phys.Rev. 176 (1968) 250
[16] L. Firlej and B. Kuchta, Stud. Surf. Science and Catalysis, 156 (2005) 689
[17] L. Firlej and B. Kuchta, J.Low Temp.Phys. 139 (2005) 591
[18] R. Etters, E. Flenner, B. Kuchta, L. Firlej and W. Przydrozny, J.Low Temp. Phys. 122 (2001) 121.
[19] R.D.Etters and B.Kuchta, Phys.Rev. B54 (1996) 12057
[20] L. Firlej, B. Kuchta, R. Etters, W. Przydróżny and E. Flenner, J.Low Temp. Phys. 122 (2001) 171
[21] M.S. Dresselhaus, Carbon Nanotubes: Synthesis, Structure, Properties, and Applications, Springer Verlag, 2001
[22] A.Mitsutake, Y. Sugita and Y. Okamoto, Biopolymers, 60 (2001) 96
[23] C. Beauvais, X Guerrault, F.-X. Coudert, A. Boutin and A. Fuchs, J.Phys.Chem. B108 (2001) 399

Studies in Surface Science and Catalysis 160
P.L. Llewellyn, F. Rodriquez-Reinoso, J. Rouqerol and N. Seaton (Editors)

# Molecular simulation study on the structure of templated porous materials obtained from different inorganic precursors

**A. Patti, A. D. Mackie and F. R. Siperstein**

Departament d'Enginyeria Química, Universitat Rovira i Virgili,
Av. Països Catalans 26, 43007 Tarragona, Spain

Monte Carlo simulations are used to determine the equilibrium phase behaviour of surfactant–inorganic oxide–solvent systems, in which the hydrophobic/hydrophilic nature of the inorganic precursor is modified to study how this change affects the final structure of the hybrid material. Lattice Monte Carlo simulations in the canonical ensemble are used to model the aggregation behaviour of the hybrid materials and to obtain the ordered mesoporous structure in a system where phase separation occurs. The model used depicts the general behaviour of the system such as the self-assembly of surfactants in complex aggregates, phase separation and the formation of ordered lyotropic liquid crystal phases. Hexagonal, lamellar, and perforated lamellar liquid crystal phases are observed at high surfactant concentrations. Ternary phase diagrams for partial and complete miscibility between the inorganic precursor and the solvent are reported.

## 1. INTRODUCTION

Self-assembled hybrid materials result from the combination of a soft organic material with an inorganic oxide through an ion exchange or liquid crystal templating mechanism [1]. In the last decade, hybrid materials have received an increasing attention mainly because of their key-role in the synthesis of mesoporous materials, first synthesized in 1992 by Beck et al. [2]. Understanding the behaviour of organic-inorganic hybrids is key to the rational design of materials with improved resistance, flexibility and other desired chemical and physical properties [3]. The large range of applications of these materials in catalysis [4], adsorption [5], and molecular separation [6] has been the driving force to improve our understanding of the synthesis mechanisms by means of experiments, theory and simulation. Such research has focused on understanding which surfactants to choose as structure-directing agents to form a given hybrid material, according to the desired pore size and geometry. Previous simulation studies that addressed the formation of ordered surfactant/inorganic structures have used simple models that captured the essential features of these systems [7-9]. When interest is in fully atomistic descriptions of the solid walls of the porous materials the structure of the surfactant phase is presumed [10].

In this work, lattice Monte Carlo simulations are used to study the phase behaviour of a ternary system in which surfactants and the inorganic component are depicted by flexible chains of connected sites on a lattice box. Under no inorganic condensation conditions (where the reaction to form a 3-dimensional silica network is avoided due to the high pH of the solution), these systems phase separate into a hybrid liquid crystal phase in equilibrium with a solvent-rich phase [11]. In simulations, the formation of a hybrid-rich phase in equilibrium with a dilute phase is observed when the interactions between silica and surfactant are stronger than the ones between solvent and surfactant [9]. In this work, we show that the nature of the inorganic component can deeply affect the range of formation of surfactant templated materials and their equilibrium structure. We analyse how changing the structure of the inorganic component with a partial hydrophobic or hydrophilic behaviour, can affect the equilibrium of the system and the order of the final structures.

## 2. MODEL

The model, although very simple, is able to describe the general behaviour of the system of interest, where the self-assembly of non ionic surfactants in complex aggregates, the phase separation and the formation of ordered lyotropic liquid crystal phases are observed. In particular, depending on the overall concentration of the system, different ordered structures can be obtained. The model surfactant, $H_4T_4$, is made up of a chain with four tail segments $T$, being the hydrophobic part, and four head segments $H$, being the hydrophilic part. Given that a head group $H$ is approximately equivalent to one oxyethylene unit (-C-C-O-) and one tail group $T$ represents about two or three $CH_2$ groups, then the $H_4T_4$ surfactant is roughly equivalent to the real surfactant $CH_3$-$(CH_2)_y$-$(O$-$CH_2$-$CH_2)_4$-$OH$, where $y$ is between 7 and 11 [12].

A two segment chain represents the inorganic precursor whose individual interaction parameters can be tuned in order to simulate a complete or partial miscibility with the solvent. The nature of the inorganic precursor can also be varied by increasing or decreasing its global hydrophobicity (or hydrophilicity). A pure silica unit was represented by $I_2$, whereas organosilica units were obtained by modifying the chain with $IT$ or $IH$, where $T$ and $H$ are equivalent to the surfactant tails and heads respectively. Solvent molecules occupy single sites on the lattice. In this model we do not take into account some features of aqueous solutions, such as hydrogen bonding. Nevertheless, we are able to identify how the modelled system is able to form liquid crystal phases.

In this work, the dimensionless temperature is defined using the head-tail interchange energy by $T^*=kT/\omega_{HT}$, where $k$ is the Boltzmann constant, $T$ is the absolute temperature and $\omega_{HT}$ is the surfactant head-tail interaction energy. The global interchange energy between different types of sites in the box is given by:

$$\omega_{ij} = \varepsilon_{ij} - \frac{1}{2}\left(\varepsilon_{ii} + \varepsilon_{jj}\right)$$

with $i \neq j$ and $\varepsilon_{ij}$ being the individual interaction energies of different types of sites. The individual values of the interaction parameters are reported in Table 1, where the soluble inorganic and the insoluble organic components are indicated with $I$ and $I'$ respectively. The surfactant–solvent interactions have been proposed in previous works by other researchers [13,14]. Two different cases are presented: in the first case, the inorganic precursor and the solvent are completely miscible ($\omega_{IS} = 0$), in the second they are as immiscible as surfactant heads and tails ($\omega_{I'S} = \omega_{HT}$), as in previous work [9].

Table 1.
Individual interaction parameters ($\varepsilon_{ij}$) between inorganic oxide and solvent

| | Complete | miscibility | | | | Partial | miscibility | |
|---|---|---|---|---|---|---|---|---|
| | I | H | T | S | | I' | H | T | S |
| I | 0 | | | | I' | -2 | | | |
| H | -2 | 0 | | | H | -3 | 0 | | |
| T | 0 | 0 | -2 | | T | -2 | 0 | -2 | |
| S | 0 | 0 | 0 | 0 | S | 0 | 0 | 0 | 0 |

## 3. SIMULATION METHOD

Lattice Monte Carlo (MC) simulations in the canonical ensemble (*NVT*) were used to simulate the aggregation behaviour of hybrid materials and to obtain the ordered mesoporous silica structures in a system where phase separation occurs. Periodic boundary conditions were applied to a fully occupied three-dimensional lattice in which the silica and surfactant chains were moved by reptation, partial regrowth or configurational bias moves [15]. A typical mix of the MC moves used is 80% reptation, 10% partial regrowth, and 10% complete regrowth. This combination, being the best compromise between CPU time required for each step and percentage of accepted movements, was chosen by analysing the equilibration for several possible combinations of MC moves. During the configurational bias moves, only 10 different directions, out of the 26 possible directions given by the selected lattice coordination number, were used to grow the chain. These 10 directions were selected randomly for each segment to be grown.

Direct interfacial simulations were performed at the reduced temperature $T^*=8.0$ in a $24\times24\times100$ box for at least $3\times10^9$ MC steps. Single phase simulations were performed in a $40\times40\times40$ box to study in more detail the structures of some liquid crystal phases, and to calculate the radial distribution functions. In such cases, because of the low solvent volume fraction (1-5%), and large number of surfactant chains, the number of MC steps was at least $70\times10^9$. To generate the initial configuration for the direct interfacial simulations, inorganic oxide and surfactant chains were randomly distributed in a concentrated region of the box where their total concentration was around 60%; in the single phase simulations the chains were allocated filling all the spaces in the lattice and the system was allowed to evolve at a very high temperature ($T^*=10^4$) for $2\times10^6$ MC steps: this created a completely random re-distribution of the chains and the initial configuration for the simulations.

After the system reached equilibrium, the ternary phase diagrams for the surfactant–inorganic–solvent system were obtained by simulating the system at different global concentrations and averaging the composition of each phase along the $z$ direction, obtained from density profiles far from the interfaces. According to the differences in the nature of the inorganic precursor, six different phase diagrams were obtained.

## 4. RESULTS AND DISCUSSION

In this section we present the results from the simulations of six different systems: $H_4T_4 - I_2$, $H_4T_4 - I'_2$, $H_4T_4 - IT$, $H_4T_4 - I'T$, $H_4T_4 - IH$ and $H_4T_4 - I'H$ solutions at $T^*=8.0$. For each ternary system the phase separation, the aggregation behaviour and the structural order are studied. All the concentrations reported are volume fractions.

The high surfactant concentration phases for both partial and complete silica/solvent miscibility are reported in Fig. 1 and Fig. 2 respectively. The dilute phases at equilibrium with the concentrated phases are not shown in the diagrams. The points in the phase diagrams are obtained by analysing the density profiles of the last configuration. To guarantee a good estimate of the equilibrium compositions, the average densities are calculated in the regions far from the interfaces. When a clear interface is not observed, namely when free micelles are present in the dilute phase, the uncertainty is considered large, otherwise the error is smaller than the size of the symbols.

When the inorganic, $I$, is completely miscible with the solvent, the size of the immiscibility gap decreases as $I_2 > IH > IT$; whereas for partially immiscible inorganic precursors, the size of the immiscibility gap decreases as $I'_2 > I'T > I'H$. For the completely miscible precursor, the driving force for the separation is the strong interaction between $I$ and $H$ sites. Modifying the inorganic precursor with $H$ or $T$ groups, the driving force for the separation decreases. In fact, for $IT$, where half of the inorganic precursor has a net repulsive interaction with the surfactant heads ($\omega_{HT} = 1$), the immiscibility region is very small.

For the partially immiscible inorganic precursor, a different trend is observed. In this case, the phase separation is due to two different reasons: the immiscibility of $I'$ with the solvent, and the strong interactions between $I'$ and $H$ sites. Modifying the inorganic precursor from $I'_2$ to $I'H$, increases its solubility in the solvent and it reduces the size of the immiscibility gap. Modifying the inorganic precursor from $I'_2$ to $I'T$, does not increase the solubility with the solvent, but it reduces the net strength of the interaction between the surfactant head and the inorganic precursor, resulting also in a smaller immiscibility gap.

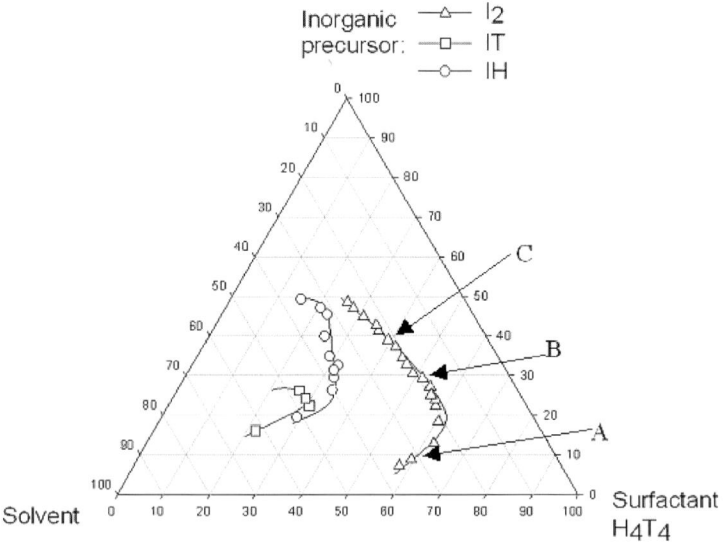

Fig. 1. Ternary phase diagram $H_4T_4$/inorganic precursor/solvent at $T^*=8.0$ for complete miscibility between the solvent and the inorganic precursor. Structures A and B are reported in Fig. 3. Lines are guides for the eyes.

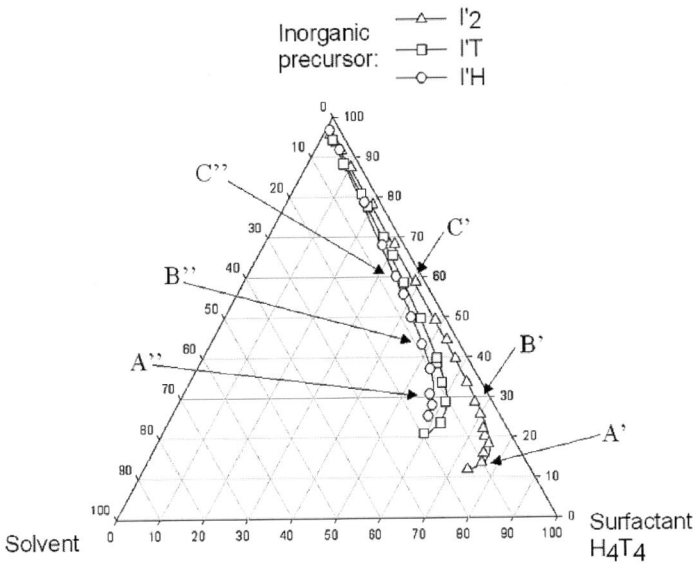

Fig. 2. Ternary phase diagram $H_4T_4$/inorganic precursor/solvent at $T^*=8.0$ for partial miscibility between the solvent and the inorganic precursor. Structures A', B', and A'', B'' are reported in Fig. 4 and in Fig. 6 respectively. Lines are guides for the eyes.

*$H_4T_4 - I_2$ and $H_4T_4 - I'_2$.*
Fig. 3 shows the final configuration for two concentrated phases in a 24×24×100 simulation box where complete miscibility between solvent and inorganic component occurs, that correspond to the points indicated on Fig. 1.

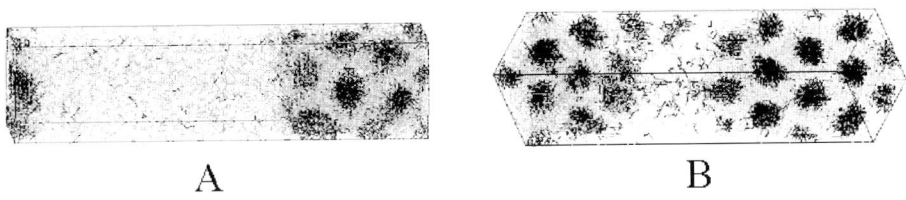

Fig. 3. Phase separation observed for two miscible silica/solvent solutions at $T^* = 8.0$ in a 24×24×100 lattice box. Global concentrations: 20% $H_4T_4$ - 20% $I_2$ (A), 40% $H_4T_4$ - 5% $I_2$ (B). Dark shading represents the surfactant tails, light shading represents the surfactant heads or the inorganic precursor. The solvent is not shown.

In boxes A and B ordered hexagonal phases are observed. Free micelles are found in the dilute phase of box B with a surfactant concentration of 15%, whereas in the lyotropic crystal phase (48% $H_4T_4$, 30% $I_2$) the surfactant self-assembles to form cylindrical aggregates. Systems with an higher inorganic content show a simple phase separation with no morphological order for the concentrated phase (point C on Fig. 1). Two immiscible silica/solvent systems are reported in Fig. 4.

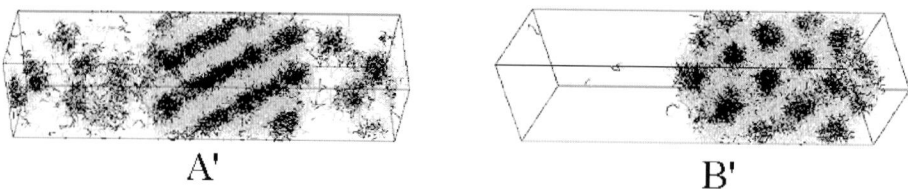

A'                                                                    B'

Fig. 4. Phase separation observed for two immiscible silica/solvent solutions at $T^*$=8.0 in a 24×24×100 lattice box. Global concentrations: 40% $H_4T_4$ - 5% $I'_2$ (A'), 30% $H_4T_4$ - 12% $I'_2$ (B'). Dark shading represents the surfactant tails, light shading represents the surfactant heads or the inorganic precursor. The solvent is not shown.

A perforated lamellar phase (76% $H_4T_4$, 14% $I_2$) at equilibrium with a dilute micellar phase (19% $H_4T_4$, 0.04% $I'_2$) is found in box A'. Lamellar and perforated lamellar crystals are observed at high surfactant concentrations (above 70%). Concentrated phases where the surfactant volume fraction lies between 60% and 70% show hexagonal structures, as observed from box B'. No morphological order is observed for those systems where the inorganic concentration in the surfactant-rich phase is more than 40% (point C' on Fig. 2).

The structure of the concentrated phase in box A was analyzed in a $40^3$ simulation box (52% $H_4T_4$, 32% $I_2$) in order to check that an hexagonal phase was formed, and a clear hexagonal structure was obtained (Fig. 5). The radial distribution function $g(r)$ is calculated for all the possible pair distributions, only the $H$-$H$, $I$-$I$ and $T$-$T$ distributions are reported in Fig. 5.

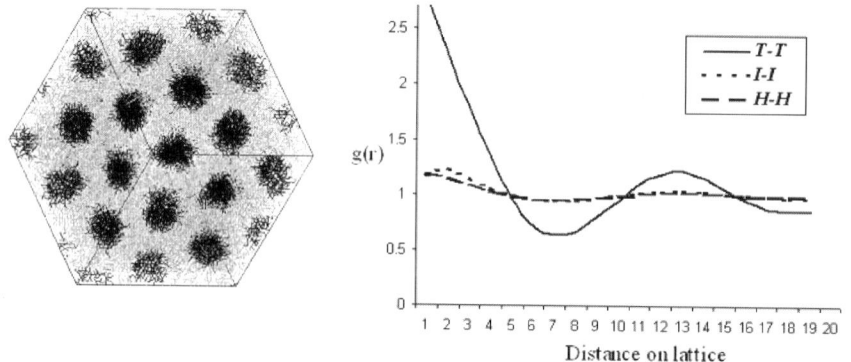

Fig. 5. Hexagonal structure formed at $T^*$=8.0 for a miscible silica/solvent (52% $H_4T_4$, 32% $I_2$) system in a $40^3$ box (left) and its radial distribution function for $T$-$T$, $I$-$I$ and $H$-$H$ distributions (right). Dark shading represents the surfactant tails, light shading represents the surfactant heads or the inorganic precursor. The solvent is not shown.

The $T$-$T$ radial distribution function indicates the periodicity of the cylinders made by the tails of the surfactant, which is approximately 12-13 lattice units. $I$-$I$ and $H$-$H$ radial distribution functions do not show this clearly, as these types of sites form a continuous "phase" throughout the box.

*$H_4T_4 - IT$ and $H_4T_4 - I'T$.*

When the inorganic precursor chains are partially modified by increasing the hydrophobicity of one of their two segments, the formation of ordered structures becomes more difficult, especially when the solvent and the inorganic component are miscible. In this case, the immiscibility gap is reduced considerably and no ordered phases were observed. For the system containing $I'T$ phase separation is observed, but without any structural order in the surfactant-rich phases because the hydrophobic segment of the inorganic precursor interacts favorably with the surfactant tails, preventing the formation of ordered phases.

*$H_4T_4 - IH$ and $H_4T_4 - I'H$.*

The $IH$ and $I'H$ precursors have a similar aggregation behaviour as $I_2$, phase separation occurs for both partial and complete miscibility, and a well-defined hexagonal order is identified at high surfactant concentrations. No ordered structures are found in the $IH - H_4T_4$ system because the highest surfactant concentration reported (around 35%) is not high enough to form complex aggregates. For $I'H - H_4T_4$ the shape of the immiscibility gap is such that hexagonal structures appear at high surfactant volume fractions. In this case, it would be interesting to analyze how lowering the temperature could affect the morphology of the hybrid-rich phase as other ordered structures should be observed.

In Fig. 6 two structures from the more representative cases of partial miscibility are reported. In boxes A'' and B'' an hexagonal ordered phase is at equilibrium with free micelles or monomers. Systems with an higher inorganic content show a phase separation where no clear morphological order exists (point C'' on Fig. 2). A determination by the radial distribution function would be needed.

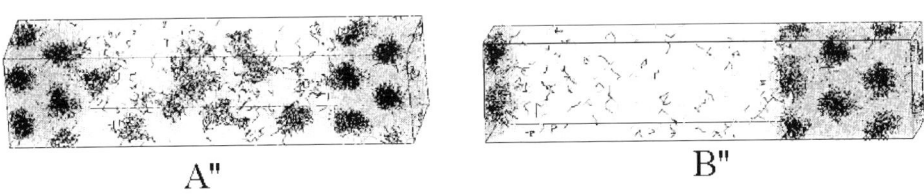

$$A'' \qquad\qquad\qquad\qquad B''$$

Fig. 6. Phase separations observed for two immiscible silica/solvent solutions at $T^*=8.0$ in a $24\times24\times100$ lattice box. Global concentrations: 30% $H_4T_4$ - 10% $I'H$ (A''), 20% $H_4T_4$ - 15% $I'H$ (B''). Dark shading represents the surfactant tails, light shading represents the surfactant heads or the inorganic precursor. The solvent is not shown.

## 5. CONCLUSIONS

A simple molecular model was used to simulate the equilibrium phase behaviour of hybrid materials for different inorganic precursors. Although some physical and chemical properties are not explicitly included in the model, it is possible to detect the general behaviour of the system such as the phase separation and the self-assembly of surfactants in hexagonal, lamellar or perforated lamellar structures.

Ternary phase diagrams were obtained for partial and complete miscibility between the solvent and the silica or organosilica precursor at the reduced temperature $T^*=8.0$. In some cases, a liquid crystal phase with high surfactant concentration is at equilibrium with a solvent-rich phase containing a small fraction of free monomers or micelles. Hexagonal phases are obtained for surfactant volume fractions between 50% and 70%, whereas lamellar or perforated lamellar structures are found for higher volume fractions. No ordered phases were observed for $H_4T_4/IT$, $H_4T_4/I'T$ and $H_4T_4/IH$ systems.

When the hydrophobic/hydrophilic nature of the inorganic precursor is changed, the surfactant concentration in the concentrated phase can decrease to an extent where no ordered structures are observed. This phenomena is observed because the surfactant heads have stronger interactions with $I_2$ than with other precursors.

## ACKNOWLEDGMENTS

This work was supported by project CTQ-2004-03346/PPQ from the Spanish Ministerio de Educación y Ciencia. Support from the program ACI2003-026 of the Generalitat de Catalunya is greatfully acknowledged. FRS acknowledges the support of the RyC program from the Spanish Ministerio de Educación y Ciencia. AP acknowledges the support of the URV graduate student fellowship.

## REFERENCES

[1]     C. Sanchez, G.J.D.A. Soler-Illia, F. Ribot, T. Lalot, C.R. Mayer, V. Cabuil, Chemistry of Materials 13 (2001) 3061.
[2]     J.S. Beck, J.C. Vartuli, W.J. Roth, M.E. Leonowicz, C.T. Kresge, K.D. Schmitt, C.T.-W. Chu, D.H. Olson, E.W. Sheppard, S.B. McCullen, J.B. Higgins, J.L. Schlenker, J. Am. Chem. Soc 114 (1992) 10834.
[3]     G.J.D. Soler-illia, C. Sanchez, B. Lebeau, J. Patarin, Chemical Reviews 102 (2002) 4093.
[4]     T. Linssen, K. Cassiers, P. Cool, E.F. Vansant, Advances in Colloid and Interface Science 103 (2003) 121.
[5]     J. Lei, J. Fan, C.Z. Yu, L.Y. Zhang, S.Y. Jiang, B. Tu, D.Y. Zhao, Microporous and Mesoporous Materials 73 (2004) 121.
[6]     B. Lee, Y. Kim, H. Lee, J. Yi, Microporous and Mesoporous Materials 50 (2001) 77.
[7]     A. Bhattacharya, S.D. Mahanti, Journal of Physics-Condensed Matter 13 (2001) 1413.
[8]     S.E. Rankin, A.P. Malanoski, F. Van Swol, Materials Research Society Symposium Proceedings 636 (2001) D1.2/1
[9]     F.R. Siperstein, K.E. Gubbins, Langmuir 19 (2003) 2049.
[10]    C. Schumacher, J. Gonzalez, M. Perez-Mendoza, P.A. Wright, N.A. Seaton, Studies in Surface Science and Catalysis 154 (2004) 386.
[11]    A. Firouzi, F. Atef, A.G. Oertli, G.D. Stucky, B.F. Chmelka, Journal of the American Chemical Society 119 (1997) 3596.
[12]    B. Fodi, R. Hentschke, Langmuir 16 (2000) 1626.
[13]    R.G. Larson, L.E. Scriven, H.T. Davis, Journal of Chemical Physics 83 (1985) 2411.
[14]    A.D. Mackie, A.Z. Panagiotopoulos, I. Szleifer, Langmuir 13 (1997) 5022.
[15]    D. Frenkel, B. Smit, Academic Press: San Diego (2002).

Studies in Surface Science and Catalysis 160
P.L. Llewellyn, F. Rodriquez-Reinoso, J. Rouqerol and N. Seaton (Editors)
© 2007 Elsevier B.V. All rights reserved

# The structure of high-pressure adsorbed fluids in slit-pores

**Suresh K. Bhatia, Thanh X. Nguyen, Kien Tran and David Nicholson**

Division of Chemical Engineering, The University of Queensland, St. Lucia, QLD 4072, Brisbane, Australia.

We present simulation results for the packing for single center Lennard Jones models of adsorbed fluids such as methane, carbon dioxide and carbon tetrachloride at high pressure in carbon slit pores. These show a series of packing transitions that are well described by a lattice density functional theory model developed in our laboratory. By contrast, simulations show that these transitions are absent for a three-center model, which provides a more adequate representation of carbon dioxide. Analysis of the simulation results shows that alternations of flat lying molecules and rotated molecules can occur in this case as the pore width is increased. The presence or absence of quadrupoles has negligible effect on these high-density structures.

## 1. INTRODUCTION

The high pressure adsorption capacity of an adsorbent for a given fluid provides an important physical performance limit for any adsorption process involving the fluid solid pair. Besides this, the capacity is a key parameter in many classical adsorption models, such as the Langmuir or Dubinin models and their many variants [1], or others based on multisite or vacancy solution theory [2-5] that have a kinetic, thermodynamic or statistical thermodynamic starting point. In such models the capacity is generally considered dependent only on temperature and independent of pore size, and determined by fitting macroscopic adsorption data, though carbons invariably possess a distribution of pore sizes. This seems reasonable for an adsorbed fluid in large pores such as macropores or mesopores where capacity is independent of the pore volume. However this is not the case for micropores whose pore widths are only a few adsorbate molecular diameters. Here the close packed configuration of the adsorbed phase can vary with pore width, passing sequentially through hexagonal and square geometries, in a similar manner to that observed experimentally when colloid suspensions are confined between parallel glass plates [6-9].

Here we report a density functional theory of the high pressure packing of Lennard Jones adsorbates in carbon slit pores, which accurately models the packing transitions with variation in pore size. The theoretical results for the pore size depenent capacity are validated by comparison with grand canonical Monte Carlo simulations, for the adsorption of methane, carbon tetrachloride and carbon dioxide. Because of the particular importance of high pressure adsorption in applications for sequestration of carbon dioxide, established as the major greenhouse contributor, we have also examined the capacity using a more representative 3-centre model of carbon dioxide. Earlier studies [10-13] with the 3-centre model have not adequately investigated the packing and microstructure of the adsorbate, about which little is known. Here we report on the microstructure observed in our simulations with this model of carbon dioxide, and compare the results with those obtained using the sngle site model.

## 2. THEORY

Density functional theory has proved to be a powerful tool for the investigation of adsorption equilibrium of single-site molecules in inhomogeneous systems, and in our laboratory has also been used to effect in the investigation of dense packing [14]. The approach minimizes the grand potential

$$\Omega[\rho] = F[\rho] + \int \rho(\mathbf{r})[v_{ex}(\mathbf{r}) - \mu]d\mathbf{r} \tag{1}$$

Here $\rho(\mathbf{r})$ is the density profile in the presence of external potential, $v_{ex}(\mathbf{r})$. Further, $\mu$ is the chemical potential of the adsorbate, and $F[\rho(\mathbf{r})]$ is the intrinsic Helmholtz free energy, decomposed in terms of an ideal gas part, $F_{id}[\rho(\mathbf{r})]$, and an excess part $F_{ex}[\rho(\mathbf{r})]$

$$F[\rho] = F_{id}[\rho] + F_{ex}[\rho] \tag{2}$$

Given suitable models for these free energies, the general procedure is to obtain the optimal profile that minimizes the grand potential for the chosen system.

In the close-packed limit the adsorbed phase is considered to be of a set of identical unit cells, which are detailed in a later section. The local density can be then be treated as the density in the unit cell. Consequently, Eq. (1). can be rewritten in terms of the unit cell potential as follows:

$$\Omega = F_{id} + F_{ex} + N(v_{ex} - \mu) = F_{id} + F_{ex} + Nv_{ex} - N\mu , \tag{3}$$

where N is the number of adsorbed molecules in a unit cell.

In our work the above general approach has been implemented for Lennard jones fluids in slit pores of carbon, using the established 10-4-3 potential of Steele [15], assuming infinitely thick pore walls. Although the latter assumption has been questioned for carbons [16], at high pressures where fluid-fluid interactions dominate this simplification is not of consequence.

Using the above fluid- fluid and fluid-solid potential models, the total interaction energy of a unit cell has been evaluated for various transitions. We considered close packed L-J particles forming a two-dimensional structure, comprising unit cells. To determine the total interaction potential for the unit cell, the following assumptions were applied
- Pairwise additivity.
- The adsorbed phase consists of an integral number of identical unit cells.
- Structural transitions of the adsorbed phase correspond to a series of following crystalline structures as the pore width increases, following

$$1\Delta \rightarrow 2\square \rightarrow 2\Delta \rightarrow 3\square \rightarrow 3\Delta....$$

The symbol $\Delta$ represents hexagonal packing and the symbol $\square$ square packing, and the numbers preceding the symbols represent the number of layers in the adsorbed phase. This sequence of crystalline structure has been studied for hard sphere systems [17, 18], as well as for frozen phases of an LJ fluid in slit-shaped pores [19, 20], and two main transition types may be identified:

1. The transitions $n\square \rightarrow n\Delta$, or rhombic transitions. In this case, only the structure within each layer changes.

2. The transitions $n\square \rightarrow (n+1)\Delta$, or buckling transitions. In this case, a new layer is formed together with structural variation of the adsorbed phase.

Based on the above assumptions, the grand potential of the unit cell corresponding to each transition has been minimised to yield unit cell parameters, and thereby the packing density.

## 3. SIMULATION

Grand canonical Monte Carlo (GCMC) simulations have been conducted to verify the predictions of the above theory for LJ fluids, and to study the capacity variation with pore size for a 3-center model of CO2. These simulations mimicking the $\mu,V,T$ ensemble used the established Metropolis sampling scheme [21] for moving (including rotation for the 3-center model), creating or deleting molecules.

Throughout a simulation, the numbers of attempted deletions and creations were kept equal to maintain microscopic reversibility. To ensure reliable values for the capacity, a range of high fugacities between 1000 and 6000 bar was investigated. At high pressure, the acceptance rate for creating or destroying particles can be low, and ten creation or deletion attempts to one move attempt were therefore performed. In the simulation, the LJ potential is truncated at 2.5 nm. For the 3-center model the potential parameters used were the same as those established by Murthy et al. [22], incorporating point charges on the atom centers. The Lorentz-Berthelot rules were used to estimate parameters for the atom-atom cross interactions. The parameters of the single LJ site CO2 were determined by matching the bulk isotherm for this fluid with that of the 3-center fluid between 0 and 2500 bar, at 308 K. This yielded $\sigma_{ff} = 0.37$nm and $\varepsilon_{ff}/k = 220$ K.

## 4. RESULTS AND DISCUSSION

### 4.1 Lennard Jones fluid

Initially computations were performed at high bulk fugacities of 1000-6000 bar, for the 1-centre models of methane, carbon tetrachloride and carbon dioxide, over a range of pore sizes and temperatures. These yielded excellent agreement between theory and simulation for all three fluids. As an example Fig. 1. depicts the results for carbon dioxide at 318 K, with the line representing the DFT predictions, matching well the simulation data given by the symbols.

As seen in the figure the packing structure and transitions considered in the theory provide a satisfactory quantitative explanation of the capacity variation with slit width. The different branches in Fig. 1. correspond to different packing structures as follows.

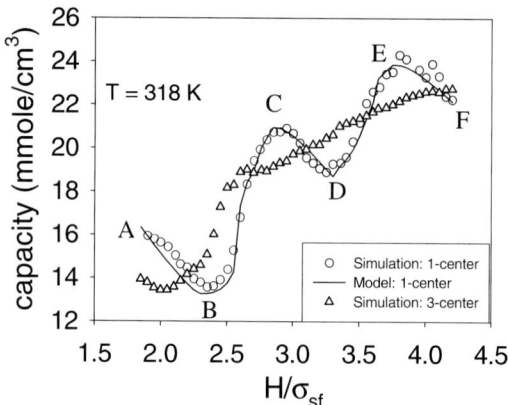

**Fig. 1. Variation of adsorbed high pressure capacity with pore size, for carbon dioxide at 318 K**

- Branch AB correspond to the monolayer with LJ molecules in 2-D hexagonal configuration. On this branch the interparticle separation is constant at $r_o = 2^{1/6}\sigma_{ff}$, and the density reduces with increase in pore size.
- On branch BC the structure gradually adjusts from hexagonal to square packing as a new layer is initiated, and at C two layers are fully established.
- On branch CD no new layer is formed, but the two layers gradually undergo a transition from square to hexagonal geometry.
- On branch DE a third layer is gradually established, with BCC unit cell structure of the packing. Each layer gradually attains square geometry.
- On branch EF a rhombic transition again occurs, with each layer attaining hexagonal geometry.

Our simulation results confirmned the above packing structure and transitions on each branch, based on the pair distributions obtained. As an example Fig. 2. depicts the pair distributions for two different pore sizes, yielding hexagonal and square geometry. In the former case the peaks at $r_o$, $\sqrt{3}\,r_o$, $2r_o$ and $\sqrt{7}\,r_o$ corresponding to the locations of increasingly distant neighbors in hexagonal configuration, while in the latter case the peaks are consistent with the locations $r_o$, $\sqrt{2}\,r_o$, $2r_o$ and $\sqrt{5}\,r_o$, typical of square consfiguration. Some deviations do occur in the latter two peaks in both cases, where the distance is comparable to or exceeds the pore width.

**4.2 3-Center Fluid**

Also superimposed in Fig. 1. are the simulation results for the 3-centre model of carbon dioxide, showing considerable difference from the LJ model results, qualitatively as well as quantitatively. For the 3-centre fluid no clear transitions are found, and the capacity increases continuously with pore size, except for small pore widths below $H = 2\sigma_{ff}$ for which a decrease is noted. Thus, the LJ model is inaccurate in this case. However, the error is generally within 5-8 percent, except in the region ABC where error upto 30 percent is evident. Consequently, since the LJ model overpredicts in some ranges of pore size, and underpredicts in others, it is likely that in the presence of a wide pore size distribution, as exists in most carbons, the error in the overall capacity is not large.

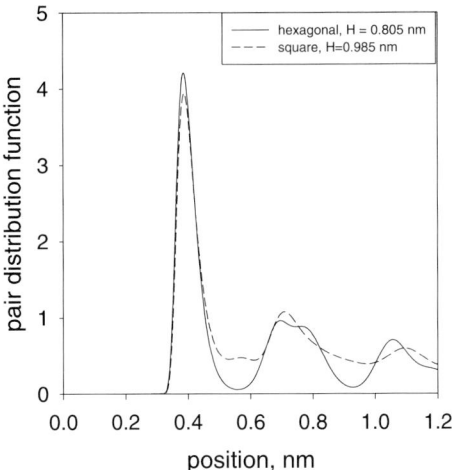

**Fig. 2. Pair distribution function of LJ carbon dioxide obtained by GCMC simulation, illustrating the different structure in hexagonal and square configuration.**

Examination of the structure of the packing of 3-centre carbon dioxide at different pore sizes revealed a clear pattern of change with respect to pore size, associated with layer formation. Both the density profiles and the profiles of molecular orientation showed a monolayer of carbon dioxide oriented parallel to the pore walls at small pore size below about H = 0.71 nm. However, with increase in pore width above 0.68 nm there is a tendency for the flat molecules to rotate, in order to permit additional molecules to adsorb. Above about 0.71 nm an additional layer is formed, with molecules near the wall tending to lie flat and those at the centre tending to rotate relative to the axis. This pattern of behaviour is repeated as the pore size is increased. Fig. 3. depicts snapshots of the structure at different pore sizes. They reveal the formation of a central relatively flat layer, followed by rotation and subsequent separation of two distinct rotated layers. This pattern is consistently followed as the pore size increases, and in this way additional layers are created. This was confirmed from the simulation results, by examination of profiles of density and the molecular orientation. Simulations with the 3–centre fluid without charges at the sites were also conducted, and yielded similar trends as given by the fluid with charges, with only a small reduction in capacity. Thus, the difference in behavior compared to the LJ fluid is clearly related to the different molecular shape represented by the 3–center fluid, and not to electrostatic effects.

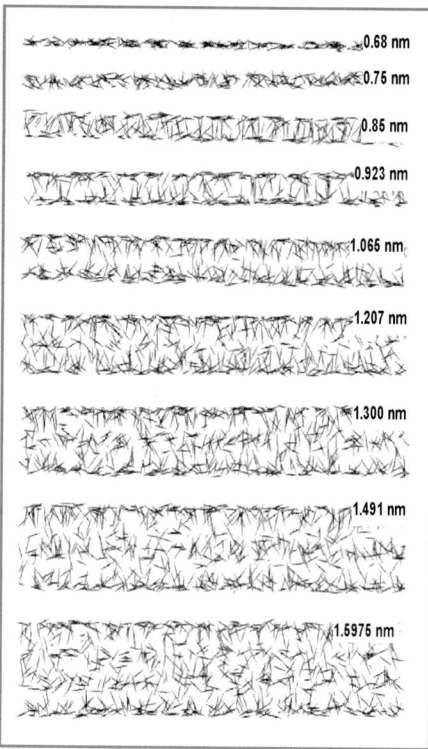

Fig. 3. Snapshots of adsorbate microstructure at different pore sizes. For clarity in viewing the (dense) packing the molecules are displayed as lines, with the red portion representing oxygen, and the central grey part representing carbon.

## 5. CONCLUSIONS

It is found that a theory considering successive continuous transitions between hexagonal and square packing successfully predicts the oscillatory variations of high pressure capacity with pore size for the LJ Fluid. However, simulations with the 3–center potential model for CO2 yield significantly different results from the LJ model for the filled pores with a capacity that increases monotonically with pore size beyond the monolayer width. At monolayer widths (pores smaller than 0.71 nm) reduction in capacity with increase in pore size occurs. Investigation of the adsorbate microstructure has shown that for the 3–center fluid packing transitions are absent, and instead rotation of flat-lying molecules occurs at the pore center to accommodate additional adsorbed molecules as pore size increases, starting from monolayer pore widths. At sufficiently large pore size, however, the central layer splits into two layers, followed by the appearance of a new layer of flat-lying molecules at the center at larger pore size. With continued increase in size this process repeats itself. The results suggest that at high pressure the single-center LJ approximation is inadequate to account for the structure of linear molecules such as $CO_2$ in narrow pores.

# REFERENCES

[1] R.T. Yang, Gas Separation by Adsorption Processes, Imperial College Press: London (1997).

[2] T. Nitta, T. Shigetomi, M. Kuro-Oka, and K. Takashi, J. Chem. Eng. Japan, 17 (1984) 39.

[3] B.P. Bering, V.V. Serpinskii, and T.S. Yakubov, Izv. Akad. Nauk SSSR Ser. Khim, 4 (1977) 727.

[4] S.K. Bhatia, and L.P. Ding, AIChE J, 47 (2001) 2136.

[5] L.P. Ding, and S.K. Bhatia, AIChE J, 49 (2003) 883.

[6] P. Pieranski, and L. Strzelecki, Phys. Rev. Lett, 50 (1983) 331.

[7] C.A. Murray, Physical Review B, 42 (1990) 688.

[8] C.A. Murray, in Bond-Orientational Order in Condensed Matter Systems, K.J. Strandburg, ed., Springer: New York (1992).

[9] J.A. Weis, D.W. Oxtoby, D.G. Grier, and C.A. Murray, J. Chem. Phys., 103 (1995) 1180.

[10] S. Samios, A.K. Stubos, N.K. Kanellopoulos, R.F. Cracknell, G.K. Papadopoulos, D. Nicholson, Langmuir, 13 (1997) 2795.

[11] D. Nicholson, Langmuir, 15 (1999) 2508.

[12] D. Nicholson, and K.E. Gubbins, J. Chem. Phys., 104 (1996) 8126.

[13] G.K. Papadopoulos, J. Chem. Phys., 114 (2001) 8139.

[14] T.X. Nguyen, S.K. Bhatia, and D. Nicholson, J. Chem. Phys., 117 (2002) 10827.

[15] W.A. Steele, Surf. Sci., 36 (1973) 317.

[16] S.K. Bhatia, Langmuir, 18 (2002) 6845.

[17] M. Schmidt, and H. Lowen, Physical Review E, 55 (1997) 7228.

[18] E. Johnson, Science, 296 (2002) 477.

[19] A. Vishnyakov, and A.V. Neimark, J. Chem. Phys., 118 (2003) 7585.

[20] K.G. Ayappa, and C.J. Ghatak, J. Chem. Phys., 117 (2002) 5373.

[21] N. Metropolis, A.W. Rosenbluth, M.N. Rosenbluth, A.N. Teller, and E.J. Teller, J. Chem Phys., 21 (1953) 1087.

[22] C.S. Murthy, S.F. O'Shea, and I.R. McDonald, Mol. Phys., 50 (1983) 531.

Studies in Surface Science and Catalysis 160
*P.L. Llewellyn, F. Rodriquez-Reinoso, J. Rouqerol and N. Seaton (Editors)*

# Monte Carlo simulation of the isosteric heats – implications for the characterisation of porous materials

Yufeng He[+] and Nigel A. Seaton

Institute for Materials and Processes, School of Engineering and Electronics, University of Edinburgh, Kenneth Denbigh Building, King's Buildings, Mayfield Road, Edinburgh, EH9 3JL, UK

The isosteric heats of adsorption of methane in BPL activated carbon and ethane in MCM-41 were obtained by Monte Carlo simulation. The simulated absolute isosteric heats were converted into their experimental excess counterparts using a thermodynamic equation, which was derived by the thermodynamic analysis of the Clausius-Clapeyron equation for the isosteric heats. The difference between absolute and excess adsorption is small at low pressure in small pores but becomes bigger as the pressure increases, and is substantial in pores with a pore size bigger than 20 Å even at low pressures. Excellent fits were obtained between experimental and simulated isosteric heats of adsorption of methane in BPL activated carbon and ethane in an MCM-41 sample. A pore size distribution model was used to relate simulation results for pores of different sizes to the experimental adsorbent. It is found that the isosteric heat is a more sensitive measure of the structure of activated carbon adsorbents than an adsorption isotherm.

---

[+] Corresponding Author. Current Address: Department of Chemical Engineering, University of Bath, Bath BA2 7AY, UK. E-mail: y.he@bath.ac.uk

## 1. INTRODUCTION

Adsorption equilibrium is characterized not only by the adsorption isotherms but also by the isosteric heats of adsorption. The isosteric heat is a direct indicator of the heterogeneity of the interaction between the adsorptive and the adsorbent. The heterogeneity of an adsorbent can be broken down into two aspects: energetic heterogeneity (due to surface irregularities and surface composition) and structural heterogeneity (due to pores of different sizes and shapes, and variable connectivity between pores). In this work, we have studied the relationship between the isosteric heat and both types of heterogeneity. The energetic heterogeneity of MCM-41 is characterized by studying the isosteric heat of ethane in a model adsorbent using Monte Carlo simulation. The structural heterogeneity of an activated carbon, BPL, is studied by a combination of Monte Carlo simulation and a pore size distribution (PSD) model of the pore structure, using methane adsorption as the adsorptive.

## 2. ABSOLUTE AND EXCESS ADSORPTION

The isosteric heats of adsorption are usually obtained experimentally either from a calorimeter or from the evaluation of pure-gas adsorption at a constant loading using the Clausius-Clapeyron equation [1]:

$$q_{st} = -R\left(\frac{\partial \ln P}{\partial (1/T)}\right)_n = RT^2\left(\frac{\partial \ln P}{\partial T}\right)_n, \tag{1}$$

where $P$ is the pressure, $n$ is the number of moles adsorbed, $R$ is the universal gas constant and $T$ is the temperature. This Clausius-Clapeyron equation method can also be used to obtain the isosteric heats from Monte Carlo simulation data. However, in this work, the approach proposed by Vuong and Monson [2] was applied to obtain the isosteric heats in a Canonical Monte Carlo (CMC) simulation since it is more efficient and requires only one simulation isotherm at the temperature of interest:

$$q_{st} = RT - \left(\frac{\partial U}{\partial N}\right)_T, \tag{2}$$

where $U$ and $N$ are the internal energy of the adsorbate and the number of molecules adsorbed, respectively.

Monte Carlo simulations generate absolute adsorption data, i.e. the actual number of molecules present in the simulated pore space, while adsorption experiments give Gibbs excess properties [3], obtained by either volumetric or gravimetric methods. Therefore, the simulation results must be converted to their excess counterparts before they can be used to analyze experimental data. The excess amount adsorbed (experimental result), $N_{ex}$, is given by:

$$N_{ex} = N_{abs} - N_b = N_{abs} - \rho_b V, \tag{3}$$

where $N_{abs}$ and $N_b$ are the simulated absolute amount adsorbed and the number of moles of gas that would be present in the pore without adsorption, at the bulk density; $V$ is the accessible volume in the model pore and $\rho_b$ is the density of the bulk gas at the same temperature and pressure. The value of $V$ is obtained by simulating the adsorption of helium in the model pore, as this is consistent with the use of helium (a very weakly adsorbing gas) to determine the experimental dead volume.

To relate the absolute and excess isosteric heats, we begin by using the Clausius-Clapeyron Equation (Equation 1) explicitly for absolute and excess adsorption:

$$q_{st\_abs} = RT^2 \left( \frac{\partial \ln P}{\partial T} \right)_{N_{abs}},$$

(4)

and

$$q_{st\_ex} = RT^2 \left( \frac{\partial \ln P}{\partial T} \right)_{Nex},$$

(5)

where $q_{st\_abs}$ and $q_{st\_ex}$ are the absolute and excess isosteric heats respectively. Applying the chain rule to the right-hand side of Equation (5), we obtain

$$\left( \frac{\partial \ln P}{\partial T} \right)_{N_{abs}} = -\frac{\left( \dfrac{\partial N_{abs}}{\partial T} \right)_{\ln P}}{\left( \dfrac{\partial N_{abs}}{\partial \ln P} \right)_{T}} = -\frac{\left( \dfrac{\partial N_{ex}}{\partial T} \right)_{\ln P} + \left( \dfrac{\partial N_b}{\partial T} \right)_{\ln P}}{\left( \dfrac{\partial N_{ex}}{\partial \ln P} \right)_{T} + \left( \dfrac{\partial N_b}{\partial \ln P} \right)_{T}}.$$

(6)

After rearranging and substituting Equation (7), we get

$$q_{st\_ex} = q_{st\_abs} + \frac{q_{st\_abs} \left( \dfrac{\partial N_b}{\partial \ln P} \right)_{T} + RT^2 \left( \dfrac{\partial N_b}{\partial T} \right)_{\ln P}}{P \left( \dfrac{\partial N_{ex}}{\partial P} \right)_{T}}.$$

(7)

$N_b$ is related to the compressibility factor of the bulk gas, $Z$, by

$$N_b = \frac{PV}{ZRT},$$

(8)

It follows that

$$\left( \frac{\partial N_b}{\partial \ln P} \right)_{T} = N_b \left( 1 - \frac{P}{Z} \left( \frac{\partial Z}{\partial P} \right)_{T} \right),$$

(9)

and

$$\left(\frac{\partial N_b}{\partial T}\right)_{\ln P} = -N_b\left(\frac{1}{T} + \frac{1}{Z}\left(\frac{\partial Z}{\partial T}\right)_P\right). \tag{10}$$

Therefore

$$q_{st\_ex} = q_{st\_abs} + \frac{N_b\left(q_{st\_abs}\left(1 - \frac{P}{Z}\left(\frac{\partial Z}{\partial P}\right)_T\right) - RT\left(1 + \frac{T}{Z}\left(\frac{\partial Z}{\partial T}\right)_P\right)\right)}{P\left(\frac{\partial N_{ex}}{\partial P}\right)_T}. \tag{11}$$

This general result allows us to convert the simulated absolute isosteric heat ($q_{st\_abs}$) into its excess counterpart heat ($q_{st\_exs}$), which can then be compared with experimental values.

For an ideal gas, $Z = 1$ and

$$q_{st\_ex} = q_{st\_abs} + \frac{N_b\left(q_{st\_abs} - RT\right)}{P\left(\frac{\partial N_{ex}}{\partial P}\right)_T}. \tag{12}$$

An alternative formulation, which is explicit in $q_{st\_abs}$, is obtained by applying the chain rule to Equation (8), and then substituting Equation (7):

$$q_{st\_abs} = q_{st\_ex} - \frac{N_b\left(q_{st\_ex}\left(1 - \frac{P}{Z}\left(\frac{\partial Z}{\partial P}\right)_T\right) - RT\left(1 + \frac{T}{Z}\left(\frac{\partial Z}{\partial T}\right)_P\right)\right)}{P\left(\frac{\partial N_{abs}}{\partial P}\right)_T}. \tag{13}$$

For an ideal gas,

$$q_{st\_abs} = q_{st\_ex} - \frac{N_b\left(q_{st\_ex} - RT\right)}{P\left(\frac{\partial N_{abs}}{\partial P}\right)_T}. \tag{14}$$

Equations (12) and (14) are equivalent, differing only in whether the equation is explicit in the absolute or excess isosteric heat.

## 3. THE ISOSTERIC HEAT AND THE PSD

The adsorption isotherm for the carbon material is related to the PSD by the adsorption integral equation:

$$n(P_i) = \int_0^\infty \rho(w, P_i) f(w) dw \qquad\qquad i = 1.....n, \qquad\qquad (15)$$

where $n(P_i)$ is the experimentally determined amount adsorbed at pressure $P_i$, $\rho(w, P_i)$ is the adsorbate density in a model pore of width $w$ at pressure $P_i$, $f(w)$ is the PSD and $n$ is the total number of experimental data points. The isosteric heat of adsorption in the adsorbent as a whole is related to the isosteric heat in individual pores and the PSD by an adsorption integral equation analogous to Equation (15):

$$q_{st}(P_i) = \frac{\int_0^\infty q_{st}(w, P_i)\rho(w, P_i) f(w) dw}{\int_0^\infty \rho(w, P_i) f(w) dw}, \qquad\qquad (16)$$

where $q_{st}(P_i)$ is the experimental isosteric heat of adsorption at pressure $P_i$, and $q_{st}(w, P_i)$ is the simulated isosteric heat of adsorption in a individual pore of width $w$ at pressure $P_i$.

## 4. MOLECULAR MODELS

The interaction between adsorbate molecules is modelled using the Lennard-Jones potential. The methane molecule is represented by a single Lennard-Jones site with parameters of $\sigma = 3.81$ Å *and* $\varepsilon/k_B = 148.2$ K [4] while ethane consists of two sites with parameters of $\sigma = 3.512$ Å *and* $\varepsilon/k_B = 139.8$ K [5]. The bond length between two sites is 2.353 Å [5]. The pores in the activated carbon adsorbent are modelled as slits. Each pore wall consists of an infinite number of structureless graphitic layers composed of Lennard-Jones, sites. Steele's 10-4-3 potential [6] was applied to account for the interaction between a methane molecule and a single semi-infinite slab of graphite. A heterogeneous, amorphous cylindrical model pore with an amorphous interfacial region, generated by an energy-minimization procedure, was used for the simulation of the adsorption of ethane in MCM-41; details of the model can be found in ref. [7].

GCMC simulations, in which the temperature, the volume of the simulation cell and the chemical potential of the adsorbate are kept constant, were carried out for the adsorption isotherms of methane and ethane in slit-shaped pores (representing pores in BPL carbon) and of ethane in cylindrical pores (representing pores in MCM-41). The absolute configurational energy of the adsorbates was obtained by a Canonical Monte Carlo (CMC) simulation, in which the number of molecules in the pore, the temperature and the volume of the simulation cell are kept constant. Details of the GCMC and CMC simulations can be found in refs. [8, 9].

## 5. RESULTS AND DISCUSSION

Fig. 1 shows the absolute isosteric heats of methane in slit-shaped pores of width 7.62 Å and 26.7 Å, along with their excess counterparts, obtained by Equation (12). For both pores, the excess isosteric heats of adsorption are larger than the absolute values. The difference between the absolute and excess values is quite small in the 7.62 Å pore, less than 1.5% when the pressure is less than 0.5 MPa; and it becomes greater than 10% when the pressure is higher than 1 MPa. In contrast, the difference between the absolute and excess isosteric heats in big pores is substantial (greater than 10% in the case of 26.7 Å pore) across the whole pressure range.

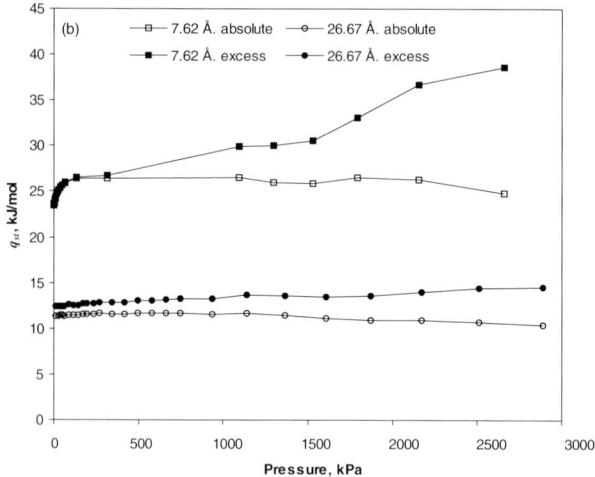

Figure 1. Absolute and excess isosteric heat of methane in slit-shaped pores at 301.4 K.

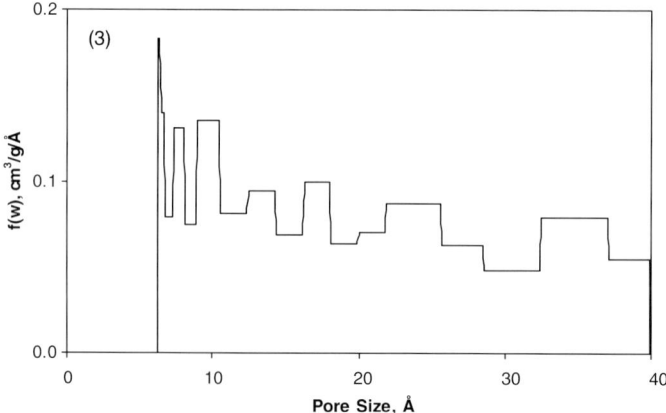

Figure 2. Pore size distributions for BPL obtained using Equation (16).

Equation (16) was used to obtain the PSD of BPL by fitting the simulated isosteric heat of adsorption of methane to experimental isosteric heat data. The PSD is shown in Fig. 2. Fig. 3 reports the fit between the simulated and experimental isosteric heats of methane. The experimental isosteric heats were obtained using the Clausius-Clapeyron equation method (Equation (1)). A very good fit between the experimental and simulated heats is observed. Fig.3 also shows the predicted excess isosteric heats based on another two PSDs obtained by the more usual process of fitting to isotherm data [10]. These two PSDs, which differ in the initial values used in the fitting process, gave excellent fits between the simulated and experimental isotherms (not shown here), however this does not guarantee a good prediction for the excess isosteric heats: both PSDs underestimate the isosteric heats at low pressures, and PSD-1 overestimates the isosteric heats at high pressures while PSD-2 gives a good estimate at high pressures [10]. The PSD obtained by fitting to isosteric heat data (shown in Figure 2) was then used to predict the adsorption of methane in the same BPL at 301.4 K and 264.6 K, and also the adsorption isotherm of ethane at 301.4 K. The results are shown in Fig. 4. Clearly, this PSD gives accurate predictions of the isotherms of methane in BPL at different temperatures with a relatively small error (mostly less than 3%), and also gives accurate predictions for ethane at the same temperature.

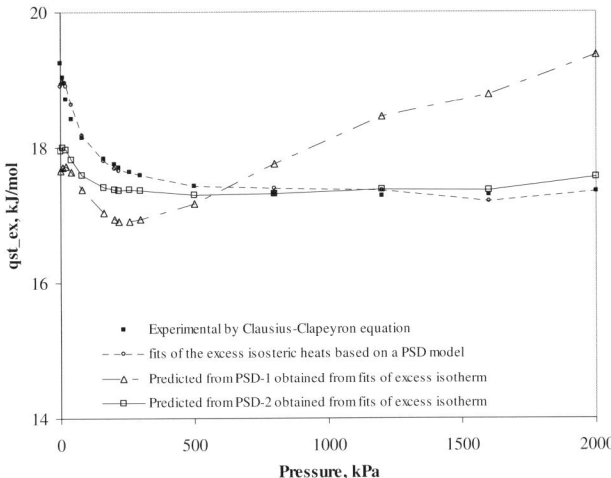

Figure 3. The excess isosteric heat of adsorption of methane based on the PSD model.

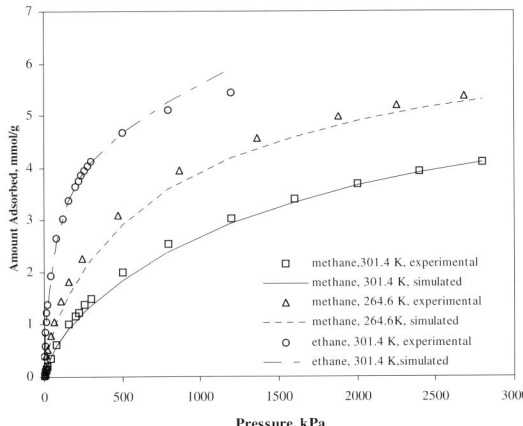

Figure 4. Prediction of isotherms based on the PSD obtained by fitting the isosteric heat.

The same method (Equations (2) and (14)) was applied to study the excess isosteric heats of ethane in MCM-41 at 273 K. Fig. 5. shows the experimental isosteric heats, obtained using the Clausius-Clapeyron equation (Equation (1)), together with the simulation excess isosteric heats. The two isosteric heats both show a decreasing isosteric heats with loading at very low pressures and fit each other quite well, which shows that our amorphous model is a realistic representation of MCM-41 and that MCM-41 is slightly energetically heterogeneous [7].

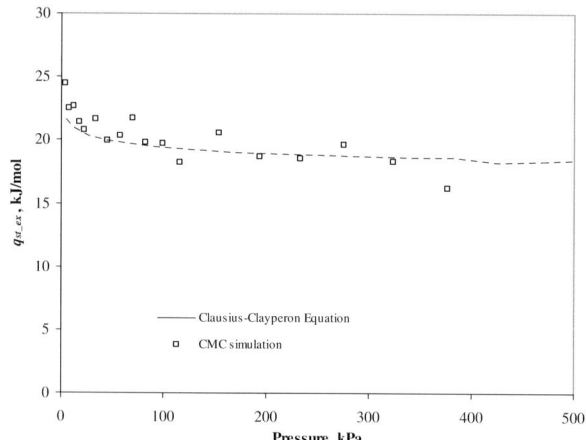

Figure 5. Excess isosteric heats of ethane in MCM-41 at 273 K.

## 6. CONCLUSIONS

A thermodynamic analysis of the Clausius-Clapeyron equation gives a method to convert the absolute isosteric heats obtained by simulation to the excess values. This method was applied to the analysis of adsorption in an activated carbon and in MCM-41, using a Monte Carlo simulation-based analysis. Excellent fits between experimental and simulation results were found for both adsorption systems.

## ACKNOWLEDGEMENT

The financial support of the U. K. Engineering and Physical Sciences Research Council and an ORS reward from the Universities UK are gratefully acknowledged.

## REFERENCES

1. D.P. Valenzuela and A.L. Myers, Adsorption Equilibrium Data Handbook, Prentice Hall, Englewood Cliffs, 1989.
2. T. Vuong and P. A. Monson, Langmuir,12 (1996) 5425.
3. S. Sircar, Ind. Eng. Chem. Res., 38 (1999) 3670.
4. J.O. Hirschfelder, C.F. Curtiss, and R.B. Bird, Molecular Theory of Gases and Liquids, John Wiley and Sons, Inc., New York, 1954.
5. J. Fischer, U. Heinbuch, and M. Wendland, Molecular Physics, 61(1987) 953..
6. W.A. Steele, The Interaction of Gases with Solid Surfaces, Pergamon Press, Oxford 1974.
7. Y. He and N.A. Seaton, Langmuir, 19 (2003) 10132.
8. M.P. Allen and D.J. Tildesley, Computer Simulation of Liquids, Clarendon, Oxford, 1987.
9. D. Frenkel and B. Smit, Understanding Molecular Simulations, Academic Press, San Diego, 1996
10. Y. He and N.A. Seaton, Langmuir, (2005) submitted.

Studies in Surface Science and Catalysis 160
P.L. Llewellyn, F. Rodriquez-Reinoso, J. Rouqerol and N. Seaton (Editors)
© 2007 Elsevier B.V. All rights reserved

# Pore size distribution in microporous carbons obtained from molecular modeling and density functional theory

**Vanessa Fierro, Gemma Bosch and Flor R. Siperstein**

Departament d'Enginyeria Química, Universitat Rovira i Virgili, Av. dels Països Catalans 26, Tarragona 43007, Spain

We present the pore size distribution (PSD) for activated carbons obtained from lignin pyrolysis and KOH activation. A comparison of the PSDs obtained from integral inversion of the adsorption isotherms generated using molecular simulation and density functional theory are compared in order to identify the differences observed when using different models to represent the adsorbate: single or two center Lennard-Jones models. Fair agreement is obtained when pores larger than 2 nm are present, despite the different models used. Only important differences are observed for porous carbons with pore sizes smaller than 1 nm.

## 1. INTRODUCTION

Gas adsorption characterization methods are commonly used to obtain information regarding the pore size distribution and surface properties of microporous materials. However, the interpretation of adsorption data may not be straightforward. Recently, several authors have made a critical comparison of the different methods available to determine the pore size distribution of microporous materials [1-4]. Among the mostly used methods to determine the PSD in microporous materials are the Dubinin-Astakhov (DA), Horvath-Kawazoe (HK) and methods based on the integral equation of the adsorption isotherm, which can be calculated using DFT or molecular simulation. Each of these methods has their own limitations and range of applicability, for example, HK and DFT based methods indicate the presence of micropores in nonporous carbons [1], with DFT performing better than HK. It has been suggested that what appears as a small amount of micropores in the PSD obtained from DFT, is the presence of surface heterogeneity in the sample, that is not considered in the DFT model.

Some authors have pointed out that DFT based methods commonly indicate that carbon based materials have a bimodal distribution of pores, where pores of an approximate width of 1nm are not present in the sample [2,4-6]. It has been suggested that this is due to artifacts in the model or in the method used for the inversion of the integral to obtain the PSD. It has been pointed out that surface energetic heterogeneity does not explain the appearance of this minimum in the PSD, which is due to packing effects [5]. In this work we examine whether the use of different models to describe the adsorbate, with different ways of packing, can explain the appearance of the minimum in the PSD.

To obtain the pore size distribution from inversion of the integral equation of an adsorption isotherm, it is necessary to generate a collection of isotherms with different pore sizes. For carbon based materials, the slit pore model with smooth graphite walls is generally adopted [7], and the experimental adsorption isotherm, $N_{exp}(P)$, is calculated as [8]:

$$N_{\exp}(P) = \sum_{H\,\min}^{H\,\max} N_s(P,H) f(H) dH \tag{1}$$

where $N_s(P,H)$ is the kernel of theoretical isotherms in model pores where excess adsorption isotherms should be used, and $f(H)$ is the pore size distribution. Although many of the shortcomings of such models have been reported in the literature, researchers are still bothered by the bimodal PSD of carbons, and the spurious minimum at approximately 1nm.

In this work, we compare the PSD of porous carbons synthesized through Kraft lignin pyrolysis, using inversion of the integral adsorption isotherm, where the integral kernel is obtained using DFT (DFT PLUS Micromeritics software) and molecular simulations. Kraft lignin is a macromolecule which is obtained in large quantities as byproduct in the conversion of wood chips to pulp for manufacturing paper. In this work, microporous carbons with a highly developed microporosity are obtained by activation of lignin with KOH, under different conditions (temperature, lignin/KOH ratio, and nitrogen flow). Due to the starting material used, and the temperature range for the pyrolysis (700-900°C), we do not expect a highly ordered material, where a bimodal PSD could be observed, but a smooth curve covering a relatively wide range of pore sizes is expected.

## 2. THEORETHICAL APPROACH

The inversion of the integral adsorption isotherm to obtain the PSD is the well-known so-called 'ill-posed' problem. Several approaches have been proposed to address this problem [3,7,8]. The DFT PLUS micromeritics software was used to determine the PSD using individual adsorption isotherms calculated by DFT. Details of this software are found elsewhere [9]. The use of this software requires the selection of the renormalization parameter, which in many cases is selected using the judgment of the researcher [4]. We used a low renormalization parameter for this work, which is comparable to what was used for the determination of the PSD from isotherms obtained using Monte Carlo simulations. Smoother curves can be obtained if a larger renormalization parameter is used.

Grand canonical Monte Carlo simulations were performed to calculate individual adsorption isotherms in carbon slit pores. The interactions potential, $U_{sf}$, of a fluid molecule at a distance $z$ from a single solid wall is described by the 10-4-3 Steele potential [10]:

$$U_{sf}(z) = 2\pi \rho_s \varepsilon_{sf} \sigma_{sf}^2 \Delta \left[ \frac{5}{2} \left( \frac{\sigma_{sf}}{z} \right)^{10} - \left( \frac{\sigma_{sf}}{z} \right)^4 - \frac{\sigma_{sf}^4}{3\Delta(0.61\Delta + z)^3} \right] \tag{2}$$

where $\rho_s$=114 nm$^{-3}$ is the density of graphite, and D=0.335 nm is the interlayer spacing in graphite. The energetic and scale parameters of the Lennard-Jones potentials, $\varepsilon_{sf}$ and $\sigma_{sf}$, are obtained using Lorentz-Berthelot mixing rules, where $\varepsilon_{ss}/k_B = 28$ K and $\sigma_{ss} = 0.34$ nm. The potential field in a micropore, where the centers of the carbon atoms are separated at a distance $H$, includes the contributions from two opposite walls:

$$U_{sf,\,pore}(z) = U_{sf}(z) + U_{sf}(H - z) \tag{3}$$

The pore width, $w$, can be calculated by substracting $\sigma_{ss}$ from the distance between the centers of carbon atoms in opposite walls of the pore, $H$.

In this work, nitrogen was described as a two-centre Lennard Jones molecule, with interaction parameters that describe well the nitrogen vapor-liquid equilibrium properties [11]. The interaction between two nitrogen molecules, $U_{2CLJQ}$, with orientations $\omega_i$ and $\omega_j$, that are separated at a distance $r_{ij}$, is given by the sum of the centre-to-centre Lennard-Jones interaction, $U_{2CLJ}$, and the quadrupole-quadrupole interactions, $U_Q$ :

$$U_{2CLJQ}\left(r_{ij},\omega_i,\omega_j,l,Q\right) = U_{2CLJ}\left(r_{ij},\omega_i,\omega_j,l\right) + U_Q\left(r_{ij},\omega_i,\omega_j,Q\right) \qquad (4)$$

where $l$ is the distance between each of the centers in a nitrogen molecule, and $U_{2CLJ}$ and $U_Q$ are given by:

$$U_{2CLJ}\left(r_{ij},\omega_i,\omega_j,l\right) = 4\varepsilon \sum_{a=1}^{2}\sum_{b=2}^{2}\left[\left(\frac{\sigma}{r_{ab}}\right)^{12} - \left(\frac{\sigma}{r_{ab}}\right)^{6}\right] \qquad (5)$$

$$U_Q\left(r_{ij},\omega_i,\omega_j,Q\right) = \frac{3}{4}\frac{Q^2}{4\pi\varepsilon_0\left|r_{ij}\right|^5}\left[1 - 5\left(c_i^2 + c_j^2\right) - 15c_i^2c_j^2 + 2\left(c - 5c_ic_j\right)^2\right] \qquad (6)$$

where $r_{ab}$ is the distance between the centers $a$ and $b$ of different nitrogen molecules; $c_i = \cos\theta_i$, and $c_j = \cos\theta_j$. The angles between the axis of the molecule and the center-to-center connection line are $\theta_i$ and $\theta_j$, and the difference in azimuthal angles of the molecules $i$ and $j$ is $\phi_{ij}$.

The Lennard-Jones interaction parameters for nitrogen are $\sigma = 0.33211$ nm and $\varepsilon/k_B = 34.897$ K. The two sites in the nitrogen molecule are separated at a distance $l = 0.10464$ nm, and the point quadrupole is Q=1.4397 DÅ.

## 3. EXPERIMENTAL METHOD

Kraft lignin (KL) was provided by Lignotech Iberica S.A. (Spain), and was presented in the form of a fine dark brown powder. The removal of the inorganic matter from KL was achieved as follows: batches of 100 g were introduced in 2 l of water, leading to black suspensions of pH 9.5, and lignin was precipitated by adding $H_2SO_4$ until the pH decreased to 1. The precipitate was gently washed with distilled water until the pH of the rinse was constant, and finally dried overnight at 105 °C. The lignin prepared this way was nearly mineral-free and was termed demineralised Kraft lignin (KL$_d$).

KOH lentils (Scharlau) were ground and physically mixed with KL$_d$ according to various KOH/ KL$_d$ mass ratios ($R$ = 1:1, 3:1, or 4.3:1). The carbonisation was carried out in a horizontal furnace and the samples were heated at a rate of 5°C/min, from room temperature up to the final carbonisation temperature ($T_{carb}$= 700, 800 or 900 °C) in different nitrogen flows ($f_{N2}$ = 200, or 800 ml/min). Samples were kept under nitrogen at the final temperature for one hour before cooling down.

During the experiments, both metallic potassium (produced by the reduction of KOH by carbon at high temperature) and KOH were partly transported in the vapour phase, and could be observed at the outlet of the reactor. Metallic potassium mixed with potassium carbonate was also present inside the crucible; therefore the latter was submitted to atmospheric humidity for two days, during which the alkaline metal slowly oxidised. Finally, the activated carbon was washed with extreme care, first with 1M HCl, and finally with distilled water until the pH of the rinse remains constant and close to 6. After drying in an oven during 24h, a very light activated carbon was obtained.

## 4. RESULTS

Measured adsorption isotherms are shown in Figs. 1-3. In all cases, the first column shows the actual capacity of the isotherms, while the second column shows the fractional coverage as a function of pressure. By inspection of the isotherms, we can say that increasing the pyrolysis temperature will increase the amount of "large" pores, as saturation capacity is reached at higher pressures. On the other hand, modifying the nitrogen flow rate has little influence in the pore size distribution, but it modifies the total pore volume. Finally, modifying the lignin to KOH ratio seems to have a strong influence in the pore size distribution up to a certain limit, because only very small pores are observed for low KOH/lignin ratios, but the pore size distribution is expected to be very similar for KOH/lignin ratios of 3 and 4.3.

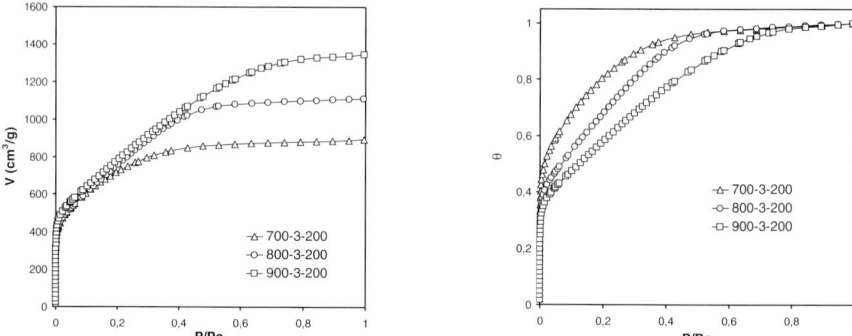

Fig 1. Adsorption isotherms of nitrogen at 77K on different carbons obtained from lignin pyrolysis at 700, 800, and 900°C, with a KOH/lignin mass ratio of 3, and nitrogen flowrate of 200 ml/min.

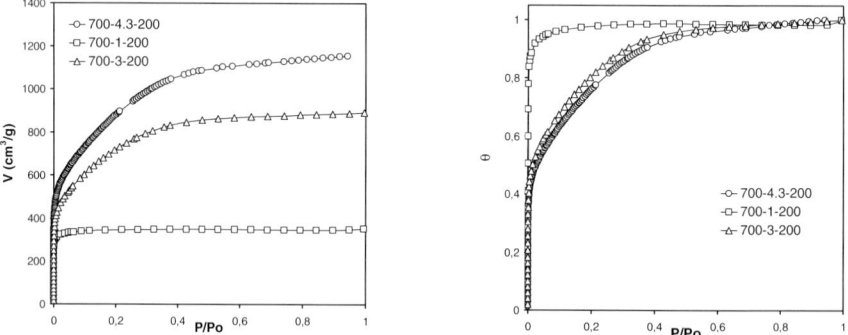

Fig 2. Adsorption isotherms of nitrogen at 77K on different carbons obtained from lignin pyrolysis at 700°C, with a KOH/lignin mass ratio between 1 and 4.3, and nitrogen flowrate of 200 ml/min.

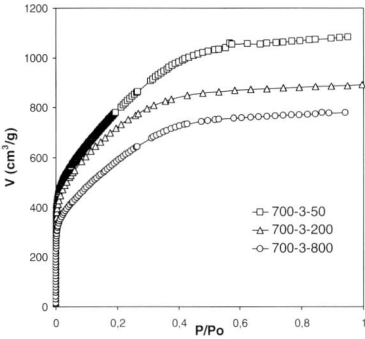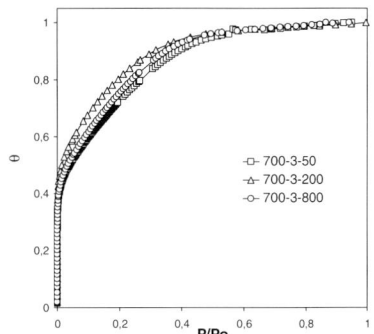

Fig 3. Adsorption isotherms of nitrogen at 77K on different carbons obtained from lignin pyrolysis at 700°C, with a KOH/lignin mass ratio of 3, and nitrogen flowrate between 50 and 800 ml/min.

Calculated pore size distributions using DFT and GCMC simulations to determine the kernel of the integral are shown in Fig. 4-6.

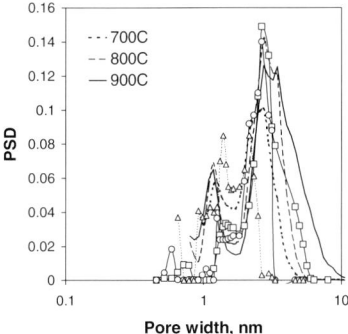

Fig 4. Pore size distributions obtained from DFT (lines) and GCMC simulations (symbols) for carbons pyrolyzed at different temperatures: 700°C (dotted line and triangles), 800°C (dashed line and circles), and 900°C (solid line and squares)

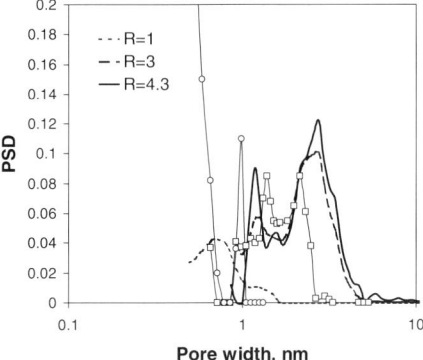

Fig 5. Pore size distributions obtained from DFT (lines) and GCMC simulations (symbols) for activated carbons with different KOH/lignin ratios: 1 (dotted line and circles), 3 (dashed line and squares), and 4.3 (solid line). PSD for R=4.3 using GCMC is not shown.

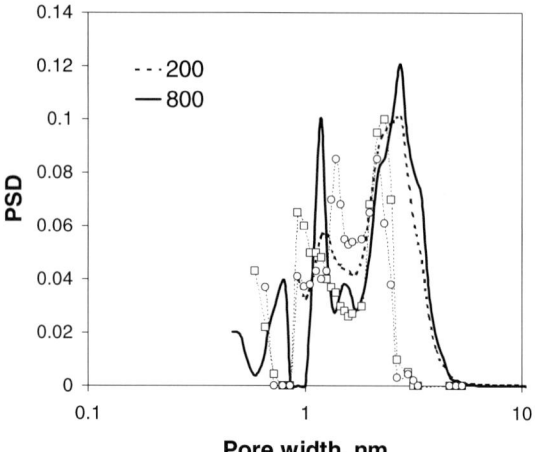

Fig 6. Pore size distributions obtained from DFT (lines) and GCMC simulations (symbols) for activated carbons at 700oC under different nitrogen flow rates: 200 ml/min (dotted line and circles) and 800 ml/min (solid line and squares).

Considering that different models are used to describe the adsorbate, namely, singles spherical particles in the DFT calculations, and two center Lennard-Jones molecules with point quadrupole in the simulations, it is interesting that both determinations show similar distributions. Probably the most important difference is observed for the carbon obtained with a low KOH/lignin ratio, which has a small total pore volume and only small pores are present. In this case, the PSD obtained from simulations predicts pores of considerably smaller size than those obtained from DFT. This may be closely related to the different models used in both methods. It must be pointed out that even when the two-center Lennard Jones with point quadurlpole model is a more realistic representation of the nitrogen molecule, in both cases the carbon is considered as a smooth wall. For very small pores, as the ones where differences are observed with the two methods, it is possible that the smooth wall approximation is not valid.

Inspection of the PSDs obtain from GCMC adsorption isotherms, it is evident that they show that there are no pores present with approximately 0.85 nm of pore width. Inspection of the structures obtained from simulations for systems of pore width, $w$, slightly smaller than 1 nm indicate that the transition from two adsorbed layers to three adsorbed layers occurs at approximately this pore width. Figure 7 shows a collection of snapshots of adsorbed systems at $P/Po=0.8$. The fact that no pores of these sizes appear in most porous carbons, is probably closely related to the particular shape of the individual adsorption isotherms when the pore is "too large" to accommodate only 2 layers of adsorbed molecules, but too small to have a third layer. Probably models that include pore-connectivity or that take into account smooth variations in size of a given pore, will not show this effect.

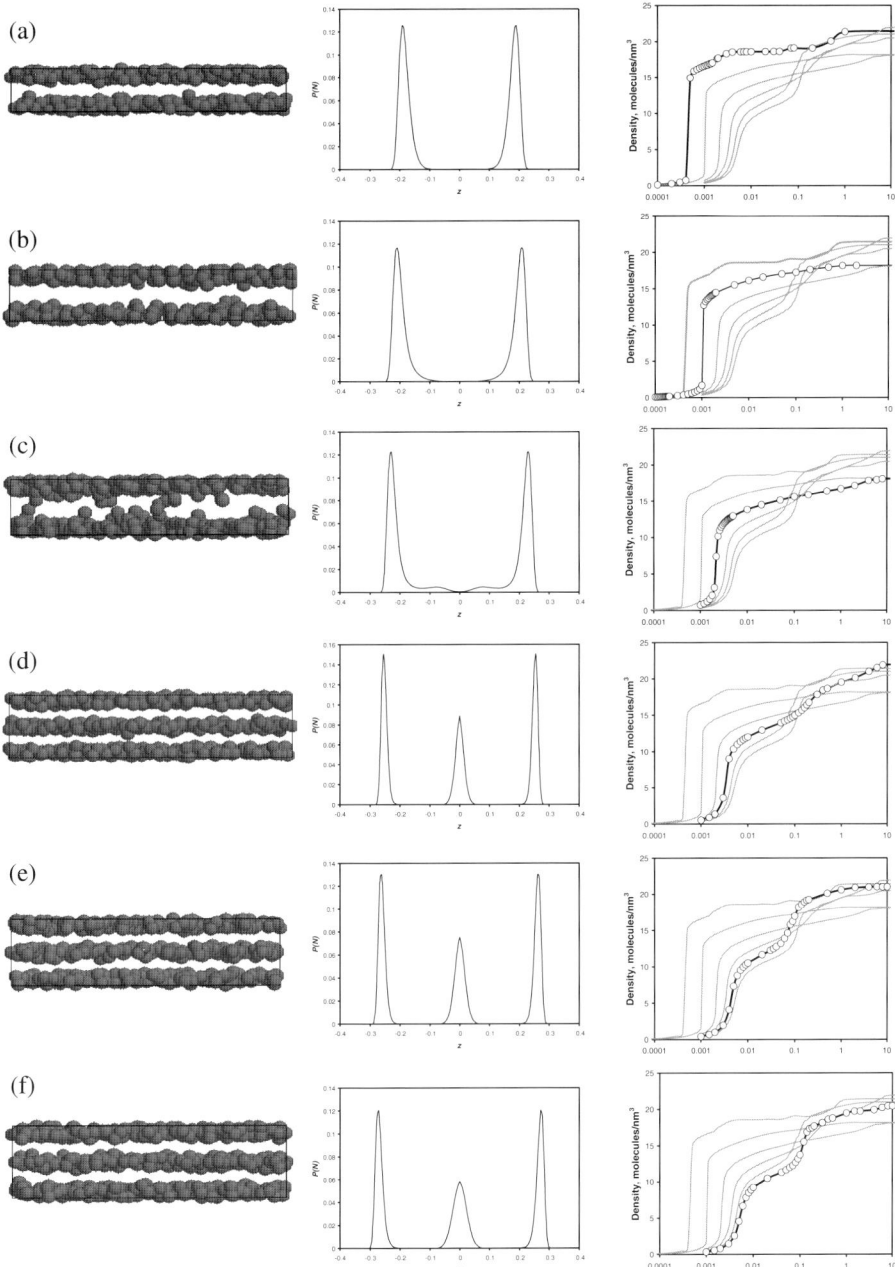

Fig. 7 Nitrogen adsorption snapshots from GCMC simulations, probability density distribution along a perpendicular axis to the solid wall at P/Po=0.8, and adsorption isotherms at 77 K for different pore sizes: (a) $w = 0.73$ nm, (b) $w = 0.79$ nm, (c) $w = 0.86$ nm, (d) $w = 0.93$ nm, (e) $w = 0.99$ nm, (f) $w = 1.06$ nm. The adsorption isotherm where circles are shown corresponds to the specific pore size.

## 5. CONCLUSIONS

In this work we show that the pore size distributions of porous carbons, obtained from lignin pyrolysis and KOH activation, determined from molecular simulation and DFT are equivalent. We suggest that the spurious minimum observed in the PSD of carbons obtained form inversion of the integral adsorption isotherms is due to artifacts of the model, such as the fact that, uniform size, infinite non-connected pores are considered. In such cases, for the pore width where transition from two to three layers occurs, the difference in shape of the calculated adsorption isotherms results in an exclusion of some pore widths at the time of the inversion of the integral. Taking into account smooth changes between pore sizes, or pore connectivity may solve this problem.

## ACKNOWLEDGMENTS

This work was supported by grant CTQ-2004-03346/PPQ from the Spanish Ministerio de Educación y Ciencia, by grant ACI2003-26 from the Generalitat de Catalunya, and the Lignocarb ALFA II 0412 FA FI. VF and FRS are grateful to the RyC program of the Spanish Ministerio de Educacion y Ciencia.

## REFERENCES

[1] M. Kruk, M. Jaroniec and J. Choma, Carbon 36 (1998) 1447.
[2] S.A. Korili and A. Gil , Adsorption 7 (2001) 249.
[3] P. Kowalczyk, A.P. Terzyk, P.A. Gauden, R. Leboda, E. Szmechtig/Gauden, G. Rychlicki, Z. Ryu, H. Rong, Carbon 41 (2003) 1113.
[4] P.A. Gauden, A.P. Terzyk, G. Rychlicki, P. Kowalczyk, M.S. Cwiertnia, and J.K. Garbacz, J. Colloid and Interface Science 273 (2004) 39.
[5] J.P. Olivier, Carbon 36 (1998) 1469.
[6] P. Kowalczyk, E.A. Ustinov, A. P. Terzyk, P.A. Gauden, K. Kaneko, G. Rychlicki, Carbon 42 (2004) 851.
[7] P.I. Ravikovitch, A. Vishnyakov, R. Russo, A.V. Neimark, Langmuir 16 (2000) 2311.
[8] N.A. Seaton, J.P.R.B. Walton, N. Quirke, Carbon 27 (1989) 853.
[9] Micromeritics DFT Plus Operator's Manual, 2003
[10] W.A. Steele, The Interactions of Gases with Solid Surfaces, Pergamon, Oxford, 1974.
[11] J. Vrabec, J. Stoll, H.. Hasse J. Phys. Chem B 105 (2001) 12126.

Studies in Surface Science and Catalysis 160
P.L. Llewellyn, F. Rodriquez-Reinoso, J. Rouqerol and N. Seaton (Editors)

# Modeling Triblock Surfactant Templated Mesoporous Silicas (MCF and SBA-15): A Mimetic Simulation Study

**Supriyo Bhattacharya[a], Benoit Coasne[b], Francisco R. Hung[a] and Keith E. Gubbins[a]**

[a] Center for High Performance Simulation and Department of Chemical and Biomolecular Engineering, North Carolina State University, Raleigh, NC 27695-7905, USA.

[b] Laboratoire de Physicochimie de la Matière Condensée (UMR CNRS 5617), Université Montpellier II, Place Eugène Bataillon, 34095 Montpellier, Cedex 5, France.

We have developed models for templated mesoporous silicas such as Mesostructured Cellular Foams and SBA-15. The first part of our work elaborates the effect of oil concentration on the pore morphology of the triblock surfactant templated mesoporous materials. Our Lattice Monte Carlo simulations mimic the synthesis process by equilibrating a mixture of triblock surfactant, oil, water and silica at a constant temperature and density. With increasing oil concentration, we find the pore geometry to change according to the sequence: cylinders → lamellae → mesocells, which is in qualitative agreement with experimental results. In the second part of our work, we develop realistic atomistic models of the SBA-15 material, starting from the mesoscale model obtained from Lattice Monte Carlo simulations. Both the pore surface heterogeneity and the micropores are derived from the mimetic simulations. The simulated TEM and pore size distribution of the model qualitatively resemble the real material.

## 2. INTRODUCTION

Templated mesoporous materials (TMMs) [1-3] are synthesized by polymerizing silica or an inorganic metal oxide around a surfactant liquid crystal mesophase and finally calcining the resulting precipitate. Detailed synthesis method depends on the material of interest [4,5]. It is possible to tune the pore size and geometry in these materials by varying the synthesis conditions such as temperature and surfactant/solvent concentrations. In this paper, we developed models of two very closely related mesoporous silicas, MCF [6,7] (Mesostructured Cellular Foams) and SBA-15 [8]. Both of these substances are synthesized using a nonionic triblock surfactant ($PEO_{20}PPO_{70}PEO_{20}$; PEO: polyethylene, PPO: polypropylene) as the template in the presence of a hydrophobic solvent (oil). The oil modifies the geometry of the liquid crystals producing small (5-30 nm) cylindrical pores (SBA-15) at low oil concentrations and very large (20-50 nm) spherical pores (MCF) at high oil concentrations. Fig. 1 shows schematics of both MCF and SBA-15. The spherical pores of MCF are connected to each other via windows. The large size of the MCF pores is suitable for adsorption and catalytic applications involving proteins and polymers. The structure of SBA-15 consists of straight cylindrical pores interconnected by micropores of diameter < 1 nm [9-11]. These micropores increase the surface roughness and provide more surface area for adsorption. The effect of micropores on the adsorption properties of SBA-15 has been extensively studied in the literature [9-13].

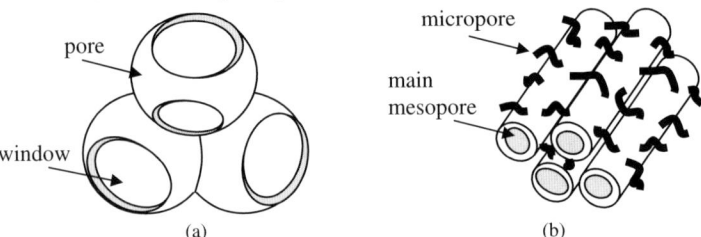

Fig. 1. Schematic diagrams of (a) MCF and (b) SBA-15

　　While experiments provide valuable information regarding the synthesis of TMMs, simulations are indispensable for understanding the complex synthesis process at the molecular level. Large system sizes and slow dynamics make it difficult to simulate the synthesis of templated mesoporous materials using atomistic simulations. The advantages of using lattice models for studying surfactant-inorganic oxide-solvent systems have been discussed elsewhere [14,15]. Lattice Monte Carlo simulations have been used extensively to study the properties of surfactant solutions [16-19] and the synthesis of mesoporous materials [14,15]. The mimetic lattice models for MCF and SBA-15 presented here were developed based on the approach by Siperstein and Gubbins. [15]. Mesoscale material models provide valuable information regarding the pore geometry and the thermodynamic principles behind synthesis. However, an atomistic pore model is necessary in order to understand the various intra-pore phenomena, such as phase transitions and chemical reactions in confinement. Several approaches have been used in modeling atomistic pores in order to study confinement phenomena [20,21]. Coasne *et al.* [21,22] developed realistic models of MCM-41 materials, starting from a mesoscale lattice model. Partly based on this approach, we develop an atomistic model for SBA-15.

## 3. SIMULATION DETAILS

In our Lattice Monte Carlo simulations, the surfactant is modeled as a chain of five hydrophobic tail units in the middle with three hydrophilic head units on both sides of the tail ($H_3T_5H_3$). Oil, water and silica are modeled as individual lattice units. Comparison between lengths of the model and the real surfactants yields a lattice unit length of 3.6 nm, which serves as a guideline for mapping the simulation and the experimental length-scales. The inter-particle interactions are modeled as square-well potentials keeping in mind the following criteria, which are analogous to the experimental situation: heads attract water, tails attract oil and the silica-head attraction is stronger than the water-head attraction. The system is equilibrated using Monte Carlo moves such as reptation, twist and chain regrowth using configurational bias. The detailed methodology has been discussed elsewhere [23]. From the equilibrated system, the mesoscale pore model is obtained by removing the surfactants, oil and water from the system.

　　In Fig. 2, we summarize the detailed steps involved in developing the atomistic pore model from the lattice model. From the lattice configuration (Fig. 2a), we isolate a pore and model its surface as a b-spline [24]. The b-spline approach serves as a reasonable interpolation to obtain a pore surface with sub-atomistic resolution < 0.1 nm. The control points in this b-spline model are the lattice surface coordinates, which allows us to retain the surface undulations and irregularities from the mimetic simulation. The atomistic pore is now generated by carving out of a cristobalite (silica) block the pore defined by the b-spline surface. Once the mesopores have been created, the micropores are carved out using the locii of the surfactant chains in the lattice configuration. The micropores appear due to surfactant chains embedded in the silica matrix. Therefore this procedure ensures that we have a realistic distribution of micropores in the model pore. The diameters of these micropores are allowed to vary between 0.6 and 1 nm and are selected randomly from a Gaussian distribution with a mean value at 0.8 nm. According to experimental observations [11], most of the micropores in SBA-15 are less than 1 nm in diameter.

After carving out the meso and the micropores, the silicon atoms which are in an incomplete tetrahedral environment are first removed. The oxygen atoms with two dangling bonds are next removed and those with one dangling bond are saturated with hydrogen atoms. This procedure ensures that the simulation box is electroneutral and that none of the silicon and oxygen atoms has dangling bonds. Finally, the pore surface was relaxed by randomly displacing all the atoms in the system by a small distance in order to create an amorphous structure. This procedure to generate atomistic silica pore surfaces in a realistic way was originally proposed by Pellenq and Levitz to model Vycor glasses [25]. To summarize, surface heterogeneities in the order of 4 nm or above in our pore model are directly obtained from the mimetic simulations. Irregularities smaller than this length-scale are incorporated by removing Si and O atoms from the pore surface and saturating the O dangling bonds with H atoms. This step can be compared to the calcination process in the experimental synthesis, where existing silicate bonds are broken and new bonds formed as the pore structure is rearranged.

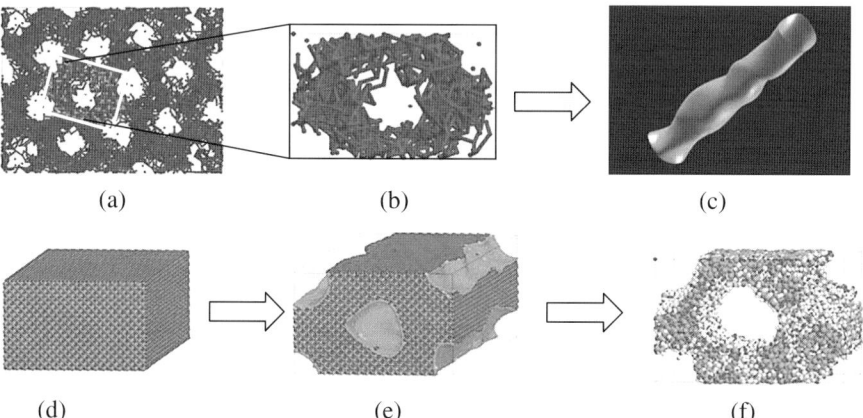

Fig. 2. Steps in developing the atomistic pore from the mesoscale model. (a) snapshot of lattice simulation box; (b) isolated pore; (c) pore surface modeled as b-spline; (d) silica block; (e) mesopores carved out; (f) model material after carving out the micropores and pore surface relaxation. White particles are the hydrogen atoms on the pore surface.

## 4. RESULTS AND DISCUSSION

In the first part of this section, we discuss the effect of oil concentration on the self-assembled structures. In the second part, we analyze the physical properties of the atomistic SBA-15 pore, such as the simulated TEM, pore size distribution, porosity and surface area.

## 4.1 Lattice Monte Carlo simulation of surfactant-oil-water-silica systems

The ternary phase diagram for the $H_3T_5H_3$-oil-water system is shown in Fig. 3a. Spherical micelles are observed at low oil and surfactant concentrations. Elongated micelles and hexagonal arrangements of cylindrical micelles are observed at medium surfactant concentrations, near the surfactant-water side of the phase diagram. Starting from the hexagonal phase on the surfactant-water side, we find the emergence of a lamellar phase as we move towards high oil concentrations. In our simulations, bicontinuous structures appeared only in one point on the phase diagram, although we believe that additional points would be found by doing extensive simulations in this zone of the phase diagram. It must be noted that the transitions between the different structures are not first order in nature. Phase separation is observed on the surfactant-oil side between a dilute surfactant solution and a surfactant rich phase. The characteristics of the phase diagram are highly dependent on the interactions of the oil with the other components in the system.

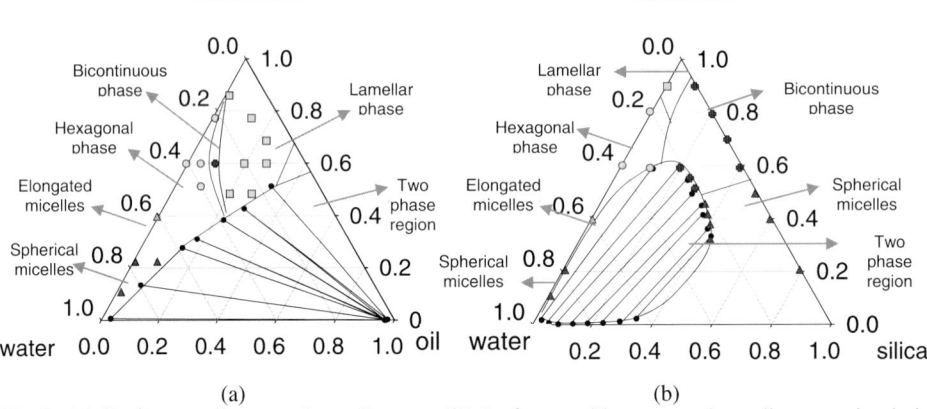

(a)                                                                                (b)

Fig. 3. (a) Surfactant-oil-water phase diagram; (b) Surfactant-silica-water phase diagram; simulation points are shown as follows: (▲) spherical micelles, (△) elongated micelles, (○) cylindrical micelles, (●) bicontinuous phase, (□) lamellar phase. Phase boundary lines are provided as a guide to the eye.

The surfactant-water-silica phase diagram is shown in Fig. 3b. Due to the stronger head-silica attraction as compared to the water-head attraction, we observe phase separation between the surfactant-rich, silica-rich and water-rich phases. Ordered phases that are observed at different regions of the phase diagram depend on the component concentrations; spherical micelles are found at low to medium surfactant concentrations, bicontinuous structures are observed at high surfactant and intermediate silica concentrations, whereas cylindrical structures are seen at high surfactant and high water concentrations.

We now investigate the self-assembly behavior by introducing oil to the surfactant-water-silica solution. The effect of oil on the surfactant-silica liquid crystal phase has been highlighted in the literature [6]. We mimicked the experimental pathway by starting from zero oil concentration on the surfactant-water-silica phase diagram. We selected this initial point in the region of the hexagonal phase because it has been reported that the silica structure is hexagonal in absence of oil [6]. The oil concentration is now raised in small increments, keeping the proportion of the other components constant. In Fig. 4, we show simulation snapshots of three different structures with increasing oil concentration. This also shows the nature of the pore structure after removal of the surfactants and solvents. At low oil concentrations, the structure is cylindrical (similar to SBA-15) and becomes lamellar as the oil level is increased. As the oil concentration is increased further, the lamellar structure transforms into spherical mesocells resembling the MCF structure, with a pore diameter almost 2.5 times that of the cylinders and the lamellae. Fig. 5 shows the experimental results from Lettow *et al.* [6]. In both the simulations and the experiments, cylindrical structures are observed at low oil levels and these structures transform into mesocellular foams once the oil concentration is increased. However, the transition from cylinders to mesocells follows separate pathways in the simulations and the experiments. As proposed by Lettow *et al.* [6], undulations occur on the surface of the cylinders prior to the formation of the mesocells. In our simulations, we were unable to find any such structure. Moreover, lamellar structures were not found in the experiments. This behavior suggests that there may be other structure-directing forces in the medium oil concentration regime, which are not being considered in the simulations. It is also possible that the lamellar structures occurred within a very small region in the experimental phase diagram and were thus undetected.

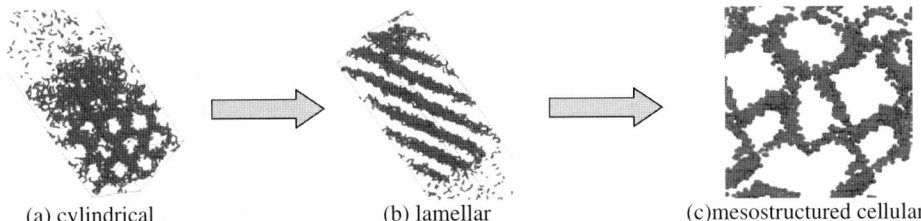

| (a) cylindrical | (b) lamellar | (c)mesostructured cellular |

Fig. 4. Simulation: Change in pore structure with increasing oil concentration (by volume). (a) 2%, (b) 9%, (c) 23%. In Fig. 4c, two periodic images are shown alongside for better visualization. All three snapshots are of the same scale. For clarity, only the silica particles are shown.

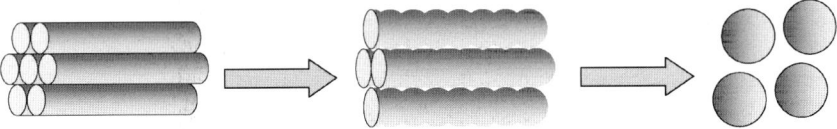

| (a) cylindrical | (b) undulating cylinder | (c)mesostructured cellular |

Fig. 5. Experiment [6]: Change in pore structure with increasing oil / surfactant mass ratios. (a). 0.00, (b) 0.21, (c) 0.50.

## 4.2 Physical properties of the pore model

We compare the pore size distributions of the generated model materials with those of the real materials. The method we used to calculate the pore size distribution has been discussed in the literature [20]. Fig. 6a shows the pore size distribution (PSD) for the model SBA-15. The pore size ranges from 7 nm to 34 nm with the most prominent peak at about 22 nm. Experimentally [6], the pore size of SBA-15 ranges from 5 nm to 30 nm, which is in good agreement with our simulations. Fig. 6b shows the pore size distribution for the model MCF. Two distinct peaks are observed, one at 30 nm, the other one at 87 nm. The peak at smaller pore size is attributed to the window diameter, while the larger peak is due to the mesocell diameter. Experimentally [6,7], the pore sizes of the cells and the windows can be computed from nitrogen or argon adsorption by assuming a cellular model as shown in Fig. 1a. The cell size is obtained from the adsorption branch of the isotherm, while the desorption branch gives the window size. The experimental PSD for a MCF is shown in Fig. 6c. We find the pore sizes in the simulations to be much larger than the experimental ones. This can be due to inaccuracies in mapping the lattice parameters with the experimental dimensions. However, the relative increase in pore diameter from cylinders to mesocells is the same in both simulation and experiment. The average mesocell diameter in the simulations is 87 nm, which is 4 times that of the cylinders (22 nm), while the average MCF mesocell diameter in the experiments is 42 nm, which is also 4 times that of SBA-15 (≈10 nm).

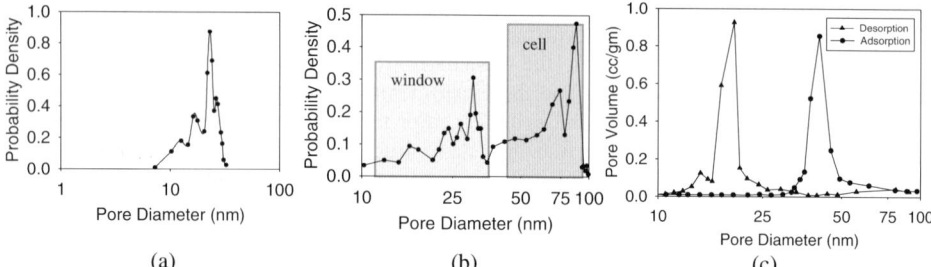

(a)                              (b)                              (c)

Fig. 6. Pore size distributions of two different Lattice Monte Carlo-generated structures. (a) cylindrical SBA-15, (b) spherical MCF, (c) Pore size distribution of MCF material obtained by nitrogen adsorption [7].

In Fig. 7a, we show high-resolution TEM for the mesoscale model of SBA-15. TEM was computed using the software package CERIUSII®. Regions of high electronic densities are denoted by various shades of grey, while the dark regions represent the pore interior. From the electron density map, it is observed that the pore walls are comprised of regions with high and low silica densities. This is further clarified in Fig. 7b, which is the enlarged version of a single pore. The region close to the pore surface has a lower wall density compared to regions away from the pore. The density fluctuations are also visible in the experimental TEM [10] of a real SBA-15 material (Fig. 7c). These density variations occur due to the numerous micro-cavities that are present in the pore walls. These cavities were previously occupied by the surfactant chains prior to their removal. The microchannels close to the pore surface are more closely packed, because the surfactant concentration is higher near the micellar center.

All the results above were for the mesoscale models of MCF and SBA-15. We now discuss the atomistic SBA-15 pore model. Fig. 8a shows the pore size distribution of the atomistic SBA-15 pore. The smaller peak at approximately 1 nm is due to the micropores, while the higher peak at 5 nm is due to the mesopores. Fig. 8b shows the experimental PSD of a SBA-15 sample with a mesopore diameter of 5.8 nm [9]. The median micropore diameter is at 1 nm in both of the plots. However, the real SBA-15 shows a wider distribution of micropore sizes compared to the model material. In the numerical model, we set the micropore diameter to vary within a small interval (between 0.6 nm and 1.0 nm), which results in a sharp micropore size distribution. In the real material, the micropore size distribution is dependent on the temperature and duration of annealing, although the relationship between the anealing conditions and the micropore size variation is poorly understood. The model mesopore diameter has a wider variation compared to the experimental material. The probable reason is, the experimental PSD measures the distribution of pore diameters averaged over a very large number of pores, whereas in the model, we measure the distribution for a single pore over a length of 15 nm. This also results in the small undulations adjacent to the main mesopore peak, which arise due to the lack of a sufficiently big statistical sample.

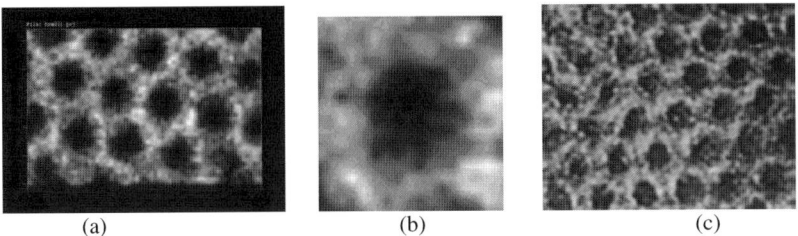

(a)        (b)        (c)

Fig. 7. TEM of (a) model SBA-15; (b) single pore cropped from (a); (c) experimental SBA-15. Reprinted in part with permission from [10]. © 2000 American Chemical Society.

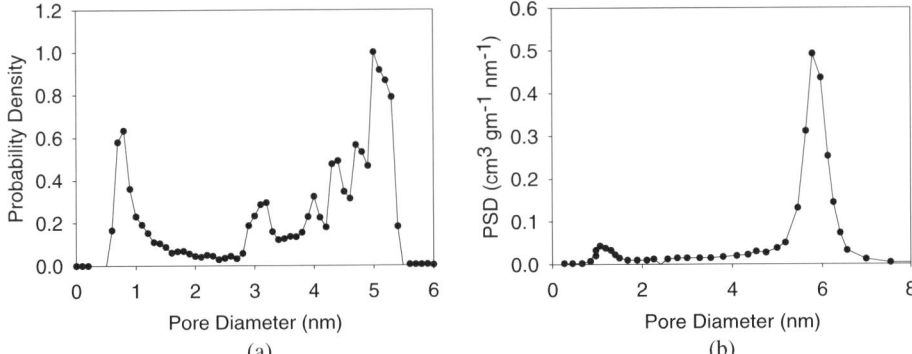

(a)             (b)

Fig. 8. Pore size distributions of (a) the atomistic SBA-15 model, (b) experimental SBA-15 [9]

The statistical properties of the atomistic pore model SBA-15 are compared with those of the real material in Table 1. These properties have been computed using the methodology described by Gelb and Gubbins [20]. The surface areas and pore volumes in both the pore model and the experimental material are for argon adsorption. Due to the presence of micropores, the SBA-15 pores have a significantly higher pore volume and surface area compared to MCM-41 pores. The parameters for our model show good agreement with the experimental values. However, the surface area of the pore model is higher than that of the experimental material. We note that the surface area in experiments may be slightly lower if some pore regions are unexplored by the adsorbate due to the bottleneck effect. The parameter wS/V, where w is the mesopore diameter, S the surface area, and V the pore volume provides the deviation of the pore geometry from a regular cylinder; wS/V is 4.0 for an open cylinder. Although the main mesopores are cylindrical, the value of wS/V for SBA-15 ranges from 5 to 10, which suggests that a significant part of its porosity is due to the micropores.

Table 1
Physical properties of experimental and atomistic SBA-15 Pores

|  | Experiment [13] | Model |
|---|---|---|
| Pore Volume | 1 cm$^3$/gm | 0.862 cm$^3$/gm |
| Mesopore Volume | 70% | 73% |
| Micropore Volume | 30% | 26% |
| Pore Surface Area | 600-900 m$^2$/gm | 1268 m$^2$/gm |
| wS/V (see text) | 5.8-10.8 | 7.65 |

## 5. CONCLUSIONS

Lattice Monte Carlo simulations were used to mimic the synthesis of surfactant templated mesoporous materials. We predicted some important features of surfactant-oil-water-silica systems, which qualitatively agree with the experiments. The simulations suggest that the pore geometry changes with increasing oil concentration in the sequence: cylinders → lamellae → mesocells. Upon the cylinder to mesocell transition, the pore diameter increases 3-4 times. The simulated TEM and pore size distributions of the model materials show qualitative agreement with experiment. Starting from the mesoscale model, we also developed atomistic models of the SBA-15 pore. The properties of this atomistic model are in agreement with those extracted from adsorption experiments. Simulations of argon adsorption in the model SBA-15 pore are currently in progress.

We would like to thank Henry Bock, Flor Siperstein and Lauriane Scanu for many helpful discussions. This work was funded by Department of Energy (DOE) under grant no. DE-FG02-98ER14847. Supercomputing time was provided by San Diego Supercomputer Center under grant NSF/MRAC CHE050047S. CERIUSII® software package and other computing resources were provided by Research Computing Services, UNC-Chapel Hill.

## REFERENCES

[1] V. Chiola, J. E. Ritsko and C. D. Vanderpool, US Patent No. 3 556 725 (1971).
[2] T. Yanagisawa, T. Shmizu, K. Kuroda and C. Kato, Bull. Chem. Soc. Jpn., 62 (1990) 1535.
[3] C. T. Kresge, M. E. Leonowicz, W. J. Roth, J. C. Vartuli and J. S. Beck, Nature 359 (1992) 710.
[4] G. J. de A. A. Soler-Illia, C. Sanchez, B. Lebeau and J. Patarin, Chem. Rev., 102 (2002) 4093.
[5] J. Y. Ying, C. P. Mehnert and M. S. Wong, Angew. Chem., Int. Ed., 38 (1999) 56.
[6] J. S. Lettow, Y. J. Han, P. Schmidt-Winkel, P. Yang, D. Zhao, G. D. Stucky and J. Y. Ying, Langmuir, 16 (2000) 8291.
[7] P. Schmidt-Winkel, W. W. Lukens Jr., P. Yang, D. I. Margolese, J. S. Lettow, J. Y. Ying and G. D. Stucky, Chem. Mater., 12 (2000) 686.
[8] D. Zhao, J. Feng, Q. Huo, N. Melosh, G. H. Fredrickson, B. F. Chmelka and G. D. Stucky, Science, 279 (1998) 548.
[9] R. Ryoo, C. H. Ko, M. Kruk, V. Antochshuk and M. Jaroniec, J. Phys. Chem. B, 104 (2000) 11465.
[10] M. Impéror-Clerc, P. Davidson and A. Davidson, J. Am. Chem. Soc., 122 (2000) 11925.
[11] S. H. Joo, R. Ryoo, M. Kruk and M. Jaroniec, J. Phys. Chem. B, 106 (2002) 4640.
[12] A. Galarneau, H. Cambon, F. Di Renzo and F. Fajula, Langmuir, 17 (2001) 8328.
[13] A. Galarneau, H. Cambon, F. Di Renzo, R. Ryoo, M. Choi and F. Fajula, New J. Chem., 27 (2003) 73.
[14] F. R. Siperstein and K. E. Gubbins, Mol. Simul., 27 (2001) 339.
[15] F. R. Siperstein and K. E. Gubbins, Langmuir, 19 (2003) 2049.
[16] R. G. Larson, L. E. Scriven and H. T. Davis, J. Chem. Phys., 83 (1985) 2411.
[17] R. G. Larson, J. Phys. II France, 6 (1996) 1441.
[18] A. Z. Panagiotopoulos, M. A. Floriano and S. K. Kumar, Langmuir, 18 (2002) 2940.
[19] S. Y. Kim, A. Z. Panagiotopoulos and M. A. Floriano, Mol. Phys., 14 (2002) 2213.
[20] L. D. Gelb and K. E. Gubbins, Langmuir, 14 (1998) 2097.
[21] B. Coasne, F. R. Hung, R. J.-M Pellenq, F. R. Siperstein and K. E. Gubbins, Langmuir, submitted (2005).
[22] B. Coasne, F. R. Hung, F. R. Siperstein and K. E. Gubbins, Ann. Chim-Sci. Mat., in press (2005).
[23] S. Bhattacharya and K. E. Gubbins, J. Chem. Phys., accepted (2005).
[24] L. Piegl and W. Tiller, The NURBS Book, Springer, 1995.
[25] R. J.-M. Pellenq and P. E. Levitz, Mol. Phys., 100 (2002) 2059.

Studies in Surface Science and Catalysis 160
P.L. Llewellyn, F. Rodriquez-Reinoso, J. Rouqerol and N. Seaton (Editors)

# Influence of temperature on water adsorption / desorption hysteresis loop in disordered mesoporous silica glass by Grand Canonical Monte Carlo simulation method

Joël Puibasset[a]\*, Roland J.-M. Pellenq[b]

[a]Centre de Recherche sur la Matière Divisée, CNRS-Université d'Orléans, 1b, rue de la Ferollerie, 45071 Orléans cedex 02, France

[b]Centre de Recherche en Matière Condensée et Nanosciences, CNRS, Campus de Luminy, case 913, 13288 Marseille cedex 09, France

\*puibasset@cnrs-orleans.fr

The water adsorption / desorption properties in a mesoporous silica glass are investigated by way of Grand Canonical Monte Carlo simulations. The SPC and PN-TrAZ potential are used to describe water-water and water-silica interactions. The numerical sample of mesoporous silica glass (pore size: 3.6 nm) was obtained by off-lattice reconstruction, known to reproduce in a realistic way the geometrical complexity of high specific surface Vycor.
The ability of the PN-TrAZ potential to describe the hydrophilic properties of silica surfaces is shown through the calculation of the adsorption isotherm and isosteric heat of adsorption at 300 K, which compare well to experimental data for Vycor. This study is extended to several temperatures, and the evolution of the hysteresis loop is examined.

## 1. INTRODUCTION

Many interesting features are exhibited by fluids adsorbed in porous materials like "capillary condensation" (appearance of a dense liquid-like state in mesoporous adsorbents for chemical potential lower than its bulk saturating value). [1-3] Despite the numerous efforts to give a comprehensive theoretical analysis of the adsorption phenomena, there are still open questions arising from the complexity of the real materials: topological and morphological disorder, as surface roughness, surface chemistry, large pore size distribution, interconnections between pores, etc. The development of molecular simulations is a powerful tool to take into account this disorder.

This molecular simulation study focuses on the adsorption of water in mesoporous Vycor-like silica glasses, which are of great interest for geophysics, pharmaceutics, industry, or environment. The adsorbent was previously constructed by off-lattice methods to mimic the micro- and meso-textural properties of real Vycor. [4-6] Whereas most of the previous studies focused on the dynamical or structural properties of confined water [7-11], this one focuses on the thermodynamic properties which are actually essential prior to a molecular dynamics study for instance (to "equilibrate" the initial configuration). The Grand Canonical Monte Carlo (GCMC) method has been used since it mimics a real adsorption experiment where the temperature and the chemical potential of the fluid (or the pressure of the gas) are fixed. The water adsorption isotherms at 300 K, 350 K, 400 K, 500 K and 650 K are calculated in this numerical mesoporous glass sample, as well as the configurational energies, the isosteric heat of adsorption, and compared to experiments when possible. The nature of the hysteresis loop observed at 300 K is also discussed.

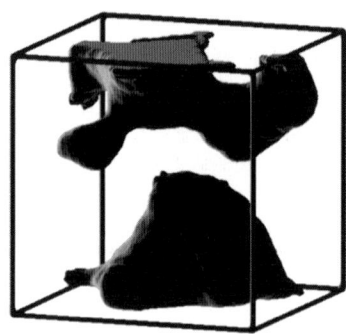

Fig. 1: 3-D representation of the numerical sample of Vycor used in this study. The porosity is in grey. The mean pore size is 3.6 nm

## 2. COMPUTATIONAL DETAILS

The atomistic Vycor-like mesoporous structure considered in this study has been generated by off-lattice reconstruction algorithm. Portions of an initial cubic crystal of cristobalite (106.95 Å of edge) are cut out by applying the off-lattice functional (shown in Fig. 1) representing the gaussian field associated to the volume autocorrelation function of the studied porous structure.[4, 5] It has been shown[6] that this numerical sample reproduces quite well micro- and meso-textural properties of real Vycor, and constitutes an improvement of the topological description of silica mesoporous glasses, because of the presence of disordered and interconnected pores. In order to have a realistic description of the surface chemistry, all silicon atoms in an incomplete tetrahedral environment are removed, and all oxygen dangling bonds are saturated with hydrogen atoms placed 1 Å from oxygen perpendicular to the surface. The porosity of the sample is 0.28, the density 1.56 g/cm$^3$, the specific surface area 210 m$^2$/g, the mean pore diameter 36 Å (from chord distribution), and OH surface density 7 OH/nm$^2$, all in agreement with experimental values: 0.28, 1.50 g/cm$^3$, 200 m$^2$/g, and around 7 OH/nm$^2$ if out-gassed around 400°C[12, 13], except for the pore size slightly smaller in our pseudo-Vycor compared to real one (40 Å). It has been shown that this numerical sample reproduces quite well micro and meso textural properties of real Vycor. [4-6]

The adsorbed water is described by the SPC model[14], because it is a fast computable model well suited for very large systems (the completely saturated system contains more than 12000 water molecules). This model reproduces well the thermodynamic and structural properties around ambient temperature, like vapor pressure, enthalpy of vaporization, and radial distribution functions.[15]

The interaction of water with Si, O and (surface) H Vycor species is assumed to remain weak, in the physisorption energy range. In this work, we have used a TrAZ form of the original PN-type potential function as reported for adsorption of rare gases and nitrogen in silicalite-1.[16][17] The PN-TrAZ potential function is based on the usual partition of the adsorption intermolecular energy restricted to two body terms only (a dispersion interaction term, a repulsive short range contribution and an induction term). The choice of this particular model to describe the water/adsorbent potential was motivated by the good degree of parameter transferability. Indeed, in a previous study[18], we found that using a set of potential parameters previously derived for adsorption in silica zeolite[16] augmented to take into account hydroxyl groups, the TrAZ model allows reproducing both low coverage experimental adsorption isotherm (amount adsorbed versus pressure at constant temperature) and isosteric heat curve with no further adjustment.

Minimal image convention is adopted to calculate all interactions. The electrostatic contribution is evaluated by summing on neutral subgroups of atoms of highest symmetry (tetrahedral silicon $SiO_4$ and surface hydroxyls OH). Implementation of Ewald summation procedure[19,20] has proven to be of little improvement, probably due to the large box size, absence of isolated charges, and negligible total dipolar moment of the substrate (due to isotropy in normal orientation on Vycor surface). Considering the polarization contributions, one has to take into account the fact that the SPC model already contains some polarization effects (treated as effective contributions in the bulk phase parameterization). It is then impossible to simply add the surface polarization contribution (the local electric field from surface and water molecules have to be added instead). However, since the surface contribution term was shown to be small, it was decided to neglect this                                                                                                    contribution.

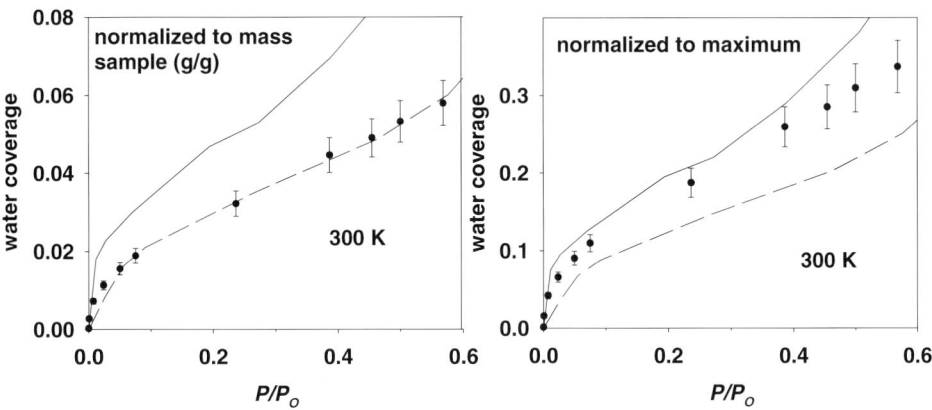

Fig 2: Water adsorption isotherms in Vycor versus relative pressure. Circles: this work. Solid line: gravimetric data by Takei et al.([39]). Dashed line: microcalorimetric data by Markova et al.([40]).

The water-water interaction calculations become excessively slow at low temperature when the substrate is almost completely filled (more than 10000 water molecules). It was then decided to implement a hierarchy method[21]. The simulation box is divided in a 3-dimensional array of 5 Å wide cells. The total interaction energy of a water molecule with the others is the sum of the individual contributions of the molecules contained in the neighboring cells (within a distance of 15 Å) and the multipole representation for the other cells. The uncertainty introduced by this method does not exceed few percent.

The adsorption isotherms are calculated with the Grand Canonical Monte Carlo (GCMC) technique by increasing the chemical potential step by step.[22] An equal number of trials for translation, rotation, and creation or destruction of molecules has been chosen for each Monte Carlo (MC) step at fixed chemical potential. The configurations are considered for visual examination and statistical calculations every block of $10^4$ Monte Carlo steps ($4.10^4$ MC trials) along the Markov chain, so that they are de-correlated. For each chemical potential value the averages are performed over approximately $10^6$ trials per molecule, or more for the highest coverage, in order to compensate the decrease in insertion acceptance level. After the complete saturation of the Vycor sample is reached for high chemical potential μ, one can start decreasing μ to acquire the desorption branch (hysteresis).

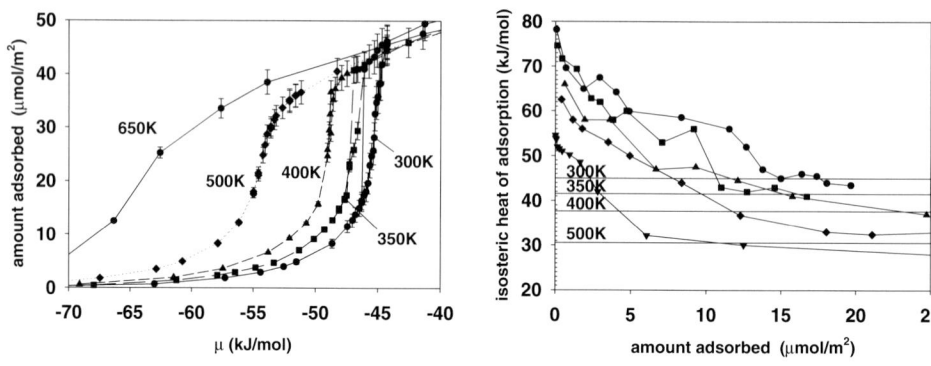

Fig 3 : Left panel: water adsorption / desorption isotherms in Vycor versus chemical potential at different temperatures. Lines are guides to the eye. Right panel: Isosteric heat of adsorption versus water coverage at different temperatures (300, 350, 400, 500 and 650K from top to bottom); symbols: this work; horizontal lines: latent heat of vaporization of bulk water.

In order to validate the molecular model, a first study was performed at room temperature (300 K). The numerical results are shown in Fig. 2 (circles), as well as two experimental data found in the literature for comparison. The first one (solid line) is a gravimetric measurement by T. Takei, et al.[23] in a 7930 Vycor sample characterized by nitrogen adsorption (pore size around 4.72 nm in diameter). The other one (dashed line) has been obtained by Markova, et al.[24] with a double twin microcalorimeter, in a 7930 Vycor glass (pore size between 4 and 7 nm). As can be seen, the choice of the normalization slightly influences the comparison because the geometric characteristics of the experimental and numerical Vycor are not exactly identical. The natural normalization (adsorbed amount divided by the geometric surface area of the sample) has been discarded due to the uncertainties in the experimental determination of the geometric surface, and replaced by two others. In Fig 2a, the adsorbed water mass has been normalized to the mass of the Vycor sample, and plotted as a function of the relative pressure (saturating vapor pressure $P_0$ =

0.044 bar at 300 K for the SPC model[25, 26]), related to the chemical potential assuming that the water vapor is ideal[25]. In Fig 2b, the amount of adsorbed water is normalized to its maximum (full saturation). This normalization is less sensitive to morphological details of the sample, but is sensitive to the choice of the saturating coverage. The calculations (circles) show a rapid increase of the adsorbed amount at low pressure (below 10% of the saturating pressure) characteristic of hydrophilic silica surfaces. The agreement between the simulation and the experimental results validates the good transferability of the potential model used in this study.

## 3. RESULTS AND DISCUSSIONS

The whole adsorption / desorption isotherms at 300 K, 350 K, 400 K, 500 K, and 650 K are plotted in Fig 3 (left panel) (normalized to the surface area of the substrate exposed to water vapor, in μmol / m$^2$). The result is given as a function of the chemical potential of water imposed by the GCMC simulation instead of pressure to avoid the introduction of uncertainties in conversion. It is important to note that this chemical potential does not contain the ideal rotational contribution $-kT(-4.09+3/2lnT)$, with $T$ in Kelvin.[27] The high temperature (650 K) isotherm is reversible (supercritical), whereas the low temperature cases present large hysteresis (type IV isotherm in the IUPAC classification). The steep rises are associated to the capillary condensation of water in the mesoporosity of the Vycor glass. These simulation results are in qualitative agreement with experimental data, with an asymmetric adsorption / desorption hysteresis characteristic of disordered and interconnected pores.[1, 28, 29]

The adsorption / desorption isotherms show that the pseudo-critical temperature (disappearance of the hysteresis loop) is lower than the bulk critical temperature since the T=500 K isotherm is reversible. An estimate of this temperature give a value between 450 and 500 K in our simulation box. This is the first estimation of the pseudo-critical temperature of water in a Vycor-like structure.

Visual inspection of molecular configurations show that the condensation occurs in the high curvature regions of the pore. A meniscus forms in the narrow necks, and finally fills up the whole pore. On the other hand, the desorption proceeds by gas phase nucleation in the large pore (formation of a bubble).

The isosteric heat of adsorption, which is experimentally accessible, measures the heat released by the adsorption of a given amount of fluid, and is numerically deduced from cross-fluctuations in energy (U) and adsorbed quantity (N) through the formula:[22]

$$q_{st} = -\frac{\langle U N \rangle - \langle U \rangle \langle N \rangle}{\langle N^2 \rangle - \langle N \rangle^2} + R T \tag{1}$$

The results are given in Fig. 3 (right panel) for different temperatures, as a function of coverage in g/g for comparison with experimental data. Generally, the low chemical potential value of the isosteric heat of adsorption gives information on the strength of the adsorbate/adsorbent interaction. The low coverage isosteric heat of adsorption at 650 K and 300 K are close to 55 kJ/mol and 80 kJ/mol respectively. The numerical results at 300 K are in good agreement with experimental values.[24] At 300 K, the curve shows that beyond one monolayer coverage (around 0.06 g/g), the isosteric heat reaches 44 kJ/mol, that is to say the latent heat of vaporization of water (horizontal line). The strong influence of the silica surface has then disappeared in the second and subsequent layers, which shows again that the surface interactions are actually relatively short range in a polar fluid.

## 4. CONCLUSION

The adsorption of water in disordered mesoporous silica glass (Vycor-like) has been analyzed by GCMC simulation method at various temperatures: 300 K, 350 K, 400 K, 500 K and 650 K. The results show a hysteresis loop in the adsorption/desorption data at low temperature, while the two highest temperature curves are reversible. The amount adsorbed at low pressure are in good agreement with published experimental data, which validates the good transferability of the interaction model used. The simulation shows that the water adsorption in Vycor-like silica is irreversible at low temperature (hysteresis). The hysteresis loop observed in the adsorption / desorption isotherm is shown to shrink with increasing temperature. Its temperature of disappearance (pseudo-critical temperature) is estimated between 450 and 500 K, a value significantly lower than the bulk critical temperature.

## ACKNOWLEDGMENT.
The Institut de Développement des Ressources en Informatique Scientifique (IDRIS-CNRS, Orsay, France) is gratefully acknowledged for the CPU grant n°051153

## REFERENCES

[1] L. D. Gelb, K. E. Gubbins, R. Radhakrishnan, et al., Reports on Progress in Physics **62**, 1573 (1999).

[2] S. J. Gregg and K. S. W. Sing, *Adsorption, Surface Area and Porosimetry* (Academic Press, New York, 1982).

[3] F. Rouquerol, J. Rouquerol, and K. S. W. Sing, *Adsorption by Powders and Porous Solids* (Academic Press, London, 1999).

[4] P. Levitz, Advances in Colloid and Interface Science **76/77**, 71 (1998).

[5] P. Levitz, G. Ehret, S. K. Sinha, et al., Journal of Chemical Physics **95**, 6151 (1991).

[6] R. J.-M. Pellenq, B. Rousseau, and P. E. Levitz, Physical Chemistry Chemical Physics **3**, 1207 (2001).

[7] S. H. Lee and P. J. Rossky, Journal of Chemical Physics **100**, 3334 (1994).

[8] E. Spohr, A. Trokhymchuk, and D. Henderson, Journal of Electroanalytical Chemistry **450**, 281 (1998).

[9] M. Rovere, M. A. Ricci, D. Vellati, et al., Journal of Chemical Physics **108**, 9859 (1998).

[10] C. Hartnig, W. Witschel, E. Spohr, et al., Journal of Molecular Liquid **85**, 127 (2000).

[11] P. Gallo, M. A. Ricci, and M. Rovere, Journal of Chemical Physics **116**, 342 (2002).

[12] H. Landmesser, H. Kosslick, W. Storek, et al., Solid State Ionics **101-103**, 271 (1997).

[13] M. J. D. Low and N. Ramasubramanian, Journal of Physical Chemistry **71**, 730 (1967).

[14] H. J. C. Berendsen, J. P. M. Postma, W. F. van Gunsteren, et al., in *Intermolecular Forces* (B. Pullman, Dordrecht: Reidel, 1981), p. 331.

[15] W. L. Jorgensen, J. Chandrasekhar, J. D. Madura, et al., Journal of Chemical Physics **79**, 926 (1983).

[16] R. J.-M. Pellenq and D. Nicholson, Journal of Physical Chemistry **98**, 13339 (1994).

[17] R. J.-M. Pellenq and D. Nicholson, Molecular Physics **95**, 549 (1998).

[18] J. Puibasset and R. J.-M. Pellenq, Journal of Chemical Physics **118**, 5613 (2003).

[19] P. Ewald, Annalen der Physik **64**, 253 (1921).

[20] D. M. Heyes, Physical Review B **49**, 755 (1994).

21  H.-Q. Ding, N. Karasawa, and W. A. Goddard III, Journal of Chemical Physics **97**, 4309 (1992).

22  D. Nicholson and N. G. Parsonage, *Computer simulation and the statistical mechanics of adsorption* (Academic Press, London, 1982).

23  T. Takei, A. Yamazaki, T. Watanabe, et al., Journal of Colloid and Interface Science **188**, 409 (1997).

24  N. Markova, E. Sparr, and L. Wadsö, Thermochimica Acta **374**, 93 (2001).

25  J. R. Errington and A. Z. Panagiotopoulos, Journal of Physical Chemistry B **102**, 7470 (1998).

26  J. Vorholz, V. I. Harismiadis, B. Rumpf, et al., Fluid Phase Equilibria **170**, 203 (2000).

27  D. Frenkel and B. Smit, *Understanding Molecular Simulation* (Academic Press, London, 1996).

28  T. Takei, K. Musaka, M. Kofuji, et al., Colloid and Polymer Science **278**, 475 (2000).

29  C. G. V. Burgess, D. H. Everett, and S. Nuttall, Pure and Applied Chemistry **61**, 1845 (1989).

Studies in Surface Science and Catalysis 160
*P.L. Llewellyn, F. Rodriquez-Reinoso, J. Rouqerol and N. Seaton (Editors)*

# Determination of pore size distribution in microporous carbons based on $CO_2$ and $H_2$ sorption data.

**M. Konstantakou**[a,b*], **S. Samios**[a,b], **Th. A. Steriotis**[b], **M. Kainourgiakis**[b], **G. K. Papadopoulos**[c], **E. S. Kikkinides**[a] **and A.K. Stubos**[b]

[a]Department of Engineering and Management of Energy Resources, University of Western Macedonia, Kastorias and Fleming Srt., 50100 Kozani, Greece

[b]National Center for Scientific Research "Demokritos", 15310 Ag. Paraskevi Attikis, Athens, Greece

[c]Department of Chemical Engineering, National Technical University of Athens, 15780 Athens, Greece

A method for the determination of pore size distribution (PSD) is applied in the case of AX-21 carbon. Adsorption isotherms of $CO_2$ (253, 298 K) and $H_2$ (77 K) up to 20 bar have been measured, while the computed isotherms resulted from GCMC simulations. The optimal PSD is determined by inverting the adsorption integral equation. Such PSDs were used to predict isotherms of other adsorbates and temperatures. Best predictions were obtained by employing the full relative pressure range PSD ($CO_2$ at 253 K). In addition, optimal PSDs were deduced from the combined use of $CO_2$ and $H_2$ data. These were found to be capable of reproducing more accurately all the available experimental isotherms.

## 1. INTRODUCTION

The reliable characterization of the pore structure is crucial for the utilization and design of improved porous systems in several applications including separation processes, removal of various pollutants and gas storage. A number of more or less established characterization methods are currently in use for mesoporous and macroporous materials, providing information on the pore size distribution (PSD), pore network connectivity, and other structural parameters of the material [1, 2]. On the contrary, the reliable assessment of microporosity (involving pores of sizes less than 2 nm) is much less advanced in terms of relating sorption properties to the underlying microstructure. The commonly used Dubinin-Radushkevich, Dubinin-Astakhov and Dubinin-Stoeckli methods employ phenomological models of adsorption based on the thermodynamic approach of Dubinin [3]. Since the mechanism(s) of adsorption in micropores is still under active debate, the limitations of these and other conventional methods used in practice (such as the MP an the Horvath-Kawazoe methods) for micropore size characterization have been repeatedly discussed in the literature (see [4, 5, 6] and references therein). As macroscopic descriptions of states of matter are invalid in micropores, improved molecular level approaches have been developed based for the micropore structure characterization. In particular, Density Functional Theory (DFT) can provide an accurate description of simple fluids in geometrically simple confined space [7, 8,

9, 10, 11, 12, 13, 14]. To capture more accurately the behavior of the adsorbates in micropores, it is often necessary to model them as non-spherical molecules with electrostatic interactions. Given the limited capabilities of DFT in this context, molecular simulation based on the Grand Canonical Monte Carlo (GCMC) technique has been established for the generation of adsorption isotherms in carbons [15, 16, 17] and the determination of PSDs [18, 19, 20, 21, 22, 23, 24, 25, 26]. A review on both methods is given in [27].

$N_2$ molecules are small enough to enter into micropores, they are considered reasonably inert and their shape is not very far from spherical. Additionally, isothermal scans at 77 K are carried out in an easy and cost efficient manner by simply using liquid $N_2$ baths, while the saturation vapor pressure of $N_2$ allows for the accurate measurement of relative pressures over a wide range (commonly $10^{-5}$ – 0.995). For these reasons, but also due to tradition, standardization issues, etc. in the majority of cases, the surface areas and PSDs of porous solids are experimentally determined from $N_2$ adsorption isotherms at 77 K. On the other hand microporous solids have pores and constrictions with dimensions comparable to the size of $N_2$ molecules. Under this extreme confinement molecular motion is in many cases stereochemically hindered, leading to the so-called activated-diffusion mechanism, where molecules need to overcome energy barriers in order to pass through constrictions and penetrate into micropores [3]. In such a case, as temperature is decreased the probability of overcoming energy barriers also decreases and thus diffusion becomes slower and equilibration time increases. It is nowadays widely accepted, that at 77 K the diffusion of nitrogen molecules into micropores is very slow. Furthermore, it was pointed out that such diffusional limitations could influence adsorption especially in ultra-micropores (pores smaller than 0.7 nm) [28]. Porous carbons usually contain a wide range of pore sizes including ultra-micropores and this might lead not only to time-consuming measurements but also to significant under estimation of the adsorption isotherm. It has been proposed [29,30] that such problems can be minimized by using high temperature $CO_2$ adsorption analysis (e.g. at 273 or 298 K). Nevertheless, as the vapor pressure of $CO_2$ at 273 K is ~35 bar (and ~65 bar at 298 K) high-pressure adsorption equipment is required to capture the contribution of all the pores of the system. It should be particularly stressed that commonly measured 273 K $CO_2$ adsorption isotherms at sub atmospheric conditions (i.e. by using conventional gas porosimeters) should be interpreted with extreme precaution. Such measurements should only be used for the pore size analysis of samples having only very fine pores (e.g. if a plateau is observed), and are bound to lead to erroneous results if the sample contains larger pores.

Much more reliable PSDs of microporous carbons could be based on probing different pore sizes by using different experimental adsorption isotherms in terms of temperatures and/or gases. Based on the arguments outlined above, instead of using $N_2$ at 77 K, high temperature $CO_2$ isotherms are recommended given that high pressure equipment is available, in order to cover the full relative pressure range. On the other hand a good approach for the analysis of very fine pores would be the use of $H_2$ isotherms at 77K [13]. Beyond the obvious practical reasons of experimental convenience, the small size of $H_2$ and the fact that 77 K is by far above its supercritical temperature ensures that even the smallest pores, possibly not accessible to other molecules, would easily be monitored, while equilibration kinetics are expected to be fast enough for reliable measurements.

In this work we report adsorption data of $CO_2$ at 253 and 298 K as well as $H_2$ at 77 K on the high surface area AX-21 activated carbon. The individual adsorption isotherms are used in order to calculate PSDs based on the GCMC approach. In a further step the individual isotherms are combined and new PSDs that respect combinations of the experimental isotherms are calculated. The reliability of the calculated PSDs (individual or combined) is examined by their ability to predict the experimental adsorption isotherms.

## 2. EXPERIMENTAL

The adsorption isotherms of $CO_2$ at 253 K and 298 K as well as $H_2$ at 77 K were measured in a pressure range 0-20 bar on the Intelligent Gravimetric Analyzer (IGA – Hiden Analytical Ltd). Prior to each run, the sample (AX-21, KOH activated carbon made by Amoco Co. and kindly provided by S. R. Tennisson, MAST Carbon) was outgassed at 623 K until no mass change was observed. The IGA reactor containing the sample was immersed in different liquid baths for the isothermal scans, namely liquid $N_2$ for 77K, NaCl/ice for 253 K and a PID controlled circulating oil bath for 298 K. While the $CO_2$ isotherms were completely reversible, this was not the case for $H_2$, which revealed a residual $H_2$ quantity after outgassing at 77K, possibly due to chemisorption. The residual amount was subtracted and two more isothermal scans were performed in order to ensure that the physically adsorbed quantity was accurately determined.

## 3. SIMULATION MODEL

The use of GCMC method for obtaining the PSD of microporous carbonaceous materials involves the following three major steps [18, 23, 24]:
1. Determination (and validation whenever possible) of a molecular model for the adsorbate – adsorbate and adsorbate – adsorbent interactions.
2. Generation of a database of GCMC sorption isotherms with respect to a specific adsorbate for a set of pore widths, pressures and temperatures.
3. Inversion of the adsorption integral equation:

$$N(p) = \int_{H\,min}^{H\,max} f(H)n(H)dH \tag{1}$$

where $N(p)$ is the experimentally measured amount of adsorbate, $n(H,p)$ is the average density of adsorbate at pressure $p$ in a pore of width $H$, and $f(H)$ is the PSD sought.
The solution of the Eq. (1) is an ill-posed problem. Depending on the form of the kernel $n(H,p)$ and the isotherm $N(p)$, there can be from zero to an infinity of solutions for $f(w)$ (detailed discussions on the methods for the solution of Eq. (1) and the application of suitable constraints to force physically sound or appealing solutions including constraints on the smoothness of $f(H)$ and the range of $H$, see [19,21,23,24] and references therein. Nonetheless, our work aims at finding useful solutions to Eq. (1) in the sense that the gas adsorption properties of microporous carbons can be reliably predicted.

### 3.1. Adsorbates

$CO_2$ is modeled as a three charged center Lennard – Jones (LJ) molecule, according to Murthy and co-workers [31] with the parameters $\varepsilon_{OO}/k_B = 75.2$ K, $\sigma_{OO} = 0.3026$ nm, $\varepsilon_{CC}/k_B = 26.3$ K, $\sigma_{CC} = 0.2824$ nm. The O-O and C-O distances of the model are 0.2324 nm and 0.1162 nm respectively. The intermolecular potential $u_{ij}$ is assumed to be a sum of the interatomic potentials between the atoms of molecules $i$ and $j$, plus the electrostatic interactions due to $CO_2$ quadrupole moment with point partial charges $q_O = -0.332e$ and $q_C = +0.664e$, i.e.

$$u_{ij}(r) = \sum_{\alpha\beta} \left\{ 4\varepsilon_{\alpha\beta} \left[ \left( \frac{\sigma_{\alpha\beta}}{r_{\alpha\beta}} \right)^{12} - \left( \frac{\sigma_{\alpha\beta}}{r_{\alpha\beta}} \right)^{6} \right] + \frac{q_\alpha q_\beta}{4\pi\varepsilon_0 r_{\alpha\beta}} \right\} \tag{2}$$

where $\varepsilon_0$ is the permittivity of vacuum.

$H_2$ was similarly treated as a two center LJ molecule with $\varepsilon_{HH}/k_B = 12.5$ K and $\sigma_{HH} = 0.259$ nm. The hydrogen – hydrogen distance in the model was taken to be the actual bond length (0.074 nm). The parameters for the two-site model were adopted from [32].

### 3.2. Adsorbent

Pore walls were treated as stacked layers of carbon atoms separated by a distance $\Delta = 0.335$ nm, and having a number density $\rho_w = 114$ atoms nm$^{-3}$ per layer. The adsorbate – wall interaction at distance $r_z$ was calculated by the 10-4-3 potential of Steele [33]:

$$u_w(r_z) = 2\pi\rho_w\varepsilon_{\alpha\beta}\sigma_{\alpha\beta}^2\Delta \left[ \frac{2}{5}\left( \frac{\sigma_{\alpha\beta}}{r_z} \right)^{10} - \left( \frac{\sigma_{\alpha\beta}}{r_z} \right)^4 - \frac{\sigma_{\alpha\beta}^4}{3\Delta(0,61\Delta + r_z)^3} \right] \tag{3}$$

The potential parameters of the solid surface are $\varepsilon_{SS}/k_B = 28.0$ K and $\sigma_{SS} = 0.340$ nm. Eq. (3) neglects the energetic inhomogeneity of the surface, nevertheless, this is not expected to affect the results significantly especially at ambient temperatures [4]. All the cross interaction potential parameters between different sites ($\alpha \neq \beta$) were calculated according to the Lorentz – Berthelot rules ($\sigma_{\alpha\beta}=(\sigma_{\alpha\alpha}+\sigma_{\beta\beta})/2$, $\varepsilon_{\alpha\beta}=(\varepsilon_{\alpha\alpha}\varepsilon_{\beta\beta})^{1/2}$). The potential energy $U_w$ due to the walls inside the slit pore model for each atom of the adsorbate molecules is given by the expression:

$$U_w = u_w(r_z) + u_w(H - r_z) \tag{4}$$

where $H$ is the distance between the carbon centers across the slit pore model (physical width). For the determination of PSDs, the corrected width $H'$ (chemical width) should be used since this is the one involved in the experimentally obtained isotherms, namely:

$$H' = H - 2z_0 + \sigma_g \tag{5}$$

where $\sigma_g$ is the root of the adsorbate – adsorbent Lennard – Jones function, and $z_0$ the root of its first derivative. If the above relation is applied in the present $H_2$ or $CO_2$ – graphite system, it is found that about 0.24 nm should be subtracted from $H$ to define $H'$ [34].

### 3.3. Adsorption isotherms

The Grand Canonical Ensemble Monte Carlo method was employed to probe the statistically important regions of the configuration space in the $(\mu, V, T)$ ensemble according to the prescription given elsewhere [35]. In this work the pore range $H=0.6 - 3.0$ nm (physical width) was divided in 25 equidistant intervals (pore groups) with 0.1 nm spacing between them. Three types of molecular moves are attempted with equal probability: (a) a compound move enabling random displacement and reorientation, with the maximum allowed displacement being adjusted so that the acceptance ratio of the move is about 20% in order to sample phase space more efficiently (b) a compound move consisting of random insertion of the center of mass of a molecule in a random orientation, by generating a unit vector distributed uniformly on the surface of a sphere centered at the origin of the Cartesian system of coordinates of the simulation box (Marsaglia algorithm [35]) and (c) a random deletion of a fluid molecule. Periodic boundary conditions have been applied in the directions other than the width of the slit. For a given simulation, the size of the box (i.e. the two dimensions other than $H$) was varied in order to ensure that sufficient particles (ca. about 500) remained in the simulation box at each pressure. Statistics were not collected over the first $3 \times 10^6$ configurations to assure adequate convergence of the simulation. The uncertainty on the computed equilibrium properties such as ensemble averages of the number of adsorbate molecules in the box and the total potential energy is estimated to be less than 4%.

## 4. MICROPORE SIZE DISTRIBUTIONS

The process of defining the optimal PSD begins by assigning an assumed initial volume, $V_j$ to every pore group $H_j$ and constructing thus a computed isotherm. Consequently, this constructed isotherm is compared to its experimental counterpart and by iterative variation of the $V_j$ matrix elements (which are subject to the constraint of non-negativity), the "optimum" micropore size distribution, namely the one providing the best fit, is selected. For this reason, the routine E04NCF of NAG library has been implemented. It is a routine solving linearly constrained linear least-squares problems based on a two-phase (primal) quadratic programming method with features to exploit the convexity of the objective function due to Gill et al. [36]. In the full-rank case, the method is related to that of Stoer [37].

The technique is applied in the case of the microporous carbon AX-21, for which the $CO_2$ at 253 K, $CO_2$ at 298 K and $H_2$ at 77 K experimental isotherms have been obtained. The resulting PSDs from the individual experimental isotherms are included in the form of histograms in Fig.1. The $H_2$ PSD reveals practically 3 classes of pores centered at around 0.5, 1.6 and 2.8 nm. At very small pore sizes ($H' < 0.8$ nm), the GCMC $H_2$ calculated isotherms are of Langmuir type in the pressure range studied, while at larger pore sizes the curvature is gradually lost and isotherms for pores with $H' > 1.5$ nm are of Henry type (i.e. straight lines). This is actually a sign that over the pressure range used, $H_2$ pore filling occurs only in very fine pores, while surface coverage is prevailing at wider pores. In practical terms the amount adsorbed from large pores is not a function of the pore size (it is practically driven by the specific surface area available) and thus during the process of inverting the adsorption integral any combination of individual pore isotherms can reproduce the experimental data. In this respect a straightforward result of our simulations is that $H_2$ sorption isotherms cannot be used for the pore size characterization of samples having pores beyond the ultramicropore region, unless of course adsorption experimental data at much larger pressures are available. The PSDs of $CO_2$ at both temperatures are in marked contrast with that of $H_2$, giving a much broader distribution of sizes spread over the entire range studied. The main differences between them are observed at the low and high limits of pore sizes considered. The 253 K PSD reveals the existence of large pores (around 2.6 nm), not present in the 298 K PSD. On the other hand the 298 K PSD shows the existence of very fine pores (0.36 nm). This can be explained if we consider the following: The experimental isotherms were carried out at exactly the same equilibration pressures and thus much lower relative pressures have been measured at 298 K. In this respect the 298 K isotherm is expected to depict better the contribution of fine pores. Following the same thought, high relative pressures have not been measured at 298 K ($p/p_0 < 0.55$), while the measurement at 253 K contains the full relative pressure range (up to $p/p_0$ 0.94). It is thus expected that the information on large pores is minimal at 298 K, and much more complete at 253 K.

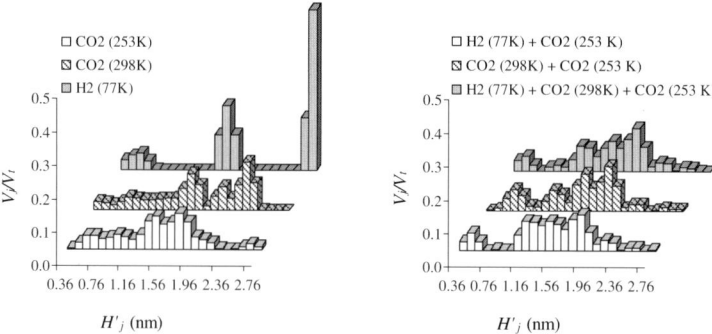

Figure 1. Pore size distributions from individual isotherms and combinations ($V_j$: volume of pores having width $H'_j$, $V_t$: total pore volume)

For the reasons outlined above the PSDs deduced from the individual isotherms have in general, limited predictive potential. This is obvious in Fig.2, where each PSD is used to calculate in a reverse manner the adsorption isotherms. Of course all the PSDs can quite accurately predict their experimental counterparts, nevertheless experimental data on different molecule and/or temperature cannot be accurately reproduced. The worst predictions are obtained by using the $H_2$ PSD. For instance, this PSD produces a step isotherm for the full relative pressure range isotherm ($CO_2$, 253 K), presumably pertaining to the filling of the three different classes of pores. A similar picture, to a much smaller extend, is also observed for the $CO_2$ 298 K PSD. As explained before, this can be attributed to the fact that both the $H_2$ and the $CO_2$ at 298 K isotherms contain information on the smaller pores and cannot thus predict the filling of larger pores at higher relative pressures. On the contrary, the resulting PSD from the full pressure range isotherm of $CO_2$ at 253 K can reasonably predict all the experimental isotherms, however it fails to capture the filling of ultramicropores by $H_2$ again due to the fact that experimental information on this pore size range is limited.

By inverting the adsorption integral equation after including different combinations of experimental and GCMC data sets, new PSDs are obtained (Fig.1). In contrast to the individual PSDs, the three "combined" PSDs are similar to each other and can more accurately reproduce all the experimental isotherms (Fig. 2). As expected, the optimal PSD is obtained from the combination of all the data. It should also be mentioned that the combination of the two "low relative pressure" isotherms ($H_2$ and $CO_2$ at 298 K) could not reproduce the $CO_2$ isotherm at 253 K (again because large pore information is missing).

In the GCMC PSD approach, the adsorption integral inversion mathematically degenerates to a minimization problem, while experimental data probe different pore sizes with varying accuracy (from fine to gross in the order $H_2$ 77 K>$CO_2$ 298 K>$CO_2$ 253 K). We can thus conclude that in order to deduce a PSD that can adequately describe the pore system, a high temperature (in the sense that equilibrium is attainable) and full relative pressure range isotherm (in order to capture the contribution of all the pore sizes) should be the basis. This set should then be improved by simultaneously using data on finer pores. For instance, the incorporation of $H_2$ data in the calculation fine-tunes the analysis by introducing small pores sizes, not detectable by $CO_2$ (Fig. 2).

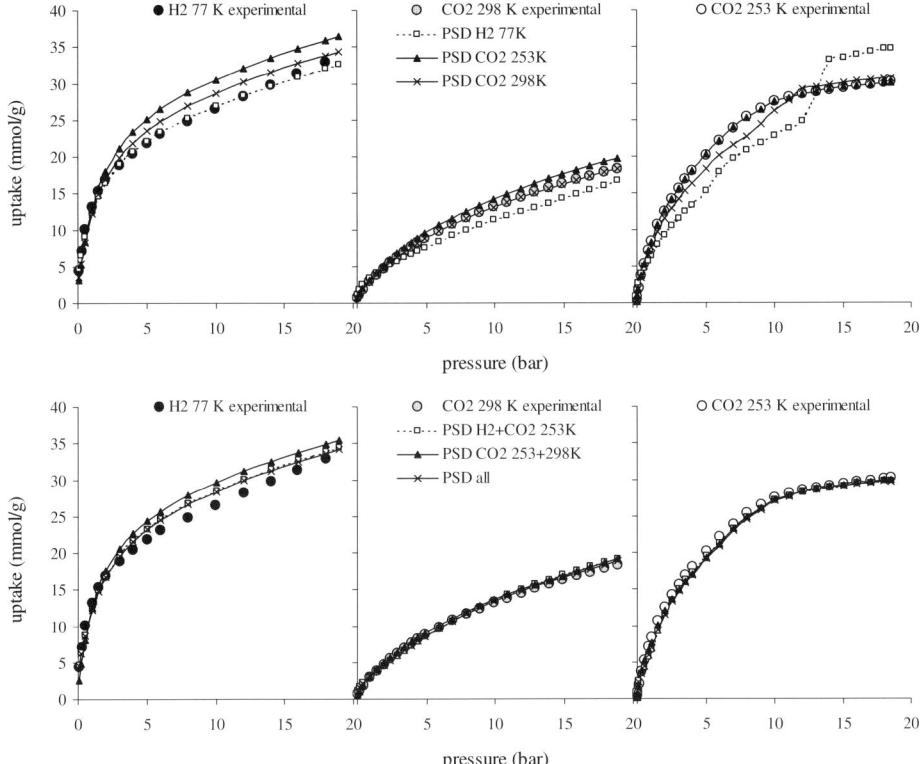

Figure 2. Experimental and calculated (from the PSDs shown in fig 1) isotherms

## 5. CONCLUSIONS

Adsorption isotherms of $CO_2$ at 253 and 298 K as well as $H_2$ at 77 K on the AX-21 activated carbon have been performed and used individually for the calculation of PSDs based on the GCMC approach. The calculated PSDs were then used in order to predict isotherms (other temperature/adsorbate). The best predictions were obtained with the full relative pressure range PSD ($CO_2$ at 253 K), however limited predictive potential was in general observed. Consequently the individual isotherms were combined and new PSDs that respect combinations of the experimental isotherms were deduced. These combined PSDs can reproduce much more accurately the experimental data. We thus propose that in order to calculate PSDs that can describe real systems in a sufficiently accurate manner, high temperature full relative pressure range isotherms (e.g. $CO_2$ at 253 K) should be combined with very low relative pressure data (e.g. $H_2$ at 77 K).

## REFERENCES

[1] K. Kaneko, J. Membrane Sci., 96 (1994) 59.
[2] Y.C. Yortsos, Experimental Methods in Physical Sciences, Academic Press, New York, 1999, pp 69-117.
[3] S.J. Gregg, K.S.W. Sing, Adsorption, Surface Area and Porosity, Academic Press, London (1982).
[4] D. Nicholson, J. Chem. Soc., Faraday Trans. 90 (1994) 181.
[5] D. Nicholson, J. Chem. Soc., Faraday Trans. 92 (1996) 1.
[6] F. Stoeckli, A. Guillot, D. Hugi-Cleary and A.M. Slasli, Carbon, 38 (2000) 938.
[7] N.A. Seaton, J.P.R.B. Walton and N. Quirke, Carbon, 17 (1989) 853.
[8] C. Lastoskie, K.E. Gubbins and N. Quirke, J. Phys. Chem., 97 (1993) 4786.
[9] P.N. Aukett, N. Quirke, S. Riddiford and S.R. Tennison, Carbon, 30 (1992) 913.
[10] P.I. Ravikovitch, S.C. Ó'Domhnaill, A.V. Neimark, F. Schuth and K.K. Unger, Langmuir, 11 (1995) 4765.
[11] K.A. Sosin and D.F. Quinn, J. Porous Mater., 1 (1995) 111.
[12] S. Scaife, P. Kluson and N. Quirke, J. Phys. Chem. B, 104 (2000) 313.
[13] J. Jagiello and M. Thommes, Carbon 42 (2004)1227.
[14] T. X. Nguyen and S. K. Bhatia, Langmuir, 20 (2004) 3532
[15] T. Ohba, D. Nicholson and K. Kaneko, Langmuir, 19, (2003) 5700
[16] D.D. Do, H.D. Do, Colloids and Surfaces A: Physicochem. Eng. Aspects 252 (2005) 7
[17] T. X. Nguyen, S. K. Bhatia and D. Nicholson, Langmuir, 21 (2005) 3187-3197
[18] S. Samios, A.K. Stubos, N.K. Kanellopoulos, R.F. Cracknell, G.K. Papadopoulos and D. Nicholson, Langmuir, 13 (1997) 2795.
[19] V.I. Gusev, J.A. O'Brien and N.A. Seaton, Langmuir, 13 (1997) 2815.
[20] M.V. Lopez-Ramon, J. Jagiello, T.J. Bandosz and N.A. Seaton, Langmuir, 13 (1997) 4435.
[21] G.M. Davies, N.A. Seaton and V.S. Vassiliadis, Langmuir, 15 (1999) 8235.
[22] S. Samios, A. Stubos, G.K. Papadopoulos, N.K. Kanellopoulos and F. Rigas, J. Colloid Interface Sci., 224 (2000) 272.
[23] P.I. Ravikovitch, A. Vishnyakov, R. Russo and A.V. Neimark, Langmuir, 16 (2000) 2311.
[24] M.B. Sweatman and N. Quirke, J. Phys. Chem. B, 105 (2001) 1403.
[25] A. Vishnyakov, P.I. Ravikovitch and A.V. Neimark, Langmuir, 15 (1999) 8736.
[26] X. Shao, W. Wang, R. Xue and Z. Shen, J. Phys. Chem. B, 108 (2004) 2970.
[27] D.D. Do, H.D. Do, Adsorption Science and Technology, 21(2003) 389
[28] F. Rodriguez-Reinoso and Linares-Solano A. In: Thrower PA, editor. Chemistry and Physics of Carbon, 21. New York: Marcel Dekker; 1988.
[29] J. Garrido, A. Linares-Solano, J.M. Martin-Martinez, M. Molina-Sabio, F. Rodriguez-Reinoso and R. Torregosa, Langmuir, 3 (1987) 76.
[30] J. Jagiello, Langmuir, 10 (1994) 2778.
[31] C.S. Murthy, S.F. O'Shea and I.R. McDonald, Mol. Phys., 50 (1983) 531.
[32] R.F. Cracknell, Phys. Chem. Chem. Phys., 3 (2001) 2091.
[33] W.A. Steele, The Interaction of Gases with Solid Surfaces, Pergamon, Oxford, 1974.
[34] K. Kaneko, R.F. Cracknell and D. Nicholson, Langmuir, 10 (1994) 4606.
[35] M. Allen and D.J. Tildesley, Computer Simulation of Liquids, Clarendon, Oxford, 1987.
[36] P.E. Gill, W. Murray, M.A. Saunders and M.H. Wright, ACM Trans. Math. Softw., 10 (1984) 282.
[37] J. Stoer, SIAM J. Numer. Anal., 8 (1971) 382.

Studies in Surface Science and Catalysis 160
P.L. Llewellyn, F. Rodriquez-Reinoso, J. Rouqerol and N. Seaton (Editors)
© 2007 Elsevier B.V. All rights reserved

# Assessment of the development of the pore size distribution during carbon activation: a population balance approach

**M V. Navarro[a], N. A. Seaton[a\*], A. M. Mastral[b], R. Murillo[b]**

[a]Institute for Materials and Processes, School of Engineering and Electronics, University of Edinburgh, EH9 3JL, UK

[b]Instituto de Carboquímica, CSIC, M Luesma Castán 4, 50015-Zaragoza, Spain

In the current study the changes produced in the pore size distribution (PSD) of lignite char during physical activation have been assessed. A mathematical model has been developed based on a population balance. In our approach, the evolution of the pore size distribution is computed as a function of time. To provide the necessary experimental input to the analysis, and to test the model, the pore size distribution is determined experimentally for samples obtained at a sequence of times. Data for the adsorption of nitrogen at 77 K and ethane at 264 K were obtained and analyzed by a combination of density functional theory (DFT) for nitrogen, and our own Grand Canonical Monte Carlo (GCMC) simulation-based method for nitrogen and for ethane.

## 1. INTRODUCTION

The range of applications of porous solids (activated carbons, zeolites, membranes) is growing rapidly and covers a wide variety of processes: separation of components, purification of feed or exhaust streams, catalytic reactions and gas storage. To produce these porous solids with optimal properties for a particular application it is necessary to understand the mechanisms involved in the formation of the solids. Although the aims of the production processes vary from solid to solid, in all cases the production process affects the physical properties (porosity, average pore size and pore size distribution) of the material. Therefore, it is important to develop an understanding of the way a production process affects the physical properties of the solid.

Activated carbons, highly porous carbon materials, can be produced from a variety of carbonaceous source materials such coals, lignite, agricultural waste or waste synthetic polymers such as tyres[1, 2]. In general, microporosity in carbons is created by the removal of carbon atoms by an activation process. However, after the evolution of the microporosity to a certain degree, further activation to increase the microporosity is accompanied by mesopore and macropore evolution[3]. The possibility of modifying the activation process to create smaller or larger pores, tailored to adsorb a group of specific molecules, make activated carbons important industrial adsorbents[4]. The activated-carbon porous structure is developed by either chemical or physical activation. Chemical activation involves the carbonisation of the raw material impregnated in a dehydrating or oxidising agent such as $H_3PO_4$ or KOH. In contrast, the physical activation process comprises one step of carbonisation followed by a partial gasification at high temperatures with an oxidant gas within the whole porous system[5]. In this process, the reaction gas penetrates the particle and attacks the carbon atoms producing the porosity development. The product gas diffuses out through the porous structure[5]. Nevertheless, in spite of numerous efforts, the activated carbon material is still poorly understood in terms of its nanostructure and especially with respect to its pore morphology and pore wall structure[6] due to its high disordered nature.

The micropore structure can be determined by several methods such as immersion calorimetry, small-angle X-ray scattering (SAXS) high resolution transmission electron microscopy (HRTEM) and gas- and liquid-phase adsorption, among which the most widely used is gas adsorption[7]. The pore structure of activated carbon is usually characterised in terms of the pore size distribution (PSD), perhaps the most important aspect of characterization of the structural heterogeneity of porous solids used in industrial applications. This PSD could be obtained as an arbitrarily chosen form such as, for instance, gamma or Gaussian distribution[8]. For a local isotherm one may choose traditional models, statistical mechanical methods such as DFT, or, most accurate for micropores, methods based on Monte Carlo simulation.

In this work, the adsorption of ethane at 264 K and nitrogen at 77 K was studied on a sequence of samples. These data were then analysed by DFT and a more accurate method for microporosity based on Monte Carlo simulations. A method based in the population balance approach was developed to follow the development of the PSD in the microporosity range. The process of PSD development has been proven to be a much more complicated process than growth of the pores.

## 2. PRODUCTION

A lignite char obtained by carbonisation in a stainless-steel, swept, fixed-bed reactor[9] was used to produce the activated carbons in the presence of $CO_2$. This solid was produced in several batch reactions of 60 g of lignite particles (diameter 0.2 - 0.5 mm) with a heating rate of 8 °C/min until 900±3 °C and this temperature was held for 3 hours to ensure effective devolatilisation. A stream of 2 l/min of $N_2$ was introduced to sweep the volatiles.

The second stage of activation with $CO_2$ was carried out in the same installation with a smaller reactor (height 27 cm). The samples with different degree of porosity development were obtained in batch reactions of 5 g of lignite char particles of 0.2-0.5 mm diameter, at 725 °C in a stream of 91% $N_2$ - 9% $CO_2$. The flowrate and the bed length used were 0.52 cm/s and 10 cm respectively to ensure a kinetically controlled regime as well as a homogeneous reaction in the whole bed. Ceramic rings were used as inert packing. The samples selected in this study were the char and the four activated carbons (AL721, AL723, AL726, AL7212) obtained at 1, 3, 6 and 12 hours of reaction. The burn-off of the samples, as a measure of activation, was defined as the change in weight of the dry sample due to the reaction of carbon.

## 3. CHARACTERISATION

The samples obtained were characterised by $N_2$ and ethane adsorption at 77 and 264 K, respectively, using an ASAP 2000 (Micromeritics) apparatus and an apparatus developed in the research group. Both gases had a purity of 99.99 %. The adsorption equilibria for pure ethane were determined using a bench-scale open-flow adsorption/desorption apparatus. A detailed description and a schematic representation of the apparatus can be found elsewhere[10]. In this study, both apparatuses were used with dry sample weights from 2 g to 0.6 g depending on the availability of the sample and its adsorption capacity. Each isotherm was measured twice. Preceding each pure-component isotherm measurement, regeneration of the sample was carried out at 150 °C under a vacuum of less than $2 \cdot 10^{-3}$ Torr for at least 3 h. Also prior to the experiment, the volumes of the charged adsorbent and the desorption chamber were determined by the expansion of helium gas.

The pore volume of the samples was divided in different pore sizes and values for the total, meso- and microporosity were obtained from the nitrogen data. The total pore volume of the sample was calculated from the $N_2$ adsorbed at a relative pressure of 0.98, the mesoporosity volume was studied by the Barret-Joyner-Halenda (BJH) method and the microporosity volume was calculated from the data obtained by the GCMC. The overall pore surface area was obtained by the Brunauer-Emmett-Teller (BET) theory. The surface areas of the mesopores and the macropores were obtained by the BJH method and GCMC respectively.

To obtain the PSD from the $N_2$ and ethane isotherm data by molecular simulation, a method based on GCMC developed by Davies et al.[11, 12] was used. The PSD is obtained from a set of model-pore isotherms and an experimental isotherm measured on the porous solid applying the adsorption integral equation:

$$N(T,P)_i = \int_0^\infty \rho(w,T,P)f(w)dw \qquad \text{i=1,....,n} \qquad (1)$$

where $N(T,P)$ is the amount adsorbed at temperature $T$ and pressure $P$, $\rho(w,T,P)$ is the "model-pore isotherm" calculated using GCMC simulation, $f(w)$ is the PSD and n is the number of adsorption measurements used in the analysis.

In this work we have modelled ethane and nitrogen as two-site molecules. The sites of ethane can be treated as being nearly spherical and nonpolar and the sites of nitrogen can be also treated in this way at 77 K, since the effect of its quadrupole at this temperature is negligible[13]. We have employed the Lennard-Jones interaction potential with the parameters given in Table 1[10, 13, 14] for those two-site molecules. In simulations of activated carbons, a slit-shaped model pore is commonly used[15] to described the pores and different number of graphite layers have been used to describe the solid. In this study 3 layers were chosen. To model the interaction between a site on an adsorptive molecule and other site in a pore wall of graphite, we employed the standard Steele 10-4-3 potential[16]. Table 1 shows the solid-fluid parameters obtained by applying the Lorentz-Berthelot combining rules to the Steele potential.

Table 1
Adsorbates and adsorbent parameters

|  | Nitrogen | Ethane | Adsorbent | Ads-Nitrogen | Ads-Ethane |
|---|---|---|---|---|---|
| $\sigma$, Å | 3.318 | 3.512 | 3.4 | 3.359 | 3.456 |
| $\varepsilon/k_B$, K | 37.8 | 139.8 | 28 | 32.5 | 62.6 |
| $r$, Å | 1.098 | 2.353 | - | - | - |
| $\Delta$, Å | - | - | 3.35 | - | - |
| $\rho_C$, Å$^{-3}$ | - | - | 0.114 | - | - |

## 4. POPULATION BALANCE METHOD

A population balance was used to follow the evolution of the PSD in a porosity development process. The general equation for a population balance is:

$$\frac{\partial n(w,t)}{\partial t} + \frac{\partial(Gn(w,t))}{\partial w} = B - D \tag{2}$$

where $n(w,t)$ is the number density function dependent on pore size, $w$, and time, $t$, $G$ is the growth rate and $B$ and $D$ are the terms due to birth and death of pores in the range studied.

In this study, as for the PSD study, a slit-shaped carbon pore has been used and only the distribution of the pore widths has been considered since the length and breadth of the pores are assumed to be semi-infinite.

The population balance deals with number of pores rather than the pore volume, in a given size range. However, if the number density function is assume to be equal to the pore volume size distribution divided by the average pore volume in each pore-size range and both breath and length are independent of time and uncorrelated with the width, then a parameter proportional to the number density function can be defined as follows:

$$n(w,t) \propto n'(w,t) = \frac{f(w,t)}{\overline{w}} \tag{3}$$

where $n'(w,t)$ is proportional to the number density function obtained from the PSD and $\overline{w}$ is the average pore size in any pore size range.

For a pore system undergoing growth, the rate of change of the number density function with respect to time and pore size is given by the equation:

$$\frac{\partial n'(w,t)}{\partial t} + \frac{\partial(Gn'(w,t))}{\partial w} = 0 \tag{4}$$

In this study $G$ is considered to be constant (reflecting a constant composition of the reactant gas[17]) and independent of the pore size.

To solve the partial differential equation defined from the population balance the discretised method of Hounslow et al.[18] was chosen.

## 5. RESULTS AND DISCUSSION

A set of four samples with different burn-off, from 16.7 to 78.9 %, was selected from the samples obtained at 725 °C. These samples were chosen to study their porosity evolution during the activation process.

Table 2
Activated lignite char samples ($T$=725 °C, $u$ = 0.52 m/s, 9 % $CO_2$).

| Ref. | %Burn-off | $S_{BET}$, $\frac{m^2}{g}$ | $V_T$, $\frac{cm^3}{g}$ | $S_{meso}$, $\frac{m^2}{g}$ | $V_{meso}$, $\frac{cm^3}{g}$ | $S_{micro}$, $\frac{m^2}{g}$ | $V_{micro}$, $\frac{cm^3}{g}$ |
|---|---|---|---|---|---|---|---|
| CHAR | 0 | 278 | 0.19 | 34.85 | 0.04 | 310.4 | 0.16 |
| AL721 | 16.7 | 459 | 0.29 | 91.92 | 0.09 | 497.0 | 0.25 |
| AL723 | 39.1 | **579** | 0.43 | 183.9 | 0.20 | 544.7 | **0.27** |
| AL726 | 58.6 | 569 | 0.52 | **204.9** | 0.24 | **547.7** | 0.26 |
| AL7212 | 78.9 | 477 | **0.56** | 185.2 | **0.26** | 429.7 | 0.21 |

To study the pore structure development, both the pore surface area and the pore volume variation in the different pore size ranges were studied using the data from the $N_2$ isotherms. The data obtained are given in Table 2. The BET surface area shows a maximum around 40% burn-off, while for $S_{meso}$ and $S_{micro}$ the maximum development is reached close to 60% burn-off. In the case of the pore volume, only $V_{micro}$ presents a maximum for the sample with an activation of 40 %; $V_{total}$ and $V_{meso}$ present a continuous increase with burn-off.

In order to study these different trends in porosity evolution depending on the range of pore size, a deeper study to characterise the PSD was carried out. DFT was applied to $N_2$ isotherms and the PSDs obtained for the samples are plotted in Fig. 1. In this Figure two main ranges of pore size can be distinguished, from 4 to 10 Å (range I) and from 10 to 50 Å (range II). Both ranges show a maximum in the mean pore sizes although this maximum appears at a different time. Porosity in range I seems to be created in the very first step of the reaction and later destroyed, hence the pore volume is decreasing with reaction time. On the other hand, the maximum pore volume for the mean pore size in the range II is reached in a second step of reaction, for the sample AL723. The samples obtained at 6 and 12 hours present similar results in this range of pore sizes.

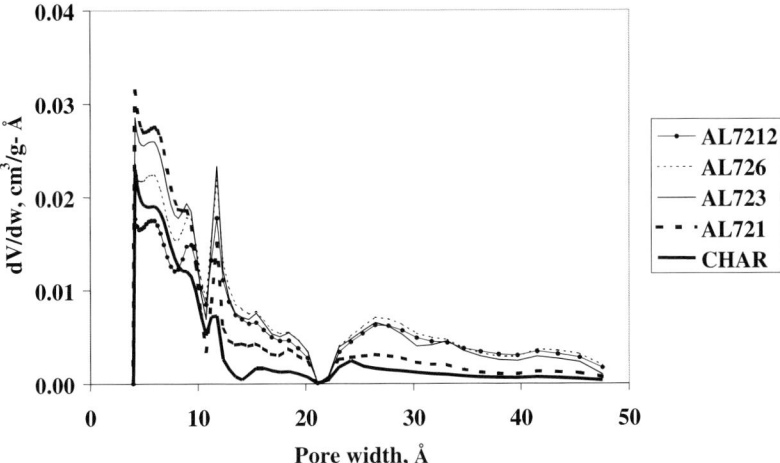

Fig. 1. PSD obtained from DFT.

The PSD obtained from the $N_2$ isotherms of the samples by means of GCMC are plotted in Fig. 2. With this method more accurate data can be obtained for pore sizes between 6 and 20 Å. In this case the evolution of the PSD with the activation reaction is focused on three main pore regions around 15, 12 and 7 Å. It is worth pointing out that the pores of 7 Å in the sample are mainly produced in the activation process, hence there is none in the PSD of the char. All the regions defined present a maximum in the mean pore size, for the region around 12 Å the maximum is for the sample with 40 % burn-off while for both regions around 7 and 15 Å, the maximum is reached at 60 % burn-off.

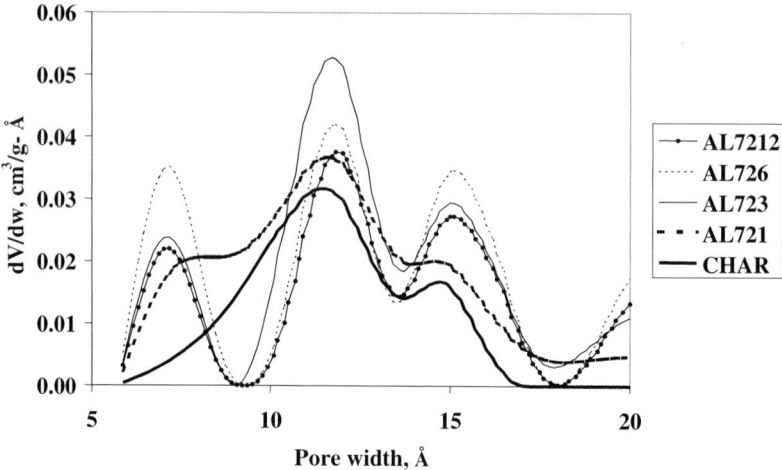

Fig. 2. PSD obtained from GCMC.

The PSDs obtained from the ethane isotherms are shown in Fig. 3. The region of pore sizes studied by the ethane, from 6 to 20 Å, can be divided into two main regions in this set of samples, above and below 13 Å as the PSD evolves differently in these two regions. In the region of small pore sizes, the maximum is reached at the beginning of the reaction for the sample with 20 % burn-off, which also shifts to slightly smaller micropore sizes. As the reaction takes place, there is a continuous decrease in the pore volume related to the mean pore size. In the case of the region from 13 to 20 Å the maximum is reached for the sample with 60% of burn-off and for a pore size of 17 Å.

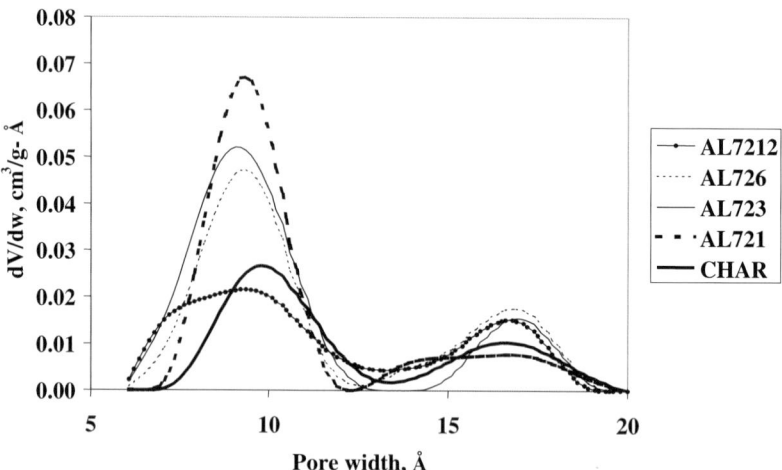

Fig. 3. Ethane PSD obtained from GCMC

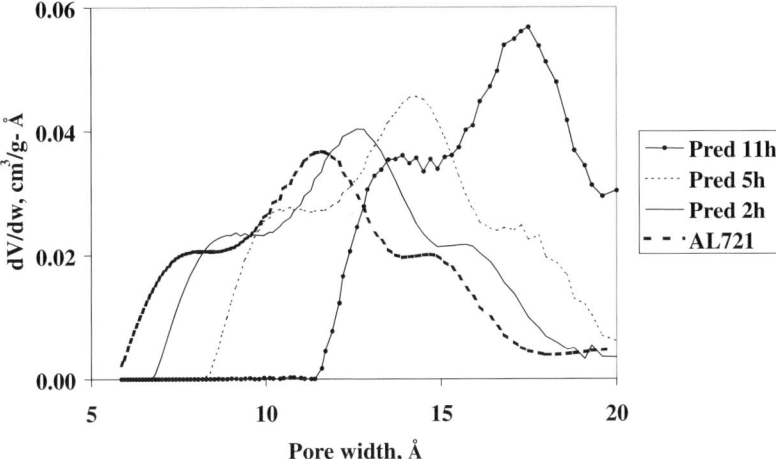

Fig. 4. PSD growth prediction with the population balance approach..

In Fig. 4 can be seen the predicted evolution of the PSD obtained from the $N_2$ isotherms of the first activated carbon produced, sample AL721, obtained from the population balance (Eq. (2) – (4)). $G$, the pore growth rate, is chosen to match the experimental rate of pore growth between samples AL721 and AL723.

Three main trends can be observed in the predicted development of the PSD: (i) the shape of three main regions remains constant, (ii) the continuously increasing height of the peak in the PSD, and (iii) the increasing mean pore size. The experimental results from $N_2$ and ethane (see Figures 2 and 3) keep the division of the PSD in main regions, three for $N_2$ and two for ethane, however, they show a manly stable mean pore size and a maximum in the pore volume related to it for each of these main pore regions.

In addition to the pore growth, other processes of porosity development during physical activation have been identified in the literature. The suggested processes include new micropore generation by selective activation and reopening of previously inaccessible pores[19] or pore coalescence due to the collapse of the pore walls[20]. Other processes are based on the structural description of the solid composed of graphitic crystallites or just graphene sheets. This graphitic crystallites are units composed by carbon and hydrogen organised in parallel layers of polyhexagonal carbon atoms whose edges are capped by hydrogen atoms[21]. Feng and Bhatia[5] observed the reopening pores which had been closed during heat treatment, by the presence of disorganised carbon which linked graphitic crystallites. In addition, to explain the apparently fixed size of some of the pores, they suggest that the crystallite structure is shrinking during gasification while retaining the same distance between crystallites. Another model, the "falling cards model"[22] deals with the evolution of the small honeycomb layers of carbon atoms, called graphene sheets, forming the activated carbons. By reaction, the links between adjacent sheets are broken, allowing some to rotate into parallel orientations increasing the number of organised regions and the porosity. One more explanation for the constant appearance of the main peaks could lie in a characterisation factor. The slit-shaped pore model has been proven to be able to resolve square or rectangular pores into a predefined set of pores which is function of the original pore size and aspect ratio (or "corner density")[23]. This suggests particular pore sizes would represent corners, walls and inner parts of the pores of the real material. While these rectangular or square pores grow, the different slit-shape pore sizes specified will grow in a different pattern, which would add complexity to the PSD development assessment.

## 6. CONCLUSIONS

The study of the pore volume and surface development of samples obtained from lignite char by $CO_2$ activation in a sequence of times established the development of the PSD. When the PSD is calculated applying a more accurate method for micropores based in GCMC, in both adsorbates, $N_2$ and ethane, three main trends are found. The number of peaks presented in the PSD is maintained for the whole range of burn-off studied: three and two for $N_2$ and ethane respectively. The mean pore sizes determined by the peaks in the PSD is also maintained mainly constant and, finally, the pore volume related to each mean pore size present a maximum for different activation time depending in the size range. However, according to the prediction obtained applying the population balance approach to pore growth, the initial division into regions is kept, but the mean pore size shift to bigger values and the size of the peak in the PSD grows continuously. Therefore, in addition to the pore growth, a range of different processes should be taken into account.

## ACKNOWLEDGMENTS

M.V. Navarro gratefully acknowledge the financial support from the MEC (Spain).

## REFERENCES
[1]    N. Nishiyama, T. Zheng, Y. Yamane, E. Egashira and K. Ueyama, Carbon, 43 (2005) 269.
[2]    D. Lozano-Castellon, D. Cazorla-Amoros, A. Linares-Solano and D.F. Quinn, Carbon, 40 (2002) 989.
[3]    R.C. Bansal, J.B. Donnet and F. Stoeckli, Active Carbon, Marcel dekker, New York, 1988.
[4]    M.B. Sweatman and N. Quirke, Journal of Physical Chemistry B, 105 (2001) 1403.
[5]    B. Feng and S.K. Bhatia, Carbon, 41 (2003) 507.
[6]    X. Py, A. Guillot and B. Cagnon, Carbon, 42 (2004) 1743.
[7]    F. Stoeckli, A. Guillot, A.M. Slasli and D. Hugi-Cleary, Carbon, 40 (2002) 383.
[8]    S. Ismadji and S.K. Bhatia, Applied Surface Science, 196 (2002) 281.
[9]    A.M. Mastral, R. Murillo, M.S. Callen and T. Garcia, Fuel Process. Technol., 60 (1999) 231.
[10]   J.H. Yun, T. Duren, F.J. Keil and N.A. Seaton, Langmuir, 18 (2002) 2693.
[11]   G.M. Davies, N.A. Seaton and V.S. Vassiliadis, Langmuir, 15 (1999) 8235.
[12]   G.M. Davies and N.A. Seaton, Langmuir, 15 (1999) 6263.
[13]   A.R. Turner and N. Quirke, Carbon, 36 (1998) 1439.
[14]   P.S.Y. Cheung and J.G. Powles, Molecular physics, 30 (1975) 921.
[15]   Y.F. Yin, B. McEnaney and T.J. Mays, Carbon, 36 (1998) 1425.
[16]   W.A. Steel, The interaction of gases with solid surfaces, Pergamon Press, Oxford, 1974.
[17]   O. Levenspiel, Chemical Reaction Engineering, Wiley, New York, 1975.
[18]   M.J. Hounslow, R.L. Ryall and V.R. Marshall, AIChE Journal, 34 (1988) 1821.
[19]   F. Rodriguez-Reinoso, Fundamental issues in control of carbon gasification reactivity, Kluwer Academic Publishers, Netherlands, 1991.
[20]   Q. Cai, Z.H. Huang, F. Kang and J.B. Yang, Carbon, 42 (2004) 775.
[21]   S. Yang, H. Hu and G. Chen, Carbon, 40 (2002) 277.
[22]   J.R. Dahn, W. Xing and Y. Gao, Carbon, 35 (1997) 825.
[23]   G.M. Davies and N.A. Seaton, Carbon, 36 (1998) 1473.

Studies in Surface Science and Catalysis 160
P.L. Llewellyn, F. Rodriquez-Reinoso, J. Rouqerol and N. Seaton (Editors)
© 2007 Elsevier B.V. All rights reserved

# Chemically modified nanoporous carbons obtained using template carbonization method

**C. O. Ania and T.J. Bandosz**

Chemistry Department, The City College of New York, New York, NY 10031,USA. E-mail: tbandosz@ccny.cuny.edu

New nanoporous carbons with extremely high mesopore volumes and a highly dispersed distribution of metals on the carbon surface have been obtained using mesoporous silica. Polystyrene sulfonic acid-based organic salts were used as carbon precursor. The precursor chemistry was modified by cation exchange with catalytically active metals (i.e., copper, nickel, cobalt), prior to carbonization. The carbons have pore sizes predominantly in the range of 10-50 nm.

## 1. INTRODUCTION

Activated carbons have been extensively used in a large number of industrial applications, due to their expanded pore network and to their ability to react with other heteroatoms [1, 2]. Recent developments in technology require porous carbons with tailorable pore sizes, specific shapes of pores, and large surface areas. Preparation of mesoporous carbons with uniform mesopores, large surface areas and large pore volumes carbons has attracted widespread attention, due to their application in emergent areas [3-5]. Although to meet such demands many novel approaches to design a pore structure have been proposed [6] only a few of them allow for full control of porosity. The template carbonization has been proven to be an excellent method for obtaining carbons with large surface areas, high porosity and controlled narrow pore size distributions in the mesopore range [6]. Many polymeric precursors including polyacrylonitrile, poly(furfuryl alcohol), poly(vinylidene chloride), and phenol resin have been applied as porous carbon precursors [7-10].

On the other hand, the knowledge of the surface chemistry of carbons is also of paramount importance. It determines the chemical stability and the reactivity in adsorptive and catalytic processes [2]. Very often, the natural chemistry of the activated carbon surface is not potent enough to enhance the specific adsorbate-adsorbent interactions or the catalytic properties of the carbon surfaces. Thus, the possibility of enhancing the physicochemical properties of carbons via modification of their surface by the incorporation of the desired heteroatoms is among current research interests in carbon science. Various methods have been described in the literature to modify carbon surface chemistry [11-14]. One of them is pyrolysis of certain organic salts containing metal cations. This method leads to carbonaceous materials with a high surface area and well-developed microporosity [15]. Moreover, when the pyrolysis of polystyrene sulfonic acid-co-maleic acid salts is carried out at 1073 K metal species are highly dispersed on the microporous carbon surface.

This work focuses on tailoring the porous structure and the metal content of carbon materials using a carbonization-templating procedure. The objective of this research is to study the effects of the templates porosity and the type of metals in the carbon precursors on the surface features of template derived carbons.

## 2. EXPERIMENTAL

### 2.1. Synthesis of mesoporous silica

The mesoporous template is synthesized in a two-step pathway using tetraethyl orthosilicate as the silica source and non-ionic surfactants following the procedure described by Boissiere et al. [16]. As a surfactant, non ionic poly(ethylene oxide) -PEO- surfactant Tergitol 15-S-N was selected. Briefly, the silica source is added under stirring to the solution of the surfactant and the pH is adjusted at around 2 with HCl. The solution is kept in a closed vessel for 18 h without stirring at room temperature. Then, a small amount of NaF is added to promote silica condensation. The mixture is aged for 3 days at two temperatures, 293 K and 333 K. The white precipitate obtained is filtrated, dried and calcined at 873 K. The samples are referred to as T30 and T60, respectively.

### 2.2. Carbon templating

Sucrose and poly-(styrene sulfonic acid co maleic acid)- sodium salt are used as carbon precursors in the template carbonization method. For the sucrose synthesis, organic phase is introduced into the silica channels by liquid impregnation of a sucrose solution according to the process described by Ryoo et al. [17]. After drying at 433 K, the organic precursor/silica composite is again impregnated with a sucrose solution and heated at 1023 K in an inert atmosphere. After two impregnation cycles, the amount of carbon deposited in the template represents 36 wt %.

In the case of polystyrene-based polymer, the silica is coated with an aqueous solution containing 16% of the organic salt, until the incipient wetness is achieved. The impregnated sample is then carbonized under nitrogen at 1023 K for 40 min. In some cases, the precursor chemistry is modified by incorporation of catalytically active metals. Previous to impregnation, cation-exchange in the organic precursor is done using $Cu(NO_3)_2$, $Co(NO_3)_2$, and $Ni(NO_3)_2$) as sources of cations. These exchanged solutions are used to impregnate the silica matrix up to the incipient wetness. After removal the silica matrix using hydrofluoric acid (48%) at room temperature, a Soxhlet washing with distilled water is carried out to remove an excess of water-soluble inorganic salts (sodium and excess of transition metal salts). The samples are referred to as T30 or T60 followed by SUC or PS M which stand either for sucrose or polymeric salt of a corresponding metal (T30-SUC or T60-PS Ni).

### 2.3. Characterization of Samples

Textural characterization was carried out by measuring the $N_2$ adsorption isotherms at 77 K. Before the experiments, the samples were outgassed under vacuum at 393 K. The isotherms were used to calculate the specific surface area, $S_{BET}$, total pore volume, $V_T$, (at $p/p_o > 0.99$) and pore size distributions. The pore size distributions were evaluated using density functional theory (DFT) [18].

X-ray diffraction (XRD) patterns at small ($2\theta = 0.5$–5) angles were obtained on a Siemens D5000 instrument operating at 40 kV and 20 mA and using Cu K$\alpha$ radiation (k = 0.15406 nm). The analyzed powdered carbon samples were spread as thin layers on a glass slide and analyzed. The structure of the silica and carbon materials was characterized by scanning electron microscopy (SEM).

## 3. RESULTS AND DISCUSSION

A mesostructured silica material (MSU-1) which is known to posses a 3-D wormhole structure was used as template in order to obtain the porous carbons [19]. The XRD patterns of the silica samples synthesized at different temperatures (i.e. 303K and 333K) show a single XRD peak, which is distinctive for this kind of silica materials [19]. The peak is shifted to lower angles as the synthesis temperature increases. Since MSU-1 silicas do not have an ordered structure, the single, broad peak in the XRD pattern reflects the pore-pore correlation distance. Thus, the silica wall thickness value can be deduced by subtracting the pore diameter value from $d$-spacing. As expected, the silica walls thickness decreased as a function of the synthesis temperature [16], varying from 1.5 nm at 303 K to 1.1 nm at 333 K. SEM micrographs presented in Fig.1 reveal that the synthesized silicas consist of particle aggregates of spherical shape within the micrometric range (2-10 micron).

Details of the porous structure of the silica samples are provided from the nitrogen adsorption isotherms. Two different aging temperatures used for the synthesis of the silica lead to development of mesostructured frameworks with different porous network [20, 21]. Raising the aging temperature of the second step in the silica synthesis results in an increase in the pore diameter. This is expected due to the changes in the amphiphilic character of the surfactant with the temperature and the fact that the silica synthesis proceeds via an assembly mechanism between the surfactant and the silica source [22]. Both silica samples have mesopores of uniform sizes. The maximum of the pore size distributions for the T30 and T60 samples are in 2.7 nm and 3.7 nm, respectively. Moreover, the pore size distributions are narrow in both cases with a full width at the half maximum (FWHM) about ~1 nm. The BET surface areas slightly decrease with an increase in aging temperature from 1014 to 980 $m^2g^{-1}$ for T30 and T60, respectively.

The morphology of the silica particles and their structural characteristics are preserved in the template-derived carbons, as presented in SEM images in Fig.1. The carbon replicas, are made up of spherical particles with diameters of 2-10 μm, regardless the carbon precursor. However, when copper was incorporated to the polystyrene carbon precursor, along with the carbon spheres, a bunch of metallic clusters appeared randomly dispersed at the surface of the carbon matrix. SEM micrographs show numerous particle aggregates of metal of a few nm in size. Taking into account the low reduction potential of the pair Cu (II)/Cu, and the reductive atmosphere during carbonization, the chemical status of copper in these aggregates is assigned to copper zero. During carbonization formation of metallic particle occurs, and the aggregates are produced as a result of migration of the reduced metals to the surface. This hypothesis is in good agreement with the finding indicating new forms of carbons created during pyrolysis in the presence of catalytic metals [15, 23]. Moreover, EDX analysis of the carbon particles loaded with metals showed that, despite the clusters, metals are also highly dispersed inside the carbon matrix. The diffraction patterns show well-dispersed metal sulfides, sulfur, and oxides agglomerates on the surface of our materials.

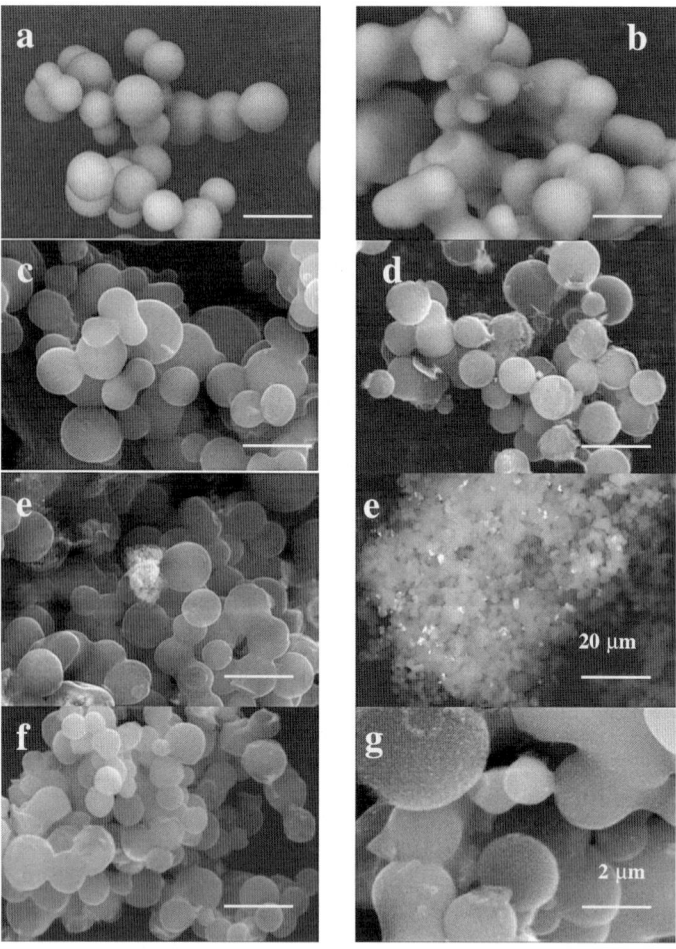

Fig.1. SEM pictures of the silicas and template-derived carbons: a) Silica T30, b) Silica T60, c) T60 suc, d) T60 Ps, e) T60 PS Cu, f) T60 Ps Ni, g) T60 Ps Co, (magnification is 5 μm, except when specified)

The XRD patterns for the template-derived carbons reveal that the samples obtained from sucrose as a carbon precursor have a diffraction pattern similar to that of the MSU-1, with a broad peak in the low angle range. On the other hand, application of a polystyrene-based salt as a carbon precursor leads to a strong alteration of the silica template. In all cases, the diffraction peak at the low angle range disappeared, indicating that the wormhole structure of the silica is not replicated and that the carbons posses a disordered and disorganized structure.

Carbons prepared from sucrose have surface areas up to $1500 \text{ m}^2 \text{ g}^{-1}$, and pore volumes up to $1 \text{ cm}^3 \text{ g}^{-1}$ (Table 1). These carbons also have micropore volumes of around $0.210 \text{ cm}^3 \text{ g}^{-1}$, which is likely related to the intrinsic microporosity of the carbon precursor. For both samples a capillary condensation occurs in the $p/p_o$ range between 0.1-0.4 and 0.3-0.6, which clearly indicates that the structural porosity of the silica framework is replicated in the carbon. The pore size distributions for these carbons are very narrow with peak maxima at 1.2 nm and 1 nm, for T30 SUC and T60 SUC, respectively. Additionally, a slight development of complementary mesoporosity (estimated from the amount of nitrogen adsorbed at $p/p_o > 0.7$) can be seen, accounting for 6 % and 7 % of the total pore volume.

The use of polystyrene-based organic salts as carbon precursor, as well as the incorporation of metals to the carbonaceous structure leads to a strong alteration of the ordered porous structure of the silica. The replicated carbons have a large mesopore system, as indicated by the presence of a large hysteresis loop and capillary condensation step at high relative pressures in the nitrogen adsorption data. All the samples show two capillary condensation steps at relative pressures between 0.2-0.4 and $p/p_0 > 0.7$. The first one, as it also appears for the silica, is associated to the structural porosity – framework of confined mesopores- resulting from the dissolution of the inorganic template. The second condensation step at relative pressures $> 0.7$ is assigned to the porosity (complementary or textural porosity) that arises from either interparticle voids due to the agglomeration of very fine particles which contains framework-confined mesopores [24] or it is the effect of the reaction of sodium with the silica during carbonization which leads to destruction of thin pore walls and formation of large pores in the template. Since the latter effect is dynamic and occurs during carbonization, the only feasible way to control it is by changing the thickness of pore walls and the content of sodium. As estimated from nitrogen adsorption, this complementary mesoporosity accounts for almost 80 % of the porosity in the samples template-derived from the T60 silica. Indeed, the pore walls in the parent silica are thinner in T60 than those in T30. For the samples obtained from T30, complementary mesoporosity represents between 65-75 % of the total porosity. An exception is for cobalt nitrate use for a carbon precursor for which the relatively small volume of large mesopores was formed (48 %). These differences might be due to the different amounts of precursors loaded in the silica matrix. The T60 silica besides, having thinner pore walls has also larger pore volumes, which leads to a higher amount of the carbon precursor infiltrated, and thus, larger contributions of complementary pores are attained.

Table 1
Structural parameters calculated from nitrogen adsorption isotherms

| Sample | $S_{BET}$ [m$^2$ g$^{-1}$] | $V_{mic}$ [cm$^3$ g$^{-1}$] (DFT) | $V_{mes}$ [cm$^3$ g$^{-1}$] (DFT) | $V_t$ [cm$^3$ g$^{-1}$] p/p$_o$<0.99 | V at p/p$_o$>0.7 [cm$^3$ g$^{-1}$] | Complementary Mesopores [%] |
|---|---|---|---|---|---|---|
| T30 PS | 997 | 0.083 | 1.754 | 2.42 | 1.57 | 65 |
| T30 PS Ni | 959 | 0.131 | 2.171 | 2.60 | 1.846 | 71 |
| T30 PS Cu | 907 | 0.073 | 2.426 | 2.87 | 2.15 | 75 |
| T30 PS Co | 820 | 0.118 | 0.989 | 1.29 | 0.623 | 48 |
| T30 SUC | 1560 | 0.15 | 0.652 | 0.99 | 0.055 | 6 |
| T60 PS | 943 | 0.081 | 2.228 | 3.61 | 2.811 | 78 |
| T60 PS Ni | 758 | 0.087 | 1.869 | 2.25 | 1.661 | 74 |
| T60 PS Cu | 661 | 0.054 | 1.949 | 2.46 | 1.912 | 78 |
| T60 PS Co | 738 | 0.073 | 1.942 | 2.32 | 1.732 | 75 |
| T60 SUC | 1382 | 0.212 | 0.861 | 1.20 | 0.083 | 7 |

The PSD presented in Fig. 2 show wide distributions with a maximum at 20 nm. Moreover, the close similarities in the pore size distributions exist, especially for the complementary mesopores. A remarkable difference in porosity is noticed for the cobalt-containing sample for which the peak related to the complementary mesopores is much narrower (maximum at 9.3 nm) and has lower intensity (48 %) than those for the nickel and copper counterparts. This is accompanied by the smaller BET surface area than those for copper and nickel or sodium counterparts. The latter effect had also been observed when the cobalt-loaded precursor was carbonized in the absence of template [16]. In that case it was attributed to the uniformity of metal dispersion achieved in the carbon precursor, as it is well known that this factor is extremely important for the formation of novel carbon entities on various supports, and to the forces that arise from the decomposition of the carbon precursor [15].

The origin of the textural porosity in the template derived carbon is complex. This behavior has been attributed in the literature to the water-vapor released during the carbonization process (due to the use of a liquid carbon precursor) that could hydrolyze the silica network [25]. However, the volume of complementary mesopores decreases significantly when a sucrose solution is used as precursor (10 % and 16 % for T30 SUC and T60 SUC, respectively). This corroborates the hypothesis proposed by Fuertes et al. [26] on the effect of the impregnation method in the porous structure of the templates. They attributed the formation of complementary mesopores to the coalescence of unfilled silica pores during carbonization, and the subsequent collapse of the porous structure when silica walls are removed [26]. Therefore, the one-step infiltration procedure with the organic salt (as opposed to the 2-step impregnation procedure of the sucrose providing a better filling of the silica pores) led to a larger number of unfilled silica pores that might coalescence during carbonization and collapse the porous structures resulting in a wider distribution of mesopores.

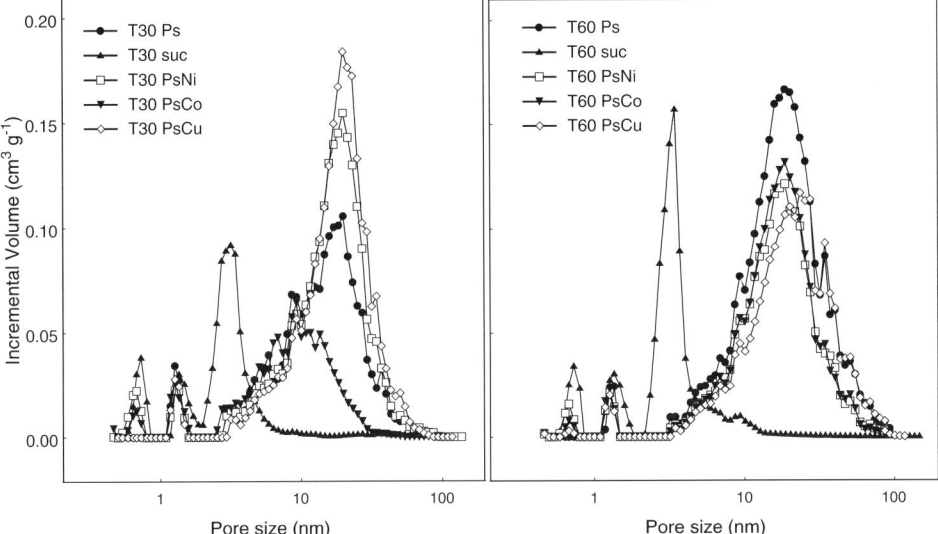

Fig.2. Pore size distributions for the carbon derived in T30 and T60 silica matrices

It is important to mention that the case of our samples, another factor contributing to porosity development arises which is associated with the carbon precursor composition. As mentioned above, during carbonization sodium salts react with the silica walls, creating wider porosity in the carbons. We refer to this process as a dynamic template effect. The fact that for the thinner pore walls silica template and the same content of sodium in the matrix the remarkably similar distributions of pore sizes are obtained support the hypothesis about the existence of this effect and its role for porosity development.

## 4. CONCLUSIONS

The results of this study show the feasibility of the template carbonization method as a route for tailoring the size of mesopores along with the surface functionalities of activated carbons with a spherical morphology. When a MSU-1 silica is used as a template and polystyrene-based organic salts as carbon precursors, the template-derived carbons have large pore volumes (up to 2.55 cm$^3$ g$^{-1}$) in the mesopore range (10-100 nm), and highly dispersed metals with a catalytic activity on the surface. The presence of reactive gases during the carbonization results in expansion of the carbonaceous structure, and thus in wide mesopore size distributions of the template-derived carbons. The dynamic effect of sodium reacting with the silica during carbonization also results in the development of high volume of mesopores.

## Acknowledgments

The financial support for this research, provided by FICYT (Fundación para el Fomento en Asturias de la Investigación Científica Aplicada y la Tecnología, Spain) and PSC CUNY (PSC CUNY 66382-0035) is gratefully acknowledged. We thank Dr. Pis from Instituto Nacional del Carbon for kindly providing SEM and XRD.

## REFERENCES

[1]   F. Derbyshire, M. Jagtoyen, R. Andrews, A. Rao, I. Martín-Gullón and E.A. Grulke, (L.R. Radovic, ed.) Chemistry and Physics of Carbon, 27, Marcel Dekker, New York 2000, p. 1.
[2]   L.R. Radovic, C. Moreno-Castilla, J. Rivera-Utrilla, (L.R. Radovic. Ed.), Chemistry and Physics of Carbon, 27, M. Dekker: New York, 2000, p. 227.
[3]   S. Yoon, J. Lee, T. Hyeon and S.M. Oh, J. Electrochem. Soc. 147 (2000) 2507.
[4]   S.H. Joo, S.J. Choi, I. Oh, J. Kwak, Z. Liu. O. Teresaki and R. Ryoo, Nature, 412 (2001) 169.
[5]   S. Han, K. Sohn and T. Hyeon, Chem. Mater. 12 (2000) 3337.
[6]   T. Kyotani, Carbon 38 (2000) 269.
[7]   L. Feng, S.H. Li, J. Zhai, Y.L. Song, L. Jiang and D.B. Zhu, Synthetic Metals, 135-136 (2003) 817.
[8]   A.B. Fuertes, Micro. Meso. Mater., 67 (2004) 273.
[9]   C.J. Meyers, S. D. Shah, S.C. Patel, R.M. Sneeringer, C.A. Bessel, N.R. Dollahon, R.A. Leising and E.S. Takeuchi, J. Phys. Chem. B105, ( 2001) 2143.
[10]  T. Kyotani, T. Nagai, S. Inoue, A. Tomita, Chem. Mater., 9 (1997) 609.
[11]  H. Oka, M. Inagaki, Y. Kaburagi and Y. Hishiyama, Solid State Ionic, 121 (1991) 157.
[12]  F. Goutfer-Wurmser, H. Konno, Y. Kabuagi, K. Oshida and M. Inagaki, Synth. Met., 118 (2002) 33.
[13]  M. Inagaki, Y. Okada, K. Miura and H. Konno, Carbon, 37 (1999) 329.
[14]  C. Zhou, J. Kong, E. Yenilmez and H. Dai, Science, 290 (2000) 1552.
[15]  D. Hines, A. Bagreev and T.J. Bandosz, Langmuir, 20 (2004) 3388.
[16]  C. Boissiere, A. Larbot, A. van der Lee, P.J. Kooyman and E. Prouzet, Chem Mater., 12 (2000) 2902.
[17]  R. Ryoo, S.H. Joo and S. Jun, J. Phys. Chem. B, 103 (1999) 7743.
[18]  J. Olivier, Carbon, 36 (1998) 1469.
[19]  S.A. Bagshaw, E. Pouzet and T.J. Pinnavaia, Science, 269 (1995) 1242.
[20]  E. Prouzet, T.J. Pinnavaia, Angew Chem Int Ed., 36 (1997) 516.
[21]  S. Alvarez and A.B. Fuertes, Carbon, 42 (2004) 423.
[22]  C. Boissiere, M.A.U. Martines, M. Tokumoto, A. Larbot and E. Prouzet, Chem. Mater., 15 (2003) 509.
[23]  H. Konno, R. Matsuura, M. Yamasaki and H. Habazaki, Synth. Met., 1215 (2002) 167.
[24]  A.B. Fuertes, Materials Letters, 58 (2004) 1494.
[25]  J. Parmentier, C. Vix-Guterl, P. Gibot, M. Reda, M. Ilescu, J. Werckmann and J. Patarin, Micro. Meso. Mater., 62 (2003) 87.
[26]  A.B. Fuertes and D.M. Nevskaia, Micro Meso. Mater., 62 (2003) 177.

Studies in Surface Science and Catalysis 160
P.L. Llewellyn, F. Rodriquez-Reinoso, J. Rouqerol and N. Seaton (Editors)

# Influence of the synthesis conditions on the pore structure and stability of MCM-41 materials containing aluminium or titanium

**M.M.L. Ribeiro Carrott, C. Galacho, F.L. Conceição and P.J.M. Carrott**

Centro de Química de Évora and Departamento de Química, Universidade de Évora, Colégio Luís António Verney, 7000-671 Évora, Portugal

A comparison of the pore structural properties of MCM-41 containing titanium, prepared at room temperature, with those of aluminosilicate grades is presented. The influence on the structural characteristics of using different metal sources and metal content is also considered. Additionally, the stability of Al-MCM-41 and Ti-MCM-41 samples, with Si/M=30, towards prolonged exposure to pure water vapour at 298K was investigated.

## 1. INTRODUCTION

Amongst the most important features of ordered mesoporous materials for their use in catalysis are the extremely high surface area, which potentially allows an efficient dispersion of active sites, and the large pore diameters that favour the diffusion of bulky molecules. However, the introduction of the heteroatoms into ordered mesoporous silicas to make them catalytically active can lead to a decrease in the quality of the pore structure as compared with the pure silica grades, the extent of which may depend on the preparation conditions. Therefore, the control of the quality of the pore structure is a relevant aspect to consider.

MCM-41 containing aluminium or titanium have been commonly used, respectively, for acid or oxidation catalysed reactions, and they are frequently obtained by grafting of active species onto the inner surface of the mesopores of a pure silica material or by its incorporation in the walls during synthesis [1]. This is usually done by a hydrothermal route, but an alternative procedure at room temperature was reported for Al-MCM-41, that allows the preparation in reduced times of well structured materials containing aluminium mainly tetracoordinated [2]. Furthermore, we have shown [3] that samples prepared by this method, even with very low Al content (Si/Al$\geq$138), had enough acidity to catalise the reaction of double bond position of 1-butene, without the need of post-synthesis cationic exchange treatments, usually required in hydrothermal synthesis to obtain the protonated forms, that cause further disruption of the pore structure. Also, we have adapted the room temperature method to obtain Ti-MCM-41 which showed negligible acidity, as compared with a Ti grafted MCM sample, which is an advantage for selectivity and activity of some redox reactions [4].

In the present study we have used other metal sources and higher metal contents and present a comparison of the effects of introduction of aluminium and titanium on the pore structural properties. Additionally, considering that a severe limitation of ordered mesoporous materials, either pure silica grades [5] as well as some hydrothermally synthesised MCM-41 containing aluminium [6] or titanium [7], is that structural changes usually occur in the presence of water vapour, a study of the stability towards water vapour at 298K is also presented.

## 2. EXPERIMENTAL

Pure silica MCM-41 samples were prepared at room temperature following a procedure previously reported [8], but with the tetraethoxysilane (TEOS) being dissolved in an equal volume of propan-2-ol (Si-p) or ethanol (Si-e) prior to the addition to the surfactant and ammonia mixture. The preparation in the presence of the alcohols had the purpose to use conditions similar to those used in metal containing samples. The titanosilicate MCM-41 samples were prepared by a similar method to that previously developed for aluminosilicate samples [2], but using always the same volume (equal to that of TEOS) of the solution in propan-2-ol of the titanium alcoxides, whose quantities were varied to achieve the required molar ratio. Titanium ethoxide, isopropoxide and n-butoxide were used as metal sources, respectively, for samples Ti-Ep-X, Ti-Pp-X, and Ti-Bp-X (with X=nominal Si/Ti). The samples containing aluminium, designated Al-Pp-X (with X=nominal Si/Al) were prepared according to the room temperature method in ref.[2] using aluminium isopropoxide in propan-2-ol. Sample Al-Be-30 was prepared in a similar way but using aluminium tert-butoxide dissolved in ethanol, as it is more soluble in this alcohol. In the synthesis of Al-P-30 and Al-B-30, the alcoxide was directly dissolved in the TEOS.

In all syntheses, hexadecyltrimethylammonium bromide was used as template and the alcohol and water were previously bi-distilled. In all cases, after stirring for 1h, the products were recovered by filtration and washed with bi-distilled water, dried at 343K and finally calcined in air at 823K (heating rate of 3Kmin$^{-1}$) for a minimum of 11h.

All samples were characterised by X-ray diffraction and nitrogen adsorption at 77K. The XRD measurements were obtained on a Bruker AXS, D8 Advance, powder diffractometer using CuK$_\alpha$ radiation. Nitrogen adsorption isotherms at 77K were determined on a CE Instruments Sorptomatic 1990. Diffuse reflectance UV-Vis spectra of the titanium containing samples were collected on a Varian Cary 5-E. The stability towards water was evaluated by measuring two consecutive water vapour adsorption-desorption isotherms at 298K, followed by XRD and nitrogen adsorption characterization. Isotherms of n-pentane and n-hexane, respectively, for Ti-Ep-30 and Al-Pp-30, were also determined before and after the exposure to water vapour. Water and hydrocarbon vapour adsorption isotherms at 298K were determined in a vacuum gravimetric apparatus using a CI Electronics MK2 vacuum microbalance with a Robal control unit and Edwards Barocel 600 capacitance manometers.

## 3. RESULTS AND DISCUSSION

### 3.1. Effect of metal introduction on the pore structural characteristics of MCM-41

It can be seen in Figs 1-4 that either pure silica sample prepared in the presence of propan-2-ol or metal containing samples, present X-ray diffraction patterns and nitrogen adsorption isotherms typical of MCM-41 materials. Practically coincident results to those of Si-P were obtained for pure silica samples prepared in the absence of alcohol and in the presence of ethanol.

In all cases, the three or four peaks observed at low diffraction angles could be indexed to only one hexagonal phase and the unit cell parameter values are presented in Table 1. Additionally, no peaks at higher angles were observed, indicating the absence of any crystalline phases containing aluminium or titanium, even for the higher metal contents used.

In most cases the nitrogen isotherms were completely reversible, except for the samples for which the desorption points are shown, and which presented hysteresis at high relative pressures after the pore filling step, indicating interparticle agglomeration. The total and external surface areas and the pore volume were estimated in the usual manner from the corresponding $\alpha_s$ plots, constructed using standard data for adsorption of nitrogen on non-porous partially hydroxylated silica, and the results are presented in Table 1. Although it is well know that the surface area is overestimated and that the hydraulic pore width approach underestimates the pore size [9,10] the values will be considered for comparative analysis between samples. It is evident from the results presented that all samples have considerably high pore volumes and surface areas, but that the exact pore structural parameters as well as the regularity and uniformity of the pore structure depend on some synthesis parameters.

In general, the introduction of titanium in the MCM-41 materials has some similar effects to those of aluminium, in the sense that some loss of pore structure quality is observed and this increases with increasing the metal content, which qualitatively agrees with the previous findings for Al-MCM-41 [2]. However, it is quite clear from the comparison done in Figs. 1 and 2 that the introduction of titanium has a less disruptive effect on the pore structure than aluminium, the difference between the two metals being more pronounced with increasing metal content. In fact, it can be seen that in particular for the nominal molar ratios of 30 and 10, the Ti-MCM-41 samples present steeper pore filling steps and better defined XRD peaks than those of Al-MCM-41 with comparable or even lower (Si/Al=15) aluminium content. Additionally, the nitrogen adsorption isotherm of the Al-Pp-15 sample presents pronounced hysteresis at high pressures, associated with secondary mesoporosity, in contrast with that of Ti-Pp-10, which is almost completely reversible.

Therefore, the results show that the previously proposed method to obtain Al-MCM-41 at room temperature can also be used with even better results for the preparation of Ti-MCM-41 materials. This synthesis can also be carried out with other titanium alcoxides giving samples with very similar pore volumes and pore uniformity as can be seen in Fig. 3(a) and Table 1, for the series prepared with Si/Ti=50. However, it can be seen that with titanium n-butoxide the regularity appears to be smaller, which can be due, at least partially, to the n-butanol formed by the hydrolysis and its higher tendency to solubilize in the micelles than the smaller chain alcohols. Although the expected amount of n-butanol formed is less than the one reported to give another coexisting hexagonal and a lamellar phase [11], it may be sufficient to cause some disordering of the hexagonal array of the micelles.

Additionally, we noticed that these three titanium alcoxides gave completely clear solutions in both alcohols, while the aluminium isopropoxide was not completely soluble and the original solution in freshly bi-distilled propan-2-ol probably contained already some polymerised species. The attempt to use other aluminium source and dissolution directly in TEOS, gave materials with very similar pore volumes and total surface areas as can be seen in Fig. 4(a) and Table 1. Small differences are observed with regard to external surface area, lower values being obtained when aluminium tert-butoxide is used as metal source. It is evident from the XRD patterns that the direct dissolution in the TEOS results in materials with slightly lower degree of structural ordering, although no significant differences are observed with regard to pore volume and uniformity.

So, it can be noted that, either with Ti or Al, the pore structural characteristics are not affected very much by the metal source used in this study, but the changes are more pronounced with increasing metal content. This can result either from the successful incorporation of the bigger metal species or from the presence of extra-structural polymerised species, which can be formed in the synthesis gel and also during the calcination step. In fact, in the original work [2] the authors reported that, although the aluminium is predominantly

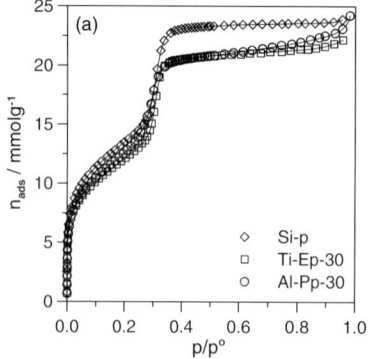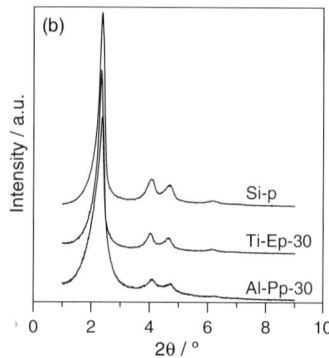

Fig. 1. (a) Nitrogen adsorption isotherms, at 77K, and (b) X-ray diffraction patterns of pure silica MCM-41 prepared in the presence of propan-2-ol and Ti- and Al-MCM-41 of Si/M=30.

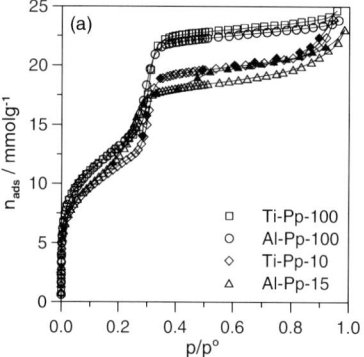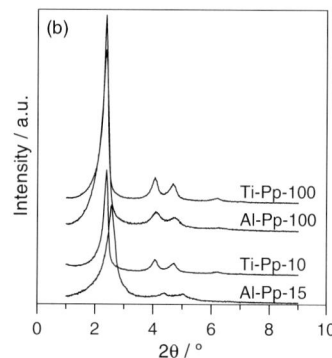

Fig. 2. (a) Nitrogen adsorption-desorption isotherms, at 77K, and (b) X-ray diffraction patterns of Ti-MCM-41 (Si/Ti=100 and 10) and Al-MCM-41 (Si/Ti=100 and 15). (empty symbols: adsorption, filled symbols: desorption)

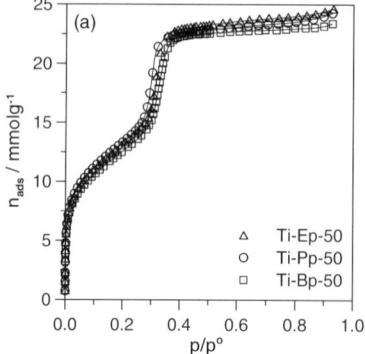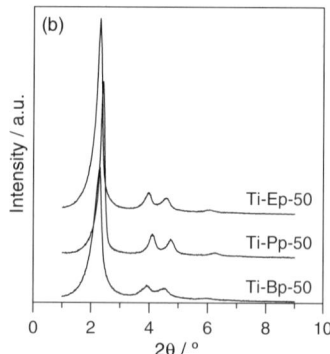

Fig. 3. (a) Nitrogen adsorption isotherms, at 77K, and (b) X-ray diffraction patterns of Ti-MCM-41 of Si/Ti=50 prepared with different titanium alcoxides.

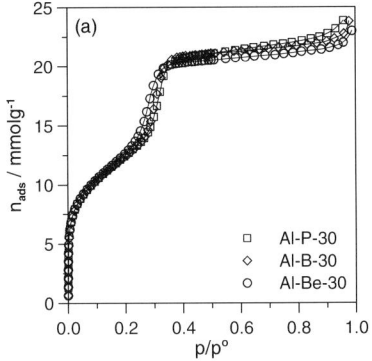 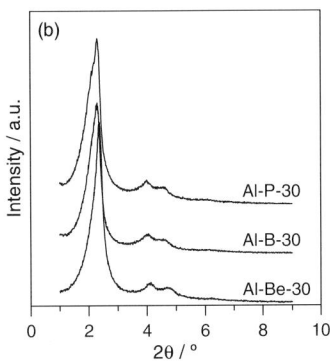

Fig. 4. (a) Nitrogen adsorption isotherms, at 77K, and (b) X-ray diffraction patterns of Al-MCM-41 of Si/Al=30 prepared with different aluminium alcoxides and conditions.

Table 1
Results of the structural characterisation by XRD and nitrogen adsorption at 77K.

| Sample | Metal Source* | Alcohol | XRD | Nitrogen adsorption at 77K[#] | | | |
|---|---|---|---|---|---|---|---|
| | | | $a_0$ (nm) | $A_s$ $(m^2g^{-1})$ | $A_{ext}$ $(m^2g^{-1})$ | $V_p$ $(cm^3(liq)g^{-1})$ | $d_p(H)$ (nm) |
| Si-p | - | propan-2-ol | 4.34 | 1063 | 17 | 0.80 | 3.06 |
| Si-e | - | ethanol | 4.33 | 1057 | 17 | 0.79 | 3.04 |
| Ti-Pp-100 | TIP | propan-2-ol | 4.34 | 1042 | 32 | 0.79 | 3.13 |
| Ti-Ep-50 | TE | propan-2-ol | 4.45 | 1033 | 33 | 0.80 | 3.20 |
| Ti-Pp-50 | TIP | propan-2-ol | 4.29 | 1053 | 33 | 0.79 | 3.10 |
| Ti-Bp-50 | TB | propan-2-ol | 4.51 | 995 | 23 | 0.78 | 3.21 |
| Ti-Ep-30 | TE | propan-2-ol | 4.36 | 955 | 28 | 0.72 | 3.11 |
| Ti-Pp-10 | TIP | propan-2-ol | 4.33 | 902 | 103 | 0.63 | 3.15 |
| Al-Pp-100 | AIP | propan-2-ol | 4.31 | 1032 | 32 | 0.77 | 3.08 |
| Al-Pp-30 | AIP | propan-2-ol | 4.32 | 976 | 71 | 0.70 | 3.09 |
| Al-Be-30 | ATB | ethanol | 4.32 | 980 | 52 | 0.69 | 2.97 |
| Al-P-30 | AIP | - | 4.42 | 975 | 78 | 0.70 | 3.12 |
| Al-B-30 | ATB | - | 4.41 | 971 | 49 | 0.71 | 3.08 |
| Al-Pp-15 | AIP | propan-2-ol | 4.04 | 920 | 108 | 0.59 | 2.91 |

*TE,TIP and TB – titanium ethoxide, isopropoxide and n-butoxide, respectively; AIP and ATB – aluminium isopropoxide and tert-butoxide, respectively.

[#] Considering $\sigma_{N2} = 0.162$ nm$^2$ and $\rho_{N2} = 0.808$ g cm$^{-3}$.

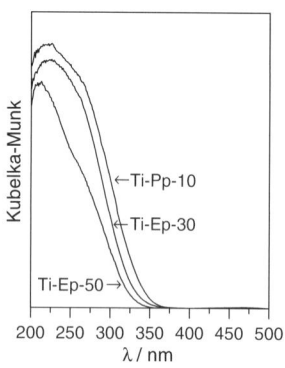

Fig. 5. DRUV-Vis spectra of
Ti-MCM-41.

in a tetrahedral environment, octahedrally coordinated aluminium species are also present in the samples prepared by this method using aluminium isopropoxide. With regard to Ti-MCM, the diffuse reflectance UV-Vis spectra shown in Fig. 5, indicate that the materials contain mainly isolated Ti (IV) incorporated in the silicate structure, as the maxima at ~212nm (Ti-Ep-50) and at ~225nm (Ti-Ep-30 and Ti-Pp-10)) can be ascribed to tetracoordinated species, namely tetrapodal ($Ti(OSi)_4$) and tripodal (such as $Ti(OH)(OSi)_3$) units [12]. For the higher Ti content samples (Ti-Ep-30 and Ti-Pp-10) it is evident that there is broadening of the band to higher wavelength. This indicates an increase of penta- and hexacoordinated species, probably due to the formation of partially polymerised species with Ti-O-Ti bonds or nanodomains of amorphous $TiO_2$-$SiO_2$, which also exist at lower titanium content but to a much lesser extent. Taking these results into account and that the introduction of Ti with Si/Ti=50 hardly changed the pore volumes and regularity, it is reasonable to consider that the extraframework species contribute to the disruption of structural properties with increasing titanium content.

### 3.2. Structural stability towards water vapour

It can be seen from Fig. 6 that the water vapour isotherms obtained in the alumino and titanosilicate samples present some similar features indicating an initial surface with low adsorption affinity which undergoes hydroxylation during the determination of the first isotherm. In both cases the pore filling step of the second isotherm occurs at lower relative pressures, the hysteresis loop is narrower and the desorption of the first and second isotherms practically coincide. Additionally, the values of volume adsorbed at p/p°=0.9 presented in Table 2 for the two Al and Ti containing samples are very similar which is consistent with the results obtained from nitrogen adsorption. Also, the higher uniformity of pore size of Ti-Pp-30 previously inferred from Fig.1 results in a pore filling step, in the water isotherm, steeper than in the case of Al-Pp-30. For this sample the pore filling occurs at lower relative pressure (~0.56 ) than for Ti-Ep-30 (~0.59) which may result from differences in surface chemistry, as expected on the basis of the presence of the different metals incorporated in the silicate structure. In fact, an estimate of the Kelvin radius ($r_K$), gives 1.8 and 2.0 nm, respectively, for Al-Pp-30 and Ti-Ep-30, indicating that at the onset of condensation the preadsorbed layer of water is different in the two samples, as the pore size was similar taking into consideration the nitrogen adsorption results.

The general shape of the water isotherms is quite similar to those previously obtained for pure silica samples [5]. However, there is an important difference: after the step the second isotherm is coincident (Al-Pp-30) or only slightly lower (Ti-Pp-30) than the first, which is in marked contrast with the behaviour of pure silica MCM-41 for which the limiting uptakes on repeat runs were significantely reduced. The comparison of the results of characterization by XRD, nitrogen and hydrocarbon adsorption, before and after exposure to water vapour shown in Figs. 7-8 and Table 2, clearly indicates that the materials still retain considerable structural ordering and high pore volumes and that structural alterations occurred mainly during the first isotherm and to a much lesser extent than with pure silica materials [5]. This is also in

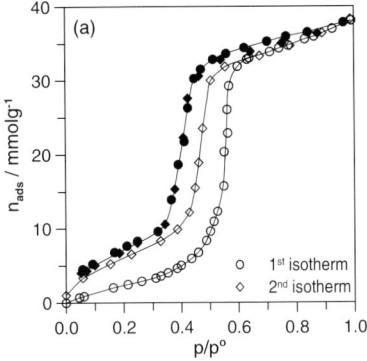 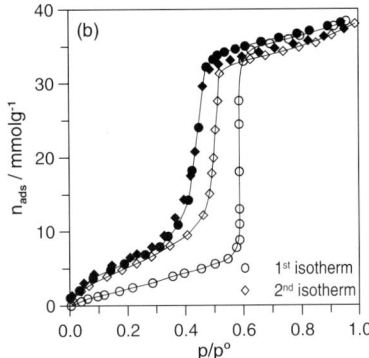

Fig. 6. Water vapour adsorption-desorption isotherms determined at 298K on: (a) Al-Pp-30 and (b) Ti-Ep-30. (empty symbols: adsorption, filled symbols: desorption)

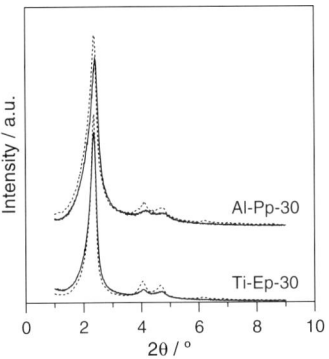
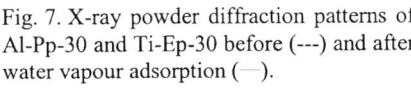

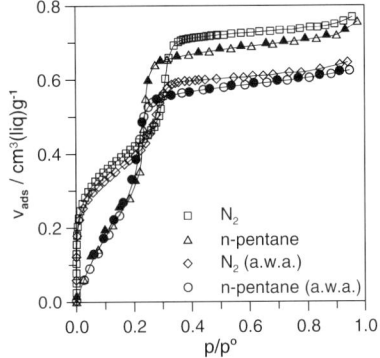

Fig. 7. X-ray powder diffraction patterns of Al-Pp-30 and Ti-Ep-30 before (---) and after water vapour adsorption (—).

Fig. 8. Nitrogen, at 77K, and n-pentane, at 298K, adsorption isotherms determined on Ti-Ep-30 before and after water vapour adsorption (a.w.a.). (empty symbols: adsorption, filled symbols: desorption).

Table 2
Results of the analysis of water vapour isotherms at 298K and of the characterisation of samples before and after water vapour adsorption (a.w.a.) by XRD, nitrogen adsorption at 77K and hydrocarbon adsorption at 298K.

| Sample | XRD | Water adsorption | | $C_n$* adsorption | Nitrogen adsorption | | | |
|---|---|---|---|---|---|---|---|---|
| | $a_o$ (nm) | $V_{0.9}(1)$ | $V_{0.9}(2)$ ($cm^3(liq)g^{-1}$) | $V_{0.9}$ ($cm^3(liq)g^{-1}$) | $A_s$ ($m^2g^{-1}$) | $A_{ext}$ ($m^2g^{-1}$) | $V_p$ ($cm^3(liq)g^{-1}$) | $d_p(H)$ (nm) |
| Al-Pp-30 | 4.32 | 0.66 | 0.66 | 0.71 | 976 | 71 | 0.70 | 3.01 |
| a.w.a. | 4.26 | | | 0.64 | 910 | 37 | 0.60 | 2.79 |
| Ti-Ep-30 | 4.36 | 0.68 | 0.66 | 0.72 | 956 | 28 | 0.72 | 3.11 |
| a.w.a. | 4.34 | | | 0.62 | 886 | 49 | 0.59 | 2.82 |

* n-hexane for Al-Pp-30 and n-pentane for Ti-Ep-30.

marked contrast with some hydrothermally synthesised Al-MCM-41 and Ti-MCM-41 whose stability was extremely low [6,7]. Furthermore, Al-Pp-30 appears to be slightly more resistent than Ti-Ep-30. It should also be noted that the reduction of pore volume results partially from the increase in mass due to the residual water present after the determination of the isotherms as well as from the pore narrowing that occurred. Taking into account that during the measurements the samples were in contact with pure water vapour for over a month, it is evident that these samples are reasonably stable. So, this study shows that the introduction of the titanium and of aluminium lead to an increase in the structural stability, which may be due to the formation of stronger Si-O-M bonds and a higher degree of condensation in the pore walls. The presence of extraframework species may also contribute to the enhancement of stability, by avoiding the interaction of water with nearby Si-O-Si bonds.

## 4. CONCLUSIONS

The results show that the room temperature synthesis method previously developed for MCM-41 containing aluminium is an adequate method for obtaining titanosilicate of different Si/Ti with highly regular MCM-41 structure, high pore volume, very uniform pore size and containing tetracoordinated titanium. Additionally, it was found that the incorporation of titanium has a less disruptive effect on the pore structure than aluminium. The increase of Ti or Al content leads to a gradual reduction of the quality of the pore structure, but no significant differences were observed with the different metal sources used.

The introduction of either titanium or aluminium by this method leads to materials that are considerably more stable towards prolonged exposure to water vapour than pure silica grades of MCM-41 as well as some hydrothermally synthesised Al and Ti-MCM-41 samples.

### Acknowledgements

The authors are grateful to the Fundação para a Ciência e a Tecnologia (FCT, Portugal) and the Fundo Europeu para o Desenvolvimento Regional (FEDER) for financial support (project no. POCTI/CTM/45859/2002).

### REFERENCES

[1] A. Taguchi and F. Schüth, Micropor. Mesopor. Mater., 77 (2005) 1.
[2] A. Matsumoto, H. Chen, K. Tutsumi, M. Grun and K. Unger, Micropor. Mesopor. Mater., 32 (1999) 55.
[3] T.N. Silva, J.M. Lopes, F.R. Ribeiro, M. Ribeiro Carrott, P.C. Galacho, M.J. Sousa and P. Carrott, React. Kinet. Catal. Lett., 77 (2002) 83.
[4] A. Tuel, Micropor. Mesopor. Mater., 27 (1999) 151.
[5] M.M.L. Ribeiro Carrott, A.J.E. Candeias, P.J.M. Carrott, and K.K. Unger, Langmuir, 15 (1999) 8895.
[6] X.S. Zhao, F. Audsley and G.Q. Lu, J. Phys. Chem. B, 102 (1998) 4143.
[7] K.A. Koyano and T. Tatsumi, Micropor. Mater., 10 (1997) 259.
[8] M. Grün, K. Unger, A. Matsumoto and K. Tsutsumi in Characterisation of Porous Solids IV, B. McEnaney, T.J. Mays, J. Rouquerol, F. Rodriguez-Reinoso, K.S.W. Sing and K.K. Unger (eds.), Royal Society of Chemistry, Cambridge, 1997, p.81.
[9] M. Kruk, M. Jaroniec and A. Sayari, J. Phys. Chem. B, 101 (1997) 583.
[10] M.M.L. Ribeiro Carrott, A.J. Candeias, P.J.M. Carrott, P.I. Ravikovitch, A.V. Neimark and A.D. Sequeira, Micropor. Mesopor. Mater., 47 (2001) 323.
[11] P. Ågren, M. Linden, J.B. Rosenholm, R. Schwarzenbacher, M. Kriechbaum, H. Amenitsch, P. Laggner, J. Blanchard and F. Schüth, J. Phys. Chem. B, 103 (1999) 5943.
[12] P. Ratnasamy, D. Srinivas and H. Knözinger, Adv. Catal., 48 (2004) 1.

Studies in Surface Science and Catalysis 160
P.L. Llewellyn, F. Rodriquez-Reinoso, J. Rouqerol and N. Seaton (Editors)
© 2007 Elsevier B.V. All rights reserved

# Effect of oxidizing agent on activated carbon cloth porosity and surface chemistry

**P. Pendleton[a], A. Badalyan[a], R. Bromball[a] and W. Skinner[b]**

[a]Center for Molecular and Materials Sciences; [b]Ian Wark Research Institute;
University of South Australia, Mawson Lakes, South Australia, 5095, Australia

FM1/250 and FM1/700 activated carbon cloth samples were chemically treated for 24 hours in 4M $NaOH$ at 373 K. Chemical activation was also carried out in 10 % (w/vol) saturated ammonia peroxydisulfate solution, and in 10 % (w/w) potassium peroxydisulfate solution at room temperatures for 24 hours. Nitrogen adsorption on chemically treated samples was measured at 77 K. All samples showed a classical Type I microporous adsorption isotherm. $NaOH$ treatment caused an increase in total micropore volume for FM1/250, but no statistically significant change for FM1/700. In contrast, a change in primary micropore volume for FM1/700 occurred, but no statistically significant change for FM1/250. The peroxydisulfate treatment decreased primary and total micropore volumes for both samples of activated carbon cloth. Both treatments caused relative increases in oxygen surface structures promoting their selective adsorption potential.

## 1. INTRODUCTION

Activated carbon cloth (ACC) is used for the removal of volatile organic compounds from effluent gas streams [1]. Specific micropore structures in ACCs make them real candidates for use in adsorption processes where a high rate of adsorption is accompanied with short contact time between adsorbent and adsorbate. It is very important for manufacturers to produce ACC with desirable pore size distributions, and be able to control the development of various pores from primary micropores to mesopores during activation processes. Adsorption selectivity towards various gases is another very important property, which can be achieved by introducing oxygen complexes onto the surface during activation [2-4].

Our recent surface analyses of powdered activated carbon [5] has been extended to the more regularly defined surfaces exhibited by ACC. These adsorbents often display a "memory" in that their physical and chemical resistance submits to various solvents, albeit to a lesser extent than their parent (un-carbonised) material. We treated two microporous ACC samples with an excessive aqueous oxidizing agent, 4M $NaOH$, and milder conditions provided by 10 % (w/vol) aqueous ammonium and potassium peroxydisulfates solutions. The oxidation differently affects the distribution of surface carbon-oxygen structures, leading to different adsorption behaviour of ACC samples.

## 2. EXPERIMENTAL

## 2.1. Materials and methods

Two samples of activated carbon cloth (ACC), FM1/250 and FM1/700, from Calgon Carbon Corporation were used. These are single, plain weave materials, with the second number designating the value of wetting heat in J/g. The samples were exposed for 24 hours (suffix '24') to 4M *NaOH* at 373 K at atmospheric pressure. Non-treated samples have suffix '0'. Other samples were suspended via a rotary mixer at room temperatures for 24 hours in 25 mL of 10% (w/vol) ammonia peroxydisulfate (APDS), of analytical grade (suffix '10 %-APDS'), in 10% (w/vol) potassium peroxydisulfate (PPDS), of analytical grade (suffix '10 %-PPDS'). After treatment, samples were washed in deionised water for 48 hours. These samples were then heated in an atmospheric oven at 398 K for 24 hours for moisture removal. Some moisture still remains in these ACC samples, but subsequently removed when the samples underwent heating in high vacuum before nitrogen adsorption.

## 2.2. Nitrogen adsorption measurements

Gaseous nitrogen adsorption measurements were carried out using an automatic manometric gas adsorption apparatus described elsewhere [6]. Before adsorption measurements, all samples were heated in the electric oven at 473 K (much lower than their decomposition temperature of >773 K) and evacuated to a pressure of 0.1 mPa for 8 hours to remove moisture and adsorbed gases. Liquid nitrogen level around the tube with ACC sample was controlled with a precision of ±0.02 mm using a modified liquid nitrogen delivery system [7]. Nitrogen adsorption was carried out at 77 K [6]. Samples reached equilibrium within 15-20 minutes. The number of experimental points (58-60) was sufficient for the evaluation of BET specific surface areas (BET-SSA), pore volumes and pore size distribution. Thermal transpiration corrections were made at pressures below 266 Pa [8]. At higher pressures, differences between corrected and non-corrected equilibrium pressures are smaller than the uncertainties in pressure measurements. Relative combined standard uncertainties (RCSUs) of the nitrogen amount adsorbed were calculated for each experimental point [9].

## 2.3. XPS analysis

X-ray Photoelectron Spectroscopy (XPS) is a widely used procedure for determination of surface functional groups. A Kratos Axis-Ultra X-ray photoelectron spectrometer (Kratos Analytical, UK) was used to determine oxygen-containing functional groups on the surfaces of the treated and untreated ACC samples. The X-ray source was monochromatic $AlK_\alpha$. (1486.6 eV) at 150W power. Atomic concentrations were calculated using core-line peak areas from survey spectra and tabulated sensitivity factors for the instrument. Survey spectra were collected in the binding energy range from 0 to 1100 eV at a pass energy of 160 eV. High resolution scans of the $C_{1s}$ were collected at a pass energy of 20 eV. These scans were curve-fitted to determine the presence of various carbon-carbon graphitic character and carbon-oxygen surface functional groups.

## 3. RESULTS AND DISCUSSION

### 3.1. Nitrogen adsorption isotherms

Fig. 1 presents nitrogen adsorption isotherms for FM1/250 samples. All adsorption isotherms showed classical Type I shape representative of a wholly microporous adsorbent. Since adsorption isotherms are almost parallel over the relative pressure range from 1.E-4 to 1.E-1, the external adsorption surface is unaffected by chemical activation. Chemical activation by

*NaOH* changes adsorption behaviour of FM1/250-24 at very low relative pressures $P/P^0 <$ $10^{-4}$. There is appreciable adsorption at relative pressures below $\approx 10^{-6}$, compared to the amount adsorbed at relative pressure of about $10^{-6} < P/P^0 < 10^{-5}$ by FM1/250-0 sample. We interpret this increase as the formation of smaller pores within the micropore structure due to chemical etching. As a reference, for comparison purposes only, the BET-SSA increased by 4.40 % after treatment in *NaOH*. In Table 1, we intentionally left one additional significant digit in the reported values for indication purposes only. Chemical treatment by 10 %-APDS and 10 %-PPDS reduced the BET SSA by 14.43 and 3.33 %, respectively.

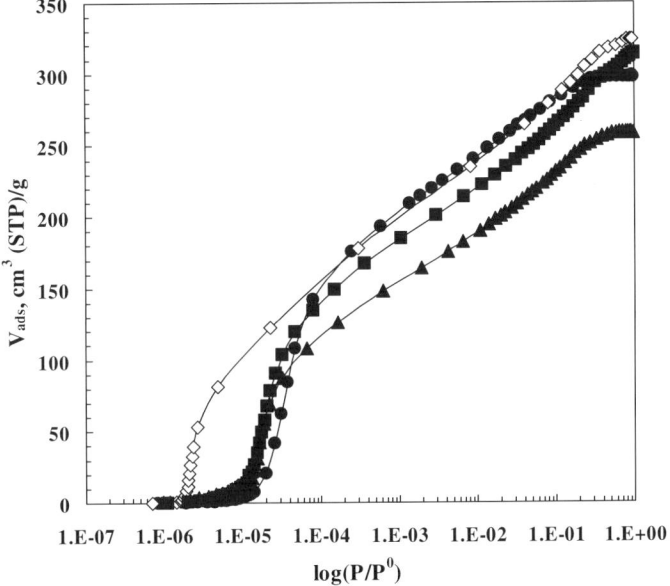

Fig. 1. Nitrogen adsorption isotherms for FM1/250 ACC. (● - FM1/250-0, ◇ - FM1/250-24, ▲ - FM1/250-10%-APDS, ■ - FM1/250-10%-PPDS).

Table 1
BET-SSA and pore volumes for FM1/250 and FM1/700 ACC

| Sample Name | $S_{BET}$, m$^2$/g | $W^\alpha_{prim}$, mL/g | $W^\alpha_{tot.}$, mL/g |
|---|---|---|---|
| FM1/250-0 | 1033.9±10.9 | 0.1927±0.0080 | 0.4595±0.0077 |
| FM1/250-24 | 1079.4±5.5 | 0.1883±0.0018 | 0.5007±0.0172 |
| FM1/250-10%-APDS | 884.7±16.3 | 0.1362±0.0043 | 0.3992±0.0859 |
| FM1/250-10%-PPDS | 999.5±10.7 | 0.1628±0.0025 | 0.4035±0.0356 |
| FM1/700-0 | 1157.7±6.7 | 0.1640±0.0014 | 0.5775±0.0354 |
| FM1/700-24 | 1213.3±9.9 | 0.1779±0.0035 | 0.5927±0.0414 |
| FM1/700-10%-APDS | 979.9±18.0 | 0.1389±0.0049 | 0.4646±0.0549 |
| FM1/700-10%-PPDS | 1066.3±18.7 | 0.1609±0.0036 | 0.5004±0.0650 |

Nitrogen adsorption isotherms for FM1/700 samples are also Type I, but again shown in Fig. 2 as $V_{ads}$ vs $log(P/P^0)$-plots to emphasize the low pressure amount adsorbed. In contrast with FM1/250-24, the *NaOH*-treated FM1/700-24 sample does not show an increased amount adsorbed, however, it does show a modest 4.80% increase in BET-SSA over FM1/700-0. Treatment with APDS and/or PPDS results in 15.36 and 7.90% reduction in BET-SSA.

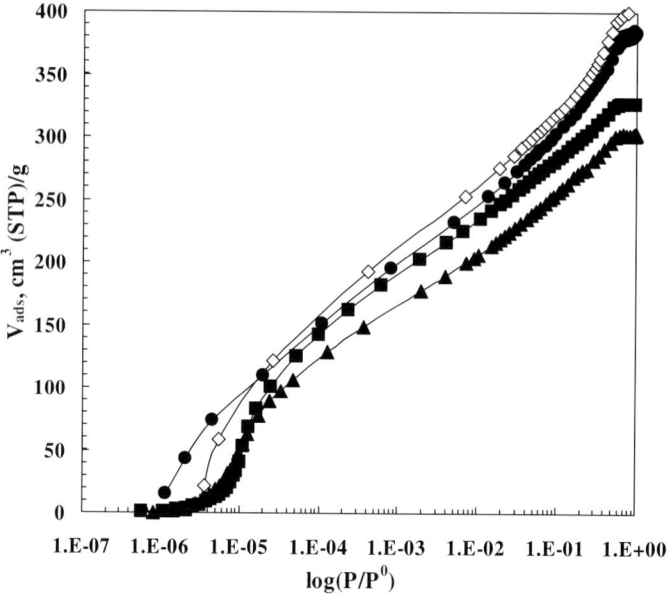

Fig. 2. Nitrogen adsorption isotherms for FM1/700 ACC. (● - FM1/700-0, ◇ - FM1/700-24, ▲ - FM1/700-10%-APDS, ■ - FM1/700-10%-PPDS).

### 3.2. Micropore volumes and pore size distribution

Primary and total micropore volumes were determined after $\alpha_S$ -uncertainty analyses [9]. A non-graphitized, non-porous carbon black [7] was used as the adsorbent standard. The increased amount adsorbed by FM1/250-24 at low pressures (in Fig.1) are equated to an increase in total micropore volume, however, the primary volume remains statistically constant. The similar treatment of FM1/700 saw no statistical change in total micropore volume, but an 8.48% increase in primary micropore volume. Treatment with PDS solutions results in a widening of the pore volume distributions as suggested by the increase in uncertainty in each primary and total micropore volume. Changes in micropore structure are further borne out by the Horwath-Kawazoe defined micropore-size distribution data in Fig. 3 and 4. Chemical activation of FM1/250 by *NaOH* shifted the maxima of the pore size distributions from ≈ 0.50 nm to ≈ 0.40 nm, thus indicating the opening of smaller micropores due to possible leaching effects. Similar treatment of FM1/700 did not significantly change pore size distribution. Treatment with either of the PDS solutions shifts the pore size distribution maxima to ≈ 0.46 nm for the both FM1/250 and FM1/700.

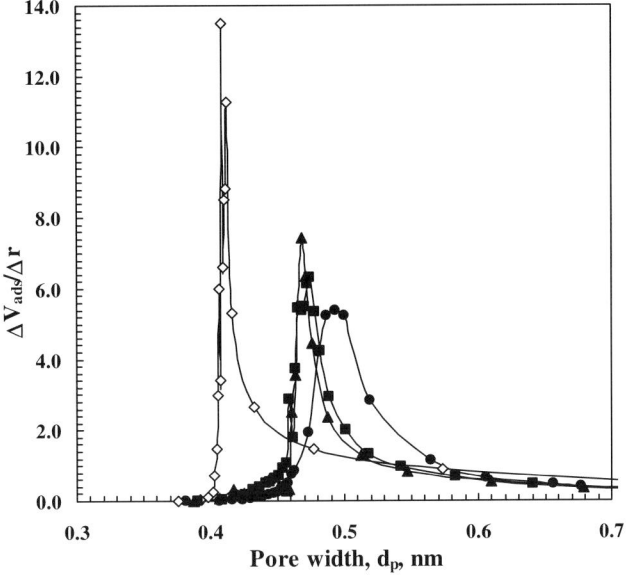

Fig. 3. Pore size distribution for FM1/250 ACC. (● - FM1/250-0, ◇ - FM1/250-24, ▲ - FM1/250-10%-APDS, ■ - FM1/250-10%-PPDS).

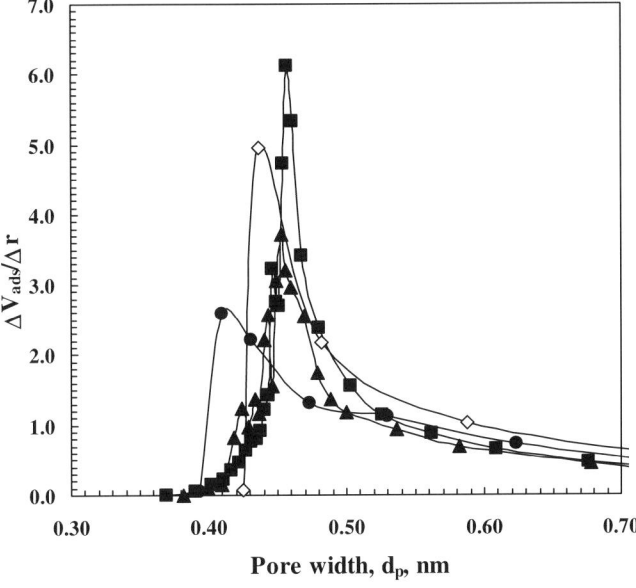

Fig. 4. Pore size distribution for FM1/700 ACC. (● - FM1/700-0, ◇ - FM1/700-24, ▲ - FM1/700-10%-APDS, ■ - FM1/700-10%-PPDS).

### 3.3. XPS results

According to Calgon, the nomenclature 250 and 700 refers to the heat of wetting (presumably water) with units of J/g [10]. These relatively high enthalpies indicate the pores are of similar dimensions to water and/or due to strong interactions with the surface functional groups. The nitrogen adsorption isotherms for both materials show a condensation over the same relative pressure ranges, indicating both materials contain pores of similar dimensions. Thus, the difference in heat of wetting is primarily due to strong water-surface functional group interactions. The type of oxygen functional groups defined via XPS are: ether ($C$-$O$-$C$) at $\approx 284.5$ eV, carbonyl ($C=O$) at $\approx 286.5$ eV, carboxylic ($O$-$C=O$) at $\approx 288.2$ eV, and anhydrate ($O$-$(C=O)$-$O$) at $\approx 290.2$ eV (see Fig. 5-8).

Both FM1/250 and FM1/700 samples show similar relative concentrations of $C$-$C$ carbon, regardless of differences during their activation manufacturing process. Slight difference in relative concentrations of ether and carbonyl groups in both non-treated ACC samples are probably due to the above activation differences.

On exposure to *NaOH*, FM1/250 exhibits relatively greater susceptibility to surface ether structure formation, a more non-polar oxygen structure than the carbonyl oxygen. In this case, surface oxidation is associated with leaching giving enhanced adsorption in the lower pressure range. A small decrease in the relative concentration of the carboxylic group is due to specific reaction with *NaOH*. In contrast, the more activated material FM1/700 experiences a relative increase in carbonyl structure and a nominal relative increase in acid and anhydride structures. These relative increases are structural oxidation effects leading to pore widening.

Treatment with APDS and PPDS solutions reduces apparent relative concentration of $C$-$C$ graphitic structure, with the effect of PPDS being more pronounced. APDS solution gave a relative increase in carbonyl content for both FM1/250 and 700, with associated nominal widening of the micropores and overall reduction in pore volume. The 10 % PPDS solution gave no appreciable relative change in oxygen content in FM1/250, but affected the FM1/700 in a manner analogous to the 10 % APDS treatment.

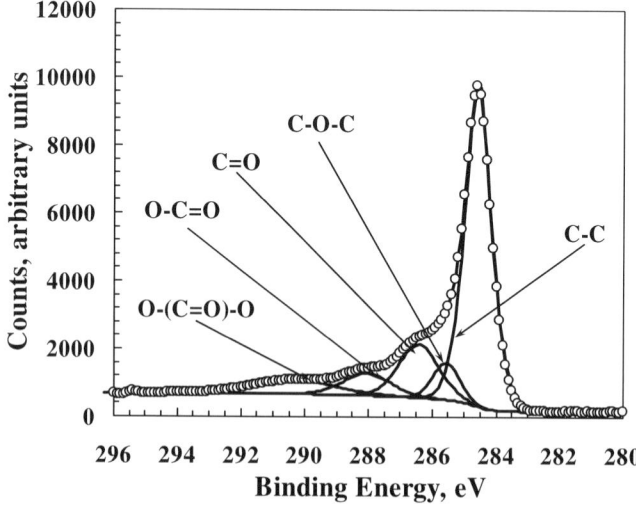

Fig. 5. Analysis of $C_{1s}$ XPS spectrum for FM1/250-0 showing types of C-O surface groups.

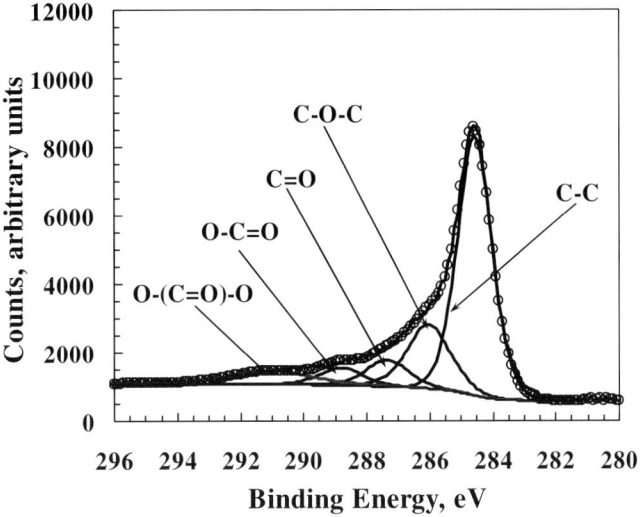

Fig. 6. Analysis of $C_{1_S}$ XPS spectrum for FM1/250-24 showing types of C-O surface groups.

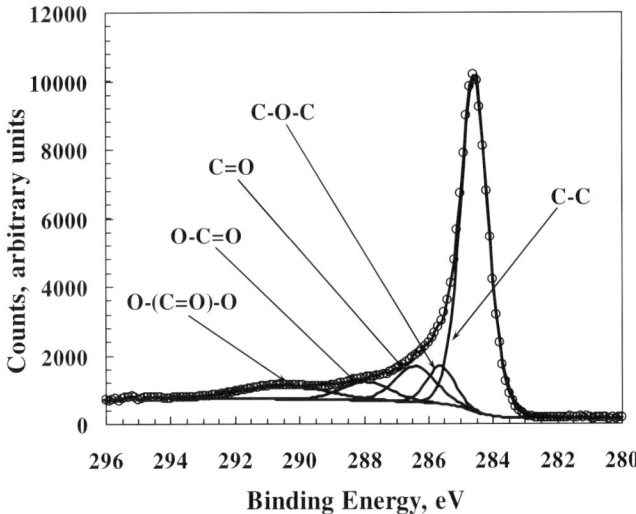

Fig. 7. Analysis of $C_{1_S}$ XPS spectrum for FM1/700-0 showing types of C-O surface groups.

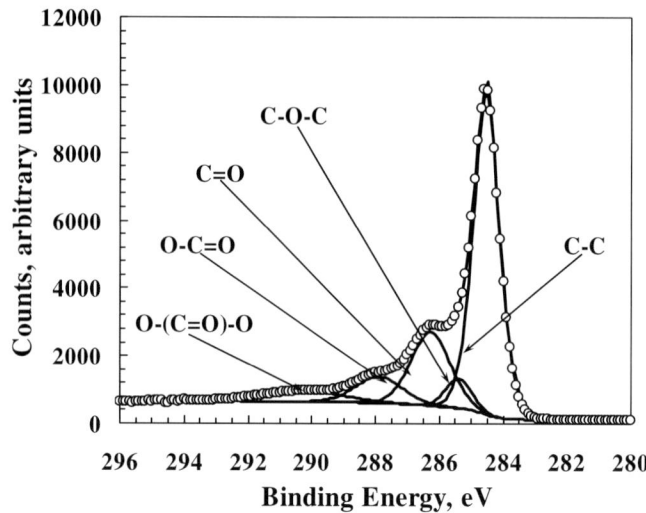

Fig. 8. Analysis of $C_{1_s}$ XPS spectrum for FM1/700-24 showing types of C-O surface groups.

## 4. CONCLUSIONS

ACC derived from polymer-based starting materials are succeptible to oxidation and leaching on contact with 4M *NaOH*. Basic *peroxydisulfate* solutions also oxidise these materials, resulting in wider micropore widths and reduced pore volumes. Each oxidising agent introduces a relatively greater oxygen content, sites which may subsequently be used for selective adsorption and/or catalyst support or binding.

## ACKNOWLEDGEMENT

We thank the Center for Molecular and Materials Sciences and the University of South Australia Research Office for their generous support of this work.

## REFERENCES

[1] M.P. Cal, M.J. Rood and S.M. Larson, J. Gas Sep. Purif., 10 (1995) 117.
[2] S.S. Barton and J.E. Koresh, JCS, Faraday Trans., 73 (1983) 1173.
[3] K. Bh. Pradhan and N.K. Sandle, Carbon, 37 (1999) 1323.
[4] C.L. Mangun, K.R. Benak, J. Economy and K.L. Foster, Carbon, 39 (2001) 1809.
[5] S.H. Wu and P. Pendleton, J. Colloid Interface Sci., 243 (2001) 306.
[6] A. Badalyan., P. Pendleton and H. Wu, Rev. Sci. Instrum., 72 (2001) 3038.
[7] A. Badalyan and P. Pendleton, Langmuir, 19 (2003) 7919.
[8] T. Takaishi and Y. Sensui, Trans. Faraday Soc., 59 (1963) 2503.
[9] A. Badalyan and P. Pendleton. In *7th International Symposium on the Characterisation of Porous Solids, COPS-7. 26-28 May 2005*. Aix-en-Provence, France. Poster 69.
[10] G. Palmgren, Calgon Carbon Corporation. Private Communication, (2003).

Studies in Surface Science and Catalysis 160
P.L. Llewellyn, F. Rodriquez-Reinoso, J. Rouqerol and N. Seaton (Editors)

# Study of the efficiency of monolithic activated carbon adsorption units.

**M. Yates[a], J.A. Martin[a], M.A. Martin-Luengo[b] and J. Blanco[a].**

[a] Instituto de Catálisis y Petroleoquímica, C.S.I.C., c/Marie Curie 2, 28049 Madrid, Spain.

[b] Instituto de Materiales de Madrid, C.S.I.C., c/Sor Juana Inés de la Cruz 3, 28049 Madrid, Spain.

Although the use of activated carbons as a purification step in the treatment of gaseous effluents from industrial plants is a well established technique, when large volumes of gas need to be treated pressure drop limitations may arise from the use of conventional adsorption beds. For these applications the conformation of the adsorption bed as honeycomb monoliths with high activated carbon content take advantage of the almost null pressure drop of these open channel structures and significantly improves the handling characteristics. In this study a commercial activated carbon was conformed as a ceramic monolith using a natural magnesium silicate as the agglomerating agent. The adsorption studies employed ortho-dichlorobenzene (o-DCB) as a probe molecule, since it can be considered as representing approximately half a molecule of tetra-chloro-dibenzene-dioxin (TCDD) the most toxic isomer of the dioxin family. It has been shown that the dynamic adsorption capacity of these units at 30°C was equivalent to the micropore volume. However, in industrial applications the adsorption step should take place at higher temperatures so that the gases exiting from the chimney rise rapidly into the upper atmosphere. Thus, the adsorption capacities of these units were determined over a range of temperatures from 30°C to 150°C at linear gas velocities between 0.3m·s$^{-1}$ to 2m·s$^{-1}$. The effects of variation in the monolith geometry: overall length versus gas linear velocity, on the dynamic adsorption capacity were studied in order to predict the optimum monolith dimensions depending on the gas flow to be treated.

Key words: Activated Carbon, Gas Purification, Honeycomb monolith, VOCs.

## 1. INTRODUCTION

In relation to price/performance, physical adsorption is one of the most important techniques to control air pollution. It is well known that the efficiency of activated carbons (ACs) as adsorbents is due to their high micropore volumes and large specific surface areas [1,2,3]. Furthermore, selectivity of ACs towards organic vapours compared to water vapour or air is advantageous [4,5,6]. For the treatment of large volumes of gas conventional adsorption beds can suffer from limitations due to pressure drop. In these applications conformation of the adsorption bed as an array of open channel monolithic honeycomb structures give rise to an almost negligible pressure drop across the adsorption bed [7]. However, at high AC contents the integrity of the adsorption unit and its mechanical strength can be causes for concern, especially if thermal regeneration of the adsorption bed is a requirement.

Results obtained previously by the authors [8,9] have shown that the static adsorption capacities of several commercially available ACs, conformed as monolith composites, could be directly related to the textural characteristics of the adsorption units and the relative vapour pressure of the molecule to be adsorbed. Of the original ACs studied in these static adsorption tests the most promising, chosen to maintain a wide selection of AC sources: coal, wood, coconut shell and peat, were selected for further study in dynamic adsorption conditions. These results confirmed that at 30°C with a contact time of one second if the textural characteristics of the adsorption units were adequate the adsorption capacity, before the efficiency of the units started to decrease, were equivalent to their micropore volumes [10]. However, for the industrial application of these adsorption units a higher effluent gas temperature may be a requirement in order to ensure that the stack gases rise quickly into the upper atmosphere. Also for the treatment of large volumes of gas the minimum dimensions of the adsorption unit should be known. Thus, in this study the dynamic adsorption capacity of the most promising AC composite from those studied previously was determined over a range of temperatures up to 150°C with linear gas velocities up to 2 ms$^{-1}$ reducing the contact time to only 0.15 s, using an air stream spiked with *o*-DCB. This molecule was chosen since it may be considered as approximately corresponding to half a molecule of tetrachloro-dibenzene dioxin (TCDD), the most toxic isomer of the dioxin family.

## 2. EXPERIMENTAL

### 2.1 Monolith preparation

The monolith composite was prepared at a ratio of 1:1 by weight of the AC (Fluesorb B from Chemviron) and a magnesium silicate clay (Pansil 100 from Tolsa SA) used as the agglomerating agent. After premixing of the dry powders a paste was formed by careful addition of water with adequate rheological properties for its successful extrusion. This dough was extruded as a honeycomb monolithic structure with parallel channels of square section at a cell density of 8 cells cm$^{-2}$ and a wall thickness of 0.9 mm using a Bonnot single screw extruder. The green ceramic body was allowed to dry at room temperature for two days in order to maintain the integrity of the monolith which was subsequently heated at 3°C min$^{-1}$ up to 150°C and maintained at this temperature for 4h. This material was subsequently used in all of the characterisation techniques and adsorption performance tests.

### 2.2 Characterisation techniques and dynamic adsorption tests

Nitrogen adsorption/desorption isotherms at -196°C were determined using a Carlo Erba 1800 Sorptomatic. The raw materials and monolith composite were outgassed overnight at 150°C to a vacuum of $\leq 10^{-2}$ Pa, ensuring a dry clean surface free from any loosely held adsorbed species. The specific surface areas ($S_{BET}$) were calculated by application of the BET equation [11] taking the area of the nitrogen molecule as 0.162 nm² [12]. Owing to the microporous nature of the AC the linear range of the BET equation was taken between relative pressures of: $p/p° = 0.02$-$0.15$. The micropore volume and external surface area, *i.e.* the area not associated with the micropores, were calculated using a *t*-plot [13] analysis, taking the thickness of an adsorbed layer of nitrogen as 0.354 nm assuming that the arrangement of nitrogen molecules in the film is hexagonal close packed [14].

Mercury intrusion porosimetry (MIP) analyses were employed to determine the pore size distribution and pore volume over the range of approximately 100 μm down to 7.5 nm diameter, utilising CE Instruments Pascal 140/240 apparatus, on samples previously dried overnight at 150°C. Approximately 0.2g of sample was accurately weighed into the sample holder that was subsequently outgassed at room temperature for five minutes to a vacuum of 0.1 kPa before filling with mercury and starting the analysis. The pressure/volume data were analysed by use of the Washburn Equation [15] assuming a cylindrical nonintersecting pore model, taking the mercury contact angle as 141° and surface tension as 484 mN m⁻¹[11]. The primary particle sizes of the raw materials were determined from analysis of the intrusion curves in the interval related to the filling of the interparticulate pore space, assuming a spherical particle geometry [16]. For the monolithic sample a single piece two channels wide and approximately 1 cm in length was employed for the measurement. Combination of the results from nitrogen isotherms and MIP leads to the characterisation of the whole pore size distribution and the total pore volume of the conformed composite material.

The mechanical strength of the monolith was determined using a Chatillon LTCM Universal Tensile Compression and Spring Tester with a test head of 1 mm diameter. The test head was positioned over one of the channel walls of the monolith composite, and the pressure slowly increased until rupture of the wall was caused; an average of ten measurements was taken to ensure the precision of the result.

Dynamic adsorption measurements were carried out in a glass reactor with an internal diameter of 2.54 cm at 94 kPa and temperatures ranging from 30°C to 150°C. The monolithic samples, of four channels 2 x 2 and between 25 to 80 cm in length, were first dried overnight at 150°C and then weighed. The monolith was then loaded into the reactor and the space between the external walls of the monolith and the internal wall of the reactor filled with silicon carbide with an average particle diameter of 0.84 mm to ensure that the gas flowed through the monolith channels. The samples were then preconditioned at 150°C in a dry air stream overnight, then subsequently cooled to the desired temperature: 30°C, 100°C or 150°C and the air stream switched to one containing either 30 ppm or 100 ppm of *o*-DCB at linear velocities between 0.321 ms⁻¹ and 2 ms⁻¹. The *o*-DCB concentration was controlled by bubbling air through the organic maintained in a thermostatic bath and then diluting this spiked stream with further air. The concentrations at the inlet and outlet were measured with a sensitivity of ±0.1 ppm by means of a Beckman FID (model 400A), the breakthrough point taken when 0.3 ppm of *o*-DCB was detected in the exit gas, indicating that the system was beginning to fail.

## 3. RESULTS AND DISCUSSION

### 3.1 Textural properties of the raw materials

The AC used in this study had a primary particle size of 90%< 12.3 µm, a specific surface area ($S_{BET}$) of 1093 $m^2g^{-1}$ of which 115 $m^2g^{-1}$ were due to the external area, with a micropore volume of 0.42 $cm^3g^{-1}$ and a mesopore volume of 0.22 $cm^3g^{-1}$. The clay binder had a much smaller particle size of only about 0.3 µm, a specific surface area ($S_{BET}$) of 149 $m^2g^{-1}$ of which 113 $m^2g^{-1}$ were due to the external area. This sample also had a small contribution of micropores of 0.02 $cm^3g^{-1}$ but a higher mesopore volume of 0.42 $cm^3g^{-1}$. The $S_{BET}$ in the case of microporous materials should be considered as an apparent surface area due to the associated micropore filling mechanism [17]. The external area and micropore volumes were calculated from analysis of the *t*-plot of the corresponding isotherms. The pore volume in pores below 50 nm was determined from the amount of gas adsorbed at a relative pressure of 0.96 on the desorption branch of the isotherm, the mesopore volumes being calculated from the difference between this value and the micropore volume.

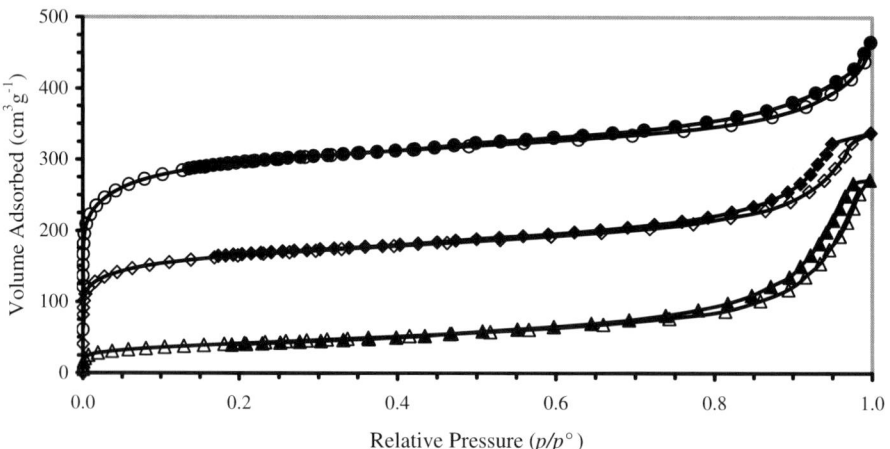

Fig. 1. Nitrogen isotherms for AC (○●), Clay (△▲) and the Monolith Composite (◊♦).

The surface area, micro and mesopore volumes of the monolith composite were very close to those expected for a purely physical mixture of the two raw materials heat treated at 150°C indicating that there was no significant chemical interaction between the two components during fabrication of the monolith, the AC only being intimately mixed with the clay binder. Thus, the monolith composite had a $S_{BET}$ of 600 $m^2g^{-1}$ with an external area of 114 $m^2g^{-1}$, a micropore volume of 0.21 $cm^3g^{-1}$ and a mesopore volume of 0.29 $cm^3g^{-1}$. From the nitrogen isotherms for the two raw materials and the monolith composite shown in Fig. 1 the type I shape of the AC was indicative of the microporous nature of this material, while the mixed type I/II shape of the clay with a H3 hysteresis loop was characteristic for a sample with a poorly defined mesoporosity that extends into the macropore range. The isotherm obtained for the composite material was also a mixed type I/II showing the characteristics of the two raw materials used in its production.

The most obvious change in the textural properties with the conformation of the monolith could be seen in the macropore region. Due to the compaction of the primary particles during the conformation of the paste, extrusion and heat treatment at 150°C the interparticulate porosity of the raw powders was reduced to a minimum. Thus, for the monolith composite the porosity in pores of less than 300 nm due to the inherent porosity of the individual primary particles was unchanged on conformation as a monolith but the porosity in larger pores was greatly reduced and shifted to narrower pores. These changes in the porosity on conformation as a monolith can be easily appreciated from the total pore size distribution curves of the two raw materials and the composite, shown in Fig. 2. In this figure the inflection in the curve at *c.* 1.8 μm for the monolith composite should be noted. This inflection is usually termed as the threshold diameter and can be considered as the limiting diameter for any transport processes into the interior of the porous structure. The position of this inflection has been shown to be directly related to the primary particle size of the AC and is of great importance during dynamic adsorption since it may cause mass transfer limitations that reduce the efficiency of the adsorption unit [10].

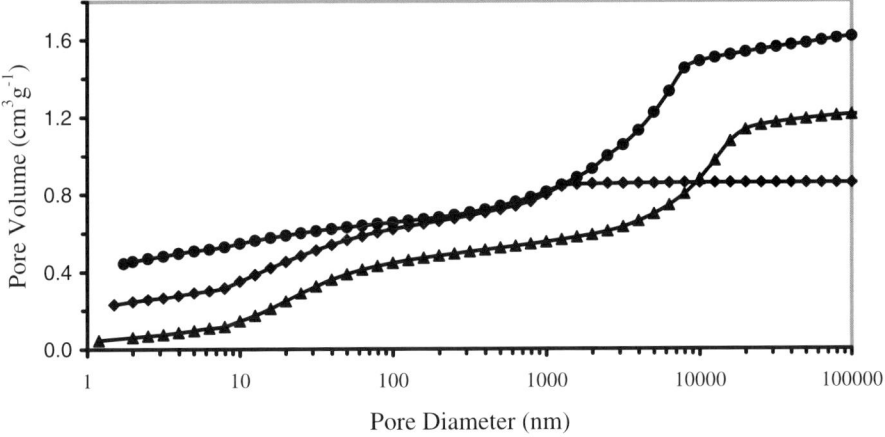

Fig. 2. Total pore volumes for the AC (●), Clay (▲) and the Monolith Composite (◆).

The mechanical strength of the monolith was found to be 10.5 MPa. From previous studies with these types of materials it had been shown that the mechanical strength could be related to either the threshold diameter or the interparticulate porosity [9]. Thus, although a wide threshold diameter was a requirement for high dynamic adsorption efficiencies, avoiding diffusion limitations, a practical limit is met due to the requirement for an adequate mechanical strength.

## 3.2 Dynamic adsorption capacities

It had been previously demonstrated that the dynamic adsorption capacities for monolithic adsorption units based on ACs were equivalent to their micropore volumes [10]. However, in order to achieve this, external surface areas of more than *c*. 100 $m^2g^{-1}$ and threshold diameters of greater than 1 μm were necessary. These experiments had been conducted with 100 ppm of *o*-DCB in a dry air stream with a linear velocity of 0.321 $ms^{-1}$ with an adsorption bed temperature of 30°C on samples of approximately 25 cm in length, giving about 0.8 s for the gas to pass through the monolith. However, due to the large volumes and high temperature of industrial effluents the effects on this optimum efficiency of much higher temperatures and linear velocities was determined. In these experiments concentrations of 30 ppm of o-DCB were used at linear velocities of 1 or 2 $ms^{-1}$ on samples of 30 to 80 cm in length at adsorption temperatures from 30°C to 150°C.

The breakthrough time was taken when 0.3 ppm of organic was detected in the exit gases, indicating the fall in the efficiency of the unit. The amount adsorbed at this point was calculated in terms of the liquid volume taking the liquid density of *o*-DCB as 1.3048 [18] in order that comparison with the textural characteristics of the adsorbent could be made. Each dynamic adsorption test was performed with fresh samples of monolith. The full results from these experiments are given in Table 1.

Table 1

Dynamic adsorption results for the AC monolith composite

| Test Number | Adsorption Temperature (°C) | *o*-DCB Concentration (ppm) | Monolith Length (cm) | Linear Velocity (m s$^{-1}$) | Contact Time (s) | Hours to Breakthrough | Amount Adsorbed (cm$^3$g$^{-1}$) | Percent Micropore Volume |
|---|---|---|---|---|---|---|---|---|
| 1 | 30  | 100 | 27.4 | 0.321 | 0.854 | 93.00 | 0.198 | 100.0 |
| 2 | 30  | 30  | 50   | 1.000 | 0.500 | 150.50 | 0.167 | 84.3 |
| 3 | 30  | 30  | 80   | 2.000 | 0.400 | 86.20 | 0.121 | 61.1 |
| 4 | 100 | 30  | 80   | 2.000 | 0.400 | 25.75 | 0.036 | 16.2 |
| 5 | 150 | 30  | 80   | 2.000 | 0.400 | 12.40 | 0.017 | 8.6 |
| 6 | 150 | 30  | 40   | 2.000 | 0.200 | 4.50 | 0.013 | 6.4 |
| 7 | 150 | 30  | 35   | 2.000 | 0.175 | 2.00 | 0.006 | 3.0 |
| 8 | 150 | 30  | 30   | 2.000 | 0.150 | 0.75 | 0.002 | 1.0 |

The first set of experiments were performed at 30°C but with increasing linear velocities (0.321 $ms^{-1}$ to 2 $ms^{-1}$) to determine the effect of reducing the contact time on the efficiency of the adsorption unit. A second set of experiments maintained the monolith length (80 cm) and linear velocity (2 $ms^{-1}$) constant but increased the adsorption bed temperature from 30°C to 150°C. Finally a series of experiments were carried out at 150°C with monoliths of various lengths (30 cm to 80 cm) at the highest velocity (2 $ms^{-1}$) to determine the importance of the contact time under these extreme conditions.

It was assumed that the measured micropore volume of the monolith composite was equivalent to the maximum volume that could be adsorbed under dynamic conditions. Thus, the adsorption efficiencies for all of the tests was calculated from the volume of o-DCB adsorbed when the breakthrough point was reached, compared to the micropore volume of the material. The effects on the dynamic adsorption efficiencies of the monolith composite from the three series of tests may be more clearly illustrated from the graph show in Fig. 3.

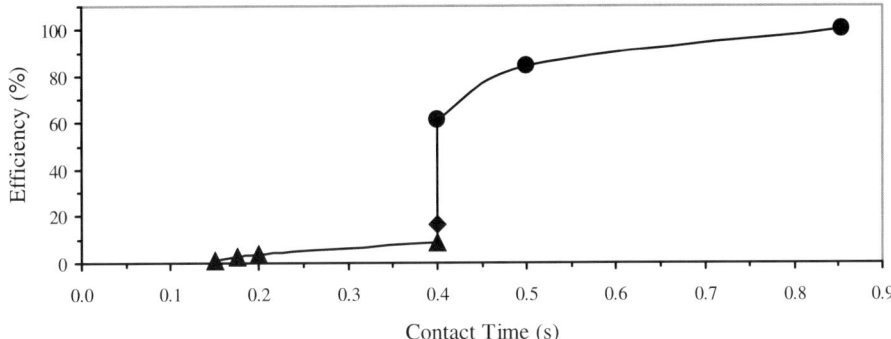

Fig. 3. The dynamic adsorption efficiency towards *o*-DCB for different contact times and temperatures: 30°C (●), 100°C (▲) and 150°C (♦).

From the results shown in Table 1 and Fig. 3 by comparing the results of Tests 1, 2 and 3 it may be seen that the efficiency of these adsorption units at 30°C fell from 100% to only *c*. 60% on reducing the contact time from c. 0.9s to only 0.4 s. This was to be expected since the adsorption process requires movement of the gas to be treated from the centre of the monolith channel to the monolith wall and then into the internal structure of the adsorbent during the passage of the gas along the open channel. Consequently the efficiency of these adsorption units could be improved by reducing the open channel width as this would reduce the distance that the gas travelled before contacting the monolith wall. This would also have the advantage of increasing the amount of adsorbent per unit volume and thus lead to smaller adsorption units.

The reduction in the adsorption capacity due to the increase in the adsorption bed temperature from 30°C to 150°C was much more severe, as may be seen by comparing the results of Tests 3, 4 and 5 where the amount adsorbed fell to only *c*. 9% at 150°C. This was also to be expected since physical adsorption is controlled by weak Van der Waals interactions between the adsorbent and the adsorbate. Moreover, previous studies had already shown that after saturation at 30°C these adsorption units could be rapidly regenerated by heating to 250°C to drive off the *o*-DCB.

Finally, working under the most extreme conditions with an adsorption bed temperature 150°C and a linear velocity of 2 ms$^{-1}$ experiments were conducted (Tests 5, 6, 7, and 8) to confirm the importance of the contact time on the overall efficiency. With the reduction of the adsorption bed length the mass transfer limitations became severe.

## 4. CONCLUSIONS

At 30°C with a contact time of approximately 1 s it has been shown that the dynamic adsorption capacity of activated carbon containing monoliths was equivalent to their micropore volumes if the measured external surface area was higher than 100 $m^2g^{-1}$ and the threshold diameter at least 1 μm. However, if the contact time is reduced due to mass transfer limitations during the gas passage along the monolith channels the overall efficiency is reduced. These results help to indicate a minimum length of the adsorption bed when monoliths with the tested geometry are used. If monoliths with narrower channels are produced presumably the same efficiency could be achieved with shorter adsorption beds. More importantly it has been shown how an increase in the adsorption bed temperature from 30°C to 150°C seriously reduced the overall efficiency of the system from about 60% down to < 10% for similar length samples.

### Acknowledgements
The authors appreciate the financial assistance received from the European Union through its project Ref. RFC-CR-04007.

### REFERENCES
[1]  M. Iley, H. Marsh and F. Rodriguez-Reinoso, Carbon, 11 (1973) 633.
[2]  M. Domingo-Garcia, I. Fernandez-Morales, F.J. López-Garzón and C. Moreno-Castilla, Langmuir, 7 (1991) 339.
[3]  F.J. López-Garzón, I. Fernández-Morales, C. Moreno-Castilla and M. Domingo-García, Stud. Surf. Sci. Cat., 120 (1998) 397.
[4]  S.J. Gregg and K.S.W. Sing, Adsorption Surface Area and Porosity, Academic Press, London (1982).
[5]  S.S. Barton, M.J.B. Evans, J. Holland and J.E. Koresh, Carbon, 22, 3 (1984) 265.
[6]  E.N.Ruddy and L.A.Carroll, Chemical Engineering Progress, July (1993) 28.
[7]  R.K. Shah and A.L. London, Dept. Mech. Eng. Stanford University CA, Tech. Rep. 75 (1971).
[8]  M. Yates, J. Blanco, P. Avila and M.P. Martin, Micro. Meso. Mater., 37 (2000) 201.
[9]  M. Yates, J. Blanco, M.A. Martin-Luengo and M.P. Martin, Micro. Meso. Mater., 65 (2003) 219.
[10] M. Yates, J. Blanco, M.A. Martín-Luengo and M.P. Martín, Stud. Surf. Sci.Cat., 144 (2002) 569.
[11] S. Brunauer, P.H. Emmett and E. Teller, J. Amer. Chem. Soc., 60 (1938) 309.
[12] J. Rouquerol, D Avnir, C.W. Fairbridge, D.H. Everett, J.H. Haynes, N. Pericone, J.D.F. Ramsay, K.S.W. Sing and K.K. Unger, Pure and Appl. Chem., 66, 8 (1994) 1739.
[13] B.C. Lippens and J.H. de Boer, J. Cat., 4 (1965) 319.
[14] B.C. Lippens, B.G. Linsen and J.H. de Boer, J. Cat., 3 (1964) 32.
[15] E.W. Washburn, Proc. Nat. Acad. Sci. U.S.A., 7 (1921) 115.
[16] R.P. Mayer and R.A. Stowe, J. Colloid Interf. Sci., 20 (1965) 893.
[17] F. Rouquerol, J. Rouquerol and K. Sing, Adsorption by powders and porous solids, Academic Press, London (1999).
[18] CRC Handbook of chemistry and physics, 59th Edition, 1978.

Studies in Surface Science and Catalysis 160
P.L. Llewellyn, F. Rodriquez-Reinoso, J. Rouqerol and N. Seaton (Editors)

# Amino functionalisation of microemulsion templated mesoporous silica foams

**S. Boskovic[a,b,c], F. Separovic[b], T.W. Turney[a], G.W. Stevens[c] M.L. Gee[b], and A. J. O'Connor[c]**

[a] CSIRO Manufacturing and Infrastructure Technology, Private Bag 33, Clayton South, Victoria 3169, Australia

[b] School of Chemistry, The University of Melbourne, Victoria 3010, Australia

[c] Department of Chemical and Biomolecular Engineering, The University of Melbourne, Victoria 3010, Australia

Large pore microemulsion templated mesoporous silica foams (MCF), with spherical cell sizes of around 40 nm, were chemically modified via a reaction in dry solvent with (3-aminopropyl)triethoxysilane (APTES) and a vapour phase reaction with (3-aminopropyl)dimethylethoxysilane (APDMES). Nitrogen sorption measurements showed that in both cases the functionalised materials retained the underlying pore structure of the unfunctionalised MCF material, with the spherical cell size of the APTES-MCF material seen to be reduced by ~ 2 nm. The nature of the chemical attachment was confirmed by $^{29}$Si MAS NMR and differed in each case. APDMES-MCF showed a single additional peak over that of the framework silica, associated with the single chemical environment of the attached silane, whereas the APTES-MCF showed multiple overlapping peaks associated with the multiple chemical environments of the attached silane. Thermogravimetric analysis revealed the loading of organic material to be higher for APTES-MCF over APDMES-MCF, consistent with the proposed attachment mechanisms.

## 1. INTRODUCTION

Since their discovery in 1992, the M41S family of mesoporous molecular sieves has shown promise for separation and immobilisation of biomolecules. The pore sizes of the original members of the M41S family probably limits their applications to smaller biomolecules, although their discovery has led to further developments into the upper end of the mesopore range. SBA-15, initially reported in 1998 [1], was obtained via a synergistic co-assembly of an amphiphilic block copolymer micellar template. It is a highly ordered hexagonal mesostructure, similarly structured to MCM-41 from the M41S family, but with bigger mesopores, some degree of microporosity and interconnection via smaller mesoporosity [2-4]. Another related material, known as messocellular foam (MCF) is obtained via the addition of 1,3,5-trimethylbenzene (TMB) to the synthesis of SBA-15 [5]. In this case the TMB and the amphiphilic block copolymer form a microemulsion, from which the MCF is templated [6]. MCF is composed of large spherical cells 22 - 42 nm in diameter, that are interconnected by windows of 10 nm in diameter to create a continual 3-dimenisonal pore system [5]. The spherical cell size of these materials can be controlled by increasing the TMB/block copolymer ratio, which increases the microemulsion droplet size, while the addition of ammonium fluoride can increases the window size from 10 to 20 nm without affecting the spherical cell size [7]. A further microemulsion templated mesoporous silica has also been reported with an analogous structure to MCF and designated as MSU-F [8].

Functionalising these types of silica with amino groups is important for potential application in both biomolecular separation and immobilisation. In the separation of proteins by techniques such as ion exchange chromatography, amino surface groups are a commonly used functionality. For immobilisation of biomolecules, the surface amino group can be used as a first functional group to which other chemical species can be attached and consequently used to immobilise biomolecules [9].

For amino functionalisation of silica, aminopropylsilanes are popular silanating agents. Most commonly, solvent phase reactions have been employed with the trialkoxysilane (3-aminopropyl)triethoxysilane (APTES), whereby the silica is refluxed in a dry solvent under an inert atmosphere [10]. For non-aqueous silylation, some water is required for the reaction to occur but the presence of water can also lead to polymerized products being deposited [11], leading to the attachment as shown in Figure 1(A). The monoalkoxysilane, (3-aminopropyl)dimethylethoxysilane (APDMES), has a greater vapour pressure than APTES, and has been shown to covalently attach to the surface of silica in a vapour phase reaction, whereby the APDMES is difunctionally adsorbed to two surface hydroxyl groups on the silica, leading to lower surface densities [11] as shown in Figure 1(B).

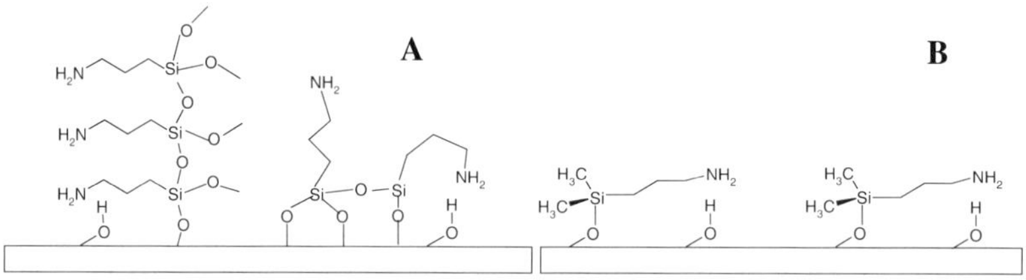

Figure 1. Schematic representation of a silica surface after functionalisation with: (A) APTES in a solvent phase reaction, and (B) APDMES in a vapour phase reaction [11].

MCF and SBA-15 were the first reported templated mesoporous materials to sequester and release proteins, taking advantage of size exclusion and ion exchange simultaneously [12]. More recently it has been shown that MCF materials have high immobilised enzyme capacities and activities compared to other templated mesoporous materials such as SBA-15 and MCM-41 [13]. In both these examples the MCF materials used were functionalised with APTES in a solvent phase reaction. In this study we report on the use of APTES in a solvent phase reaction and APDMES in a vapour phase reaction to functionalise large pore MCF materials and provide detailed characterisation of the resulting materials via nitrogen sorption, $^{29}$Si MAS NMR and thermogravimetric analysis (TGA).

## 2. MATERIALS AND METHODS

### 2.1 Materials

(3-aminopropyl)dimethylethoxysilane (APDMES, United Chemical Technologies), (3-aminopropyl)triethoxysilane (APTES, Aldrich, 99%), Pluronic triblock copolymer $EO_{20}PO_{70}EO_{20}$ (P123, BASF), tetraethoxysilane (TEOS, Fluka, 99 %), 1,3,5-

trimethylbenzene (TMB, Ajax) and ammonium fluoride (NH₄F, Aldrich, 97%), were used as received. Hydrochloric acid (1.6 M) was made from a 35 weight % solution (Merck, AR) with distilled water, and toluene (Merck, 99%) was initially dried on 4Å molecular sieves, distilled over Na metal wires and stored on 4Å molecular sieves prior to use.

## 2.1 Synthesis

Two MCF samples, labelled MCF-A and MCF-B were synthesised via a previously reported method [7]. For each sample, 4 g of P123 was dissolved in 150 mL of 1.6 M HCl in a beaker with stirring. The resulting solution was then heated at 37-40°C prior to the addition of 11.4 mL of TMB and 46 mg of NH₄F. The mixture was then allowed to stir for a further 30 minutes followed by the addition of 8.8 g of TEOS. After 20 hours at 37-40°C, the mixture was transferred to a Teflon coated autoclave and aged under static conditions at 100°C for 24 hours. The mixture after cooling was then filtered on a Buchner funnel and pumped to dry for 2 days, after which it was calcined at 500°C in air for 8 hours to give the MCF materials.

## 2.2 Functionalisation
### APTES

MCF-A was functionalised with APTES in a solvent phase reaction via a previously reported method for MCM-41 mesoporous silica [10]. Initially, 0.5 g of MCF-A was dispersed in 50 mL of dry toluene in a round bottom flask under nitrogen, which was followed by the addition of 4.4 g of APTES. The mixture was then refluxed for 18 hours. After cooling to ambient temperature, the sample was filtered on a Buchner funnel, washed with dry toluene and isopropanol and heated overnight in a vacuum oven at 100°C. The resulting material is subsequently referred to as APTES-MCF-A

### APDMES

MCF-B was functionalised with APDMES in a vapour phase reaction previously reported method for colloidal silica [11]. Initially ~ 0.5 g of MCF-B was degassed under vacuum (< 0.02 atm) at ~350°C for 8 hours. The sample was then allowed to cool to room temperature while the vacuum was maintained. After cooling, MCF-B was isolated from the vacuum system and the APDMES vapour was introduced and the silylation reaction was allowed to occur over a period of 2 days. At the end of this reaction period, the reactor was evacuated to remove the reaction by-product (ethanol) and any unreacted APDMES. The resulting material is subsequently referred to as APDMES-MCF-B

## 2.3 Characterisation

Nitrogen sorption isotherms were recorded at 77 K, using a Micromeritics ASAP 2000. Prior to analysis, the samples were degassed overnight at 150°C under vacuum. BET surface areas were calculated using adsorption data recorded in the relative pressure range, 0.05 - 0.2. The total pore volume was derived from the amount of gas adsorbed at a relative pressure of ~0.99. Pore sizes were determined by the BdB-FHH method [14]. This method allows the determination of spherical cell size from the adsorption branch and window size from the desorption branch of the isotherm.

Nuclear magnetic resonance (NMR) analysis was performed on a Varian 300 NMR spectrometer operating at 59.60 MHz for ²⁹Si. Acquisition and processing of data were performed using Varian software. The spectra were recorded by magic angle spinning (MAS) at ~ 6 kHz to enhance resolution in a 5 mm zirconia rotor, using a 3.5 µs π/2 pulse and 60 s recycle time, with a proton decoupling power of ~ 70 kHz, a spectral width of 40 kHz and a line broadening of 100 Hz. Typically, ~ 1500 scans were collected for each spectrum.

Thermogravimetric analysis (TGA) was completed on a Pyris 1 Perkin-Elmer TGA under flowing nitrogen. Approximately 2-4 mg of each sample was used. Initially, the sample was held for 2 minutes at 30°C followed by heating to 700°C at a rate of 10°C/minute.


## 3. RESULTS AND DISCUSSION

The nitrogen sorption isotherms and corresponding BdB-FHH pore size distributions are shown in Figure 2 for APTES-MCF-A and APDMES-MCF-B along with the unfunctionalised MCF samples, MCF-A and MCF-B. A summary of the values determined is contained in Table 1. It can be seen that for the unfunctionalised MCF samples, MCF-A and MCF-B, the shape and form of the isotherm and the parameters of BET surface area and total pore volume determined, are very similar, allowing comparisons to be readily made between APTES-MCF-A and APDMES-MCF-B.

Table 1
Nitrogen sorption parameters for MCF

|  | $N_2$ Sorption Parameters | | | | |
|---|---|---|---|---|---|
| sample | $S_{BET}$ ($m^2$/g) | $C_{BET}$ | $V_t$ ($cm^3$/g) | $d_{cell}$ (nm) | $d_{window}$ (nm) |
| MCF-A | 570 | 114 | 2.6 | 41 | 18 |
| APTES-MCF-A | 288 | 48 | 1.6 | 38 | 16 |
| MCF-B | 560 | 119 | 2.5 | 38 | 18 |
| APDMES-MCF-B | 347 | 33 | 2.0 | 38 | 18 |

$S_{BET}$: BET surface area; $C_{BET}$: BET constant; $V_t$: total pore volume; $d_{cell}$: cell diameter (BdB-FHH method); $d_{window}$: window diameter (BdB-FHH method).

For both APTES-MCF-A and APDMES-MCF-B, the form of the nitrogen sorption isotherm is maintained after functionalisation, with the step observed in the isotherm at a relative pressure of ~ 0.9, associated with capillary condensation within the pores, seen to be relatively unchanged. The BdB-FHH pore size distributions also show similar narrow forms before and after functionalisation. Both results indicate that the underlying pore structure of the MCF materials has been maintained after functionalisation.

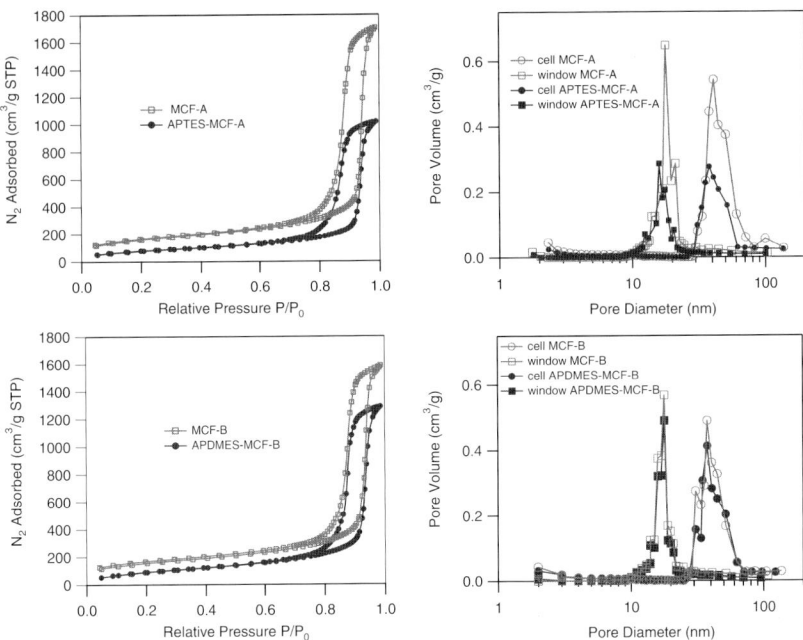

Figure 2. Nitrogen sorption isotherms (left) and corresponding BdB-FHH pore size distributions (right) for APTES-MCF-A (upper) and APDMES-MCF-B (lower).

From the $N_2$ sorption isotherms it is also apparent that the amount of nitrogen that can be adsorbed has decreased, more so for APTES-MCF-A than for APDMES-MCF-B. For a given amount of material, functionalisation will result in an increase in the material's weight, which will give rise to reduced specific material properties, such as the BET surface area and total pore volume. This reduction is greater for APTES-MCF-A which experiences a drop of 2 to 3 nm in both the spherical cell and window size distributions, whereas these remain unchanged for APDMES-MCF-B after functionalisation. The reduction in pore size in APTES-MCF-A is most likely due to the formation of an APTES polymerized layer within the pore [11].

For both APTES-MCF-A and APDMES-MCF-B the percentage reduction in the BET surface area is seen to greater than that seen in the total pore volume. The BET surface area is sensitive to the surface roughness, as seen by the nitrogen molecule probe in inorganic-organic hybrid materials such as these, and this is a likely contributing factor [15]. However, we have recently shown that MCF materials are also microporous [16], as in other mesoporous materials templated from amphiphilic block copolymers containing poly (ethylene oxide) components [2-4, 17, 18], which could be responsible for a significant contribution to the BET surface area. In the same study we were able to show for a sample prepared similarly to APTES-MCF-A, that functionalisation with APTES results in the blocking of these micropores by the polymer layer and hence results in a greatly reduced BET surface area.

Solid-state $^{29}Si$ MAS NMR spectra of MCF-A, APTES-MCF-A and APDMES-MCF-B are shown in Figure 3. The unfunctionalised MCF-A, shows what appears to be three distinct peaks, which can be assigned as surface geminal groups ($Q^2$), surface vicinal silanol groups

($Q^3$) and framework silica ($Q^4$), respectively, at ~ -96, -106 and -113 ppm, according to previously reported data for SBA-15 and our work with MCM-48 mesoporous silica [19, 20].

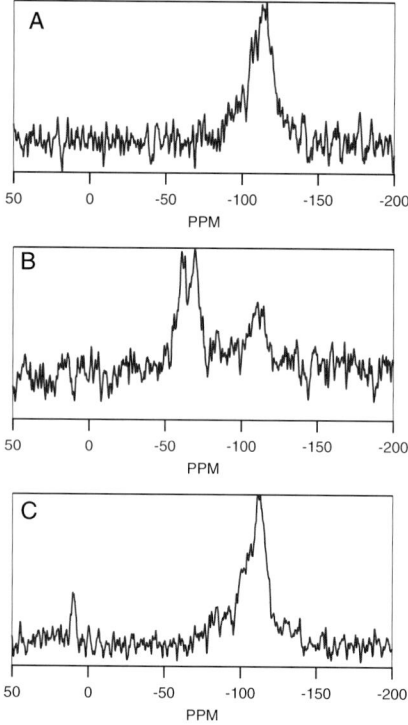

Figure 3. $^{29}$Si MAS NMR spectra of: (A) MCF-A, (B) APTES-MCF-A, and (C) APDMES-MCF-B.

For APTES-MCF-A it appears as if two additional overlapping peaks are present at ~ -61 and -69 ppm. An organotrialkoxy silane such as APTES can condense forming multiple Si – O – Si bonds [21] (Figure 4). In this case the resonances can be assigned as being due to two Si – O – Si bonds (-61 ppm) and three Si – O – Si bonds (-69 ppm), based on previously reported data for a porous silica gel functionalised with APTES in a similar manner [21].

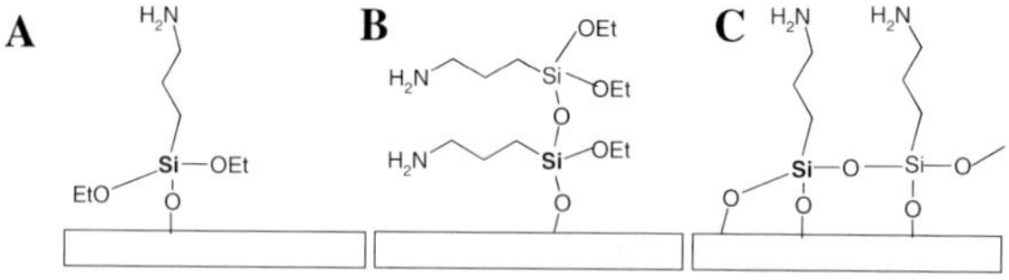

Figure 4. APTES condensed on a silica surface with: (A) one, (B) two, and (C) three Si-O-Si bonds.

The APDMES-MCF-B spectrum shows a single additional peak not present in the spectrum of the unfunctionalised silica, at ~ 10 ppm. This can be attributed to the attached aminosilane groups, which are present in only one distinct chemical environment (Figure 1B) as opposed to those in APTES functionalised silica. Previously, for MCM-48 templated mesoporous silica functionalised in a similar manner, we have been able to use a method reported for trimethylsilane groups [22], to deconvolute the NMR spectra and calculate the surface concentrations of aminopropylsilane groups [19]. However, for the spectra obtained in this case, deconvoloution was not attempted due to the low signal to noise ratio.

Thermogravimetric weight loss curves are shown for MCF-A, APTES-MCF-A and APDMES-MCF-B in Figure 5. The weight loss profile of the MCF is significantly altered by attachment of either APTES or APDMES. For the unfunctionalised MCF the weight loss curve is seen to be relatively flat indicating that there is no appreciable condensation of silanol groups occurring on the surface [23]. For the APTES-MCF-A, the curve is much like that reported previously for similarly functionalised MCM-41 mesoporous silica; the step at 100°C can be attributed to the loss of physically adsorbed water and the slope between 400 and 700°C to the decomposition of the bonded aminopropylsilane [10]. Interestingly, APTES-MCF-A contains more adsorbed water than MCF-A, This is likely to be due to the significant increase in the number of polar surface functionalities that can hydrogen bond to water (i.e. $NH_2$) upon functionalisation with APTES, as illustrated in Figure 1(A). For APDMES-MCF-B, no clear step can be seen associated with the loss of physically adsorbed water, while the slope from 200 to 600°C associated with the decomposition of the aminopropylsilane is seen to be less pronounced than that for APTES-MCF-A.

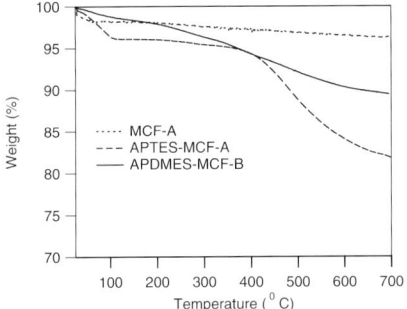

Figure 5. TGA weight loss curves for MCF-A, APTES-MCF-A and APDMES-MCF-B.

In a comparison of the organic loadings, the final weight loss of the APTES-MCF-A is almost twice that of the APDMES-MCF-B. It has been shown previously that the height of the step where the decomposition of the attached silane is occurring is proportional to the carbon content present in the sample [10], which again shows a reduced amount present for the APDMES-MCF-B. The reduced loading is consistent with the attachment mechanisms whereby APTES forms a polymerized layer [11] and APDMES is difunctionally adsorbed to two surface hydroxyl groups on the silica, leading to lower surface densities [11].

For applications in the separation or immobilisation of biomoleules in large mesopore materials, such as the MCF considered here, the solvent phase reaction with APTES would offer certain advantages over the vapour phase reaction with APDMES due the higher loadings. For other smaller pore materials, where any further pore size reduction could limit applications, the vapour phase reaction with APDMES could be the preferred option.

## 4. CONCLUSIONS

MCF was chemically modified via a reaction in dry solvent with APTES and a vapour phase reaction with APDMES. The functionalised materials were characterised via nitrogen sorption, $^{29}Si$ MAS NMR and thermogravimetric analysis. Nitrogen sorption results showed that in both cases the functionalised MCF retained the underlying pore structure of the unfunctionalised material, with the spherical cell size of the APTES-MCF material being slightly reduced. The nature of the chemical attachment was confirmed by [29]Si MAS NMR and was seen to be different in each case: APDMES-MCF showed a single peak (additional to that of the framework) associated with the environment of the attached silane, whereas the APTES-MCF showed several additional overlapping peaks associated with the multiple chemical environments of the attached silane. TGA revealed the loading of organic material to be higher for APTES-MCF than APDMES-MCF, consistent with the attachment mechanisms of each silane.

## ACKNOWLEDGEMENT
S.B. is grateful for the grant of an Australian Postgraduate Award (APA) and CSIRO Scholarship. The authors acknowledge access to infrastructure from the Particulate Fluids Processing Centre, a special research centre of the Australian Research Council (ARC), and ARC Discovery Grant DP0451387. Dr I. Burgar is thanked for running the [29]Si MAS NMR.

## REFERENCES
[1]     D. Y. Zhao, J. L. Feng, Q. S. Huo, N. Melosh, G. H. Fredrickson, B. F. Chmelka and G. D. Stucky, Science, 279 (1998) 548.
[2]     A. Galarneau, H. Cambon, F. Di Renzo, R. Ryoo, M. Choi and F. Faluja, New J. Chem., 27 (2003) 73.
[3]     M. Kruk and M. Jarioniec, Chem. Mater., 12 (2000) 1961.
[4]     R. Ryoo, C. Hyun Ko, M. Kruk, V. Antochshuk and M. Jaroniec, J. Phys. Chem. B., 104 (2000) 11465.
[5]     P. Schmidt-Winkel, W. W. J. Lukens, D. Zhao, P. Yang, B. F. Chmelka and G. D. Stucky, J. Am. Chem. Soc., 121 (1999) 254.
[6]     P. Schmidt-Winkel, J. Glinka and G. D. Stucky, Langmuir, 16 (2000) 356.
[7]     P. Schmidt-Winkel, W. W. J. Lukens, P. Yang, D. I. Margolese, J. S. Lettow and G. D. Stucky, Chem. Mater., 12 (2000) 686.
[8]     S. S. Kim, T. R. Pauly and T. J. Pinnavaia, Chem. Comm., (2000) 1661.
[9]     H. H. Weetall, Appl. Biochem. Biotechnol., 41 (1993) 157.
[10]    C. P. Jaroniec, M. Kruk, M. Jaroniec and A. Sayari, J. Phys. Chem. B., 102 (1998) 5503.
[11]    L. D. White and C. P. Tripp, J. Colloid Interface Sci., 232 (2000) 400.
[12]    Y. J. Han, G. D. Stucky and A. Butler, J. Am. Chem. Soc., 121 (1999) 9897.
[13]    P. H. Pandya, R. V. Jasra, B. L. Newalkar and P. N. Bhatt, Microporous Mesoporous Mat., 77 (2005) 67.
[14]    W. W. J. Lukens, P. Schmidt-Winkel, D. Zhao, J. Feng and G. D. Stucky, Langmuir, 15 (1999) 5403.
[15]    M. Kruk and M. Jaroniec, Chem. Mater., 13 (2001) 3169.
[16]    S. Boskovic, A. J. Hill, T. W. Turney, M. L. Gee, G. W. Stevens and A. J. O'Connor, Prog. Solid State Chem., submitted, (2005).
[17]    C. G. Goltner, B. Smarsly and M. Antonietti, Chem. Mater., 13 (2001) 1617.
[18]    B. Smarsly, S. Polarz and M. Antonietti, J. Phys. Chem. B., 105 (2001) 10473.
[19]    A. Daehler, S. Boskovic, M. L. Gee, A. J. O'Connor, F. Separovic and G. W. Stevens, J. Phys. Chem. B., accepted, (2005).
[20]    M. W. McKittrick and C. W. Jones, Chem. Mater., 15 (2003) 1132.
[21]    E. J. R. Sudholter, R. Huis, G. R. Hays and C. M. Alma, J. Colloid Interface Sci., 103 (1985) 554.
[22]    B. H. Wouters, T. Chen, M. Dewilde and P. J. Grobet, Microporous Mesoporous Mat., 44-45 (2001) 453.
[23]    C. P. Jaroniec, P. K. Gilpin and M. Jaroniec, J. Phys. Chem. B., 101 (1997) 6861.

Studies in Surface Science and Catalysis 160
P.L. Llewellyn, F. Rodriquez-Reinoso, J. Rouqerol and N. Seaton (Editors)

# Effect of activation process on resin based activated carbons

**F.K. Katsaros[a]\*, Th.A. Steriotis[a], A.K. Stubos[b], N.K. Kanellopoulos[a] and S.R. Tennison[c]**

[a] NCSR Demokritos, Institute of Physical Chemistry, 153 10 Ag. Paraskevi Attikis, Greece
[b] NCSR Demokritos, Institute of Nuclear Technology – Radiation Protection, 153 10 Ag. Paraskevi Attikis, Greece
[c] MAST Carbon Ltd., Henley Park, Guildford, Surrey GU3 2AF, U.K.

A series of porous carbon powders, prepared by carbonisation and activation of a commercial phenolic resin, is studied, in terms of pore structure, using Nitrogen and Carbon Dioxide adsorption isotherms at 77 K and 194.5 K, respectively. The activation process leads to an increase in the porosity, while the sizes of the pores developed are independent of the activation, up to 30 % weight loss. The main structure is created during carbonisation of the polymeric precursor, which acts as a structure template, while activation (<30% weight loss) pertains mainly to the removal of amorphous material, which is developed during the initial pyrolysis and blocking the porous network. Additionally, carbon membrane analogues are studied by adsorption and permeation techniques and the activation mechanism, as well as structural and diffusion characteristics are resolved. Finally, the optimum conditions of activation for the membranes are determined.

## 1. INTRODUCTION

Activated Carbon (AC) is the most widely used porous material. Due to its low cost, large surface area and mainly its easily tailored properties, in terms of both surface chemistry and pore structure, AC is abundantly used in a wide range of industrial applications. Numerous diverse processes involve AC in both powder and membrane form, including protection against toxic gases and vapours [1], removal of organic materials (SOC and NOM) and metal ions from water [2,3,4], storage and refinement of natural gas [5,6] etc. AC is produced by physical or chemical treatment of a precursor, the final properties of the product depending strongly on the starting material.

The "physical" AC preparation process involves two separate stages, the first being carbonisation, i.e. pyrolysis of the raw material at temperatures usually between 873-1073 K, in an inert atmosphere, where the pore network is developed and the non-carbon species are essentially removed. In the subsequent step of activation, the substrate is heated under an oxidizing agent (Carbon dioxide, air, water) at about 1073-1173 K. This process aims to improve the characteristics of the pore structure, e.g. increase of pore volume and surface area, tuning of the pore size etc. This step, which is a heterogeneous solid-gas reaction, involves the gasification of the more reactive carbon from the lattice of the solid substrate. Higher temperatures (>1273 K) lead to decrease in the size of the pores. The most wide used oxidizing medium is Carbon dioxide, which exhibits better kinetic behaviour and leads to a process that can be easily controlled. A crucial parameter of the process is the degree of activation, (i.e. the % weight loss of the substrate during heating), which for a certain system (substrate - oxidizing medium) depends on the time and the temperature of treatment. It is generally accepted that small degree of activation leads to improved microporous structures, while weight loss greater than 30% can cause the formation of mesopores and macropores. Finally, activation to a higher degree directs to the decrease of pore volume and surface area, due to collapse of the substrate's pore structure [7,8].

In this work, a series of porous carbon powders is studied using Nitrogen and Carbon Dioxide isotherms at 77 K and 194.5 K, respectively. The pore structure characteristics are deduced and a mechanism of the activation process is proposed. Composite carbon membranes analogues are also studied using both adsorption and permeation techniques. Structural and diffusion characteristics are derived as well as the activation process mechanism of these composite membranes. Finally, the optimum conditions of activation of the membranes are determined.

## 2. EXPERIMENTAL

### 2.1 Activated Carbon Resol Powders

The five AC powders used in the present study were produced after carbonisation of ground, crosslinked, fully cured, polymer (resol) precursor at 1073 K under Nitrogen atmosphere for 60 minutes [9]. One of the samples is used without further treatment (Sample P1), while the other four samples, prior to their characterisation, were activated at 1073 K, under $CO_2$ flow, to different degrees (weight loss), as follows: Sample P2 - 6.5%, Sample P3 – 10%, Sample P4 – 16% and Sample P5 –30% weight loss, respectively.

The effect of the degree of activation on the porous structural properties, like pore volume, porosity, specific surface area and pore size distribution, is studied using gas porosimetry. Adsorption isotherms of $N_2$ at 77 K and $CO_2$ at 194.5 K were performed on the aforementioned samples by means of QUANTACHROME AUTOSORB-1 gas porosimeter automatic volumetric analyser with Krypton upgrade. Initially the samples were outgassed at 523 K for about 72 hours, under high vacuum ($10^{-3}$ Pa).

### 2.2 Carbon Membrane Analogues

The carbon membranes, which were produced through carbonisation and activation of phenol-formaldehyde resins, consist of a cylindrical, macroporous novolac resin substrate (granular novolac J1048, porosity 40 %, pore size 5 μm), and a 40 μm microporous resol layer [10]. The carbonisation process took place by heating of the raw material at 1073 K, under nitrogen atmosphere, while activation process was occurred by heating at 1073 K, using $CO_2$ flow, through the internal side of the membrane, for different reaction times, as follows: sample M1 for 15 minutes, sample M2 for 60 minutes and sample M3 for 90 minutes, respectively.

Parts of the membranes were used for i) Nitrogen adsorption isotherms at 77 K on AUTOSORB-1 porosimeter and ii) permeability experiments. To this end, membrane samples were connected to a specially designed permeability apparatus [11,12] and permeability experiments were carried out for different gases (He, $N_2$, $CO_2$) at 273 K and 308 K.

## 3. RESULTS AND DISCUSSION

### 3.1 Carbon Resol Powders

The adsorption isotherms of $N_2$ at 77 K (Fig. 1) and $CO_2$ at 194.5 K (Fig. 2) performed on the powder samples are of type I according to IUPAC classification. All the adsorption isotherms exhibit a sharp increase in the adsorbed volume, at low relative pressures, while at higher pressures ($0.2 < p/p_0 < 0.9$), the slope of the isotherm is almost zero and a "plateau" appears, indicating that the samples are purely microporous. The enhanced uptake of $N_2$ at relative pressures close to 1.0 ($p/p_0 > 0.95$) can be attributed to interparticle condensation. The adsorption data were interpreted on the basis of BET [13], Langmuir [14] and Dubinin – Radushkevich methods [15] and the specific surface area of the samples is calculated (Table 1), using the value 0.162 $nm^2$/molecule for $N_2$.

As the degree of activation increases, an incremental tendency in total pore volume, $V_p$, is monitored. Activation up to 6.5% weight loss leads mainly to a gain in micropore volume. Further activation (10%) affects, not only the micropore, but the mesopore volume (0.027 instead of 0.007 $cm^3 \ g^{-1}$) as well. Higher degree of activation does not produce significant change, effecting micropores only, while stronger treatment (>30%), leads to an increase in micropore volume. Same results on pore volume are also derived from $CO_2$ isotherms at 194.5 K (Fig. 2). In addition, the difference between $V_p$ and $V_{0.95}$ (V at $p/p_0=0.95$) is not significant. This fact can be associated with the different adsorption mechanisms, namely surface coverage for $CO_2$ and micropore filling for $N_2$.

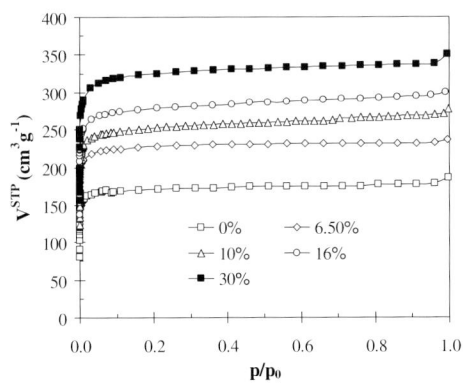

Fig. 1. $N_2$ adsorption isotherms at 77 K for AC powders

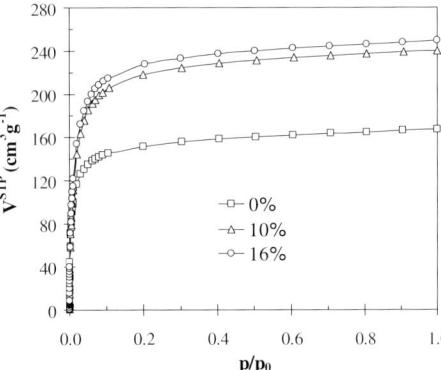

Fig. 2. $CO_2$ adsorption isotherms at 195 K for AC powders

Table 1 - Data calculated from $N_2$ adsorption at 77 K (powder samples).

|  | 0 % | 6.5 % | 10 % | 16 % | 30 % |
|---|---|---|---|---|---|
| $A_{BET}$ $(m^2 g^{-1})$ | 688.51 | 913.54 | 1007.0 | 1100.2 | 1284.8 |
| $A_{lanqmuir}$ $(m^2 g^{-1})$ | 767.30 | 1010.6 | 1162.0 | 1269.4 | 1463.7 |
| $A_{DR}$ $(m^2 g^{-1})$ | 745.34 | 998.18 | 1096.4 | 1219.8 | 1435.4 |
| Porosity (%) | 36.7 | 42.8 | 46.4 | 48.9 | 53.2 |

The micropore volumes calculated depend on the degree of activation and the adsorptive used ($N_2$, $CO_2$). The relatively low micropore volume of unactivated sample is attributed to the constrictions, formed during carbonisation from amorphous material. These constrictions hinder the diffusion of $N_2$ into the micropores at 77 K. In contrast, $CO_2$ molecules, though similar in dimension with $N_2$, can diffuse into the micropores, due to much higher experimental temperature. This hindrance phenomenon, which is pure kinetic effect, is closely connected with the existence of ultra-micropores and small degree of activation (<5%). Stronger activation (up to ~ 15%), leads to widening of the pores and partial withdrawal of the kinetic hindrance of $N_2$ molecules, so that the micropore volumes, calculated for $N_2$ and $CO_2$, are almost identical. Thus, the pore system is rather homogeneous and is characterised by the existence of ultra- and super-micropores. Activation degrees ranging between 15 and 30% produce materials, for which the micropore volume calculated from $N_2$ isotherms is larger than the corresponding $CO_2$ volume. These materials are heteroporous and are characterised by a wide pore size distribution [16].

By applying Dubinnin-Astakhov analysis for slits [17] on the $N_2$ isotherms, the micropore size distribution can be deduced. The psd curves range from 0.7 nm to around 1.8 nm, the

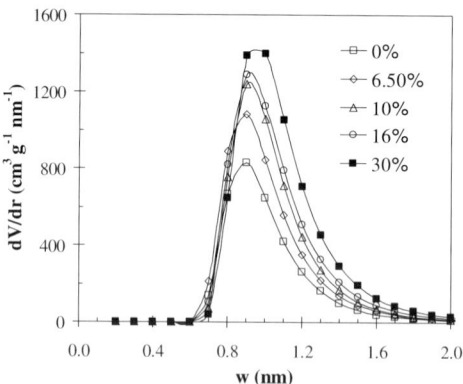

Fig. 3. Pore size distribution (DA) of carbon powder samples from $N_2$ porosimetry

maximum being around 0.9-1.1 nm, depending on the sample. From the analysis, the DA constant, $n_{DA}$, is approximately equal to 2. This value corresponds to a relatively homogeneous porous system with a narrow psd. As the degree of activation increases, the peak is shifted to higher values, while the height of the curve increases as well. The increment in the pore volume without a corresponding shift to higher dimensions implies that during activation the carbon pore structure does not change. This fact is associated mainly with the removal of amorphous carbon from the pore system, making the already existing pores accessible. In a smaller scale, new pores are developed and widening of the existing pore system occurs. These results are in accordance with the change in the volume of different classes of pores (micro-, and meso-) from the adsorption isotherms, and the deviations of the calculated micropore volume from $N_2$ and $CO_2$ adsorption data.

### 3.2 Carbon Membranes

*3.2.1 Adsorption Isotherms*

The Nitrogen adsorption isotherms at 77 K on carbon membranes are shown in (Fig. 4). An incremental trend in total pore volume, $V_p$, with the activation time is observed. Activation time 15 minutes brings about changes in both the micropore and the mesopore volume (0.02 instead of 0.01 cm$^3$ g$^{-1}$). On the contrary, activation time, between 60 to 90 minutes, leads to a notable gain in the micropore volume only (rise from 0.200 to 0.255 and finally to 0.355 cm$^3$g$^{-1}$). At this point, it must be noted that no macropores are observed during the activation process for all the samples.

Using the DA analysis on the $N_2$ isotherms the pore size distribution (psd) can be deduced (Fig. 5). The psd curves show an increase in the height of the peek as well as a small shift to higher radius. The increment in the pore volume, related to the widening of the pore structure of the membranes during activation, follows the same mechanism as in the case of the powders.

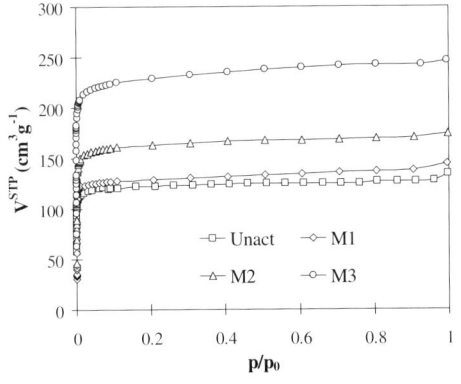

Fig. 4. N$_2$ adsorption isotherms at 77 K for the membranes of type A

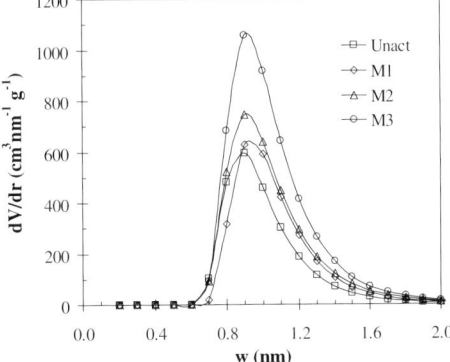

Fig. 5. DA psd of carbon membranes obtained from N$_2$ adsorption at 77 K.

In physical adsorption, the unconstrained flow of the oxidating medium into the pore system, leads to a more effective removal of amorphous carbon from the internal side of the pores. It is expected then, that the activation process will be promoted through large pores. During the activation from the surface layer, the presence of mesopores of the skin causes a non-uniform treatment and the formation of large channels throughout the length of the surface layer. In this case, the activation process is localised in regions near these large channels. On the other hand, the pore network of the substrate, it is expected to facilitate the flow of $CO_2$ and to cause milder and more symmetrical activation. Therefore, the activation of the membranes, in the present study, is taking place through $CO_2$ flow from the internal side of the cylinder (substrate), while there is flow of He from the external side (micropore layer).

### 3.2.3 Permeability Measurements

The Helium permeabilities (Fig. 6) remain stable and independent of the inlet pressure, in the area 0-10 kPa, for all samples. It is also observed that the permeability increases with the activation time. This can be attributed to, either the widening of the effective pore size, due to the removal of amorphous carbon from the pores, or to the formation of large channels throughout the surface layer during activation. These aspects are in accordance with the results obtained form static techniques (Adsorption isotherms).

Additionally, the permeabilities of He, $N_2$ and $CO_2$ on M1 sample (Fig. 7) remain stable and independent of the applied pressure (inlet pressure). This fact is evidence that gas flow through the membrane is described by the Knudsen equation. By processing the permeability data according to the molecular weight of the penetrating gases, divergences between the calculated and the theoretical values arise. The variations between experimental and theoretical values (for molecular stream) are greater for $CO_2$. Similar results are also obtained for the M2 membrane. In this case, the decrease in $CO_2$ permeability is much smaller than the corresponding in M1 sample.

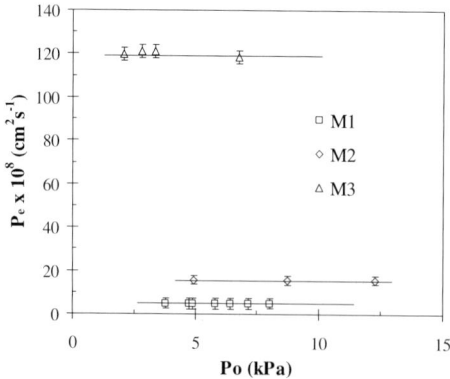

Fig. 6. Helium integral permeability experiments at 273 K

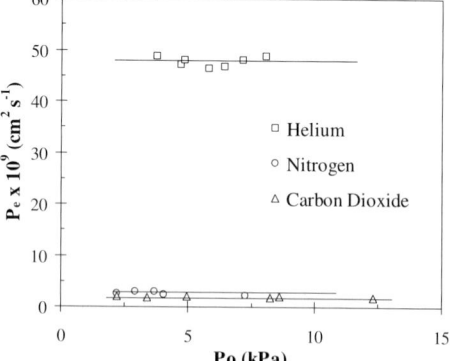

Fig. 7. Integral permeability experiments at 273 K for M1 sample

Table 2 – Carbon membrane permeability data ($\times 10^8$ cm$^2$ s$^{-1}$)

|  | Helium | Nitrogen | Carbon Dioxide |
|---|---|---|---|
| M1 | 4.78 | 0.26 | 0.19 |
| M2 | 15.6 | 1.14 | 0.84 |
| M2 (308 K) | 38.2 | - | - |
| M3 | 120 | 10.6 | 8.12 |

Helium permeance experiments on M2 sample at 273 K and 308 K indicate that the increase in permeability, in cm$^2$ s$^{-1}$ units (Table 2), is not proportional to the square root of the temperature, as defined by molecular flow (Knudsen regime). The higher value of the permeability ratio, against the temperature square root ratio, suggests that helium flux cannot be described by the Knudsen approximation, but by the corresponding of activated diffusion. In this case, the presence of ultra micropores and the constrictions in the pores, hinder the molecular motion. As the temperature rises, the kinetic energy of the molecules increases and they can overcome the energy barrier of the diffusion. The phenomenon of activated diffusion is very common on micropore system and it can be expressed by an Arrhenius relation:

$$D = D_\infty e^{-E_e/RT} \qquad \text{Eq. 1}$$

where D: the diffusion coefficient at temperature T, $D_\infty$: the diffusion coefficient of non activated diffusion and $E_a$: the activation energy. The activation energy, for a certain porous system, mainly depends on the molecular kinetic diameter, while the nature of the surface as well as the shape of the pores plays a minor role [18].

The effect of constrictions on gas flow, which appears at both the entrance and the way through the pores, is more intensive as the gas molecular kinetic diameter is larger. Consequently, the lower values of N$_2$ and CO$_2$ permeability, in regard to He values, can be attributed to the differences in molecular sizes ($\sigma_{Nitrogen}$=0.3798 nm, $\sigma_{Carbon\ Dioxide}$=0.3941 nm, $\sigma_{Helium}$=0.2551 nm) [19]. Since the activation time increases, the effect of hindrance attenuated, due to the removal of amorphous carbon. In this case, higher N$_2$ and CO$_2$ permeability values, in regard to Helium value, are observed. Additionally, the higher kinetic energy of gas molecules, at the 308 K, surpasses effectively the energy barrier of entrance and pass through the pores, leading to a gain in permeability. However, at constant temperature, the flow of a certain gas takes place according to molecular stream mechanism.

Based on the abovementioned flow mechanism, and relating the maximum pressure, where the permeability is independent of the mean pressure, with the mean free path (molecular stream flow takes place where the r/λ<0.05), it is possible to calculate the maximum size of a continuous path of the membrane or in other words, an average size for defects. According to permeability data, the M2 sample seems to be is defect free, having pores less than 15 nm. On the other hand the M3 sample has pores larger than 400 nm. This result is in agreement with the high permeability value of the M3 membrane in regard to M2. Thus, it can be concluded that the optimum activation time is ranged between 30-60 minutes since extensive activation time (90 minutes) causes partial destruction of the surface layer.

## 4. CONCLUSIONS

The activation process leads to an increase of the porosity while the pore sizes developed are independent of activation up to 30 % weight loss. The main structure is created during carbonisation of the polymeric precursor. The amorphous material, which blocks the pore network, is easily removed during activation, rendering the already existing pores accessible. The same mechanism is also observed during the activation of carbon membrane analogues. In this case, the presence of different types of pores leads to asymmetric activation. The optimum activation time is 30-60 minutes while activation is more efficient when using $CO_2$ as oxidizing agent, flowing from the inner side (substrate) of the membrane.

Additionally we may conclude that low-pressure integral permeability is a quick and easy to use tool for crack (defect) detection and leads to detailed structural characterization of asymmetric membranes. Furthermore, permeability experiments give a rough but reliable measure of the maximum pore size of the structure under study.

## References

[1] W.M.T.M Reimerink, Adsorption and its application in industry and environmental protection», Vol. 1, Application in Industry, A. Dabrowski, Elsevier, 751 (1999)

[2] T.Lebeau, C.Lelievre, H.Buisson, D.Clevet, L.W. van de Venter, P. Cote, Desalination 117 (1998) 219

[3] Y. Matsui, Y. Fukuda, T. Inoue and T. Matsushita, Water Res. 37 (2003) 4413

[4] V.López-Ramón, C. Moreno-Castilla, J. Rivera-Utrilla and L. R. Radovic, Carbon 41 (2003) 2020

[5] S.Sircar, T.C.Golden, M.B. Rao, Carbon, 34 (1996) 1

[6] K. Inomata, K. Kanazawa, Y. Urabe, H. Hosono and T. Araki, Carbon, 40 (2002) 87

[7] A.Ahmadpour, D.D. Do, Carbon, 34 (1996) 471

[8] J. A. Maciá-Agulló, B. C. Moore, D. Cazorla-Amorós and A. Linares-Solano, Carbon 42 (2004) 1367

[9] F. K. Katsaros, Ph.D. Thesis, National Kapodistrian, University of Athens (2000).

[10] F.K.Katsaros, T.A. Steriotis, A.K. Stubos, A. Ch. Mitropoulos, N.K.Kanellopoulos, S. Tennison, Micropor. Mat., 8 (1997) 171

[11] T.A.Steriotis, F.K.Katsaros, A. Ch.Mitropoulos, A.K.Stubos, N.K. Kanellopoulos, Meas. Sci. Technol., 8 (1997) 168

[12] T.A.Steriotis, F.K. Katsaros, A. Ch. Mitropoulos, A.K. Stubos, P. Galiatsatou, N.K. Kanellopoulos, Rev. Sci. Instrum., 67 (1996) 2545

[13] S.Brunauer, P.H.Emmett, E. Teller, J. Amer. Chem. Soc., 60 (1938) 309

[14] S.J.Gregg, K.W.S. Sing, Adsorption, Surface area and Porosity. Academic Press London (1982)

[15] M.M.Dubinnin, Chem. Rev., 60 (1960) 235

[16] J.Carrido, A. Linares-Solano, J.M. Martin-Martinez, M. Molina-Sabio, F. Rodriguez-Reinoso, R. Torregros, Lanqmuir, 3 (1987) 76

[17] M.J.G.Janssen, C.W.M. van Oorschot, The characterisation of Zeolites by Gas Adsoprtion, in Zeolites: Fact, Figures, Future. P.A Jocobs and R.A. Van Santen Elsevier (1989)

[18] J. Karger, D.M. Ruthven, Diffusion in Zeolites and other Microporous Solids. John Wiley & Sons Inc., NY (1991)

[19] R.C. Reid, J.M. Prausnitz, B.E. Poling Properties of Gases and Liquids. McGraw-Hill International Editions, Chemical Engineering Series, Fourth Edition (International) (1988).

Studies in Surface Science and Catalysis 160
P.L. Llewellyn, F. Rodriquez-Reinoso, J. Rouqerol and N. Seaton (Editors)
607

# Highly microporous carbons prepared by activation of kraft lignin with KOH

**V. Fierro[a], V. Torné-Fernández[a] and A. Celzard[b]**

[a]  Departament d'Enginyeria Química, Universitat Rovira i Virgili, Campus Sescelades, Av. dels Països Catalans 26, 43007 Tarragona, Spain.

[b]  Laboratoire de Chimie du Solide Minéral, UMR CNRS 7555, Université Henri Poincaré, 54506 Vandoeuvre-lès-nancy Cédex, France.

Highly microporous carbon materials with high apparent surface areas (up to ~ 3000 m$^2$ g$^{-1}$) were obtained by heat treatment of mixtures of demineralised kraft lignin (KL$_d$) and KOH. The effects of five parameters: temperature of activation (500-900 °C), KOH/KL$_d$ ratio (1-5), time of activation (0.5-2h), heating rate (5 and 10 °C min$^{-1}$) and nitrogen flow rate (200-800 cm$^3$min$^{-1}$) on carbon yield, surface area, pore volume and pore size distribution were investigated. An increase in the activation degree of KL$_d$ produced a gradual enhancement in the volume of total micropores. Highly activated samples also presented noteworthy mesoporosity. Too high activation temperature resulted in the burn-off of carbon structures and widening of micropores to mesopores.

## 1. INTRODUCTION

The term lignin refers to a group of phenolic polymers accounting for the strength and the rigidity of the vegetal cell walls. The objective of any chemical pulping process is to remove enough lignin to separate cellulosic fibres from each other, producing a suitable pulp for the manufacture of paper and other related products. In terms of industrial chemical modification of lignin, the kraft pulping process is the main one. The kraft method produces black liquor, a residue composed of lignin (30 - 40 %) and other inorganic compounds, which is used as in-house fuel for the recovery of both energy and residual inorganic matter. Several alternatives to combustion have been considered. One of the main possible applications of by-product kraft lignin (KL) consists in preparing activated carbons. Recently, the chemical activation of KL impregnated with H$_3$PO$_4$ was reported [1,2]. The activated carbons produced were essentially microporous with surface areas as high as 1300 m$^2$/g.

The literature evidences a growing interest in alkaline hydroxide activation process, and KOH has been found to be one of the most effective compounds for that purpose [3-7]. High surface areas and pore volumes are reported for lignocellulosic materials, carbons and chars activated by KOH. However, controlling the mean pore size and the pore size distribution is necessary for using such materials in a given application. The present study shows the possibility of producing highly microporous carbons by activation of KL$_d$ with KOH. The effects of five experimental parameters: activation temperature, KOH/KL$_d$ ratio, time of activation, heating rate and nitrogen flow rate on surface area and pore size distribution were investigated.

## 2. EXPERIMENTAL

### 2.1. Demineralisation of KL

KL was supplied by Lignotech Iberica S.A. (Spain), and was presented in the form of a fine dark brown powder. The removal of the inorganic matter from KL was achieved as follows: batches of 100 g were introduced in 2 l of water, leading to dark brown suspensions of pH 9.5, and lignin was precipitated by adding $H_2SO_4$ until the pH decreased to 1. The precipitate was gently washed with distilled water until the pH of the rinse was constant, and finally dried overnight at 105 °C. The lignin prepared this way was nearly mineral-free and was termed demineralised Kraft lignin ($KL_d$).

### 2.2. Preparation of carbons

KOH lentils (Scharlau) were ground and physically mixed with $KL_d$ according to various KOH/ $KL_d$ mass ratios (R=1:1, 2:1, 3:1, 4:1 or 5:1). The carbonisation was carried out in a horizontal furnace and the samples were heated (r = 5 or 10 °C/min) from room temperature up to the final carbonisation temperature ($T_{carb}$= 500, 600, 700, 800 or 900 °C) in different nitrogen flows ($f_{N2}$ = 200, 400, 600 or 800 ml/min). Samples were kept at the final temperature for different carbonisation times ($t_{carb}$ = 0.5, 1 or 2 h) before cooling down under nitrogen.

During the experiments, both metallic potassium (produced by the reduction of KOH by carbon at high temperature) and KOH were partly transported in the vapour phase, and could be observed at the outlet of the reactor. Metallic potassium mixed with potassium carbonate was also present inside the crucible; therefore the latter was submitted to atmospheric humidity for two days, during which the alkaline metal slowly oxidised. Finally, the activated carbon was washed with extreme care, first with 1M HCl, and finally with distilled water until the pH of the rinse remains constant and close to 6. After drying in an oven during 24 h, a very light activated carbon was obtained.

### 2.3. Characterisation of lignin and carbons

*Proximate and ultimate analyses.* Elemental analysis of C, H, S and N content in lignins and activated carbons was done using a Carlo Erba EA-1108 instrument. Oxygen was calculated by difference. The proximate analysis was carried out by thermogravimetrical analysis in a Perkin-Elmer TGA 7 microbalance equipped with a 273–1273K programmable temperature furnace following the weight losses at 110°C/air (moisture), 900 °C/non-oxidising atmosphere (volatile matter), 900 °C/air (fixed carbon); ash content was obtained by difference.

*SEM studies.* The surface morphology of KL and $KL_d$ was studied by scanning electron microscopy (SEM) with a JEOL JSM-6400. The microscope was equipped with an energy dispersive X-ray (EDX) microanalyser that was used for observing the dispersion of the mineral matter in KL and $KL_d$.

*Surface area and porosity.* Surface area and porosity were determined from the corresponding nitrogen adsorption–desorption isotherms obtained at 77 K with an automatic instrument (ASAP 2020, Micromeritics). The samples were previously outgassed at 523 K for several hours. $N_2$ adsorption data for $P/P_0$ from $10^{-5}$ to 0.99 (in a set of values previously fixed) were analysed according to : (i) the BET method [8] for calculating the specific surface area, $S_{BET}$; and (ii) the $\alpha_S$ method [9] (using Carbopack F Graphitised Carbon Black as reference material    [10] ) for calculating the micropore volume, $V_{\alpha\ micro}$, and the supermicropore volume, $V_{\alpha\ super}$. The total pore volume, $V_{0.99}$, was calculated from nitrogen adsorption at a relative pressure of 0.99.

## 3. RESULTS AND DISCUSION

Table 1 shows the proximate and ultimate analyses of KL and KL$_d$. KL has a high ash content (11.1 % on dry ash-free (daf basis) which is nearly removed (0.2 % on daf basis) after the treatment with $H_2SO_4$. The high S content (2.2 %) in KL is due to both the Kraft or sulphate process, which consists in a treatment with NaOH and $Na_2S$ to separate the cellulose from the other wood constituents, and to organically bound sulphur (up to 1.5 %) [11]. XRD analysis of lignin showed that Na is found combined with S and C inside the phase $Na_2CO_3$·2 $Na_2SO_4$. Lignin demineralisation produced a decrease of S content (down to 0.5 %) and of O content (from 33.3 to 27.8 %). Analysis by SEM-EDX showed that S and Na are uniformly distributed in the polymeric matrix before and after the demineralisation treatment.

Table 1. Proximate and ultimate analyses of KL and KL$_d$ (wt. %)

|  | Proximate Analysis (wt %, dry basis) | | | Ultimate Analysis (wt %, daf) | | | | |
|---|---|---|---|---|---|---|---|---|
|  | Fixed Carbon | Volatile matter | ash | C | H | N | S | O$^*$ |
| KL | 36.4 | 52.5 | 11.1 | 59.5 | 5.1 | 0.1 | 2.2 | 33.3 |
| KL$_d$ | 39.7 | 60.1 | 0.2 | 65.8 | 5.9 | 0.0 | 0.5 | 27.8 |

* Estimated by difference

The effect of the experimental parameters considered in this study is discussed below.

### 3.1. Effect of the temperature of activation

Figure 1 a) shows the variation of carbon yield with the temperature of activation. Increasing the activation temperature produces the decrease of the carbon yield due: (i) to the pyrolysis of lignin up to 600 °C; (ii) to the activation by KOH and $K_2CO_3$ that starts at 450-500°C. The dissociation of the two phenomena, pyrolysis and activation, is impossible because lignin have already reacted with KOH in some extent before pyrolysis finishes.

Activation with KOH involves the oxidation of C and the production of K metal, $H_2$ and $K_2CO_3$ according to the main reaction :

$$6 \text{ KOH} + 2 \text{ C} \leftrightarrow 2 \text{ K} + 3 \text{ H}_2 + 2 \text{ K}_2CO_3 \tag{1}$$

However, reaction (1) is certainly not the only one, since various molecules like CO, $CO_2$, $H_2$ and $H_2O$ originating from the thermal decomposition of lignin are also present. Thus, $K_2CO_3$ produced by reaction of KOH with $CO_2$ also acts as an efficient activating agent. Hayashi and coworkers [12] showed that high surface areas of nearly 1700 and 2000 $m^2/g$ can be obtained by activation of KL with $K_2CO_3$ for R=2 at 700 and 800 °C, respectively. Moreover, carbons prepared by $K_2CO_3$ activation showed higher surfaces than those prepared by KOH activation at temperatures higher than 600°C. Hayashi has also worked with different raw materials: husks [13], nutshells [14] or formaldehyde resins [15] reaching very high surfaces even at 700°C. McKee [16] studied the gasification of graphite powder by a serie of alkali metal salts and found that $K_2CO_3$ was reduced in inert atmosphere by carbon as follows:

$$K_2CO_3 + 2 \text{ C} \leftrightarrow 2 \text{ K} + 3 \text{ CO} \tag{2}$$

The decomposition of $K_2CO_3$ to $CO_2$ and $K_2O$ could also lead to activation by the two latter products, according to :

$$C + CO_2 \leftrightarrow 2 \text{ CO} \tag{3a}$$

$$2 \text{ K}_2O + C \leftrightarrow 4 \text{ K} + CO_2 \tag{3b}$$

$$K_2O + C \leftrightarrow 2\,K + CO \tag{3c}$$

Nevertheless, $CO_2$ and $K_2O$ individually are not expected to be activating agents until high temperatures (above 800 °C) are reached. Hence, the other most efficient activant is rather the water vapour evolved from lignin pyrolysis, and giving the following reaction :

$$H_2O + C \leftrightarrow H_2 + CO \tag{4}$$

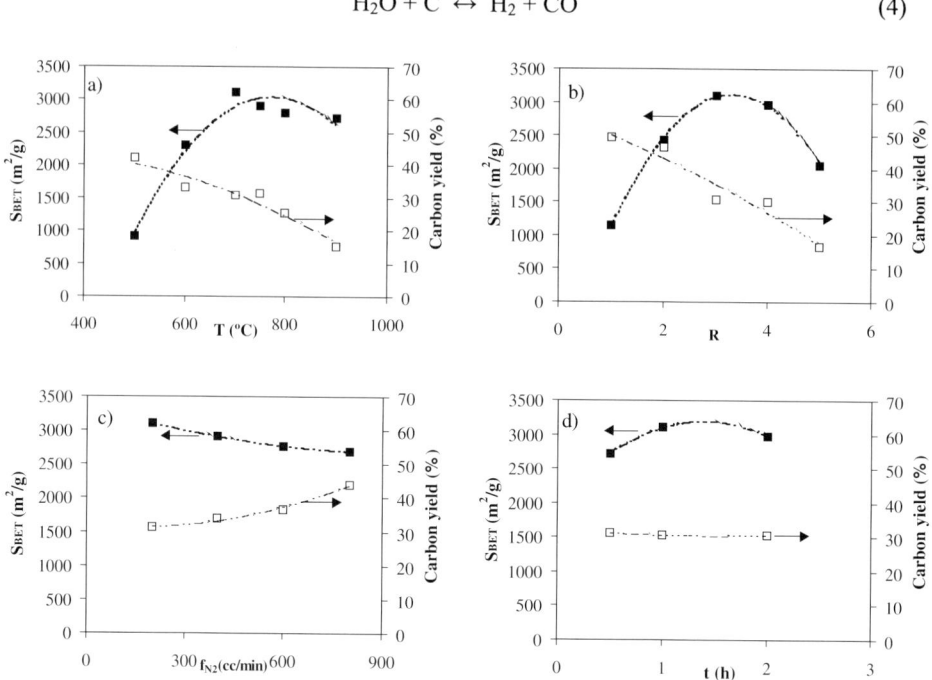

Figure 1. Variation of the $S_{BET}$ (■) and carbon yield (□) with: a) temperature of activation (R= 3, $f_{N2}$= 200 cc/min, r= 5°C/min, $t_{carb}$= 1h); b) R (T= 700 °C, $f_{N2}$= 200 cc/min, r= 5°C/min, $t_{carb}$= 1h); c) $N_2$ flow (T= 700 °C, R= 3, r= 5 °C/min, $t_{carb}$= 1 h); d) activation time (T= 700 °C, R= 3, $f_{N2}$= 200 cc/min, r= 5 °C/min).

Figure 2 a) shows the adsorption-desorption isotherms of $N_2$ at 77 K of the activated carbons prepared at 500, 600, 700, 800 and 900 °C. The carbon prepared at 500 °C presents an extensive plateau in the range of medium to high relative pressures, indicating an essentially microporous character. As carbonization temperature increases the knee of the isotherms widens and the width of the plateau decreases, indicating a widening of the pores. Thus, the material prepared at 900 °C shows a hysteresis loop, evidencing a well-developed mesoporosity. The decrease of carbon yield is accompanied by an increase of both surface area and microporosity up to the temperature of 750 °C, above which these properties decrease as seen in Figure 1 a).

Figure 3 a) shows the total pore, miropore and ultramicropore volumes. The total pore volume always increases with activation temperature but at temperatures higher than 750°C there is a widening of micropores to create mesopores. Figure 3 b) shows the corresponding pore size distributions calculated by application of the Horwatz-Kawazoe method; the maximum is always centered in the micropore region but, as the temperature increases the contribution of wider pores is more important, in agreement with the results of Figures 2a) and 3a).

### 3.2. Effect of the KOH/KL$_d$ weight ratio (R)

At constant temperature (700 °C), the value of R has a marked effect on both the carbon yield and the BET surface area $S_{BET}$. Figure 1b) shows that there is a linear decrease of carbon yield with R whereas a maximum of the $S_{BET}$ can be observed at R values around 3. Figure 2 b) shows that an increase of R from 1 to 3 produces a great enhancement of N$_2$ adsorption capacity at 77 K but higher values of R reduce it. The micropore volumes are also reduced for R ≥ 3 (see Figure 3 c) with the concomitant increase of the fraction of wider pores (see Figure 3 d).

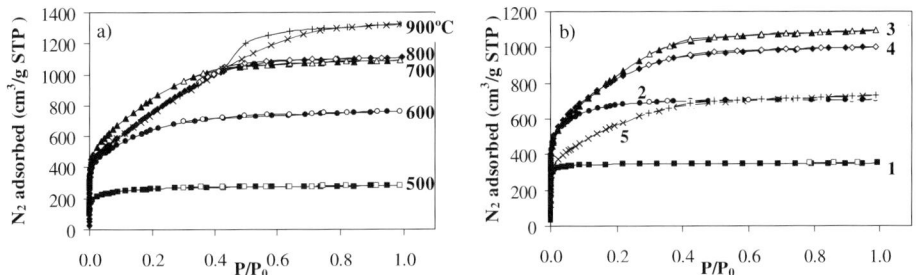

Figure 2. Adsorption-desorption isotherms of N$_2$ at 77 K on the activated carbons derived from KL$_d$ : a) effect of T and b) effect of R (open symbols and x: adsorption isotherms; full symbols and +: desorption isotherms).

### 3.3. Effect of the N$_2$ flow

The flow of N$_2$ removes the gaseous reaction products but also a part of the activating agent as it was observed in this study. Figure 1 c) shows that the increase in the N$_2$ flow rate from 200 to 800 cc/min produces an increase of carbon yield because the activating agent is increasingly swept out.

Such a decrease of the activation efficiency may be explained both by a lower contact time of KOH vapour with the solid matter, and by the removal of other possible activating agents : CO$_2$ (either as such, or as K$_2$CO$_3$ after its reaction with KOH), and H$_2$O.

$S_{BET}$ and the porosity, in the whole pore diameter range, decrease with increasing N$_2$ flow rate as it is shown in Figures 1c) and 3 e) respectively. These results are different from those found by Linares-Solano and coworkers who found that N$_2$ flow enhances the activation of the carbon [6]. However, it is obvious that the different nature of the precursor has an effect on the activation process and so on the activated carbon produced. These results, apparently opposed, could not be contradictory. When pyrolysis and activation take place simultaneously, as for biomass precursors, the co-activation by water vapour evolved during pyrolysis could be important as suggested above.

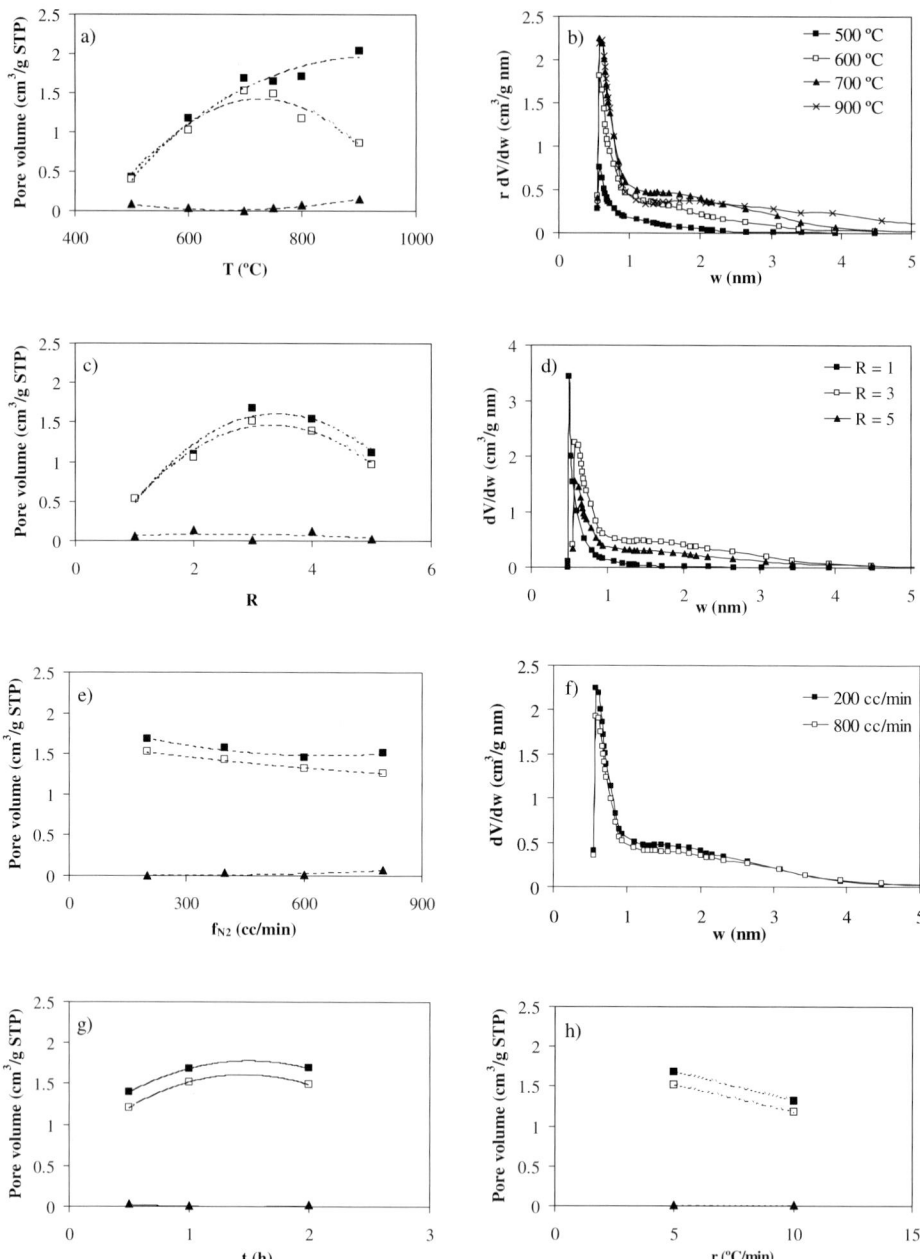

Figure 3. Variation of the $V_{0.99}$ (■), $V_{\alpha\ micro}$ (□) and $V_{\alpha\ ultra}$ (▲) with: a) temperature of activation ($R= 3$, $f_{N2}= 200\ cc/min$, $r= 5°C/min$, $t_{carb}= 1h$); c) R ($T= 700\ °C$, $f_{N2}= 200\ cc/min$, $r= 5°C/min$, $t_{carb}= 1h$); e) $N_2$ flow rate ($T= 700\ °C$, $R= 3$, $r= 5\ °C/min$, $t_{carb}= 1\ h$); g) activation time ($T= 700\ °C$, $R= 3$, $f_{N2}= 200\ cc/min$, $r= 5\ °C/min$); h) heating rate ($T= 700\ °C$, $R= 3$, $f_{N2}= 200\ cc/min$, $t_{carb}= 1h$). Evolution of the

pore size distribution of the carbons with: b) temperature of activation (experimental conditions as in Fig. 3 a); d) R (experimental conditions as in Fig. 3 c).

Figure 3 f) shows the PSD of two activated carbons prepared with a $N_2$ flow of 200 or 800 cc/min. Increasing the $N_2$ flow does not change the PSD but decreases the pore volume ($S_{BET}$ and $V_{0.99}$ decrease). Since the size of the pores is the same, finding a lower pore volume should be related to a lower number of pores. Therefore, the effect of the $N_2$ flow is very different from that of R, even is the latter is more noticeable.

### 3.4. Effect of the activation time

The duration of activation at a given temperature does not seem to affect neither the carbon yield nor the elemental composition, since differences in both analyses are very small. Figure 1 d) and 3 g) show that there is an optimum in the activation time, between 1 and 2 h, to get the highest surface area and microporosity. The $S_{BET}$ decreases from 3105 to 2990 m$^2$/g when increasing activation time from 1 to 2h respectively.

### 3.5 Effect of the heating rate

An increase of the heating rate from 5 to 10 °C/min produces a lowering of the $S_{BET}$ from 3105 to 2493 m$^2$/g (see figure 3h). Actual technical limitations do not allow using higher heating rates but our results agree with previous findings [6]. During the heating process the hydroxide melts, then, it is reasonable that a lower heating rate allows a longer contact time between carbon and liquid hydroxide, and hence a better impregnation, before the reaction temperature is reached.

## 4. CONCLUSIONS

The present exploratory study evidenced the possibility of preparing highly microporous active carbons from demineralised Kraft lignin, using KOH in suitable experimental conditions. The most relevant parameters were found to be activation temperature and mass ratio KOH/lignin, while the other ones (flow of inert gas, duration time, heating rate) were found to have minor effects within the corresponding range of values investigated. Thus, the best materials (surface area ~ 3000 m$^2$/g, micropore volumes ~ 1.5 cm$^3$/g) were obtained at 700 °C and KOH/KL$_d$ = 3. Such results are very close to those already reported for anthracites, which are known to lead to very good adsorbents when prepared in similar conditions [17]. Furthermore, modifying the experimental conditions easily leads to a range of active carbons, from almost purely microporous to mesoporous. Hence, even if the detailed mechanisms are still unclear, chemical activation now appears to be a valuable (rapid, simple and cheap) process for the valorization of lignin.

**Acknowledgements**

This research was made possible in part by financial support from MCYT (project PPQ2002-04201-CO02), DURSI (2001SGR00323 and 2002AIRE) and ALFA Program (project ALFA II 0412 FA FI). V. Fierro acknowledges the MCYT and the Universitat Rovira i Virgili (URV) for the financial support of her 'Ramón y Cajal' research contract. V. Torné-Fernández acknowledges the URV for her PhD grant.

## REFERENCES

[1]  V. Fierro, V. Torné-Fernández, D. Montané and J. Salvadó, 'Activated Carbons Prepared from Kraft Lignin by Phosphoric Acid Impregnation', In proceedings of Carbon'03, Oviedo (Spain) 2003.
[2]  V. Fierro, V. Torné-Fernández, D. Montané and A. Celzard, *Thermochim. Acta*, 433 (2005) 153.

[3]   A. Ahmadpour and D.D. Do, Carbon, 34 (1996) 471.
[4]   T. Otowa, Y. Nojima, T. Miyazaki, Carbon, 35 (1997) 1315.
[5]   C. Liang, Z.Wei, Q. Xin and C. Li, Appl. Catal. A, 208 (2001) 193.
[6]   D. Lozano-Castelló, M.A. Lillo-Ródenas, D. Cazorla-Amorós and A. Linares-Solano, Carbon, 39 (2001) 741.
[7]   E. Frackowiak and F. Beguin, Carbon, 40 (2002) 1775.
[8]   S. Brunauer, P.H. Emmett and E. Teller, J. Chem. Am. Soc. 60 (1938) 309.
[9]   K.S.W. Sing, Carbon, 27 (1989) 5.
[10] M. Kruk, Z.J. Li, M. Jaroniec, W.R. Betz, Langmuir, 15 (1999) 1435.
[11] C.H. Hoyt, D.W. Goheen, in: K.V Sarkanen, C.H. Ludwig (Eds.), Lignins, Wiley-Interscience, New York, 1971, Chapter 20, p. 833.
[12] J. Hayashi, A. Kazehaya, K. Muroyama, A.P. Watkinson, Carbon, 38 (2000) 1873.
[13] J. Hayashi, T. Horikawa, K. Muroyama , V.G. Gomes, Micropor. Mesopor. Mat., 55 (2002) 63.
[14] J. Hayashi, T. Horikawa, I. Takeda, K. Muroyama, F.N. Ani Carbon, 40 (2002) 2381.
[15] J. Hayashi, M. Uchibayashi, T. Horikawa, K. Muroyama, V.G. Gomes, Carbon, 40 (2002) 2747.
[16] D.W. McKee, Carbon, 20 (1982) 59.
[17] A. Celzard and V. Fierro, Energy and Fuels, 19 (2005) 573.

Studies in Surface Science and Catalysis 160
*P.L. Llewellyn, F. Rodriquez-Reinoso, J. Rouqerol and N. Seaton (Editors)*
© 2007 Elsevier B.V. All rights reserved

# Preparation and Characterization of Nanoporous Ternary Mixed Cerium Oxides

**Christothea Attipa, and Charis R. Theocharis**

Porous Solids Group, Department of Chemistry, University of Cyprus, P.O. Box 20537, 1678, Nicosia, Cyprus

We report the synthesis and characterisation of ceria samples containing two heteroatoms, namely vanadium and copper or iron. For certain compositions, the samples contained both micropores and mesopores. High BET surface areas were obtained in relation to pristine ceria.

## 1. INTRODUCTION

We have been interested in the synthesis and study of porous ceria, including heteroatom containing ceria, because of the importance of these solids in several catalytic applications[1,2,3]. Previous work in these laboratories has shown that ceria or metal-doped ceria prepared from dilute aqueous solutions by ammonia precipitation, is mesoporous[4-6]. The mesoporosity of these solids is susceptible to changes in synthesis conditions such as pH and concentration of the cerium and other cations precursors used. In this contribution we present results from the synthesis of cerium (IV) oxide precipitated from aqueous solution in the presence of different concentrations of vanadium (V) and copper (II) or vanadium (V) and iron (II) ions. We chose to study the synthesis and properties of ternary mixed oxides with the view of preparing phases with interesting catalytic properties.

The aim of these studies was to examine the effect of the heteroatoms on the surface texture and the porosity of ceria, as well as the chemistry of the surface in terms of surface groups present and surface acidity (or basicity), as well as on bulk properties such as crystal structure and oxygen storage capacity. Vanadium was chosen because of the multiplicity of oxidation states in which it can be present. In previous studies, copper as a guest ion was shown to increase the BET surface area of ceria in certain concentrations. Iron was chosen because of the similarity of the electrode potential of the $Fe^{3+}/Fe^{2+}$ and $Ce^{4+}/Ce^{3+}$ pairs. Further, vanadium, copper and iron oxides have proven catalytic properties either in pure form or mixed with other phases.

## 2. EXPERIMENTAL

The mixed cerium oxide samples were prepared by homogenous co-precipitation from aqueous solutions, using 1M ammonia solution as base. The metal precursors used were cerium (IV) ammonium nitrate, ammonium metavanadate (V) and copper (II) nitrate or iron (II) nitrate. The concentrations of all the solutions used were 0.01M throughout. The appropriate solutions were mixed in the appropriate ratios, to achieve the desired results. The precipitate was left to stand in the mother liquor for 24h, centrifuged and dried at 373K for 24h.

Nitrogen adsorption isotherms were measured at 77K from a starting pressure of 0.15Pa, after outgassing at 383K for 24h, using an ASAP2010 Micromeretics apparatus. The crystalline structure of the samples was determined by X-ray diffraction using Cu Kα radiation employing a Shimadzu XRD-6000 diffractometer. The thermal behaviour of the samples was measured by both thermogravimetry (Shimadzu apparatus) and differential scanning calorimetry (DSC model Q100). Surface areas were calculated from the adsorption data using the BET equation, whilst the pore size distribution was estimated using DFT methods employing a slit-shaped pore model.

Four sets of solids were prepared, namely $V_xCu_xCe_{1-2x}O_2$ and $V_xFe_xCe_{1-2x}O_2$, where x was less or equal to 0.45, for which each sample contained equal amounts of each heteroatom and varying proportions of Ce(IV). The second set of samples had compositions $V_xCu_{0.4-x}Ce_{0.6}O_2$ and $V_xFe_{0.4-x}Ce_{0.6}O_2$ with x less or equal to 0.4. Each of these samples contained 60 mole percent of Ce(IV) and varying amounts of the heteroatoms totalling 40 mole percent.

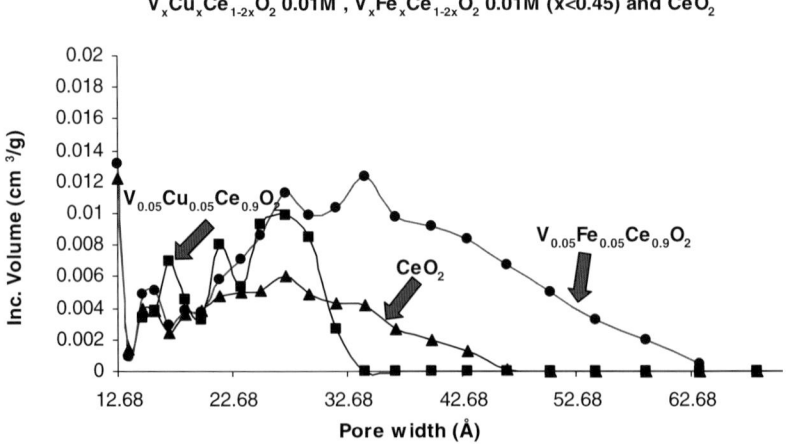

Figure 1 DFT plots for $V_{0.05}Cu_{0.05}Ce_{0.9}O_2$, $V_{0.05}Fe_{0.05}Ce_{0.9}O_2$

## 3. RESULTS AND DISCUSSION

Table 1 contains the pertinent surface texture results for the samples discussed in this presentation. Figure 1 contains the DFT plots for $V_{0.05}Cu_{0.05}Ce_{0.9}O_2$, $V_{0.05}Fe_{0.05}Ce_{0.9}O_2$. The latter was chosen because as it can be seen from Table 1 it exhibits the highest BET surface area of all samples in the relevant series, while the former sample is the equivalent where iron has replaced copper.

Table 1
Samples prepared with their $S_{BET}$ values, total pore volume (V) and average pore diameter d

| SAMPLE | $S_{BET}$ $m^2 g^{-1}$ | V $/cm^3 g^{-1}$ | d/ nm |
|---|---|---|---|
| $CeO_2$ (100%) | 232 | 0.13 | 2.2 |
| $V_{0.05}Cu_{0.05}Ce_{0.90}O_2$ | 285 | 0.15 | 2.1 |
| $V_{0.1}Cu_{0.1}Ce_{0.80}O_2$ | 299 | 0.19 | 2.6 |
| $V_{0.2}Cu_{0.2}Ce_{0.60}O_2$ | 242 | 0.14 | 3.6 |
| $V_{0.3}Cu_{0.3}Ce_{0.40}O_2$ | 211 | 0.13 | 2.6 |
| $V_{0.35}Cu_{0.35}Ce_{0.3}O_2$ | 175 | 0.14 | 3.2 |
| $V_{0.4}Cu_{0.4}Ce_{0.20}O_2$ | 122 | 0.14 | 2.6 |
| $V_{0.45}Cu_{0.45}Ce_{0.1}O_2$ | 133 | 0.15 | 4.4 |
| $V_{0.1}Cu_{0.3}Ce_{0.60}O_2$ | 250 | 0.15 | 2.4 |
| $V_{0.3}Cu_{0.1}Ce_{0.60}O_2$ | 261 | 0.17 | 2.6 |
| $V_{0.4}Ce_{0.60}O_2$ | 192 | 0.12 | 2.4 |
| $Cu_{0.4}Ce_{0.60}O_2$ | 245 | 0.14 | 2.2 |
| $V_{0.15}Cu_{0.25}Ce_{0.60}O_2$ | 265 | 0.15 | 2.4 |
| $V_{0.25}Cu_{0.15}Ce_{0.60}O_2$ | 267 | 0.16 | 2.4 |
| $V_{0.05}Cu_{0.35}Ce_{0.60}O_2$ | 294 | 0.17 | 2.2 |
| $V_{0.35}Cu_{0.05}Ce_{0.60}O_2$ | 233 | 0.15 | 2.4 |
| $V_{0.01}Fe_{0.01}Ce_{0.98}O_2$ | 251 | 0.14 | 2.2 |
| $V_{0.02}Fe_{0.02}Ce_{0.96}O_2$ | 263 | 0.16 | 2.4 |
| $V_{0.03}Fe_{0.03}Ce_{0.94}O_2$ | 235 | 0.13 | 2.2 |
| $V_{0.04}Fe_{0.04}Ce_{0.92}O_2$ | 263 | 0.15 | 2.2 |
| $V_{0.05}Fe_{0.05}Ce_{0.90}O_2$ | 330 | 0.22 | 2.6 |
| $V_{0.1}Fe_{0.1}Ce_{0.80}O_2$ | 304 | 0.21 | 2.8 |
| $V_{0.2}Fe_{0.2}Ce_{0.60}O_2$ | 290 | 0.21 | 2.8 |
| $V_{0.3}Fe_{0.3}Ce_{0.40}O_2$ | 244 | 0.22 | 3.6 |
| $V_{0.35}Fe_{0.35}Ce_{0.30}O_2$ | 247 | 0.25 | 4 |
| $V_{0.4}Fe_{0.4}Ce_{0.2}O_2$ | 280 | 0.25 | 3.6 |
| $V_{0.45}Fe_{0.45}Ce_{0.1}O_2$ | 281 | 0.24 | 3.4 |
| $V_{0.35}Fe_{0.05}Ce_{0.60}O_2$ | 247 | 0.23 | 3.8 |
| $V_{0.05}Fe_{0.35}Ce_{0.60}O_2$ | 251 | 0.16 | 2.6 |
| $V_{0.1}Fe_{0.3}Ce_{0.60}O_2$ | 275 | 0.20 | 2.8 |
| $V_{0.3}Fe_{0.1}Ce_{0.60}O_2$ | 252 | 0.20 | 3.2 |
| $V_{0.15}Fe_{0.25}Ce_{0.60}O_2$ | 256 | 0.21 | 3.2 |
| $V_{0.25}Fe_{0.15}Ce_{0.60}O_2$ | 260 | 0.21 | 3.2 |

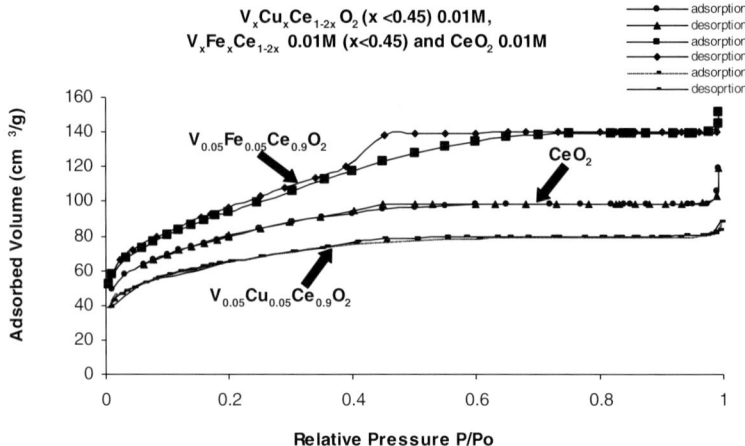

Figure 2 Nitrogen adsorption isotherms for samples $V_{0.05}Cu_{0.05}Ce_{0.9}O_2$, $V_{0.05}Fe_{0.05}Ce_{0.9}O_2$

It can be seen from Figure 1 that both these samples have the majority of their pores well within the mesopore range, but also possess considerable porosity in the microporous range. This is the first example of substituted ceria exhibiting this type of porosity, which is reflected in the relatively high BET surface area. It should be noted that pristine ceria does not possess microporosity. The low values for microporous pore volume for this sample must be, presumably, an artifact of the model used. It can also be seen from Figure 1 that the iron containing sample $V_{0.05}Fe_{0.05}Ce_{0.9}O_2$ has a wide pore size range in the mesopore range, whilst $V_{0.05}Cu_{0.05}Ce_{0.9}O_2$, has a much more narrow one.

From Figure 2, it can be seen that these samples exhibit a type IV isotherm with a well-defined plateau at high partial pressures. Sample $V_{0.05}Cu_{0.05}Ce_{0.9}O_2$ has lower pore volume as well as apparent surface area, whereas $V_{0.05}Fe_{0.05}Ce_{0.9}O_2$ has considerably higher total pore volume as well as apparent surface area.

From Table 1, it can be seen that for some copper containing ternary ceria samples the apparent surface area was superior to that for the pristine sample when heat-treated and outgassed under similar conditions, whilst for others the apparent surface area was lower. For the ternary samples containing iron, the situation was that for all compositions the samples exhibited apparent surface areas well above the value for the pristine sample.

Figure 3 contains the nitrogen adsorption isotherms for $V_{0.05}Cu_{0.35}Ce_{0.6}O_2$, $V_{0.05}Fe_{0.35}Ce_{0.6}O_2$, and, for comparison, for pristine ceria. Both these samples exhibited apparent surface areas comparable to those for pristine ceria, as it can be seen also from Table 1. On the contrary, the iron-containing sample had considerably higher total pore volume than the pristine sample.

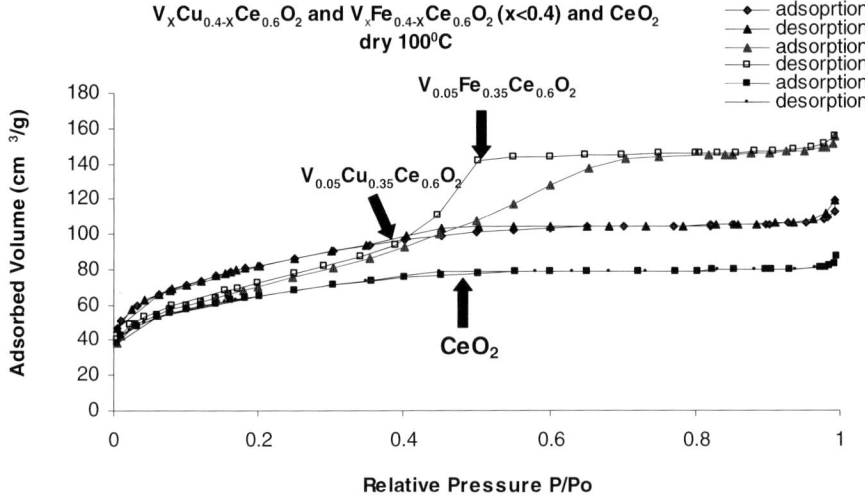

Figure 3 Nitrogen Adsorption Isotherms for $V_{0.05}Cu_{0.35}Ce_{0.6}O_2$, $V_{0.05}Fe_{0.35}Ce_{0.6}O_2$, and pristine ceria

Specifically, sample $V_{0.05}Fe_{0.35}Ce_{0.6}O_2$ had a total pore volume of 0.23 $cm^3g^{-1}$ as compared to 0.13 $cm^3g^{-1}$ for the pristine sample and 0.15 $cm^3g^{-1}$ for sample $V_{0.05}Cu_{0.35}Ce_{0.6}O_2$. In similar fashion to the previous set of samples, the isotherms were again of type IV. DFT analysis indicated that a small amount of microporosity is present in these samples. In general, with one exception, copper-containing samples had higher BET surface areas than the equivalent one containing iron. One exception, is the sample set out in Figure 2. Similarly, for both the copper-containing and iron-containing samples, highest BET surface area values were obtained for samples containing 5 to 20 mole percent of vanadium. The variation of BET surface area values with cerium content was rather unclear. For the copper-containing series there was no clear pattern of dependence upon cerium content, but maximum surface area was obtained for 90 mole percent cerium. For the iron-containing series, a maximum apparent surface area was obtained for 80 mole percent cerium content.

X-ray diffractometry has shown that upon synthesis all samples had peaks attributable to cerium (IV) oxide (fluorite structure). However, upon calcination to 773K further crystallization took place, and in addition to ceria, $V_2O_5$, $CeVO_4$ was present as well as CuO and $Cu_2O$ for the copper-containing samples, and FeO and $Fe_2O_3$ for the iron-containing solids. The phases were identified from the literature [7-9]. The absence of any peaks other than those of ceria in the as-prepared samples was surprising given that for certain samples that was the minority phase present. However, it does not imply that a solid solution was prepared possessing that structure. Rather, although it is likely that at least part of the sample consisted of of a solid solution, it is more likely that the other oxides are present as poorly crystalline or microcrystalline phases. That phase changes occur, as well as loss of water and ammonia during heat treatment, it is also observed using thermogravimetry, as well as calorimetry. The fact that $CeVO_4$ is one of the phases present suggests that mixed oxide phases are indeed present in the as-prepared samples. It is noteworthy that all hetero-atoms were present in two oxidation states, V +4 and +5, Cu in +1 and +2, and Fe +2 and +3. This augurs well for the usefulness of these solids in catalysis.

## REFERENCES

[1] Bunluesin T, Gorte RJ and Graham GW, **1998**, *Applied. Catal. B-Environmental*,15,107.

[2] Shen WJ and Matsumura Y, **2000**, *Phys Chem Chem Phys,* 2, 1519.

[3] Overbury SH, Mullins DR, and Huntley DR, **1999**, *J Catal* , 186, 296.

[4] Pashalidis I and Theocharis CR, **2000**, in (Eds K.K. Unger, G. Kreysa and J.P. Baselt), "*Studies in Surface Science and Catalysis*", Elsevier Science Publishers, 128, 643.

[5] Kyriakou G, Paschalidis I, and Theocharis CR, **2002**, *Fundamentals of Adsorption, 7* (Editors: K.Kaneko, H. Kanoh and Y. Hanzawa), 201.

[6] Theocharis CR, Christophidou M, and Pashalidis I., **2002**, "*Studies in Surface Science and Catalysis*", Elsevier Science Publishers, 144, 75

[7] R. Cousin, D. Coursot, E.A. Aad, S. Capelle, J.P. Amoureux, M. Dourbin, M. Guelton, A. Aboukais, *Colloids and Surface A*, 158 (1999) 43-49.

[8] U.O. Krasovec, B. Orel, A. Surca, N. Bukovec, R. Resfeld, *Solid State Ionics*, 118 (1999) 192-214.

[9] G. R. Rao, H.R. Saho, B.G. Mishra, *Colloids and Surfaces A: Physicochem. Eng. Aspects,* 220 (2003) 261-269.

Studies in Surface Science and Catalysis 160
P.L. Llewellyn, F. Rodriquez-Reinoso, J. Rouqerol and N. Seaton (Editors)

# Preparation and dynamic adsorption properties of activated carbons with tailored micro- and mesoporosity

**P.A. Barnes[a], E.A. Dawson[a], M W Smith[b], J.L. Ward[b] & H.M. Williams[a]**

[a] Department of Chemical & Biological Sciences, University of Huddersfield, HD1 3DH, United Kingdom

[b] Physical Sciences Department, Dstl, Porton Down, Salisbury, Wiltshire SP4 0JQ, United Kingdom

## 1. INTRODUCTION

This research aims to study the influence of pore size distribution on the adsorption by activated carbons of Volatile Organic Chemicals (VOCs) from a dry or humid flowing air stream. Carbons containing various proportions of micro- and mesoporosity have been prepared from phenolic resin. Initially, highly mesoporous carbons were prepared using a templating technique. Subsequently these materials were activated in carbon dioxide to increase the proportion of microporosity. The resulting carbons have been characterised using nitrogen adsorption at 77K, together with water and pentane adsorption at 295K. The dynamic adsorption properties in dry and humid conditions have been assessed at 295K using pentane and octane vapours.

## 2. EXPERIMENTAL

### 2.1. Preparation
Mesoporous carbons were prepared using a templating method based on that described by Han et al [1]. Two silica gel templates were used in this study, the first (obtained from Sigma-Aldrich) denoted "100Å" and the second by a Crosfield (now INEOS Silicas) code, SD1503.

Carbons were prepared by dissolving SMD 30207 phenolic resin (Schenectady Europe Ltd., ref. SMD30207) in methanol (1 part resin to either 1 or 2 parts of methanol by mass) and adding hexamine (Aldrich 99%) in a ratio of 10:1 by mass of resin. The resulting solutions were added to the silica gel in varying quantities and the solvent allowed to evaporate in air. The samples were crosslinked in air at 200°C and then pyrolysed at 500°C under flowing nitrogen for 6.5 hours. To remove the silica gel template, the samples were heated twice in a concentrated sodium hydroxide (NaOH) solution (ca. 4M). Finally, the samples were washed with 0.2M hydrochloric acid (HCl). Atomic absorption spectroscopy was used to confirm that no significant residual levels of sodium were present (<25µg of sodium per gram of carbon).

Microporosity was introduced into the carbons through activation in carbon dioxide at 900°C, with the activation time being varied. All carbons were formed as fine powders and were of similar, but not necessarily identical, particle sizes.

Sample codes used in this study are derived as follows: the first two letters indicate the precursor (PR = phenolic resin for all samples described here). Following this is the template used (either 100 for "100Å" or SD for SD1503); following the divider is the activation time in hours (with "0" denoting the unactivated templated carbon). Thus PRSD/2 denotes a phenolic resin carbon templated using SD1503 silica gel, then activated for 2 hours in carbon dioxide at 900°C.

## 2.2. Characterisation

### 2.2.1. Nitrogen adsorption at 77K

Nitrogen adsorption isotherms were measured at 77K, using an automated instrument (Omnisorp 100cx, Beckman Coulter Electronics Ltd). Adsorbent samples (approximately 0.1g) were out-gassed at 200°C to a residual pressure of $10^{-4}$ mbar for at least 6 hours prior to the measurement of isotherms. Specific surface areas were derived from the isotherms using the BET equation [2]. Total pore volumes were obtained using the Gurvitsch Rule and the micropore volumes were determined using the $\alpha_s$ method [2]. Mesopore volumes were subsequently determined by subtraction.

### 2.2.2. Water adsorption at 295K

The polarity of the carbons was assessed through the measurement of water adsorption isotherms using an automated instrument (CISORP95, C I Electronics Ltd). Approximately 50mg of dry carbon was placed in an open container suspended from a microbalance. The sample was heated to 60°C (the upper limit of the instrument) in dry air for 60 minutes then sequentially equilibrated at various relative humidities ranging from 5 to 90% at 22°C. The water uptake, as a percentage of the dry adsorbent weight, was calculated for each step of the isotherm using the recorded weight data.

### 2.2.3. VOC adsorption at 295K

Adsorption isotherms for pentane, octane and methanol were determined at 295K using an Intelligent Gravimetric Analyser (Hiden Isochema Ltd). The use of this instrument to obtain adsorption isotherms for organic chemicals has previously been described in detail by Fletcher and Thomas [3]. Approximately 40mg of carbon was placed in an open container and outgassed at <0.1mbar for 3 hours at 130°C. The carbon was then cooled to 295K and sequentially exposed to a pre-programmed range of partial pressures of the adsorptive. At each point, equilibrium was deemed to have been reached when the uptake reached 99% of the asymptotic value predicted by the software.

### 2.2.4. Dynamic adsorption testing at 295K

Carbon samples (0.1g) were packed in 4.5mm diameter tubes and mounted in a test apparatus maintained at 295K. Test atmospheres were prepared by vaporising a known quantity of either pentane or octane into a known volume of air at the required humidity, contained in a Lamofoil bag. The test concentration was 2000mg m$^{-3}$ throughout. The test atmosphere was drawn through the sample at 50cm$^3$min$^{-1}$. For humid testing, the carbons were placed in an environmental chamber at RH80% and 295K for 48 hours prior to testing. During tests, the effluent airstream was monitored using capillary gas chromatography (Varian CP-3800, CP-Sil 5CB column, 15m x 0.25mm with 0.25µm film thickness, 15ml min$^{-1}$, 50°C), internally calibrated prior to the test using the test vapour mixture. Tests were continued until the influent and effluent concentrations were equal.

As the carbons were not of identical particle sizes, comparison has predominantly been carried out on the basis of equilibrium adsorption capacities, with breakthrough times being shown as a guide only.

## 3. RESULTS

### 3.1. Nitrogen adsorption at 77K

Fig. 1 illustrates nitrogen adsorption / desorption isotherms at 77K for the templated and 8 hour activated carbons.

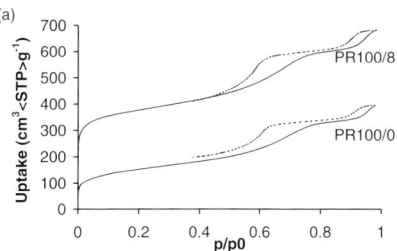 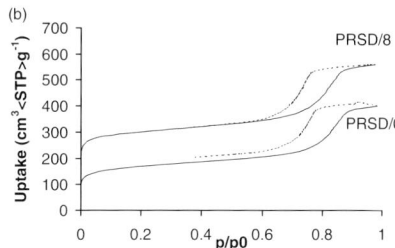

Fig. 1. Nitrogen adsorption (solid line) / desorption (dotted line) isotherms at 295K for (a) PR100 series and (b) PRSD series carbons

The BET surface areas, micropore volumes and mesopore volumes derived from the isotherm data are listed in Table 1, as are the burn-off levels resulting from the activation.

Table 1
BET surface areas, micropore volumes and mesopore volumes derived from nitrogen adsorption data

|         | Burn-off (%) | Specific surface area $(m^2 \, g^{-1})$ | Micropore volume $(cm^3 \, g^{-1})$ | Mesopore volume $(cm^3 \, g^{-1})$ |
|---------|---------|---------|---------|---------|
| PR100/0 | 0       | 546     | 0.07    | 0.45    |
| PR100/2 | 26      | 753     | 0.20    | 0.43    |
| PR100/8 | 50      | 1307    | 0.37    | 0.60    |
| PRSD/0  | 0       | 592     | 0.15    | 0.45    |
| PRSD/2  | 21      | 653     | 0.21    | 0.45    |
| PRSD/8  | 39      | 1031    | 0.35    | 0.51    |

### 3.2. Water adsorption at 295K
Throughout this and the following sections, volumetric uptakes have been derived from mass uptakes based on the assumption of liquid density within the pores.

Fig. 2 illustrates water adsorption / desorption isotherms at 295K for PR100 and PRSD series carbons.

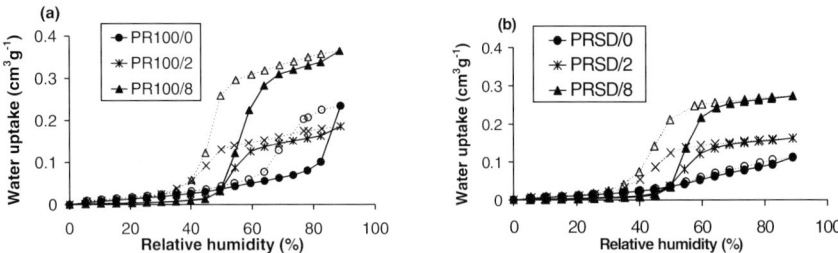

Fig. 2. Water adsorption (solid line) / desorption (dotted line) isotherms at 295K for (a) PR100 series and (b) PRSD series carbons

### 3.3. VOC adsorption at 295K
Fig. 3 illustrates pentane adsorption / desorption isotherms at 295K for PR100 and PRSD series carbons.

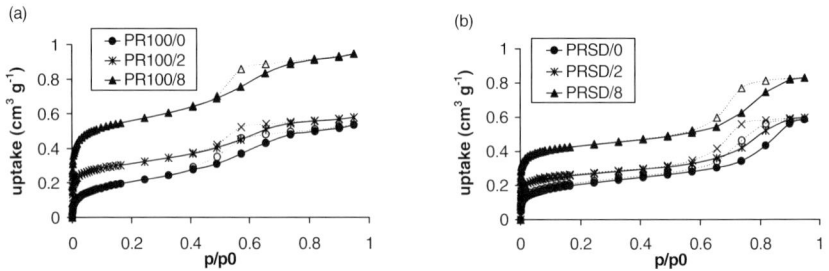

Fig. 3. Pentane adsorption (solid line) / desorption (dotted line) isotherms at 295K for (a) PR100 series and (b) PRSD series carbons

### 3.4. Dynamic adsorption testing at 295K
Table 2 lists breakthrough times for pentane through each of the carbons. This table also compares capacities determined from breakthrough data with those obtained from the relevant point on the adsorption branch of the isotherm (*ca.* $p/p0 = 0.001$).

Table 2
Breakthrough times and uptakes from dynamic testing using 2000mg m$^{-3}$ pentane, compared
with uptakes from adsorption isotherms at p/p0=0.001 and 0.95

|  | Breakthrough time (minutes) | Dynamic uptake (cm$^3$g$^{-1}$) | Isotherm uptake at p/p0 = 0.001 (cm$^3$g$^{-1}$) | Isotherm uptake at p/p0 = 0.95 (cm$^3$g$^{-1}$) |
|---|---|---|---|---|
| PR100/0 | 13 | 0.05 | 0.04 | 0.54 |
| PR100/2 | 60 | 0.12 | 0.13 | 0.58 |
| PR100/8 | 132 | 0.23 | 0.24 | 0.95 |
| PRSD/0 | 48 | 0.09 | 0.09 | 0.59 |
| PRSD/2 | 64 | 0.14 | 0.14 | 0.60 |
| PRSD/8 | 118 | 0.22 | 0.23 | 0.83 |

Tables 3 and 4 compare breakthrough times and dynamic uptakes in dry or humid airstreams
for pentane and octane.

Table 3
Effect of humidity on breakthrough times and uptakes from dynamic testing using 2000mg
m$^{-3}$ pentane

|  | Dry air (<15% RH) | | Humid air (80% RH) | |
|---|---|---|---|---|
|  | Breakthrough time (minutes) | Dynamic uptake (cm$^3$g$^{-1}$) | Breakthrough time (minutes) | Dynamic uptake (cm$^3$g$^{-1}$) |
| PR100/0 | 13 | 0.05 | 9 | 0.02 |
| PR100/2 | 60 | 0.12 | 35 | 0.08 |
| PR100/8 | 132 | 0.23 | 33 | 0.09 |
| PRSD/0 | 48 | 0.09 | 23 | 0.05 |
| PRSD/2 | 64 | 0.14 | 23 | 0.07 |
| PRSD/8 | 118 | 0.22 | 27 | 0.11 |

Table 4
Effect of humidity on breakthrough times and uptakes from dynamic testing using 2000mg
m$^{-3}$ octane

|  | Dry air (<15% RH) | | Humid air (80% RH) | |
|---|---|---|---|---|
|  | Breakthrough time (minutes) | Dynamic uptake (cm$^3$g$^{-1}$) | Breakthrough time (minutes) | Dynamic uptake (cm$^3$g$^{-1}$) |
| PR100/0 | 102 | 0.18 | 36 | 0.10 |
| PR100/2 | 219 | 0.45 | 202 | 0.33 |
| PR100/8 | >380 | >0.54 | - | - |
| PRSD/0 | 137 | 0.22 | 73 | 0.16 |
| PRSD/2 | 190 | 0.31 | 116 | 0.22 |
| PRSD/8 | - | - | - | - |

## 4. DISCUSSION

The nitrogen isotherms show that the carbons contain both micro- and mesoporosity. The templated carbons are predominantly mesoporous, whilst subsequent activation increases mainly the microporosity. The location and shape of the hysteresis loops suggest that the PRSD carbons contain larger mesopores than the PR100 carbons; pore size distribution analysis (not shown here) using the BJH method [2] suggests these to be centred around ca. 8nm and 5nm respectively.

Analysis by water adsorption / desorption shows that the increase in microporosity in the activated samples results in a significant change to the isotherm at relative humidities above about 50%. At high humidities, it is interesting to note that the PR100/0 carbon shows a higher uptake at 90% humidity than the PR100/2 carbon, despite the higher porosity of the latter sample. The reason for this is not known, although it has been replicated in other sample series not included here. The instrument does not allow accurate measurement at relative humidities above 90%.

Alcaniz-Monge *et al* have previously reported [4] that the microporosity is filled at a relative humidity of around 82%. Fig 4 illustrates the correlation between micropore volumes derived from nitrogen adsorption at 77K and water uptakes at 80% relative humidity at 295K.

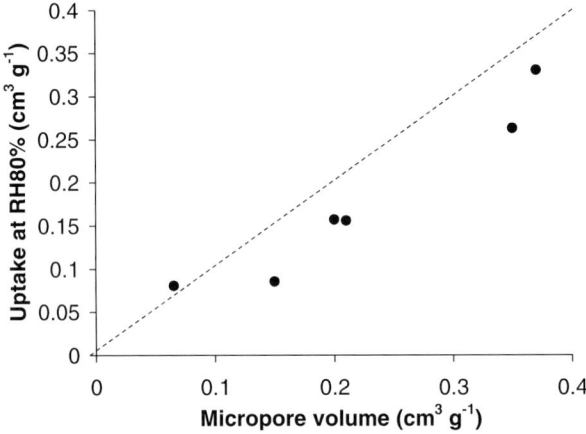

Fig. 4. Relationship between micropore volumes derived from nitrogen adsorption at 77K and water uptakes at 80% relative humidity at 295K (the dotted line represents y=x)

The data show that the water uptakes, calculated on the assumption of liquid density in the pores, tend to be slightly lower than the micropore volumes determined from nitrogen adsorption. However it has been reported that the density of water adsorbed in microporous carbons is less than the liquid value [5]. Taking this into account suggests that for dynamic testing of carbons pre-equilibrated to a relative humidity of 80%, the micropores will be filled with water. Study of the filling of the mesoporosity by water is limited by the experimental range of the apparatus, although it is apparent from the isotherms that the water uptakes between 80% and 90% relative humidity vary significantly from carbon to carbon, despite the high levels of mesoporosity in all samples.

Table 2 shows, not surprisingly, a clear correlation between the breakthrough times for pentane vapour and the level of porosity in the samples. The uptakes derived from the breakthrough curves show an excellent match to those read from the adsorption isotherm; comparison with the uptakes at p/p0 = 0.95 show how little of the microporosity is utilised at this test concentration (a result of the low partial pressure). Lillo-Ródenas *et al* have recently shown [6] a correlation between benzene adsorption at 200ppmv and the amount of narrow microporosity as determined by $CO_2$ adsorption at 273K; it is assumed that only the narrower micropores are occupied by pentane in these tests.

Table 3 shows that exposure to water vapour both before and during the test has a significant impact on both the breakthrough time and capacity for pentane. It is noticeable from this data that the introduction of higher levels of microporosity no longer has a significant effect on the breakthrough time. Comparison is drawn here on an equal mass basis; for similar comparison on an equal volume basis, the most highly activated samples would show worse performance than the intermediate samples.

Similar data for octane breakthrough show that the presence of water vapour has far less effect. This is partly due to the higher partial pressure of octane (testing having been carried out at constant concentration), which is reflected in the higher uptakes seen during adsorption from dry air. Being less volatile than pentane, octane is therefore able to displace pre-adsorbed water from the carbon more readily and will compete for a greater range of the microporosity than pentane. Therefore additional microporosity is beneficial in this case. These observations are consistent with the findings of Russell and LeVan [7], who reported that water has a greater influence on the adsorption of lighter alkanes, and that the effect is greatest at low alkane partial pressure or high relative humidity.

## 5. CONCLUSIONS

Carbons containing controlled levels of micro- and mesoporosity have been prepared using a combination of templating and activation. Analysis by nitrogen adsorption shows that the templated carbons are predominantly mesoporous. Subsequent activation of the templated carbons increases the micro:mesopore ratio.

Adsorption of pentane from a dry flowing airstream reveals that the dynamic uptakes are consistent with the uptake from adsorption isotherms. Comparison with the isotherm data shows that only a limited amount of the microporosity is utilised. Increasing the micropore volume results in an increase in breakthrough time and uptake.

When in competition with water vapour, the breakthrough time and capacity for pentane are both significantly reduced. Introduction of high levels of microporosity can reduce performance (especially if comparison is drawn on an equal volume basis), as a result of the high quantities of water present.

The effect of water vapour is less pronounced for octane than for pentane, as a result of its lower volatility and hence higher partial pressure. This enables octane to more readily displace water from the micropore volume, thus limiting the reduction in breakthrough time and capacity. In this case, additional microporosity does provide a beneficial effect.

## REFERENCES

[1]  S.J. Han, M. Kim and T. Hyeon, Carbon 41 (2003) 1525

[2]  F. Rouquerol, J. Rouquerol, and K. Sing, Adsorption by powders & porous solids, Academic Press London, 1999

[3]  A.J. Fletcher and K.M. Thomas, Langmuir 15 (1999) 6908

[4]  J. Alcañiz-Monge, A. Linares-Solano, B. Rand, J.Phys.Chem. B 105 (2001) 7998

[5]  J. Alcañiz-Monge, A. Linares-Solano, B. Rand, J.Phys.Chem. B 106 (2002) 3209

[6]  M.A. Lillo-Ródenas, D. Cazorlo-Amorós, A. Linares-Solano, Carbon 43 (2005) 1758

[7]  B.P. Russell, M.D. LeVan, Ind.Eng.Chem.Res. 36 (1997) 2380

Studies in Surface Science and Catalysis 160
*P.L. Llewellyn, F. Rodriquez-Reinoso, J. Rouqerol and N. Seaton (Editors)*

# Preparation of functionally graded alumina ceramic materials with controlled porosity

J. Andertová, J. Havrda, R. Tláskal

Department of Glass and Ceramics, Institute of Chemical Technology Prague, CZ - 166 28, Prague 6, Technická 5, Czech Republic, e-mail: Jana.Andertova@vscht.cz

Functionally graded ceramic materials with controlled porosity are the object of research in the surgical applications of field of bioinert ceramic materials. The use of ceramics in medical applications (implants) depends on the possibility of adjusting their properties to those of bone tissue. The great differences in properties of ceramic materials and bone tissues lead to the study of utilization of functionally graded materials for preparation of biomaterials. Preparation of bioinert alumina ceramics with gradient of porosity by slip casting to the porous molds was studied. The sol-gel transition of AlO(OH) was used to stabilize pore-generating agent with a particle size of 150–190 μm and the preparation of a reproducible aqueous suspension of $\alpha$-$Al_2O_3$ for the slip-casting of bodies with defined porosity. The physical and mechanical properties of one-component and composite bodies with layers of variable controlled porosity were determined. The dependence of the determined properties on the value and type of porosity is reported.

## 1. INTRODUCTION

Present trends in research of the bioinert ceramics are characterized by an effort to harmonize ceramic and bone tissue properties. It was proved that interaction of the implant material properties with those of the bone might cause problems with the shock transport to the skeleton causing pain in people with a lower threshold of pain. The bioinert implants must also meet certain functional requirements. These include the ability to transfer load and stress distribution as well as demand for so called iso-elasticity, particularly for intensive stressed parts of human body. The differences in values of the elasticity modulus (Young's modulus) of ceramic materials and the bone tissue is disadvantage the applied sintered; in fact non-porous, ceramic materials on basis of alumina or zirconia. The Young's modulus values for these materials are from 380 to 200 GPa, i.e. higher by one order compared with the bone tissue (7-30 GPa). Through the porosity, pore size and shape, the Young's modulus values for ceramic materials can be regulated. The Young's modulus values bottom limit or porosity upper limit are given by the requirements for the implants mechanical properties. These requirements predetermine that the mechanically stressed functional parts of the implants must be continually prepared from the sintered non-porous ceramics for the required mechanical properties. One of the possible approaches for a solution, i.e. for elimination of sharp interface between the implant and bone is the introduction of the FGM; "functionally graded materials" into the proposal of their preparation. The idea behind of this effort is a fact that chemical composition or microstructure in the body volume can be controlled with aim to reach the gradient of properties. The functionally graded ceramic materials are prepared with the aim to reach a situation that within the interface implant-bond will be material of similar properties [1- 5]. The objective of the work is the preparation of functionally graded alumina, i.e. layered ceramics with controlled porosity gradient. The model system can be characterised as a layered structure with different, continually changing porosity of individual layers where single layers differ by the pores content. The method of pouring the alumina aqueous suspensions into the porous block was used as a method for preparation of alumina bodies with different porosity. The work was divided into two parts. In the first part of this work the alumina ceramics with different constant porosity were prepared. In the second part functionally graded alumina ceramic with variable porosity was prepared, i.e. functionally graded alumina ceramic with gradual change of porosity and thus also properties in individual layers [6- 8].

## 2. MATERIALS AND METHODS

The work was carried-out with the powder $\alpha$-$Al_2O_3$ (AKP 15-Sumito Chemical Co., Ltd., Japan), d=0,6-0,8.$10^{-6}$m. Thin walled balls (150-190.$10^{-6}$m) were used as a pore-generating agent on the basis of $Al_2O_3$. As a pore-generating stabilization and sedimentation prevention of pore generating agent the boehmit gel AlO(OH)(Dispersial sol P2) was applied. For suspension stabilisation Sokrat 32A(CHZ Sokolov) was used as a defllocculant. The rheological properties were tested on a rotational viscometer (RV1, Haake, Germany) equipped with coaxial cylinders sensors (shear rate range 0-1000 $s^{-1}$). Flow behavior was recorded for each type of suspension. The apparent viscosity $\eta$ was determined from flow curves (at a shear rate 50 $s^{-1}$). The ceramic bodies were prepared by the method of slip casting the suspensions into plaster moulds. The prepared bodies were low-and high temperature processed, i.e. dried and fired. The physical (bulk density, porosity) and mechanical properties (Young's modulus) of bodies were tested. Bulk density was determined by the double weighing method (Archimedes´ method). Porosity $P$ is the volume of open and closed pores and cavities in the sample / total volume of sample (including all pores and cavities) ratio, Eq.1.

$$P = \frac{V_0 + V_n}{V_a} .100 = \left(1 - \frac{\rho_v}{\rho}\right).100 \quad [\%] \qquad (1),$$

where $V_o$ is volume of open pores, $V_n$ is volume closed pores and $V_a$ is total volume of sample, $\rho_v$ and $\rho$ are sample bulk and ultimate density respective. For the optical microscopic analysis of the bodies' surface the image analysis (LUCIA, Laboratory Imaging, Czech Republic) was used. The elastic modulus E (Young's modulus) was measured by the resonant frequency method using a resonant frequency tester (Erudite, CNC ELECTRONICS, UK).

## 3. EXPERIMENT

### 3.1. Preparation of basic alumina suspensions for slip casting

The basic aqueous suspensions were prepared by mixing alumina powder, deflocculant Sokrat 32A together and the rheological parameters were determined. The solid phase content has ranged from 60% wt – 85 wt%, deflocculant content 0,2 wt%-0,5 wt% (wt% based on dry solid mass). Based on the dependence of the apparent viscosity $\eta$ on the deflocculant content added, optimum amount of deflocculant 2,3 wt% was determined (minimum value of viscosity). From the rheological curves of suspensions with liquid content in range $C \in$ $<0,39m^3m^{-3}; 0,5m^3m^{-3}>$ the coagulation concentration $C_k = 0,403$ corresponding to the rheological transition from the viscous to the viscoplastic material was expressed; character of established dependence is shown in Fig. 1.

### 3.2 Preparation of suspension with pore–generating agent

The effect of the addition of 5 wt% pore-generating agent to suspension with 76 wt % solid mass content was initially evaluated. Particular attention was paid to the kinetic stability of the suspension. It was found that after a certain time (~minutes) the suspension's kinetic stability decreased and the pore-generating agent stayed on the surface. Changing the content of the pore-generating agent, or changing the solid phase in the suspension did not increase the stability of the suspension. The kinetic stability of the suspension was also unaffected by a change in the deflocculant content. Based on this, the original adaptation of the suspension in which the suspension is derived from the addition of sol and formation of sol-gel transition was used. Studying the interaction of the pore-generating agent surface with the suspension, it was discovered that good wetting of the pore-generating agent surface occurred with the gel - see Fig.2. The effect of adding AlO(OH) on the rheological properties of the suspension, initially without the pore-generating agent and later with the pore-generating agent (range 0 wt%.–7 wt%) was next investigated. From these rheological measurements, the dependence of the apparent viscosity $\eta$ of the system on the AlO(OH) solid mass content C in the suspension was evaluated - see in Fig.3. Flow curves of suspensions with the AlO(OH) content above 3 wt% indicate a certain degree of time-dependence.

The suspension flowing properties were assessed using the time of the suspension outflow (25ml) from the outflow viscometer. Dependence of the outflow time for suspensions with 76 wt % solid phase content on the pore - generating agent content is shown in Fig.4, where $S_{pga}$ is the pore-generating agent content. The results show, that pore-generating agent over 5 wt% content negatively affects the suspension flow properties.

To find out the kinetic stability of the pore-generating agent distribution in the suspension volume, suspensions with variable AlO(OH) content were evaluated. For the stability evaluation, the modified Andreasen's method was employed. The suspension with the pore-generating agents (range 3wt% - 5 wt %) was poured into a cylinder ($15.10^{-3}$m diameter with height level of $8.10^{-2}$m). After 1 and after 3 hours, $3,5 .10^{-3}$l of suspension

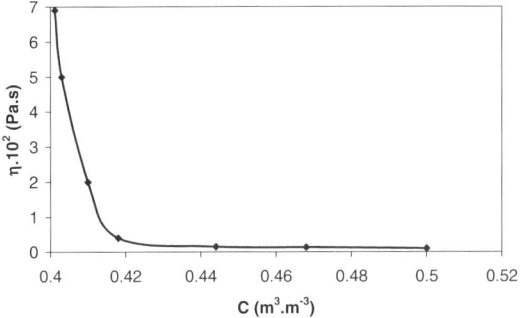

Fig. 1 Dependence of apparent viscosity $\eta$ on the liquid concentration C in the suspension

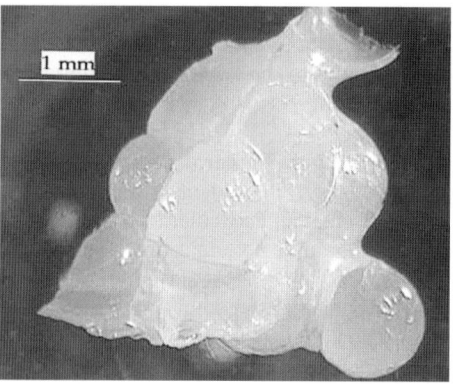

Fig. 2 Wetting of pore-generating agent with AlO(OH) gel

content was withdrawn from different levels of the suspension and weighed. Next the pore-generating agent was separated from the suspension and weighed. Using this data the stability of the pore-generating agent distribution in the total suspension volume was evaluated. The results of the individual experiments for suspensions with 5 wt% pore-generating agent are presented in Table 1.

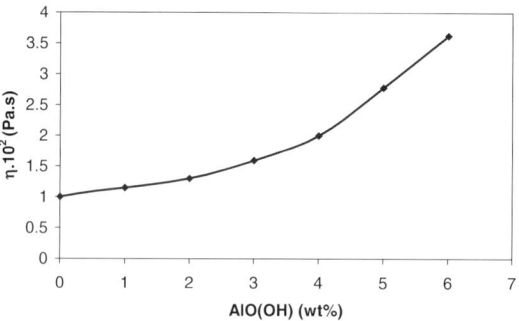

Fig.3 Dependence of apparent viscosity $\eta$ on the solid mass content of AlO(OH), suspensions without the pore-generating agent

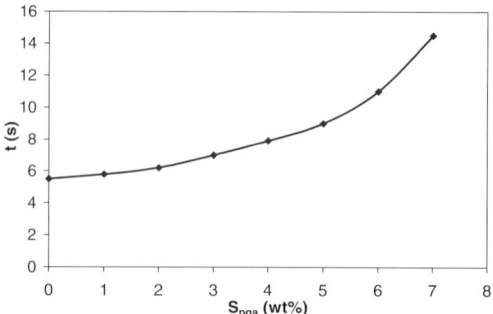

Fig.4 Dependence of the outflow time for suspensions with 75wt% solid phase content on the pore-generating agent content

Table 1
Content of the pore-generating agent in the different layers of the suspensions (3 hour after preparation of the suspension)

| α–Al$_2$O$_3$ / AlO(OH) | Upper layer | Central layer | Bottom layer |
|---|---|---|---|
| wt %/ wt % | [g/g]* | [g/g]* | [g/g]* |
| 75 / 1 | 0.0558 | 0.0561 | 0.0555 |
| 74 / 2 | 0.0466 | 0.0469 | 0.0504 |
| 73 / 3 | 0.0589 | 0.0765 | 0.0834 |
| 72 / 4 | 0.1051 | 0.0734 | 0.0522 |
| 71 / 5 | 0.0915 | 0.0690 | 0.0852 |
| 76 / 3 | 0.0558 | 0.0534 | 0.0543 |
| 76 / 4 | 0.0508 | 0.0496 | 0.0513 |
| 76 / 5 | 0.0615 | 0.0634 | 0.0643 |

*The pore-generated agent content in 1g of suspension

Based on these experiments, the optimal parameters of the final suspension was established: 76 wt% of alumina mixture, 3 wt % of Al(OOH), 2,3 wt% of deflocculant and range 0 wt% - 5 wt% (maximally 7wt%) of the pore-generating agent. The time of the kinetic stability of the pore - generating agent in the suspension is 3 hours.

### 3.3. Preparation of alumina bodies with different, constant porosity

Using the final suspension testing bodies were prepared with content of pore-generating agent ranging from 0 wt% - 7 wt% by the method of slip casting to the porous (gypsum) mould. After the low-temperature (105°C) and high- temperature (2 °C/min, 1570°C, dwell 2 h ) treatment, the bulk density, porosity and Young's modulus were determined. The mean values of tested properties are given in Table 2.

Using the optical microscope to analyse the surface character of bodies, the number of pores, the pore diameters and pore distribution were assessed. Surface character of bodies with different pore content is shown in Fig.5. For the determination of pore characteristics, i.e. determination of number of pores accrued by pore-generating agents, the pore distributions on surface of bodies without pore-generating agent was evaluated. This value was subtracted from the value obtained from those bodies prepared with the agent. Surface contribution and median value of pore diameter in bodies with content of pore-generating agent ranging from 0 wt%-7 wt % is presented in Table 3.

Table 2
The physical and mechanical properties of bodies with constant growing porosity

| Content of pore-generating agent | Bulk density $\rho_v$ | Porosity $P$ | Young's modulus E |
|---|---|---|---|
| [wt%] | [g.cm$^{-3}$] | [%] | GPa |
| 0 | 3,80 | 5,6 | - |
| 1 | 3,51 | 12,4 | 310 |
| 2 | 3,39 | 15,0 | 275 |
| 3 | 3,13 | 21,8 | - |
| 5 | 2,92 | 27,0 | 226 |
| 6 | 2,82 | 29,5 | - |
| 7 | 2,40 | 39,0 | 187 |

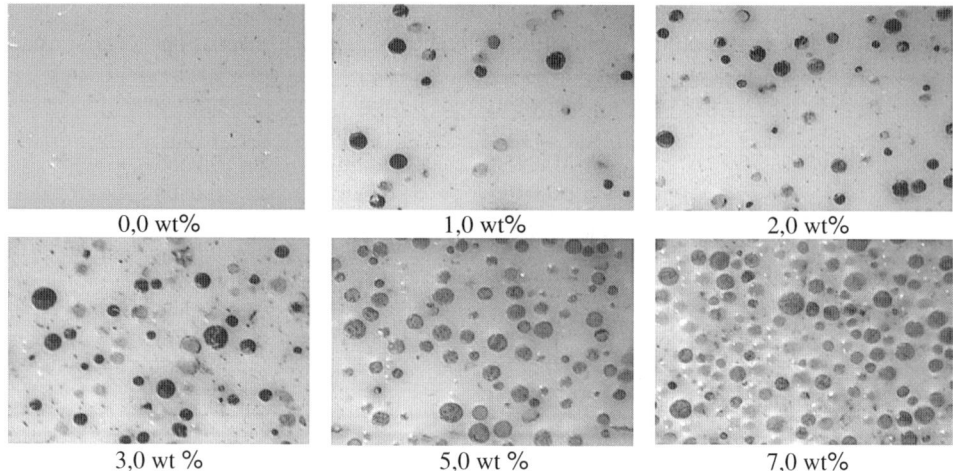

| 0,0 wt% | 1,0 wt% | 2,0 wt% |
| 3,0 wt % | 5,0 wt % | 7,0 wt% |

Fig.5. Micrograph of surface character of samples with different porosity. Bottom: pore-generating agent content in wt%

Specific characteristics of prepared bodies of different porosity were necessary for the preparation of bodies with porosity gradient and this was the following stage of this work.

### 3.4. Preparation of functionally graded alumina ceramic

Using the slip casting method, two layered bodies, $2.10^{-3}$m thickness, were prepared. Each layer was of a different porosity. In the first experiment, the pore-generating agent content was constant in one layer (0,0 wt%) and variable in second layer (0 wt%-7 wt%). In the second experiment the content of pore-generating agent of both layers was variable (0,0 wt% -6,8 wt%). These bi-layered bodies were then studied. It was proved that the successful preparation is conditioned mainly by perfection of the joint without any defect between individual layers and by harmonization of the length changes of the jointed layers. In Fig.6 is showed effect of the stress, acting on the interface between the layers in the bi-layered bodies consisting of layer without the pore-generating agent and layer with continually growing content of the pore-generating agent. Different porosity value in individual layers generates a stress for reason of difference of the coefficient of thermal expansion $\alpha$ in individual layer. The generating stress causes deformation of the samples. Based on the length

Table 3

Pore characteristics obtained from bodies' surface micrograph analysis

| Pore-generating agent content | Surface content of pores | Median of pore diameter |
|:---:|:---:|:---:|
| [wt %] | [%] | [μm] |
| 0 | 0,73 | 6 |
| 1 | 6,64 | 92 |
| 2 | 9,27 | 85 |
| 3 | 10,90 | 80 |
| 5 | 20,52 | 100 |
| 6 | 21,24 | 94 |
| 7 | 36,05 | 97 |

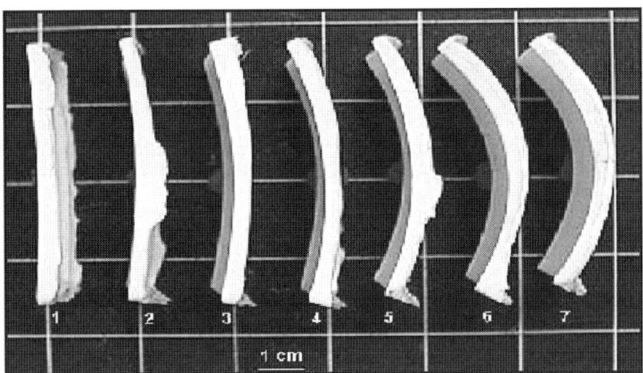

Fig.6. Fired bi-layered bodies with growing content of pore-generating agent in second layer
1 – (0,0-0,3)[*]; 2 – (0,0-1,2)[*]; 3 – (0,0-1,7)[*]; 4 – (0,0-2,7)[*]; 5 – (0,0-3,8)[*]; 6 – (0,0-5,7)[*]; 7 –
(0,0-6,8)[*];[*)] numbers in brackets give the pore-generating agent content in the layer (in wt %)

changes measuring within temperature intervals 20 °C-1570 °C, the non-linear model of
temperature and porosity dependence of coefficient of thermal expansion $\alpha$ was determined.
From this non-linear model and from value of deformation of bi-layered bodies, the critical
difference of coefficient of thermal expansion $\Delta\alpha_{lim} = 0{,}2 \cdot 10^{-6}$ $K^{-1}$ between adjacent layers
was determined [9]. The value of $\alpha_{lim}$ was used for prediction of optimal conditions for
preparation of defect-free bi-layered alumina ceramics with layers of controlled porosity. It
was found that the bi-layered bodies consisting of layers with content of 0,0 wt% and 0,3 %wt
of pore-generating agent have an acceptable difference of values of coefficient of thermal
expansion ($\Delta\alpha_{lim} \sim 0{,}2 \cdot 10^{-6}$ $K^{-1}$) and their deformation is identical as those of fired bodies
without the pore-generated agent. In addition it was found, that the deformation of bi-layered
bodies with $\Delta\alpha_{lim} \sim 0{,}2 \cdot 10^{-6}$ $K^{-1}$ could be caused also by the imperfect technology of body
preparation.

For investigation of physical and mechanical properties of the functionally graded
alumina ceramics with variable porosity the set of samples was prepared. The values of tested
properties are given in Table 4.

Interface character between layers with different porosity and character of pore shape in
the bi-layered samples is demonstrated in Fig 7.

| 0,0 wt%- 0,3 wt%, | 4,4 wt%- 5,0 wt%, | 6,2 wt%- 6,8 wt% |

Fig.7. Character of interface between layers of different growing porosity in bi-layered
samples. Bottom: pore-generating content in each layer (black line-joint of layers)

Table 4
Properties of functionally graded ceramics with variable porosity

| The pore-generating agent content | Porosity $P$ | Young's modulus $E_\parallel$*) | Young's modulus $E_\perp$*) | Bulk density $\rho_v$ |
|---|---|---|---|---|
| [wt%] | [%] | [GPa] | [GPa] | [g.cm$^{-3}$] |
| AlO(OH) free alumina | 0,8 | 378,6 | 378,6 | |
| 0,0-0,0 | 4,3 | 349,2 | 349,2 | 3,83 |
| 0,0-0,5 | 5,7 | 340,3 | 339,9 | 3,77 |
| 0,5-1,0 | 9,6 | 326,2 | 326,2 | 3,62 |
| 1,0-1,5 | 11,9 | 302,2 | 300,9 | 3,52 |
| 1,5-2,0 | 13,2 | 281,0 | 281,0 | 3,47 |
| 2,0-2,6 | 15,5 | 275,0 | 274,9 | 3,38 |
| 2,6-3,2 | 18,0 | 259,5 | 259,0 | 3,28 |
| 3,2-3,8 | 20,4 | 238,4 | 238,1 | 3,19 |
| 3,8-4,4 | 22,3 | 223,7 | 223,5 | 3,11 |
| 4,4-5,0 | 26,9 | 217,0 | 217,0 | 2,93 |
| 5,0-5,6 | 27,5 | 214,7 | 214,7 | 2,90 |
| 5,6-6,2 | 27,9 | 201,2 | 200,4 | 2,88 |
| 6,2-6,8 | 29,6 | 186,7 | 186,7 | 2,82 |

*) The values of Young's modules were calculated from values determined for each layer, $E_\perp = E_1E_2/(E_2V_1+E_1V_2)$, $E_\parallel = E_1V_1/E_2V_2$, where $E_1, E_2$ are Young's modules and $V_1, V_2$ volume content each phase, $E_\parallel$, $E_\perp$ are Young's modules determined for equidistantly a perpendicularly applied stress respectively [10].

## 4. DISCUSSIONS AND CONCLUSION

Based on the experimental results described above the following conclusions can be made:

The optimal composition of kinetically stable suspensions for slip casting of bodies with variable porosity is 76 wt% of $\alpha$-Al$_2$O$_3$; 2,3 wt% of deflocculant; 3 wt% of AlO(OH) and the pore-generating agent content in the range 0wt % - 5wt %. The addition of the pore-generating agent to the suspension can change real porosity approximately in 4 to 32 %. The defect-free character of the interface as noted on micrographic examination of the bi-layered bodies confirmed the good jointing of the layers. Pore size values, 160-250 μm, and the spherical shape of the pores confirm the appropriateness of the selected pore-generating agent for creation of spherical pores with controlled size for bone tissue ingrowths. From a non-linear model and from deformation of composite bodies the critical difference of coefficient of thermal expansion $\Delta\alpha_{lim} = 0,2.10^{-6}$ K$^{-1}$ between two layers was determined. It is possible to prepare alumina ceramics with non-porous and porous layer, which-in spite of the still existing difference exhibit a Young's modulus closer to that of bone. By using of functionally graded alumina ceramic with increasing porosity, it is possible to approximate the physical and mechanical properties of the material to that of bone tissue properties.

## ACKNOWLEDGEMENT

This study was part of research programme MSM 6046137302 Preparation and research of functional materials and material technologies using micro- and nanoscopic methods

## REFERENCES

[1] L.L.Hench: *Bioceramics: From Concept to Clinic*, J.Am.Ceram.Soc. 74, 7, (1991), 1487-1510.
[2] M.Vallet-Regí: *Ceramics for medical applications*, J.Chem.Soc. Dalton Trans, 97-108, (2001)
[3] E.Dorre, W.Dawihl, V.Krohn, G.Altmeyer, M.Semlitsch: *"Do Ceramic Components of Hip Joints Maintain their Strength in Human Bodies?"* in Ceramics in Surgery, ed. P. Vinenzi (Elsevier, Amsterdam 1983) 61
[4] L.L.Hench: *Bioceramics*, J.Am.Ceram.Soc. 81, 1705-28, (1998)
[5] Y.T.Pei, V.Ocelík, J.Th.M. De Hosson: *Interfacial adhesion of laser clad functionally graded materials*, Mat.Sci Eng. A342 (2003), 192-200
[6] P.Kolařík: *PhD Thesis*, (in Czech) ICT Prague (1995)
[7] R.Tláskal: *PhD Thesis*, (in Czech) ICT Prague (2005)
[8] J.Andertová, R.Tláskal, J.Havrda: *Study of suspensions rheological behaviour for preparation of functionally graded ceramic materials*, Proceeding of CHISA, Prague (2004), (full version on CD) 48
[9] J. Andertová, R Tláskal, J. Havrda, I. Zedníková: Functionally graded alumina ceramic materials-heat treatment of bodies prepared by slip casting method, Proceeding of IX Conference & Exhibition of European Ceramic Society, (full version on CD), Portoroz, Slovenia (2005)
[10] R.W.Davidge: *Mechanical Behaviour of Ceramics*, Cambridge University Press Cambridge (1976)

Studies in Surface Science and Catalysis 160
P.L. Llewellyn, F. Rodriquez-Reinoso, J. Rouqerol and N. Seaton (Editors)

639

# Preparation of Mesoporous Ceria in the Presence of Non-Aqueous Phases

**Aphrodite Tillirou, and Charis R. Theocharis**

Porous Solids Group, Department of Chemistry, University of Cyprus, P.O.Box 20537, 1678 Nicosia, Cyprus

In this paper we present the use of aniline in methanol solution as the precipitating agent in the synthesis of mesoporous ceria. Different reaction conditions were used including elevated temperature and ageing time, whereas the use of a templating matrix was also examined.

## 1. INTRODUCTION

There has been much recent interest in the preparation of various inorganic solids (notably aluminium phosphates) with controlled porosity in the microporous and mesoporous range [1], in ways analogous to that used in the synthesis of aluminosilicate zeolites. For the synthesis of mesoporous solids, the synthetic paths make use of long chain surfactant molecules. At the University of Cyprus there is a long-standing interest in the synthesis and study of porous ceria [2-4], because of the importance of that solid in several catalytic applications, and notably in automobile exhaust catalysts. In this presentation we discuss the use of aniline as the precipitating base instead of ammonia, which has hitherto been used.

The use of aniline as precipitating agent delayed the onset of precipitation, thus affording the synthesis of ceria under sol-gel conditions enabling better control of the process. The effect of the difference in pH during synthesis was also examined, on both the surface characteristics of the product, as well as the nature of surface groups present, and consequently surface acidity. Non-aqueous solvents, specifically methanol and ethanol were also employed.

## 2. EXPERIMENTAL

Ceria was prepared in a sol-gel procedure. Equal amounts of aqueous cerium (IV) ammonium nitrate solution and methanolic aniline solution were mixed. Concentrations in the ranges $0.001 < [Ce^{4+}] < 0.1 M$ (mol $l^{-1}$) and $0.1 < [C_6H_5NH_2] < 5M$ were used. The resulting gelatinous precipitate was allowed to age for different lengths of time in the reaction liquor. The reaction was carried out at room temperature and at 323K. The samples were centrifuged off at the end of the ageing period, dried at 373K for 24h and calcined for 673K for 2h in air.

Nitrogen adsorption isotherms were measured at 77K from a starting pressure of 0.15Pa, after outgassing at 383K for 24h, using an ASAP2010 Micromeretics apparatus. The crystalline structure of the samples was determined by X-ray diffraction using Cu Kα radiation employing a Shimadzu XRD-6000 diffractometer. The thermal behaviour of the samples was measured by both thermogravimetry (Shimadzu apparatus) and differential scanning calorimetry (DSC model Q100). Surface areas were calculated from the adsorption data using the BET equation, whilst the pore size distribution was estimated using DFT methods employing a slit-shaped pore model.

## 3. RESULTS AND DISCUSSION

Table 1 summarizes some samples prepared, the concentrations of reactants employed and the BET surface area, total pore volume at $p/p^o=0.995$ (V), average pore diameter (d), and crystallite size. The last parameter was estimated from X-ray diffractogrammes, and confirmed with transmission electron microscopy. Table 2 shows the effect of synthesis conditions on the properties of the samples. Table 2 refers to syntheses using 0.01M cerium and 2M aniline solutions, after calcination for 2h at 673K.

Table 1
Synthesis and Characterization of Samples after Calcination at 673K for 2h

| $Ce^{4+}$ / mol $l^{-1}$ | Aniline/ mol $l^{-1}$ | $S_{BET}$ / $m^2g^{-1}$ | V/ $cm^3g^{-1}$ | d/ nm | Crystallite size / nm |
|---|---|---|---|---|---|
| 0.1 | 1 | 11 | 0.05 | 18.8 | 28 |
| 0.05 | 1 | 14 | 0.05 | 15.4 | 25 |
| 0.01 | 1 | 29 | 0.03 | 4.6 | 14 |
| 0.1 | 2 | 28 | 0.08 | 11.9 | 16 |
| 0.01 | 2 | 56 | 0.06 | 4.5 | 11 |
| 0.005 | 2 | 106 | 0.09 | 3.5 | 9 |
| 0.001 | 2 | 116 | 0.13 | 4.6 | 7 |

Table 2
Effect of Synthesis Conditions on Surface Texture

| Synthesis Conditions | $S_{BET}$ / $m^2g^{-1}$ | V/ $cm^3g^{-1}$ | d/ nm |
|---|---|---|---|
| Room temperature, ageing for 2h | 84 | 0.06 | 3.6 |
| 323K, ageing for 2h | 110 | 0.09 | 4.0 |
| Room temperature, ageing for 24h | 56 | 0.09 | 6.7 |
| Room temperature for 24h, in the presence of 1g $l^{-1}$ humic acid | 79 | 0.1 | 5.0 |

It can be seen from Table 1 that lowering the concentration of the cerium precursor tends to lead to samples with increased surface area, due to a lowering of particle size. This was in agreement with observations previously made in relation to the use of ammonia. On the other hand, there appears to be an advantage in increasing the concentration of the aniline solutions used, at least up to 2M. Table 2 displays the effect of reaction temperature as well as of ageing in the reaction mixture on the surface texture of the ceria samples. It is evident that ageing leads to crystallite growth and therefore BET surface area reduction. On the other hand, ageing appears to lead to an increase in the average pore diameter of the samples, as well as in total pore volume. In Table 2 we also present the data for a sample prepared in the presence of humic acid. It is obvious that under this synthesis regime, humic acid acts as a templating agent, in a manner similar to the behaviour displayed when ammonia was used [5]. On the contrary, it does not appear that aniline itself behaves strongly as a template, but its behaviour appears to be mainly that of a base.

Figure 1 shows the pore size distribution for a series of ceria samples prepared in the presence of differing amounts of aniline. It is evident that the effect of precursor cerium ions concentration on both pore volume and on the pore size distribution is much more important than that of aniline concentration, although the latter does appear to influence $S_{BET}$ as discussed earlier. It is clear from Figure 1 that the sample prepared from a 0.1M cerium solution has very ill-defined pores with a very widely spread size distribution and very low total pore volume. When the precursor concentration was increased to 0.01M this led to a more narrow size distribution, but still relatively low total pore volume. On the other hand, at even lower cerium concentrations, there is a more marked increase in pore volume, but whereas the pore size distribution shows a marked maximum value, there is a fairly wide distribution of pore diameter values, from 2nm to 7,5nm for 0.005M and 10nm for 0.001M precursor concentration. It should be noted that aniline concentration was kept constant.

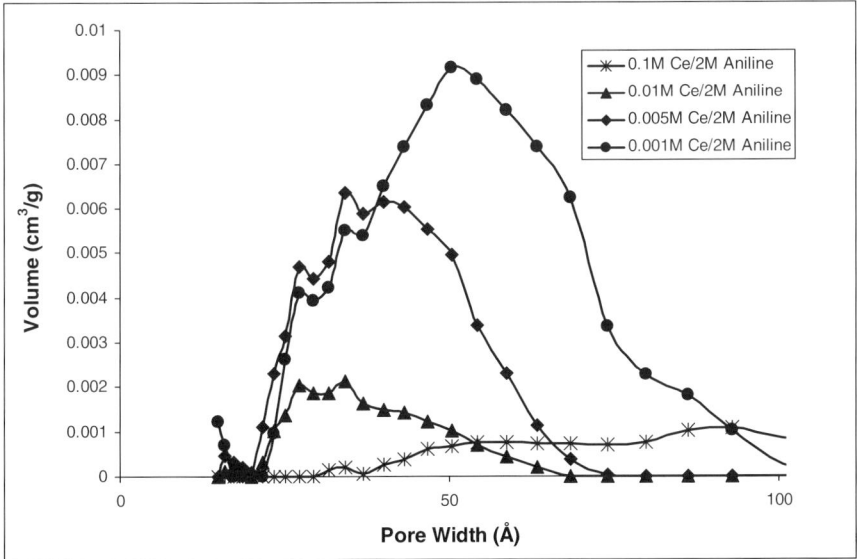

Figure 1 DFT analysis of pore size distribution for a series of ceria samples

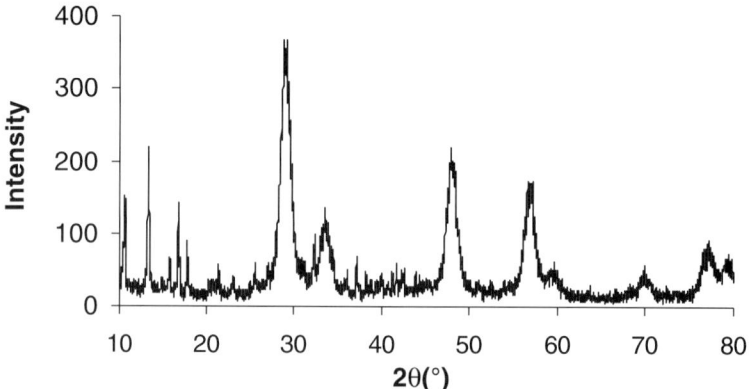

Figure 2 X-ray Diffractogramme for sample 0.05M Ce$^{+4}$/1M Aniline after drying at 373K for 24h

Figure 2 contains the X-ray diffractogramme for ceria prepared from 0.05M cerium solution and 1M aniline. The major peaks present were those of the poorly crystalline fluorite ceria phase previously identified [see, for example, 6]. However, contrary to what obtains for other samples presented in this paper, extra peaks have appeared in the diffractogramme for this sample, in the 2Θ region 10 to 20°.

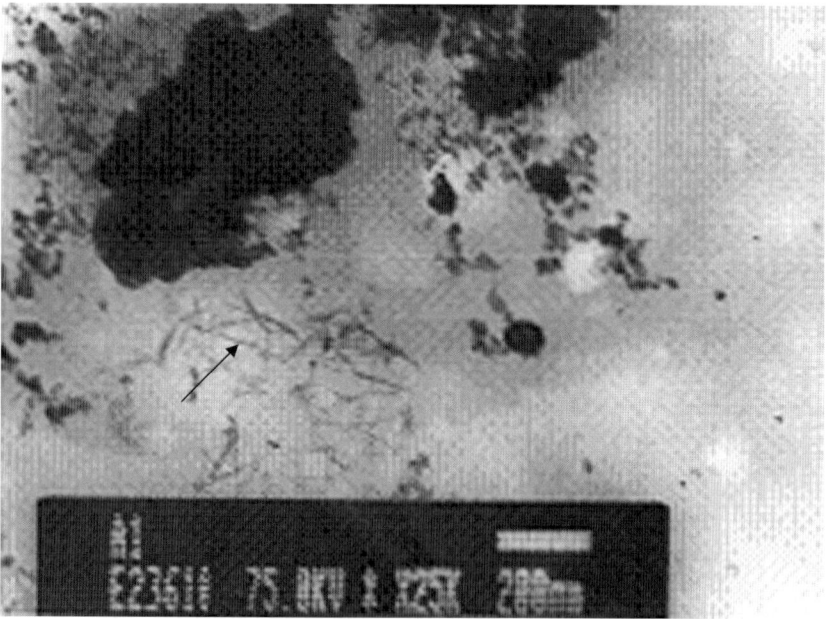

Figure 3 Transmission electron micrograph of Ceria prepared from a solution containing 0.05M cerium ions. The arrow indicates the hitherto unidentified second phase.

Figure 3 shows a transmission electron micrograph of the same sample, in which needle-like crystallites can be identified, not present in samples that did not present the extra peaks in the X-ray diffractogramme. It is clear that the two are correlated, and it is suggested that the new phase represents an aniline-cerium (IV) complex that precipitates out under the specific synthesis condition. Calcination of the sample leads to elimination of the phase. This phase, presumably, does not contribute significantly to the overall surface area of the sample due to its relatively large particle size, and may, in part, explain the values in Table1.

Figure 4 shows the nitrogen adsorption isotherms for various ceria samples prepared using 2M aniline solutions. It can be seen that whereas for precursor cerium concentrations of 0.1M and 0.01M the isotherms are of type II in the IUPAC classification, whereas for the sample prepared from a 0.005M solution, the isotherm was of type IV. These changes are reflected in the pore size distribution graphs presented in Figure 1 on a previous page. It is clear that the dilution of the reaction solution leads to a change of charge density on the surface of the crystallites, thus affecting the mode and strength of aggregation.

Although it was stated above that the templating effect of aniline does not appear to be a strong one, it must nevertheless have some influence, given the evidence for complexation of Ce(IV) by aniline, leading to the precipitation of a complex. In addition, FTIR spectra of samples prepared in the presence and absence of aniline present differences in both the OH stretch peaks at about 3400 cm$^{-1}$ as well the Ce-O-Ce bands at 1300 cm$^{-1}$.

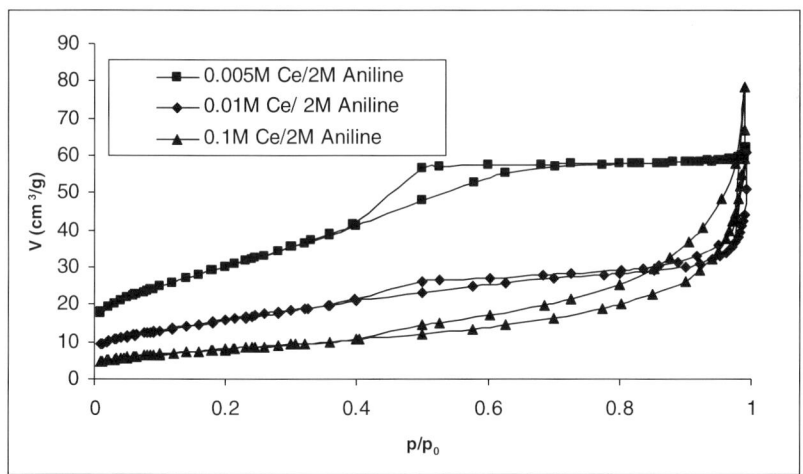

Figure 4 Nitrogen adsorption isotherms for ceria samples prepared using 2M aniline solutions

It appears that the additional phase, attributed to a cerium (IV) –aniline complex is present in samples prepared from cerium (IV) solutions with metal ion concentrations 0f 0.05M and above. Samples prepared from a 0.01M Ce (IV) solution that presented the smallest particle size and thus largest specific surface area, did not contain the second phase.

It was necessary to use methanol or ethanol as solvent for aniline, given the low solubility of that base in water. Preliminary results had indicated that use of methanol gave samples of higher BET surface area than ethanol. Ceria synthesis using 2-, 3-, and 4-methyl pyridine was also carried out, for which ethanol as solvent gave the best results, in terms of the apparent surface area of the samples. This suggests that the polarity of the solvent has an effect on the precipitation process, presumably due to changes in the solvation sphere of the ions in solution. The changes observed in the properties of ceria synthesised in the presence of aniline compared to ammonia should reflect differences in the synthesis pH, but also the incorporation, to some extent, of the organic molecule coordinated to the cerium ions in the resultant solid.

## REFERENCES

[1] J.M. Thomas and W.J. Thomas, "Heterogeneous Catalysis", VCH, Weinheim, Germany, (1997)
[2] I. Pashalidis, and C. R. Theocharis, in "Studies in Surface Science and Catalysis", Elsevier Science Publishers, 128, 643- 652, (2000)
[3] I. Pashalidis, C.R. Theocharis, K. Eliotou and M. Pittaki, 1st International Conference on Chemical Sciences and Industry, South-East European Countries, Halkidiki, Greece, June 1998.
[4] I. Pashalidis and C.R. Theocharis, Second European East West Workshop on Chemistry and Energy, Sintra, Portugal, March 1995
[5] G. Kyriakou, I. Paschalidis and C.R. Theocharis, Fundamentals of Adsorption, 7 (Editors: K.Kaneko, H. Kanoh and Y. Hanzawa), 201-207, (2002)
[6] C.R. Theocharis, M. Christophidou and I. Pashalidis, "Studies in Surface Science and Catalysis", Elsevier Science Publishers, 144, 75-82, (2002)

Studies in Surface Science and Catalysis 160
*P.L. Llewellyn, F. Rodriquez-Reinoso, J. Rouqerol and N. Seaton (Editors)*

# Preparation and Characterization of Nanoporous Solids With Composition $Ce_xMn_{1-x}O_{2-y}$ With x Values 0 to 1

**Evroulla Hapeshi, and Charis R. Theocharis***

Porous Solids Group, Department of Chemistry, University of Cyprus, P.O. Box 20537, 1678, Nicosia, Cyprus

A series of mixed ceria-mangania solids were synthesised covering the whole concentration range from 0% to 100% cerium. It was shown that at high cerium concentrations a single phase was obtained, whereas at low concentrations the samples consisted of two phases. The behaviour of manganese oxide under heat treatment was modified in the presence of cerium ions, both in the nature of oxide phases (and oxidation states) produced at various temperatures, as well as the temperature at which the phase change occurred. The surface properties and texture of the solids prepared were characterised.

## 1. INTRODUCTION

We have been interested in the synthesis and study of porous ceria, because of its importance in several catalytic applications, and notably in automobile exhaust catalysts [1,2,3]. Work so far, has indicated that ceria or metal-doped ceria prepared from dilute aqueous solutions by ammonia precipitation, is mesoporous [4-6]. The surface texture of these solids (BET surface area, size and shape of pores, total pore volume) is susceptible to changes in synthesis conditions such as pH and concentration of the cerium and other cation precursors used. For several cations used, a lowering of the BET surface area of ceria was obtained. However, for several metallic cations, including manganese, a significant increase in BET surface area was obtained.

Manganese oxide is itself a proven catalyst, in oxidation-reduction reactions, such as, for example, the oxidation of methane and CO. We have chosen to study this particular system because of the multiple oxidation states of manganese, and the number of different phases that can be synthesised by varying the heat treatment. In this presentation we discuss the properties of mixed ceria-mangania (III) solids over the range between 0% and 100% manganese content.

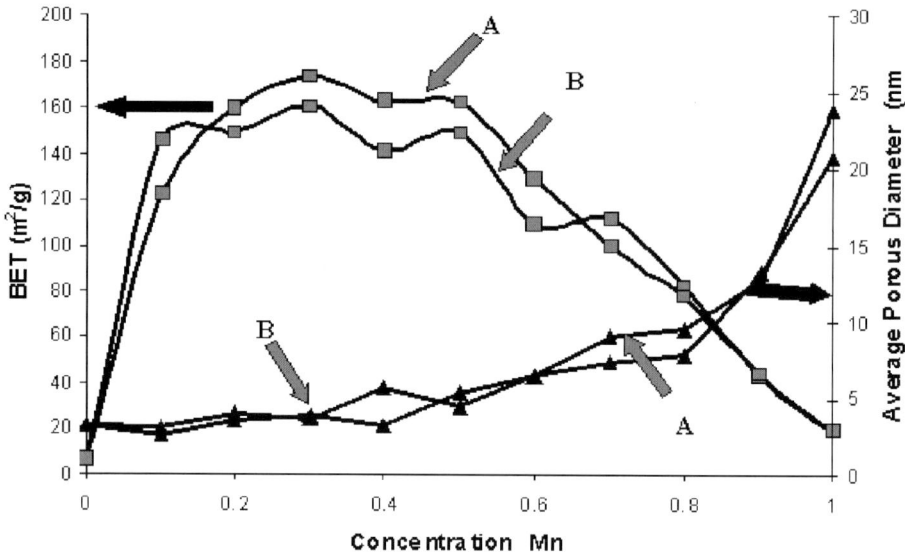

Figure 1 BET surface area and average pore diameter for samples prepared using manganese chloride or nitrate as precursor. A: $MnCl_2 4H_2O$, B: $Mn(NO_3)_2 4H_2O$

## 2. EXPERIMENTAL

The mixed cerium-manganese oxide samples were prepared by homogenous co-precipitation from aqueous solutions, using 1M ammonia solution as base. The metal precursors used were cerium (IV) ammonium nitrate and hydrated manganese (II) nitrate or chloride 0.1M solutions mixed in the appropriate ratios. The precipitate was left to stand in the mother liquor for 24h, centrifuged and dried at 473K for 24h.

Nitrogen adsorption isotherms were measured at 77K from a starting pressure of 0.15Pa, after outgassing at 383K for 24h, using an ASAP2010 Micromeretics apparatus. The crystalline structure of the samples was determined by X-ray diffraction using Cu Kα radiation employing a Shimadzu XRD-6000 diffractometer. The thermal behaviour of the samples was measured by both thermogravimetry (Shimadzu apparatus) and differential scanning calorimetry (Quantachrome DSC model Q100). Thermogravimetric analysis was carried out in air with a heating rate of 10K min[-1]. Surface areas were calculated from the adsorption data using the BET equation, whilst the pore size distribution was estimated using DFT methods employing a slit-shaped pore model.

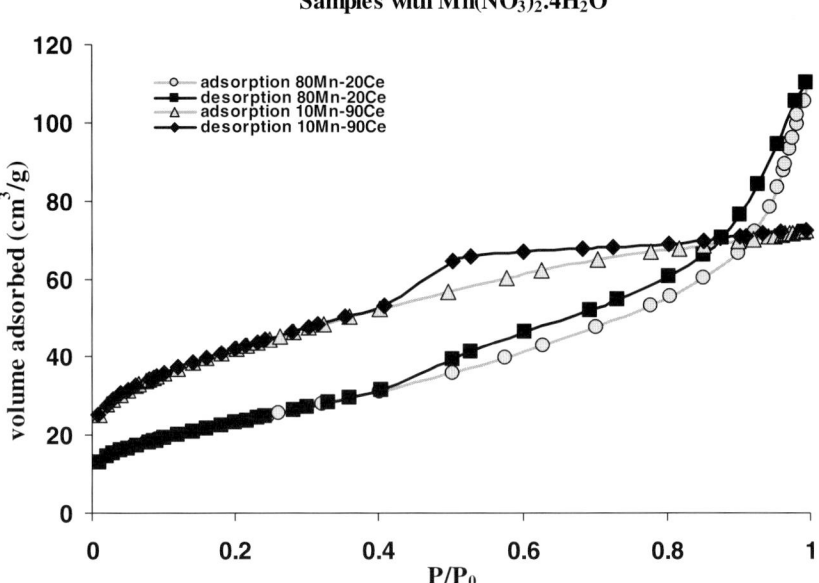

Figure 2 Nitrogen adsorption isotherms for two mixed ceria-mangania oxides.

## 3. RESULTS AND DISCUSSION

Figure 1 summarizes the BET surface area and average pore diameter results for all samples prepared. It can be seen that the effect of changing the manganese precursor on these properties is not significant. Conversely, there is a significant effect upon both parameters of the quantity of manganese present. The average pore diameter increases monotonically with manganese content, whereas the apparent surface area shows a maximum for the composition $Ce_{0.7}Mn_{0.3}O_2$. This is a significant increase in comparison to the data for pristine ceria [4]. Two different types of isotherms were observed, as shown in Figure 2: for low manganese content (10 to 30 percent), type IV isotherms in the IUPAC classification were obtained with H2 hysteresis loops, whereas for higher content (50 to 90 percent), a type II isotherm with an H3 type hysteresis loop was obtained.

Figure 3 shows the thermogravimetric trace for pure manganese oxide, indicating that after an initial weight loss, there is a weight increase at 650K due to oxygen uptake, and the increase in oxidation state of manganese from 2.8 to 3.0, corresponding to a phase change from $Mn_3O_4$ to $Mn_2O_3$. This was confirmed from the literature [7-9].

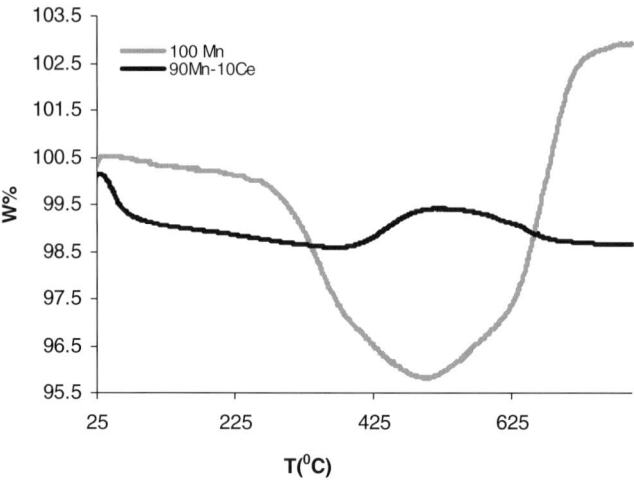

Figure 3 Thermogravimetric plots for manganese oxide ($Mn_3O_4$) and $Mn_{0.9}Ce_{0.1}O_2$.

The synthesis of $Mn_3O_4$ under the conditions used here was confirmed from the literature [10]. Further temperature increase would change the oxidation state to 3.2, accompanied by weight loss. In Figures 3 and 4 the thermogravimetric trace for sample $Mn_{0.9}Ce_{0.1}O_2$ is shown. An increase in weight is observed at 750K. At the same temperature, an endothermic event, was observed by scanning calorimetry, shown in Figure 4. X-ray diffractometry has shown that in sample $Mn_{0.9}Ce_{0.1}O_2$ two phases are present, $Mn_3O_4$, and the fluorite-like $CeO_2$. It is suggested that the difference in temperature at which oxidation occurs must be due to the presence of some cerium ions in the $Mn_3O_4$ phase.

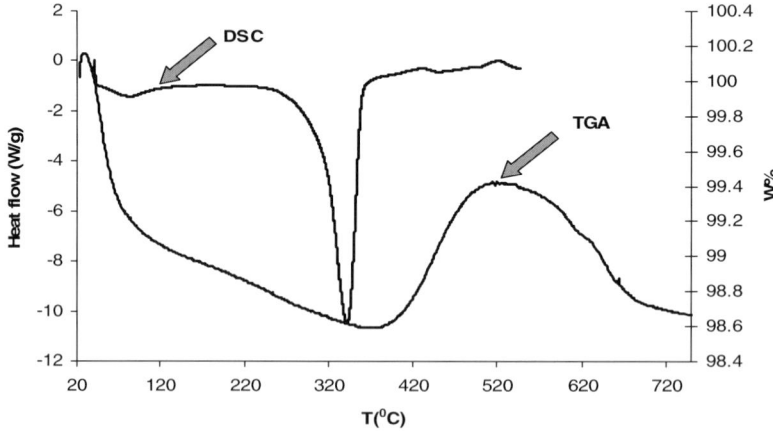

Figure 4  DSC and TGA traces for sample $Mn_{0.9}Ce_{0.1}O_2$.

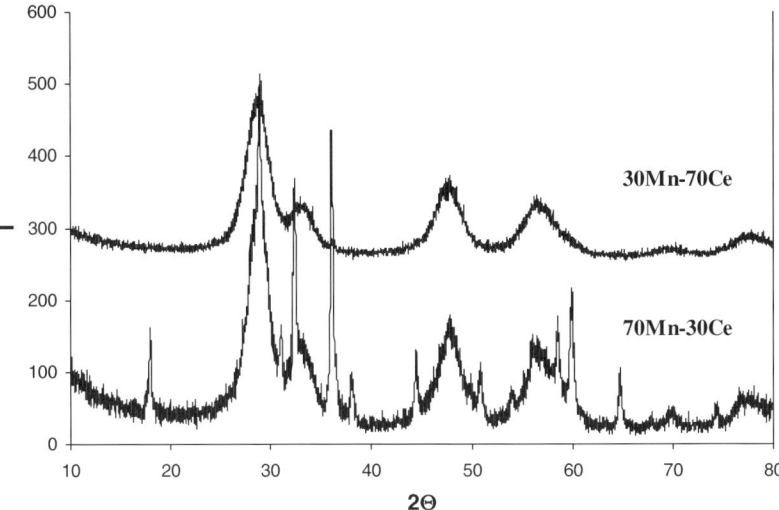

Figure 5 XRD traces for samples $Mn_{0.3}Ce_{0.7}O_2$ and $Mn_{0.7}Ce_{0.3}O_{0.2}$.

Figure 5 shows the X-ray diffractograms for samples $Mn_{0.3}Ce_{0.7}O_2$ and $Mn_{0.7}Ce_{0.3}O_2$. It can be seen that in the former sample the only peaks present were those hitherto assigned to cerium (IV) oxide. The same was observed for samples containing less that 30 mole percent manganese. Conversely, sample $Mn_{0.7}Ce_{0.3}O_{0.2}$, in common with all phases with metal content 40 mole percent or higher manganese contained two phases, one corresponding to ceria, and the second to $Mn_3O_4$. $Mn_3O_4$ contains manganese ions in oxidation states of +2 and +3, whereas in $Mn_2O_3$ only in the +3 state. In a pure manganese oxide phase oxidation occurs by oxygen absorption, resulting in the weight increase observed by thermogravimetry. However, in the mixed oxide, oxidation of Mn (II) to Mn (III) can be driven by the reduction of Ce (IV) ions to Ce (III), resulting in oxygen uptake occurring at a higher temperature, presumably subsequent to the conversion of Ce (IV) to Ce (III) reaching equilibrium.

As mentioned above, for manganese content 30 percent or less of metal content, mixed ceria-mangania solids consisted of a single phase corresponding to that of ceria (fluorite). As manganese content was increased, a second phase was developed, with a structure previously identified [10] as corresponding to that of $Mn_3O_4$. It is suggested that this phase must contain a certain amount of cerium (III or IV) ions. The particle size of the manganese-rich phase was larger than that of the cerium-rich phase, resulting in sharper peaks for the former in the X-ray diffractograms than for the latter.

The change in the shape of the nitrogen adsorption isotherm from type IV to type II with manganese content can be correlated with the presence of two phases, and the, at least partial, separation of the two oxides. This would lead to a lowering of non-stoichiometry of the samples, and presumably, a change in particle surface charge, and thus in the aggregation regime.

## REFERENCES

[1] Bunluesin T, Gorte RJ and Graham GW, 1998, Applied. Catal. B-Environmental,15,107.

[2] Shen WJ and Matsumura Y, 2000, Phys Chem Chem Phys, 2, 1519.

[3] Overbury SH, Mullins DR, and Huntley DR, 1999, J Catal , 186, 296.

[4] Pashalidis I and Theocharis CR, 2000, in (Eds K.K. Unger, G. Kreysa and J.P. Baselt), "Studies in Surface Science and Catalysis", Elsevier Science Publishers, 128, 643.

[5] Kyriakou G, Paschalidis I, and Theocharis CR, 2002, Fundamentals of Adsorption, 7 (Editors: K.Kaneko, H. Kanoh and Y. Hanzawa), 201.

[6] Theocharis CR, Christophidou M, and Pashalidis I., 2002, "Studies in Surface Science and Catalysis", Elsevier Science Publishers, 144, 75

[7] M. I. Zaki, M. A. Hasan, L. Pasuputely, K. Kumari, Thermochimica Acta 311 (1998) 97-103

[8] C. M. Julien, M. Massot, C. Poinsignon, Spectrochimica Acta Part A 60(2004) 689-700

[9] J. B. Macstre, E. F. Lopez, J. M. Gallardo-Amores, R. Ruano Casero, V. Sánchez Escribano, E. P. Bernal, International Journal of Inorganic Material 3 (2001) 889-899

[10] E. R. Stobbe, B. A. de Boer, J. W. Geus, Catalysis Today 47 (1999) 161-167

Studies in Surface Science and Catalysis 160
P.L. Llewellyn, F. Rodriquez-Reinoso, J. Rouqerol and N. Seaton (Editors)
651

# Thermal stability of ion exchange and adsorption properties of titania gels prepared from titanous chloride and hydrogen peroxide

A. R. Ramadan, N. Yacoub and J. Ragai

Department of Chemistry, the American University in Cairo,
113 Kasr El Aini Street, P.O. Box 2511, Cairo, Egypt

A comparative study of the uptake of the cations, $Cu^{2+}$, $Ni^{2+}$, $Co^{2+}$ and $Ca^{2+}$ by hydrous titanium oxides unheated and heat-treated at $400°C$ was carried out. The hydrous oxides were prepared at different pH values, using titanous chloride as starting material and hydrogen peroxide as oxidizing agent. Similarly to the unheated oxides, the heat-treated samples were found to be amphoteric in nature and exhibited an isolelectric point of     ~6.6. The total ion exchange capacity was found to generally decrease with heat treatment, so did the selectivity to a number of cations. Characterization of the samples by infrared and nitrogen adsorption studies were also carried out.

## 1. INTRODUCTION

Hydrous titanium oxides are important ion exchangers with different selectivities to different ions, exhibiting certain desirable catalytic and adsorptive properties [1-6]. One particular advantage to these hydrous oxides is their good thermal and radiation stabilities. Studies undertaken by a number of workers have shown that hydrous titanium oxide has ion exchange characteristics that are highly dependent on preparative conditions[7-8], and that the titania surface displayed amphoteric characteristics [ 9-10].

The present work investigates the thermal stability of the ion exchange properties of hydrous titanium oxides prepared at different pH values from 1 to 10, using titanous chloride as starting material and hydrogen peroxide as oxidizing agent. This is carried out by investigating the uptake of the cations $Cu^{2+}$, $Ni^{2+}$, $Co^{2+}$, and $Ca^{2+}$ by the oxides before and after subjecting them to heat treatment at $400°C$, a temperature of significance for a number of industrial applications.

## 2.  EXPERIMENTAL

### 2.1  Reagents and apparatus

The titanous chloride, $H_2O_2$, and ammonium hydroxide were obtained from Fischer (20% stabilized solution, 100 volume, 29.8% solution, respectively). Sodium hydroxide was obtained from Aldrich (99.99% pure), hydrochloric acid obtained from Fischer (38% concentrated), and sodium chloride obtained from Riedel (99.5% pure). The solutions of metal cations $Cu^{2+}$, $Ni^{2+}$, $Co^{2+}$ and $Ca^{2+}$ were prepared from the corresponding nitrates obtained from Fischer (98.9% pure), Analar (A.R.), Aldrich (99% pure), and Fischer (99.8% pure), respectively.

Infrared studies were carried out using a 337 Perkin-Elmer double beam grating spectrophotometer, and x-ray diffraction was carried out using a Philips x-ray diffraction analyzer system, PW 1840, using a Nickel-filtered Cu $K_\alpha$ radiation. A Thermolyne 48000 furnace was used for sample heating.

### 2.2  Sample preparation

The preparation of the hydrous titanium dioxide samples was described in detail elsewhere [11]. $H_2O_2$ was added to the titanous chloride followed by the addition of the ammonium hydroxide until the required pH values were obtained. Six samples were thus prepared at pH values of 1.26, 2.68, 4.47, 6.08, 8.18 and 9.83. The precipitated samples were then filtered, washed free of chloride ions , air dried at room temperature then redispersed in 250 ml of distilled water for 48 hours, refiltered, and finally air dried at room temperature. Subsequently, part of each sample was transformed into the hydrogen form using the column method of Inoue and Yamazaki [12] to yield samples TH1, TH2, TH4, TH6, TH8 and TH9, corresponding to pH values of 1.26, 2.68, 4.47, 6.0 , 8.18 and 9.83 respectively .

These samples were then heat-treated in air at 400°C for a period of two hours and denoted TH1(400), TH2(400), TH4(400), TH6(400), TH8(400) and TH9(400), with the water content, determined by weight percentage loss after heating to 800°C for 6 hours, being 7.2, 6.8, 6.0, 7.2, 7.7 and 1.6, respectively.

### 2.3  Surface and related characterization of the hydrous titania gels

Surface characterization was carried out using low temperature nitrogen adsorption. Samples were outgassed overnight at room temperature to residual pressures of approximately $25 \times 10^{-4}$ torr, then measurements carried out at 77K using a conventional volumetric technique. The gas pressures were measured on a mercury manometer. Infrared spectroscopy was used for further characterization, with the solid samples being prepared in the form of KBr pellets.

### 2.4  Determination of the ion exchange capacity

The ion exchange capacity was determined by the uptake of the gels of NaCl. 0.1 grams of the gels were equilibrated with 20ml of solutions maintained at a constant ionic strength of 0.1M using NaCl. This was carried out at different pH values ranging from 2-12, adjusted by adding suitable volumes of 0.1M HCl or 0.1M NaOH. The mixtures were kept in closed containers with intermittent shaking to reach equilibrium, with any drifts in pH due to ion exchange and liberation of $H^+$ and/or $OH^-$ corrected by the addition of suitable amounts of the HCl or the NaOH. The amounts of chloride and sodium ions adsorbed were then determined as the difference between initial and final concentrations in the solutions. Chloride was determined titrimetrically using More's method, and sodium determined by flame photometry.

## 2.5 Determination of the distribution coefficients ($K_d$) for the cations

The solutions of metal cations $Cu^{2+}$, $Ni^{2+}$, $Ca^{2+}$ and $Co^{2+}$ were prepared from the corresponding nitrates. 0.1gram of the gels were equilibrated with 10ml of 0.01M metal ion solutions at an ionic strength of 0.1M adjusted by NaCl. This was carried out at different pH values within a range limited by the precipitation of the $Ni^{2+}$, $Ca^{2+}$ and $Co^{2+}$ above a pH of ~6.2 and of $Cu^{2+}$ above a pH of ~ 5.3. These preparations resulted in the following solutions:

- $Cu^{2+}$: four solutions of pH values of 2.5, 3.1, 3.9 and 5.3.
- $Ni^{2+}$: four solutions of pH values of 3.0, 4.1, 4.9 and 6.0
- $Ca^{2+}$: four solutions of pH values of 3.0, 4.0, 4.8 and 6.2
- $Co^{2+}$: four solutions of pH values of 3.0, 4.0, 4.9 and 6.2

The mixtures were kept at 25°C in tightly stoppered containers. The distribution ratio was calculated using the formula:

$$K_d = (C_i - C_f) (v/m) / C_f$$

where $C_i$ and $C_f$ are respectively the initial and final concentrations of the metal ion, "v" the volume of the solution in ml and "m" is the mass of gel in grams. The concentrations of the metal ions were determined by complexometric titration [13].

## 3. RESULTS AND DISCUSSION

### 3.1 Infrared results

Infrared studies indicate that the presence of the ammonium ion (band at 1400 $cm^{-1}$) persists after heat treatment to 400°C in all samples, except TH8(400) and TH9(400). Moreover, interstitial water is also found in all heat treated samples (band in the region between 1600 $cm^{-1}$ and 1633 $cm^{-1}$, corresponding to the bending mode of molecular water[14]), as well as chemisorbed water (band at 3470 $cm^{-1}$[15]), though to a lesser extent than in the unheated oxides. The band at 3150 $cm^{-1}$, attributable to adsorbed ammonium ions [15], and found in the unheated samples of TH6 and TH8, is not detected after heat treatment.

The hydroxo complexes of titanium are manifested by bands at 970 $cm^{-1}$ for the bending mode of a bridging OH group and 1080 $cm^{-1}$ for a bending mode of a terminal OH group [14,16]. In this respect, the 970 $cm^{-1}$ band present in the unheated hydrous oxides TH2, TH4, TH6 and TH9, disappeared upon heat treatment, so that none of the heat treated oxides display the presence of bridging OH groups. The 1080 $cm^{-1}$ band persists in the heat treated samples TH2(400), TH4(400), TH6(400), TH8(400) and TH9(400), decreasing significantly, particularly for the latter two oxides.

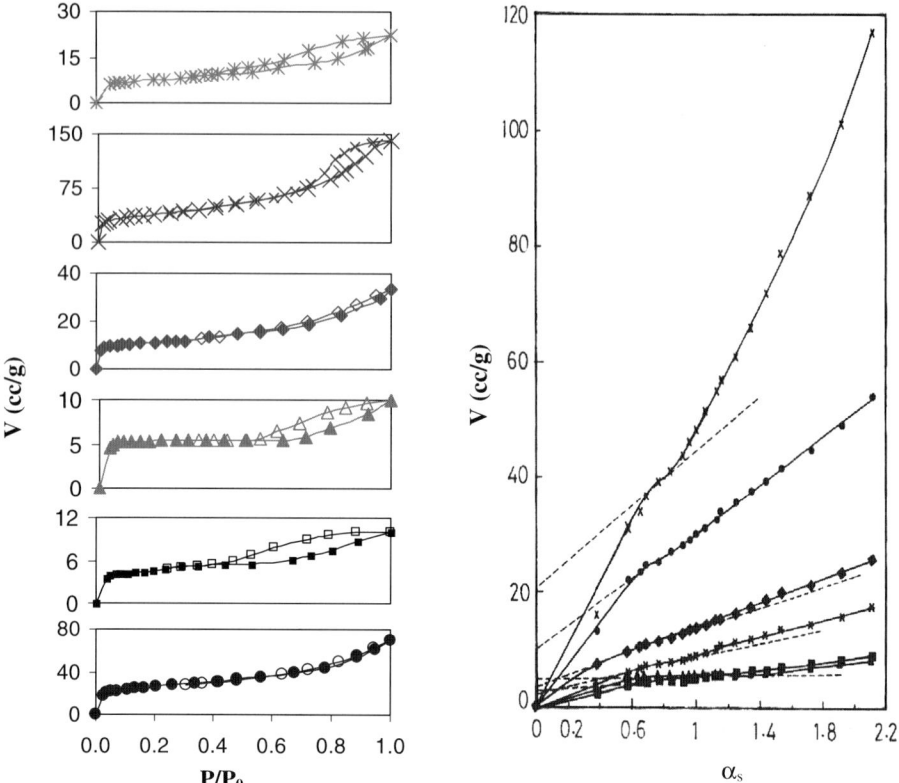

Figure 1. Adsorption isotherms and $\alpha_s$ plots for the TH(400) series. Data points relate to the following samples: ●: TH1(400); ■: TH2(400); ▲: TH4(400); ◆: TH6(400); ✗: TH8(400); ✱: TH9(400).

### 3.2 Adsorption measurements

Representative nitrogen adsorption isotherms for the TH(400) series of hydrous oxides and the corresponding $\alpha_s$ plots are shown in figure 1. $\alpha_s$ is defined as (amount adsorbed)/(amount adsorbed at $P/P^o = 0.4$) [17]. Details of the construction of the $\alpha_s$ plots, as well as that of the calculation of the surface areas $S_{BET}$ and $S_S$, the pore volumes and the effective micropore volumes are given elsewhere [17,18]. Table 1 presents these values for the TH and TH(400) series of gels.

Interpretation of the results of the adsorption studies carried out on the unheated TH series is presented elsewhere [11] Heat-treatment at 400°C results in most cases in a decrease in the extent of surface area relative to the unheated gels. Considered in conjunction with the infrared results depicting the disappearance of the band of the bending mode of bridging OH groups, this could possibly indicate the conversion of the hydroxo bridges to oxo bridges between contiguous titanium ions with heat treatment. The shape of the isotherms for the heat treated samples are suggestive of a more open gel structure (displaying some Type II and IV

Table 1
Analysis of the nitrogen adsorption data for the $TiO_2$ unheated and heat-treated samples

| Sample | $V_m$ (ml/g) | BET C-const | $S_{BET}$ (m²/g) | $S_s$ (m²/g) | Total pore volume (ml/g) | Micropore volume (ml/g) | Porosity |
|--------|------|------|------|------|------|------|------|
| TH1 | 60.4 | 184 | 263.0 | 266.0 | 0.1654 | 0.1440 | micro |
| TH2 | 3.5 | 12575 | 15.3 | 15.0 | 0.0072 | 0.0064 | micro |
| TH4 | 8.9 | 690 | 38.7 | 39.5 | 0.0203 | 0.0185 | micro |
| TH6 | 1.6 | 531 | 6.8 | 7.5 | 0.0036 | 0.0028 | micro |
| TH8 | 73.5 | 317 | 320.0 | 328.0 | 0.2186 | 0.1980 | micro + meso |
| TH9 | 32.9 | 2736 | 143.0 | 144.0 | 0.0758 | 0.0600 | micro |
| | | | | | | | |
| TH1(400) | 20.9 | 584 | 91.0 | 89.2 | 0.0952 | 0.0166 | micro |
| TH2(400) | 3.7 | 200000 | 16.1 | 16.8 | 0.0147 | 0.0046 | micro |
| TH4(400) | 4.9 | 481 | 21.1 | 21.2 | 0.0133 | 0.0078 | micro + meso |
| TH6(400) | 9.1 | 2779 | 39.5 | 40.0 | 0.0445 | 0.0076 | micro + meso |
| TH8(400) | 30.7 | 597 | 134.0 | 137.0 | 0.2075 | 0.0270 | micro + meso |
| TH9(400) | 6.1 | 885 | 26.7 | 26.5 | 0.0306 | 0.0076 | micro + meso |

character as suggested in the BDDT classification). The $\alpha_s$ plots indicate the prevalence of microporosity after heat treatment and the appearance of mesoporosity for some oxide samples.

### 3.3 The ion exchange capacity

The ion exchange capacities of the heat-treated samples indicate that heat treatment does not alter the amphoteric nature of the gels, which still exhibit similar isoelectric points to the unheated gels at pH ~6.6. However, heat treatment is found to significantly decrease the cation exchange capacity of the gels, while having a more limited effect on the anion exchange capacity. This can be attributed to the more significant loss of the bridging OH groups, acting as cation exchange sites, with heating, which is reflected in the disappearance of the corresponding IR band.

For the anion exchange capacity, of concern to the work presented here, the decrease is least significant for samples TH4(400) and TH6(400). Samples of lower and higher pH of preparation exhibited more significant decreases. This decrease can be understood in terms of the loss of terminal OH groups, acting as anion exchangers, with heating. The difference in the decrease of the anion exchange capacities for the different samples is reflected in the IR spectra, and can generally be attributed to the structural differences of these oxides resulting from different preparative pH values. Representative results for the change in the ion exchange capacity are depicted in figure 2.

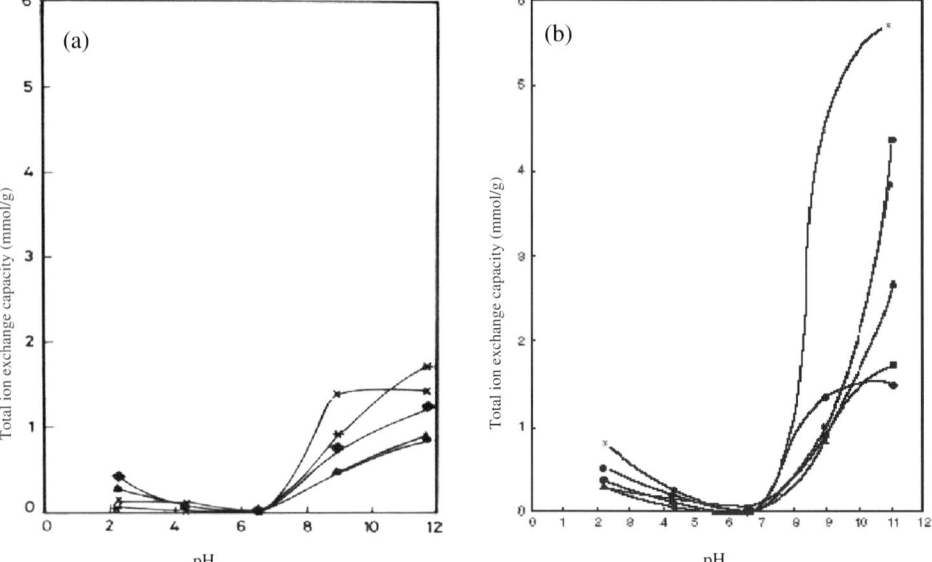

Figure 2. Ion exchange capacity for the TH(400) series (a) and the TH series (b). Data points relate to the following samples: ●: TH1(400); ■: TH2(400); ▲: TH4(400); ◆: TH6(400); ✗: TH8(400); ✱: TH9(400).

### 3.4 Selectivity measurements

The uptake of the metal cations under investigation took place at pH values below the isoelectric point, and cannot, therefore, be explained in terms of a simple cation exchange mechanism. Uptakes in the case of the cations $Cu^{+2}$, $Co^{+2}$ and $Ni^{+2}$ are assumed to be due to complex formation with one or more $H_2O$ molecules in the hydration sphere of the aquo ion being exchanged by the gel OH groups [19]. The latter would be acting as a bridge between the metal ion and the titanium ion. As for the $Ca^{2+}$ cations, their uptake is suggested to take place through an electrostatic attraction between these hydrated cations in solution and the hydrous titanium oxides [12].

Details of the selectivity measurement results for the unheated samples TH1 to TH9 are presented elsewhere [11] Heat treatment at 400°C results in a decrease of the $K_d$ values for the cations investigated for samples TH2(400), TH4(400) and TH6(400), which can be explained in terms of the loss of complexed OH groups as supported by the infrared studies. However, the $K_d$ values were found to increase upon heat treatment for samples at higher preparative pH values (TH8(400), TH9(400)). This could possibly be attributed to the increase in the pore size with heating, as reflected in the surface studies, which would facilitate the uptake of the cations from solution [1].

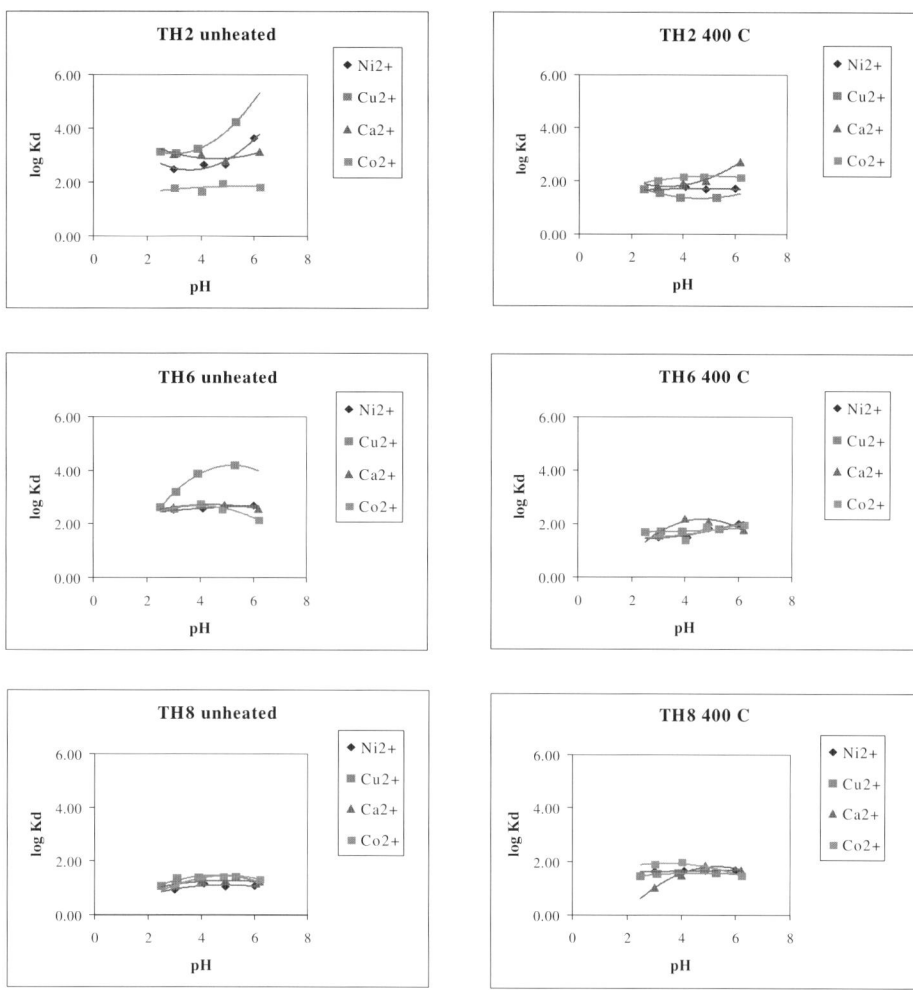

Figure 3. Selective uptake of $Cu^{2+}$, $Ni^{2+}$, $Co^{2+}$ and $Ca^{2+}$ for the unheated and heat treated gels of preparative pH values of 2.68, 6.08 and 8.18.

The trend for the uptake of the cations $Cu^{+2} > Ni^{+2} > Co^{+2}$, with $Ca^{2+}$ selectivity falling between $Cu^{+2}$ on the one hand and the $Ni^{+2}$ and $Co^{+2}$ on the other for the unheated hydrous oxides is found to change to $Cu^{+2} < Ni^{+2} < Co^{+2}$ with $Ca^{2+}$ selectivity falling between $Ni^{+2}$ and $Co^{+2}$ upon heat treatment for the gels prepared at pH values of 2.68 and 4.47. The trend is unchanged with heat treatment for the other gels. This could possibly be interpreted by the relative prominence of two processes affecting the uptake of the cations from solution: the diffusion process of the hydrated cations to the OH sites of the gels, and the surface hydrolysis process resulting in the attachment of these cations onto the gels through the OH groups. For the unheated samples, results indicate that the uptake seems to be controlled by the hydrolysis process with the selectivity of uptake in agreement with the first hydrolysis constants for the cations [20]. As heat treatment is carried out, and the presence of complexed OH groups becomes limited as supported by the infrared studies, two scenarios seem to arise. For samples TH2(400) and TH4(400), the diffusion of the hydrated cations to these fewer OH groups seems to predominate. The trend of selective uptake is found to change and could be seen as more in association with the size of the cations. For samples TH6(400), TH8(400), and TH9(400), the hydrolysis process seems to predominate. The trend of selective uptake is found to reflect the trend of the first hydrolysis constant of the cations. The observations for TH6(400), TH8(400), and TH9(400) could possibly be attributed to the observed increase in mesoporosity, which would increase the accessibility of the fewer OH groups of the gels to the cations in solution, allowing the hydrolysis process to have the more significant effect on the selectivity of uptake.

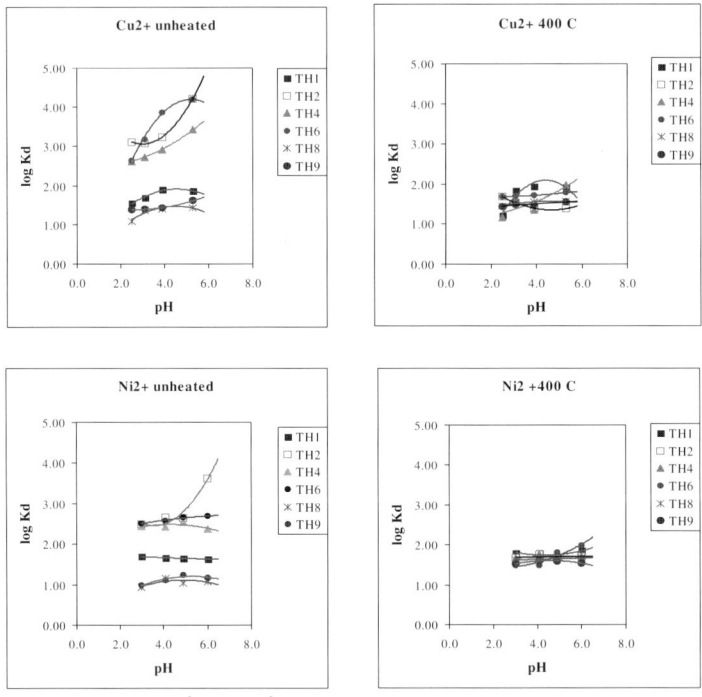

Figure 4. Selective uptake of $Cu^{2+}$ and $Ni^{2+}$ cations for the unheated and heat treated gels.

## 4. CONCLUSIONS

Heat treatment at 400°C of hydrous oxides of titanium prepared using titanous chloride as starting material and $H_2O_2$ as oxidizing agent does not affect the *amphoteric* properties of the gels nor their isoelectric points.

A significant decrease in the cation exchange capacity for the heat-treated hydrous oxides, and a more limited decrease in their anion exchange capacity, are observed. These are concomitant with an observed decrease in the presence of both bridging and terminal OH groups, with the decrease of the former being more significant. An associated decrease in surface area with heat treatment is also observed.

Observed trends for the uptake of the cations by the heat treated gels appear to be associated to both the availability of terminal OH groups, as well as the porosity of the gels. Upon heat treatment and the associated reduction in the presence of complexed OH groups, the trend of the selective uptake of the cations is found to be more in correlation with the sizes of the hydrated cations for some of the hydrous oxide samples. For these samples, the diffusion of the hydrated cations to the limited OH groups of the gels appear to predominate the uptake process. For other hydrous oxides the trend of the selective uptake of cations remains unchanged from the unheated samples, and is in correlation with the first hydrolysis constant for these cations. For these samples, an increase in mesoporosity is observed with heating and it seems to improves the accessibility of the limited OH groups of the gels to the hydrated cations from solution, counter effecting the reduction in the presence of these OH groups. For these samples, the hydrolysis of the cations appears to predominate the uptake process, similar to the cases of unheated gels.

## ACKNOWLEDGEMENT

The authors would like to express their thanks to the American University in Cairo (Office of Graduate Studies and Research) for the grant awarded to them which enabled the present work to be undertaken.

## RRFERENCES

1- N. Yacoub, J. Ragai and S.A. Selim, J. Mater. Sci., 26 (1991) 4937.
2- T. A. Egerton and I. R. Tooley, J. Mater. Chem., 12 (2002) 1111.
3- T. Fesionowski, J. Pigm. Resin Technol., 30 (2001) 287.
4- G. Uday Chand, D. Manotosh, D. Sushanta and B. Subhas Chandra, Water, Air, Soil Pollut., 143 (2003) 245.
5- E. I. Shabana and M.I. El-Dessouky, J. Radioanal. Nucl. Chem., 253 (2002) 281.
6- T. J. Gardner and L. I. Mc Laughlin, J. Mater. Res. Soc. Proc., 432 (1997) 249.
7- Y. Komatsu, Y. Fujiki and T. Sasaki, Bull. Chem. Soc. Jpn., 58 (1985) 97.
8- H. Hayashi, T. Iwasaki, Y. Onodera and Y. Fujiki, Bull. Chem. Soc. Jpn., 62 (1989) 371.
9- M. Sugita, M. Tsuji, and M. Abe, Bull. Chem. Soc. Jpn, 63 (1990), 559.
10- H. Kita, N. Henmi, K. Shimazu and K. Tanabe, J. Chem. Soc. Faraday Trans. I, 77 (1981) 2451.
11- N. Yacoub, A. R. Ramadan and J. Ragai, Adsorpt. Sci. Technol., 23 (2005) 215.

12- Y. Inoue and H. Yamazaki, Bull. Chem. Soc. Jpn., 60 (1987) 891.

13- T. S. West, Complexometry with EDTA and Related Reagents, BDH Chemicals Ltd., Poole,1969.

14- K. Nakamoto in: Infrared and Raman Spectra of Inorganic and Coordination Compounds, John Whiley and Sons, New York,1978, pp. 228 229.

15- N. D. Parkyns in: Chemisorption and Catalysis, Hepple Publications, London,1970, pp. 150 172.

16- J. Ragai, J. Chem. Technol. Biotechnol., 32 (1982) 998.

17- K. S. W. Sing in: D. H. Everett and R. H. Ottewill (eds.), International Symposium on Surface Area Determinations, Butterworths, London, 1970.

18- S. J. Gregg and K. S. W. Sing, Adsorption, Surface Area and Porosity, Academic Press, London, 1982, pp. 94 95.

19- R. O. James, P. J. Stiglich and T.W. Healy, Faraday Discuss. Chem. Soc., 59 (1975) 142.

20- J. E. Huheey, Inorganic Chemistry: Principles of Structure and Reactivity, Harper and Row, New York, 1983, p. 295.

Studies in Surface Science and Catalysis 160
*P.L. Llewellyn, F. Rodriquez-Reinoso, J. Rouqerol and N. Seaton (Editors)*

# In-situ SAXS on Transformations of Mesoporous and Nanostructured Solids

Peter Laggner[a], Marlene Strobl[a], Philipp Jocham[a], Peter M. Abuja[b] and Manfred Kriechbaum[a]

[a]Institute of Biophysics and X-Ray Structure Research, Austrian Academy of Sciences, Schmiedlstrasse 6, A-8042 Graz, Austria

[b]HECUS X-Ray Systems GmbH, Reininghausstrasse 13a, A-8020 Graz, Austria

## 1. INTRODUCTION

Pore size and inner surface are important characteristics of solid materials and determine many important properties, such as mechanical, chemical and thermal stability, compactability, solubility, dissolution rate and sorption properties. Many of the available methods for the determination of these inner structure parameters in solids require extensive sample pre-treatment, which is often not desirable as it will lead to significant alterations of the structures of interest. Moreover, the methods are time-consuming and thus do not permit to follow the inner structure development over time, which is of interest in production and use of nanoporous materials. The non-invasive measurement of inner surface by small-angle X-ray scattering (SAXS), is a convenient alternative to sorption methods, allowing inner surface measurements within minutes, and without sample pre-treatment. In this paper we give a brief overview about the principles of this method, and present examples to illustrate the applicability of SAXS for the determination of inner surface and pore size.

## 2. THEORY

SAXS measurements typically cover length scales between 1 and 100 nm, which according to Bragg's law

$$n \lambda = 2 \, d \, \sin \theta$$

($\lambda$, the X-ray wavelength, typically 0.154 nm for Cu-$K_\alpha$; d, the real-space distance; $2\theta$, the scattering angle) correspond to scattering angles between 0.8 and 80 mrad. For random heterophase systems, the scattering curves, i.e. the angular dependence of the scattering intensity I(q) (where $q = 4\pi \sin \theta/\lambda$ is the modulus of the scattering vector), contain direct information on the inner surface of the system, provided that the geometry of the structure is non-fractal – i.e. Euclidean [1]. The relevant parameters are the limiting decay coefficient, k, towards large angles (the 'Porod coefficient'), and the 'invariant' Q. The k-value is defined by the following equations

$$\lim_{q \to \infty} I(q) = k / q^4$$

and

$$\lim_{q \to \infty} \widetilde{I}(q) = \widetilde{k} / q^3$$

(The tilde in the second equation indicates from slit-length smeared scattering data obtained from line-collimation cameras, as used in this study). For two-phase systems, $k$ is proportional to the total interfacial area, S, between the two phases.

The invariant, which is directly proportional to the total volume, V, of the scattering system,

is given by the following integral:

$$Q = \int_0^\infty I(q)q^2 \cdot dq$$

or

$$\tilde{Q} = \int_0^\infty \tilde{I}(q)q \cdot dq$$

From the ratio of k and q (or their slit-smeared analogs) it is therefore possible to calculate, without any absolute-scale measurement, the specific inner surface, $S_i$, surface per volume, of such a two-phase system by the following equations:

$$S_i = \pi \cdot \phi(1-\phi) \cdot \frac{k}{Q}$$

or

$$S_i = 4 \cdot \phi(1-\phi) \cdot \frac{\tilde{k}}{\tilde{Q}}$$

The pore volume fraction, $\phi$, can be determined by other methods, such as He-pycnometry, or using the X-ray absorption at a certain length scale [2], in combination with the skeletal density and chemical composition. From the inner surface, the average chord length through any of the two phases, i.e. the characteristic length scale of the system comparable to the radius of gyration in the scattering from particles, is obtained by the equations

$$\bar{l} = 4 \cdot \phi(1-\phi) \cdot \frac{1}{S_i}$$

or

$$\bar{l} = \frac{4}{\pi} \cdot \frac{Q}{k} \quad \text{and} \quad \bar{l} = \frac{\tilde{Q}}{\tilde{k}}$$

The average pore diameter, $l_p$, and wall thickness, $l_s$, are related to average chord length by by

$$\bar{l}_p = \frac{\bar{l}}{1-\phi} \quad \text{and} \quad \bar{l}_s = \frac{\bar{l}}{\phi}$$

## 3. EXPERIMENTAL

All experiments were performed on a HECUS System3 SWAXS camera (HECUS X-Ray Systems GmbH, Graz, Austria), with slit-block collimation, mounted on a Seifert ID-3003 X-ray generator (Seifert, Ahrensburg, Germany), equipped with a X-ray tube with copper anode, operated at 50 kV and 40 mA. To suppress the contribution from $Cu$-$K_\beta$ radiation, Ni-filtering was used. Position-sensitive gas detectors (MBraun-Hecus PSD 50M) with real-time readout were used to monitor the scattering curves. The samples were investigated in a thermostatted quartz capillary of 1 mm inner diameter at 293K. Measuring times were 300 sec.

To avoid texture artefacts, the capillaries were rotated during the measurements using the SpinCap accessory (HECUS XRS, Graz, Austria). This precaution proved advantageous not only with microcrystalline powder specimens, but also with mesoporous reaction systems in liquid suspension, to prevent sedimentation effects on the SAXS intensities.

Certified reference materials (alumina), PM-102, -103, and -104 were obtained from the Deutsche Bundesanstalt für Materialforschung (BAM), Berlin, Germany. The certified values for pore size and inner surface for these materials had been determined using the Brunauer-Emett-Teller (BET) gas sorption method.

## 4. RESULTS AND DISCUSSION

### *SAXS of alumina standards (BAM)*

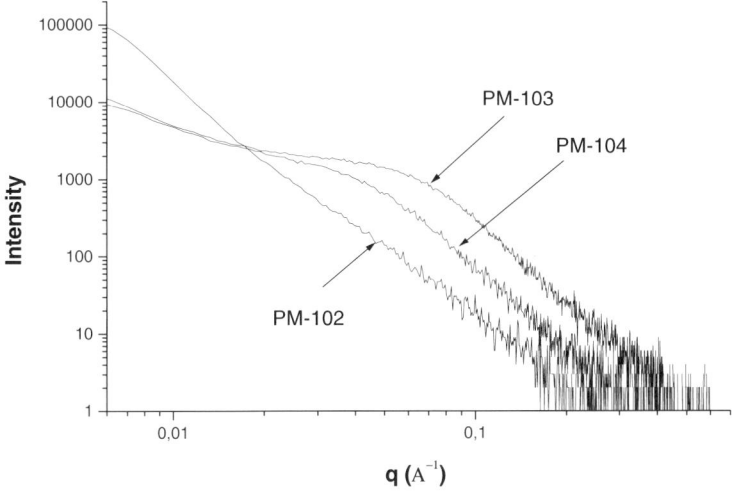

Fig. 1: SAXS scattering curves of the BAM alumina standards (Exposure times: 300 sec)

To validate the of values for specific inner surface, three certified standard samples of alumina, ranging between BET surfaces of 5,41 and 156 $m^2$/g, were investigated by SAXS. A representative set of scattering curves from the alumina samples is shown in Figure 1. The region of $q > 0.09$ $Å^{-1}$ was used to determine the slit-smeared k-values, and for the determination of Q the whole experimental range was used and extrapolated to large angles according to the $q^{-3}$ decay. The measuring times of 300 sec were found to provide sufficiently good counting statistics to determine the specific inner surface to a precision of better than +/- 3 %. This shows that SAXS is by at least one order of magnitude faster than the BET method. It should be noted that the use of line-collimation provides a distinct advantage over point-collimation in this respect, because the $q^{-3}$ decay leads to a significantly better counting statistics in the low-intensity part of the curves than a $q^{-4}$ decay, and therefore allows shorter exposure times.

The numerical results results for the specific inner surface values are compiled in Table 1. There is generally good agreement with the BET values, however, the SAXS values are consistently slightly higher. The most likely reason for this is, that SAXS is also sensitive to closed pores which are not accessible to gas sorption in the BET method, and to the nanostructured surface of the particles. Also the fact, that it is not necessary to exhaustively dry the samples for the SAXS measurements, may contribute to the differences.

Table 1: Comparison of inner surface values by BET and SAXS of BAM standards.

| BAM Standard | BET value ($m^2$/g) | SAXS value ($m^2$/g) |
|---|---|---|
| PM-102 | 5.41 | 6.1 |
| PM-104 | 79.8 | 97.4 |
| PM-103 | 156.0 | 161.4 |

*Monitoring Inner Surface in M41S Silicas in Presence of Swelling Agents*

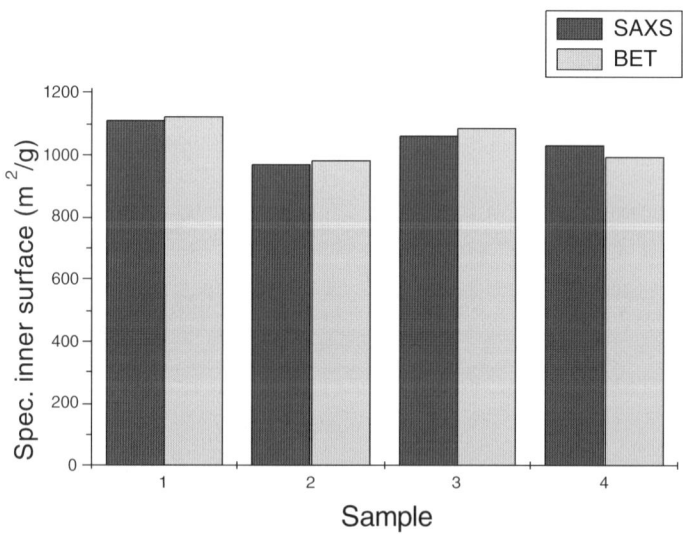

Fig. 2: Comparison of BET and SAXS derived values for inner surface for M41S silicas.
1 ... Mesitylene
2 ... TiPB (tri-isopropylbenzene) 20 %
3 ... TiPB 40 %
4 ... TiPB 80 %

Mesoporous M41S materials were synthesized from gels according to [3] alone and in the presence of various organic agents (mesitylene, triisopropylbenzene (TiPB))   leading to different micellar sizes. After calcination at 813 K the products exhibited different pore sizes accordingly. Pore sizes and inner surface were determined both by BET and by SAXS (Fig. 1) and showed excellent agreement.

The production process of the M41S involved the formation of an organic-inorganic mesophase, consisting of a sodium silicate gel in which hexadecyltriammonium chloride and dodecyltriammonium chloride mixed micelles (with or without incorporated swelling agents) were embedded. The mesophases were heated for 96 h at 373 K. Every 24h the pH of the gel was adjusted to 10.6 by addition of acetic acid. During this time the inner surface formation was followed by SAXS by taking aliquots (Fig. 2). During the acidification steps the reduction of inner surface is clearly visible.

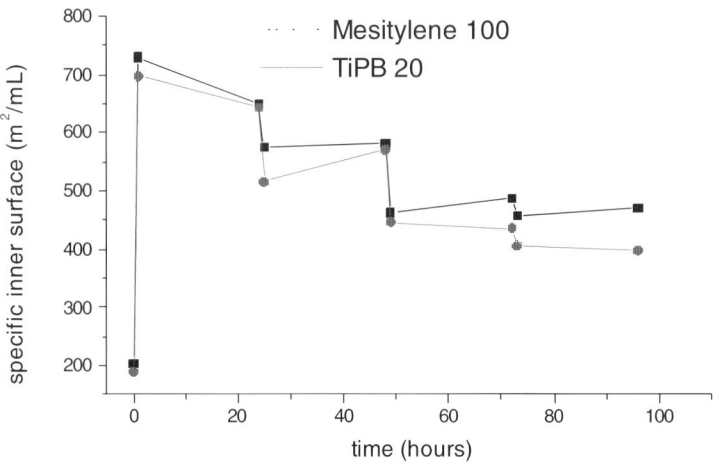

Fig. 3: Development of inner surface of M41S silicas containing mesitylene and TiPB during 96 hours as monitored by SAXS. The acidification steps are at 24, 48 and 72 h.

## 5. CONCLUSION

The use of SAXS as a tool for inner surface measurement is particularly advantageous if, as shown in the M41S example, preconditioning of the sample by drying is impossible or undesirable, e.g. because the interface between two phases in the wet state needs to be monitored. Besides delivering results nearly identical to other methods, like BET, SAXS is not only faster, but also it is not limited to samples to the dry state. Moreover, not only porous systems – i.e. solid/air interfaces - can be investigated, but any two-phase system, provided there is a difference in electron density between the two phases. This is particularly valuable in investigations on liquid, heterophase reaction systems, where the specific inner surface can provide important clues to the kinetics and mechanisms of structure growth and decay.

## REFERENCES

[1] Porod, G. Kolloid-Z. 124 (1951) 83 – 144.
[2] Spalla, O., Lyonnard, S., Testard, F. J. Appl. Cryst. 36 (2003) 338 – 347.
[3] Lüchinger M., Pirngruber, G. D., Lindlar, B., Laggner, P., Prins, R. Micropor. Mesopor. Mater. 79 (2005) 41-52.

Studies in Surface Science and Catalysis 160
*P.L. Llewellyn, F. Rodriquez-Reinoso, J. Rouqerol and N. Seaton (Editors)*
© 2007 Elsevier B.V. All rights reserved

# Confinement effects on freezing of binary mixtures

**Benoit Coasne[a,b], Joanna Czwartos[c], Keith E. Gubbins[a], Francisco R. Hung[a] and Malgorzata Sliwinska-Bartkowiak[c]**

[a] Center for High High Performance Simulation and Department of Chemical and Biomolecular Engineering, North Carolina State University, Raleigh, NC 27695-7905, USA

[b] Laboratoire de Physicochimie de la Matière Condensée, CNRS (UMR 5617) and University Monpellier 2, Place Eugène Bataillon, 34095 Montpellier Cedex 05, France

[c] Institute of Physics, Adam Mickiewicz University, Umultowska 85, 61-614 Poznan, Poland

We report molecular simulations and experimental measurements of the freezing and melting of mixtures confined in nanopores. Dielectric relaxation spectroscopy was used to determine the experimental phase diagram of mixtures confined in activated carbon fibers. Grand Canonical Monte Carlo simulations combined with the parallel tempering technique were used to model the freezing of several Lennard – Jones mixtures in graphite slit pores. The effect of confinement is discussed for mixtures having a simple solid solution or an azeotropic solid – liquid phase diagram. We also investigate how the competition between the wall – fluid and fluid – fluid interactions affects the freezing temperature of the confined system. The structure of the crystal phase in the simulations is also investigated by means of positional and bond-orientational pair correlation functions and bond-order parameters.

## 1. INTRODUCTION

Many experiments and molecular simulations of the freezing of fluids confined in nanoporous solids have been reported [1]. This effort is devoted to the understanding of the effect of confinement, surface forces, and reduced dimensionality on the thermodynamics of fluids. These works are also of practical interest for applications involving confined systems (lubrication in nanotechnologies, synthesis of nano-structured materials, phase separation, etc). Beside the abundant literature for pure fluids in nanopores, few studies [2-7] have focused on the freezing of confined mixtures. As in the case of pure substances, the pore width $H$ and the ratio of the wall/fluid to the fluid/fluid interactions (parameter $\alpha$ [8]), play an important role in the phase behavior of the mixture. The ratio of the wall/fluid interaction for the two species is also a key parameter in describing freezing of these systems.

In this paper, we review our experimental and simulation work [5-7] and include new results on the solid – liquid phase behavior of mixtures confined in nanopores. Dielectric relaxation spectroscopy (DRS) was used to study the experimental phase diagram of $CCl_4/C_6H_{12}$ mixtures confined in activated carbon fibers (ACF). Grand Canonical Monte Carlo (GCMC) simulations with the parallel tempering technique were used to model the freezing of Lennard-Jones mixtures in slit pores. Mixtures having a simple solid solution or an azeotropic solid-liquid phase diagram were considered. We also investigate the effect of the ratio $\alpha$ of the wall-fluid to the fluid-fluid interactions on the freezing temperature of the confined system. The structure of the crystal phase is also discussed using both positional and bond-orientational pair correlation functions and bond-order parameters.

## 2. EXPERIMENTAL AND SIMULATION METHODS

### 2.1. Experiment

ACF with a pore width of 1.2 nm was used to study freezing of confined $CCl_4/C_6H_{12}$ mixtures. Pores in this material, which are approximately of a slit geometry, are expected to accommodate two layers of $CCl_4$ or $C_6H_{12}$ since the reduced pore width is $H^* \sim 2.4$. DRS was performed using a parallel plate capacitor of empty capacitance $C_0 = 69.1$ pF. The capacitance $C$ and the tangent loss $\tan(\delta)$ (where $\delta$ is the angle by which current leads the voltage) of the filled sample were measured at different temperatures using a SI 1260 impedance/gain phase analyzer in the frequency range 10 Hz – 10 MHz. The real and imaginary parts of the dielectric permittivity $\varepsilon^* = \varepsilon' - i\varepsilon''$ are related to C and $\delta$, $\varepsilon' = C/C_0$ and $\varepsilon'' = \tan(\delta)/\varepsilon'$ [9]. Melting can be monitored in DRS by a large increase in $\varepsilon'$. The sample was introduced in the capacitor as a suspension of ACF filled with the mixture in the bulk mixture. As a result, the measurements yield an effective permittivity that has contributions from the bulk and confined mixtures. Full details regarding the experiments can be found in Ref. [7].

### 2.1. Molecular simulation

The phase diagram of the bulk and confined mixtures AB were determined using the Gibbs-Duhem integration (GDI) technique [10,11] and GCMC simulations [12], respectively. The GDI consists of determining the phase coexistence conditions by integrating the Clapeyron equation at constant pressure. Such a method allows one to find the relation $T(\xi_B)$ that describes the solid/liquid coexistence temperature when the fugacity fraction for species B, $\xi_B$, varies from 0 to 1. At each coexistence condition, Monte Carlo simulations in the $NPT\xi_B$ ensemble are performed to estimate the enthalpies and mole fractions for the liquid and solid phases. The GDI and $NPT\xi_B$ Monte Carlo algorithms used in this work are similar to those developed by Hitchcock and Hall [7,11].

The GCMC technique consists of simulating a system having a constant volume $V$ (the pore with the confined phase) in equilibrium with a fictitious reservoir of particles imposing its chemical potentials $\mu_A$, $\mu_B$ and its temperature $T$. We combined the GCMC simulations with a parallel tempering technique to improve the sampling of phase space [12,13]. The input parameters, $\mu_A(T)$ and $\mu_B(T)$ at $P = 1$ atm, were determined using the equation of state for Lennard – Jones mixtures of Johnson *et al.* [14]. Starting with well-equilibrated configurations, we performed GCMC simulations with the parallel tempering technique. 16 replicas were used in each run and the temperature difference between two successive replicas is $\Delta T = 3$ K. After equilibration, density profiles, order parameters, and correlation functions were averaged in a second run. Full details regarding the methods can be found in Refs. [5,7].

The fluid/fluid interactions were modeled using Lennard-Jones potentials with parameters that reproduce properties of the bulk liquids. The cross-species A/B parameters were calculated using the Lorentz-Berthelot combining rules [15]. The slit pore was described as an assembly of two structureless parallel walls. Periodic boundary conditions were applied in the directions parallel to the pore walls. The fluid/wall interaction was calculated using the Steele '10-4-3' potential [16]. The fluid/wall parameters, $\varepsilon_{w/X}$, $\sigma_{w/X}$ (X = A or B) were determined by combining the wall/wall and fluid/fluid parameters using the Lorentz-Berthelot rules with the values $\varepsilon_{ww}/k_B = 28$ K and $\sigma_{ww} = 0.34$ nm for the carbon wall.

Strong layering of the confined system was observed due to the interaction with the attractive pore walls. The structure of the confined mixture was investigated by calculating for each layer $i$ the 2D bond-order parameters $\Phi_{n,i}$ ($n = 4$ and 6 for a square and triangular structure, respectively). $\Phi_{n,i}$ was determined as the average value of the local order parameter $\Psi_{n,i}(r)$, which measures the bond order at a position $\mathbf{r}$ of a particle in the layer $i$ [17]:

$$\Phi_{n,i} = \frac{\left| \int \Psi_{n,i}(\mathbf{r}) \, d\mathbf{r} \right|}{\int d\mathbf{r}} \quad \text{with} \quad \Psi_{n,i}(\mathbf{r}) = \frac{1}{N_b} \sum_{k=1}^{N_b} \exp(in\theta_k) \quad (1)$$

where $\theta_k$ are the bond angles between the particle and each of its $N_b$ nearest neighbors. $\Phi_{n,i}$ is close to 1 for a crystal layer having a triangular ($n = 6$) or a square ($n = 4$) structure and close to 0 for a liquid layer. We also monitored the 2D positional and bond-orientational pair correlation functions, $g_i(r)$ and $G_{n,i}(r) = <\Psi_{n,i}^*(0) \, \Psi_{n,i}(r)>$. The latter measures the correlations between the local bond-order parameter $\Psi_{n,i}(r)$ at two positions separated by a distance r.

## 3. RESULTS AND DISCUSSION

### 3.1. Solid solution mixtures

We first discuss the freezing of solid solution mixture Ar/Kr confined in a graphite slit pore having a width $H = 1.44$ nm. Density profiles (not shown) show that the slit pore accomodates two contact and one inner layers. The confined mixture always has a larger Kr mole fraction than the bulk. This result is in agreement with previous works for confined mixtures [18,19], which showed that the mole fraction of the component having the strongest fluid/wall interaction is increased compared to the bulk. The bond-order parameter, $\Phi_6$, for the confined layers of a mixture in equilibrium with a bulk mixture $x_{Kr} = 0.05$ is shown in Fig.1(a) as a function of the temperature $T$. $\Phi_6$ sharply increases upon freezing at $T \sim 136$ K, which reveals that the confined layers undergo a liquid to crystal phase transition. $\Phi_6$ varies from $\sim 0.08$ in the liquid region up to $\sim 0.80$ in the crystal region. This latter value suggests that the crystals layers have a hexagonal structure (triangular symmetry) with, however, some defects. The Kr mole fraction of the confined mixture is shown as a function of the temperature in Fig. 1(b). $x_{Kr}$ sharply increases at $T = 136$ K from 0.57 in the liquid phase up to 0.67 in the solid phase. This analysis provides a first set of crystal/liquid coexistence conditions: $T$, $x_{Kr}$(liquid), and $x_{Kr}$(crystal).

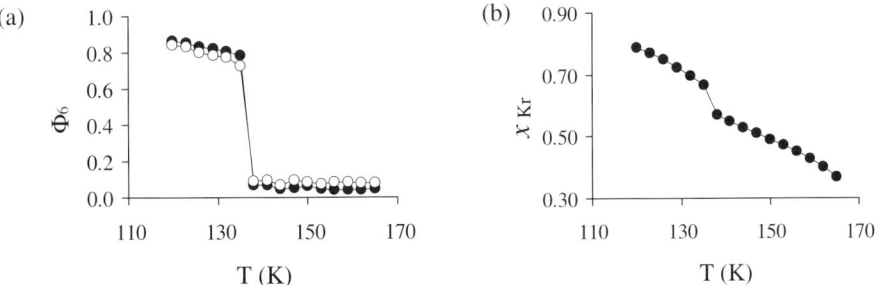

Fig. 1. (a) $\Phi_6$ versus $T$ for Ar/Kr mixtures in a slit pore $H = 1.44$ nm: ($\bullet$) contact, ($\circ$) inner layers. (b) $x_{Kr}$ versus $T$ for Ar/Kr mixtures in a slit pore $H = 1.44$ nm.

The freezing temperature for the confined mixture, $T_f = 136$ K, is much larger than the freezing point of a bulk mixture having the same composition, $T_f^{bulk} \sim 111$ K [11]. The sharp increase observed in the 2D bond-order parameter $\Phi_6$ and the Kr mole fraction suggest that freezing of the confined layers is a first order transition; however, free energy calculations are required to confirm in a rigorous way the nature of the liquid to crystal transition. Thanks to the use of the parallel tempering technique in which both the liquid and crystal phases are simulated in the same run, we did not perform calculations for the melting process. In our previous work [5,6], it has been shown that simulations of melting and freezing phenomena give similar results, provided that the parallel tempering method is used and a significant fraction of swap moves are accepted [13]. Moreover, Hung *et al.* [20] showed for pure fluids that this technique gives the same results as free energy calculations.

In-plane 2D positional g(r) and orientational $G_6(r)$ pair correlation functions are shown in Figs. 2 and 3 for the contact and inner layers of Ar/Kr confined mixtures in equilibrium with the bulk mixture $x_{Kr} \sim 0.05$. At $T = 129$ K, the confined layers appear as 2D hexagonal crystals with long-range positional order as can be seen from the features of the g(r) function for this temperature; the amplitude between the first and second peaks is close to 0, the second peak is split into two secondary peaks, and the third peak has a shoulder on its right side. Moreover, the $G_6(r)$ function at this temperature has a constant average value as expected for a hexagonal crystal layer with long-range orientational order. At $T = 144$ K, the confined layers exhibit a liquid-like behavior as revealed by the g(R) functions, which are characteristic of a phase having short-range positional order. This result is confirmed by the exponential decay observed in the $G_6(R)$ function; such a decay is typical of 2D liquid phases, which have short-range orientational order. Analysis of the in-plane 2D pair correlation functions corroborate the results shown in Fig. 1(a) for the order parameter $\Phi_6$; the transition temperature between the crystal and liquid phases was found to be $T_f \sim 136$ K. For all compositions studied in this work, it seems that freezing of the confined layers involves a direct phase transition between a 2D-crystal and a 2D-liquid. This result departs from previous works for confined fluids in which the existence of a hexatic phase between the crystal and liquid phases was reported [21]. Such an intermediate phase is expected according to the KTHNY theory for 2D melting [17]. The stability of the hexatic phase depends on the size of the system, so that the existence of such an intermediate phase cannot be ruled out or confirmed in the present work unless a scaling size analysis is performed [21].

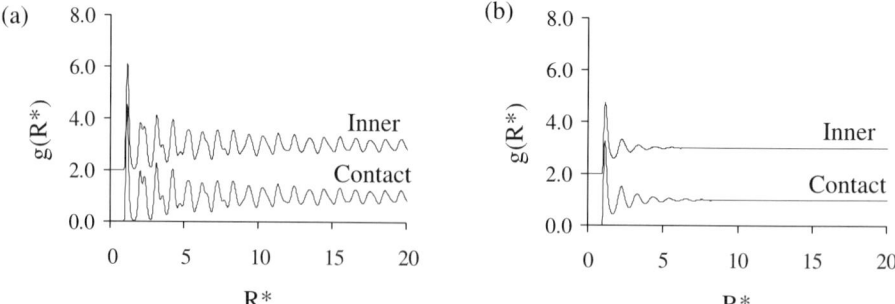

Fig. 2. In-plane 2D pair correlation functions g(R) for the contact and inner layers of Ar/Kr mixtures in a slit pore $H = 1.44$ nm at (a) $T = 129$ K and (b) $T = 144$ K. For the sake of clarity the g(R) function for the inner layer has been shifted by +2. R* is the reduced distance with respect to $\sigma_{Ar}$.

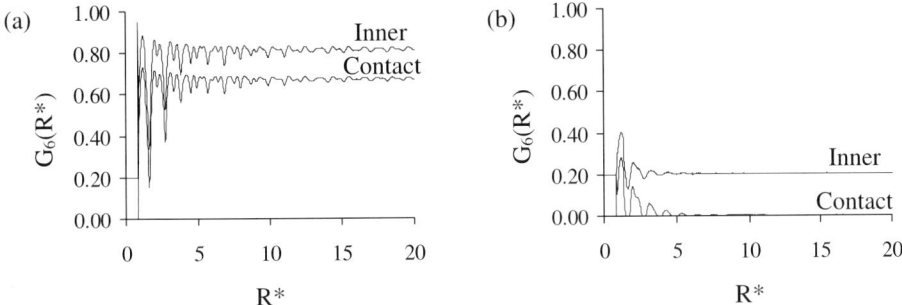

Fig. 3. In-plane 2D pair correlation functions $G_6(R)$ for the contact and inner layers of Ar/Kr mixtures in a slit pore $H = 1.44$ nm at (a) $T = 129$ K and (b) $T = 144$ K. For the sake of clarity the $G_6(R)$ function for the inner layer has been shifted by +0.2. $R^*$ is the reduced distance with respect to $\sigma_{Ar}$.

The phase diagram $(T, x_{Kr})$ for Ar/Kr mixtures confined in the 1.44 nm slit pore is shown in Fig. 4(a). The solid/liquid coexistence conditions were determined for different compositions following the analysis described above. Results are compared with the phase diagram obtained for bulk Lennard-Jones Ar/Kr mixtures [11]. The phase diagram for the confined mixture has the same shape as the bulk, but the solid/liquid coexistence lines are shifted to larger temperatures. In agreement with previous works on confined fluids [1,8], this increase in the freezing temperature can be explained by the fact that the ratio of the wall/fluid to the fluid/fluid interactions is larger than 1 for Ar and Kr ($\alpha_{Ar} = 1.93$, $\alpha_{Kr} = 1.78$ [5]). We also report in Fig. 4(a) the phase diagram for Ar/Kr mixtures in the 1.44 nm slit pore after an arbitrary reduction of the Ar/wall interaction so that $\alpha_{Ar}$ is lower than 1 ($\alpha_{Ar} = 0.80$) [6]. In this case, we found that the freezing temperature of the confined system is larger than the bulk for mixtures rich in Kr. In contrast, the freezing temperature for the confined mixture is lower than the bulk for mixtures rich in Ar. Configurations of the contact layer of the confined mixture are shown in Fig. 4(b-c). In each case, the same structure was observed for the inner layer. For all mixtures, the crystal layers have a triangular symmetry, which corresponds to a centered hexagonal structure; this structure is usually observed for strongly attractive pores ($\alpha > 1$) as it corresponds to a dense close-packing of the atoms at the pore wall. In contrast, a square crystal is found for pure Ar in the case of the mixture with $\alpha_{Ar} = 0.80$; such a symmetry is observed for pore sizes where a change in the number of layers occurs [1].

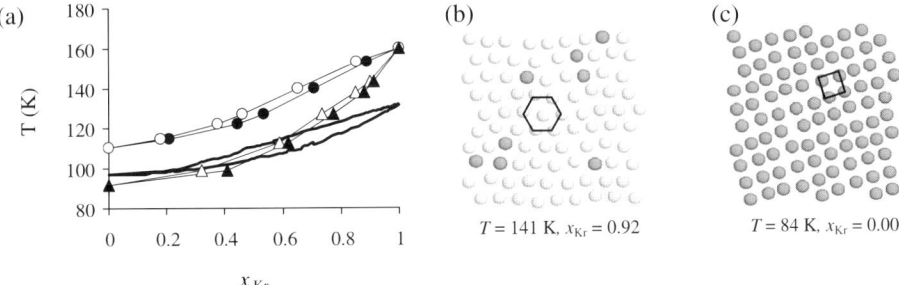

Fig. 4. (a) (○●): $(T, x_{Kr})$ phase diagram of Ar/Kr mixtures in a pore $H = 1.44$ nm. Open and closed symbols are the liquid and crystal coexistence lines, respectively. (△▲) same as circles but after reduction of the interaction parameter $\varepsilon_{Ar/W}$ from 62.9 K down to 26.1 K ($\alpha_{Ar} = 0.80$). The thick lines are the phase diagram for the bulk (from [11]). (b),(c) Triangular and square crystals of Ar/Kr mixture in the $H = 1.44$ nm pore (with $\alpha_{Ar} = 0.80$). Gray and white spheres are Ar and Kr atoms, respectively.

### 3.2. Azeotropic mixtures

We now present experiments and molecular simulations for azeotropic mixtures confined in slit pores. Measurements of the dielectric constant $\varepsilon' = C/C_0$ allow the investigation of melting phenomena, as the polarizability of the liquid and solid phases are significantly different [9]. The capacitance curve, $C$, as a function of the temperature, $T$, is shown in Fig. 5(a) for a $CCl_4/C_6H_{12}$ mixture with $x_{C6H12} = 0.4$ confined at $P = 1$ atm in ACF with a pore width $H = 1.2$ nm. Melting for both the bulk and the confined mixtures is observed as the sample consists of a suspension of filled ACF in the bulk mixture. A first large increase in the capacitance is observed at $T = -25.0$ $^0$C; this corresponds to the melting temperature of the bulk mixture for $x_{C6H12} = 0.4$. Such an increase at $T = -25.0$ $^0$C indicates that the bulk crystal mixture starts melting i.e. the system reaches the crystal coexistence line. At T = -23.3 $^0$C, the transformation of the bulk crystal into the liquid mixture is complete, i.e. the system reaches the liquid coexistence line, and the capacitance of the system decreases as expected for a liquid phase [9]. This analysis provides both the liquid and crystal coexistence temperatures for bulk $CCl_4/C_6H_{12}$ mixture having a molar composition $x_{C6H12} = 0.4$. At a much higher temperature, T = 12 $^0$C, a second increase in the capacitance of the system is observed. The sudden change at this temperature, which does not correspond to any known transition temperature for a bulk $CCl_4/C_6H_{12}$ mixture with $x_{C6H12} = 0.4$, is believed to represent the melting of the material confined within the ACF. As in the case of the bulk mixture, the crystal and liquid coexistence temperatures for this molar composition were estimated from the temperatures where the capacitance starts increasing and where the capacitance reaches a maximum, respectively.

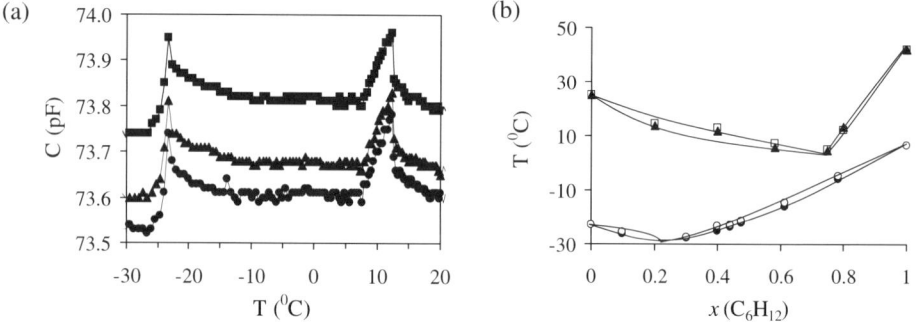

Fig. 5. (a) $C$ versus $T$ curve at $P = 1$ atm for a $CCl_4/C_6H_{12}$ mixture with $x_{C6H12} = 0.4$ confined in ACF ($H = 1.2$ nm): (●) $\omega = 100$ kHz, (▲) $\omega = 600$ kHz, and (■) $\omega = 1$ MHz. The signal is for both the bulk and confined mixtures as the sample is a suspension of filled ACF in the bulk mixture. (b) ($T$, $x_{C6H12}$) phase diagram at $P = 1$ atm for $CCl_4/C_6H_{12}$ mixture: (○●) bulk, (□▲) in ACF with a pore width $H = 1.2$ nm. Open and closed symbols are the liquid and crystal coexistence lines, respectively.

The experimental process described above was repeated for different mole fractions in order to obtain the solid/liquid phase diagram for confined $CCl_4/C_6H_{12}$ mixtures (see Fig. 5(b)). The phase diagrams for the confined mixture is of the same type as that for the bulk, i.e. azeotropic, but the solid/liquid coexistence lines are located at higher temperature. The shift in the coexistence conditions is consistent with previous works, which showed that the freezing temperature for systems confined in strongly attractive pores is increased compared to the bulk [1,8]. The relative increase in the freezing temperature is 1.19 for pure $CCl_4$ and 1.12 for pure $C_6H_{12}$. The larger shift for $CCl_4$ can be explained by the larger $\alpha$ value for this fluid ($\alpha_{CCl4}$ = 1.93, $\alpha_{CH4}$ = 1.76 [7]). We note that the increase in freezing temperature as a function of $x_{C6H12}$ cannot be discussed quantitatively as only the global mole fraction of the sample (bulk and confined) is known. For the same reason, the location of the azeotrope for the confined mixture $x_{0,C6H12} \sim 0.75$, which is larger than the bulk azeotrope $x_{0,C6H12} \sim 0.23$, cannot be discussed.

In order to complement our experimental investigation, we performed GCMC simulations for Ar/CH$_4$ azeotropic mixtures confined in a slit pore of a width H = 1.02 nm. Such a size was chosen because it corresponds to a similar reduced pore size to that used in the experiments, i.e. $H^* \sim 3$. Density profiles of the confined mixture are shown in Fig. 6(a) for two different temperatures; the confined mixture has a layered structure, composed of two symmetrical layers. In order to obtain the phase diagram for the confined mixture, we performed simulations for different compositions of the bulk mixture. For each run, the liquid/solid coexistence was determined following the analysis described in section 3.1; freezing was monitored through the changes of $x_{CH4}$ and $\Phi_6$ with temperature, and the structure of the confined mixture was studied using both in-plane 2D g(r) and G$_6$(r) functions. The solid/liquid phase diagram for Ar/CH$_4$ mixtures in the H = 1.02 nm pore is shown in Fig. 6(b). The confined mixture has the same type of phase diagram as the bulk mixture, but the liquid and crystal coexistence lines are located at higher temperatures. Again, the larger freezing temperature for the confined mixture compared with the bulk can be explained by the fact that both $\alpha_{Ar}$ and $\alpha_{CH4}$ are larger than 1 ($\alpha_{Ar}$ = 2.14, $\alpha_{CH4}$ = 2.16 [7]). The larger increases in the freezing temperatures in the case of the simulations, $T_f^*/ T_f^{*,bulk} \sim 1.5 - 1.6$, compared to the experiments, $T_f^*/ T_f^{*,bulk} \sim 1.1 - 1.2$, can be explained by the larger values of the $\alpha$ parameters for the mixtures considered in the simulations [7]. The azeotrope for the confined Ar/CH$_4$ mixture is located at $x_{CH4}$ = 0.20 and T = 120 K; the crystal for $x_{CH4}$ < 0.20 is richer in Ar than the liquid, while the crystal for $x_{CH4}$ > 0.20 is richer in CH$_4$ than the liquid. This situation is similar to that observed for the bulk where freezing involves an increase (decrease) in $x_{CH4}$ for mole fractions above (below) the azeotrope. The simulations for Ar/CH$_4$ mixtures confined in slit pores are in general agreement with the experiments for CCl$_4$/C$_6$H$_{12}$ in ACF; the phase diagram for the confined mixture is of the same type as that for the bulk, but the liquid/crystal coexistence is shifted to higher temperatures.

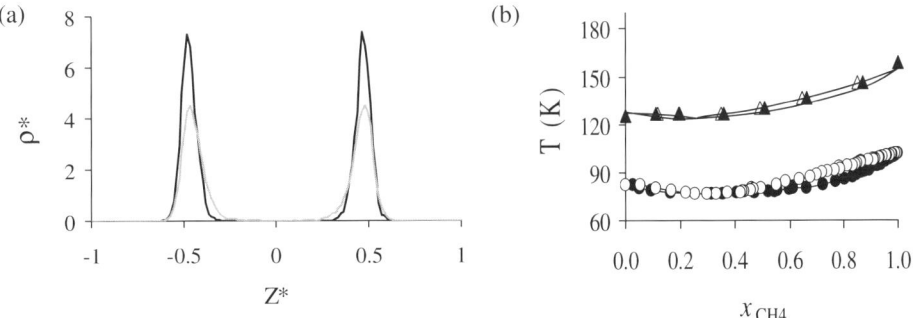

Fig. 6. (a) Density profiles in reduced units $\rho^* = \rho\sigma_{Ar}^3$ of Ar/CH$_4$ mixture with $x_{CH4} \sim 0.9$ confined in a slit pore H = 1.02 nm: T = 137 K (black line) and T = 163 K (grey line). Z$^*$ is the distance from the pore center in reduced units with respect to $\sigma_{Ar}$. (b) (T, $x_{CH4}$) phase diagram at P = 1 atm for Ar/CH$_4$ mixtures: (○●) bulk, (△▲) in a slit pore H = 1.02 nm. Open and closed symbols denote the liquid and solid coexistence lines, respectively. The lines are provided as a guide to the eye.

## 4. CONCLUSIONS

Experiments and simulations of the freezing of mixtures confined in nanopores are reported. Mixtures having either a solid solution or an azeotropic phase diagram were considered. In both cases, the phase diagram of the confined mixture is of the same type as that for the bulk. Depending on the ratio $\alpha$ of the wall/fluid to the fluid/fluid interactions, the freezing temperature of the confined mixture is increased ($\alpha > 1$) or decreased ($\alpha < 1$) compared to the bulk. Further studies are needed to corroborate our results. Differential scanning calorimetry should confirm the transition temperatures found in this study, while X-ray diffraction would allow us to determine the structure of the confined phases. Although the use of the parallel tempering technique greatly reduces the risk of being trapped in a metastable state, we plan to combine our simulations with free energy calculations to confirm our findings.

This work was supported by grants from the Petroleum Research Fund (ACS), KBN (PO3B 0114) and NATO (PST.CLG.978802). This research used supercomputing time from HPC-NCSU, SDSC (NSF/MRAC CHE050047S) and NERSC (DOE DE-FGO2-98ER14847).

## REFERENCES

[1] For recent reviews, see L. D. Gelb, K. E. Gubbins, R. Radhakrishnan and M. Sliwinska-Bartkowiak, Rep. Prog. Phys., 62 (1999) 1573; H. K. Christenson, J. Phys.: Condens. Matter, 13, (2001) 95; C. Alba-Simionesco, B. Coasne, G. Dosseh, G. Dudziak, K. E. Gubbins, R. Radhakrishnan and M. Sliwinska-Bartkowiak, J. Phys.: Condens. Matter, to be published (2005).
[2] R. R. Meyer, J. Sloan, R. E. Dunin-Borkowski, A. I. Kirkland, M. C. Novotny, S. R. Bailey, J. L. Hutchison and M. L. H. Green, Science 289 (2000) 1324.
[3] M. Wilson, J. Chem. Phys., 116 (2002) 3027.
[4] B. Cui, B. Lin and S. Rice, J. Chem. Phys., 119 (2003) 2386.
[5] B. Coasne, J. Czwartos, K. E. Gubbins, F. R. Hung and M. Sliwinska-Bartkowiak, Mol. Phys., 102 (2004) 2149.
[6] B. Coasne, J. Czwartos, K. E. Gubbins, F. R. Hung and M. Sliwinska-Bartkowiak, Adsorption, 11 (2005) 301.
[7] J. Czwartos, B. Coasne, F. R. Hung, K. E. Gubbins and M. Sliwinska-Bartkowiak, Mol. Phys., in press (2005).
[8] R. Radhakrishnan, K. E. Gubbins and M. Sliwinska-Bartkowiak, J. Chem. Phys., 116 (2002) 1147.
[9] M. Sliwinska-Bartkowiak, J. Gras, R. Sikorski, R. Radhakrishnan, L. D. Gelb and K. E. Gubbins, Langmuir, 15 (1999) 6060; A. Chelkowski, Dielectric Physics, Elsevier, New York, 1980.
[10] M. Mehta and D. A. Kofke, Chem. Eng. Sci., 49 (1994) 2633.
[11] M. R. Hitchcock and C. K. Hall, J. Chem. Phys., 110 (1999) 11433.
[12] D. Frenkel and B. Smit, Understanding Molecular Simulation, Academic Press, New York, 2002.
[13] Q. Yan and J. J. de Pablo, J. Chem. Phys., 111 (1999) 9509.
[14] J. K. Johnson, J. A. Zollweg and K. E. Gubbins, Mol. Phys., 78 (1993) 591.
[15] J. S. Rowlinson, Liquids and Liquid Mixtures, Butterworth Scientific, London, 1982.
[16] W. A. Steele, Surf. Sci., 36 (1973) 317.
[17] B. I. Halperin D. R. and Nelson, Phys. Rev. Lett., 41 (1978) 121; D. R. Nelson and B. I. Halperin, Phys. Rev. B, 19 (1979) 2457; K. J. Strandburg, Rev. Mod. Phys., 60 (1988) 161.
[18] R. F. Cracknell, D. Nicholson and N. Quirke, Mol. Phys., 80 (1993) 885.
[19] Z. Tan and K. E. Gubbins, J. Phys. Chem., 96 (1992) 845.
[20] F. R. Hung, B. Coasne, K. E. Gubbins, E. E. Santiso, F. R. Siperstein and M. Sliwinska-Bartkowiak, J. Chem. Phys., 122 (2005) 144706.
[21] R. Radhakrishnan, K. E. Gubbins and M. Sliwinska-Bartkowiak, Phys. Rev. Lett., 89 (2002) 076101. R. Radhakrishnan, K. E. Gubbins and M. Sliwinska-Bartkowiak, Phys. Rev. B (2005) submitted.

Studies in Surface Science and Catalysis 160
*P.L. Llewellyn, F. Rodriquez-Reinoso, J. Rouqerol and N. Seaton (Editors)*
© 2007 Elsevier B.V. All rights reserved

# New equipment for characterization of nanofiltration membranes

## V. Milisic[a], M. Mietton-Peuchot[a] and T. Courtois[b]

[a]Laboratoire de Génie des Procédés et Environnement, Université Victor Segalen Bordeaux 2, 351 cours de la Libération, 33405 Talence cedex, France

[b]GEPS Sarl, Parc scientifique Unitec 1, 2 Allée Georges Brus, 33600 Pessac, France

## 1. INTRODUCTION

Nanofiltration is a membrane technique enabling the separation at ion's and organic monomer's scale. The high separation performances of the technique are due to the complex chemical, electrical and physical interactions between solvent, solutes and membrane material. It allows to foreseen an important industrial developments on condition that those mechanisms are controlled and that the tools for selection of equipment and operating conditions are available.

The researchers of Laboratory for Process Engineering and Environment initiated, since 1995 a huge research program on nanofiltration gathering more that ten academic and industrial partners from different countries [1, 2, 3].

Finally, the results of those studies are materialized by:
- numerical code *NanoCEP* : **NanoC**(himie)**E**(lectrique)**P**(hysique) that enables to determine the characteristics of nanofiltration membrane the most adapted for a given separation, as well as operating conditions [4,5,6, 7],
- database containing the characteristics of mostly all commercially available nanofiltration membranes necessary for numerical code input,
- apparatuses able to characterize the membranes, i.e. to measure those data, *S.P.I.R.E.* (Streaming Potential Isoelectric point Retention rate Electrometer) and *Fluid-Fluid Porometer* [8].

The commercial versions of those instruments are developed by GEPS Company, already well introduced in porous media characterisation and separation studies. The last activity will be significantly supported by the extension of numerical code to ultra and microfiltration applications, at present in progress.

Since *NanoCEP* code simulates as well influence of physical parameters (dimensions of solutes to be separated, dimensions of pores and concentration/polarization layer) as electrostatic charges of both, solutes and membrane, this adaptation is relatively simple to do. In the present article, after a presentation of two apparatuses, some results of measurements will be shown and discussed.

## 2. MEASURING PROCEDURE AND EXAMPLES OF RESULTS

The nanofiltration membranes, organic or ceramic, are complex media composed of several layers, often of different nature, each made for a specific purpose: higher mechanical resistance, better permeate draining, more efficient separation [9,10, 11]. The separation efficiency is determined by thin active layer at membrane surface [12, 13].

### 2.1. Streaming Potential Isoelectric point Retention rate Electrometer (S.P.I.R.E)

The apparatus schematically presented in Fig. 1 is designed for two complementary measurements:

- streaming potential – characterizing the nature and the density of electric charges.
- retention rate – quantifying the selectivity of a membrane towards given solutes at different pH.

The first measurement is based on electrokinetic effect due to the movement of a continuous phase over a solid surface that produces an electrical potential. In function of electrodes positions, it could be measured along the membrane or across the membrane, too [13, 14, 15].

The retention rate $(R=1-Cp/Cr)$ is measured according to the classical procedure by measuring concentrations in the permeate $(Cp)$ and the retentate $(Cr)$ [16]. The test is entirely automatic and could be carried out with multi-component solution as feed at different pH and transmembrane pressures.

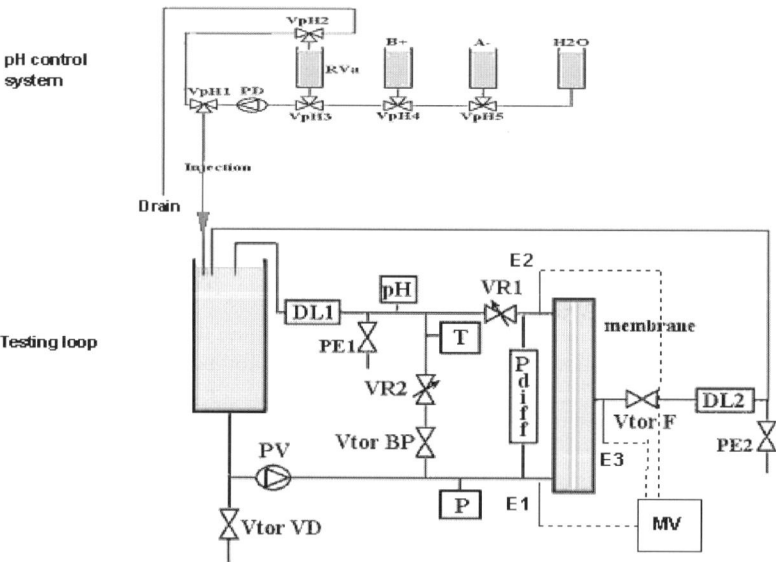

Fig. 1. S.P.I.R.E: PV – pump, DL - flow-meters, VR – electric valves for regulation of pressure, Vtor – electric valves, PE – sampling valves, pH - pH-meter, P- relative pressure transducer, Pdiff – differential pressure transducer, E – platinum or Ag/AgCl electrodes, MV - micro-voltmeter

The Fig. 1 shows the scheme of S.P.I.R.E. In the case of cross membrane streaming potential determination, the electrical potential is measured between the retentate and permeate membrane sides and is named "filtration potential". Along the membrane, it is measured, without permeation, between the inlet and the outlet of the module.

In practice, the cross membrane filtration potential is modified by a very brief jump of transmembrane pressure and, in consequence, of the permeate rate, which creates a jump of filtration potential, too (Fig. 2). For a given pH, that procedure is repeated with different pressure jump values. The streaming potential is obtained by dividing filtration potential jump value by transmembrane pressure jump value. The same test done at different pH, enables to determine the isoelectric point of a membrane (Fig. 3) and to have an idea on electrostatic charges density and sign.

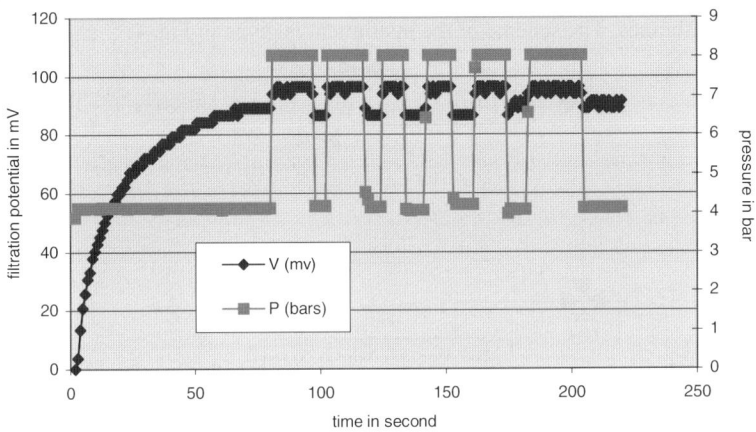

Fig. 2. Row results of streaming potential measurements

Parc scientifique Unitec 1, 2 Allée Georges Brus, 33600 Pessac, France

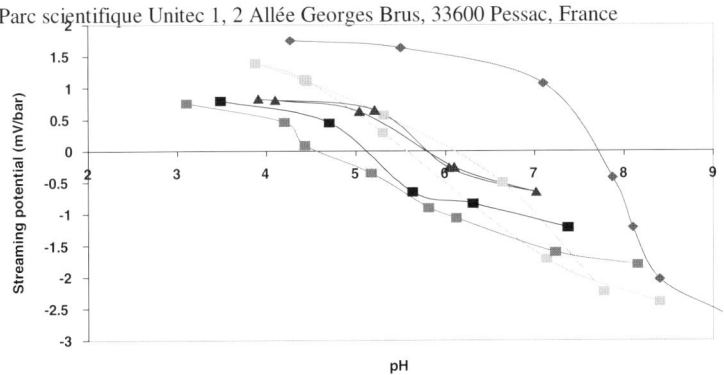

Fig. 3: Streaming potential vs. pH for

## 2.2. Fluid-fluid porometer

The measurements carried out by using that apparatus enable to obtain following information:

- bubble point up to 20 bar,
- drop point up to 20 bar,
- gas diffusion,
- gas permeability,
- solvent permeability,
- contact angle,
- interfacial tension between two immiscible solvents and
- pore diameter distribution from 2 nm to 50 μm.

All of those are very useful as well in R&D as in quality control. The last data are very precious for better understanding of transport mechanisms inside the active layer of nanofiltration membranes. The scheme of the instrument is given in Fig. 4.

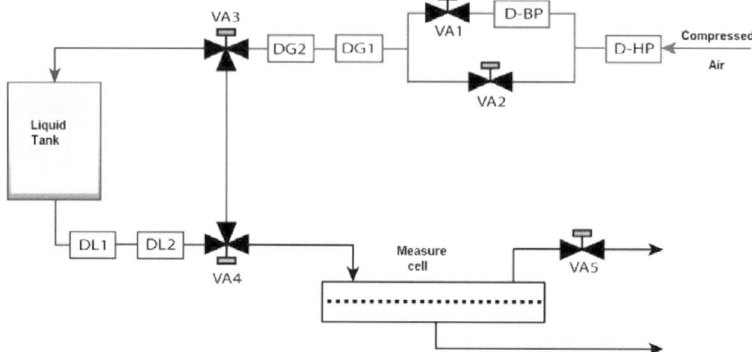

Fig. 4. Fluid-fluid porometer: D-HP – high pressure regulator, D-BP – low pressure regulator DL – solvent flow-meters, DG – gas flow-meters, VA – electric valves

The principle of pore size distribution measurement is shown in the Fig. 5. At first, the porous matrix, in occurrence nanofiltration membrane is filled with a specific "wetting fluid". The latter is then pushed out, by successively increasing pressure, by another immiscible fluid (gas or liquid).

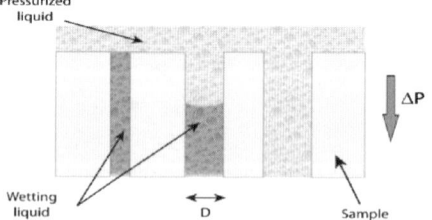

Fig. 5. Principle of measurement – pressurized immiscible fluid is liquid

The "wetting fluid" is removed from pores according to the capillary law [15, 16]:

$$\Delta P = \frac{4\sigma \cos\theta}{D} \qquad (1)$$

where $\sigma$ is liquid surface tension and $\theta$ contact angle between liquid and membrane materiel [17, 18].

The large pores are firstly concerned, then with pressure increase, the smaller ones. For each pressure, the permeate rate ($Qp$) is measured as shown in Fig. 6 (porometry curve). Once the "wetting liquid" replaced, the result is validated by a second pressure test (permeation curve) where the pressurized fluid is filtered across the membrane. The permeability curve ($Qp/\Delta P$) is then derived in function of pressure i.e. pore diameter (Eq. 1) [19,20]. An example of the results so obtained is given in Fig. 7.

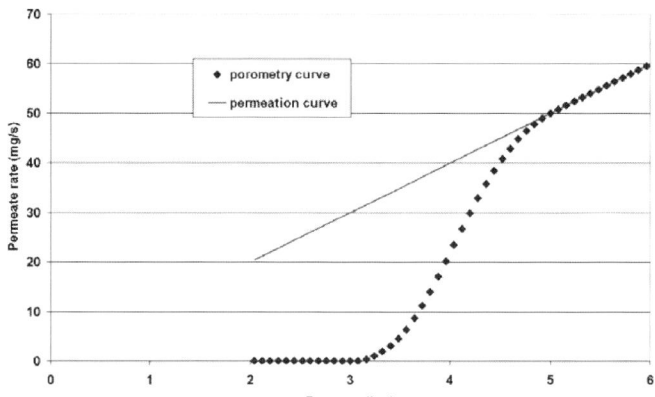

Fig. 6. Measurements of permeate rate v.s. pressure

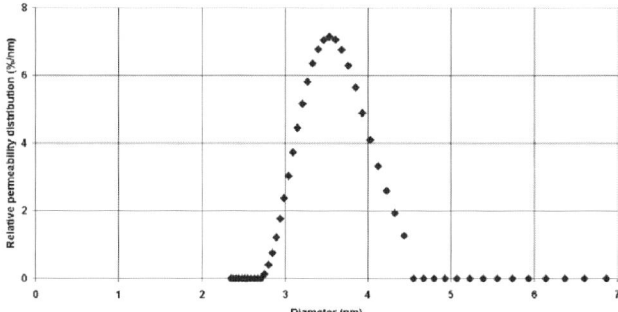

Fig. 7. Pore size distribution curve

## 3. CONCLUSION

In the frame of integrated approach to the nanofiltration, an extensive characterisation of membranes is essential for adequate choice of membrane. The results of multi-annual fundamental research enabled to define and design new equipment for membrane characterization.

After validation at laboratory and pilot scale the apparatuses are commercialized.

In addition, the data obtained by using those characterisation methods constitute the base for conception and optimisation of nanofiltration units by means of numerical code *NanoCEP*.

The same type of developments is in current for ultra and microfiltration applications: membrane characterisation device coupled with data-base and numerical code.

## REFERENCES

[1] M. Boughenou, V. Milisic et M. Mietton-Peuchot, 5ième Congrès Français de Génie des Procédés, Lyon, 9(1995), 339.
[2] M. Mietton-Peuchot, C. Peuchot, T. Courtois, L'Eau l'Industrie les Nuisances, 201 (1997), 26.
[3] M. Mietton-Peuchot, T. Courtois, Euromembrane'97, University of Twente, Netherlands, (1997), 285.
[4] V. Milisic, M. Hamachi, 3$^{rd}$ Nanofiltration and Applications Workshop, Lappeenranta, Finland, (2001), 75.
[5] S. Chevalier, Modélisation mathématique des mécanismes de séparation en nanofiltration. Ph.D, Université Bordeaux I, (1999).
[6] V. Milisic, S.Chevalier, Euromembrane '99, Leuven, Belgium, (1999), 397.
[7] S. Chevalier, V. Milisic, B. Chretien and C. Baumgartner, 8$^{th}$ World filtration congress, Brighton, GB, (2000), 661.
[8] T. Courtois, 1$^{st}$ Nanofiltration & Applications Workshop, Shawinigan, Canada, (1997), 60.
[9] M. Mietton-Peuchot, 3$^{rd}$ Nanofiltration and Applications Workshop, Lappeenranta, Finland, (2001), 82.
[10] J. Hernandez, I. Calvo, P. Pradanos and L. Palacio, T.S.Sørensen (eds), Surface Chemistry and Electrochemistry of Membranes, Stockholm, 1999.
[11] Y. Lee, J. Jeong, I.J. Young and W.H. Lee, J. Membrane Sci, 130 (1997), 149.
[12] Z. BRAHIMI, Incidence de la matière organique sur le transfert de sels simples et métalliques en nanofiltration, Ph.D.,INP de Toulouse, (2000).
[13] M. Boughenou. Contribution à la compréhension des phénomènes de transfert des solutés en nanofiltration : caractérisation des membranes et application aux composés toxiques, Ph.D., ENSICG Toulouse, (1997).
[14] W. R. Bowen et H. Mukhtar , Journal of Membrane Science, 112 (1996), 263.
[15] M. Nyström, M. Lindström et E. Matthiasson, J. Colloids and Surfaces, 36 (1989), 297.
[16] J. M. M. Peeters . Characterization of nanofiltration membranes, Ph.D., University of Twente, 1997.
[17] N. Le Bolay, M.Mietton-Peuchot, A. Ricard, 3ème Congrès Français de Génie des Procédés, Compiègne, (1991), 127.
[18] M.G. Liu, M. M.Mietton-Peuchot, R. Ben Aim, ICOM 90, Chicago, (1990), 234.
[19] M.G.Liu, M. Mietton-Peuchot, R. Ben Aim, I.C.I.M., Montpellier, (1991), 421.
[20] M.G.Liu, M.Mietton-Peuchot, R. Ben Aim, 3ème Congrès de Génie des Procédés, Compiègne, (1991), 186.

Studies in Surface Science and Catalysis 160
P.L. Llewellyn, F. Rodriquez-Reinoso, J. Rouqerol and N. Seaton (Editors)
© 2007 Elsevier B.V. All rights reserved

# The porous structure of biodegradable scaffolds obtained with supercritical $CO_2$ as foaming agent

S. Blacher[a], C. Calberg[b], G. Kerckhofs[c], A. Léonard[a], M. Wevers[c], R. Jérôme[b], J.-P. Pirard[a]

[a]Laboratoire de Génie Chimique, [b]CERM, University of Liège, B-4000 Liège, Belgium.

[c] Metallurgy and Materials Engineering, Katholieke Universiteit Leuven, B-3001 Leuven, Belgium

## ABSTRACT

Poly(ε-caprolactone) foams were prepared, via a batch process, by using supercritical $CO_2$ as foaming agent. Their porous structure was characterized through helium and mercury pycnometry, scanning electron microscopy (SEM) and X-ray microtomography observations coupled with image analysis. The pore size distributions obtained by these two latter techniques show that the pore structure is more homogeneous when the foaming process is performed under a high $CO_2$ saturation pressure (higher than 250 bars).

## 1. INTRODUCTION

Porous polymeric materials are very attractive because they show high flexibility of generating morphologies to meet specific applications. In particular, porous biodegradable polymer matrices are widely used in biomedical applications such as tissue engineering and guided tissue regeneration [1]. These matrices provide a temporary support for cell seeding and growth. They are also used to deliver growth factors to the growing cells. Among biodegradable polymers, poly(ε-caprolactone) (PCL) meets all the chemical requirements for tissue engineering [2]. The techniques reported for generating porous PCL foams include freeze-drying and classical foaming processes. However, the main drawback of these techniques is that they respectively require organic solvents and chemical blowing agents in their fabrication process. Residues of these chemicals left in the polymer after expansion may be harmful to the transplanted cells. Therefore, methods using supercritical $CO_2$ (sc $CO_2$) as foaming agent are often preferred, because $CO_2$ is known as a chemically inert and non-toxic gas [3]. Moreover, as a result of their compressed state, supercritical fluids (SCF) are highly suited to the generation of polymer foams. Solvent free approaches have thus been developed wherein a polymer is saturated with sc $CO_2$ at high pressure, followed by rapid depressurization at constant temperature. These methods take advantage of the large depression of the glass transition temperature found for many polymers in the presence of sc $CO_2$, which means that amorphous polymers may be kept in the viscous state at relatively low temperature. In the present work, we explore the preparation and the characterization of neat PCL foams prepared by using sc $CO_2$ as blowing agent. The same batch process was used in other works that allows the comparison of the results [4, 5, 6]. The obtained foams are made of an isotropic network of pores with a size distribution that depends on the experimental parameters, such as the pressure and temperature of saturation, the depressurization profile and the composition of the polymer formulation.

The texture characterization of these highly porous materials is a major issue in relation with their potential applications. Mercury porosimetry is traditionally used to characterize this kind of materials. However, the use of this technique is not appropriated in the present case. Indeed, on one hand, mercury porosimetry is limited to a maximum pore size of 75 μm and, on the other hand, depending of the parameters used in the foaming process, the porous structure can be so open that mercury would not intrude the sample but would flow through the porous structure. Moreover, it has been shown recently that anisotropic poly(L-lactide-co-ε-caprolactone) foams generated by freeze-drying shrink under the high pressure required for Hg intrusion [7,8]. As shown recently [9] X-ray microtomography constitutes a promising characterization technique for highly porous materials

In order to determine the porosity and the pore size distribution of the PCL foams, image analysis was performed on 2D images obtained by scanning electron microscopy (SEM) at high magnification and 3D images obtained by X-ray microtomography at low magnification. It is shown that the pore structure of the studied foams strongly depends on the $CO_2$ saturation pressure used for the foaming process.

## 2. MATERIAL AND METHODS

PCL used in this study was supplied by SOLVAY (CAPA 650, Mw ~ 50.000). Before foaming, PCL is moulded into sheets of 25 mm diameter and 3 mm thickness at 120°C during 10 min. Basically, foaming was processed in three steps. The samples were first saturated with $CO_2$ at high pressure and low temperature (40°C) and then kept under these conditions during 2 hours. Subsequently, $CO_2$ was slowly released from the autoclave within 12 min. at ambient temperature. The $CO_2$ pressure of saturation was changed to produce the following four different foams: A1(150 bars), A2 (200 bars), A3 (250 bars) and A4 (300 bars).

The apparent specific volume, $V_s$, which is the reverse of the bulk density, was measured by mercury pycnometry. After placing a foam sample of weight $w_s$ in a pycnometer, it is completely filled with mercury and weighed ($w_{sl}$). $V_s$ is calculated according to the expression $V_s = (w_l - w_{sl} + w_s)/w_s \rho_{Hg}$ where $w_l$ is the weight of the pycnometer filled with mercury and $\rho_{Hg}$ is the density of mercury (13.5 g cm$^{-3}$). The specific volume of the solid foam skeleton, $V_{sk}$, is measured by helium pycnometry (AccuPyc 1330, Micrometrics). The foam porosity is then calculated as $\varepsilon = (V_s - V_{sk})/V_s$ and the specific pore volume, $\rho_p$, is then calculated as the difference between $V_s$ and $V_{sk}$.

In order to observe their internal pore structure, foams are cut with a razor blade. The sections are mounted on an aluminum stub with a carbon adhesive and then coated with platinum (120 sec, Argon atmosphere). Scanning electron microscopy (SEM) is performed with a Jeol JSM-840A SEM operating at an accelerating voltage of 20kV.

Table1
Porosity data: Specific volume of the sample ($V_s$), specific pore volume ($\rho_p$), and foam porosity ($\varepsilon$), measured by pycnometry. $D_{eq}$: Equivalent average pore diameter determined from SEM image analysis ($D_{eq}$). Foam porosity determined by X-ray microtomography ($\delta$).

| Sample | $CO_2$ pressure (bar) | $V_s$ (cm³/g) | $\rho_p$ (cm³/g) | $\varepsilon$ | $\delta$ | $D_{eq}$ (mm) |
|--------|------------------------|----------------|-------------------|----------------|-----------|----------------|
| A1 | 150 | 2.47 | 1.60 | 0.65 | 0.62 | 0.18±0.08 |
| A2 | 200 | 2.87 | 2.00 | 0.70 | 0.61 | 0.21±0.10 |
| A3 | 250 | 2.83 | 1.97 | 0.70 | 0.60 | 0.15±0.07 |
| A4 | 300 | 2.96 | 2.10 | 0.71 | 0.60 | 0.15±0.09 |

The SEM micrographs are digitized on a matrix of 1024x1024 pixels with 256 gray levels. Image analysis is performed using the software 'Aphelion3.6' from Aphelion. Five images of different areas of the same foam are analyzed.

Microtomography was performed with the Philips HOMX 16 with AEA Tomohawk system. Detailed information about this device can be found in [10] . The X-ray source operates at 72 kV and 0.23 mA. The detector is a 1024x1024 10-bit range. The spatial resolution (voxel size) for the different samples is: 0.02x0.02x0.02mm for A1, 0.02x0.02x0.02mm for A2, 0.013x0.013x0.013mm for A3 and 0.015x0.015x0.015mm for A4. 3D images are obtained by stacking a series of 2D binary cross section images using ANT (Skycan) software.

## 3. RESULTS

### 3.1. SEM observations and image analysis
The influence of the $CO_2$ saturation pressure on the pore structure of PCL foams is shown in Figures 1a-d. A1 foam is composed of two populations of pores: ultramacropores with diameters larger than 1mm and macropores with diameters in the range 50-500 μm (Figure 1a). For the A2 foam, only macropores are present whose diameters are spread over a larger range (Fig. 1b). Finally for samples A3 and A4 (Fig. 1c,d ), in which the foaming process has been performed under the highest $CO_2$ saturation pressures (250 and 300 bars, respectively) the pore structure is rather homogeneous.

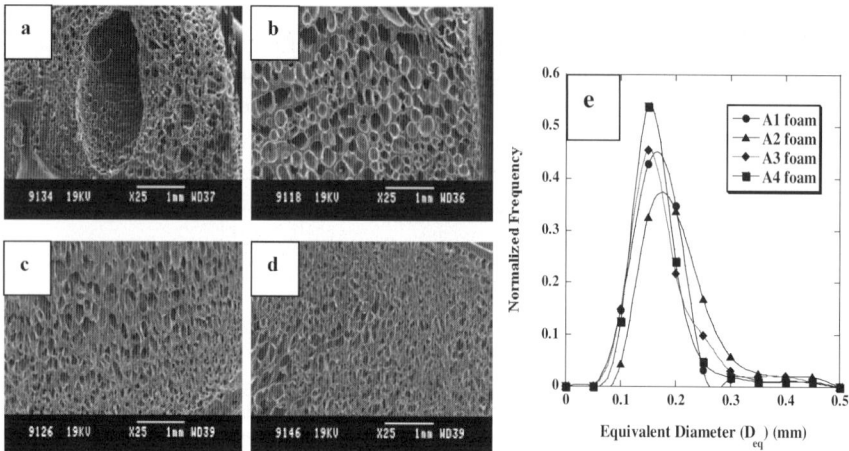

Figure 1. SEM micrographs for (a) A1, (b) A2, (c) A3 and (d) A4 foams. (e) Pores size distribution obtained by SEM image analysis

In order to quantify these observations, the pore size distribution was calculated using image analysis techniques. The used image analysis algorithms are described elsewhere [8] and enable the statistical distribution of the equivalent diameter of the pores $D_{eq}$ to be assessed. For the sake of comparison, only macropores smaller than 500 μm were evaluated, i.e. ultramacropores observed only in the A1 foam were not considered.

For all samples, a peak is observed in the distribution at $D_{eq} \approx 0.18$mm but the shape of the distribution is different according to the foaming conditions. The $D_{eq}$ distribution is narrow for foam A1. For foam A2, the distribution spreads towards large values, which could result from a lowering of the size of the ultramacropores observed in A1. Finally, the distribution becomes narrow again for samples A3 and A4 (Fig. 1e). This trend is reflected in the evolution of the mean equivalent diameters and of the standard errors, as presented in Table 1.

### 3.2. X-ray microtomography and image analysis

The 3-dimensional structure of foams was investigated by X-ray microtomography. Typical cross sections of samples A1 and A4 are shown in Figures 2a and b. As the cross-section images present a poor contrast, "tophat" and "bottom-hat" filters [11] were applied together to enhance the contrast of images. The used procedure consists in adding the original image to the tophat-filtered image, and then subtracting the bottom-hat-filtered image. Then, the resulting image is binarised using Otsu's method [12] (Fig. 2c and d).

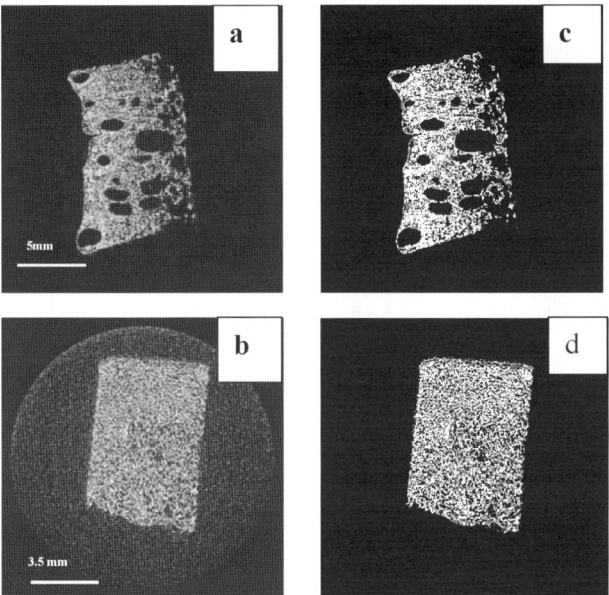

Figure 2. Tomographic cross section images for (a) A1 and (b) A4 foams and their corresponding binary images (c) and (d).

With this method, the threshold level is chosen automatically so as to maximize the interclass variance and to minimize the intraclass variance of the thresholded black (pores) and white (polymer) pixels. Finally, from sets of 100 binary cross sections for each foam, 3D binary images were reconstructed (Fig. 3a-b).

From the 3D processed binary images, the porosity $\delta$, defined as the fraction of voxels of the objects that belong to its pores was measured. It was found that $\delta$ is almost constant for all samples (Table 1).

The observation of cross-sections and 3D foams images indicates that the polymer matrix A1 (Figures 2a and 3a) presents a small pores structure, in which a small number of large pores (diameter >1mm) are dispersed. For foam A2, the largest pores disappeared and the pore structure seems denser. Finally, A3 and A4 (Fig. 2b and 3b) present a compact structure.

As the 3D images of foams present a continuous and rather disorder pore structure (Fig. 3) in which it is not possible to assign to each pore a precise geometry, a standard granulometry measure cannot be applied. Then, to quantify the larger pore sizes, we calculated the opening size distribution [9] which allows assigning a size to both continuous and individual particles. When an opening transformation is performed on a binary image with a structuring element (SE) of size $\lambda$, the image is replaced by the envelope of all SEs inscribed in its objects. For the sake of simplicity, spheres of increasing radii $\lambda$ (approximated by octahedra) were used. When an image is opened by a sphere whose diameter is smaller than the smallest features of its objects, it remains unchanged. As the size of the sphere is

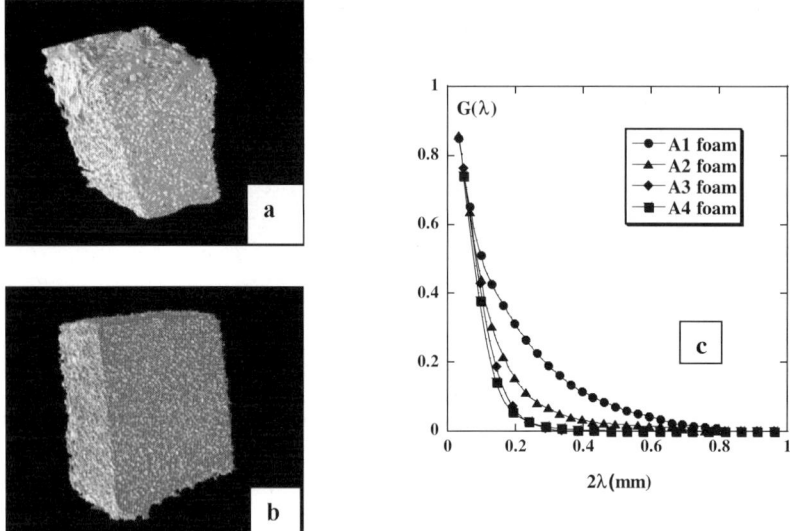

Figure 3. 3D images reconstruction for foams A1 (a) and A4(b).(c) Opening pore size distribution of the studied foams.

increased, larger parts of the objects are removed by the opening transformation Therefore opening can be considered as equivalent to a physical sieving process. This procedure was applied to the reversed 3D images of the foams, i.e. to the 3D images in which pores correspond to white measurable voxels and the support to black voxels. Figure 3e plots the volume of the porous network, $G(\lambda)$ normalized by it initial volume, as a function of the size of the sphere. The comparison of the $G(\lambda)$ distribution for the four foams (Fig. 3e) indicates that pore size distribution becomes narrower and is also shifted towards smaller sizes when the $CO_2$ pressure of saturation used in the foaming process increases.

## 4. DISCUSSION

The pore structure is a key characteristic which must be determined in function of the application. In most cases an homogeneous structure is required. However, in the field of the tissue engineering, foams porosity ideally consists in a bimodal pore distribution in which pores larger than 10 µm are essential for sustaining cell infiltration, whereas pores smaller than 10 µm contribute to cell attachment and create a large surface area for the growth of tissue layer [1,2]. In this work the influence of the sc $CO_2$ saturation pressure during the foaming process on the pore structure of PCL foams was studied. To achieve this goal, three independent methods were used: helium and mercury pycnometry, SEM, and X-ray microtomography, these two latter techniques being coupled with image analysis. These techniques have the advantage of being non-destructive for the pore structure and of enabling information at different scales to be extracted. In order to test the ability of the used methodology to characterize macro- and ultramacro- porous textures, foaming conditions were selected to obtain pores larger than 10 μm.

The porosity of foams evaluated using pycnometry ($\varepsilon$) and X-ray microtomography ($\delta$) agrees well (Table 1). However, the $\varepsilon$ values increase with the sc CO2 saturation pressure whereas the $\delta$ values remain almost constant.. This difference can be attributed to the combination of those three features: (a) the resolution of the X-ray microtomograph is not large enough to clearly distinguish the walls from the pores, due the low thickness of the walls and/or small pore sizes. In this case, it is impossible to discriminate between the gray levels of the those pores and of the polymer matrix which results, at this scale, in a single blurred texture, (b) PCL has a low X-ray attenuation coefficient, which implies to work at relatively low energy level. In our experimental conditions, this leads to a loss of focus stability and to worse detector accuracy, (c) as a result of the poor quality of the images, processing must be performed to enhance the contrast between pores and walls before image binarisation. As this processing presents a statistical character, some pixels belonging to pores could be considered as part of the walls and vice-versa, leading to some errors in the quantification of total porosity.

Image analysis of SEM images and of X-ray microtomograms show that the pore size distributions become narrower and is also shifted towards smaller sizes when the foaming $CO_2$ saturation pressure increases. In particular for sample A1, a continuous distribution in which large pores (~0.8mm) coexisting with small pores (~0.15 mm) is observed.

These results find their origin in the foaming mechanism. The number and size of the formed bubbles is determined by the competition between the rates of bubble nucleation and growth. It is well known from the homogeneous nucleation theory [13] that when the magnitude of the pressure drop induced by the reactor depressurization increases, the energy barrier for nucleation decreases. This leads to an increased nucleation rate, and hence to smaller bubbles. The presence of ultramacropores may be explained by an effect of the temperature. As the temperature drop resulting from the gas expansion is lower for low saturation pressures, the actual temperature following depressurization is higher for A1 than for the other samples. In this case, bubbles have much time to grow. This is prone to favor the coalescence of bubbles. It must be noticed that the pore structure is a key characteristic which must be determined in function of the application. In most cases an homogeneous structure is required. However, in the field of the tissue engineering, foams porosity ideally consists in a bimodal pore distribution in which pores larger than 10 μm are essential for sustaining cell infiltration, whereas pores smaller than 10 μm contribute to cell attachment and create a large surface area for the growth of tissue layer [14,15].

## ACKNOWLEDGEMENTS

The authors thank the Ministry of the "Communauté française de Belgique", Belgium (Action de Recherche Concertée 00/05-265), the Ministry of "Région Wallonne" (DGTRE), Belgium, in the frame of the W.D.U. program for their financial support. A. Léonard is grateful to the FNRS (National Fund for Scientific Research, Belgium) for a Postdoctoral Researcher position.

## REFERENCES

[1]   D.M. Liu, V. Dixit (eds) Porous materials for tissue engineerig. Uetikon-Zuerich, Trans Tech Publications Ltd., 1997.

[2] Maquet V,Blacher S,Pirard R,Pirard J-P,Jerome R. 2000;16:10463 J B.M. R. ,Appl Biomater., 43 (1998) 291.
[3] A. Cooper,. Adv. Mater, 15 (2003) 1049.
[4] Qun Xu, X. Ren, Y. Chang, J. Wang, L. Yu, K. Dean, J. Appl. Polym. Sci., 94 (2004) 593.
[5] S. Cotugno, E. Di Maio, G. Mensitieri, S. Iannace, G.W. Roberts, R.G. Carbonell, H.B. Hopfenberg, Ind. Eng. Chem. Res. 44 (2005) 1795.
[6] F. Stassin , Ph D Thesis, University of Liège, 2005.
[7] J. Tija, P.J. Moghe. Biomed. Mater.Res., Appl. Biomater. 43 (1998) 291.
[8] V. Maquet, S. Blacher, R. Pirard, J.-P. Pirard , M. N. Vyakarnam, R. Jérôme, J.B.M.R. Appl Biomater., 66 (2003) 199.
[9] S. Blacher, A. Léonard, B. Heinrichs, N. Tcherkassova, F. Ferauche, M. Crine, P. Marchot, E. Loukine, J.P. Pirard, Colloid Surface A 241 (2004) 201.
[10] http://www.mtm.kuleuven.ac.be/Research/Equipment/Mechanical/MCT.html
[11] J. Serra, Image Analysis and Mathematical Morphology, vol. 1, Academic Press, New York, 1982.
[12] N. Otsu. IEEE Trans. Syst. Man Cybern. 9 (1979) 62.
[13] O.Olabisi, L.M. Robeson, M. T Shaw, Polymer-Polymer Miscibility, Academic Press: NewYork, 1979.

Studies in Surface Science and Catalysis 160
P.L. Llewellyn, F. Rodriquez-Reinoso, J. Rouqerol and N. Seaton (Editors)
© 2007 Elsevier B.V. All rights reserved

# Detection of specific electronic interactions at the interface aromatic hydrocarbon-graphite by immersion calorimetry

**B. Bachiller-Baeza[1], A. Guerrero-Ruiz[2], and I. Rodríguez-Ramos[1]**

[1] Instituto de Catálisis y Petroleoquímica, CSIC, C/ Marie Curie, 2, Cantoblanco 28049, Madrid, Spain. b.bachiller@icp.csic.es

[2] Departamento de Química Inorgánica y Técnica, Facultad de Ciencias, UNED, C/ Senda del Rey, 9, 28040 Madrid, Spain. aguerrero@ccia.uned.es

High surface area graphites with different amounts of oxygen surface groups were prepared, and characterized by temperature programmed desorption and thermogravimetry. The oxidation treatments with $H_2O_2$ and $HNO_3$ resulted in a partial destruction of the carbon structure and in an increase of the content of the oxygen-containing groups, more evident in the sample treated with $HNO_3$. The enthalpies of immersion of the samples in methylcyclohexane, toluene and mesitylene have proved that the surface chemistry of the graphites has a great influence on the adsorption of aromatic molecules. For all the samples the enthalpies follow the order methylcyclohexane < toluene < mesitylene. On the other hand, while the enthalpy of immersion in methylcyclohexane is independent on the content of oxygen surface groups, the areal enthalpies for the aromatic molecules, toluene and mesitylene, increase as the density of oxygen groups does. Moreover, this effect is more evident in the case of mesitylene.

## 1. INTRODUCTION

One of the most important applications of the carbon materials is as adsorbents. These materials can differ in the structural and chemical characteristics of their surfaces. And, the understanding on how these factors control their behaviour is essential. One important aspect is the presence of functional groups on the carbon surface, usually oxygen-containing groups. Oxygen groups can be introduced after an oxidizing treatment of the material, and several studies have dealt with the characterization of the surface chemistry by different methods: titration methods, temperature programmed desorption, XPS, etc [1, 2]. The existence of oxygen-containing functional groups with acidic, neutral or basic character has been shown. On the other hand, the surface chemistry is known to influence on the adsorption of polar molecules and other organic molecules in aqueous medium, mainly due to the formation of H-bonds. But, the specific mechanism by which the adsorption of non-polar organic compounds takes place is still ambiguous. Several mechanisms have been proposed in the literature where carbon basal planes, unpaired electrons located at the edges of the terminal graphene layers, and the heterogeneous surface groups play an important role in the adsorption process. Basically, it is assumed that the interaction of non-polar molecules with carbon is probably limited to non-specific forces due to the non-polar character of the carbon surface. However, in the case of aromatic molecules the scenario is more complex, in particular for substituted rings as phenol, aniline and derivatives. Although the adsorption of phenol and substituted phenols has been extensively studied, numerous discrepancies still exit considering the nature of the adsorption site. The interaction between the $\pi$ electrons of the aromatic ring and the electron-rich regions located in the basal planes has been claimed [3]. It cannot be ruled out that the basic carbonyl oxygen groups acting as electron donors interact with the aromatic ring forming donor-acceptor complexes [4]. More recently, Castillejos-Lopez et al. suggested the formation of a charge-transfer complex among the aromatic electrons of the organic molecule and the lateral faces of a graphite surface [5].

In general, immersion calorimetry has been used to characterize the surface area and pore structure of the carbonaceous materials by using compounds with different molecular sizes [6, 7]. But, it can be also a useful technique in providing information about the energy of interaction between the molecules of a liquid and a solid surface. Different authors have applied immersion calorimetry mainly to the characterization of acidic and basic groups of carbonaceous materials through the determination of the enthalpies of neutralization of the groups with solutions of different acidity [8, 9]. And also, the enthalpy of immersion of carbonaceous materials into water has been correlated with their oxygen content [10]. However, calorimetric studies of the adsorption of organic molecules and the influence of the surface chemistry on that are very scarce. It seems that immersion calorimetry could be a complement of the traditional determination of the adsorption isotherms of aromatics from organic and water solutions.

Consequently, in order to gain some insight about the mechanism of adsorption of organic molecules on the carbon surface, we have applied immersion calorimetry to the adsorption of three different organic molecules, methylcyclohexane (MCH), toluene (TOL) and mesitylene (MES). Trying to avoid the additional effects of the microporosity, we have studied as adsorbents high surface area graphites with different content of oxygen surface groups. The experiments have been performed in order to detect the possible modifications of the electronic properties of the graphite surfaces as a consequence of the presence of functional oxygen groups chemisorbed on them.


## 2. EXPERIMENTAL

The high surface area graphite used as a starting material in this study and denoted H1 was purchased from Lonza Ltd. The graphite was subjected to two different oxidation treatments. An aliquot of the initial support was immersed in an aqueous solution of $H_2O_2$ 6N (50 ml·g$^{-1}$ of support), and was stirred for 48 h at room temperature. Then, the sample $H1_{ox1}$ was washed with deionised water, filtered and finally dried at 383K under air. A second fraction was oxidized in concentrated aqueous $HNO_3$ (60%). The suspension (10 ml·g$^{-1}$ of support) was heated at 353 K until reaching total evaporation of the liquid (c.a. 24 h), and the process repeated a second time. Then, the sample $H1_{ox2}$ was washed with deionised water and dried at 383 K. Also, H1 was treated under inert atmosphere of He in a Carbolite furnace at 723 K, (H2), and at 1173 K (H3) to remove, partially or completely, the oxygen surface groups.

The surface area of the samples was determined from the $N_2$ adsorption isotherm at 77 K by applying the BET equation, measured in an automatic Micromeritics ASAP-2000.

The amount of oxygen groups present on the surface was determined by thermogravimetric analysis in a CI Electronics microbalance (MK2-MC5). The sample was treated under flowing He for 2 h and then heated at a 10 K·min$^{-1}$ rate up to 1023 K.

The chemical nature of these functional groups was evaluated by TPD-MS experiments under vacuum in a conventional volumetric system connected to a BALZERS QMG 421 C mass spectrometer. The sample was evacuated for 30 min at room temperature and ramped to 1173 K at a 10 K·min$^{-1}$ rate.

The enthalpies of immersion of the samples in methylcyclohexane, toluene and mesitylene were determined at 303 K with an isothermal calorimeter of the Tian-Calvet type, Setaram C-80. Prior to experiments the samples were outgassed at 383 K for 10 h. The corrections corresponding to the energy of bulb breaking and to the energy of liquid vaporization have been considered.

## 3. RESULTS AND DISCUSSION

The N$_2$ adsorption isotherms for sample evacuated at 1173 K, H3, and for oxidized samples, H1$_{ox1}$ and H1$_{ox2}$, are displayed in Fig. 1. It can be seen that all the isotherms showed an hysteresis loop of type B which is generally associated to slit-shaped mesopores. Moreover, while the outgassing treatment does not affect the graphite structure, both oxidation treatments resulted in a reduction in the BET area, more evident in the case of the sample treated with HNO$_3$. It appears that the mesoporous structure of the carbon was destroyed. Similar modifications have been observed in activated carbons after oxidizing treatments with H$_2$O$_2$ or HNO$_3$, the oxidation with HNO$_3$ being more aggressive [11].

Table 1
Surface areas, oxygen surface groups evolved during heat treatment and enthalpies of immersion at 303 K into different liquids

| Graphite | $S_{BET}$ (m$^2$·g$^{-1}$) | Amount evolved (mmol·g$^{-1}$) | | [O] (mmol·g$^{-1}$) | $-\Delta H_{imm}$ (J·g$^{-1}$) | | |
|---|---|---|---|---|---|---|---|
| | | CO$_2$ | CO | | MCH | TOL | MES |
| H3 | 298 | 0.00 | 0.25 | 0.25 | 24 | 30 | 56 |
| H2 | 305 | 0.07 | 0.89 | 1.03 | - | 38 | 51 |
| H1 | 299 | 0.41 | 1.07 | 1.89 | 23 | 40 | 53 |
| H1$_{ox1}$ | 241 | 0.59 | 1.07 | 2.25 | 18 | 30 | 50 |
| H1$_{ox2}$ | 69 | 0.77 | 1.43 | 2.97 | 5 | 14 | 23 |

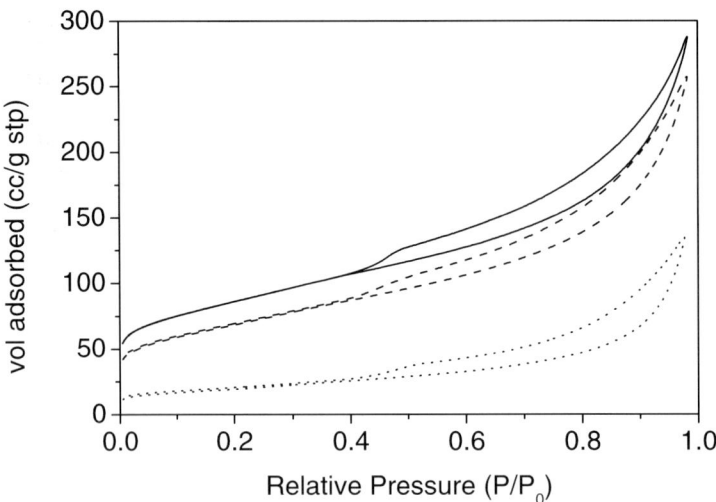

Fig. 1. Isotherms for $N_2$ adsorption at 77 K. (—) H1, (---) $H1_{ox1}$, (·····) $H1_{ox2}$.

The modification in the surface chemistry upon thermal and oxidation treatments is clearly assessed by the TPD profiles of $CO_2$ and CO shown in Fig. 2A and Fig. 2B respectively. Both H1 and $H1_{ox1}$ samples display a similar $CO_2$ desorption profile, very wide and with a maximum rate around 525 K. On the other hand, the sample treated with $HNO_3$, $H1_{ox2}$, presents a profile with two well-differentiated peaks centred at 500 K and 670 K. This result indicates that two types of chemical species or two energetically different species are present on the surface. The simultaneous evolution of $H_2O$ possibly evidence the creation of anhydride groups during the TPD run, which are more stable than the carboxylic acid groups and desorb as $CO_2$ and CO at higher temperatures. The absence of a $CO_2$ desorption peak for H2 indicates that the thermal treatment at 723 K decompose all the oxygen surface complexes that evolve as $CO_2$, mainly the carboxylic acids. Regarding the CO profiles, again H1 and $H1_{ox1}$ showed a wide band that starts at 373 K and with maximum at 840 and 740 K respectively. The difference in the temperature maximum can be due to a different relative concentration of the possible species desorbing in the range of temperatures, including the anhydride groups formed from the carboxylic acid during the experiment. In the case of sample H2, a single and well-defined peak appears at 860 K due to phenolic or lactonic groups,. The maximum desorption rate in the CO profile for sample $H1_{ox2}$ is at 920 K probably due to carbonyl groups. But, it also showed a shoulder at 750 K, which can be associated with the second peak at around 670 K of the $CO_2$ profile and be assigned to the anhydride group decomposition.

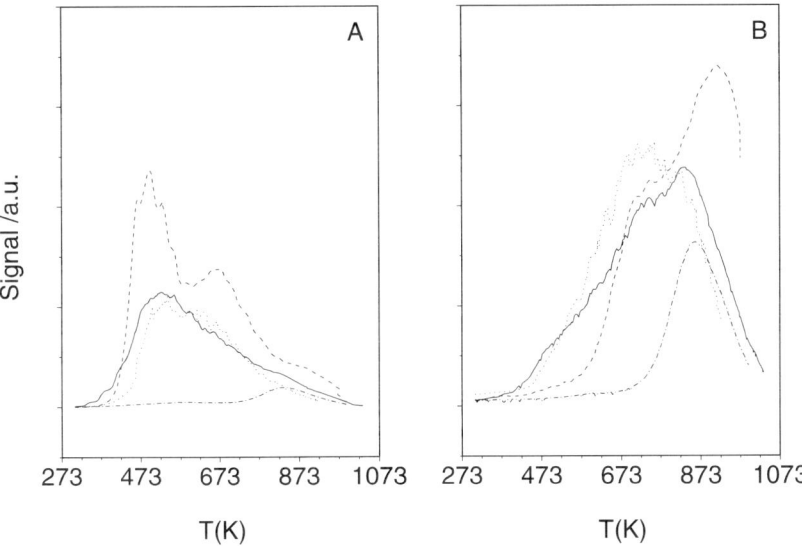

Fig. 2. Temperature-programmed desorption profiles for (A) $CO_2$ and (B) CO. (—) H1, (····) H1$_{ox1}$, (---) H1$_{ox2}$, (-··-) H2.

Table 1 contains the amount of groups per gram of sample evolving as $CO_2$ and CO calculated from the thermogravimetric measurements. The thermal treatments reduced the amount of oxygen surface groups, where H2 only have groups evolving as CO. On the other hand, the oxidized samples differ in the type and amount of groups created. While the treatment with $H_2O_2$ hardly produces an increase in the amount of $CO_2$ evolving groups and maintains the amount of groups evolving as CO, the $HNO_3$ treatment results in an increase of both CO and $CO_2$-evolving groups as it was shown in the temperature-programmed profiles.

The enthalpies of immersion of methylcyclohexane, toluene and mesitylene for the series of graphite samples are also displayed in Table 1. The sample treated at 1173 K, H3, showed a progressive increase of enthalpy values in the order methylcyclohexane < toluene < mesitylene. Methylcyclohexane is a non-aromatic molecule, and then the interactions are basically non-specific. But, in addition to the non-specific dispersion forces, the aromatic molecules can specifically interact with graphite surface. Thus, we could tentatively interpret the variations in immersion heats for toluene and mesitylene as due to modifications in the $\pi$-$\pi$ interaction between the aromatic ring and the electron-rich regions located in the graphene layers, i.e. the basal planes. The higher value obtained for mesitylene comparing with that for toluene could be due to the two additional methyl groups present on the molecule. These groups activate in a greater extent the aromatic ring by donating electrons, which would result in an increase of the specific interaction. The same tendency that it was found for H3 is observed for the other samples, the enthalpies of immersion for a sample follow the order methylcyclohexane < toluene < mesitylene.

However, comparing the enthalpy of immersion per gram for the same compound different behaviours are observed depending on the graphite sample. Particularly, a marked decrease in the values is observed for sample $H1_{ox2}$. Problems in the accessibility of the molecules to the porous structure of the HSAG, which is in the range of mesoporous, must be ruled out. Thus, the effect would be caused by two factors, differences in the surface area and/or surface electronic structure. Consequently, it seems more appropriate to compare the areal enthalpies ($J \cdot m^{-2}$) of immersion, and in Fig.3 the areal enthalpies of immersion as a function of the amount of surface oxygen ([O] $\mu mol \cdot m^{-2}$) are displayed. We can see that the areal enthalpies of immersion of methylcyclohexane for the studied samples are constant and non-dependent on the amount of surface oxygen groups. This indicates again that the interaction of the molecule with the carbon surface is via non-specific dispersion forces, and that the enthalpy is proportional to the surface area. Moreover, the enthalpies for toluene and mesitylene immersion varied with the amount of oxygen-containing groups, increasing as the density of these surface species This latter indicates that the interactions of these molecules with the carbon sites in the graphite (basal planes or edges of the crystallites) is different than that of methylcyclohexane. Therefore, it reflects the influence of these oxygen complexes on the adsorption of aromatic molecules.

A strong effect of the surface acidity on the adsorption of benzene and xylene from heptane and cyclohexane solutions has been suggested previously [12], where adsorption capacity is inversely related to the surface oxygen concentration, and consequently, to surface polarity. Also, the adsorption capacity for phenanthrene gas was higher for activated carbons with low concentration of surface oxygen groups [13]. For benzene the presence of $CO_2$-type groups are detrimental for the adsorption whereas the CO-type groups favour it [13]. In the present case, we have studied toluene and mesitylene, two hydrophobic and non-H bonding molecules, whose behaviour should be similar to that reported for benzene and xylene. Since the major fraction of surface for H3 is non-polar, the adsorption capacity for the more hydrophobic mesitylene should be higher than that for toluene, which would explain the differences in the enthalpies of immersion of Table 1 for these two molecules. For $H1_{ox2}$, the highest oxidized graphite and with a significant polar surface fraction, the lower values of enthalpy would corroborate the reduction in capacity of adsorption for the non-polar molecules. But, as shown in Fig.3 the specific interaction of the aromatic molecule with the adsorption site is higher than for methylcyclohexane. The oxygen surface groups appear as modifiers of the electronic properties of the graphite surfaces withdrawing electrons from the graphene layers and creating positive holes in the conductive $\pi$-band. This effect would increase the interaction between the aromatic adsorbates and the basal planes of the graphite. Moreover, the two additional methyl groups present on mesitylene are activating groups, which should give a higher interaction. Both effects work in the same direction, and the interaction of this molecule with carbon surface is expected to be higher. Other possibility is that in the presence of oxygen centers on the carbon surface, the interaction with the carbon sites is different, and preferential sites according to their position in the carbon structure (basal planes, edges of the crystallites, etc) could exist.

In conclusion, we have proved that the surface chemistry of higher surface area graphites has an important role in the adsorption of aromatic compounds, and that the immersion calorimetry can be used as a sensitive tool capable of identifying the type of surface states.

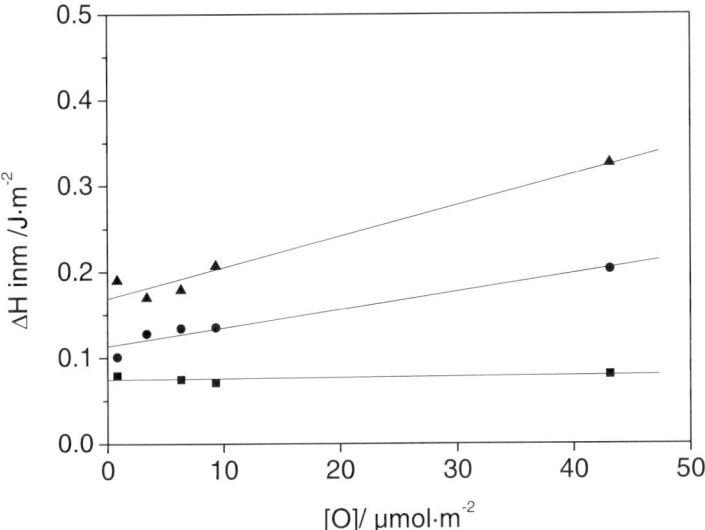

Fig. 3. Areal enthalpy of immersion for adsorption of (■) methylcyclohexane, (●) toluene and (▲) mesitylene.

## ACKNOWLEDGEMENT

The authors acknowledge the financial support from the Ministerio de Ciencia y Tecnología (Spain) (project MAT 2002-04189-C02-01 and 02). BBB also gratefully acknowledge financial support from the Ministerio de Ciencia y Tecnología in the Ramon & Cajal Program.

## REFERENCES

[1]  H.P. Bohem, Adv. Catal., 16 (1966) 179.
[2]  L.R. Radovic, C. Moreno-Castilla, J. Rivera-Utrilla, Chem. Phys. Carbon, 27 (2000) 227.
[3]  R.W. Coughlin, F.S. Ezra, Environ. Sci. Technol., 2 (4) (1968) 291.
[4]  J.S. Mattson, H.B. Mark, M.D. Malbin, W.J. Weber and J.C. Critten, J. Colloid. Interface Sci., 3(1) (1969) 116-30.
[5]  E. Castillejos-López, D.M. Nevskaia, V. Muñoz, I. Rodríguez-Ramos and A. Guerrero-Ruiz, Langmuir, 20(4) (2004) 1013-15.
[6]  R. Denoyel, J. Fernandez-Colinas, Y. Grillet and J. Rouquerol, Langmuir, 9 (1993) 515.

[7]  H.F. Stoeckli and F. Kraehenbuehl, Carbon, 19 (1981) 353.
[8]  M.V. Lopez-Ramon, F. Stoeckli, C. Moreno-Castilla and F. Carrasco-Marin, Carbon, 37 (1999)1215.
[9]  S.S. Barton, M.J.B. Evans, E. Halliop and J.A.F. MacDonald, Carbon, 35 (1997) 1361.
[10] F. Stoeckli  and A. Lavanchy, Carbon, 38 (2000) 475.
[11] C. Moreno-Castilla, M.A. Ferro-García, J.P. Joly, I. Bautista-Toledo, F. Carrasco-Marín and J. Rivera-Utrilla, Langmuir, 11 (1995) 4386.
[12] F. Ahnert, H.A. Arafat and N.G. Pinto, Adsorption, 9 (2003) 311.
[13] T. García, R. Murillo, D. Cazorla-Amorós, A.M. Mastralñ and A. Linares-Solano, Carbon, 42 (2004) 1683.

Studies in Surface Science and Catalysis 160
P.L. Llewellyn, F. Rodriquez-Reinoso, J. Rouqerol and N. Seaton (Editors)

# Characterization and modelling of argillaceous porous medium by compressional and shear acoustic waves

J. Riffaud [a,b], A. Cerepi [a], J. Marrauld [a]

[a] Institut EGID-Bordeaux 3, Université Michel de Montaigne, 1, allée Daguin, 33607, Pessac, Cedex, France, Tel : (33) 05 57 12 10 11, Fax : (33) 05 57 12 10 01

[b] TOTAL, Centre de Pau, Avenue Larribau, 64018, Pau, cedex, France

## 1. INTRODUCTION

Modelling of argillaceous minerals elastic properties in porous medium shows a great interest for the understanding of argillaceous sediments mechanical behaviour. This behaviour description remains baldy known, because of the great diversity and complexity of the studied material [1-4]. Up to now, we only know some experimental modelling [1-4] and the boundaries proposed by Reuss and Voigt and by Hashin-Shtrikman [5] in order to describe the clays elastic moduli. This behaviour depends on the clays' mineralogy and porosity. With clays porosity estimation as a function of effective pressure for each clay, it is possible to model the argillaceous formation elastic behaviour. First, this article shows the theoretical background of the present experimental modelling. Second, a new approach to model clays and shales elastic behaviours. Finally, results and discussion are meant to give a proposed modelling.

## 2. THEORITECAL BACKGROUND

The prediction of effective porosity and elastic moduli in mixed phases depends on volume fractions and the elastic modulus of each phase as well as their geometrical combination. With volume fractions specification, it is only possible to predict the mixed upper and lower elastic moduli bounds.

### 2.1. Elastic behaviour boundaries

The upper bound of n-phases mixture effective modulus is given by Voigt relation [6] as:

$$M_V = \sum_{i=1}^{N} fiMi$$

where fi is volume fraction of component i, and Mi the elastic modulus of component i. The lower bound of n-phases mixture effective modulus is given by Reuss relation [6] as:

$$\frac{1}{M_R} = \sum_{i=1}^{N} \frac{fi}{M_i} \tag{1}$$

The best known bounds modelling, without information about the geometrical combination are given by Hashin-Shtrikman [5] relation as:

$$K^{HS+-} = K_1 + \frac{f_2}{(K_2-K_1)^{-1} + f_1(K_1+\frac{4}{3}\mu_1)^{-1}} \quad (2)$$

$$\mu^{HS+-} = \mu_1 + \frac{f_2}{(\mu_2-\mu_1)^{-1} + \frac{2f_1(K_1+2\mu_1)}{5\mu_1(K_1+\frac{4}{3}\mu_1)}} \quad (3)$$

where $K_i$ is the the the bulk modulus; the $\mu_1$ is the shear modulus; $f_1$ is the volume fraction of the phase i. Upper and lower bounds are obtained by interchanging components 1 and 2 (figure 1). Generally, the upper bound is obtained when the stiffest component is the 1 in the expressions, and the lower bound when the softest component is termed 1. Differences between upper and lower depend on the difference existing between elastic moduli of the various components. Usually, with a homogeneous mixture, mixture elastic modulus is considered as the average of upper and lower bounds. This is a good approximation when elastic moduli are not too much different. But, when mixture component moduli are too different like in a mineral / pore fluid mixture, a great difference between upper and lower bounds prevents you from having a reliable estimation. For argillaceous sediment, the basic composition system is a mineral matrix with its porosity (free and bound water), with the another mineral constituents (quartz; heavy minerals,...) floating in this matrix. In order to predict volume fraction of a different phase, it is necessary to know porosity variation with burial and effective pressure. We will not consider here diagenetic mineralogical transformations within argillaceous matrix.

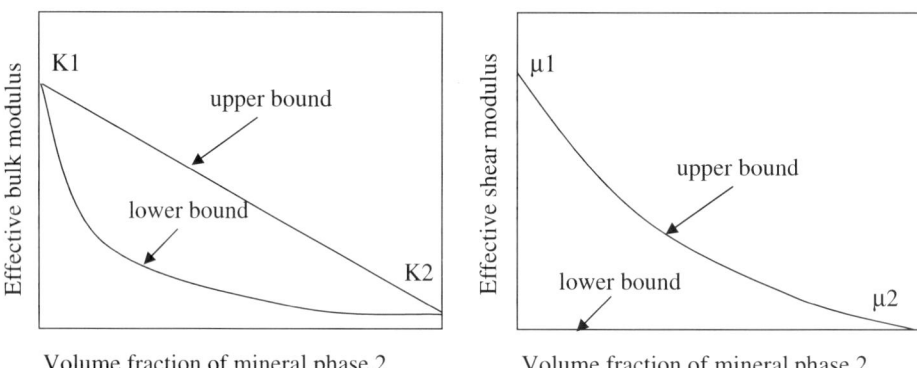

Fig.1. Hashin-Shtrikman boundaries for elastic moduli.

## 2.2. Clay porosity evolution as a function of effective pressure.

Many authors have observed a progressive modification of the shales density with burial. Athy [7] and Korvin [8] have shown that it exists a density exponential decreasing with the burial increase.

$$\rho(z) = \rho_\infty + (\rho_0 - \rho_\infty)e^{-kz} \quad (4)$$

With $\rho_\infty$ the density of clay for an infinite depth; $\rho_0$ the pore fluid density; k a constant; z the depth. If we consider a unit clay cube, we have the expression:

$$V_{clay}=1, \quad V_{clay\,min} + V_{fluid} = 1, \quad V_{clay}+\Phi=1 \quad (5)$$

And for a given depth, we have:

$$\rho_{clay}(z) = \rho_{claymin} + (\rho_{fluid} - \rho_{claymin})\Phi(z) \qquad (6)$$

Korvin [8] have shown that in a normally compacted sequence, clay density increases exponentially with burial depth, with effective pressure. We obtain the following relation:

$$\Phi = e^{-kz} \qquad (7)$$

With a description of effective pressure as a linear function of burial, it is possible to express clays porosity as:

$$\Phi_{Clay} = \alpha \bullet e^{-\beta Peff} \qquad (8)$$

With $P_{eff}$ representing the difference between the geostatic pressure and fluid pressure multiplied by Biot coefficient, and $\alpha$ and $\beta$ being constants.

## 2.3. Clay porosity from density tool.

Shales density is the average of each shale component density balanced by their volume fractions. It is possible to express measured density with density tools as :

$$\rho_{measured} = \rho_{Minerals} * V_{Minerals} + \rho_{water} * \Phi_{effective} + \rho_{clay} * V_{clay} \qquad (9)$$

And it is possible to break up the clay fraction as a clay mineral fraction and a clay porosity fraction.

$$\rho_{clay} * V_{clay} = \rho_{claymin} * (V_{clay} - \Phi_{clay}) + \rho_{water} * \Phi_{clay} \qquad (10)$$

It is thus possible to express clay porosity as:

$$\Phi_{Clay} = \frac{\rho_{measured} - \rho_{Mineral} + V_{Clay} * (\rho_{Mineral} - \rho_{Claymin}) + \Phi_{effective} * (\rho_{Mineral} - \rho_{Water})}{\rho_{Water} - \rho_{Claymin}} \qquad (11)$$

With combination of both clays porosity expressions, it is possible to estimate the coefficients $\alpha$ and $\beta$.

## 3. RESULTS AND DISCUSSION

Figure 2 shows the clay compaction evolution and the water drainage, causing density increase. We observe that clay is initially put in suspension in water, and then arranges itself progressively to raise a laminated structure. The ultimate stage of compaction tends to form a mineral block of a much reduced porosity, in which anisotropy tends to disappear.

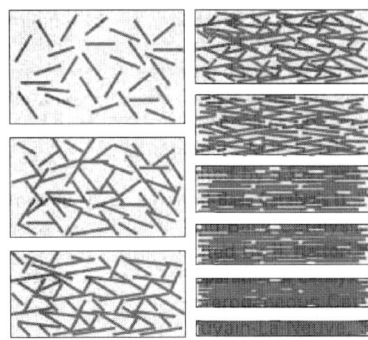

Fig.2. Clay compaction evolution with burial increase.

### 3.1. Clay elastic moduli evolution versus effective pressure
Taking into account of geometrical structure, and Reuss and Voigt formulations, it is possible to express clays bulk and shear moduli.

#### 3.1.1. Bulk modulus
In order to obtain clay bulk modulus law, we consider first, a media composed by a fluid and clay hydrated layers. The clay laminated structure and clay layers / fluid alternation imply considering Reuss formula to express the modulus.

$$\frac{1}{K_{clay}} = \frac{\Phi_{clay}}{K_{fluid}} + \frac{(1-\Phi_{clay})}{K_{claymin}} \tag{12}$$

with $K_{clay}$ the clay bulk modulus; $K_{fluid}$ the fluid bulk modulus; $K_{claymin}$ the hydrated clay layer bulk modulus. The hydrated clay layer corresponds to the clay mineral with its bounded water.

$$K_{clzy} = \frac{1}{\frac{1}{K_{clay\,min}} + \left(\frac{1}{K_{fluid}} - \frac{1}{K_{clay\,min}}\right) \bullet \Phi_{clay}} \tag{13}$$

With the expression of $\Phi_{clay}$ as a function of effective pressure, we obtain :

$$K_{clay} = \frac{1}{\frac{1}{K_{clay\,min}} + \left(\frac{1}{K_{fluid}} - \frac{1}{K_{clay\,min}}\right) \bullet \alpha e^{-\beta Peff}} \tag{14}$$

#### 3.1.2. Shear modulus
It is possible to express the shear modulus as a combination of Reuss function and Voigt function, with the same system:

$$\mu_{clay} = \mu_{min} \bullet e^{(-8+2\bullet\Phi_{clay})\bullet\Phi_{clay}} \tag{15}$$

### 3.2. P and S wave propagation velocities versus effective pressure
With $K_{clay}$ and $\mu_{clay}$ expressions, it is possible to describe shales elastic moduli. Indeed, clay is never in a pure state in nature and is often met associated with other minerals, such as quartz or silts, whose presence will modify the elastic moduli of the clay / minerals mixture. But,

as is explained at the beginning of this paper, we need to make a hypothesis about the structural combination of clay and other minerals, in order to model shales (clay and other minerals) elastic moduli. We will make the hypothesis that we have quartz floating in clay matrix. For that, we use a Reuss average to model it. Effective porosity, normally close to zero, is combined with clay volume, in order to model elastic moduli. We obtain elastic shales laws as:

$$\frac{1}{K}=\frac{V_{clay}+\Phi_{eff}}{K_{clay}}+\frac{1-\Phi_{eff}-V_{clay}}{K_{quartz}} \tag{16}$$

$$\frac{1}{\mu}=\frac{V_{clay}+\Phi_{eff}}{\mu_{clay}}+\frac{1-\Phi_{eff}-V_{clay}}{\mu_{quartz}} \tag{17}$$

With these moduli, velocities are described as:

$$V_P=\sqrt{\frac{K+\frac{4}{3}\mu}{\rho}} \tag{18} \qquad V_S=\sqrt{\frac{\mu}{\rho}} \tag{19}$$

### 3.3. Hydrated argillaceous minerals elastic moduli.

In order to use beforehand established laws, it is advisable to know elastic moduli $K_{claymin}$ and $\mu_{claymin}$ of hydrated argillaceous mineral. Several studies [1, 2, 9, 10] have proposed elastic moduli for dry clays. From these results and laboratory studies have been estimated hydrated argillaceous mineral elastic moduli. These moduli are specified in table 1. Moreover, it is necessary to know the upper porosity $\alpha$ associated with each sort of studied clay. We will use for that, Chilingar and Knight [11] studies. These measurements on three pure clays (Montmorillonite, Kaolinite and Illite) have showed, that it exist a great impact of mineralogy on clay porosity. However, measured porosities made by Chilingar seem higher than other realised on several oil fields around the world. We will consider only porosity measurements for lower effective pressures. A synthesis of values resulting from this various work is presented in table 1. Various values of elastic modulii are expressed in GigaPascal. A chart of the K and $\mu$ elastic modulii for each sort of clay is presented in figure 3. The bulk modulus of the fluid considered is 2.5 GPa.

Table I : Porosity values and elastic moduli values related to three pure clays.

| MINERAL | $\Phi_{max}$ | $K_{mindry}$ | $K_{min}$ | $\mu_{mindry}$ | $\mu_{min}$ |
|---|---|---|---|---|---|
| KAOLINITE | 0.5 | 47.6 | 20 | 24.9 | 9.2 |
| ILLITE | 0.54 | 43.2 | 18.9 | 23.7 | 8.7 |
| MONTMORILLONITE | 0.6 | 29.7 | 14.1 | 16.5 | 6.4 |

### 3.4. Poisson's ratio evolution.

Poisson's ratio v is expressed like:

$$v=\frac{3K-2\mu}{6K+2\mu} \tag{20}$$

So, with K and $\mu$ clay moduli modelling, it is possible to determine Poisson's ratio for a given depth. Poisson's ratio is 0.5 in fluid and about 0.3 in clays. Figure 4 presents the PR modelling for kaolinite as a function of clay porosity. The values of PR are well close to 0.3 when porosity tends to zero and 0.5 when porosity tends to 1. Moreover, we observed that PR increases rapidly when porosity increases.

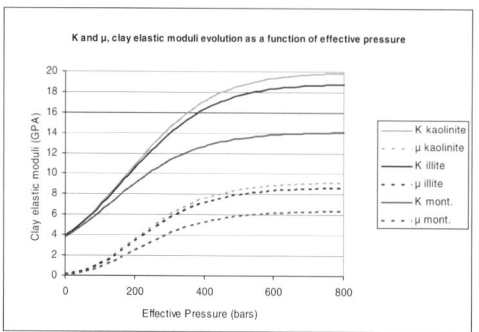 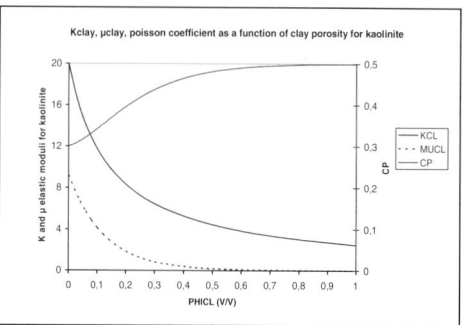

Fig.3.  Evolution of argillaceous elastic moduli K   Fig.4. Kclay, µclay and Poisson's ratio as a function
        and µ versus effective pressure.                    of clay porosity for kaolinite.

## 4. EXAMPLE OF P AND S VELOCITIES MODELLING IN ARGILLACEOUS SEDIMENTS

### 4.1. Using spectral gamma ray tools to estimate clay mineralogy

To use the good values for mineral elastic moduli, it is advisable to have a good reliability about the clay mineralogy for a given depth. Clay mineralogy determination could be done with two methods: X-Ray diffraction on plugs, and gamma ray spectroscopy. The first one has a good reliability but require to have plug, and the second, diagraphic, is less reliable but gives information on all the well. The second method gives an idea about the most dominant clay present at the point of measurement. We will use, to have a good reliability, results of gamma ray spectroscopy validated by X-Ray diffraction on plugs. The results of this interpretation will be presented in figure 5.

#### 4.1.1. Determination of the clay porosity evolution law

Clay porosity is estimated with the method presented in the section *"Clay porosity calculation with density tool"* The results of this modelling are presented in figure 6. The porosity law determined as a function of effective pressure is :

$$\Phi_{clay}=0.45 \bullet e^{-0.008 \bullet Peff} \tag{21}$$

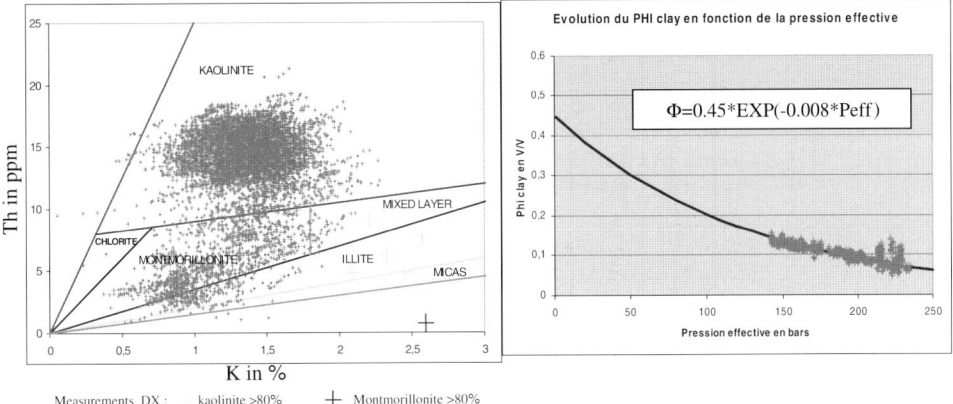

Fig. 5. Thorium versus Potassium diagram in clays for clay mineralogy definition.

Fig. 6. Clay porosity evolution in function of effective pressure.

### 4.2. Velocities modelling with mineralogy

P and S velocities calculation is realised with the process described in the second section "results and discussion" of this paper. We will determine clay elastic moduli with: i) rock petrophysical parameters from a classical log study (Gamma ray, Neutron, density, ...), which give us the volume fraction of each component; ii) clay mineralogical parameters from XRD study and spectrometry gamma ray tool results; iii) Laws established and presented in this paper. The results of modelling are presented in figure 7. It is necessary to take into account only velocities in argillaceous continuums zone where clay volume is sufficient (VCL>0.6-0.7) to have a structure with floating quartz in clay matrix.

### 5. CONCLUSION

The methods presented in this article allow to model with a good reliability P and S wave propagation velocities. The results presented in figure 6 show that modelled velocities are close to measured velocities with sonic tool. The average mistake is lower than 5% in shales continuums. Sometimes, modelling is not in compliance with measured velocities. These mistakes have three reasons. The first is the relative validity of each phase volumes fractions; the second is the presence of washout due to the bad quality of the drilled zone. These zones are viewed by the caliper tool; and finally, the third reason are the approximations made during the mathematic modelling.

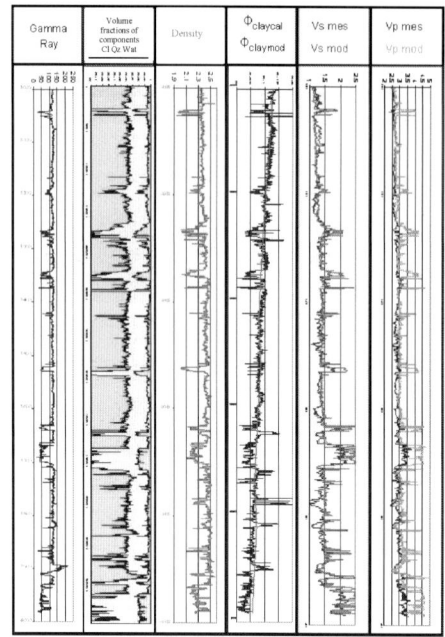

Fig. 7. P and S wave propagation velocities modelling results in shales.

## REFERENCES

[1]  C. Tosaya, and A. Nur, 1982. Effects of diagenesis and clays on compressional velocities in rocks. Geophys. Res. Lett., 9, 5-8.

[2]  J.P. Castagna, M.L. Batzle, and R.L. Eastwood, 1985. Relationships between compressional-wave and shear-wave velocities in clastic silicate rocks, Geophysics, 50, 571-581.

[3]  D.H. Han, 1986. Effects of Porosity and Clay Content on Acoustic Properties of Sandstones and Unconsolidated Sediments. Ph.D. dissertation, Stanford University.

[4]  D. Han, A. Nur, and D. Morgan, 1986. Effects of porosity and clay content on wave velocities in sandstones. Geophysics, 51, 2093-2107.

[5]  Z. Hashin and S. Shtrikman, 1963. A variational approach to the elastic behaviour of multiphase materials. J. Mech. Phys.Solids, 11, 127-140.

[6]  A. Reuss, 1929. Berchnung der Fliessgrenzen von Mischkristallen auf Grund der Plastizitätsbedingung für Einkristalle, Zeitschrift für Angewandte Mathematik und Mechanic, 9, 49-58.

[7]  L.F. Athy, 1930, Density, porosity, and compaction of sedimentary rocks: AAPG Bull., v. 14, p. 1-24.

[8]  G. Korvin, 1984. Shale compaction and statistical physics. – Geophysical Journal of the Riyal Astronomical Society, 78, 35 – 50.

[9]  Vanorio, Prasad and Nur. 2003. Elastic properties of dry clay mineral aggregates, suspensions and sandstone.

[10] Wang, Z. 1998. Elastic properties of Solid Clays. S.E.G.

[11] Chilingar, G. V., and Knight, Larry. 1960. Relationship between pressure and moisture content of kaolinite, illite and montmorillonite clays. Bull. Am. Assoc. Petroleum Geologists, v. 44, no. 1, p. 101-106..

Studies in Surface Science and Catalysis 160
*P.L. Llewellyn, F. Rodriquez-Reinoso, J. Rouqerol and N. Seaton (Editors)*

# Modelisation and circulation of fluids in geological porous systems. Images analyzing and mercury porosimetry

**S. Galaup, R. Burlot, A. Cerepi, L. Wang, M. Dai**

Institut EGID-Bordeaux 3, Université Michel de Montaigne, 1, allée Daguin, 33607, Pessac Cedex, France; Tel : (33) 05 57 12 10 00; fax : (33) 05 57 12 10 01 ;: mailto : galaup@egid.u-bordeaux.fr.

The principal object of this article is the description and the development of a software which makes it possible to simulate in real time the circulation of a fluid in a porous rock. We obtain not only rock parameters (porosity, radius of pores connection), but we also observe the progression of the fluid in the rock (phenomenon of deflation). In the second time we explain the first results which we obtained with these simulations then we will compare these results 2D with the data 3D of porosity per mercury injection to allow us to establish a link between porosity image 2D and the data 3D.

## 1. INTRODUCTION

Today, the knowledge of the deep supsoil profits of an unprecedented increase in the number of data geophysics and their quality. In particular, measurements of the rocks characteristics for the knowledge of the geometrical and pétrophysics properties in the reservoirs rocks is of an essential importance. The geometry of the porous space and the capacity of petroleum to circulate through this space are going to condition the possibilities of fluid recovery in the rock reservoir [1-4]. Among the measures implementations, the method of Hg-injection porosimetry is one of the most typical [5-6]. It allows to know the porosity and the distribution of the radius average of pores connection. But the inconvenience of the porosimetry is not to be able to observe the fluid progress in the rock. It is so difficult to understand indeed the rocks characteristics as for example the relaxation of the fluid in the rocks.

The subject of this study is to simulate the fluid circulation in a porous rock by using the physics laws associated with the capillarity phenomena. With this simulation, we obtain not only the parameters of rock (porosity, radius of connection of the pores), but we can also observe the progression of fluids in the rock. The simulations which we led relaying on images of many geological carbonate porous systems obtained in scanning electron microscopy (SEM) in the BSE mode. These images are pretreated to obtain a binary image representing the porous space and the matrix. Then, starting from the largest entry of pore at the borders of the image, we will propagate the front of the fluid through porous space. During the fluid propagation and by image treatments, information on the porosity, capillary pressure and pores diameters will be calculated.

## 2. MATERIALS AND METHODS

To carry out the algorithm of this software, we have uses like language the Visual C++ 6.0 on Microsoft Windows.

The images which we have draft are carried out on a scanning electron microscope with a Philips 515 M in the Back Scattered Electron mode. This mode makes it possible to better distinguish the various phases present on the sample with in particular the porous system (filled of resin) which appears with a strong black contrast [7]. The growth to which we have works depend on the type of porosity which we want to determine (micro or macroporosity). To compare these measurements 2D with measurements 3D, we have to proceed on the same samples measurements of mercury posimetry. The method of mercury porosimetry is a method of indirect volume measurement (3D) largely used to measure porosity and the distribution of the average radii of pores connection [8]. The porosimeter used (Autopore III) can reach pressures of 33,000 psia (228Mpa). This apparatus can detect pores whose diameter of access varies between 360 mm for largest to 0,005 mm for finest

In all study we will use like samples of Oligocene limestones with coming from the Bordeaux area.

## 3. THEORY

All the macroscopic properties of the rocks depend on the microstructure and thus of porosity. It is clear that microscopic information is only partially accessible and from there comes the importance given to certain macroscopic parameters, which are measurable like porosity, specific surface, the curves of imbibition and drainage, the formation factor (electric conductivity) and the permeability. We will approach here only the concepts essential for comprehension of calculations.

### 3.1.Porosity

The porosity of a rock is defined by the fraction of volume V of rock which is not occupied by a solid phase. The volume of solid is noted Vs and porosity is noted $\phi$.The rock being made of solid and from pores, it results from it that

$$\phi = \frac{V - V_s}{V} = \frac{V_p}{V} = \frac{pore\ volum}{total\ volum}$$

(1)

### 3.2. Capillary pressure

Capillary fluid flow through porous media has been the focus of many studies during this last decade because of its importance as a process in nature and in different porous materials such as oil filled and water wet reservoirs (sandstone or carbonate), soils and fractured rocks. Many authors have proposed different analyses, experimental results and numerical simulations that have improved the understanding of the displacement process during a one or two-phase drainage [8-13]

We want to determine capillary flow dynamics in porous media at the pore-scale. The theoretical basis of porosimetry is defined by Laplace's law which explained that the work required to expand a non-wetting fluid surface (mercury) of principal radii of curvature $R_1$ and $R_2$ is equal to the work done to the concave side of the surface. Washburn (1921) linked to Laplace's equation by using a capillary model where the porous medium is assimilated to a bundle of cylindrical capillary tubes

The Laplace's equation translates the relation existing between the pressure $P_C$ which one must apply so that the fluid penetrates in a capillary of radius R. The capillary pressure is the difference in pressure through a surface separating two non-miscible fluids, inside a capillary. This parameter depends on the structure of the pores.

The Laplace's equation of leads to : $P_c = P_0 - P_1 = \dfrac{2\gamma}{r_m} = 2\gamma\dfrac{\cos\theta}{R}$ (2)

where Pc is the capillary pressure, $P_0$ pressure of the gas (air here) in the vicinity of the meniscus, and $P_1$ pressure of liquid close to the interface, $\gamma$ the interfacial tension; $\theta$ the angle between mercury meniscus and pore wall (fig.1). Measurements of contact angle are not easy. Some of the methods are examined by Jakson, Rao and Craig [14-16].

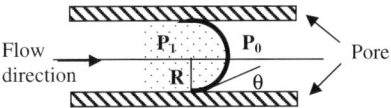

Flow direction

**fig. 1**. Model of the average pore radii $R$ of a cylindric capillary

### 3.3. Dynamics of fluid–air capillary flow at the pore-scale

The simplest equation to calculate the dynamics of Fluid–air capillary flow at the pore-scale is attributed to Washburn. It is usually considered to be rigorous for the case of capillary penetration into a uniform capillary tube or bundle of uniform capillary tubes. By combining with the Hagen–Poiseuille equation, the capillary velocity ($v$) is given by the following :

$$\overline{v} = \frac{dx}{dt} = \frac{dP}{dx}\frac{R^2}{8\eta} \qquad \text{with (2), we obtained} \qquad \overline{v} = \frac{R\gamma\cos\theta}{4x\eta} \qquad (3)$$

where $x$ is the length of fluid penetration; $dP$ is the pressure drop between the meniscus and the bulk liquid; $R$ the tube radius; $\eta$ the fluid viscosity; $\theta$ the contact angle and $\gamma$ is the interfacial tension. This equation relates the rate of meniscus advance at a given length to other physical properties. The capillary velocity is proportional to the raduis R While integrating, we obtained:

$$x^2 = \frac{Rt\gamma\cos\theta}{2\eta}$$

(4)

The position of the meniscus increase with $\sqrt{t}$ and $\sqrt{R}$. If there is a distribution of different radii capillary, the liquid will not get in at the same velocity in all radii. Sorbie et al. [17] have re-examined the basis of the Washburn equation and have extended this equation.

### 4. FLUID PROGRESSION SIMULATION IN POROUS SURFACE IMAGE

Simulation will consist in showing during various stages the fluid progression in porous space and to show the evolution of porosity image and the cumulative curve of the surface of pore. One wants to simulate what occurs when one forces the fluid to penetrate in porous space by applying an increasing pressure. The image parameter which will use to us to carry out simulation is the diameter capillary (associated with pressure).
It is admitted that the fluid penetrate in porous surface by the four faces of image. One will detect the first enter corresponding to the lowest pressure then seeking the diameter connection to be able to trace the fluid meniscus one. Finally we will build the propagation of

the fluid (with an associated coding color for each pressure) in the capillary tubes of the original image

### 4.1. Image treatments

For our problems, the first part of this software gathers all the functions of image processing relating to the improvement of the images of scanning electronic microscopy in order to determine all the porous system as well as possible and to simulate the propagation of the fluid in this porous system. We can distinguish various stages:
- Scanning Electron Microscopy images filtered
- contour detection and thus of the porous network by thresholding of the image
- post-processing of this binary image (mathematical morphology)

The first stage of this work is to treat the images in gray level obtained by scanning electron microscopy in BSE mode. An example of electron microscopy image obtained on Oligocene limestone sample is given on the figure 2a. Various filters of pre-treatments studied by many authors [18-19] are establish in this software (Average, Median, Gaussian).

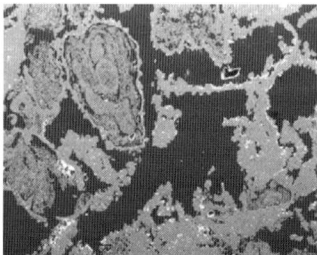 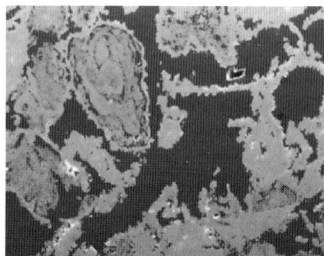

**fig.2b**. : Filtered Image

To obtain the best results, we also established an iterative filter presented by Mo. Dai [20]. It smoothes the noise, that it is additive or multiplicative and preserves discontinuities specific to the contents of the image. The method consists in convoluting the image with a very small average mask whose coefficients are of the 0 and of the 1. Criterion measuring the continuity of the image to each iteration is a threshold which varies like Gaussian. This last property allows an automatic stop of the iteration. The results of this filter on the preceding image east give on the figure 2b.

### 4.2. Threshold detection and binary process

The example of the preceding image shows an image composes of two objects, the rock and the pores. In this case, the histogram is constitute of two quite distinct modes. The thresholding will consist has to separate these two phases automatically. One determines thresholds which will constitute the limits of the various classes. Within the precise framework of the total thresholding (even thresholding for all the image), the identification of the thresholds can be done by seeking local minima or by approximating the modes by the Gaussian ones. The research of the local minima and the passages by zero require the calculation of first and second derivative function. To reduce calculations, one has thus to use the operator of Shen [21].

Before carrying out the detection of the thresholds, the histogram can undergo improvements either by smoothing, or by approximation. This part of the program is fundamental because it determines all porosity present on the image and also the microporosity which on certain image is represented only by some pixels. The results of the automatic control is watch on the binary image.

The small isolated black areas are regarded as a noise of image. Usually to reduce it, we use mathematical morphology with like major disadvantage to destroy contours. We developed a method of search automatic for area isolated in order to remove them. The secondary treatments are lead from these cleaned images. The last stage of these treatments relates to the detection of capillary contours. For this operation, we have uses a simple filtering of the Laplacien type.

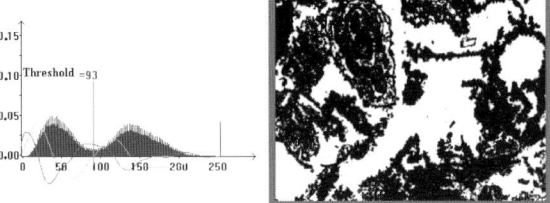

**fig.4.** : Histogram thresholding and binary Image

### 4.3. Invasion of the capillaries by a fluid

For equation (1), we have :
$$P_c = 2\gamma \frac{\cos\theta}{R} = 4\gamma \frac{\cos\theta}{D}$$

In our study we took the mercury with: $\gamma = 485$ dynes/cm and $\theta = 130°$.

### 4.3.1. Calculation of the capillary diameter

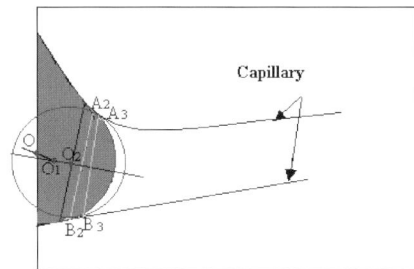 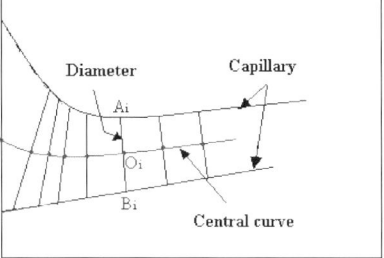

**fig. 5.** Calcul of capillary diameter **a:** sucession of tangent circle **b:** central curve

To make penetrate a fluid in the pores of the rock compared to capillary tubes systems, it is necessary to exert a pressure P on this fluid higher than capillary pressure $P_C$. Consequently, the measurement of the pressure of injection of a fluid makes it possible to calculate the diameter of the pores connection whose section is comparable with a circle. The detection of the capillary diameter is of primary importance. To calculate this diameter, we must used the central curve of this capillaries (fig. 5). For the calculation of this median instead of making a skeleton process of the capillary which is very sensitive to the least defect of contour, we preferred develops a method which is based on the calculation of a succession of tangent circles has the border of the capillaries as the figure shows it.

We redo the operation over the entire length of the capillary and one obtains the central curve of the capillaries by connecting the whole of the centers of the successive circles. With this curve, we can calculate the capillary diameters.

### 4.3.2. Detection of the capillary entry

As we work on a 2D image in of the porous surface which we let us regard as a cut of the 3D carrot (sample test for the mercury porosimetry), we must determine which are the capillaries of entry in edge of images or the invasion of the fluid will be done in priority. This detection is obtained by marking all the capillaries then by determining some the various diameters then while returning these values in chained lists.

### 4.3.3. Visualization of the propagation of the fluid

To make progress the fluid inside the porous system, we must apply to this fluid a gradual pressure. Any moment during the visualization of the fluid propagation we must be able to determine which are the capillaries which will fill. They are imperative to classify the various capillary diameters and thus the various pressures corresponding to the progression of the invasion of the system by the fluid an example is given on the figure 6.

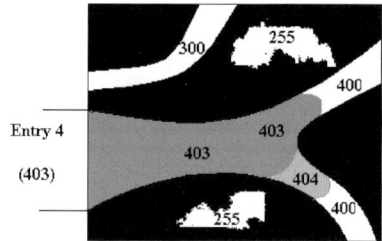

**fig.6**. : Fluid propagation

With each new value, one calculates from the pressure, the surface of filling and thus the new fluid meniscus. The result of the propagation of the fluid is displayed in real time. One associates the various pressures (and thus the diameter of the capillaries) a code of color (of the red for the largest diameters with the yellow for smallest). An example of visualization is given on the following figure 7.

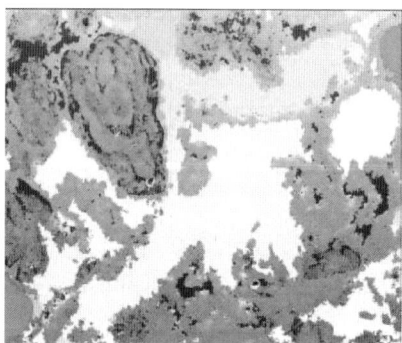

**fig.7**. : Fluid propagation in limestone

It is interesting to observe the propagation of this fluid in particular all that relates to the phenomena of deflation and trapping of air in multiple connections. Thus, one can show how the pores connectivity led to a residual saturation.

## 5. RESULTS

Series of images which we treated come from samples of Oligocene limestones whose properties are known and from which various porosities were measured in addition in mercury porosimetry. To compare these two methods, we took mercury as fluid for simulation. Several types of porosities were analyzed, the goal of this article being before all the presentation of simulation, we will present here only most outstanding. The curves of porosity below were simulated from images obtained on the sample whose curve of mercury porosimetry is given on the figure 8. On this histogram we observe that the distribution of the connections radii is bimodal with a principal peak is around 15-20 µm whereas the other presents a ray around 1 µm. Global porosity is 42.7 %.

Series of images on which we simulated the circulation of the fluid were obtained with an enlarging of x 406. To analyze the characteristics of the porosity curves obtained, we have to smooth the histograms by the Gaussian's filters (see figure 9)

We find well on the whole of the porosity curves the two modes of connections, on the other hand we do not find the same values exactly. On the whole of the curves (10 images) we obtain average values of the connections radii of 2 and of 25 µm. Although to obtain a good representation the number of treated curves is insufficient, these

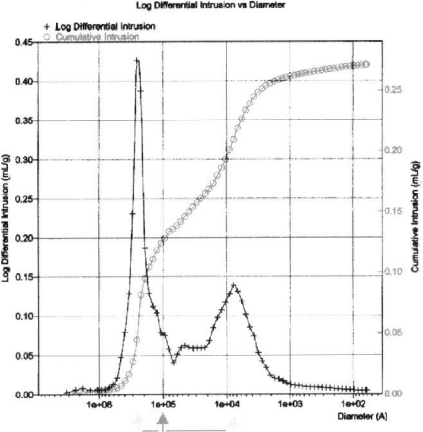

**fig.8.** : Mercury porosimetry curve

values are in concord with the porosimetry mercury. Several treatments were carried out with different samples of porosity and one always obtains good agreements for the connections radii.

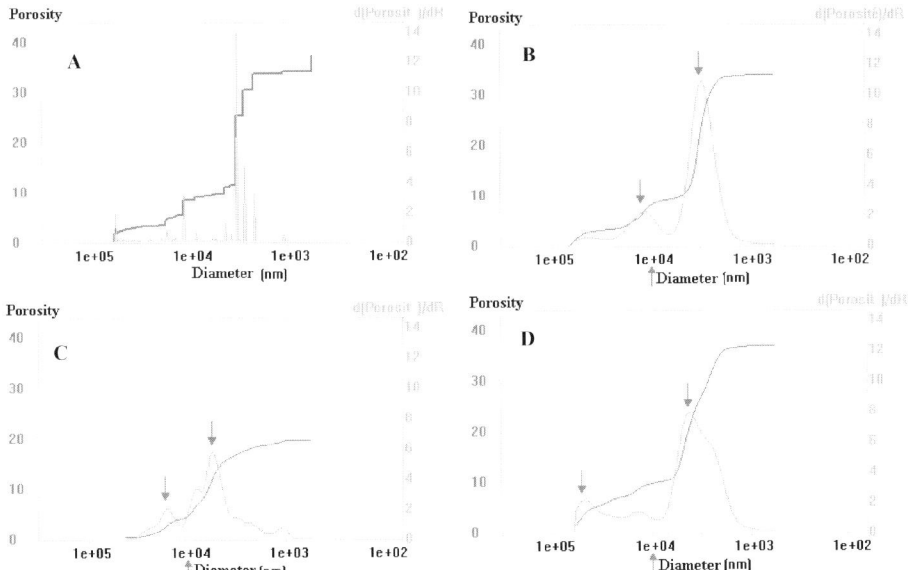

Taking into account the growths used for the images acquisition, the detection of the diameters of connection starts to $10^3$ nm and stops $10^5$ nm. To carry out the comparison between the data image and the data petrophysic, it would be necessary to have images for 2 different growths to visualize the macroporosity (R > 50 nm) and the mesoporosity (2 < R < 50 nm) of the sample.

With regard to porosity image we have to determine a porosity of 31.6 % whereas that obtained with porosity mercury is 42.7 %. Porosity calculated is thus lower than that measured. In addition to the fact of the statistics which is certainly insufficient, the problem comes owing to the fact that it is hard to extend the results 2D to results in 3D

Indeed when one proceeds to a 2D section of the porous system, it frequently sometimes happens to cut capillaries which is in another plan that the plan of the section 2D. These capillaries are found completely closed whereas in the 3D system theses capillaries could be connected. This problem involves under evaluation in the calculation of porosity image

## 6. CONCLUSION AND PROSPECTS

This modeling is satisfactory in addition to the possibility of reaching the physical parameters of the rock, one can observe in real time the progression of does not matter that it fluid in the porous system. By taking care to work on a great number of images, the modeling of the circulation of the fluid in the porous network enabled us to determine with a good precision the radii of connected analyzed rocks. These values are in conformity with those obtained in porosimetrie mercury. The problem rests in fact on the evaluation of the total porosity which under is evaluated with the software. That comes from the errors induced by the representativeness 2D. We planned to improve this software by the possibility starting from cuts seriated to rebuild an pseudo-image in 3D of the sample in order to model the invasion of the fluid on a volume 3D. That will allow to determine with more precision total porosity.

## REFERENCES

[1] M.R.E Bos, Log Anal., (1982) 17– 21,
[2] F.S Anselmetti, S. Luthi, G.P Eberli, AAPG Bull. 82 (10) (1998) 1815– 1836.
[3] T.T. Mowers, D.A. Budd, AAPG Bull. 80 (3), (1996) 309– 322.
[4] N.C Wardlaw, M. Mckellar, L. Yu, Carbonates Evaporites 3, (1988) 1 –15.
[5] C. Tsakiroglou , V. N. Burganos , J. Jacobsen, AICHE J. 50 (2) , (2004) 489 - 510
[6] R. Vocka and M. A. Dubois, Phys. Rev. E 62, (2000) 5216–5224
[7] A. Cerepi, R. Burlot, S. Galaup, J.P. Barde, C. Loisy, L. Humbert, Studies in Surface Science and Catalysis, Elsevier, (2002) vol. 144, 483-490.
[8] A. Cerepi, L. Humbert, R. Burlot, Colloids and Surfaces A : Phys. Eng. Aspects 206 (2002) 425–444
[9] E. Aker, K.K. Maloy, Phys. Rev. E 58 (2) (1998) 2217–2226.
[10] E. Aker, K.K. Maloy, A. Hansen, G.G. Batrouni, Trans. Porous Media 32 (1998) 163–186.
[11] E. Aker, K.J. Maloy, A. Hansen, Phys. Rev. E 61 (3) (2000) 2936–2946.
[12] G.N. Constantinides, A.C. Payatakes, AICHE J. 42 (3) (1996) 69–382.
[13] R.K. Lenormand, E. Touboul, C. Zircone, J. Fluid Mech. 189 (1988) 165–187.
[14] M. Jackson, M, Blunt, J. of Petroleum Science and Engineering 39, (2003) 231-246
[15] D. N. Rao and R. S. Karyampudi, J. of Adhesion Science and Technology 16 (5) (2002), 581-598
[16] F.F. Craig, SPE Monograph 3 (1970)
[17] K.S. Sorbie, Y.Z. Wu, S.R. McDougall, J. Colloids Interf. Sci. 174 (1995) 289–301.
[18] D. Marr, Proc. R. Soc. London, B, 1980, vol 207, pp 187-217
[19] I.T. Young, L.J van Vliet, Signal Processing, 44, (1995), 139-151.
[20] M. Dai, Image Anal Stereol; 20 (l1), (2001) 245-250.
[21] J.Shen, CVGIP : Graphical Model and Image Processing, 54, (2) (1992), 121-124

Studies in Surface Science and Catalysis 160
P.L. Llewellyn, F. Rodriquez-Reinoso, J. Rouqerol and N. Seaton (Editors)

# Electrical behaviour of saturated and unsaturated geological carbonate porous systems

A. Cerepi, C. Loisy, R. Toullec, R. Burlot, S. Galaup, M. Schmutz

Institut EGID-Bordeaux 3, Université Michel de Montaigne, 1, allée Daguin, 33607, Pessac, Cedex, France, Tel : (33) 05 57 12 10 11, Fax : (33) 05 57 12 10 01

## I. INTRODUCTION

The electrical conductivity of a porous solid depends on different parameters such as the spatial distribution of the constituent minerals and pore space, saturation distribution, mineralogical content, wettability and temperature. Transport properties such as the electrical conductivity and permeability depend on porosity, but are also strongly sensitive to the micro-structure of the porous system, the connectivity of the pore space and their micro-geometry [1-6]. Different electrical parameters of porous solids are studied: the formation resistivity factor, the cementation factor, the resistivity index and the saturation exponent. The frequency dispersion is explained by the chargeability factor which depends strongly on the sedimentary rock texture, the brine/gas saturation of porous solids and salinity of brine [8-10]. We report here an experimental investigation of geological carbonate porous texture effects on their electrical behaviour at different frequencies.

The pore space characterization and the understanding of the flow in the pore space are important for a measurable electrical field. Electrokinetic phenomena in a porous media such as electroosmosis (fluid-flow induced by applied electric fields) and streaming potential are known to exist in brine-saturated porous media. Electrokinetic phenomena in a physical system have been known since Klingenberg's 1941 work on the permeability. They are very interesting to understand several electrical properties such as possible indicators of imminent earthquakes, mapping subsurface fluid flow, for the study of hydrothermal activity and volcanic activity [11-12]. From that it is possible to characterize the permeability, the effective throat radius and the electric potential [11-12]. We also show here the dependence of the streaming potential coupling coefficient in two-phase flow conditions. The relative electrokinetic coupling coefficient scales with the reduced water saturation at capillary pressure equilibrium. The streaming potential coupling coefficient becomes dependent on the surface conduction. A new model tries to explain these different observations [12]. Pengra and Wong [13] have shown that permeability is rigorously related to the streaming potential and electroosmotic pressure, and that these can be easily measured by a low-frequency AC technique.

The paper is composed of five sections. Section 2 contains the theoretical principles of the complex resistivity, the electrical behaviour of a saturated and unsaturated porous medium and electrofiltration phenomena. Section 3 gives the physical properties of different samples. Section 4 shows the experimental setting. In section 5 we give some results and discussions. At last, section 6 is meant to give the conclusion.

## II. THEORETICAL BACKGROUND

### II.1. Classic saturated and unsaturated electrical parameters

Archie [5] defined a parameter known as the formation factor, F which links the resistivity of the sample at 100 % water saturation $R_o$ with the resistivity of water $R_w$ :

$$F = \frac{R_o}{R_w} = \frac{1}{\Phi^m} \tag{1}$$

where m is known as the "cementation factor"; this factor is usually found to lie in the range 1.3-2.5; $\Phi$ is the porosity.

The second Archie equation relates the resistivity of the sample to water saturation [5]:

$$IR = \frac{R_t}{R_o} = S_w^{-n} \tag{2}$$

where $I_R$ is the resistivity index; $R_t$ is the resistivity of the sample at a given water saturation, $S_w$, and $R_o$ is the resitivity of the sample at 100% water saturation. n is an empirical parameter called the Archie saturation exponent. Typically, n is approximately 2 for water-wet sandstone samples.

The empirical parameter of the chargeability factor (*M*) explained the maximum frequency dispersion normalized to its maximum value at high frequencies. The chargeability is defined as follows [7]:

$$M = \frac{R}{R + R_s} \tag{3}$$

where R is the resistivity at high frequencies (100 kHz); $R_s$ is the resistivity at the low frequencies (100 Hz).

### II.2. Electrokinetic phenomena

Electrokinetic phenomena arise from movement of ions in the electric double layer under a pore pressure gradient. In the case of a steady-state fluid circulation and for a saturated porous media, a linear relation exists between the electrical potential difference $\Delta V$ and the pressure difference $\Delta P$. This ratio is called the electrokinetic coupling coefficient [11-12] :

$$C_{sat} = \Delta V / \Delta P = (\varepsilon \zeta) / (\eta \sigma_f) \tag{4}$$

with $\zeta$ the zeta potential as the electrical potential on the shear plane; $\sigma_f$ as the electric conductivity of the circulating brine; $\eta$ as the shear viscosity of the circulating brine; $\varepsilon = \varepsilon_{water} \varepsilon_0$ is the electric permittivity of the fluid with $\varepsilon_{water}$ the relative dielectric constant of the fluid and $\varepsilon_0 = 8.8410^{-12}$ F/m the dielectric constant of vacuum [11].

In the case of partial saturation $S_w$ medium a new equation has been proposed and used by Guichet *et al.* [11]. This equation tries to relate the electrokinetic coupling coefficient to the partial water saturation and is used for m = n = 2. Guichet *et al.* [11], Revil and Cerepi [12] have generalized their results to any values of m and n :

$$C(Sw \leq 1) = \frac{\varepsilon\zeta/\eta\sigma f}{S_w^n \left(1 + m\left(\frac{F}{S_w^n} - 1\right)\frac{(\sigma s/\sigma f)}{Sw}\right)} \qquad (5)$$

with $S_w$ as the effective brine saturation defined as following :

$$S_w = \frac{S_{wi} - S_{wi}}{1 - S_{wi}} \qquad \text{as} \qquad S_w > S_{wi} \qquad \text{and} \qquad S_w = 0 \qquad \text{as} \qquad S_w < S_{wi} \qquad (6)$$

where $S_w$ is water saturation and $S_{wi}$ is the irreductible water saturation (no brine flow below this saturation in porous medium);

## III. SAMPLE DATA BASE

Table 1 shows different petrographical and petrophysical characteristics (mineralogy, texture, diagenetic phases and pore types) of different studied samples. Petrophysical characteritics are performed from the mercury porosimetry, water permeability and porosity. The 13 studied dolomite samples were chosen for this study because of their high textural heterogeneity due to a variety of depositional environments and different diagenetic evolutions. Detailed petrographic analyses of the porous microstructures were made from thin-sections in conventional optical microscopy (OM) and Scanning Electron Microscopy (SEM).

The muddy textures (mudstone-wackestone with mud-supported / grains supported texture) are characterized by two primary pore-types: intramatrix (within carbonate mud) and intragranular of small size ($<$1-2 µm) (within carbonate grains). Their pore networks are monomodal, with low porosity-low permeability trends. In this lithofacies the reservoir properties are mainly controlled by muddy matrix size.

The crystal carbonate facies is a secondary dolomitic texture characterized by the dominance of well-developed dolomite crystals. The most common porosity types in the dolomitized parts are: i) moldic and vuggy porosity; ii) intercrystalline matrix porosity; iii) intracrystalline porosity due to the selective dissolution of crystalline dolomite; iv) karst and fracture porosity. The pore structures vary from monomodal to trimodal with an increase of porosity and permeability according to the development of vugs and karsts. In this facies, the reservoir properties are mainly controlled by the crystal size.

## IV. EXPERIMENTAL SETTING AND PROCEDURES

The experimental setting is composed of a loan Mk2 "LLC" Modular Universal Stackable $P_c$&$I_R$ system developed to obtain different electrical measurements and capillary pressure curves e.g., Fig. 1. The apparatus contains multi-potential electrodes along the sample length which allows us to identify the electrical behaviour heterogeneity of each sample. The sample plugs are 3.8 mm in diameter varying in length from 3 to 8 cm. The electrical cell is connected with an impedencemeter HP which allows us to make different frequencies for resistivity measurements from 100 Hz to 100 kHz for a voltage of 1000 mV. The two end metal plates serve as current electrodes and two additional electrodes around the samples measure the voltage e.g., Fig. 1. The volume of the sample whose resistivity is measured, is situated between the two additional voltage electrodes. In order to study the frequency dependence of partially saturated rocks, the desaturation technique using semi-permeable capillary diaphragms (ceramic membranes) has been used. The brine used for all the $I_R$

experiments was 5 g/l NaCl. The main advantages of this method are the reduction of capillary end effects and the uniform saturation distribution along the core length. Core plugs were drilled in parallel with the orientation of the stratification. Then, the samples were oven dried for 48 hours at a temperature of 50 °C. The dried samples were saturated with brine under 0.001 mbar vacuum for 24 hours. We gradually increased the capillary pressure until the atmospheric pressure. Then, the effective water porosity was obtained by weighing. Once the sample is saturated at 100%, we assemble it with the jacket and the electrode pins into the pressure tube (confining pressure 29,5 bars max). Resistivities were measured at different saturations and different frequencies.

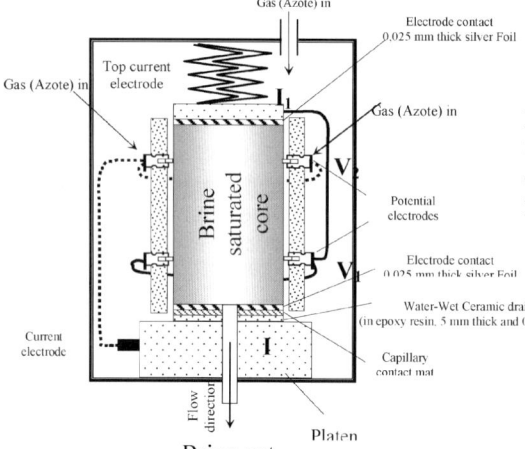

Fig. 1. Experimental Multi - electrode core flow cell configuration for IR-n, F-m determination and electrokinetic phenomena.

Table 1 : Petrographic and Petrophysical characteristics of different studied samples.

F: formation factor; m : cementation factor; n : saturation exponent; $\Phi_w$ : water porosity; $k_w$ : water perméability (mD); $S_{wi}$ : irreducible water saturation; M-W: mudstone-wackestone; C: crystal carbonate; G: grainstone; IM: intramatrix porosity; iX: intracrystalline porosity ; IX: intercrystalline porosity; V: vuggy porosity; K: karstic porosity ; F: fracturing porosity;

| Samples | Textures | Pore-types | Electric Parameters | | | Chargeability | | $S_{wi}$ | $\Phi_w$ (%) | $k_w$ (mD) |
| | | | F | Rw@5g/l NaCl | m | Factor $M$ | n | | | |
|---|---|---|---|---|---|---|---|---|---|---|
| 1-C | M-W | IM, F | 35.56 | 1.07 | 2.05 | 0.388 | 2.9 | 0.51 | 17.5 | 27.43 |
| 2-C | M-W | IM, iX, V | 8.7 | 1.07 | 1.68 | 0.491 | 1.2 | 0.21 | 27.5 | 9.25 |
| 3-C | C | IX, F, K, V | 20.6 | 1.07 | 2.48 | 0.366 | 1.87 | 0.36 | 31.1 | 203.225 |
| 4-C | C | IX, F, K | 9.54 | 1.07 | 2.14 | 0.478 | - | - | 34.9 | 90.34 |
| 5-D | C | iX, F | 23.28 | 1.07 | 1.98 | 0.411 | 2.6 | 0.6 | 20.4 | 142.9 |
| 6-D | C | K, F, iX, V | 19.67 | 1.07 | 2.07 | 0.447 | - | - | 23.8 | 185.5 |
| 7-D | C | K, F, iX, V | 13.77 | 1.07 | 1.99 | 0.423 | 2.4 | 0.87 | 26.8 | 303.4 |
| 8-D | M-W | IM | 36.97 | 1.07 | 2.31 | 0.5 | - | - | 21,0 | 0.7 |
| 9-D | C | K, F, iX, V | 28.2 | 1.07 | 2.27 | 0.483 | 3.4 | 0.73 | 22.9 | 169.96 |
| 3-GO01 | G | V,K,IM | 21.75 | 1.07 | 1.93 | 0.49 | 2.7 | 0.42 | 20.3 | 48.4 |
| 24-GO01 | G | IG, V | 19.67 | 1.07 | 1.55 | 0.19 | 4.21 | 0.62 | 14.6 | - |
| 35-GO01 | G-C | V, IG | 52.12 | 1.07 | 2.12 | 0.38 | 2.55 | 0.54 | 15.5 | - |
| 39-GO01 | C | IM, IX, IG | 96.14 | 1.07 | 2.49 | 0.32 | 3.5 | 0.4 | 15.9 | 23.8 |

## V. RESULTS AND DISCUSSIONS

### V.1. Classic electrical parameters

More than thirteen carbonate samples are studied for classic electric parameters in saturated and unsaturated geological carbonate porous systems. Table 1 gives different measurements obtained from the electrical setting.

The cementation factor (m) and the structural parameter (a) were interpreted for each sample from a logF-log$\Phi$ curve using Eq(1). The values of m obtained from this method range from 1.55 to 2.49. The dolomitic crystal texture shows the highest value of m (1.98 to 2.49) and F (9.54 to 96.14) while the mudstone-wackestone texture has low values of m (1.62 to 1.98) and F (5.47 to 1.98). The high Archie cementation factor m of dolomitic porous systems is due to its high cementation degree.

Table I and Figure 3 give the relationship between $\log I_R$ - logSw and the saturation exponent n whose values range between 1.2 and 3.5. The highest values of n are obtained for the dolomite crystal carbonate (n = 3.5) while the lowest values of n are obtained for the mudstone-wackestone texture (n = 1.2).

The values of the chargeability factor $M$ range between 0.19 and 0.5, and they depend on the brine/gas saturation of the carbonate porous system, carbonate textures and the salinity of brine in the porous medium. The highest value of the chargeability factor is obtained in M-W texture. The lowest value M is reached in the dolomitic crystal carbonate texture.

The capillary pressure versus brine saturation curves for four dolomite samples is shown in Figure 2. For thirteen studied carbonate samples the irreducible water saturations $S_{wi}$ (no water flow zone) range from 0.21 to 0.87 (table 1).

### V.2. Electrokinetic phenomena

The electrokinetic phenomena are performed on four samples (table 1, fig. 4, 5, 6). In the case of a saturated carbonate porous system, the potential variations of four electrodes as functions of the imposed pressure variation are shown in e.g., fig.5. A negative linear regression is observed which gives the value of the coupling coefficient $C_m(S_w=1)$. The coupling coefficient varies from $C_m(S_w=1) = -1.981$ mV/bars to $C_m(S_w=1) = -15.72$ mV/bars.

In the case of an unsaturated carbonate porous system, the first results are shown in Figure 4. The measured streaming potential decreases proportional to the driving pressure in the carbonate porous system for a steady state system and an unsaturated state ($S_w<1$). The measured coupling coefficients range from $C_m(S_w<1) = 0.1$ mV/bars to $C_m(S_w<1) = 1000$ mV/bars, depending on the water saturation, the texture of porous medium and the electrical conductivity. Figure 4 shows an increase of the measured coupling coefficient (streaming – pressure ratio) observed with an increase in the water saturation. For a water saturation varing between 1 and 0.4 the coupling coefficient (at $S_w<1$) decreases from 1000 mV/bars to 0.1 mV/bars.

Figure 6 shows the normalized measured coupling coefficient $C_m(S_w < 1) / C_m(S_w = 1)$ versus brine saturation. We observe that the relative streaming potential coupling coefficient decreases while the water saturation decreases. The streaming potential coupling coefficient falls to zero when the water saturation reaches the irreducible water saturation $S_{wi}$.

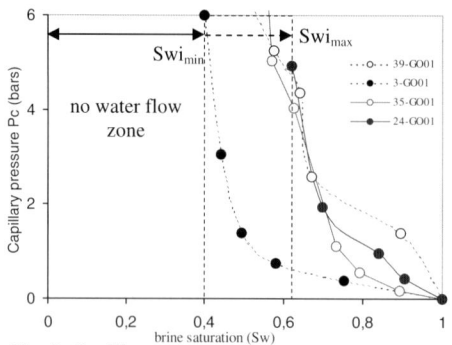

Fig. 2. Capillary pressure curve versus water saturation for four carbonate porous systems

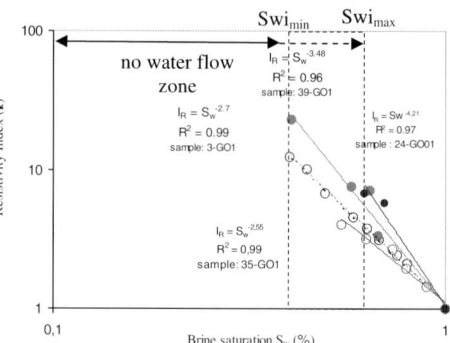

Fig. 3. Resistivity Index versus brine saturation for four carbonate porous systems

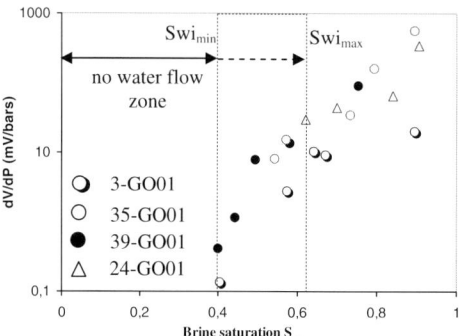

Fig. 4. Variation of the measured coupling coefficient versus water saturation for four carbonate porous systems

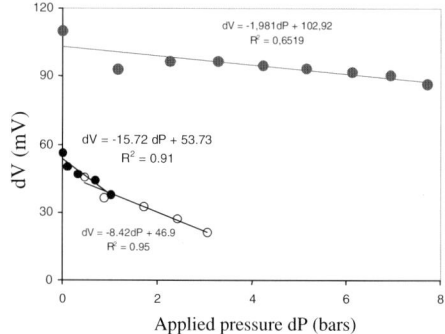

Fig. 5. Streaming potential differences versus applied pressure differences in steady state and saturated state for three carbonate porous systems.

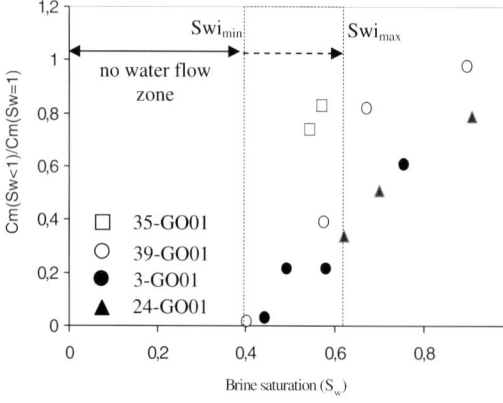

Fig. 6. Normalized coupling coefficient $C_m(S_w<1)/C_m(S_w=1)$ versus water saturation for four carbonate porous systems.
$Swi_{min}$: Minimum irreducible water saturation;
$Swi_{max}$: Maximum irreducible water saturation.

## VI. CONCLUSION

More than thirteen carbonate samples are studied to measure classic electric parameters in saturated and unsaturated geological carbonate porous systems. The values of the cementation factor m range from 1.55 to 2.49. The dolomitic crystal texture shows the highest value. The electrical behaviour of the carbonate porous system in a unsaturated medium is studied from the saturation exponent n. The highest values of n are obtained for a dolomite crystal carbonate (n = 3.5) while the lowest values of n are obtained for a mudstone-wackestone texture (n = 1.2). The values of the chargeability factor $M$ range between 0.19 and 0.5, and they depend on the brine/gas saturation of the carbonate porous system, carbonate textures and the salinity of brine in a porous medium.

The electrokinetic phenomena are performed on four samples. The measured streaming potential decreases proportional to the driving pressure in a carbonate porous system for a steady state system and an unsaturated state. The relative streaming potential coupling coefficient decreases while the water saturation decreases. The streaming potential coupling coefficient falls to zero when the water saturation reaches the irreducible water saturation $S_{wi}$.

## REFERENCES

[1] Bryant, S. & Pallat, M., 1996. Predicting formation factor and resistivity index in simple sandstones, J. Pet. Sci. Eng, 15, pp. 169-179.
[2] Katz, A.J. & Thompson, A.H., 1986. Quantitative prediction of permeability in porous rock, Physical review B, vol. 34, N° 11, pp. 8179 – 8181.
[3] Worthington, P.F., 1997, Petrophysical estimation of permeability as a function of scale, from Lovell, M.A. & Hervey, P.K. (eds), in Developments in Petrophysics, Geological Society Special Publication, N° 122, pp. 159-168.
[4] Worthington, P.F., Pallat, N., and Toussaint-Jackson, J.E., 1989. Influence of microporosity on the evaluation of Hydrocarbon saturation, SPEFE (June 1989), 203 ; Trans, AIME, 287.
[5] Archie, G.E., 1942. The electrical resistivity log as an aid in determiny some reservoir characteristics. AIME Petroleum Technology, pp. 1-8.
[6] Saner, S.Abdulaziz Al-Hasthi, Maung Than Htay, 1996, Use of tortuosity for discriminating electro-facies to interpret the electrical parameters of carbonate reservoir rocks, Journal of Petro. Science & Engineering, 16, pp. 237-249.
[7] Denical, P.S. & Jing, X.D. 1998. Effects of water salinity, saturation and clay content on the complex resistivity of sandstone samples, In : Harvey, P.K. & Lovell, M.A., (eds) Core-log Integration, Geological Society, London, Special Publications, 136, pp. 147-157.
[8] Da Rocha, B. R. P. & Habashy, T.M., 1997. Fractal geometry, porosity and complex resistivity : from hand specimens to field data, from Lovell, M.A. & Harvey, P.K., in Developments in Petrophysics, Geology Society Special Publication, N° 122, pp. 287-297.
[9] Swanson, R.G., 1981. Visualing pores and nonwetting phase in porous rock. Journal of Petroleum Technology, vol. 31, pp. 10-18.
[10] Diederix, K.M., 1982, Anomalous relationship between resistivity index and water saturations in the Rotliegend sandstone (The Netherlands), SPWLA Annual Logging Symposium, Corpus Christi, Texas, 6-9 July.
[11] Guichet, X., Jouniaux, L., Pozzi, JP., 2003. Streaming potential of a sand column in partial saturation conditions, Journal of Geophysical Research, vol. 108, No. B3, 2141.
[12] Revil, A., Cerepi, A., 2004. Streaming potentials in two-phase flow conditions, Geophysical Research letters, vol. 31, L11605, 1-4.
[13] Pengra, D.B., Wong, P., 1999. Low-frequency AC electrokinetics, Colloids and Surfaces A: physicoshemical and Engineering Aspects, 159, 283-292.

# Author Index

## A

Aboukaïs, A. ; 209
Abuja, P. M. ; 661
Albiniak, A. ; 407
Almazán-Almazán, M. C. ; 185
Andertová, J. ; 629
André, G. ; 71, 375
Ania, C. O. ; 319, 559
Arenillas, A. ; 319
Attipa, C. ; 615
Ávila, P. ; 233

## B

Bachiller-Baeza, B. ; 689
Baldayan, A. ; 383, 575
Bandosz, T. J. ; 559
Barnes, P. A. ; 621
Barrande, M. ; 33
Belkhiri, S. ; 447
Bell, R. G. ; 335
Bellarosa, L. ; 365
Belmabkhout, Y. ; 113
Benaliouche, F. ; 447
Beurroies, I. ; 33, 311
Bhatia, S. K. ; 63, 503
Bhattacharya, S. ; 527
Bichara, C. ; 169
Biggs, M. J. ; 79, 257
Blacher, S. ; 193, 681
Blanco, J. ; 583
Borówka, A. ; 431, 471
Bosch, G. ; 519
Boskovic, S. ; 591
Bossuot, C. ; 265
Boucheffa, Y. ; 447
Boulet, P. ; 311
Bowden, A. A. ; 311
Bromball, R. ; 575
Brosset, C. ; 249
Brouwer, S. ; 145
Burlot, R. ; 713, 705
Buts, A. ; 79, 257

## C

Cai, Q. ; 79, 257
Calberg, C. ; 681
Canet, X. ; 201, 209, 225
Carrott, P. J. M. ; 567
Celzard, A. ; 607
Cerepi, A. ; 697, 705, 713
Chen, B. ; 311
Cherif, M. F. ; 447
Chudek, J. A. ; 177
Coasne, B. ; 1, 153, 527, 667
Cohaut, N. ; 407
Conceição, F. L. ; 567
Cordero, S. ; 303
Coulomb, J.-P. ; 71, 375
Courtois, T. ; 675
Coveney, P. V. ; 311
Czwartos, J. ; 667

## D

Dąbrowski, A. ; 479
Dai, M. ; 705
Dawson, E. A. ; 621
De Weireld, G. ; 113, 201, 209
Denoyel, R. ; 1, 33
Domingo-Garciá, M. ; 185
Dudarko, O. A. ; 479
Düren, T. ; 161

## E

Enke, D. ; 455
Evans, J. R. G. ; 311

## F

Felipe, C. ; 303
Fernández-Morales, I. ; 185
Fierro, V. ; 241, 519, 607
Findenegg, G. H. ; 17
Firlej, L. ; 487
Fischer, H. ; 295

# STUDIES IN SURFACE SCIENCE AND CATALYSIS

**Advisory Editors:**
B. Delmon, Université Catholique de Louvain, Louvain-la-Neuve, Belgium
J.T. Yates, University of Pittsburgh, Pittsburgh, PA, U.S.A.

726

728

734